T0228942

WRITING REACTION MECHANISMS IN ORGANIC CHEMISTRY

THIRD EDITION

WRITING REACTION MECHANISMS IN ORGANIC CHEMISTRY

THIRD EDITION

KENNETH A. SAVIN
Eli Lilly and Company
Butler University

AMSTERDAM • BOSTON • HEIDELBERG • LONDON
NEW YORK • OXFORD • PARIS • SAN DIEGO
SAN FRANCISCO • SINGAPORE • SYDNEY • TOKYO
Academic Press is an imprint of Elsevier

Academic Press is an imprint of Elsevier
225 Wyman Street, Waltham, MA 02451, USA
525 B Street, Suite 1800, San Diego, CA 92101-4495, USA
32 Jamestown Road, London NW1 7BY, UK
The Boulevard, Langford Lane, Kidlington, Oxford OX5 1GB, UK

Third edition 2014

Copyright © 2014, 2000, 1992 Elsevier Inc. All rights reserved.

No part of this publication may be reproduced or transmitted in any form or by any means, electronic or mechanical, including photocopying, recording, or any information storage and retrieval system, without permission in writing from the publisher. Details on how to seek permission, further information about the Publisher's permissions policies and our arrangements with organizations such as the Copyright Clearance Center and the Copyright Licensing Agency, can be found at our website: www.elsevier.com/permissions.

This book and the individual contributions contained in it are protected under copyright by the Publisher (other than as may be noted herein).

Notices
Knowledge and best practice in this field are constantly changing. As new research and experience broaden our understanding, changes in research methods, professional practices, or medical treatment may become necessary.

Practitioners and researchers must always rely on their own experience and knowledge in evaluating and using any information, methods, compounds, or experiments described herein. In using such information or methods they should be mindful of their own safety and the safety of others, including parties for whom they have a professional responsibility.

To the fullest extent of the law, neither the Publisher nor the authors, contributors, or editors, assume any liability for any injury and/or damage to persons or property as a matter of products liability, negligence or otherwise, or from any use or operation of any methods, products, instructions, or ideas contained in the material herein.

ISBN: 978-0-12-411475-3

Library of Congress Cataloging-in-Publication Data

Savin, Kenneth.
  Writing reaction mechanisms in organic chemistry. – Third edition / Kenneth Savin.
    pages cm
  Previous edition by Audrey Miller.
  ISBN 978-0-12-411475-3
  1. Organic reaction mechanisms–Textbooks. I. Title.
  QD251.2.M53 2014
  547'.139–dc23
                          2014005928

British Library Cataloguing in Publication Data
A catalogue record for this book is available from the British Library

For information on all Academic Press publications
visit our web site at store.elsevier.com

This book has been manufactured using Print On Demand technology.

Working together
to grow libraries in
developing countries

www.elsevier.com • www.bookaid.org

Contents

Acknowledgments for the Third Edition   vii

1. Introduction—Molecular Structure and Reactivity   1

2. General Principles for Writing Reaction Mechanisms   55

3. Reactions of Nucleophiles and Bases   93

4. Reactions Involving Acids and Other Electrophiles   161

5. Radicals and Radical Anions   237

6. Pericyclic Reactions   293

7. Oxidations and Reductions   355

8. Additional Problems   433

Appendix A   481
Appendix B   483
Appexdix C   485
Index   493

Please find the companion website at http://booksite.elsevier.com/9780124114753.

Contents

Acknowledgments for the Third Edition   vii

Introduction—Molecular Structure and Reactivity   1

Formal Principles for Writing Reaction Mechanisms   37

Reactions of Nucleophiles and Bases   93

Reactions Involving Acids and Other Electrophiles   171

Radicals and Radical Anions   207

Pericyclic Reactions   249

Oxidations and Reductions   335

Additional Problems   443

Appendix A   481
Appendix B   483
Appendix C   485
Index   495

Please find the companion website at http://booksite.elsevier.com/9780124114753

Acknowledgments for the Third Edition

For the third edition of this text, the focus on the *how* of writing organic reaction mechanisms remains the foremost objective. The book has been expanded with a new chapter focused on oxidation and reduction mechanisms as well as new material throughout the text. Although oxidation and reduction reactions were considered for previous versions of the text, it was decided that this important and yet often under represented topic should be included to better equip the reader. This new chapter is set up to allow students to see how our understanding of mechanisms has developed and apply what they have learned in the earlier chapters of the book to a greater number of situations and more sophisticated systems. We have added new problems throughout the text to update and provide illustrative examples to the text that will aid in identifying key situations and patterns. The oxidations chapter also allows us to touch on other mechanistic topics including organometallics, stereochemistry, radiolabeling, and a more philosophical view of the mechanistic models we are applying. This new chapter, as well as the changes to the previous chapters, has all been done with consideration to the ultimate length of the book and the goal of keeping it portable and reasonable in length.

Additional references for the examples, problems, and key topics have been expanded with an eye toward the practical application of the concepts to yet to be encountered challenges.

I am indebted to the authors of the first two editions of this book, Philippa Solomon and Audrey Miller, for the original conceptual architecture and content. I have tried to hold to the original philosophy and organizational design of the material from the previous versions. I feel it is presented in the best way possible for a text used as a "teaching book".

I am grateful for the help I received from the reviewers who took the time to read and improve the text. Their suggestions go far beyond the grammatical corrections, but are expressive of a group of individuals who are committed to learning and have a bias for doing chemistry, not just talking about it. In particular I would like to thank Alison Campbell for her contributions to the discussions around metals, Doug Kjell who reviewed and suggested problems, LuAnne McNulty for looking at the text from both the perspective of a student as well as from the standpoint of the professor, and Andrea Frederick and Nick Magnus for their key discussion around the order in which the material is presented, how oxidation number should be described, and how to draw connections to topics that the students have already been exposed to.

I would, of course, also like to thank my family. My wife Lisa and my boys Zach and Cory, for their support and prodding through all the long evenings and weekends spent in developing the manuscript and for being tolerant of the time together we have missed as a result of this effort.

Introduction—Molecular Structure and Reactivity

Reaction mechanisms offer us insights into how molecules react, enable us to manipulate the course of known reactions, aid us in predicting the course of known reactions using new substrates, and help us to develop new reactions and reagents. In order to understand and write reaction mechanisms, it is essential to have a detailed knowledge of the structures of the molecules involved and to be able to notate these structures unambiguously. In this chapter, we present a review of the fundamental principles relating to molecular structure and of the ways to convey structural information. A crucial aspect of structure from the mechanistic viewpoint is the distribution of electrons, so this chapter outlines how to analyze and depict electron distributions. Mastering the material in this chapter will provide you with the tools you need to propose reasonable mechanisms and to convey these mechanisms clearly to others.

1. HOW TO WRITE LEWIS STRUCTURES AND CALCULATE FORMAL CHARGES

The ability to construct Lewis structures is fundamental to writing or understanding organic reaction mechanisms. It is particularly important because lone pairs of electrons frequently are crucial to the mechanism but often are omitted from structures appearing in the chemical literature.

There are two methods commonly used to show Lewis structures. One shows all electrons as dots. The other shows all bonds (two shared electrons) as lines and all unshared electrons as dots.

Writing Reaction Mechanisms in Organic Chemistry
http://dx.doi.org/10.1016/B978-0-12-411475-3.00001-4

Copyright © 2014 Elsevier Inc. All rights reserved.

A. Determining the Number of Bonds

HINT 1.1

To facilitate the drawing of Lewis structures, estimate the number of bonds.

For a stable structure with an even number of electrons, the number of bonds is given by the equation:

$$(\text{Electron Demand} - \text{Electron Supply})/2 = \text{Number of Bonds}$$

The electron demand is two for hydrogen and eight for all other atoms usually considered in organic chemistry. (The tendency of most atoms to acquire eight valence electrons is known as the octet rule.) For elements in group IIIA (e.g., B, Al, Ga), the electron demand is six. Other exceptions are noted, as they arise, in examples and problems.

For neutral molecules, the contribution of each atom to the electron supply is the number of valence electrons of the neutral atom. (This is the same as the group number of the element when the periodic table is divided into eight groups.) For ions, the electron supply is decreased by one for each positive charge of a cation and is increased by one for each negative charge of an anion.

Use the estimated number of bonds to draw the number of two-electron bonds in your structure. This may involve drawing a number of double and triple bonds (see the following section).

B. Determining the Number of Rings and/or π Bonds (Degree of Unsaturation)

The total number of rings and/or π bonds can be calculated from the molecular formula, bearing in mind that in an acyclic saturated hydrocarbon the number of hydrogens is $2n + 2$, where n is the number of carbon atoms. Each time a ring or π bond is formed, there will be two fewer hydrogens needed to complete the structure.

HINT 1.2

On the basis of the molecular formula, the degree of unsaturation for a hydrocarbon is calculated as $(2m + 2 - n)/2$, where m is the number of carbons and n is the number of hydrogens. The number calculated is the number of rings and/or π bonds. For molecules containing heteroatoms, the degree of unsaturation can be calculated as follows:

Nitrogen: For each nitrogen atom, subtract 1 from n.
Halogens: For each halogen atom, add 1 to n.
Oxygen: Use the formula for hydrocarbons.

This method cannot be used for molecules in which there are atoms like sulfur and phosphorus whose valence shell can expand beyond eight.

EXAMPLE 1.1. CALCULATE THE NUMBER OF RINGS AND/OR π BONDS CORRESPONDING TO EACH OF THE FOLLOWING MOLECULAR FORMULAS

a. $C_2H_2Cl_2Br_2$

There are a total of four halogen atoms. Using the formula $(2m+2-n)/2$, we calculate the degree of unsaturation to be $(2(2)+2-(2+4))/2=0$.

b. C_2H_3N

There is one nitrogen atom, so the degree of unsaturation is $(2(2)+2-(3-1))=2$.

C. Drawing the Lewis Structure

Start by drawing the skeleton of the molecule, using the correct number of rings or π bonds, and then attach hydrogen atoms to satisfy the remaining valences. For organic molecules, the carbon skeleton frequently is given in an abbreviated form.

Once the atoms and bonds have been placed, add lone pairs of electrons to give each atom a total of eight valence electrons. When this process is complete, there should be two electrons for hydrogen; six for B, Al, or Ga; and eight for all other atoms. The total number of valence electrons for each element in the final representation of a molecule is obtained by counting each electron around the element as one electron, even if the electron is shared with another atom. (This should not be confused with counting electrons for charges or formal charges; see Section 1.D.) The number of valence electrons around each atom equals the electron demand. Thus, when the number of valence electrons around each element equals the electron demand, the number of bonds will be as calculated in Hint 1.1.

Atoms of higher atomic number can expand the valence shell to more than eight electrons. These atoms include sulfur, phosphorus, and the halogens (except fluorine).

HINT 1.3

When drawing Lewis structures, make use of the following common structural features.

1. Hydrogen is always on the periphery because it forms only one covalent bond.
2. Carbon, nitrogen, and oxygen exhibit characteristic bonding patterns. In the examples that follow, the R groups may be hydrogen, alkyl, or aryl groups, or any combination of these. These substituents do not change the bonding pattern depicted.
   (a) Carbon in neutral molecules usually has four bonds. The four bonds may all be σ bonds, or they may be various combinations of σ and π bonds (i.e., double and triple bonds).

$$
\begin{array}{ccc}
& R & R \\
& | & \\
R-\overset{\displaystyle |}{\underset{\displaystyle |}{C}}-R & \text{or} & R : \overset{\displaystyle \cdot\cdot}{C} : R \\
& R & R \\
R-C{\equiv}C-R & \text{or} & R : C ::: C : R
\end{array}
$$

There are exceptions to the rule that carbon has four bonds. These include CO, isonitriles (RNC), and carbenes (neutral carbon species with six valence electrons; see Chapter 4).

(b) Carbon with a single positive or negative charge has three bonds.

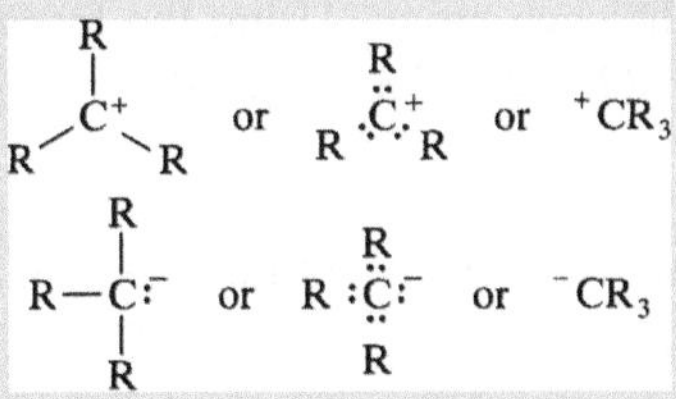

(c) Neutral nitrogen, with the exception of nitrenes (see Chapter 4), has three bonds and a lone pair.

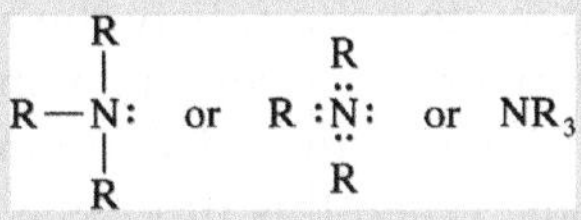

(d) Positively charged nitrogen has four bonds and a positive charge; exceptions are nitrenium ions (see Chapter 4).

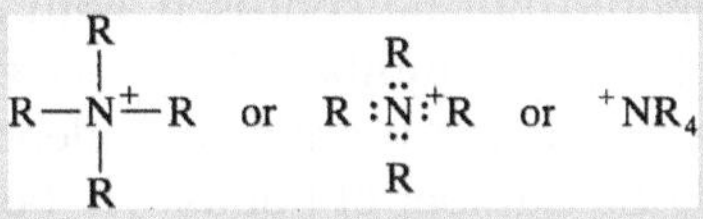

(e) Negatively charged nitrogen has two bonds and two lone pairs of electrons.

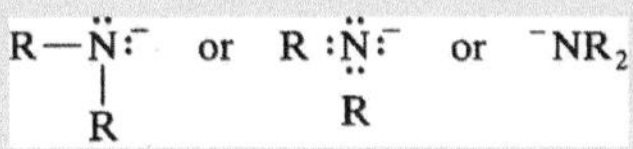

(f) Neutral oxygen has two bonds and two lone pairs of electrons.

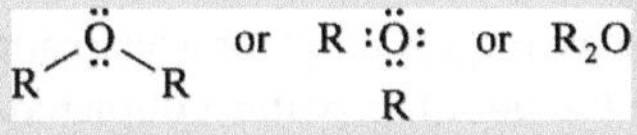

(g) Oxygen–oxygen bonds are uncommon; they are present only in peroxides, hydroperoxides, and diacyl peroxides (see Chapter 5). The formula, RCO_2R, implies the following structure:

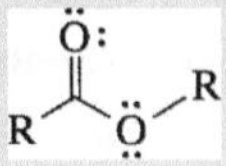

(h) Positive oxygen usually has three bonds and a lone pair of electrons; exceptions are the very unstable oxenium ions, which contain a single bond to oxygen and two lone pairs of electrons.

$$
\overset{R}{\underset{R}{\overset{|}{\underset{}{\overset{+}{O}}}}}\diagdown R \quad \text{or} \quad R:\overset{R}{\underset{R}{\overset{\cdot\cdot}{O}:^{+}}} \quad \text{or} \quad R_3O^{+}
$$

3. Sometimes a phosphorus or sulfur atom in a molecule is depicted with 10 electrons. Because phosphorus and sulfur have *d* orbitals, the outer shell can be expanded to accommodate more than eight electrons. If the shell, and therefore the demand, is expanded to 10 electrons, one more bond will be calculated by the equation used to calculate the number of bonds. See Example 1.5.

In the literature, a formula often is written to indicate the bonding skeleton for the molecule. This severely limits, often to just one, the number of possible structures that can be written.

EXAMPLE 1.2. THE LEWIS STRUCTURE FOR ACETALDEHYDE, CH₃CHO

	Electron Supply	Electron Demand
2C	8	16
4H	4	8
1O	6	8
	18	32

The estimated number of bonds is $(32 - 18)/2 = 7$.

The degree of unsaturation is determined by looking at the corresponding saturated hydrocarbon C_2H_6. Because the molecular formula for acetaldehyde is C_2H_6O and there are no nitrogen, phosphorus, or halogen atoms, the degree of unsaturation is $(6 - 4)/2 = 1$. There is either one double bond or one ring.

The notation CH_3CHO indicates that the molecule is a straight-chain compound with a methyl group, so we can write

$$CH_3-C-O$$

We complete the structure by adding the remaining hydrogen atom and the remaining valence electrons to give

$$CH_3-\overset{}{\underset{\underset{H}{|}}{C}}=\overset{\cdot\cdot}{\underset{\cdot\cdot}{O}}$$

Note that if we had been given only the molecular formula C_2H_6O, a second structure could be drawn

A third possible structure differs from the first only in the position of the double bond and a hydrogen atom.

This enol structure is unstable relative to acetaldehyde and is not isolable, although in solution small quantities exist in equilibrium with acetaldehyde.

D. Formal Charge

Even in neutral molecules, some of the atoms may have charges. Because the total charge of the molecule is zero, these charges are called formal charges to distinguish them from ionic charges.

Formal charges are important for two reasons. First, determining formal charges helps us pinpoint reactive sites within the molecule and can help us in choosing plausible mechanisms. Also, formal charges are helpful in determining the relative importance of resonance forms (see Section 5).

HINT 1.4

To calculate formal charges, use the completed Lewis structure and the following formula:

Formal Charge = Number of Valence Shell Electrons – (Number of Unshared Electrons
+ Half the Number of Shared Electrons)

The formal charge is zero if the number of unshared electrons, plus the number of shared electrons divided by two, is equal to the number of valence shell electrons in the neutral atom (as ascertained from the group number in the periodic table). As the number of bonds formed by the atom increases, so does the formal charge. Thus, the formal charge of nitrogen in $(CH_3)_3N$ is zero, but the formal charge on nitrogen in $(CH_3)_4N^+$ is +1.

Note: *An atom always "owns" all unshared electrons.* This is true both when counting the number of electrons for determining formal charge and in determining the number of valence electrons. However, in determining *formal charge*, an atom "owns" half of the bonding electrons, whereas in determining the *number of valence electrons*, the atom "owns" *all the bonding electrons*.

EXAMPLE 1.3. CALCULATION OF FORMAL CHARGE FOR THE STRUCTURES SHOWN

(a)

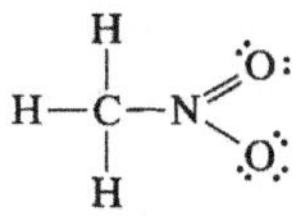

The formal charges are calculated as follows:

Hydrogen
   1·(no. of valence electrons) − 2/2·(2 bonding electrons divided by 2) = 0
Carbon
   4·(no. of valence electrons) − 8/2·(8 bonding electrons divided by 2) = 0
Nitrogen
   5 − 8/2·(8 bonding electrons) = +1

There are two different oxygen atoms:

Oxygen (double bonded)
   6 − 4·(unshared electrons) − 4/2·(4 bonding electrons) = 0
Oxygen (single bonded)
   6 − 6·(unshared electrons) − 2/2·(2 bonding electrons) = −1.

(b)

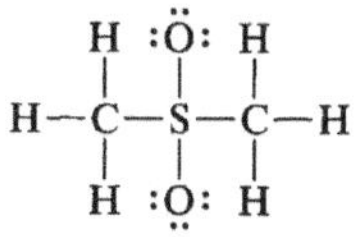

The calculations for carbon and hydrogen are the same as those for part (a).

Formal charge for each oxygen:
   6 − 6 − (2/2) = −1
Formal charge for sulfur:
   6 − 0 − (8/2) = +2

EXAMPLE 1.4. WRITE POSSIBLE LEWIS STRUCTURES FOR C_2H_3N

	Electron Supply	Electron Demand
3H	3	6
2C	8	16
1N	5	8
	16	30

The estimated number of bonds is $(30 − 16)/2 = 7$.

As calculated in Example 1.1, this molecular formula represents molecules that contain two rings and/or π bonds. However, because it requires a minimum of three atoms to make a ring, and since hydrogen cannot be part of a ring because each hydrogen forms only one bond, two rings are not possible. Thus, all structures with this formula will have either a ring and a π bond or two π bonds. Because no information is given on the order in which the carbons and nitrogen are bonded, all possible bonding arrangements must be considered.

Structures **1-1** through **1-9** depict some possibilities. The charges shown in the structures are formal charges. When charges are not shown, the formal charge is zero.

Structure **1-1** contains seven bonds using 14 of the 16 electrons of the electron supply. The remaining two electrons are supplied as a lone pair of electrons on the carbon, so that both carbons and the nitrogen have eight electrons around them. This structure is unusual because the right-hand carbon does not have four bonds to it. Nonetheless, isonitriles such as **1-1** (see Hint 1.3) are isolable. Structure **1-2** is a resonance form of **1-1**. (For a discussion of resonance forms, see Section 5.) Traditionally, **1-1** is written instead of **1-2**, because both carbons have an octet in **1-1**. Structures **1-3** and **1-4** represent resonance forms for another isomer. When all the atoms have an octet of electrons, a neutral structure like **1-3** is usually preferred to a charged form like **1-4** because the charge separation in **1-4** makes this a higher energy (and, therefore, less stable) species. Alternative forms with greater charge separation can be written for structures **1-5**–**1-9**. Because of the strain energy of three-membered rings and cumulated double bonds, structures **1-6** through **1-9** are expected to be quite unstable.

It is always a good idea to check your work by counting the number of electrons shown in the structure. The number of electrons you have drawn must be equal to the supply of electrons.

EXAMPLE 1.5. WRITE TWO POSSIBLE LEWIS STRUCTURES FOR DIMETHYL SULFOXIDE, $(CH_3)_2SO$, AND CALCULATE FORMAL CHARGES FOR ALL ATOMS IN EACH STRUCTURE

	Electron Supply	Electron Demand
2C	8	16
6H	6	12
1S	6	8
1O	6	8
	26	44

According to Hint 1.1, the estimated number of bonds is $(44 - 26)/2 = 9$. Also, Hint 1.3 calculates 0 rings and/or π bonds. The way the formula is given indicates that both methyl groups are bonded to the sulfur, which is also bonded to oxygen. Drawing the skeleton gives the following:

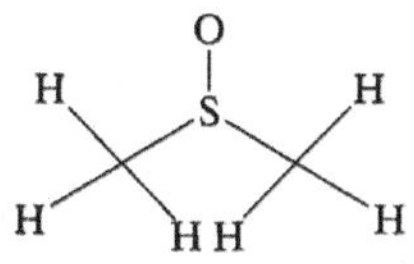

The nine bonds use up 18 electrons from the total supply of 26. Thus there are eight electrons (four lone pairs) to fill in. In order to have octets at sulfur and oxygen, three lone pairs are placed on oxygen and one lone pair on sulfur.

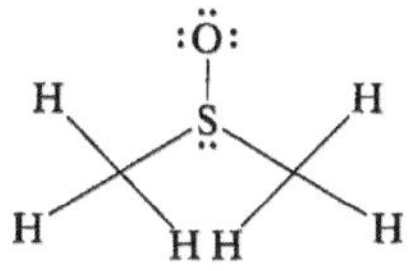

1-10

The formal charge on oxygen in **1-10** is −1. There are six unshared electrons and $2/2 = 1$ electron from the pair being shared. Thus, the number of electrons is seven, which is one more than the number of valence electrons for oxygen.

The formal charge on sulfur in **1-10** is +1. There are two unshared electrons and $6/2 = 3$ electrons from the pairs being shared. Thus, the number of electrons is five, which is one less than the number of valence electrons for sulfur.

All the other atoms in **1-10** have a formal charge of 0.

There is another reasonable structure, **1-11,** for dimethyl sulfoxide, which corresponds to an expansion of the valence shell of sulfur to accommodate 10 electrons. Note that our calculation of electron demand counted eight electrons for sulfur. The 10-electron sulfur has an electron demand of 10 and leads to a total demand of 46 rather than 44 and the calculation of 10 bonds rather than 9 bonds. All atoms in this structure have zero formal charge.

1-11

Hint 1.3 does not predict the π bond in this molecule, because the valence shell of sulfur has expanded beyond eight. Structures **1-10** and **1-11** correspond to different possible resonance forms for dimethyl sulfoxide (see Section 5), and each is a viable structure.

Why do we not usually write just one of these two possible structures for dimethyl sulfoxide, as we do for a carbonyl group? In the case of the carbonyl group, we represent the structure by a double bond between carbon and oxygen, as in structure **1-12**.

1-12   1-13

In structure **1-12**, both carbon and oxygen have an octet and neither carbon nor oxygen has a charge, whereas in structure **1-13**, carbon does not have an octet and both carbon and oxygen carry a charge. Taken together, these factors make structure **1-12** more stable and therefore more likely. Looking at the analogous structures for dimethyl sulfoxide, we see that in structure **1-10** both atoms have an octet and both are charged, whereas in structure **1-11**, sulfur has 10 valence electrons, but both sulfur and oxygen are neutral. Thus, neither **1-10** nor **1-11** is clearly favored, and the structure of dimethyl sulfoxide is best represented by a combination of structures **1-10** and **1-11**.

Note: No hydrogen atoms are shown in structures **1-12** and **1-13**. In representing organic molecules, it is assumed that the valence requirements of carbon are satisfied by hydrogen unless otherwise specified. Thus, in structures **1-12** and **1-13**, it is understood that there are six hydrogen atoms, three on each carbon.

Also, to avoid possible confusion, when nitrogen or oxygen is bonded to hydrogen it is shown in the structure *explicitly*.

HINT 1.5

When the electron supply is an odd number, the resulting unpaired electron will produce a radical, that is, the valence shell of one atom, other than hydrogen, will not be completed. This atom will have seven electrons instead of eight. Thus, if you get a 1/2 when you calculate the number of bonds, it represents a radical in the final structure.

As a quick check it may be easier to check each atom individually to be sure that the octet rule is met. This can be a faster, yet reliable, method for identifying charges and placement of unpaired electrons.

PROBLEM 1.1

Write Lewis structures for each of the following and show any formal charges.

a. $CH_2{=}CHCHO$

b. $NO_2^+ BF_4^-$

c. Hexamethylphosphorous triamide, $[(CH_3)_2N]_3P$

d. $CH_3N(O)CH_3$

e. CH_3SOH (methylsulfenic acid)

Lewis structures for common functional groups are listed in Appendix A.

2. REPRESENTATIONS OF ORGANIC COMPOUNDS

As illustrated earlier, the bonds in organic structures are represented by lines. Often, some or all of the lone pairs of electrons are not represented in any way. The reader must fill them in when necessary. To organic chemists, the most important atoms that have lone pairs of electrons are those in groups VA, VIA, and VIIA of the periodic table: N, O, P, S, and the halogens. The lone pairs on these elements can be of critical concern when writing a reaction mechanism. Thus, you must remember that lone pairs may be present even if they are not shown in the structures as written. For example, the structure of anisole might be written with or without the lone pairs of electrons on oxygen:

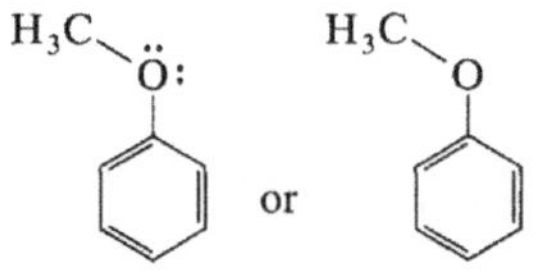

Other possible sources of confusion, as far as electron distribution is concerned, are ambiguities you may see in literature representations of cations and anions. The following illustrations show several representations of the resonance forms of the cation produced when anisole is protonated in the *para* position by concentrated sulfuric acid. There are three features to note in the first representation of the product, **1-14**. (1) Two lone pairs of electrons are shown on the oxygen. (2) The positive charge shown on carbon means that the carbon has one less electron than neutral carbon. The number of electrons on carbon = (6 shared electrons)/2 = 3, whereas neutral carbon has four electrons. (3) Both hydrogens are drawn in the *para* position to emphasize the fact that this carbon is now sp^3 hybridized. The second structure for the product, **1-15-1**, represents the overlap of one of the lone pairs of electrons on the oxygen with the rest of the π system. The electrons originally shown as a lone pair now are forming the second bond between oxygen and carbon. Representation **1-15-2**, the kind of structure commonly found in the literature, means exactly the same thing as **1-15-1**, but, *for simplicity, the lone pair on oxygen is not shown.*

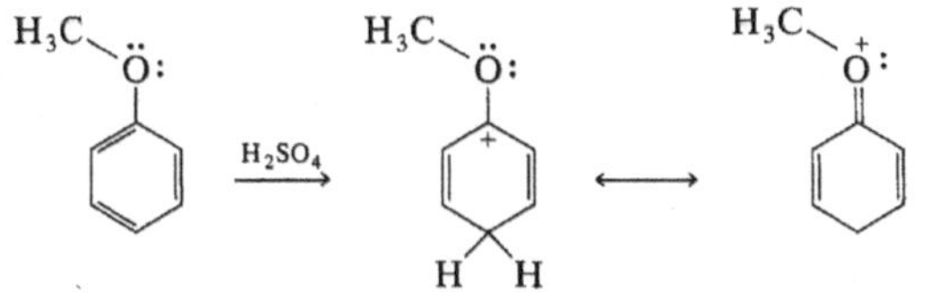

Similarly, there are several ways in which anions are represented. Sometimes a line represents a pair of electrons (as in bonds or lone pairs of electrons), sometimes a line represents a negative charge, and sometimes a line means both. The following structures represent the anion formed when a proton is removed from the oxygen of isopropyl alcohol.

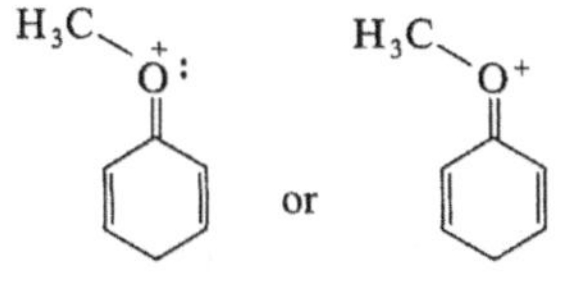

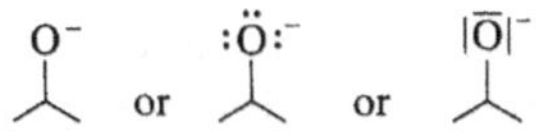

All three representations are equivalent, although the first two are the most commonly used. A compilation of symbols used in chemical notation appears in Appendix B.

3. GEOMETRY AND HYBRIDIZATION

Particular geometries (spatial orientations of atoms in a molecule) can be related to particular bonding patterns in molecules. These bonding patterns led to the concept of hybridization, which was derived from a mathematical model of bonding. In that model, mathematical functions (wave functions) for the s and p orbitals in the outermost electron shell are combined in various ways (hybridized) to produce geometries close to those deduced from experiment.

The designations for hybrid orbitals in bonding atoms are derived from the designations of the atomic orbitals of the isolated atoms. For example, in a molecule with an sp^3 carbon atom, the carbon has four sp^3 hybrid orbitals, which are derived from the combination of the one s orbital and three p orbitals in the free carbon atom. The number of hybrid orbitals is always the same as the number of atomic orbitals used to form the hybrids. Thus, combination of one s and three p orbitals produces four sp^3 orbitals, one s and two p orbitals produce three sp^2 orbitals, and one s and one p orbital produce two sp orbitals.

We will be most concerned with the hybridization of the elements C, N, O, P, and S, because these are the atoms, besides hydrogen, that are encountered most commonly in organic compounds. If we exclude situations where P and S have expanded octets, it is relatively simple to predict the hybridization of any of these common atoms in a molecule. By counting X, the number of atoms, and E, the number of lone pairs surrounding the atoms C, N, O, P, and S, the hybridization and geometry about the central atom can be determined by applying the principle of valence shell electron pair repulsion (VSEPR) to give the following:

1. If $X + E = 4$, the central atom will be sp^3 hybridized and the ideal geometry will have bond angles of 109.5°. *In exceptional cases, atoms with $X + E = 4$ may be sp^2 hybridized.* This occurs if sp^2 hybridization enables a lone pair to occupy a p orbital that overlaps a delocalized π electron system, as in the heteroatoms of structures **1-30** through **1-33** in Example 1.12.
2. If $X + E = 3$, the central atom will be sp^2 hybridized. There will be three hybrid orbitals and an unhybridized p orbital will remain. Again, the hybrid orbitals will be located as far apart as possible. This leads to an ideal geometry with 120° bond angles between the three coplanar hybrid orbitals and 90° between the hybrid orbitals and the remaining p orbital.
3. If $X + E = 2$, the central atom will be sp hybridized and two unhybridized p orbitals will remain. The hybrid orbitals will be linear (180° bond angles), and the p orbitals will be perpendicular to the linear system and perpendicular to each other.

The geometry and hybridization for compounds of second row elements are summarized in Table 1.1.

TABLE 1.1 Geometry and Hybridization in Carbon and Other Second Row Elements

Number of atoms + lone pairs $(X + E)$	Hybridization	Number of hybrid orbitals	Number of p orbitals	Geometry (bond angle)	Example[a]
4	sp^3	4	0	Tetrahedral (109.5°)	
3	sp^2	3	1	Planar (120°)	
2	sp	2	2	Linear (180°)	

[a]The geometry shown is predicted by VSEPR theory, in which orbitals containing valence electrons are directed so that the electrons are as far apart as possible. An asterisk indicates a hybridized atom.

EXAMPLE 1.6. THE HYBRIDIZATION AND GEOMETRY OF THE CARBON AND OXYGEN ATOMS IN 3-METHYL-2-CYCLOHEXEN-1-ONE

The oxygen atom contains two lone pairs of electrons, so $X + E = 3$. Thus, oxygen is sp^2 hybridized. Two of the sp^2 orbitals are occupied by the lone pairs of electrons. The third sp^2 orbital overlaps with a sp^2 hybridized orbital at C-2 to form the C—O σ bond. The lone pairs and C-2 lie in a plane approximately 120° from one another. There is a p orbital perpendicular to this plane.

C-2 is sp^2 hybridized. The three sp^2-hybridized orbitals overlap with orbitals on O-1, C-3, and C-7 to form three σ bonds that lie in the same plane approximately 120° from each other. The p orbital, perpendicular to this plane, is parallel to the p orbital on O-1 so these p orbitals can overlap to produce the C—O π bond.

Carbons 3, 4, 5, and 8 are sp^3 hybridized. (The presence of hydrogen atoms is assumed.) Bond angles are approximately 109.5°.

Carbons 6 and 7 are sp^2 hybridized. They are doubly bonded by a σ bond, produced from hybrid orbitals, and a π bond produced from their p orbitals.

Because of the geometrical constraints imposed by the sp^2-hybridized atoms, atoms 1, 2, 3, 5, 6, 7, and 8 all lie in the same plane.

PROBLEM 1.2

Discuss the hybridization and geometry for each of the atoms in the following molecules or intermediates.

a. $CH_3—C\equiv N$
b. $PhN{=}C{=}S$
c. $(CH_3)_3P$

4. ELECTRONEGATIVITIES AND DIPOLES

Many organic reactions depend on the interaction of a molecule that has a positive or fractional positive charge with a molecule that has a negative or fractional negative charge. In neutral organic molecules, the existence of a fractional charge can be inferred from the difference in electronegativity, if any, between the atoms at the ends of a bond. A useful scale of relative electronegativities was established by Linus Pauling. These values are given in Table 1.2, which also reflects the relative position of the elements in the periodic table.

TABLE 1.2 Relative Values for Electronegativities[a]

H	B	C	N	O	F
2.1	**2.0**	**2.5**	**3.0**	**3.5**	**4.0**
2.53	2.2	2.75	3.19	3.65	4.00
	Al	**Si**	**P**	**S**	**Cl**
	1.5	**1.8**	**2.1**	**2.5**	**3.0**
	1.71	2.14	2.5	3.96	3.48
					Br
					2.8
					3.22
					I
					2.5
					2.78

[a]*The boldface values are those given by Linus Pauling in* The Nature of the Chemical Bond, *3rd ed.; Cornell University Press: Ithaca, NY, 1960. p. 93. The second set of values is from Sanderson, (1983); J. Chem. Educ.* **1983**, *65, 112.*

The larger the electronegativity value, the more electron attracting the element. Thus, fluorine is the most electronegative element shown in the table.

HINT 1.6

Carbon, phosphorus, and iodine have about the same electronegativity. Within a row of the periodic table, the electronegativity increases from left to right. Within a column of the periodic table, electronegativity increases from bottom to top.

From the relative electronegativities of the atoms, the relative fractional charges can be ascertained for bonds.

EXAMPLE 1.7. RELATIVE DIPOLES IN SOME COMMON BONDS

$$\overset{\delta^+ \ \ \delta^-}{C-O} \quad \overset{\delta^+ \ \ \delta^-}{C=O} \quad \overset{\delta^+ \ \ \delta^-}{C-Br} \quad \overset{\delta^+ \ \ \delta^-}{C-N}$$

In all cases the more electronegative element has the fractional negative charge. There will be more fractional charge in the second structure than in the first, because the π electrons in the second structure are held less tightly by the atoms and thus are more mobile. The C—Br bond is expected to have a weaker dipole than the C—O single bond because bromine is not as electronegative as oxygen. You will notice that the situation is not so clear if we are comparing the polarity of the C—Br and C—N bonds. The Pauling scale would suggest that the C—N bond is more polar than the C—Br bond, whereas the Sanderson scale would predict the reverse. Thus, although attempts have been made to establish quantitative electronegativity scales, electronegativity is, at best, a qualitative guide to bond polarity.

PROBLEM 1.3

Predict the direction of the dipole in the bonds highlighted in the following structures.

a. 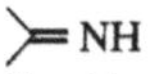 $\rangle$=NH
b. Br—F
c. CH_3—$N(CH_3)_2$
d. CH_3—$P(CH_3)_2$

---

5. RESONANCE STRUCTURES

When the distribution of valence electrons in a molecule cannot be represented adequately by a single Lewis structure, the structure can be approximated by a combination of Lewis structures that differ only in the placement of electrons. Lewis structures that differ only in the placement of electrons are called resonance structures. We use resonance structures to show the delocalization of electrons and to help predict the most likely electron distribution in a molecule.

A. Drawing Resonance Structures

A simple method for finding the resonance structures for a given compound or intermediate is to draw one of the resonance structures and then, by using arrows to show the movement of electrons and draw a new structure with a different electron distribution. This movement of electrons is formal only; that is, no such electron flow actually takes place in the molecule. The actual molecule is a hybrid of the resonance structures that incorporates some of the characteristics of each resonance structure. Thus, resonance structures themselves are not structures of actual molecules or intermediates but are a formality that help to predict the electron distribution for the real structures. *Resonance structures, and only resonance structures, are separated by a double-headed arrow.*

Note: Chemists commonly use the following types of arrows:

• A double-headed arrow links two *resonance* structures

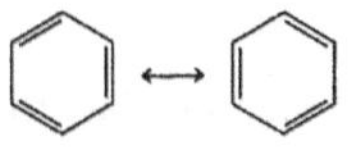

• Two half-headed arrows indicate an *equilibrium*

- A curved arrow indicates the movement of an *electron pair* in the direction of the arrowhead

$$H-\overset{..}{\underset{..}{O}}:^{\frown}+ H-\overset{..}{\underset{..}{CH_2CH}} \overset{:\overset{..}{O}}{\underset{\parallel}{}} \;\rightleftharpoons\; H-\overset{..}{\underset{..}{O}}: + :CH_2CH \overset{\overset{..}{O}:}{\underset{\parallel}{}}$$

- A curved half-headed arrow indicates the movement of a *single electron* in the direction of the arrowhead

$$(CH_3)_3C-\overset{..}{\underset{..}{O}}-\overset{..}{\underset{..}{O}}-C(CH_3)_3 \;\longrightarrow\; (CH_3)_3C-\overset{..}{\underset{..}{O}}\cdot + \cdot\overset{..}{\underset{..}{O}}-C(CH_3)_3$$

Chapter 5 describes this and other one-electron processes.

A summary of symbols used in chemical notation appears in Appendix B.

EXAMPLE 1.8. WRITE THE RESONANCE STRUCTURES FOR NAPHTHALENE

First, draw a structure, **1-16**, for naphthalene that shows alternating single and double bonds around the periphery. This is one of the resonance structures that contribute to the character of delocalized naphthalene, a resonance hybrid.

1-16   1-17

Each arrow drawn within **1-16** indicates movement of the π electron pair of a double bond to the location shown by the head of the arrow. This gives a new structure, **1-17,** which can then be manipulated in a similar manner to give a third structure, **1-18.**

1-17   1-18

Finally, when the forms have been figured out, they can be presented in the following manner:

How do you know that all possible resonance forms have been written? This is accomplished only by trial and error. If you keep pushing electrons around the naphthalene ring, you will continue to draw structures, but they will be identical to one of the three previously written.

What are some of the pitfalls of this method? If only a single electron pair in **1-17** is moved, **1-19** is obtained. However, this structure does not make sense. At the carbon labeled 1, there are five bonds to carbon; this is a carbon with 10 electrons. However, it is not possible to expand the valence shell of carbon. Similar rearrangement of other π bonds in **1-16, 1-17,** or **1-18** would lead to similarly nonsensical structures.

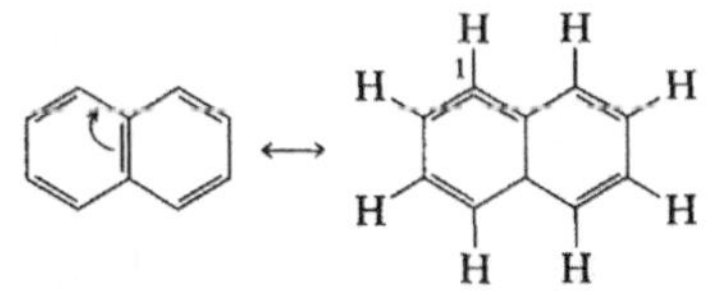

A second possibility would be to move the electrons of a double bond to just one of the terminal carbons; this leads to a structure like **1-20**. However, when more than one neutral resonance structure can be written, doubly charged resonance structures, like **1-20** and **1-21**, contribute an insignificant amount to the resonance hybrid and are usually not written.

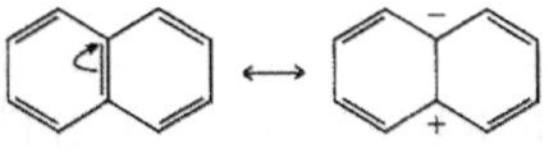

EXAMPLE 1.9. WRITE RESONANCE FORMS FOR THE INTERMEDIATE IN THE NITRATION OF ANISOLE AT THE PARA POSITION

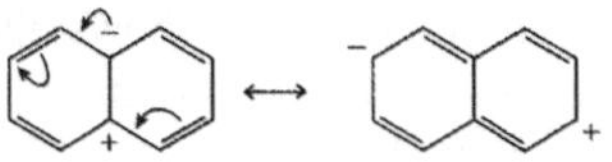

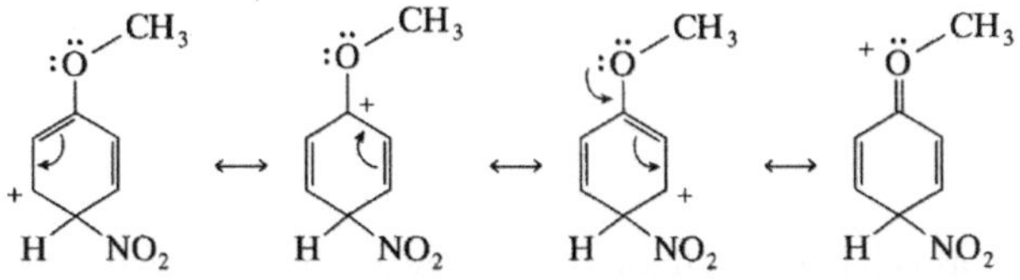

There are actually twice as many resonance forms as those shown because the nitro group is also capable of electron delocalization. Thus, for each resonance form written previously, two resonance forms can be substituted in which the nitro group's electron distribution has been written out as well:

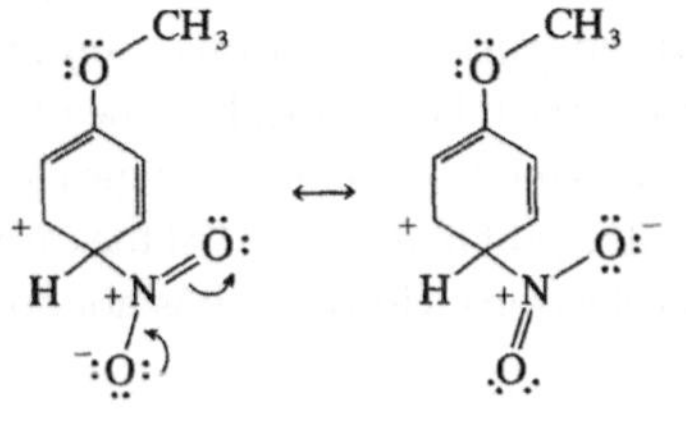

Because the nitro group is attached to an sp^3-hybridized carbon, it is not conjugated with the electrons in the ring and is not important to their delocalization. Thus, if resonance forms were being written to rationalize the stability of the intermediate in the nitration of anisole, the detail in the nitro groups would not be important because it does not contribute to the stabilization of the carbocation intermediate.

Note: When an atom in a structure is shown with a negative charge, this is usually taken to imply the presence of an electron pair; often, a pair of electrons and a negative sign are used interchangeably (see Section 2). This can sometimes be confusing. For example, the cyclooctatetraenyl anion (Problem 1.4e) can be depicted in several ways:

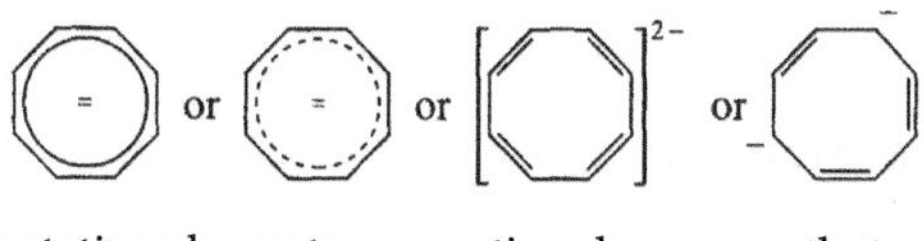

Notice that every representation shows two negative charges, so that we can be sure of the fact that this is a species with a double negative charge. In general, a negative charge sign drawn next to an atom indicates the presence of an electron pair associated with that atom. For some of the representations of the cyclooctatetraenyl anion, however, it is not clear how many electrons are in the π system (there is no ambiguity about the electrons in the σ bonds). In a situation like this, there is no hard and fast rule about how to count the electrons, based on the structural representation. To reach more solid ground, you need to know that cyclooctatetraene forms a relatively stable aromatic dianion with 10 π electrons (see Section 6). Fortunately, these ambiguous situations are not common.

PROBLEM 1.4

Draw resonance structures for each of the following.

a. Anthracene

b.

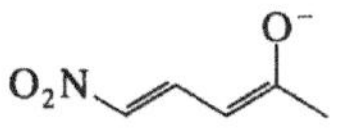

c. PhCH$_2^+$

d.

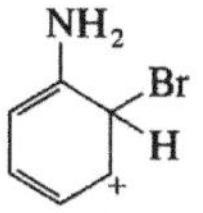

e.

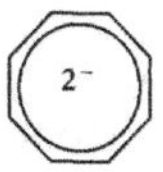

f.

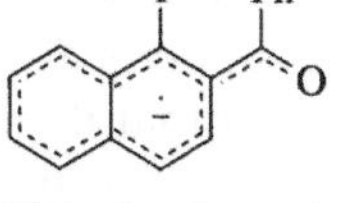

This is the anion radical of 1-iodo-2-benzoylnaphthalene. The dashed lines indicate a delocalized π system. The symbol "Ph" stands for a phenyl group.

PROBLEM 1.5

Either *p*-dinitrobenzene or *m*-dinitrobenzene is commonly used as a radical trap in electron transfer reactions. The compound that forms the most stable radical anion is the better trap. Consider the radical anions formed when either of these starting materials adds an electron and predict which compound is commonly used.

B. Rules for Resonance Structures

1. All the electrons involved in delocalization are π electrons or, like lone pairs, they can readily be put into *p* orbitals.
2. Each of the electrons involved in delocalization must have some overlap with the other electrons. This means that if the orbitals are oriented at a 90° angle, there will be no overlap. Generally, better overlap is afforded as the orbital alignment approaches a 0° angle.
3. Each resonance structure must have the same number of π electrons. Count two for each π bond; only two electrons are counted for a triple bond because only one of the π bonds of a triple bond can overlap with the conjugated π system. Also, when a π system carries a charge, count two for an anion and zero for a positive charge.
4. The same number of electrons must be paired in each structure. Structures **1-22** and **1-23** are not resonance structures because they do not have the same number of paired electrons. In **1-22,** there are two pairs of π electrons: a pair of electrons for the π bond and a pair of electrons for the anion. In **1-23,** there is one pair of π electrons and two unpaired electrons (shown by the dots).

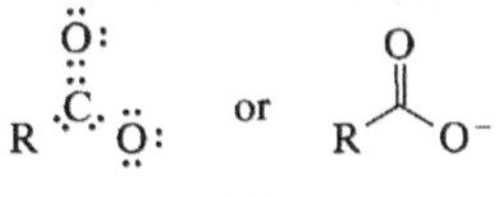

1-22

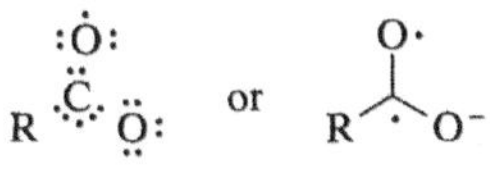

1-23

5. All resonance structures must have identical geometries. Otherwise they do not represent the same molecule. For example, the following structure (known as Dewar benzene) is not a resonance form of benzene because it is not planar and has two less π electrons. Because molecular geometry is linked to hybridization, it follows that hybridization is also unchanged for the atoms in resonance structures. (**Note:** If it is assumed that the central bond in this structure is a π bond, then it has the same number of electrons as benzene. However, in order for the *p* orbitals to overlap, the central carbon atoms would have to be

much closer than they are in benzene, and this is yet another reason why Dewar benzene is an isolable compound rather than a resonance form of benzene.)

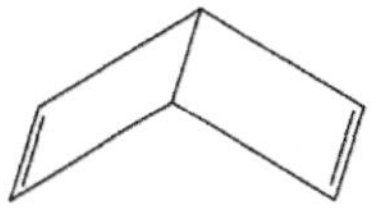

1-24

6. Resonance structures that depend on charge separation are of higher energy and do not contribute as significantly to the resonance hybrid as those structures that do not depend on charge separation.

 much more important than

 much more important than

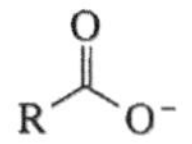

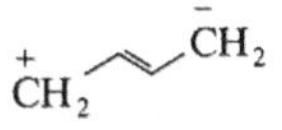

7. Usually, resonance structures are more important when the negative charge is on the most electronegative atom and the positive charge is on the most electropositive atom.

 contributes more than

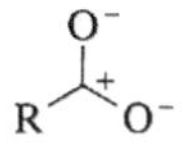

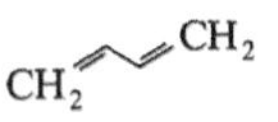

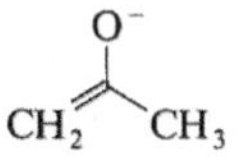

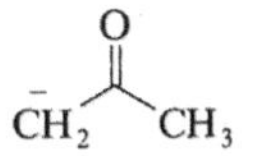

In some cases, aromatic anions or cations are exceptions to this rule (see Section 6).

In the example that follows, **1-26** is less favorable than **1-25**, because the more electronegative atom in **1-26**, oxygen, is positive. In other words, although neither the positive carbon in **1-25** nor the positive oxygen in **1-26** has an octet, it is especially destabilizing when the much more electronegative oxygen bears the positive charge.

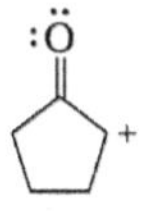

 contributes more than

1-25 **1-26**

Electron stabilization is greatest when there are two or more structures of lowest energy. The resonance hybrid is more stable than any of the contributing structures.

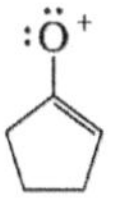

PROBLEM 1.6

a. Do you think delocalization as shown by the following resonance structures is important? Explain why or why not.

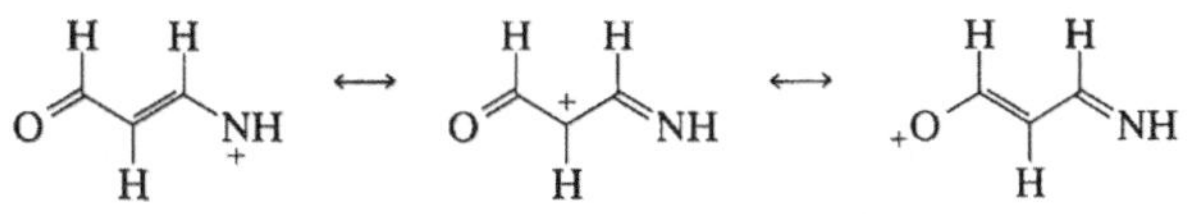

b. If the charges were negative instead of positive, would your answer be different? Explain.

PROBLEM 1.7

Write Lewis structures for each of the following and show any formal charges. Also, draw all resonance forms for these species.

a. CH_3NO_2
b. PhN_2^+
c. CH_3COCHN_2 (diazoacetone)
d. N_3CN
e. $CH_2 = CHCH_2^- Li^+$

6. AROMATICITY AND ANTIAROMATICITY

A. Aromatic Carbocydes

Certain cyclic, completely conjugated, π systems show unusual stability. These systems are said to be aromatic. Hückel originally narrowly defined aromatic compounds as those completely conjugated, monocyclic carbon compounds that contain $(4n + 2)$ π electrons. In this designation, n can be 0, 1, 2, 3, ..., so that systems that contain 2, 6, 10, 14, 18, ..., π electrons are aromatic. This criterion is known as Hückel's rule.

EXAMPLE 1.10. SOME AROMATIC COMPOUNDS THAT STRICTLY OBEY HÜCKEL'S RULE

In these systems, each double bond contributes two electrons, each positive charge on carbon contributes none, and the negative charge (designation of an anion) contributes two electrons. If the first and last two structures did not have a charge on the singly bonded atom, they would not be aromatic because the π system could not be completely delocalized. That is, if the cyclopropenyl ring is depicted as uncharged, then there are two hydrogens on the carbon with no double bond. This carbon is then sp^3 hybridized and has no p orbital available to complete a delocalized system.

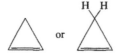

Hückel's rule has been expanded to cover fused polycyclic compounds because when these compounds have the requisite number of electrons, they also show unusual stability.

EXAMPLE 1.11. SOME AROMATIC FUSED RING SYSTEMS

1-27   1-28   1-29

Structures **1-27** and **1-28**, which contain 10 conjugated π electrons, are examples of Hückel's rule with $n = 2$. Structure **1-29** obeys Hückel's rule with $n = 4$. In fused ring systems, only the electrons located at the periphery of the structure are counted when Hückel's rule is applied (see Problem 1.8e for an example).

B. Aromatic Heterocycles

Hückel's rule can also be extended to heterocycles, ring systems that incorporate noncarbon atoms. In heterocyclic compounds, a lone pair of electrons on the heterocyclic atom may be counted as part of the conjugated π system to attain the correct number for an aromatic system.

EXAMPLE 1.12. SOME AROMATIC HETEROCYCLES

1-30   1-31   1-32   1-33   1-34

These examples illustrate how the lone pairs of electrons are considered in determining aromaticity. In each of the examples, the carbons and the heteroatoms are sp^2 hybridized, ensuring a planar system with a p orbital perpendicular to this plane at each position in the ring. In **1-30**, two electrons must be contributed by the nitrogen to give a total of six π electrons. Thus, the lone pair of electrons would be in the p orbital on the nitrogen. In **1-31**, one of the lone pairs of electrons on the oxygen is also in a p orbital parallel with the rest of the π system in order to give a six π electron aromatic system. Thus, the other lone pair on oxygen must be in an sp^2-hybridized orbital, which, by definition, is perpendicular to the conjugated π system and therefore cannot contribute to the number of electrons in the overlapping π system. The considerations concerning the sulfur in **1-32** are identical to those for oxygen in **1-31**, so this is a six π electron system. Compound **1-33** is similar to **1-30** with overlap of the additional fused six-membered ring, making this an aromatic 10 π electron system. In sharp contrast to the other examples, a totally conjugated six π electron system is formed in **1-34** with the contribution of only one electron from the nitrogen. The lone pair of electrons will then be in an sp^2-hybridized orbital perpendicular to the aromatic six π electron system. Thus, these two electrons are not part of the delocalized system (compare depictions **1-30a** and **1-34a**). In conclusion, *the heteroatom is hybridized in such a way that one or two of its electrons can become part of an aromatic system.*

EXAMPLE 1.13. A DIFFERENT VIEW OF THE AROMATIC HETEROCYCLES

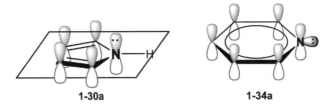

1-30a   1-34a

These views of the pyrrole and pyridine show the p orbitals. The pyrrole nitrogen p orbital is perpendicular to the plane and aligned with the other p orbitals and is thus able to participate in the aromatic system. For the pyridine, one orbital is again perpendicular to the plane and aligned with the other p orbitals and participates in the aromatic system. The second orbital on the nitrogen, representing the nitrogen lone pair, is in the plane of the ring atoms and is not aligned to participate in the aromatic system.

C. Antiaromaticity

Are conjugated systems that contain $4n$ (4, 8, 12, 16, …) π electrons also aromatic? These systems actually are destabilized by delocalization and are said to be antiaromatic.

EXAMPLE 1.14. SOME ANTIAROMATIC SYSTEMS

The first three examples contain four π electrons, whereas the last one contains eight π electrons. All are highly unstable species. On the other hand, cyclooctatetraene, **1-35,** an eight π electron system, is much more stable than any of the preceding compounds. This is because the π electrons in cyclooctatetraene are not delocalized significantly: the eight-membered ring is bent into a tublike structure and adjacent π bonds are not parallel.

1-35

PROBLEM 1.8

Classify each of the following compounds as aromatic, antiaromatic, or nonaromatic.

a.

b.

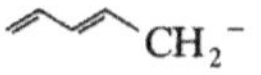

c.

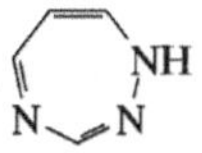

d.

e.

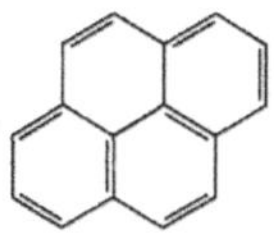

f.

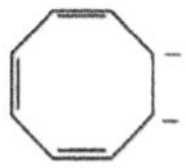

g.

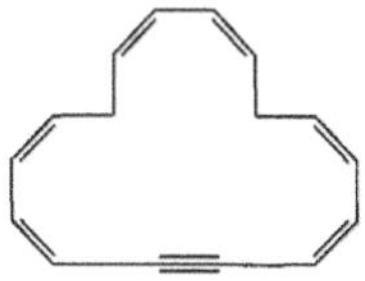

7. TAUTOMERS AND EQUILIBRIUM

Tautomers are isomers that differ in the arrangement of single and double bonds and a small atom, usually hydrogen. Under appropriate reaction conditions, such isomers can equilibrate by a simple mechanism.

Equilibrium exists when there are equal rates for both the forward and reverse processes of a reaction. Equilibrium usually is designated by half-headed arrows shown for both the forward and reverse reactions. If it is known that one side of the equilibrium is favored, this may be indicated by a longer arrow pointing to the side that is favored.

EXAMPLE 1.15. AN ACID−BASE EQUILIBRIUM

$$CH_3CO_2H + H_2O \rightleftharpoons CH_3CO_2^- + H_3O^+$$

EXAMPLE 1.16. TAUTOMERIC EQUILIBRIA OF KETONES

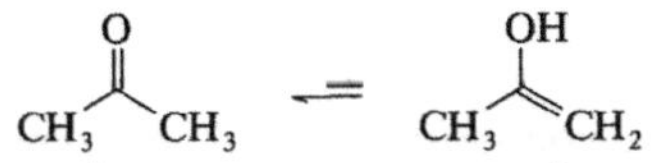

The keto and enol forms of aldehydes and ketones represent a common example of tautomerism. The tautomers interconvert by an equilibrium process that involves the transfer of a hydrogen atom from oxygen to carbon and back again.

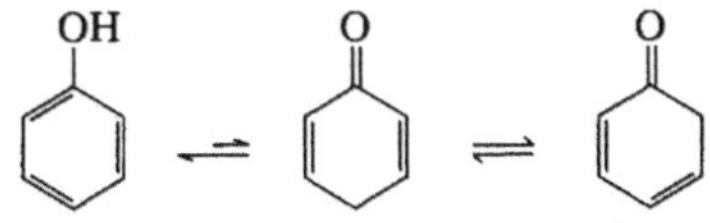

HINT 1.7

To avoid confusing resonance structures and tautomers, use the following criteria:

1. *Tautomers* are readily converted *isomers*. As such they differ in the placement of a double bond and a hydrogen atom. The equilibration between the isomers is shown with a pair of half-headed arrows.
2. *Resonance structures* represent *different π bonding patterns,* not different chemical species. Different resonance structures are indicated by a double-headed arrow between them.
3. All resonance structures for a given species have identical σ bonding patterns (with a few unusual exceptions) and identical geometries. In tautomers, the σ bonding pattern differs.

EXAMPLE 1.17. TAUTOMERISM VS RESONANCE

Compounds **1-36** and **1-37** are tautomers; they are isomers and are in equilibrium with each other.

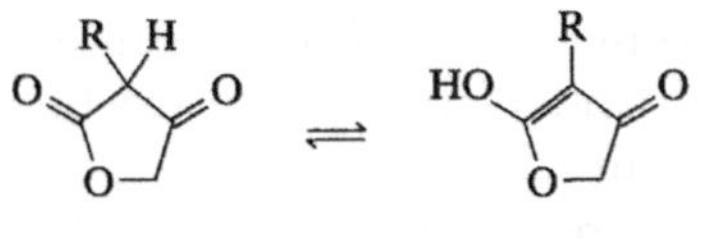

On the other hand, **1-38, 1-39,** and **1-40** are resonance forms. The hybrid of these structures can be formed from **1-36** or **1-37** by removing the acidic proton.

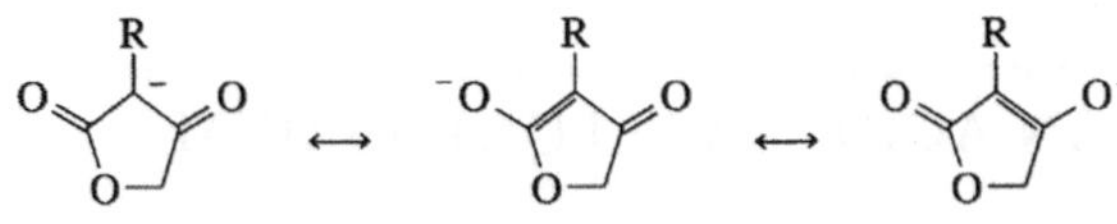

Note that **1-38, 1-39,** and **1-40** have the same atoms attached at all positions, whereas the tautomers **1-36** and **1-37** differ in the position of a proton.

PROBLEM 1.9

Write tautomeric structures for each of the following compounds. The number of tautomers you should write, in addition to the original structure, is shown in parentheses.

a.

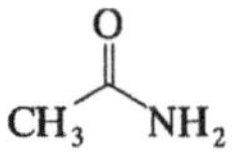

$$\underset{CH_3 \quad NH_2}{\overset{O}{\parallel}}$$ (2)

b. CH_3CHO (1)

c. $CH_3CH{=}CHCHO$ (1)

d.

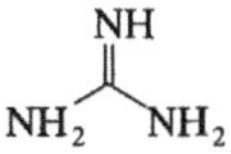

$$\underset{NH_2 \quad NH_2}{\overset{NH}{\parallel}}$$ (2)

e. 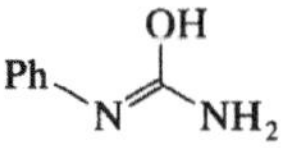 Ph$\underset{N}{\diagdown}\overset{OH}{\diagup}NH_2$ (5)

PROBLEM 1.10

For each of the following sets of structures, indicate whether they are tautomers, resonance forms, or the same molecule.

a.

b.

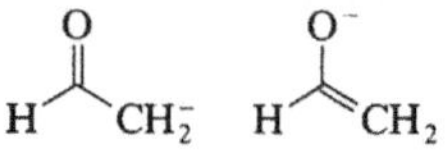

$$\underset{H \quad CH_2^-}{\overset{O}{\parallel}} \quad \underset{H \quad CH_2}{\overset{O^-}{\parallel}}$$

c.

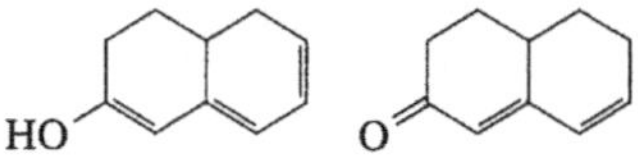

PROBLEM 1.11

In a published paper, two structures were presented in the following manner and referred to as resonance forms. Are the structures shown actually resonance forms? If not, what are they and how can you correct the picture?

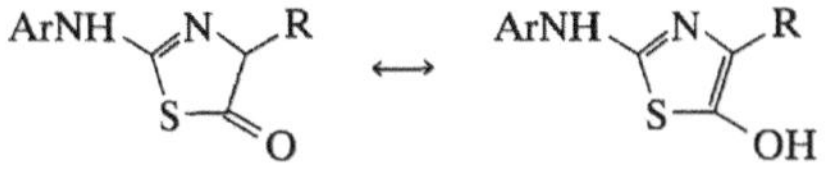

---

8. ACIDITY AND BASICITY

A Brønsted acid is a proton donor. A Brønsted base is a proton acceptor.

$$CH_3CO_2H + CH_3NH_2 \rightleftharpoons CH_3CO_2^- + CH_3\overset{+}{N}H_3$$
$$\text{acid} \qquad \text{base} \qquad \text{conjugated base} \quad \text{conjugated acid}$$

If this equation were reversed, the definitions would be similar:

$$CH_3\overset{+}{N}H_3 + CH_3CO_2^- \rightleftharpoons CH_3NH_2 + CH_3CO_2H$$
$$\text{acid} \qquad \text{base} \qquad \text{conjugated base} \quad \text{conjugated acid}$$

In each equation, the acids are the proton donors and the bases are proton acceptors.

There is an inverse relationship between the acidity of an acid and the basicity of its conjugate base. That is, the more acidic the acid, the weaker the basicity of the conjugate base, and vice versa. For example, if the acid is very weak, like methane, the conjugate base, the methyl carbanion, is a very strong base. On the other hand, if the acid is very strong, like sulfuric acid, the conjugate base, the HSO_4^- ion, is a very weak base. Because of this reciprocal relationship between acidity and basicity, most references to acidity and basicity use a single scale of pK_a values, and relative basicities are obtained from the relative acidities of the conjugate acids.

Table 1.3 shows the approximate pK_a values for common functional groups. The lower the pK_a value, the more acidic the protonated species. Any conjugate acid in the table will protonate a species lying below it and will be protonated by acids listed above it. Thus, a halogen acid (pK_a −10 to −8) will protonate any species listed in this table, whereas a carboxylic acid (pK_a of 4–5) would be expected to protonate aliphatic amines (pK_a values of 9–11), but not primary and secondary aromatic amines (pK_a values of −5 and 1, respectively).

Appendix C contains a more detailed list of pK_a values for a variety of acids. Especially at very high pK_a values, the numbers may be inaccurate because various approximations have to be made in measuring such values. This is often why the literature contains different pK_a values for the same acid.

TABLE 1.3 Typical Acidities of Common Organic and Inorganic Substances

Group	Conjugate acid[a]	Conjugate base[a]	Typical pK_a
Halogen acids	HX (X=I, Br, Cl)	X⁻	−10 to −8
Nitrile	$R-C\equiv\overset{+}{N}-H$	$RC\equiv N$	−10
Aldehyde, ketone	$R-\overset{\overset{+}{O}-H}{\underset{\|}{C}}-R'$	$R-\overset{O}{\underset{\|}{C}}-R'$	−7
Thiol, sulfide	$R-\overset{H}{\underset{+}{S}}-R'$	$R-S-R'$	−7 to −6
Phenol, aromatic ether	Ph$-\overset{H}{\underset{+}{O}}-R'$	Ph$-O-R'$	−7 to −6
Ester, acid	$R-\overset{\overset{+}{O}-H}{\underset{\|}{C}}-OR'$	$R-\overset{\overset{+}{O}-H}{\underset{\|}{C}}-OR'$	−7 to −6
Sulfonic acid[b,c]	$R-\overset{O}{\underset{O}{\overset{\|}{\underset{\|}{S}}}}-O-H$	$R-\overset{O}{\underset{O}{\overset{\|}{\underset{\|}{S}}}}-O^-$	−5 to −2
Alcohol, ether	$R-\overset{H}{\underset{+}{O}}-R'$	$R-O-R'$	−3 to −2
H_2O	$H_2\overset{+}{O}-H$	H_2O	−1.7
Amide	$R-\overset{\overset{+}{O}-H}{\underset{\|}{C}}-NR_2$	$R-\overset{O}{\underset{\|}{C}}-NR_2$	−1
Carboxylic acid[b,c]	$R-\overset{O}{\underset{\|}{C}}-O-H$	$R-\overset{O}{\underset{\|}{C}}-O^-$	3–5
Aromatic amine	Ph$-\overset{H}{\underset{+}{N}}-R'_2$	Ph$-NR'_2$[d]	4–5
Pyridine	pyridine-$\overset{+}{N}-H$	pyridine[d]	5
Alkylamine	$R-\overset{H}{\underset{+}{N}}-R'_2$	$R-N-R'^{[d]}_2$	9–11
Phenol[c]	Ph$-O-H$	Ph$-O^-$	9–11
Thiol[c]	$R-S-H$	$R-S^-$	9–11

(Continued)

TABLE 1.3   Typical Acidities of Common Organic and Inorganic Substances—cont'd

Group	Conjugate acid[a]	Conjugate base[a]	Typical pK_a
Sulfonamide[c]	$R-\overset{\overset{O}{\|\|}}{\underset{\overset{\|\|}{O}}{S}}-\overset{H}{\underset{}{N}}-H$	$R-\overset{\overset{O}{\|\|}}{\underset{\overset{\|\|}{O}}{S}}-\overset{-}{N}-H$	10
Nitro[c]	$R-\overset{H}{\underset{}{C}H}-NO_2$	$R-\overset{-}{C}H-NO_2$	10
Amide	$R-\overset{\overset{O}{\|\|}}{C}-\overset{H}{\underset{}{N}}-R$	$R-\overset{\overset{O}{\|\|}}{C}-\overset{-}{N}-R$	15–17
Alcohol	$R-O-H$	$R-O^-$	15–19
Aldehyde, ketone	$R-\overset{H}{\underset{R'}{C}}-\overset{O}{\overset{\|\|}{C}}-R'$	$R'-\overset{-}{\underset{R}{C}}-\overset{O}{\overset{\|\|}{C}}-R'$	17–20
Ester	$R-\overset{H}{\underset{R'}{C}}-\overset{O}{\overset{\|\|}{C}}-OR$	$R-\overset{-}{\underset{R'}{C}}-\overset{O}{\overset{\|\|}{C}}-OR$	20–25
Nitrile	$R-\overset{H}{\underset{R'}{C}}-C\equiv N$	$R-\overset{-}{\underset{R'}{C}}-C\equiv N$	25
Alkyne	$R-C\equiv C-H$	$R-C\equiv C^-$	25
Amine	R_2N-H	R_2N^-	35–40
Alkane	R_3C-H	R_3C^-	50–60

[a]Abbreviations: R = alkyl; R' = alkyl or H.
[b]Dissolves in 5% $NaHCO_3$ solution.
[c]Dissolves in 5% NaOH solution.
[d]Dissolves in 5% HCl solution.

HINT 1.8

It is helpful to think of the pK_a as the pH of a solution in which the acid is 50% ionized.

Weak acids with a high pK_a require a high pH (strong alkali) to lose their protons. Acids with a pK_a greater than 15.7 cannot be deprotonated in aqueous solution; those with a pK_a less than 15.7 will be soluble (to some extent) in 5% NaOH solution. (The value 15.7 is used because we are considering the pK_a of water. The pH scale is based on the ion product of water, $[H^+][OH^-]$, which equals 10^{-14}. The pK_a value is based on an H_2O concentration of 55 mol/l.) Bases whose conjugate acids have a pK_a less than 0 (i.e., negative) will not be protonated in aqueous solution because water, being a stronger base, is protonated preferentially. Bases whose conjugate acids have a positive pK_a are basic enough to dissolve (to some extent) in 5% HCl.

PROBLEM 1.12

For each of the following pairs, indicate which is the strongest base. For b and d use resonance structures to rationalize the relative basicities.

a. $H\overline{C}{=}CH_2$, $\overline{C}{\equiv}CH$

b. $\overline{C}H_2CON(Et)_2$, $[(CH_3)_2CH]_2\overline{N}$

c. CH_3O^-, $(CH_3)CO^-$

d. *p*-nitrophenolate, *m*-nitrophenolate

PROBLEM 1.13

For each of the following compounds, indicate which proton is more likely to be removed when the compound is treated with base and rationalize your answer. Assume that equilibria are involved in each case.

a. $CH_3COCH_2COCH_3$

b. $H_2NCH_2CH_2OH$

c.

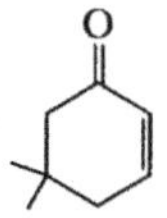

d.

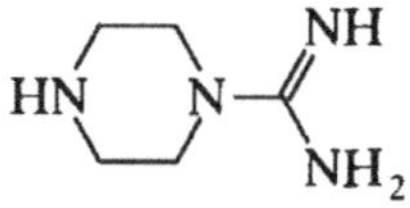

EXAMPLE 1.18. CALCULATING THE EQUILIBRIUM CONSTANT FOR AN ACID–BASE REACTION IN ORDER TO PREDICT WHETHER THE REACTION IS LIKELY TO PROCEED

$$Br^- + EtOH \rightleftharpoons HBr^- + EtO^-$$

For this reaction, the equilibrium constant is

$$K_a = \frac{[HBr][EtO^-]}{[Br^-][EtOH]}$$

Using the values listed in Appendix C, this equilibrium constant can be calculated by an appropriate combination of the equilibrium constant for the ionization of ethanol and the equilibrium constant for the ionization of HBr. The equilibrium constant for the ionization of ethanol is

$$10^{-15.9} = \frac{[EtO^-][H^+]}{[EtOH]}$$

The equilibrium constant for the ionization of HBr is

$$10^9 = \frac{[Br^-][H^+]}{[HBr]}$$

If the equilibrium constant for ethanol ionization is divided by that for HBr ionization, the equilibrium constant for the reaction of bromide ion with ethanol is obtained:

$$K_a = \frac{10^{-15.9}}{10^9} = \frac{[HBr][EtO^-][H^+]}{[H^+][Br^-][EtOH]} = \frac{[HBr][EtO^-]}{[Br^-][EtOH]}$$

K_a equals $10^{-24.9}$. Thus, it can be concluded that bromide ion does not react with ethanol.

PROBLEM 1.14

Using the values listed in Appendix C, calculate the equilibrium constants for each of the following reactions and predict which direction, forward or reverse, is favored.

a. $CH_3CO_2Et + EtO^- \rightleftharpoons CH_2CO_2Et + EtOH$

b.

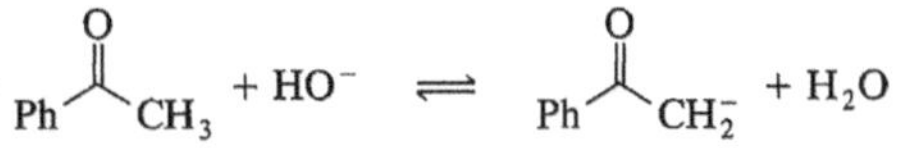

$$\underset{Ph}{\overset{O}{\|}}\!\!\!\!\!\!\!\!\! \underset{CH_3}{\quad} + HO^- \rightleftharpoons \underset{Ph}{\overset{O}{\|}}\!\!\!\!\!\!\!\!\! \underset{CH_2^-}{\quad} + H_2O$$

c. $PH_3CH + {}^-N(i-Pr)_2 \rightleftharpoons PH_3C^- + HN(i-Pr)_2$

Note: $i\text{-Pr} = -CH(CH_3)_2$.

9. NUCLEOPHILES AND ELECTROPHILES

Nucleophiles are reactive species that seek an electron-poor center. They have an atom with a negative or partial negative charge, and this atom is referred to as the nucleophilic atom. Reacting species that have an electron-poor center are called electrophiles. These electron-poor centers usually have a positive or partial positive charge, but electron-deficient species can also be neutral (radicals and carbenes, see Chapter 5). Table 1.4 lists common nucleophiles and Table 1.5 lists common electrophiles.

Reactivity in reactions involving nucleophiles depends on several factors, including the nature of the nucleophile, the substrate, and the solvent.

TABLE 1.4   Common Nucleophiles[a]

F^-, Cl^-, Br^-, I^-

$ROH, RO^-, RO{-}O^-$

$[R{-}\overset{O}{\underset{}{C}}{\sim}\overset{\bullet}{O}]^-$, $[Ar{-}\overset{O}{\underset{}{C}}{\sim}\overset{\bullet}{O}]^-$

$RSH, RS^-, ArSH, ArS^-$

$RSR', [\overset{\bullet\bullet}{S}{-}\overset{\bullet\bullet}{S}]^{2-}$ (disulfide)

$RNH_2, RR'NH, RR'R''N, ArNH_2$, etc.

H_2NNH_2 (hydrazine)

$[\overset{\bullet\bullet}{N}{=}N{=}\overset{\bullet\bullet}{N}]^-$ (azide)

$[\overset{\bullet}{N}{=}C{=}O]^-$ (isocyanate)

(phthalimidate)

$R_3\overset{\bullet}{P}$ (phosphine)

$(RO)_3\overset{\bullet}{P}$ (phosphite)

$Li\overset{\bullet}{Al}H_4, Na\overset{\bullet}{B}H_4, LiEt_3\overset{\bullet}{B}H$

$RMgX, ArMgX$

$RCuLi$

$RC{\equiv}C^-$

---

[a]*Asterisk indicates a nucleophilic atom.*

A. Nucleophilicity

Nucleophilicity measures the ability of a nucleophile to react at an electron-deficient center. It should not be confused with basicity, although often there are parallels between the two. Whereas nucleophilicity considers the *reactivity* (i.e., the rate of reaction) of an electron-rich species *at an electron-deficient center* (usually carbon), basicity is a measure of the *position of equilibrium* in reaction with a *proton*.

Table 1.6 shows nucleophiles ranked by one measure of nucleophilicity. These nucleophilicities are based on the relative reactivities of the nucleophile and water with methyl bromide at 25 °C. The nucleophilicity n is calculated according to the Swain–Scott equation:

$$\log\frac{k}{k_o} = sn$$

where k is the rate constant for reactions with the nucleophile, k_o is the rate constant for reaction when water is the nucleophile, $s = 1.00$ (for methyl bromide as substrate), and n is the relative nucleophilicity. The larger the n value, the greater the nucleophilicity.

TABLE 1.5   Common Electrophiles[a]

$\overset{\cdot\cdot}{Zn}Cl_2$, $\overset{\cdot}{Al}Cl_3$, $\overset{\cdot}{B}F_3$

$\overset{\cdot}{P}Br_3$

$\overset{\cdot}{S}OCl_2$

$-\overset{|}{\underset{|}{\overset{\cdot}{C}}}-X \; (X = Cl, Br, I)$

$-\overset{|}{\underset{|}{\overset{\cdot}{C}}}-O-SO_2R \; (R = p\text{-tolyl}, CF_3, CH_3)$

$-\overset{X}{\underset{|}{\overset{|}{\overset{\cdot}{C}}}}-CO_2H, \quad -\overset{X}{\underset{|}{\overset{|}{C}}}-CO_2R, \quad -\overset{X}{\underset{|}{\overset{|}{C}}}-\overset{O}{\overset{||}{C}}-R$

$-\overset{O}{\overset{||}{\underset{\cdot}{C}}}-OR, \quad -\overset{O}{\overset{||}{C}}R$

$\overset{\cdot}{C}H_2\overset{\cdot}{N_2^{\circ}}$ (diazomethane)

$H-\overset{\cdot\cdot}{O}-\overset{\cdot\cdot}{O}-H$ (hydrogen peroxide)

epoxide structure (epoxide)

$\overset{\cdot}{N}=\overset{*}{O}^b$ (generated from HNO_2)

[a] Asterisk indicates an electrophilic atom.
[b] To react as an electrophile, CH_2N_2 (diazomethane) must first be protonated to form the methyl diazonium cation:

$$H_2\overset{-}{C} = \overset{+}{N} = N \overset{H^+}{\longrightarrow} H_3C-\overset{*}{N} \equiv N.$$

Thus, Table 1.6 shows that the thiosulfate ion ($S_2O_3^{2-}$, $n = 6.4$) is more nucleophilic than iodide (I^-, $n = 5.0$).

B. Substrate

The structure of the substrate influences the rate of reaction with a nucleophile, and this effect is reflected in the s values defined in the previous section. For example, methyl bromide and chloroacetate both have s values of 1.00, so they react at the same rate. On the other hand, iodoacetate, with an s value of 1.33, reacts faster than methyl bromide, whereas benzyl chloride, with an s value of 0.87, reacts more slowly.

C. Solvent

Nucleophilicity is often solvent dependent, but the relationship is a complex one and depends on a number of different factors. Ritchie and coworkers have measured

TABLE 1.6 Nucleophilicities Toward Carbon[a]

Nucleophile	n	Nucleophile	N
$S_2O_3^{2-}$	6.4	Pyridine	3.6
SH^-	5.1	Br^-	3.5
CN^-	5.1	PhO^-	3.5
SO_3^{2-}	5.1	$CH_3CO_2^-$	2.7
I^-	5.0	Cl^-	2.7
$PhNH_2$	4.5	$HOCH_2CO_2^-$	2.5
SCN^-	4.4	SO_4^{2-}	2.5
OH^-	4.2	$ClCH_2CO_2^-$	2.2
$(NH_2)_2CS$	4.1	F^-	2.0
N_3^-	4.0	NO_3^-	1.0
HCO_3^-	3.8	H_2O	0.0
$H_2PO_4^-$	3.9		

[a]From Wells, (1963).

solvent-dependent relative nucleophilicities, N_+, in various solvents, using the equation

$$\log\frac{k_o}{k_{H_2O}} = N_+$$

where k_n is the rate constant for reaction of a cation with a nucleophile in a given solvent, and k_{H_2O} is the rate constant for reaction of the same cation with water in water. Some N_+ values are given in Table 1.7. Note that nucleophilicity is greater in dipolar aprotic solvents like dimethyl sulfoxide and dimethylformamide than in protic solvents like water or alcohols. For this reason, dimethyl sulfoxide is often used as a solvent for carrying out nucleophilic substitutions.

TABLE 1.7 Relative Nucleophilicities in Common Solvents[a]

Nucleophile (solvent)	N_+	Nucleophile (solvent)	N_+
$H_2O(H_2O)$	0.0	PhS^- (CH_3OH)	10.51
CH_3OH (CH_3OH)	1.18	PhS^- $[(CH_3)_2SO]$	12.83
CN^- (H_2O)	3.67	N_3^- (H_2O)	7.6
CN^- (CH_3OH)	5.94	N_3^- (CH_3OH)	8.85
CN^- $[(CH_3)_2SO]$	8.60	N_3^- $[(CH_3)_2SO]$	10.07
CN^- $[(CH_3)_2NCHO]$	9.33		

[a]From Ritchie (1975).

Sometimes relative nucleophilicities change in going from a protic to an aprotic solvent. For example, the relative nucleophilicities of the halide ions in water are $I^- > Br^- > Cl^-$, whereas in dimethylformamide, the nucleophilicities are reversed, i.e., $Cl^- > Br^- > I^-$.

PROBLEM 1.15

In each of the following reactions, label the electrophilic or nucleophilic center in each reactant. In a, b, and c, different parts of acetophenone are behaving as either a nucleophile or an electrophile.

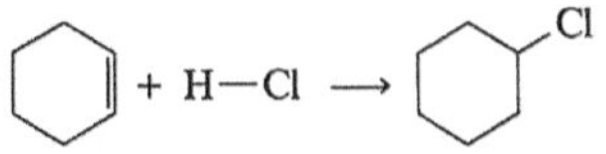

ANSWERS TO PROBLEMS

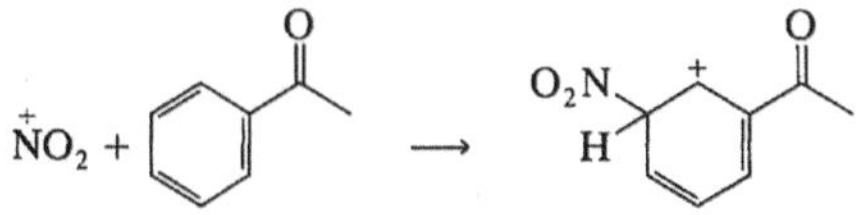

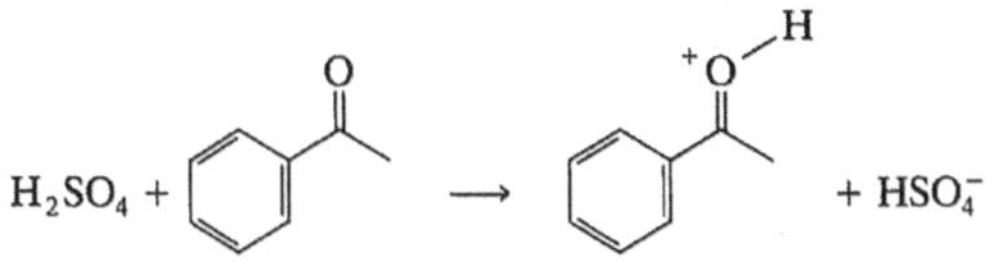

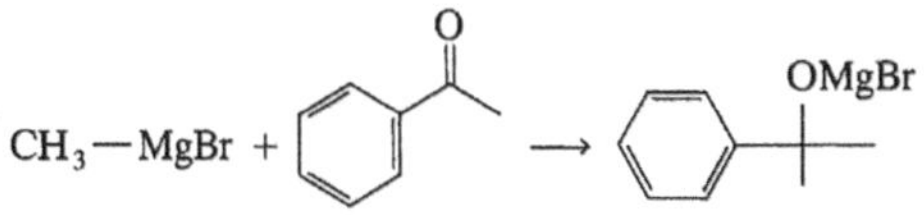

Problem 1.1

For all parts of this problem, the overall carbon skeleton is given. Therefore, a good approach is to draw the skeleton of the molecule with single bonds and fill in extra bonds, if necessary, to complete the octet of atoms other than hydrogen.

a. Electron supply $= (3 \times 4)(C) + (1 \times 6)(O) + (4 \times 1)(H) = 22$. Electron demand $= (3 \times 8)(C) + (1 \times 8)(O) + (4 \times 2)(H) = 40$. Estimate of bonds $= (40 - 22)/2 = 9$. This leaves two bonds left over after all atoms are joined by single bonds; thus, there is a double bond (as shown) between the CH_2 and CH groups and a double bond between the second CH and the oxygen to give the following skeleton:

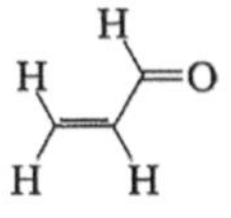

Calculation of the number of rings and/or π bonds also shows that two π bonds are present. The molecular formula is C_3H_4O. The number of hydrogens for a saturated hydrocarbon is $(2 \times 3) + 2 = 8$. There are $(8 - 4)/2 = 2$ rings and/or π bonds.

A total of 18 electrons are used in making the nine bonds in the molecule. There are four electrons left (from the original supply of 22); these can be used to complete the octet on oxygen by giving it two lone pairs of electrons.

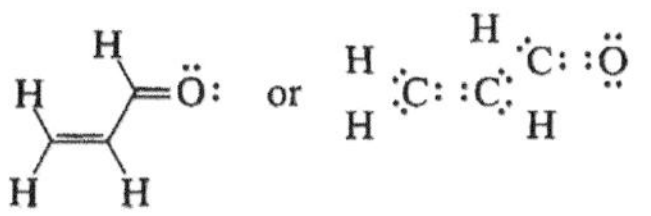

$$\text{(structures shown)}\qquad \text{or}\qquad \text{(structure shown)}$$

b. The way the charges are written in the structure indicates that the NO_2^+ and BF_4^- are separate entities. For NO_2^+, the electron demand is $3 \times 8 = 24$. Electron supply is $(1 \times 5)(N) + (2 \times 6)(O) - 1$ (positive charge) $= 16$. Estimate of bonds $= (24-16)/2 = 4$. A reasonable structure can be drawn with two double bonds. The remaining eight electrons (electron supply − electrons used for bonds) are used to complete the octets of the two oxygens to give the structure shown.

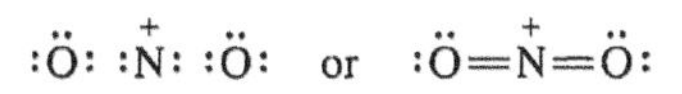

$$:\!\ddot{O}\!:\ \overset{+}{:}\!\ddot{N}\!:\ :\!\ddot{O}\!: \quad \text{or} \quad :\!\ddot{O}\!\!=\!\!\overset{+}{N}\!\!=\!\!\ddot{O}\!:$$

An alternative structure is much less stable. It has two adjacent positively charged atoms, both of which are electronegative.

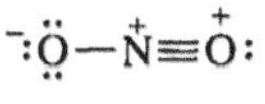

$$^{-}\!:\!\ddot{O}\!-\!\overset{+}{\ddot{N}}\!\!\equiv\!\!\overset{+}{O}\!:$$

By using an electron demand of six for B, we calculate that the electron demand for the atoms in BF_4^- is $(4 \times 8)(F) + 6(B) = 38$. The number of bonds predicted is then $(38 - 32)/2 = 3$. Because there are five atoms, clearly we need at least four bonds to join them together. In a neutral molecule, the electron demand for boron is six. In this negative ion, a reasonable move would be to assign an octet of electrons to B so that we can form four bonds and join all the atoms. These bonds use up eight electrons, and we can use the remaining 24 electrons to complete the octets around the four F atoms.

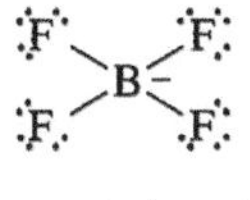

$$\begin{array}{ccc} :\!\ddot{F} & & \ddot{F}\!: \\ & \!\!B\!\!^{-}\!\! & \\ :\!\ddot{F} & & \ddot{F}\!: \end{array}$$

Although the rules presented in Chapter 1 for obtaining Lewis structures work most of the time, there are situations in which these approximations are not applicable. Boron is an element that displays unusual bonding properties in a number of its compounds.

c. Electron supply $= (6 \times 4)(C) + (18 \times 1)(H) + (3 \times 5)(N) + (1 \times 5)(P) = 62$. Electron demand $= (10 \times 8) + (18 \times 2) = 116$. The number of bonds predicted is $(116 - 62)/2 = 27$. After the bonds are placed in the molecule, there are eight electrons left $(62-2(27))$, which are used to complete the octets of phosphorus and nitrogen. All charges are zero.

d. Electron supply $= (2 \times 4)(C) + (6 \times 1)(H) + (1 \times 5)(N) + (1 \times 6)(O) = 25$. Electron demand $= (4 \times 8) + (6 \times 2) = 44$. Estimate of bonds $= (44 - 25) = 9.5$. This means that there must be nine bonds and an odd electron somewhere in the molecule. There are two likely structures **(1-41** and **1-42)** that agree with the carbon skeleton given:

I-4I   I-42

Because of the chemistry of the nitroxide functional group, the odd electron is usually written on oxygen, as in **1-41,** even though this means that the most electronegative element in the molecule is the one that does not have an octet. None of the atoms in **1-41** have a formal charge. In **1-42** there is a +1 charge on nitrogen and a −1 charge on oxygen. If you did not have an odd electron in your structure, either you had too many electrons, as in structure **1-43,** or you had too few electrons.

I-43

Structure **1-43** has an overall negative charge, which is placed on the oxygen. Because this compound is not neutral, it would not be stable in the absence of a counterion. If you did not put enough electrons in the structure, it will be positively charged. You should always count the number of electrons in your finished Lewis structure to make sure that you have exactly the number calculated for the electron supply.

Unlike the sulfur in dimethyl sulfoxide (Example 1.5), nitrogen and oxygen do not have *d* orbitals. Thus, it is not possible for this compound to have a double bond between nitrogen and oxygen, because this will leave either nitrogen or oxygen with more than eight electrons.

e. Electron supply $= (1 \times 4)(C) + (4 \times 1)(H) + (2 \times 6)(S \text{ and } O) = 20$. Electron demand $= 24(C, S, O) + 8(H) = 32$. Estimate of bonds $= (32 - 20)/2 = 6$. The molecular formula, CH_4OS, suggests that there are no rings or π bonds when sulfur has an octet.

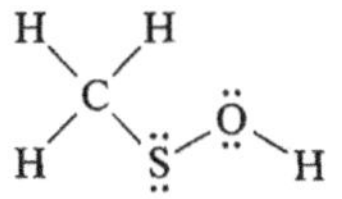

1-44

After insertion of the bonds, the eight electrons left over are used to complete the octets of sulfur and oxygen by filling in two lone pairs on each atom. There are no charges on any of the atoms in **1-44.**

The way the structure is given in the problem indicates that the hydrogen is on oxygen. Another possible structure, which used to be written for sulfenic acids, puts the acidic hydrogen on sulfur rather than oxygen:

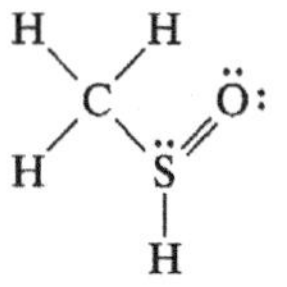

1-45 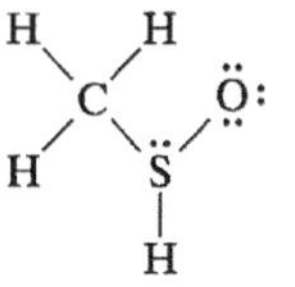   **1-46**

In **1-45** there are seven bonds because the sulfur has expanded its valence shell to accommodate 10 electrons. In **1-46,** the sulfur has a charge of +1 and the oxygen has a charge of -1.

Problem 1.2

It is always a good idea to write a complete Lewis structure before deciding on hybridization so that lone pairs are not missed.

a. The CH_3 carbon is sp^3 hybridized. The carbon is attached to four distinct groups, three hydrogens and one carbon, so $X = 4$. The other carbon and the nitrogen are both sp hybridized because $X + E = 2$. The sp-hybridized carbon is attached to the CH_3 carbon and to the nitrogen. The nitrogen is attached to carbon and also has a lone pair. The C—C—N skeleton is linear because of the central sp-hybridized atom. The H—C—H bond angles of the methyl group are approximately 109°, the tetrahedral bond angle.

b. All the carbons in the phenyl ring are sp^2 hybridized. This ring is also conjugated with the external π system, so the entire molecule is planar. The nitrogen and sulfur are sp^2 hybridized. Nitrogen is bonded to the phenyl group and the external carbon and also contains a lone pair, so $X + E = 3$; the sulfur is attached to carbon and has two lone pairs, so $X + E = 3$. The external carbon is sp hybridized. It is attached to only two atoms, the nitrogen and the sulfur, so $X + E = 2$. The N⦌C⦌S group is linear. Note, however, that the PheN⦌C bond angle is 120°.

c. For phosphorus $X + E = 4$ (three ligands and one lone pair); thus, the carbons and the phosphorus are all sp^3 hybridized with approximate tetrahedral bond angles.

Problem 1.3

a. 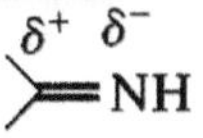 $\overset{\delta^+ \quad \delta^-}{\diagup\diagdown}=NH$

b. $\overset{\delta^+ \quad \delta^-}{Br-F}$

c. $\overset{\delta^+ \quad \delta^-}{H_3C-N(CH_3)_2}$

d. $\overset{\delta^- \quad \delta^+}{H_3C-P(CH_3)_2}$

Problem 1.4

a.

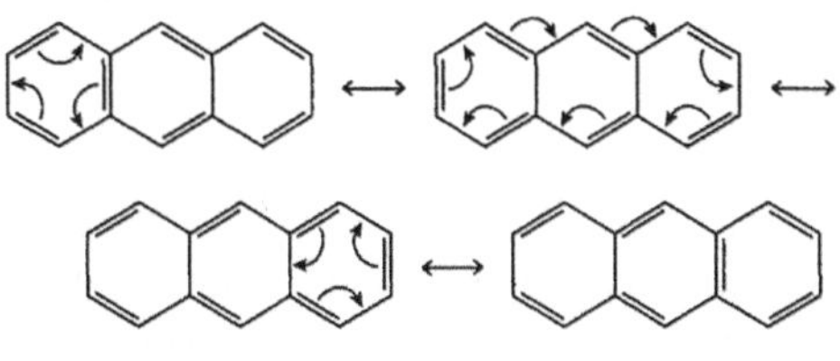

1-47

Is the resonance form **1-48** different from those written above? The answer is no.

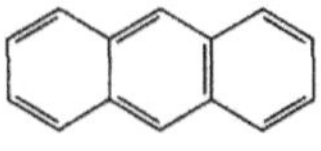

1-48

This represents exactly the same electron distribution as **1-47**. Although the double bond between the two right-hand rings is written to the right of the σ bond in **1-47** and to the left in **1-48**, the double bond is still between the same two carbons. Thus, **1-47** and **1-48** are identical.

Would the following redistribution of electrons lead to a correct resonance form? The answer is no.

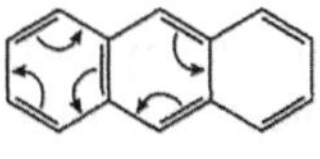

If a double bond is written at every single bond to which an arrow is drawn, we obtain structure **1-49**:

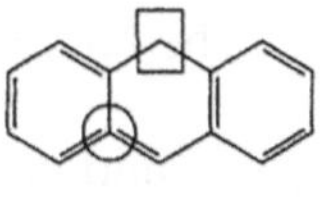

1-49

In **1-49**, the circled carbon has 10 electrons, which is impossible for carbon. Also, the boxed carbon does not have enough electrons. (This carbon is attached only to one hydrogen, not two.) Thus, this is an impossible structure.

b. The nitro group is conjugated with the π system. Hence, there are two forms for the nitro group whenever the second negative charge is located on an atom external to the nitro group.

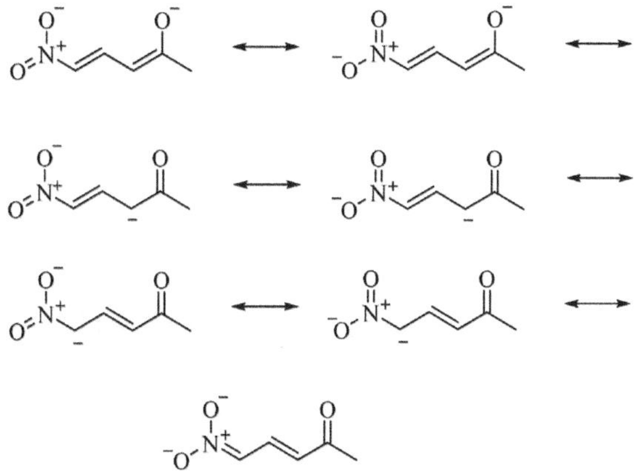

A structure like **1-50** would be incorrect because it represents a different chemical species; in fact, this compound has one less hydrogen at the terminal carbon and an additional negative charge. Remember that all resonance forms must have identical geometries.

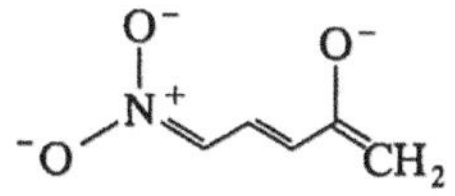

1-50

c.

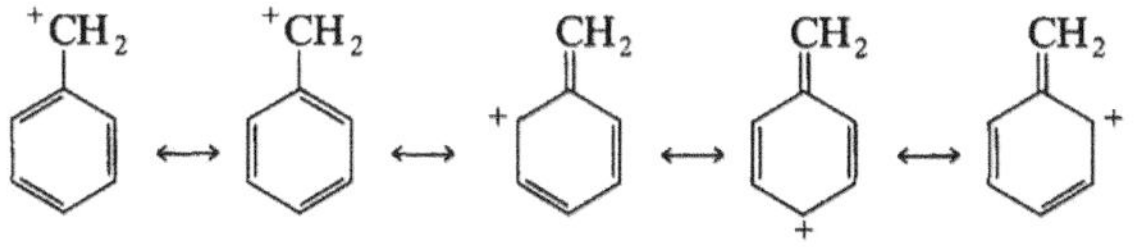

d.

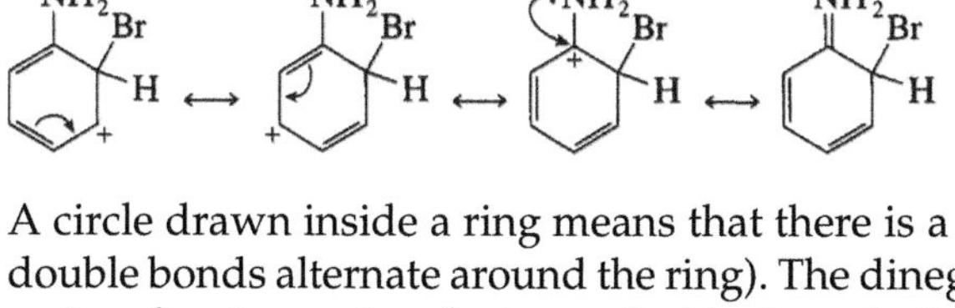

e. A circle drawn inside a ring means that there is a totally conjugated π system (single and double bonds alternate around the ring). The dinegative charge means that this conjugated system has two extra electrons, that is, two electrons have been added to the neutral, totally conjugated system (cyclooctatetraene). To draw the first structure, start with the parent compound. From it we can derive the highly unstable resonance structure with a negative and positive charge. This structure makes very little contribution to the actual structure of cyclooctatetraene and is shown purely for the purpose of "electron bookkeeping". We now add two electrons to form the dianion.

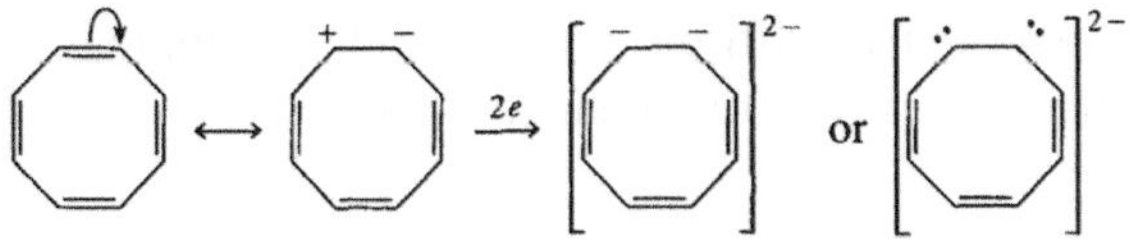

The resonance forms drawn here are a few of the many possible forms that can be drawn:

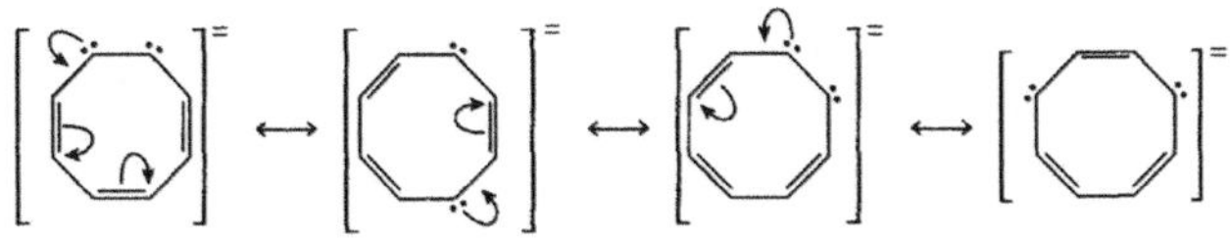

f. To figure out the first structure, from which the others can be derived by "electron pushing", examine the structure one step at a time. The dotted lines indicate that the ring system of the parent compound consists of alternating single and double bonds. From the parent compound, the radical anion can be derived by the addition of an electron (e.g., from a reducing agent such as sodium). Draw one of the resonance structures for the conjugated π system of the parent compound and then add one electron to form the radical anion. (Note the use of a half-headed arrow to show the movement of the single electron.) Use of a stepwise approach helps to make sure that you have the correct number of double bonds, charges, and electrons.

There are many resonance forms possible. Just a few are shown here. Notice that the aromatic Ph group also presents opportunities for drawing additional resonance structures.

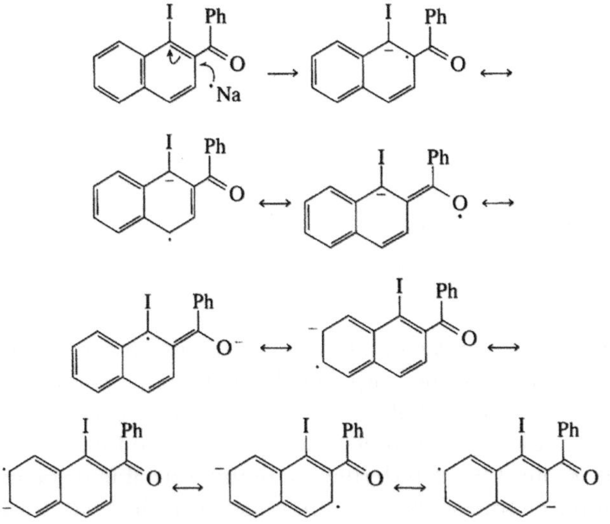

Problem 1.5

Once again, start by drawing the structure of the conjugated parent compound and then add an electron to form the radical anion. Only a few of the possible resonance forms are drawn. Nonetheless, it can be seen that the anion and radical can be delocalized onto both nitro groups simultaneously for the p-dinitrobenzene, and this leads to more possible resonance forms. Because there is more delocalization in the intermediate from the *para* compound, it should be easier to transfer an electron to p-dinitrobenzene, and hence, it should be a better radical trap.

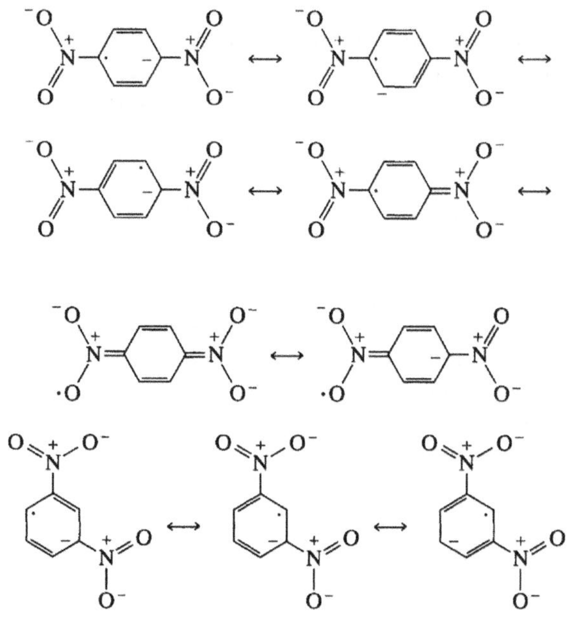

Problem 1.6

a. The resonance forms with a positive charge on nitrogen or oxygen do not contribute significantly to stabilization of the positive charge, because the nitrogen and oxygen do not have an octet of electrons when they bear the positive charge. The fact that oxygen or nitrogen does not have an octet is a critical point. When oxygen and nitrogen do have an octet, they can contribute significantly to a resonance hybrid in which they are positively charged. For example, the 1-methoxyethyl carbocation is stabilized significantly by the adjacent oxygen:

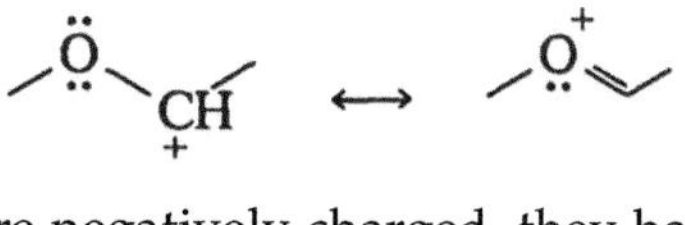

b. When nitrogen and oxygen are negatively charged, they have an octet of electrons. A negative charge on these atoms, both more electronegative than carbon, is significantly stabilizing. Thus, this delocalization would be very important.

Problem 1.7

a. Electron supply $= (1 \times 4)(C) + (3 \times 1)(H) + (1 \times 5)(N) + (2 \times 6)(O) = 24$. Electron demand $= (4 \times 8) + (3 \times 2) = 38$. Estimate of bonds $= (38-24)/2 = 7$. By using Hint 1.2, one ring or one π bond is expected. Two possible skeleton structures are **1-51** and **1-52**.

1-51   1-52

The combination of a highly strained three-membered ring and a weak O—O bond makes **1-52** unlikely. When you see the grouping NO_2 in a molecule, you can assume that it is a nitro group. Filling in the five lone pairs of electrons gives **1-53** or the resonance structure **1-54**.

1-53   1-54

In a neutral molecule, the sum of the formal charges must equal 0. Thus, if oxygen is given a formal charge of -1, some other atom in the molecule must have a formal charge of $+1$. Calculation shows that this atom is the nitrogen.

Structure **1-55** is incorrect because the nitrogen has 10 electrons and nitrogen does not have orbitals available to expand its valence shell.

1-55

b. Electron supply $= (6 \times 4)(C) + (5 \times 1)(H) + (2 \times 5)(N) - 1 = 38$. Electron demand $= (8 \times 8) + (5 \times 2) = 74$. Estimate of bonds $= (74 - 38)/2 = 18$. This cation does not have an odd electron. One electron was removed from the electron supply because the species is positively charged. The resonance structures that contribute most to the resonance hybrid are **1-56** and **1-57**.

1-56   1-57

Resonance forms **1-58** and **1-59** contribute much less than **1-56** and **1-57** to the resonance hybrid because the positively charged nitrogen in these forms does not have an octet of electrons.

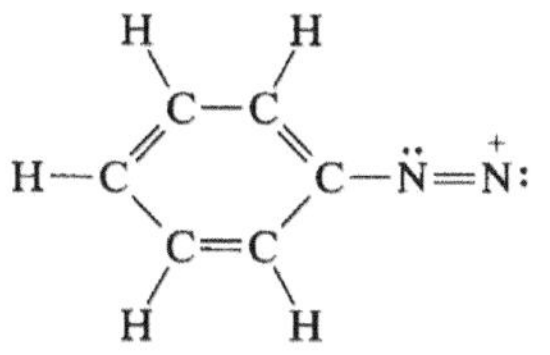

1-58 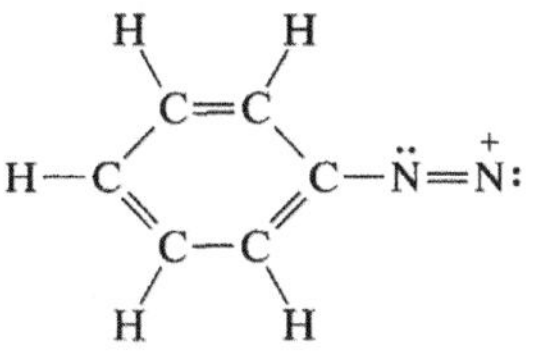   1-59

c. Electron supply $= (3 \times 4)(C) + (4 \times 1)(H) + (2 \times 5)(N) + (1 \times)(O) = 32$. Electron demand $= (6 \times 8) + (4 \times 2) = 56$. Estimate of total bonds $= (56-32)/2 = 12$. Several possible structures are shown. All are resonance forms.

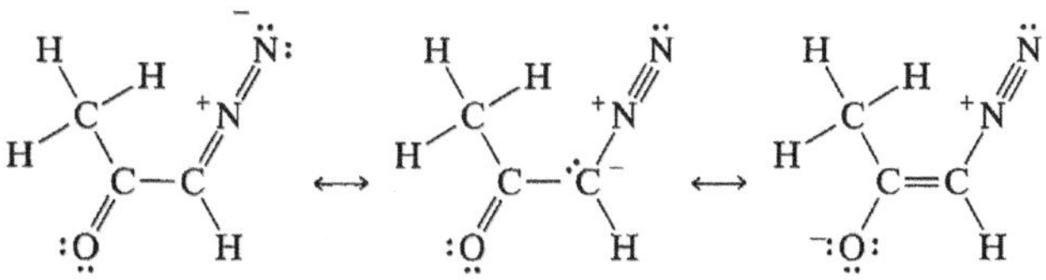

A resonance structure that will not contribute as significantly to the hybrid is **1-60**.

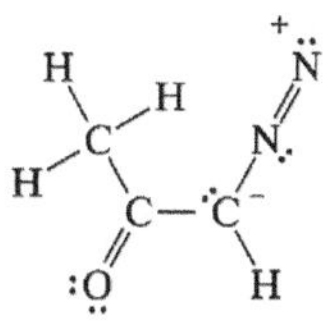

1-60

In this structure, the positively charged nitrogen does not have an octet of electrons, so that it will be less stable than the previous structures in which positively charged nitrogen does have an octet. Notice also that the number of bonds in **1-60** is one less than that estimated.

You might have interpreted the molecule's skeleton to be that shown in resonance structures **1-61** through **1-63**. However, none of these structures are expected to be stable. Structure **1-61** has a large number of charged atoms, **1-62** has an uncharged carbon with only six valence electrons (a carbene, see Chapter 4), and **1-63** has a carbene carbon and a nitrogen that has only six electrons. The first answers drawn for this problem are much more likely.

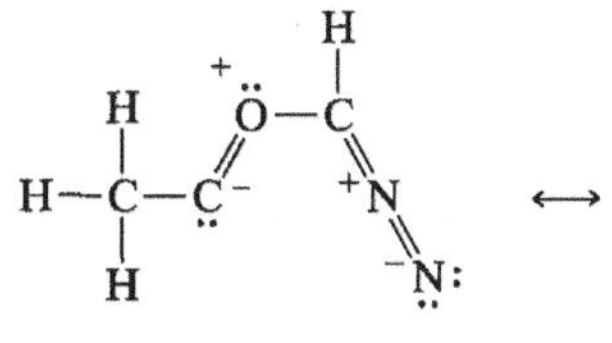

1-61

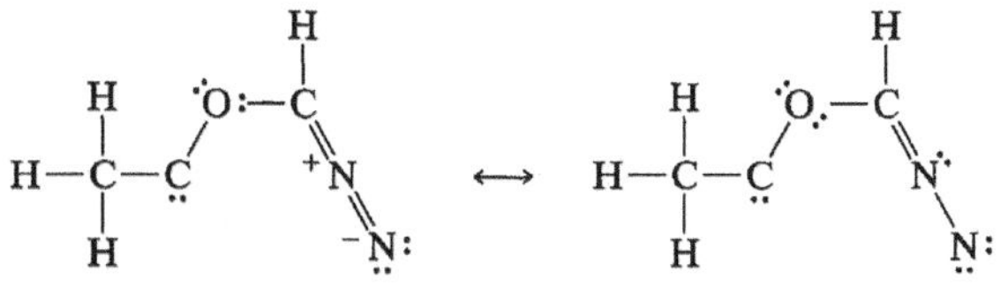

1-62   1-63

d. Electron supply $= (1 \times 4)(C) + (4 \times 5)(N) = 24$. Electron demand $= 5 \times 8 = 40$. Estimate of bonds $= (40-24)/2 = 8$. Because they are resonance structures, all the following are correct:

$$^-\!:\ddot{N}\!=\!\overset{+}{N}\!=\!\ddot{N}\!-\!C\!\equiv\!N: \quad\longleftrightarrow\quad \ddot{N}\!\equiv\!\overset{+}{N}\!-\!\ddot{N}\!-\!C\!\equiv\!N:^- \quad\longleftrightarrow$$

$$\ddot{N}\!\equiv\!\overset{+}{N}\!-\!N\!=\!C\!=\!\ddot{N}:^-$$

1-64

In a neutral molecule, the sum of the formal charges must equal 0. Thus, if one nitrogen is given a formal of $+1$, some other atom in the molecule must have a formal charge of -1. Another acceptable structure is **1-65**.

$$\begin{matrix}\ddot{N}\\ \| \\ \ddot{N}\end{matrix}\!\!>\!\!\ddot{N}\!-\!C\!\equiv\!N: \qquad \ddot{N}\!=\!\overset{*}{\ddot{N}}\!=\!\ddot{N}\!-\!C\!\equiv\!N:$$

1-65   **1-66**

On the other hand, **1-66** is unacceptable. The starred nitrogen has 10 electrons, which is not possible for any element in period 2 of the periodic table. Also, the terminal nitrogen next to this nitrogen has only six electrons instead of an octet. Other structures, **1-67** and **1-68**, in which one of the nitrogens does not have an octet, are less stable than the resonance structure **1-64**.

$$:N\!=\!\ddot{N}\!-\!\ddot{N}\!-\!C\!\equiv\!N: \qquad :\ddot{N}\!-\!\ddot{N}\!=\!\ddot{N}\!-\!C\!\equiv\!N:$$

1-67   **1-68**

Structures **1-69** and **1-70**, possible alternatives to a linear nitrogen array for the azide group, are not acceptable. In **1-69**, there is a nitrogen with 10 electrons, and in **1-70**, the nitrogen at the top does not have an octet of electrons. In addition, the electron-deficient nitrogen is located next to a positively charged nitrogen, an extremely unstable electron arrangement.

$$\begin{matrix}\overset{..}{\overset{+}{N}}\\ \| \\ N\!-\!C\!\equiv\!N:\\ \| \\ :N^-\end{matrix} \quad\longleftrightarrow\quad \begin{matrix}:\ddot{N}\\ \backslash \\ \overset{+}{N}\!-\!C\!\equiv\!N:\\ / \\ :N^-\end{matrix}$$

1-69   **1-70**

e. Electron supply for the allyl anion $= (3 \times 4)(C) + (5 \times 1)(H) + 1 = 18$. Electron demand $= (3 \times 8) + (5 \times 2) = 34$. Estimate of bonds $= (34 - 18)/2 = 8$.

$$\begin{matrix}&&H&&\\ &&|&&\\ H&&C&&H\\ \backslash&&\|&&/\\ &C&&\overset{-}{C}:&\\ |&&&&|\\ H&&&&H\\ \end{matrix} \quad\longleftrightarrow\quad \begin{matrix}&&H&&\\ &&|&&\\ H&&\overset{-}{C}&&H\\ \backslash&&/\ \ \backslash&&/\\ :C&&&&C&\\ |&&&&|\\ H&&&&H\\ \end{matrix}$$

$$\text{Li}^+ \qquad\qquad \text{Li}^+$$

The lithium cation sits equidistant from the two concentrations of negative charge at either end of the linear π system.

Problem 1.8

a. Thiophene is aromatic. The sulfur is sp^2 hybridized. One of the lone pairs of electrons is in an sp^2-hybridized orbital, and the other lone pair is in a p orbital parallel to the p orbitals on each of the carbons. Thus, this is a six π electron system.

b. The pentadienyl anion is nonaromatic. Although the structure contains six π electrons that are conjugated, it is not cyclic.

c. This heterocycle is antiaromatic or nonaromatic. If the NH nitrogen is sp^2 hybridized, the lone pair of electrons on the nitrogen can be in a p orbital that overlaps with the other six p electrons in the ring. This would be an antiaromatic system. However, because of the destabilization that delocalization would cause, it is unlikely that good overlap of the nitrogen lone pair electrons with the π system would occur.

d. Cycloheptatriene is nonaromatic. The carbon at the top of the structure is sp^3 hybridized. Thus, the delocalization of the six π electron system is interrupted by this carbon. For aromaticity, the compound must have the correct number of electrons and those electrons must be completely delocalized.

e. Pyrene is aromatic. This compound contains 16 π electrons, so, strictly speaking, it does not obey Hückel's rule and you might have expected this compound to be antiaromatic. However, if a compound has fused six-membered rings, all of which are totally conjugated, it exhibits the properties we associate with aromaticity and is usually considered to be aromatic. We can rationalize this aromaticity by looking at the alternative resonance structure that follows:

In this structure, the π bonds on the periphery form a fully conjugated loop. We can consider that the molecule has two noninteracting π systems: the 14 π electron system of the periphery, which is aromatic according to Hückel's rule, and the highlighted two π electron system of the central double bond.

f. The cyclooctatetraenyl dianion is aromatic. This is a completely conjugated, cyclic, 10 π electron system. The ring is planar. However, the situation is not so simple. It has been calculated that the energy reduction, due to aromaticity, is not large enough to compensate for the electron–electron repulsion energy. Apparently, such anionic compounds exist primarily because the metallic counterions are strongly solvated by the reaction solvents.

g. This macrocyclic alkyne is aromatic. The completely conjugated system contains 14 π electrons. Only two of the π electrons of the triple bond are counted, because the other two π electrons are perpendicular to the first two and cannot conjugate with the rest of the π system.

Problem 1.9

a.

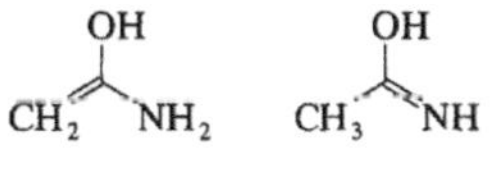

b. CH_2=CHOH

c. CH_2=CH—CH=CHOH

d.

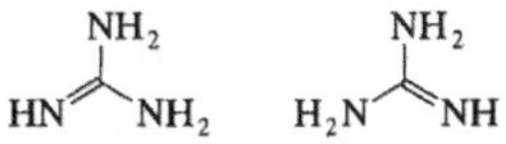

e.

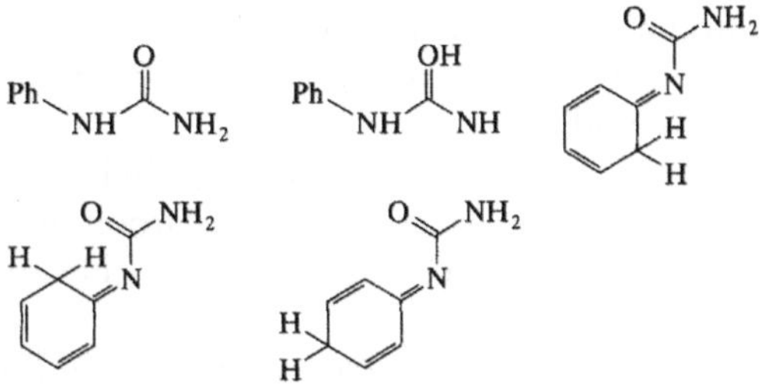

There are two more possible structures in which there is isomerism about the imine nitrogen:

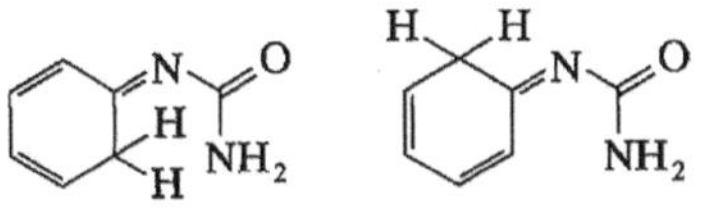

Problem 1.10

a. The first structure is the same compound as the last; the second is the same as the fourth. The first, second, and third structures meet the basic requirement for tautomers: interconversion involves only movement of a double bond and one hydrogen atom. However, most chemists would not call them tautomers because the allylic proton that moves is not very acidic. Its pK_a can be estimated from the pK_a of the allylic proton of 1-propene: 47.1–48.0 (from Appendix C). Thus, the intermediate anion necessary for the interconversion of these isomers would be formed only under extremely basic conditions. When the equilibration is this difficult to effect, the different isomers usually are not called tautomers. On the other hand, if another double bond were added to the structure, the compound would be very acidic (from Appendix C, the related cyclopentadiene has a $pK_a = 18.1$ in DMSO (dimethyl sulfoxide), which is more acidic than the α protons in acetone, $pK_a = 19.2$). Thus, most chemists would call compounds **1-71** and **1-72** tautomers.

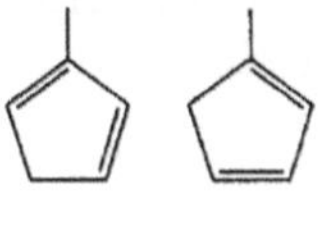

1-71   1-72

The structures in the problem are not reasonable resonance forms, because conversion from one structure to another involves movement of a hydrogen atom and rehybridization of two carbon atoms. Therefore, if the double bond were moved in either direction to the next ring location, it would leave one carbon with only three bonds (carbon 5 in the right-hand structure following) and place five bonds on another (carbon 2 in the right-hand structure following). These other structures would have extremely high energies and contribute nothing to a resonance hybrid.

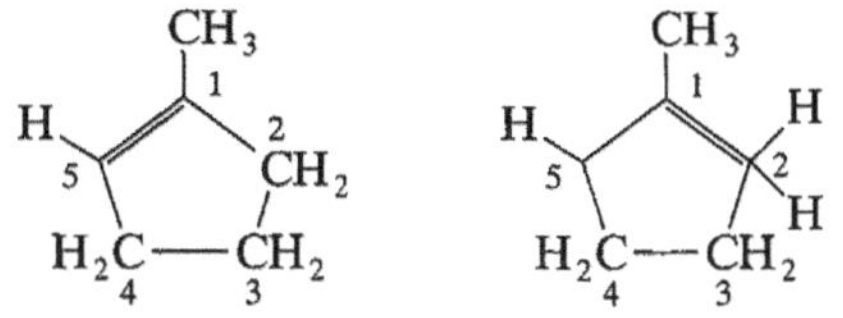

b. These two structures are resonance forms. They represent the simplest enolate ion that can exist.
c. These compounds are tautomers. They differ in the placement of a double bond and a proton, *and* they could readily be in equilibrium with each other.

Problem 1.11

The two structures are tautomers because they differ in the placement of a double bond and a proton, and they could readily be in equilibrium with each other. If the resonance arrow is changed to $\rightleftharpoons$, the picture will be correct.

Problem 1.12

a. $CH_2{=}CH^-$ is more basic than $HC{\equiv}C^-$. That is, according to Appendix C, ethene is less acidic than ethyne, so the conjugate base of ethane is more basic than the conjugate base of ethyne.
b. $[(CH_3)_2CH]_2N^-$ is more basic than $^-CH_2CON(Et)_2$. The diisopropylamide anion cannot be stabilized by resonance, because the N is bonded only to sp^3-hybridized carbons. On the other hand, $^-CH_2CON(Et)_2$ is stabilized by resonance. The negative charge is delocalized on both carbon and oxygen:

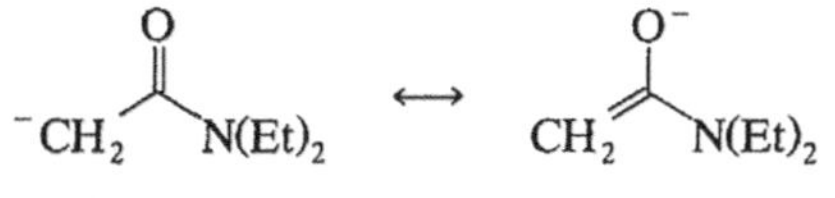

This anion is more stable than the diisopropylamide ion and, thus, diisopropylamide anion is the more basic.
c. $(CH_3)_3CO^-$ is more basic than CH_3O^-. The *t*-butoxide anion is not as well solvated as methoxide and, thus, is more basic. Appendix C also indicates that methanol is more acidic than *t*-butyl alcohol, so the relative basicity of the conjugate bases is the reverse.

d. The *m*-nitrophenolate anion is less stable than the *p*-nitrophenolate anion and, therefore, more basic. The resonance forms for *m*-nitrophenolate are as follows:

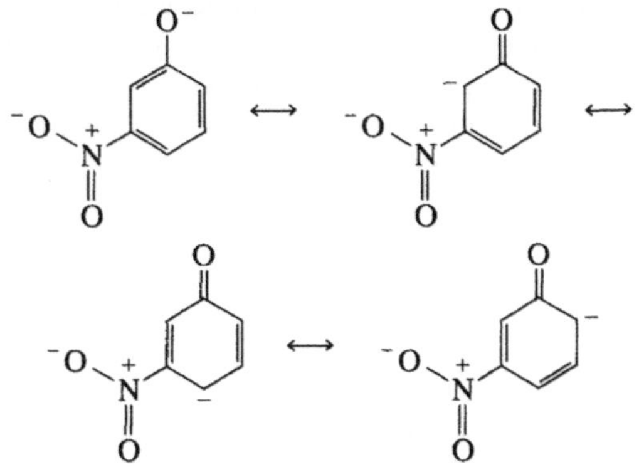

The resonance forms for *p*-nitrophenolate are as follows:

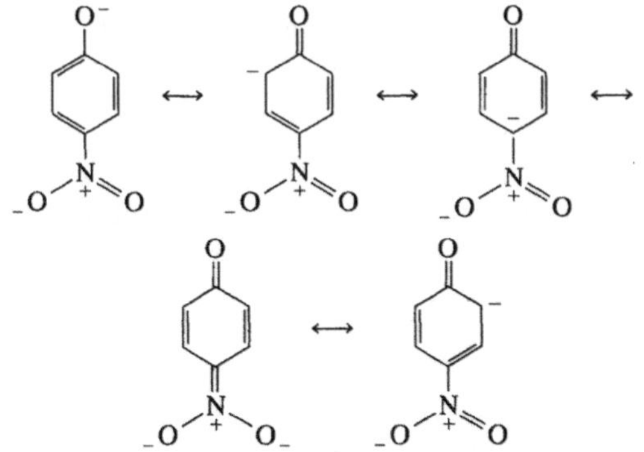

The *p*-nitrophenolate anion is more stable because the negative charge can be delocalized onto the nitro group. Because it is more stable, it will be less basic. Note that in the drawings, the other Kekulé form of the ring and the other form of the nitro group were omitted where appropriate. Because these forms also stabilize the starting materials, they do not explain a difference in energy between the starting phenol and product phenolate. It is the size of this energy difference that determines the acidity of the starting material.

Problem 1.13

a. The carbanion produced by removal of a proton on the CH_2 (methylene) group is the one stabilized most effectively by resonance because it is conjugated with two carbonyl groups.

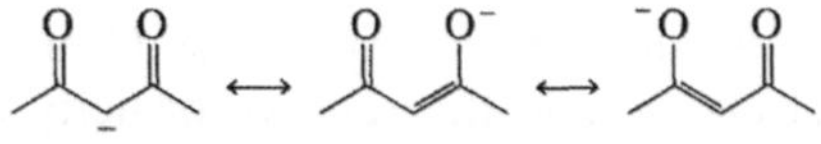

Removal of a proton from a methyl group will give a carbanion that is stabilized by conjugation with only one carbonyl group.

b. The proton will be removed from the oxygen, because this is the most electronegative element in the molecule and can stabilize the negative charge most effectively.

c. Removal of a proton from the γ position gives the most stabilized anion because more delocalization is possible.

Removal of the α proton on the sp^3-hybridized carbon gives a less stabilized anion because the charge is located on only two centers, not three.

d. The proton will be removed from the exocyclic amino nitrogen to form the resonance-stabilized anion, **1-73**.

1-73

The neutral molecule has three different types of acidic hydrogens attached to nitrogen atoms. Two of these are attached to amino nitrogens and one is attached to an imine group. To compare the acidity of the various hydrogen atoms, we need to compare the stability of the different anions formed by loss of the different protons. Ordinarily, by analogy to the acidity of carbon acids ($HC \equiv CH > H_2C = CH_2 > CH_3 - CH_3$), we might expect the imine to be more acidic than either of the amines. However, the exocyclic amine proton actually is the most acidic because the anion **1-73**, formed by loss of the exocyclic amine proton, can be represented by two resonance forms of equal energy in which the charge is distributed over two nitrogen atoms. Resonance stabilization is not available to either anion **1-74**, formed by loss of the imine proton, or anion **1-75**, derived by loss of the endocyclic amine proton.

1-74   1-75

Other factors being equal, an anion with lone pairs in an sp^2 orbital is less basic than an anion with lone pairs in an sp^3 orbital. In general, the more s character in the orbital forming the bond to hydrogen, the more acidic the proton. A good way to remember this is to note the high acidity of acetylene ($HC\equiv CH$, $pK_a = 28.8$) compared with ethylene ($H_2C=CH_2$, $pK_a = 44$) and ethane (CH_3CH_3, $pK_a \approx 50$).

Problem 1.14

a. $K_a = 10^{-30.5}/10^{-15.9} = 10^{-14.6}$

This value tells us that there is very little ester anion present at equilibrium.

b. $K_a = 10^{-24.7}/10^{-15.7} = 10^{-9}$

This value is only an approximation, because the two acidities are not measured in the same solvent. Nonetheless, the value indicates that equilibrium favors the starting material to a large extent.

c. $K_a = 10^{-30.6}/10^{-35.7} = 10^{5.1}$

In this case the reaction goes substantially to the right. By using the alternative pK_a of 39 for the amine, the answer would be $10^{8.4}$.

Problem 1.15

a. $^+NO_2$ is the electrophile. Writing a Lewis structure for this species ($:\ddot{O}::\overset{+}{N}::\ddot{O}:$) indicates that the nitrogen is positive and will be the atom that reacts with the nucleophile. In acetophenone, the π electrons of the ring act as the nucleophile. In keeping track of electrons, it is helpful to think of the nucleophile as the electron pair of the π bond to the carbon where the nitrogen becomes attached:

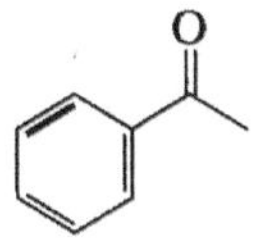

b. The electrophile is the proton of sulfuric acid that is transferred to the oxygen of acetophenone. Thus, the oxygen of acetophenone acts as the nucleophile in this reaction. The product shown would be a direct result of a lone pair of electrons on the oxygen acting as a nucleophile. One could also show the π electrons of the $C=O$ group acting as the nucleophile. This would give the following structure:

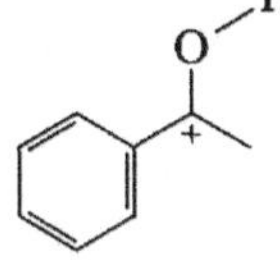

This structure and the structure drawn in the problem are the same, because they are both resonance forms that contribute to the same resonance hybrid.

This reaction could also be described as an acid–base reaction, where the acid is the proton of sulfuric acid and the base is the oxygen of the carbonyl group.

c. This bond of the Grignard reagent is a highly polarized covalent bond, so that the carbon bears a negative charge. This nucleophilic carbon of the Grignard reagent becomes attached to the electrophilic carbon of the carbonyl group in acetophenone.

d. The electron pair, constituting the π bond in cyclohexene, is the nucleophile, and a proton from HCl is the electrophile.

References

Ritchie, C. D. *J. Am. Chem. Soc.* **1975,** *97,* 1170–1179.

The Nature of the Chemical Bond, 3rd ed.; Cornell University Press: Ithaca, NY, 1960; p. 93. The second set of values is from Sanderson, R. T. *J. Am. Chem. Soc.* 1983, *105,* 2259-2261; *J. Chem. Educ.* 1983, 65, 112.

Wells, P. R. *Chem. Rev.* **1963,** *63,* 171–219.

2

General Principles for Writing Reaction Mechanisms

In writing a reaction mechanism, we give a step-by-step account of the bond (electron) reorganizations that take place in the course of a reaction. These mechanisms do not have any objective existence; they are merely our attempt to represent what is going on in a reaction. These *models* have been built over many years based on the results of many experiments and have been successfully applied to a broad spectrum of transformations.

Although experiments can suggest that some mechanisms are reasonable and others are not, for many reactions there is no evidence regarding the mechanism, and we are free to write whatever mechanism we choose, subject only to the constraint that we conform to generally accepted mechanistic patterns.

The purpose of this book is to help you figure out a number of pathways for a new reaction by showing you some of the steps that often take place under a particular set of reaction conditions. This chapter is devoted to some general principles, derived from the results of many experiments by organic chemists, which can be applied to writing organic mechanisms. Subsequent chapters will develop the ideas further under more specific reaction conditions.

It is often difficult to predict what will actually happen in the course of a reaction. If you were planning to run a reaction that had never been done before, you would plan the experiment on the basis of previously run reactions that look similar. You would assume that the steps of bond reorganization that take place in the new reaction are analogous to those in the reactions previously run. However, you might find that one or more steps in your reaction scheme give unanticipated results. In other words, although a number of general ideas about the course of reactions have been developed on the basis of experiments, it is sometimes difficult to choose which ideas apply to a particular reaction. Working through the problems in this book will help you develop the ability to make some of those choices. Nonetheless, often you will conclude that there is more than one possible pathway for a reaction.

It should be noted that through the application of structure similarity searches in reaction databases it is a simple task to find examples of similar transformations, including the conditions and outcomes of those experiments. The power that network and database technology has brought to the modern day chemist cannot be overstated. The change in access to information (the ease of access, the volume of information, and the diversity of sources and forms of information) has resulted in a significant transformation in the way chemistry is taught and the way chemists work.

Writing Reaction Mechanisms in Organic Chemistry
http://dx.doi.org/10.1016/B978-0-12-411475-3.00002-6

Copyright © 2014 Elsevier Inc. All rights reserved.

1. BALANCING EQUATIONS

HINT 2.1

It can be assumed, unless otherwise stated, that when an organic reaction is written, the products shown have undergone any required aqueous workup, which may involve acid or base, to give a neutral organic molecule (unless salts are shown as the product). In other words, when an equation for a reaction is written in the literature or in an examination, an aqueous workup usually is assumed and intermediates, salts, etc., are not shown.

From the viewpoint of organic chemistry, an equation is usually considered to be balanced if it accounts for all the carbon atoms and is balanced with respect to charges and electrons. Ordinarily, no attempt is made to account for the changes in the inorganic species involved in the reaction.

HINT 2.2

Check that equations are balanced. First, balance all atoms on both sides of the equation, and then balance the charges. Be aware that when equations are written in the organic literature, they are frequently not balanced.

EXAMPLE 2.1. BALANCING ATOMS

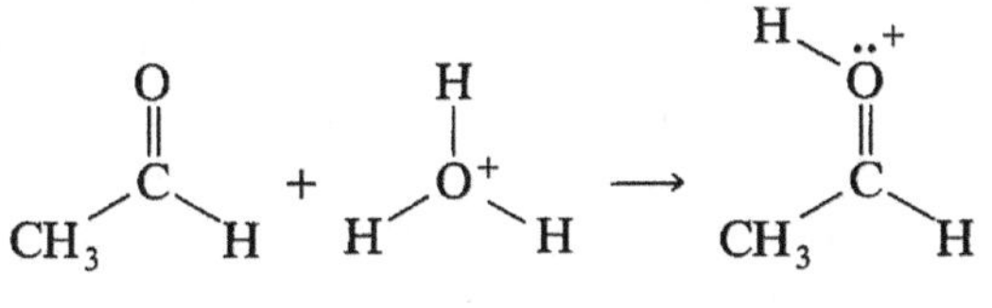

In this equation, the carbon atoms balance but the hydrogen and oxygen atoms do not. The equation is balanced by adding a molecule of water to the right-hand side.

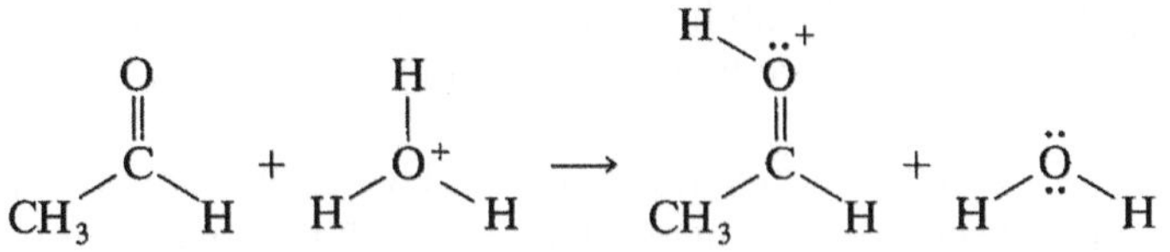

The equation is balanced with regard to charge: the positive charge on the left balances the positive charge on the right.

EXAMPLE 2.2. BALANCING CHARGES

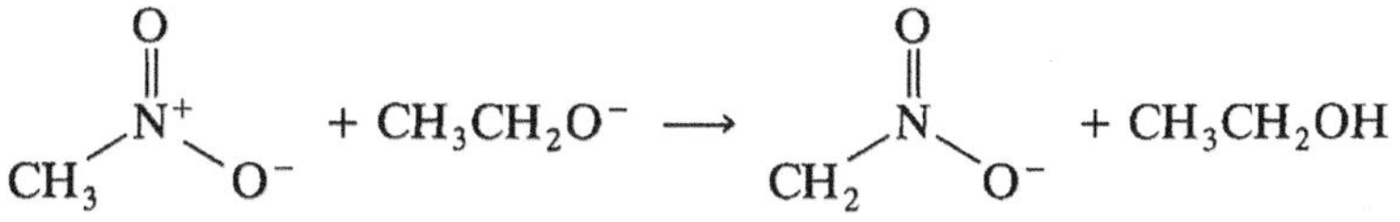

In this equation, the carbon, hydrogen, nitrogen, and oxygen atoms balance. At first glance, the charges also appear to balance because there is a single net negative charge on each side of the equation. However, the right-hand side of the equation contains an incorrect Lewis structure in which there is an electron-deficient carbon and the formal charge on nitrogen is omitted. The equation is balanced correctly by adding a negative charge on carbon and a positive charge on nitrogen.

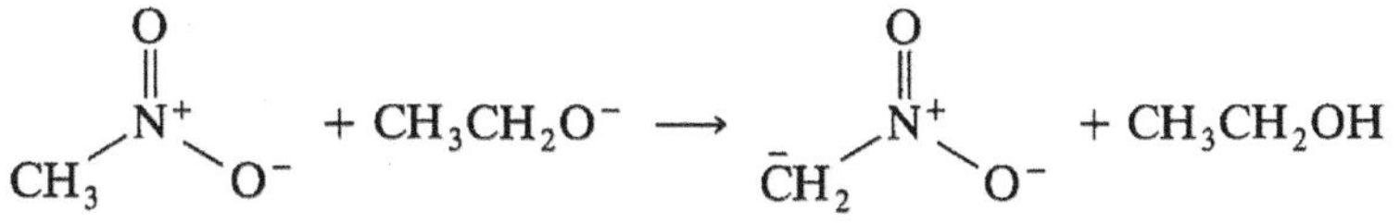

HINT 2.3

An easy way to check to see that the charges are balanced in the molecule is to make sure that the atoms that are representatives of period 2 of the periodic table (B, C, N, O, C, F) meet the octet rule on both sides of the equation and then apply the appropriate charges.

PROBLEM 2.1

In the following steps, supply the missing charges and lone pairs. Assume that no molecules with unpaired electrons are produced

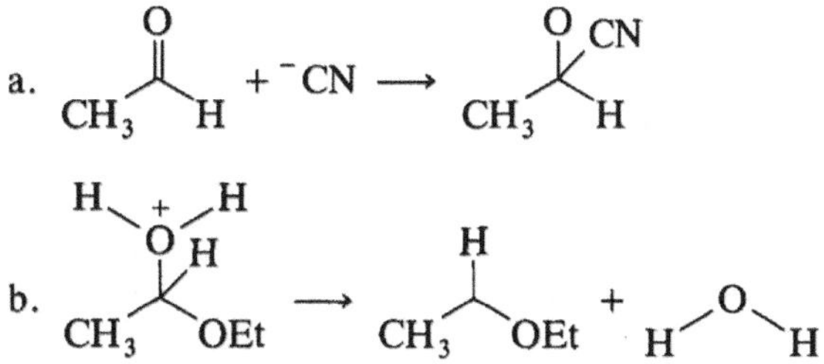

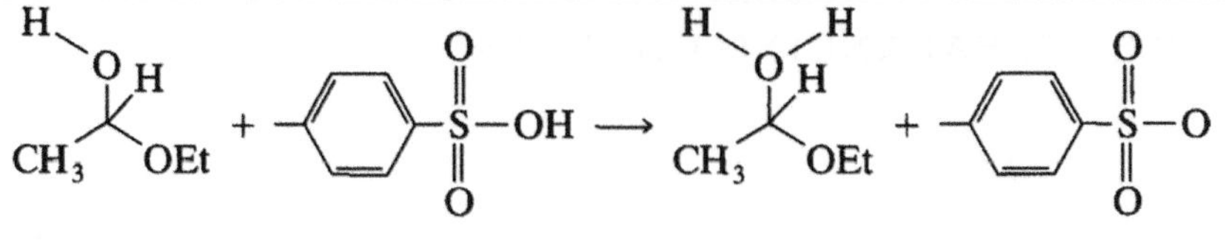

2. USING ARROWS TO SHOW MOVING ELECTRONS

In writing mechanisms, bond-making and bond-breaking processes are shown by curved arrows. The arrows are a convenient tool for thinking about and illustrating what the actual electron redistribution for a reaction may be.

HINT 2.4

The arrows that are used to show the redistribution of electron density are drawn from a position of high electron density to a position that is electron deficient. Thus, arrows are drawn leading away from negative charges or lone pairs and toward positive charges or the positive end of a dipole. In other words, they are drawn leading away from nucleophiles and toward electrophiles. Furthermore, it is only in *unusual* reaction mechanisms that two arrows will lead either away from or toward the same atom.

EXAMPLE 2.3. USING ARROWS TO SHOW REDISTRIBUTION OF ELECTRON DENSITY

The following equations show the electron flow for the transformations in Examples 2.1 and 2.2.

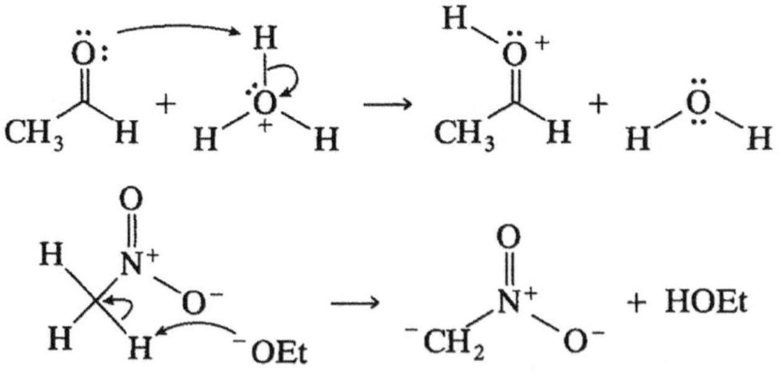

For an understanding of why the neutral oxygen in the first equation reacts with a proton of the hydronium ion rather than the positively charged oxygen, see Hint 2.10.

HINT 2.5

When you first start drawing reaction mechanisms, rewrite any intermediate structure before you try to manipulate it further. This avoids confusing the arrows associated with electron flow for one step with the arrows associated with electron flow for a subsequent step. As you gain experience, you will not need to do this. It will also be helpful to write the Lewis structure for at least the reacting atom and to write lone pairs on atoms such as nitrogen, oxygen, halogen, phosphorus, and sulfur. You may also find it helpful to include hydrogens that are at or near the reacting centers in your depictions of the mechanism. If not to better understand the reaction process, the hydrogens can also act as markers to allow perspective on what has changed in the process.

PROBLEM 2.2

For the following reactions, supply the missing charges and then use curved arrows to show the bond breaking and bond making for each step.

a.

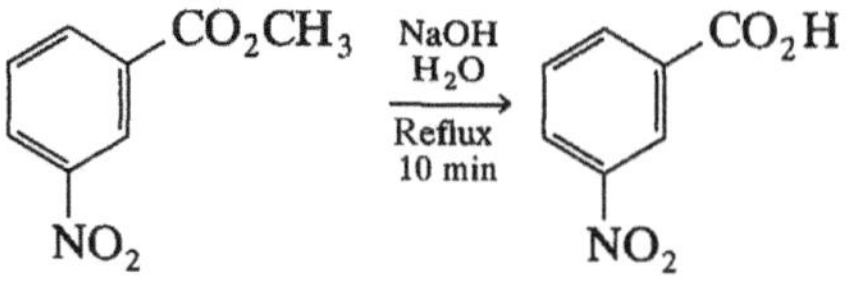

95%

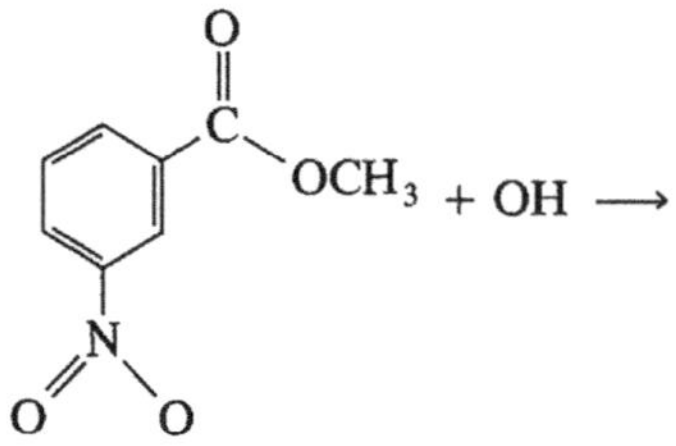

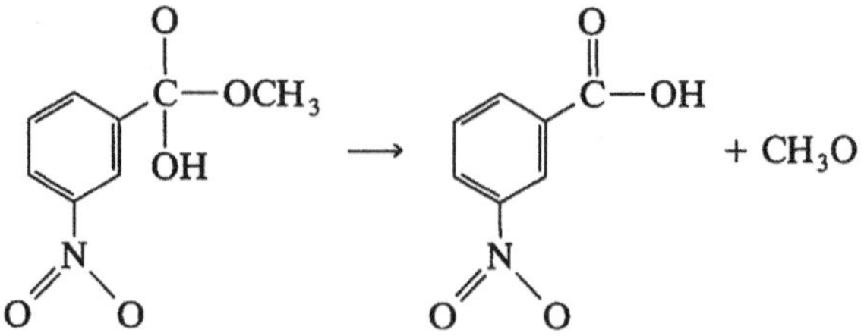

b.

HCl
H₂O
Room temperature

75%

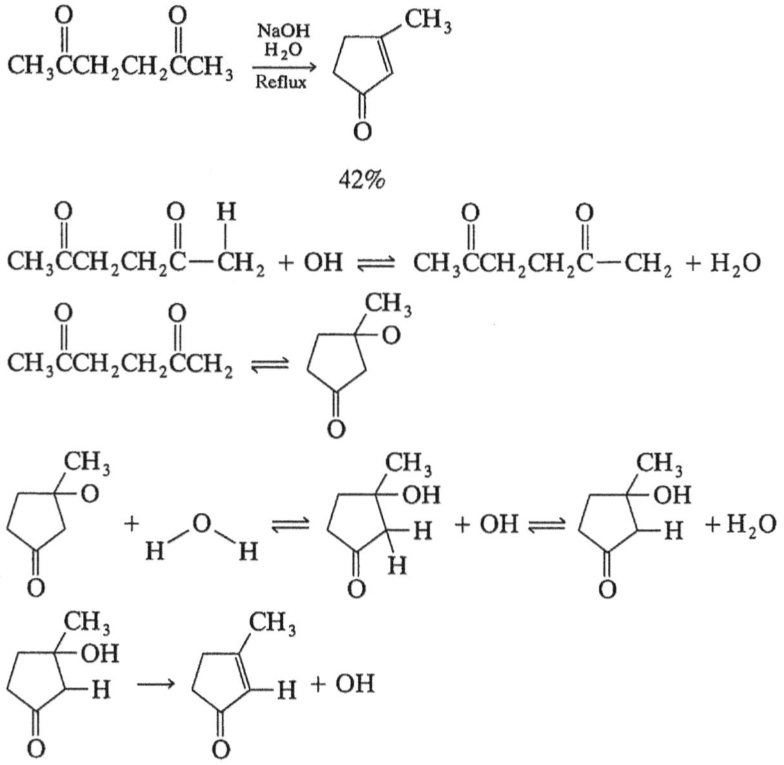

3. MECHANISMS IN ACIDIC AND BASIC MEDIA

HINT 2.6

If a reaction is run in a strongly basic medium, any *positively* charged species must be a weak acid. If a reaction is run in a strongly acidic medium, any *negatively* charged species must be a weak base. In a weak acid or base (like water), both strong acids and strong bases may be written as part of the mechanism.

We can look at what this means in some specific situations. Using the pK_a values listed in Appendix C, we find that in a strongly basic solution like 5% aqueous sodium hydroxide (pH ~ 14), the only protonated species would be those with a $pK_a > 12$ (e.g., guanidine, pK_a 13.4). In strongly acidic solutions like 5% aqueous hydrochloric acid (pH ~ 0), the only anions present would be those whose conjugate acid had a $pK_a < -2$ (e.g., $PhSO_3H$, pK_a −2.9). In a solution closer to neutrality (e.g., 5% $NaHCO_3$, pH = 8.5), we would find positively charged guanidine and aliphatic amino groups (pK_a = 13.4 and 10.7, respectively), as well as neutral aromatic amines (pK_a ~ 4), and anions such as acetate and 2,4-dinitrophenolate (pK_a = 4.7 and 4.1, respectively).

EXAMPLE 2.4. WRITING A MECHANISM IN A STRONG BASE

The following mechanism for the hydrolysis of methyl acetate in a strong base is consistent with the experimental data for the reaction.

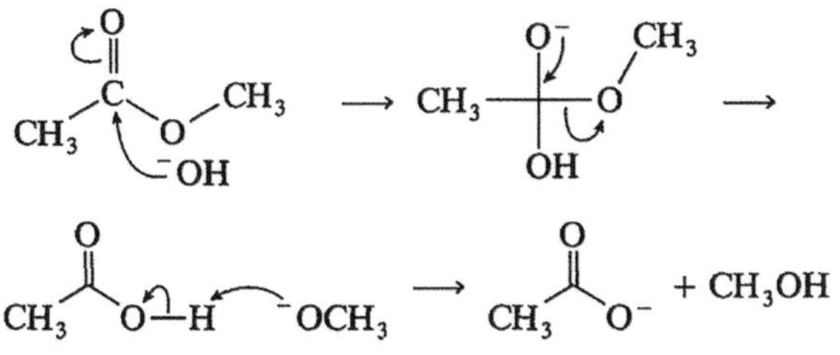

As suggested in Hint 2.6, all the charged species in this mechanism are negatively charged because the reaction occurs in strong base. Thus, the following steps would be incorrect for a reaction in base because they involve the formation of ROH_2^+ (the intermediate) and H_3O^+, both of which are strong acids.

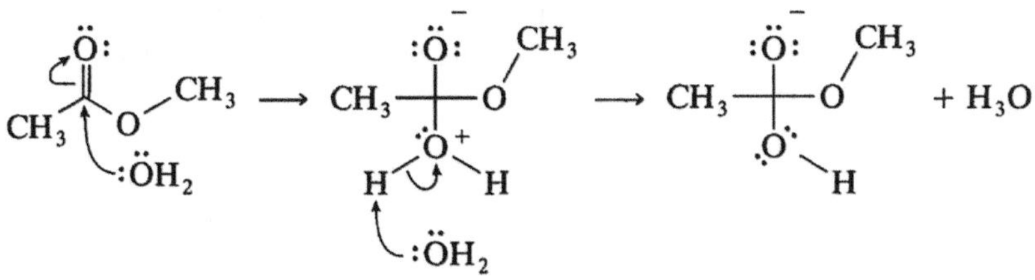

Another way of looking at this is to realize that, in aqueous base, hydroxide ion has a significant concentration. Because hydroxide is a much better nucleophile than water, it will act as the nucleophile in the first step of the reaction.

EXAMPLE 2.5. WRITING A MECHANISM IN STRONG ACID

The following step is consistent with the facts known about the esterification of acetic acid with methanol in strong acid.

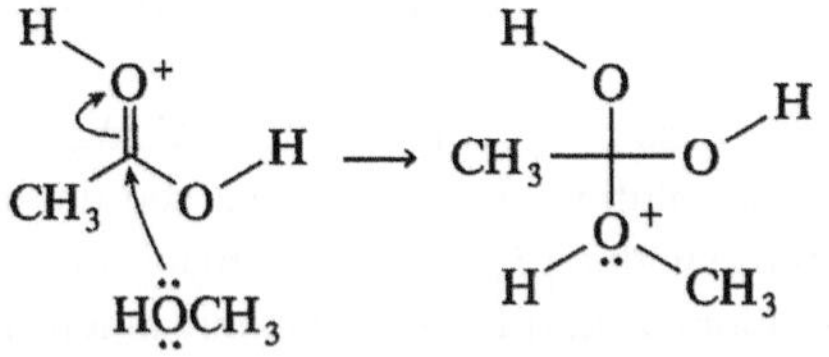

Because the fastest reaction for a strong base, CH_3O^-, in acid is protonation, the concentration of CH_3O^- would be negligible and the following mechanistic step would be highly improbable for esterification in acid:

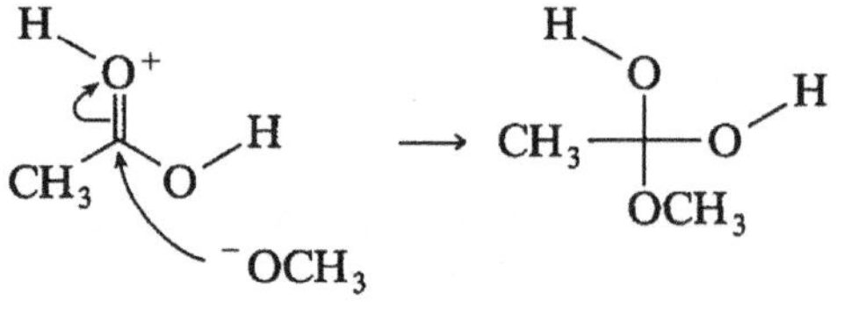

EXAMPLE 2.6. WRITING A MECHANISM IN A WEAK BASE OR WEAK ACID

When a reaction occurs in the presence of a weak acid or a weak base, the intermediates do not necessarily carry a net positive or negative charge. For example, the following mechanism often is written for the hydrolysis of acetyl chloride in water. (Most molecules of acetyl chloride probably are protonated on oxygen before reaction with a nucleophile, because acid is produced as the reaction proceeds.)

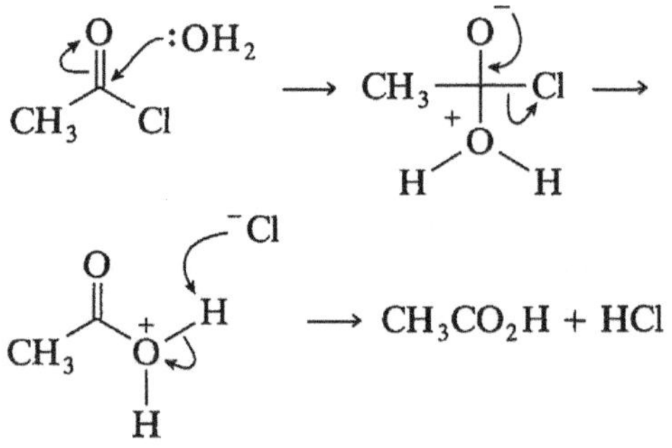

In the first step, the weak base water acts as a nucleophile. In the second step, the weak base chloride ion is shown removing a proton. This second step also could have been written with water acting as the base. Notice that in this example most of the lone pairs have been omitted from the Lewis structures. Reactions in the chemical literature often are written in this way.

EXAMPLE 2.7. STRONG ACIDS AND BASES AS INTERMEDIATES IN THE TAUTOMERIZATION OF ENOLS IN WATER (NEUTRAL CONDITIONS)

The first step is usually described as a proton transfer from the enol to a molecule of water. However, when arrows are used to show the flow of electrons, the arrow must proceed from the nucleophile to the electrophile.

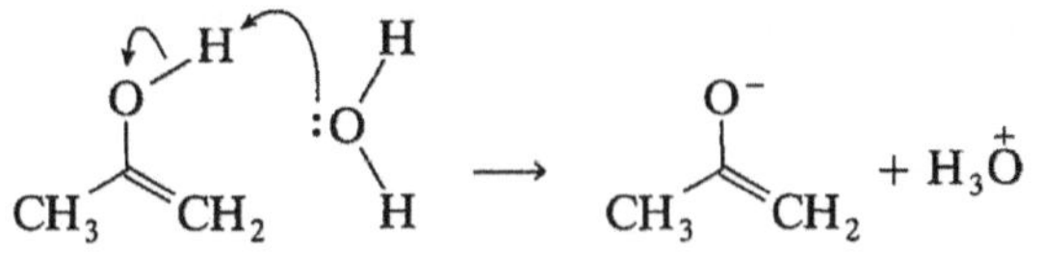

2-1

This step produces a strong acid, hydronium ion, and a strong base, the enolate anion **2-1**. This anion is a resonance hybrid of structures **2-1** and **2-2**.

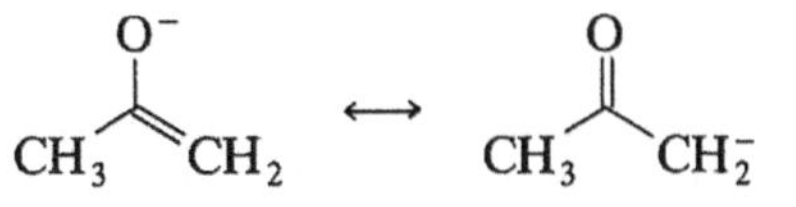

2-1   **2-2**

The hybrid can remove a proton from the hydronium ion to give the ketone form of the tautomers. Although not strictly correct (because resonance structures do not exist), such reactions commonly are depicted as arising from the resonance structure that bears the charge on the atom that is adding the proton.

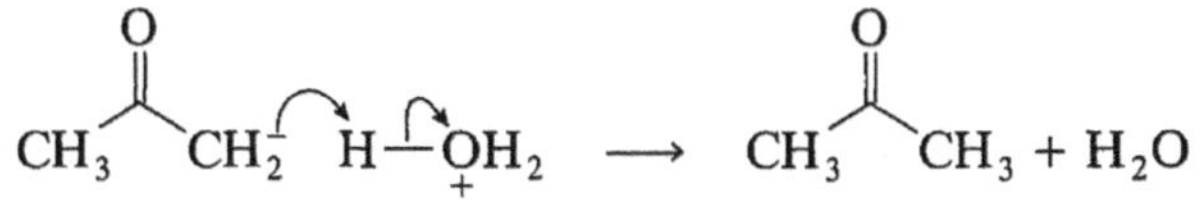

2-2

Because enolization under neutral conditions produces both a strong acid and a strong base, the reaction is very slow. Addition of a very small amount of either a strong acid or a strong base dramatically increases the rate of enolization.

HINT 2.7

Approach writing the mechanism of a chemical reaction in a logical fashion. For example, if the reagent is a strong base, look for acidic protons in the substrate, and then look for a reasonable reaction for the anion produced. If the anion formed by deprotonation has a suitable leaving group, its loss would lead to overall elimination. If the anion formed is a good nucleophile, look for a suitable electrophilic center at which the nucleophile can react (for further detail, see Chapter 3).

For mechanisms in acid, follow a similar approach and look for basic atoms in the substrate. Protonate a basic atom and consider what reactions would be expected from the resulting cation.

HINT 2.8

When writing mechanisms in acid and base, keep in mind that protons are removed by bases. Even very weak bases like HSO_4^-, the conjugate base of sulfuric acid, can remove protons. Protons do not just leave a substrate as H^+ because the bare proton is very unstable! Nonetheless, the designation $-H^+$ is often used when a proton is removed from a molecule. (Whether this designation is acceptable is up to individual taste. If you are using this book in a course, you will have to find out what is acceptable to your instructor.) A corollary is that when protons are added to a substrate, they originate from an acid, that is, protons are not added to substrates as freely floating (unsolvated) protons.

PROBLEM 2.3

In each of the following reactions, the first step in the mechanism is removal of a proton. In each case, put the proton most likely to be removed in a box. The pK_a values listed in Appendix C may help you decide which proton is most acidic.

a.

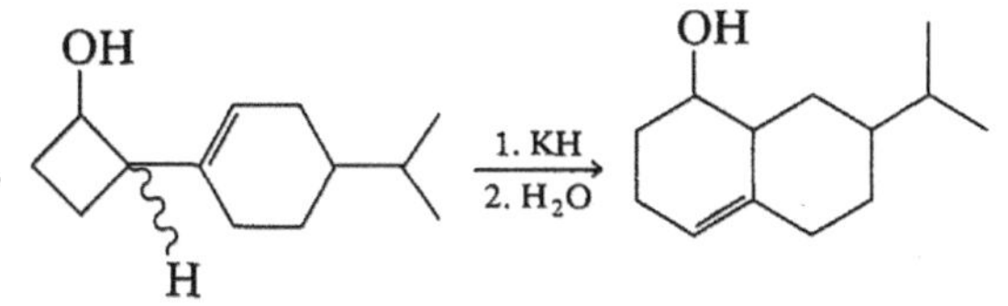

The wavy bond line means that the stereochemistry is unspecified.

b.

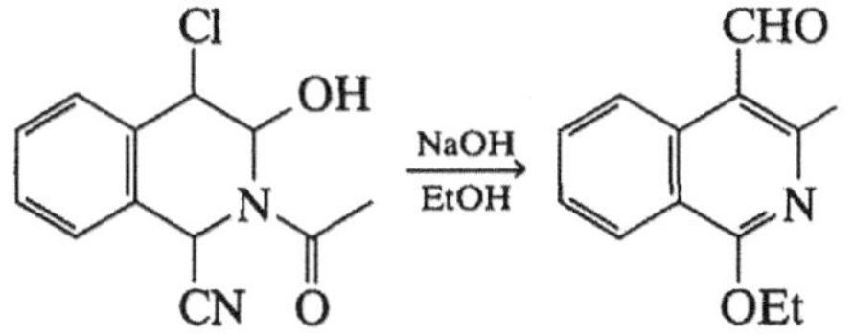

c.

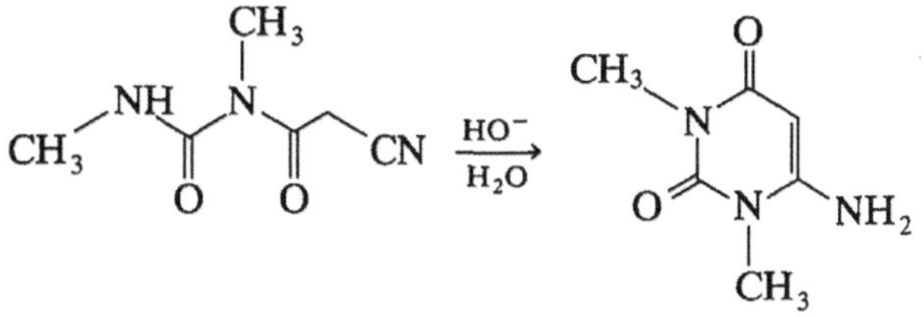

In this example, removal of the most acidic proton does not lead to the product. Which is the most acidic proton and which is the proton that must be removed in order to lead to the product?

d.

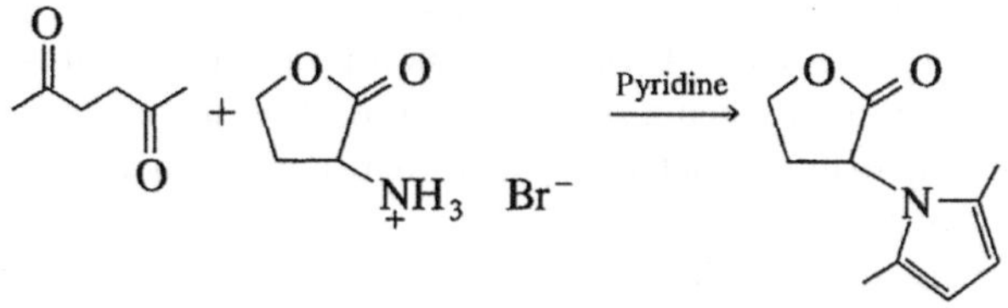

Source: Cohen and Bhupathy, (1983); Kirby et al., (1979); Kirby et al., (1985); Bernier et al., (1985); Jaffe, (1985).

PROBLEM 2.4

For each of the following reactions, the first step in the mechanism is protonation. In each case, put the atom most likely to be protonated in a box.

a.

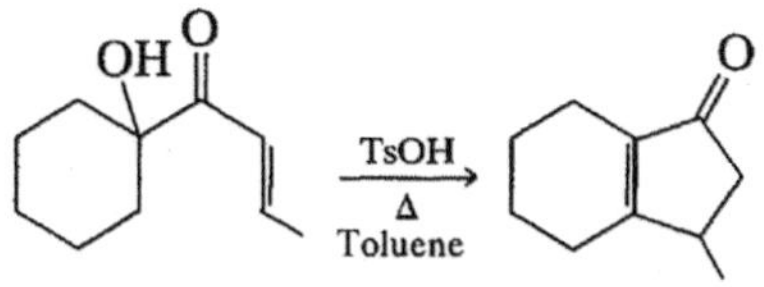

TsOH = *p*-toluenesulfonic acid

b.

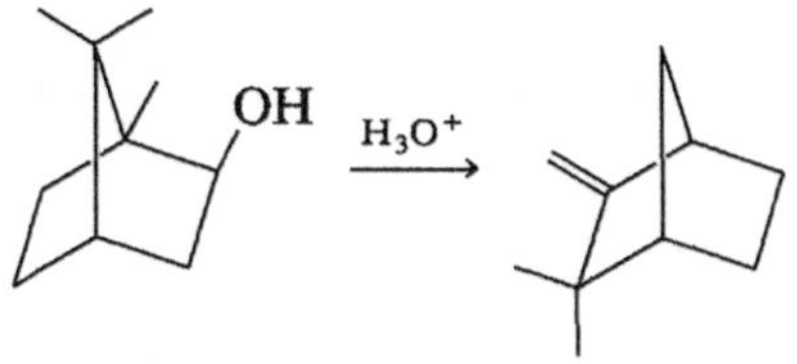

c.

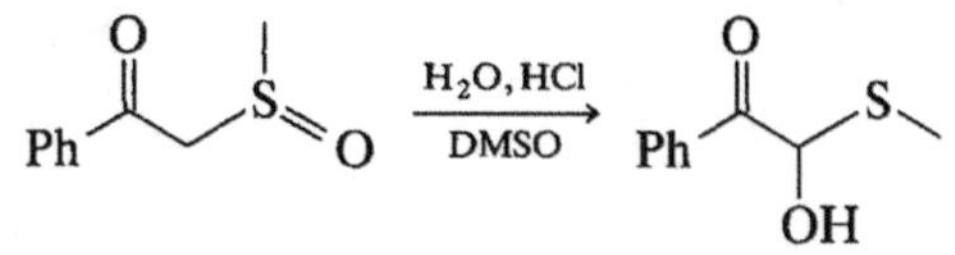

76–81%

Source: Jacobson and Lahm, (1979); House, 1972.

HINT 2.9

When a mechanism involves the removal of a proton, removal of the most acidic proton does not always lead to the product. An example is Problem 2.3c, in which removal of the most acidic proton by base does not lead to the product. (A mechanism for this reaction is proposed in the answer to Problem 2.3c.) Similarly, when a mechanism involves protonation, it is not always protonation of the most basic atom that leads to the product. Such reactions are called unproductive steps. When equilibria are involved, they are called unproductive equilibria.

4. ELECTRON-RICH SPECIES: BASES OR NUCLEOPHILES?

HINT 2.10

A Lewis base, that is, a species with a lone pair of electrons, can function either as a base, abstracting a proton, or as a nucleophile, reacting with a positively charged atom (usually carbon). Which of these processes occurs depends on a number of factors, including the structure of the Lewis base, the structure of the substrate, the specific combination of base and substrate, and the solvent.

EXAMPLE 2.8. ABSTRACTION OF AN ACIDIC PROTON IN PREFERENCE TO NUCLEOPHILIC ADDITION

Consider the reaction of methylmagnesium bromide with 2,4-pentanedione. This substrate contains carbonyl groups that might undergo nucleophilic reaction with the Grignard reagent. However, it also contains very acidic protons (see Appendix C), one of which reacts considerably faster with the Grignard reagent than the carbonyl groups. Thus, the reaction of methylmagnesium bromide with 2,4-pentanedione leads to methane and, after aqueous workup, the starting ketone.

EXAMPLE 2.9. NUCLEOPHILIC SUBSTITUTION IN PREFERENCE TO PROTON ABSTRACTION

$$
\underset{\substack{\| \\ O}}{Ph-C-OCH_2CH_2Cl} \xrightarrow[\substack{87\ ^\circ C,\ 24\ h}]{\substack{NaI \\ Methyl\ ethyl\ ketone}} \underset{\substack{\| \\ O}}{Ph-COCH_2CH_2I} + NaCl
$$

$$
80\%
$$

$$
\underset{\substack{\| \\ O}}{Ph-C-OCH_2CH_2-Cl} \longrightarrow \underset{\substack{\| \\ O}}{PhC-OCH_2CH_2-I} + Cl^-
$$

With iodide ion (I^-), a good nucleophile that is a weak base, substitution is the predominant reaction.

Source: Ford-Moore, (1964).

EXAMPLE 2.10. COMPETITION BETWEEN SUBSTITUTION AND PROTON ABSTRACTION

$$
\underset{\substack{| \\ Br}}{CH_3-CH-CH_3} \xrightarrow[\substack{CH_3CH_2OH}]{\substack{CH_3CH_2O^-Na^+}} \underset{\substack{| \\ OCH_2CH_3}}{CH_3-CH-CH_3} + CH_3-CH=CH_2
$$

$$
\sim 50\% \qquad \sim 50\%
$$

$$
CH_3CH_2O^- \quad \underset{\substack{| \\ Br}}{CH_3-CH-CH_3} \longrightarrow \underset{\substack{| \\ OCH_2CH_3}}{CH_3-CH-CH_3} + Br^-
$$

$$
CH_3CH_2O^- \quad H \quad \underset{\substack{| \\ Br}}{CH_3-CH-CH_2} \longrightarrow CH_3-CH=CH_2 + CH_3CH_2OH + Br^-
$$

In the reaction of isopropyl bromide with sodium ethoxide in ethanol, the dual reactivity of sodium ethoxide is apparent. Ethoxide ion can act as a nucleophile, displacing bromide ion from carbon to produce isopropyl ethyl ether, or it can remove a proton, with simultaneous loss of bromide ion, to produce propene.

If you know the product of a reaction, usually it is not too difficult to determine whether an electron-rich reagent is acting as a base or as a nucleophile. Predicting the course of a reaction can be a more difficult task. However, as you work through a number of examples and problems, you will start to develop a feel for this as well.

5. TRIMOLECULAR STEPS

Trimolecular steps are rare because of the large decrease in entropy associated with three molecules simultaneously assuming the proper orientation for reaction.

HINT 2.11

Avoid formulating mechanisms involving trimolecular steps. Instead, try to break a trimolecular step into two or more bimolecular steps.

EXAMPLE 2.11. BREAKING A TRIMOLECULAR STEP INTO SEVERAL BIMOLECULAR STEPS

When mechanisms for the following reaction are considered,

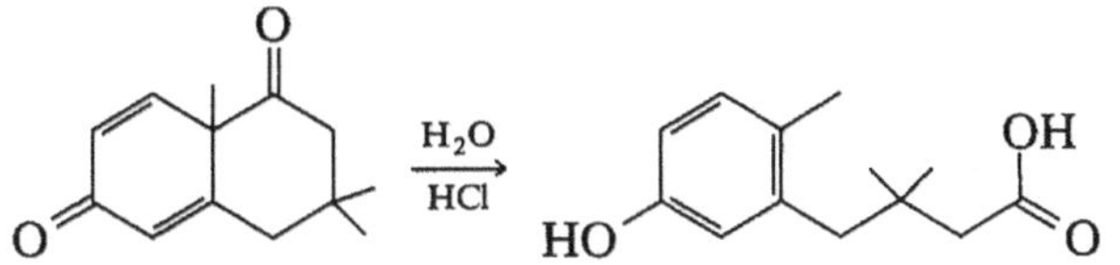

one of the steps could be written as a trimolecular reaction:

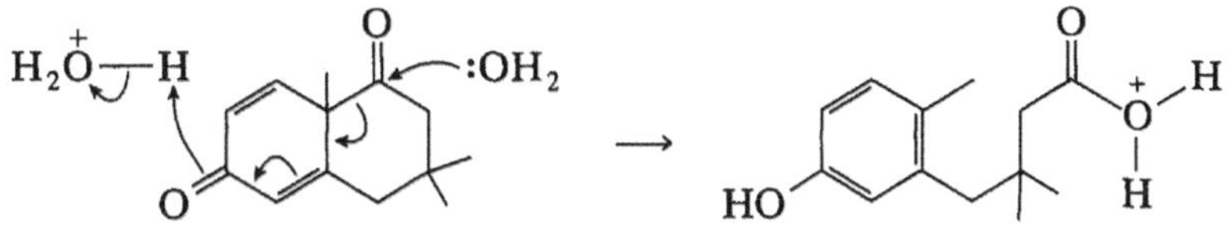

Note that, in the reaction that produces intermediate **2-3**, only one lone pair of electrons is shown on the water molecules. This follows the common practice of selectively omitting lone pairs from Lewis structures and showing only the lone pairs actually taking part in the reaction. Intermediate **2-3** can lose a proton to water to give the product.

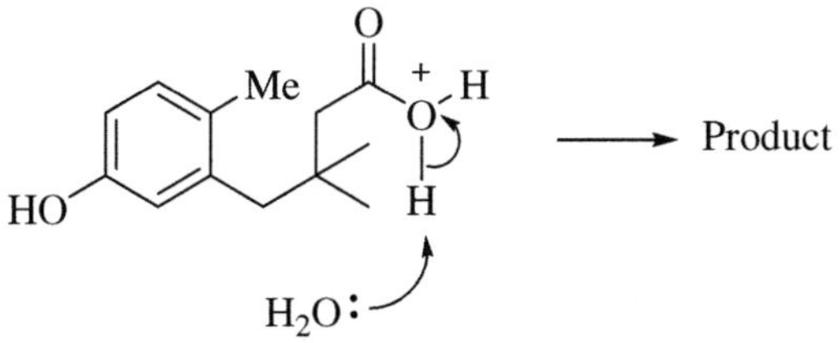

However, we can also write another mechanism, which avoids the trimolecular step:

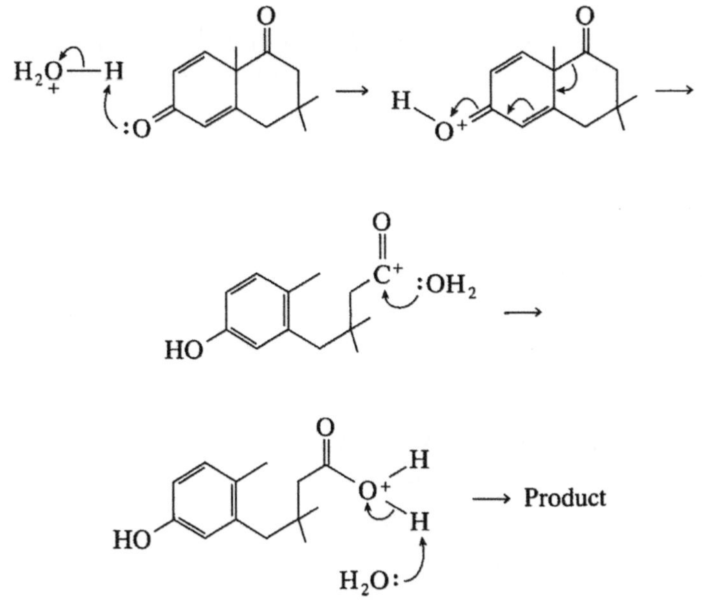

See Problem 4.11a for an alternative mechanism for this reaction.

6. STABILITY OF INTERMEDIATES

Any intermediate written for a reaction mechanism must have reasonable stability. For example, second row elements (e.g., carbon, nitrogen, and oxygen) should not be written with more than eight valence electrons, although third row elements like sulfur and phosphorus can, and do, expand their valence shells to accommodate 10 (occasionally more) electrons. In addition, positively charged carbon, nitrogen, and oxygen species with only six valence electrons are generally formed with difficulty. Although carbocations (six electrons) are high-energy intermediates that are encountered in many reactions, the corresponding positively charged nitrogen and oxygen species with six electrons are rare, especially for oxygen (for an example of an electron-deficient nitrogen species, the nitrenium ion, see Chapter 4.

HINT 2.12

Nucleophilic reaction cannot occur at a positively charged oxygen or nitrogen that has a filled valence shell. Only eight electrons can be accommodated by elements in the second period of the periodic table. However, third period elements, like sulfur and phosphorus, can (and do) expand their valence shells to accommodate 10 (occasionally more) electrons (see Hint 2.3).

EXAMPLE 2.12. NITROGEN CANNOT ACCOMMODATE MORE THAN EIGHT ELECTRONS IN THE VALENCE SHELL

The following step is inappropriate because, in the product, nitrogen has expanded its valence shell to 10 electrons.

To avoid this situation, the π electrons in the double bond could move to the adjacent carbon, giving an internal salt called an ylide. Although they are rather unstable, ylides are intermediates in some well-known reactions.

If a neutral nucleophile reacted with nitrogen in a similar manner, there would be three charges on the product. The two positive charges on adjacent atoms would make this a very unstable intermediate.

With the preceding reagents, a more appropriate reaction would be

HINT 2.13

In writing a mechanism, avoid intermediates containing positively charged nitrogen or oxygen ions with less than eight electrons. These species are rare and have high energy because of the high electronegativities of oxygen and nitrogen.

EXAMPLE 2.13. HOW TO AVOID WRITING MECHANISMS WITH ELECTRON-DEFICIENT, POSITIVELY CHARGED OXYGEN AND NITROGEN SPECIES

Take a look at the following mechanistic steps:

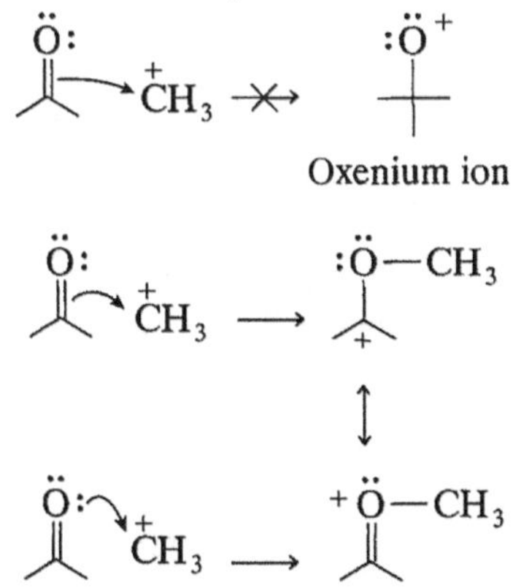

Oxenium ion

If the electrophile bonds to the carbon, the process generates an oxenium ion, a highly unstable species. If the electrophile bonds to the oxygen, the process generates a resonance-stabilized carbocation. Note that if we depict the bond being formed by the bonding electron pair between carbon and oxygen, we obtain the product with the electrophile bonded either to carbon or to oxygen. If we use the lone pair of electrons on oxygen, we obtain the product in which the electrophile is bonded to oxygen. It does not matter which pair of electrons we use as long as we draw correct Lewis structures and obtain intermediates that have reasonable stability (for another example of this, see the answer to Problem 2.5b). Our choice of mechanism is based not on which electrons we choose to "push", but on the stability of the intermediate formed.

The situation with nitrogen is analogous. The nitrenium ion is highly unstable, and the carbonium ion is resonance stabilized.

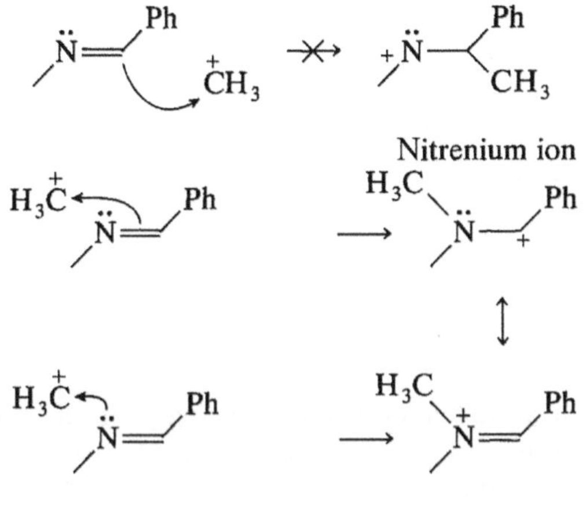

Nitrenium ion

The preceding reaction steps present another difficulty, namely, attack by the methyl cation. Primary carbocations that lack stabilizing groups are highly unstable, and the methyl cation is the least stable of the carbocations. In fact, even in "superacid" (FSO_3H-SbF_5), no primary carbocation is stable enough to be detected. Consequently, a mechanism that invokes such a species should be looked upon with suspicion.

7. DRIVING FORCES FOR REACTIONS

HINT 2.14

A viable reaction should have some energetic driving force. Examples include formation of a stable inorganic compound (a salt for example); formation of a stable double bond or aromatic system; formation of a stable carbocation, anion, or radical from a less stable one; and formation of a stable small molecule (see Hint 2.15). The progress of the reaction is made even more favorable if the small molecule lost is a gas that is released and escapes from the reaction mixture.

A reaction may be driven by a decrease in enthalpy, an increase in entropy, or a combination of the two. Reactions driven by entropy often involve the formation of more product molecules from fewer starting molecules. Reactions that form more stable bonds are primarily enthalpy driven. When writing a mechanism, constantly ask the following questions: Why would this reaction go this way? What is favorable about this particular step?

A. Leaving Groups

When a reaction step involves a nucleophilic substitution, the nature of the leaving group often is a key factor in determining whether the reaction will occur. In general, *leaving group ability is inversely related to base strength*. Thus, H_2O is a much better leaving group than OH^-, and I^- is a better leaving group than F^-. A list of common leaving groups appears in Table 3.1.

If the reaction involves a poor leaving group, then a very good nucleophile will be necessary to induce the reaction to occur, as the next example illustrates.

EXAMPLE 2.14. A RATIONALE FOR THE INVOLVEMENT OF DIFFERENT LEAVING GROUPS IN THE ACID- AND BASE-PROMOTED HYDROLYSIS OF AMIDES

Ammonia is the leaving group in the acid-promoted hydrolysis of amides. Amide ion, $^-NH_2$, is the leaving group in the base-promoted hydrolysis. The difference can be explained by the driving force of the intramolecular nucleophile relative to the ability of amide ion or ammonia to act as a leaving group. In acid, the intramolecular nucleophile is the oxygen of one of the hydroxyl groups of the tetrahedral intermediate.

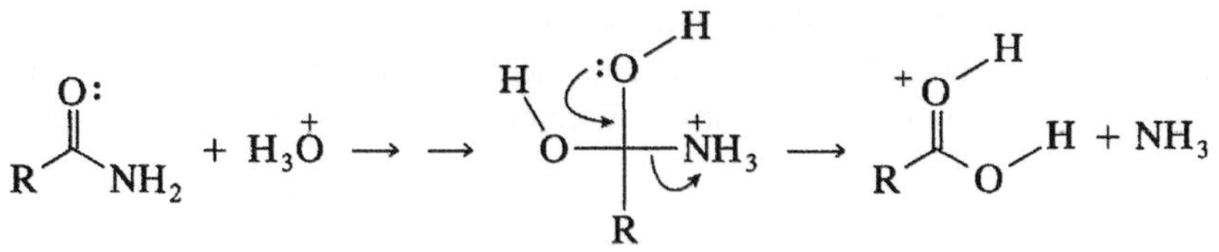

In base, the intramolecular nucleophile is the oxyanion of the tetrahedral intermediate:

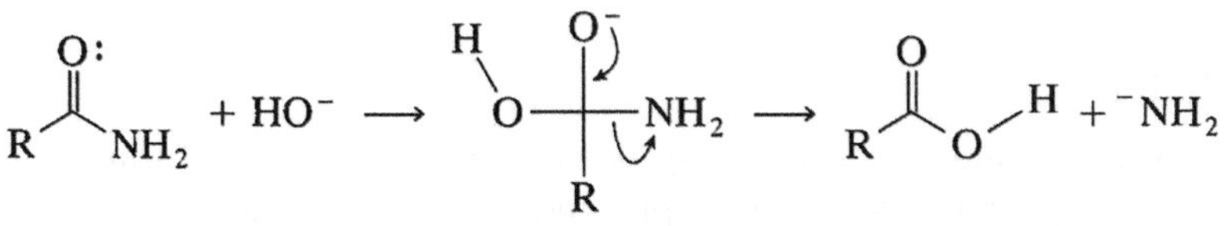

  The hydroxyl group is not a strong enough nucleophile, even intramolecularly, to drive the loss of an amide ion, so that under acidic conditions, the nitrogen must be protonated in order to form a sufficiently good leaving group. The leaving group then becomes ammonia, a better leaving group than amide anion. In base, the oxyanion formed by reaction of the original amide with a hydroxyl ion is a strong enough nucleophile to drive the loss of an amide ion.

B. Formation of a Small Stable Molecule

  Formation of a small stable molecule can be a significant driving force for a reaction because this involves a decrease in enthalpy and an increase in entropy.

HINT 2.15

  A frequent driving force for a reaction is formation of the following small stable molecules: nitrogen, carbon monoxide, carbon dioxide, water, and sulfur dioxide.

EXAMPLE 2.15. LOSS OF CARBON MONOXIDE IN THE THERMAL REACTION OF TETRAPHENYLCYCLOPENTADIENONE WITH MALEIC ANHYDRIDE

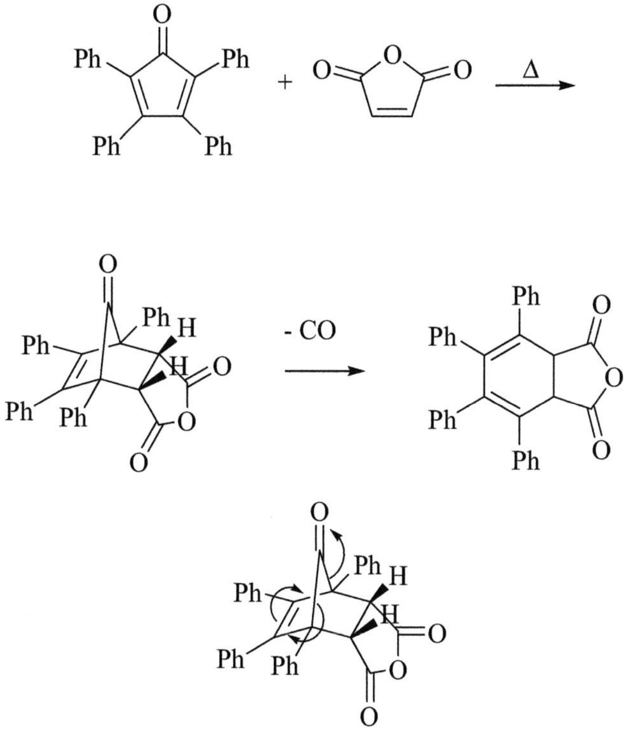

8. STRUCTURAL RELATIONSHIPS BETWEEN STARTING MATERIALS AND PRODUCTS

Numbering of the atoms in the starting material and the product can help you determine the relationship between the atoms in the starting material and those in the product.

HINT 2.16

Number the atoms of the starting material in any logical order. Next, by looking for common sequences of atoms and bonding patterns identify atoms of the product that correspond with atoms of the starting material and assign to the atoms of the product the corresponding number of the atoms from the starting material. Then, *using the smallest possible number of bond changes*, fill in the rest of the numbers.

EXAMPLE 2.16. USING A NUMBERING SCHEME WHEN WRITING A MECHANISM

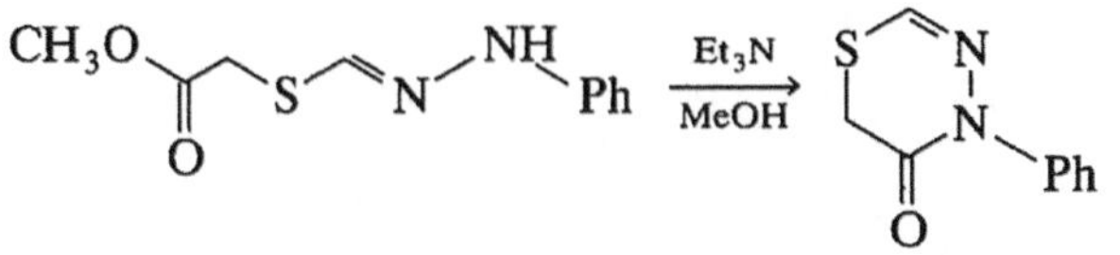

Numbering of the atoms in the starting material and product makes it clear that nitrogen-1 becomes attached to carbon 6.

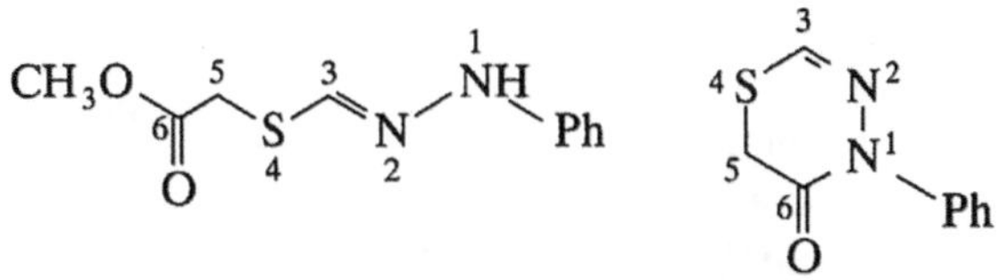

With this connection established, we can write the mechanism as follows:

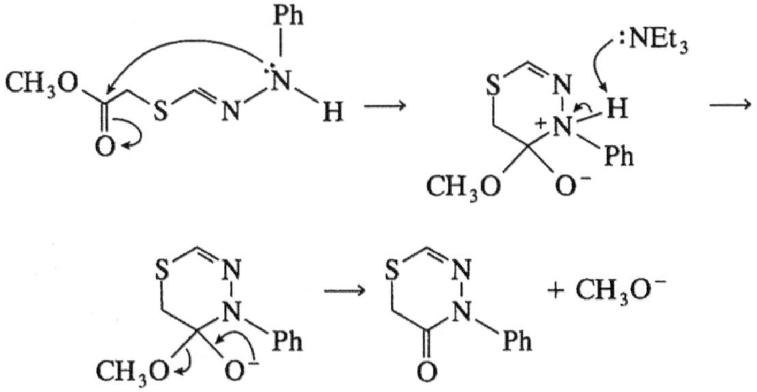

The other products, methoxide ion and triethylammonium ion, would equilibrate to give the weakest acid and weakest base (see Appendix C).

$$CH_3O^- + H\overset{+}{N}Et_3 \rightleftharpoons CH_3OH + NEt_3$$

EXAMPLE 2.17. USING A NUMBERING SCHEME TO DECIDE WHICH BONDS HAVE BEEN FORMED AND WHICH BROKEN

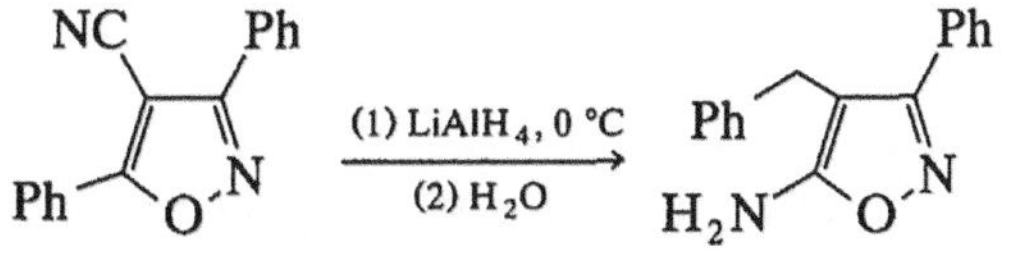

First, consecutively number the atoms in the starting material. In this example, the atoms in the product can be numbered by paying close attention to the location of the phenyl groups and nitrogens:

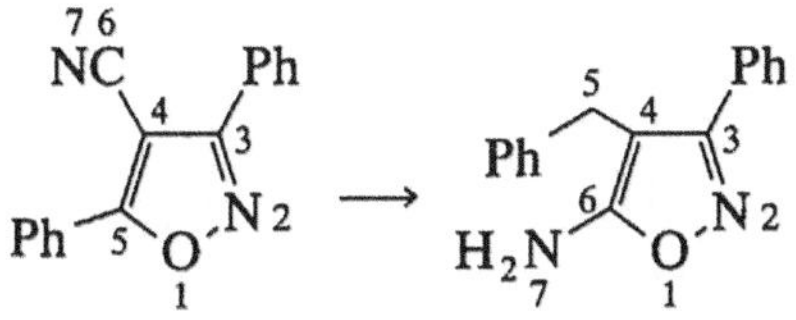

Without having to write any mechanistic steps, the numbering scheme allows us to decide that the bond between C-5 and O-1 breaks and that a new bond forms between O-1 and C-6. This numbering scheme gives the least possible rearrangement of the atoms when going from starting material to product. This information is invaluable when writing a mechanism for this reaction.

This example is derived from Alberola et al. (1984); a mechanism is suggested in this paper.

9. SOLVENT EFFECTS

Usually, the primary function of a solvent is to provide a medium in which reactants and products can come into contact with one another and interact. Accordingly, solubility dictates the choice of solvent for many organic reactions. However, the nature of the solvent can influence the mechanism of a reaction, and sometimes the choice of solvent dictates the pathway by which a reaction proceeds. In terms of effect on the mechanism, interactions of polar solvents with polar reagents are the most important. Accordingly, solvents can be divided into three groups:

1. Protic solvents, e.g., water, alcohols, and acids.
2. Polar aprotic solvents, e.g., dimethylformamide (DMF), dimethyl sulfoxide (DMSO), acetonitrile (CH_3CN), acetone, sulfur dioxide, and hexamethylphosphoramide (cyclopentyl methyl ether, N-formyl morpholine, and dimethylisosorbide are "greener" alternatives).
3. Nonpolar solvents, e.g., chloroform, tetrahydrofuran, 2-methyltetrahydrofuan, ethyl ether, toluene, carbon tetrachloride.

Interactions between a polar solvent and a charged species are stabilizing. Protic solvents can stabilize both anionic and cationic species, whereas polar aprotic solvents stabilize only cationic species. Thus, protic solvents favor reactions in which charge separation occurs in the transition state, the high-energy point in the reaction pathway. In nucleophilic substitution reactions, the pathway where two charged species are formed (i.e., S_N1 reaction) is favored in protic solvents, whereas the pathway with a less polar transition state (i.e., S_N2 reaction) is favored in nonprotic solvents.

The source is *Jessop, (2011)*.

EXAMPLE 2.18. THE INFLUENCE OF SOLVENT ON BASICITY

Chloride ion generally is a moderate nucleophile and a weak base. However, in the following dehydrohalogenation reaction, chloride ion functions as a base, removing a proton to bring about elimination of the elements of HCl.

In this reaction, chloride functions as a base because the reaction is carried out in the polar aprotic solvent DMF. In polar aprotic solvents, cations are stabilized by solvent interaction, but anions do not interact with the solvent. The "bare" chloride anion functions as a base because it is not stabilized by solvent interaction.

For other examples of the effect of solvent on reaction mechanism, see Examples 4.14 and 4.15.

Source: Warnhoff et al., (1963).

HINT 2.17

Use the combination of reagents and solvents specified as a guide to the mechanism. Ionic reagents and polar solvents point to ionic mechanisms. The absence of ionic reagents and use of a nonpolar solvent may suggest a nonionic mechanism.

10. A LAST WORD

The fourteenth-century English philosopher William of Occam introduced the principle known as Occam's razor. A paraphrase of this principle which can be applied to writing organic reaction mechanisms is expressed in Hint 2.18.

HINT 2.18

When more than one mechanistic scheme is possible, the simplest is usually the best.

PROBLEM 2.5

For each of the following transformations, number all relevant nonhydrogen atoms in the starting materials, and number the same atoms in the product.

a.

(See Problem 3.20b for further exploration of this reaction.)

b.

(See Example 4.10 for further exploration of this reaction.)

PROBLEM 2.6

In each of the following problems, an overall reaction is given, followed by a mechanism. For each mechanism shown, identify inappropriate steps, give the number of any applicable hint, and explain its relationship to the problem. Then write a more reasonable mechanism for each reaction.

a.

b. 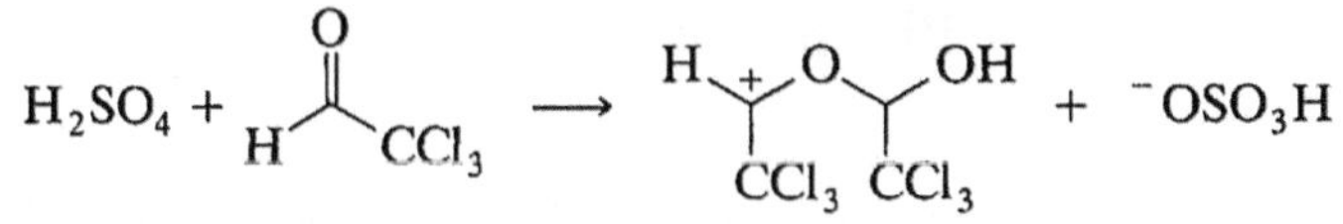 H_2SO_4 + [structure] $\longrightarrow$ [structure] + $^-OSO_3H$

2-5

[reaction scheme] $\longrightarrow$ **2-5**

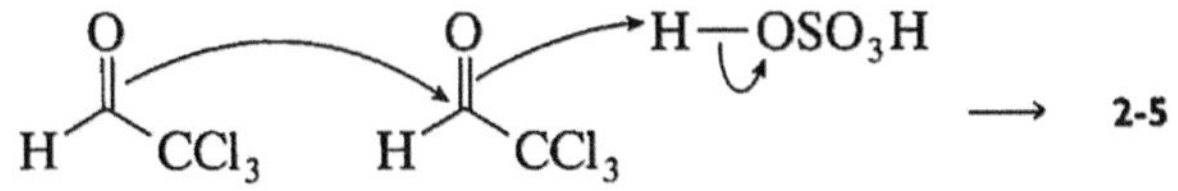

c. 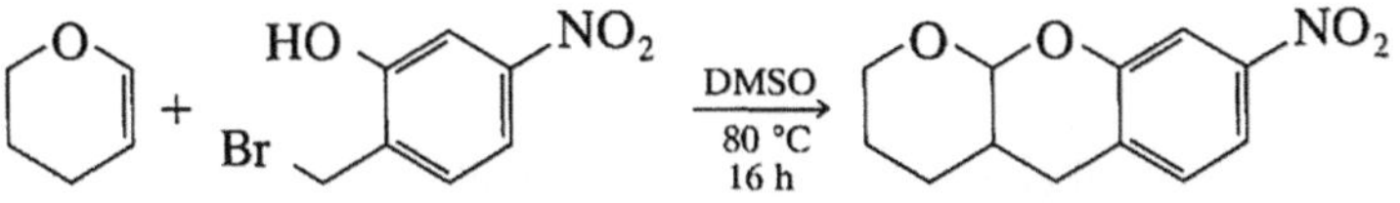 [structure] + [structure] $\xrightarrow[\substack{80\ °C \\ 16\ h}]{DMSO}$ [structure]

2-6

[reaction scheme] $\longrightarrow$ **2-6**

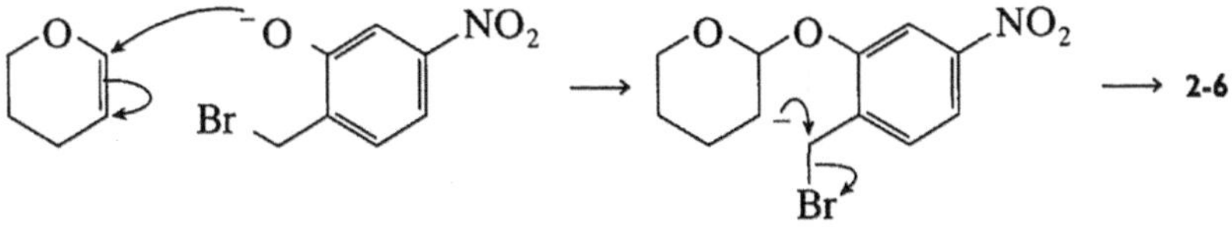

d. 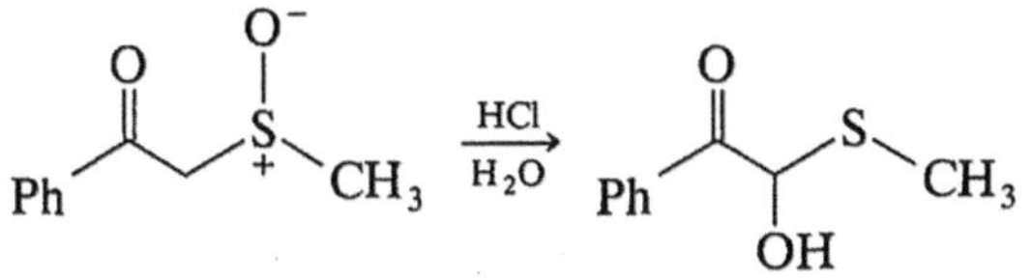 [structure] $\xrightarrow[H_2O]{HCl}$ [structure]

2-7

[reaction scheme] $\longrightarrow$

[reaction scheme]

H_3O^+ + [structure] $\longrightarrow$ **2-7**

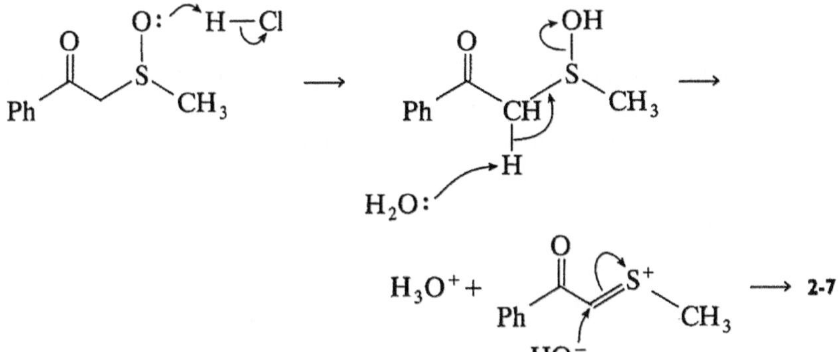

Source: Kirby and Martin, (1983); Kunieda et al., (1983).

ANSWERS TO PROBLEMS

Problem 2.1

a.

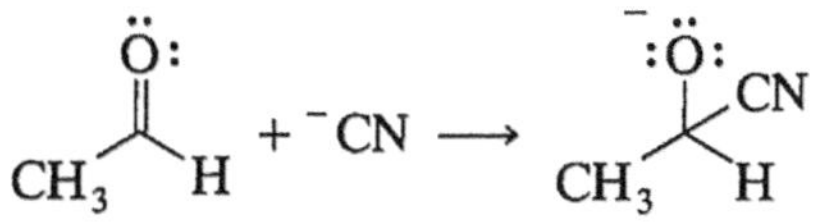

b.

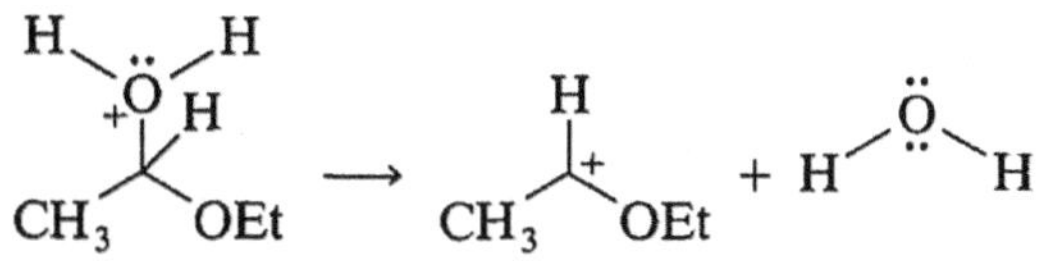

c.

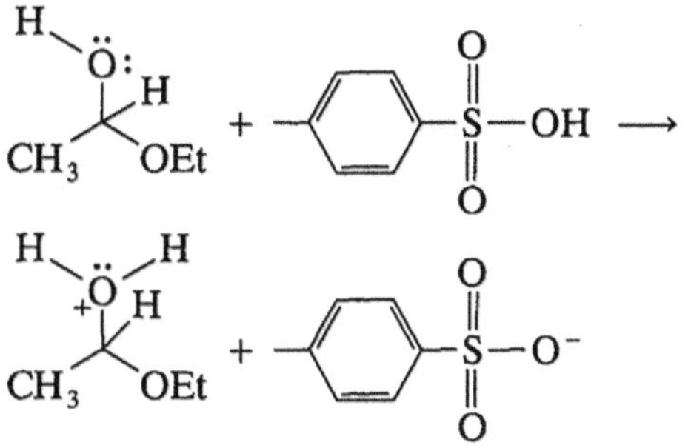

Problem 2.2

a.

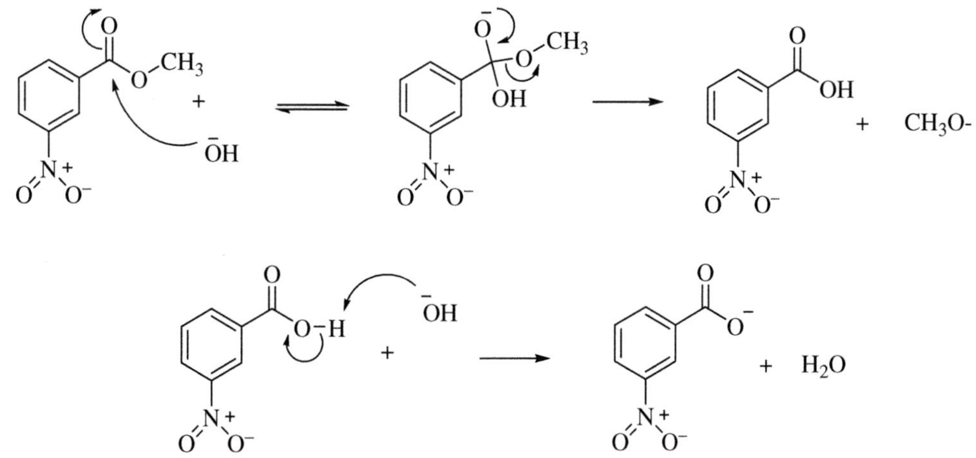

b.

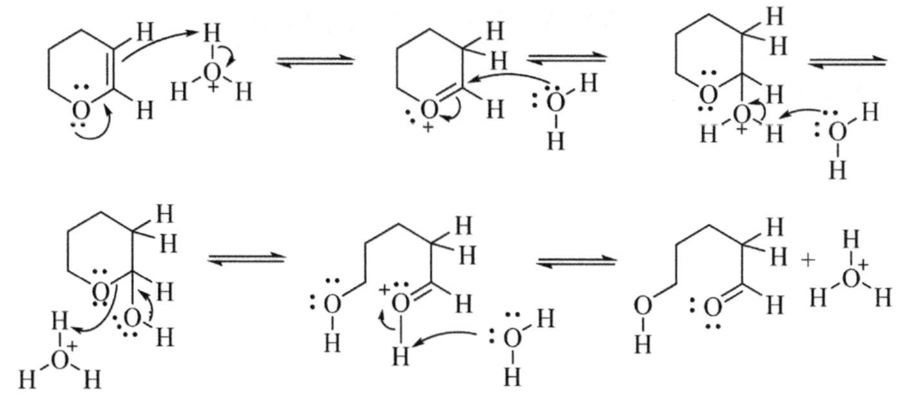

The following transformation is also plausible:

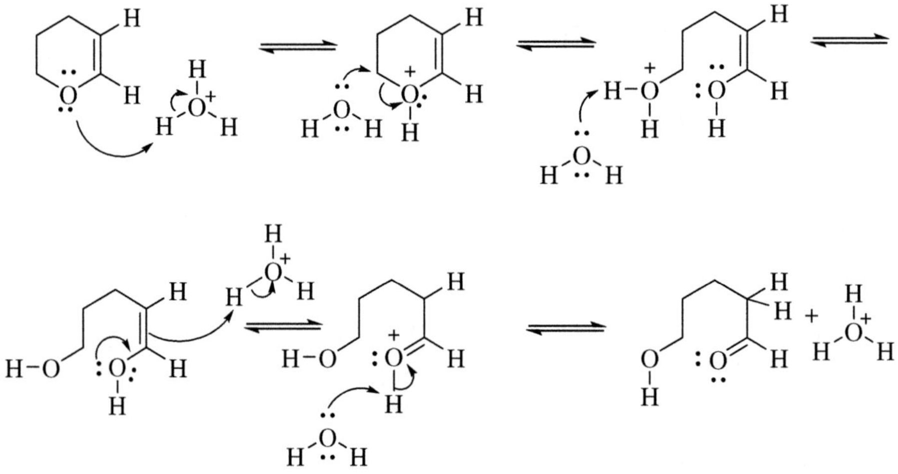

c.

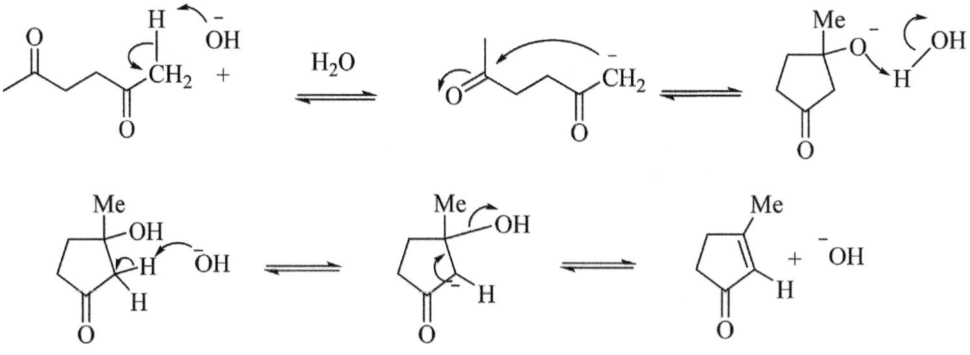

Problem 2.3

a.

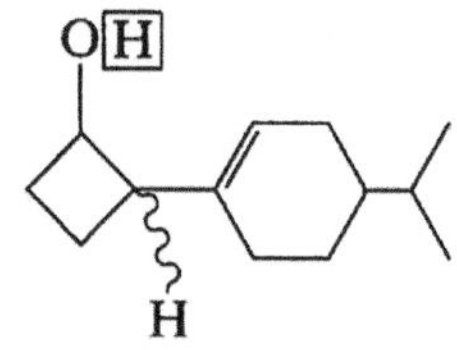

Note: There is some question about the mechanism for this rearrangement. See the paper cited for details.

b.

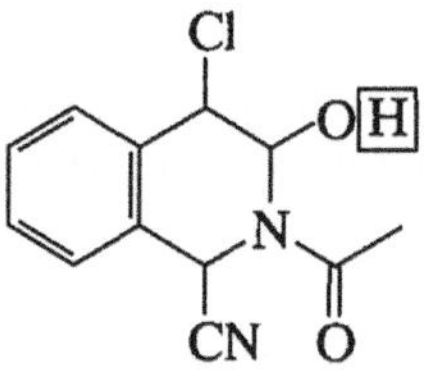

c. Removal leads to product

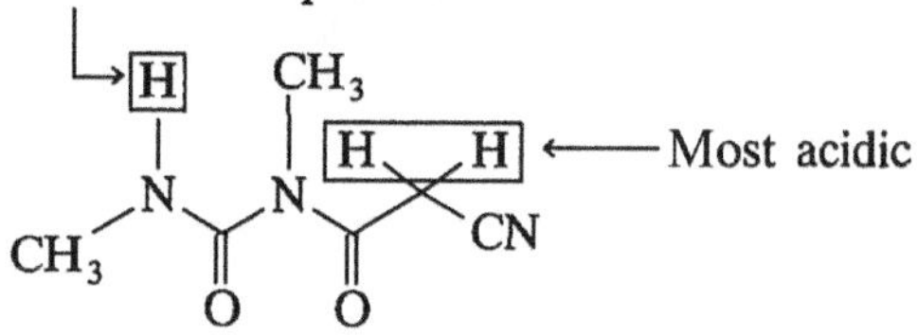

On the basis of acidity, the protons in this molecule can be divided into three groups: the methyl hydrogens, the amide hydrogen, and the methylene hydrogens. Of these, the methyl hydrogens are the least acidic, and the methylene hydrogens are the most acidic. We can estimate the acidity of the protons by examining the stabilities of the anions that result when different hydrogens are removed from the molecule. Removal of a proton from either methyl group gives an anion stabilized only by the inductive effects of nitrogen; no resonance forms are possible. Removal of hydrogen from the amide nitrogen gives an anion stabilized by two resonance forms, one with negative character on nitrogen and one with negative character on oxygen:

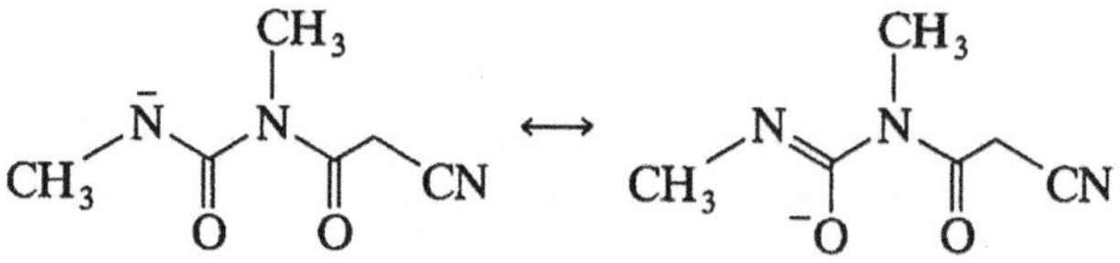

Finally, removal of hydrogen from the methylene group gives an anion with three resonance forms in which the negative charge is on nitrogen, oxygen, or carbon:

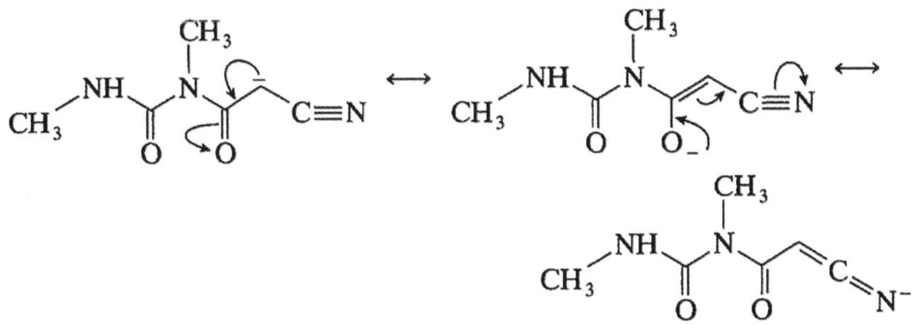

Thus, the anion formed by loss of a proton from the methylene group should be more stable. We can also compare the acidities of the various protons by using the pK_a values listed in Appendix C.

Acid	Conjugate base	pK_a
CH_3CNH_2 (C=O)	$CH_3C\overset{-}{N}H$ (C=O)	25.5
$NCCH_2COCH_3$ (C=O)	$NC\overset{-}{C}HCOCH_3$ (C=O)	12.8
$CH_3COCH_2CH_3$ (C=O)	$\overset{-}{C}H_2COCH_2CH_3$ (C=O)	30
$CH_3CN(CH_2CH_3)_2$ (C=O)	$\overset{-}{C}H_2CN(CH_2CH_3)_2$ (C=O)	34.5

Using these values, there are several ways to compare the acidity of the protons in the boxes. The pK_a of the amide protons in acetamide is 25.5. The pK_a of the amide proton in this compound should be somewhat greater, because the second amido nitrogen will reduce the resonance stabilization of the anion formed when the amide proton is removed. The precedent for the other boxed protons is the pK_a of the methylene protons of methyl cyanoacetate, 12.8. Because the carbonyl group in the given compound is an amide rather than an ester, the pK_a of its protons will be somewhat higher, but not nearly as high as the value for the amide proton.

We can arrive at the same conclusion in a different way using an alternative analysis based on the pK_a values in Appendix C. The pK_a of the protons on the α carbon of ethyl acetate is 30, whereas the pK_a of those of methyl cyanoacetate is 12.8. Therefore, a cyano group lowers the pK_a of the α carbon protons by $30-12.8 = 17$ units. The pK_a of the α carbon protons of an N,N-disubstituted amide, N,N-diethylacetamide, is 34.5. If we assume that the cyano group enhances the acidity of the methylene protons in the given compound by the same amount, their pK_a would be $34.5-17 = 16.5$. Thus, the estimated pK_a of the methylene protons is at least 9 pK units lower than that of the amide proton, a factor of 1 billion.

The mechanism for the reaction could be written as follows. First, the proton is removed by base:

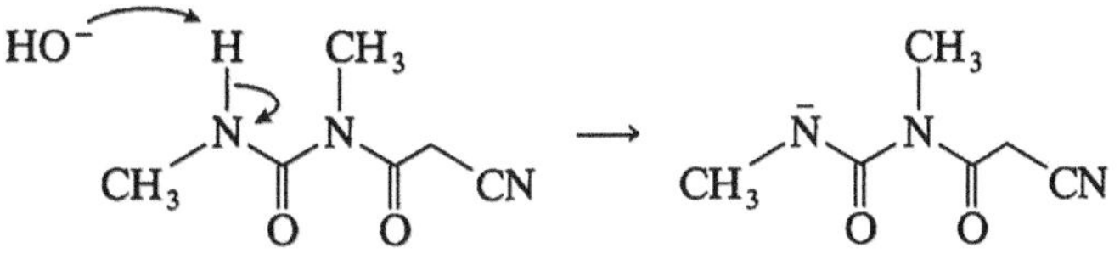

2-8

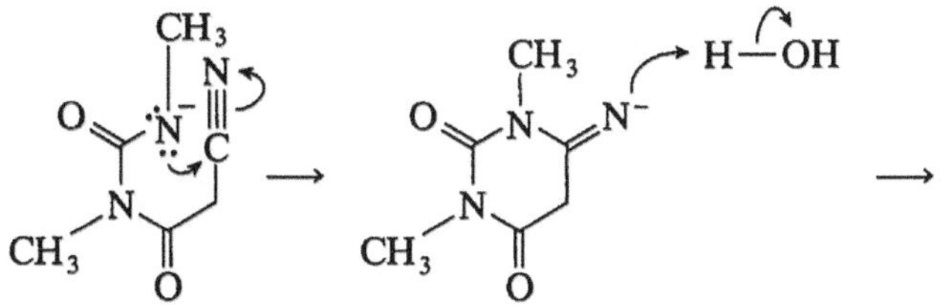

2-8

The resulting anion, **2-8**, can be written in a conformation that makes it clear how the subsequent cyclization takes place.

Now the intermediate tautomerizes to give the final product.

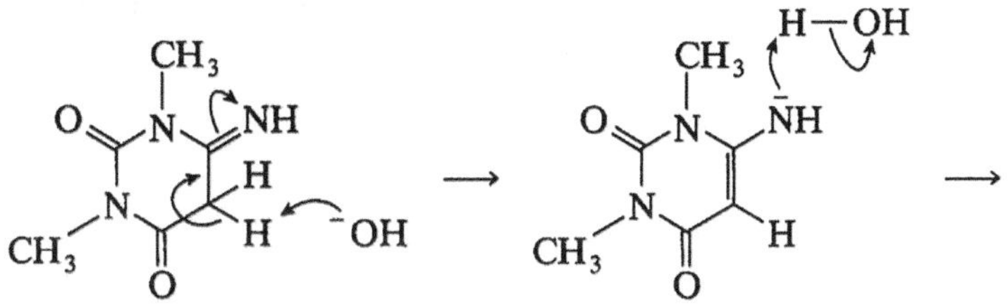

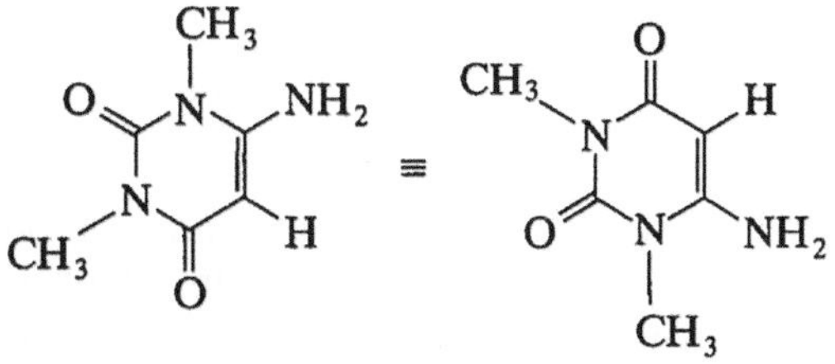

d.

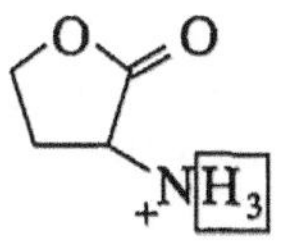

Problem 2.4

a.

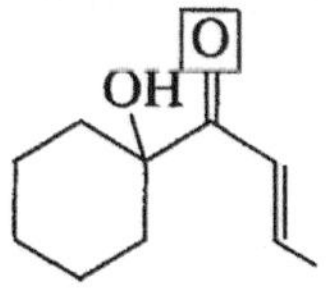

Table 1.3 and Appendix C indicate that carbonyl oxygens are more basic than the oxygen of an alcohol. The protonated carbonyl is stabilized by resonance, whereas the protonated alcohol is not. In this case, the protonated carbonyl is further stabilized by delocalizations of charge onto the C=C double bond.

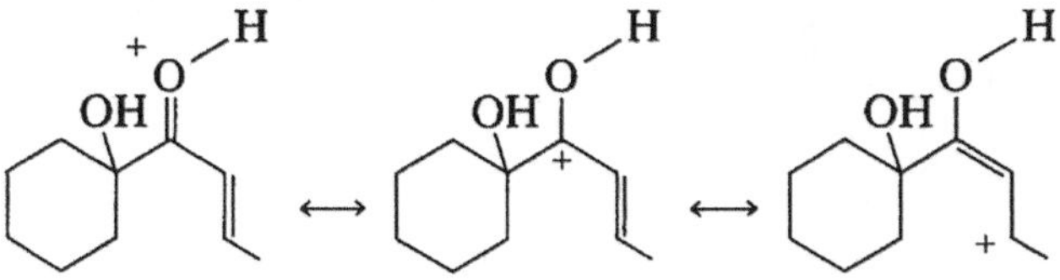

In the paper cited (Jacobsen and Lahm, 1979), the mechanism proposed suggests that the starting material is dehydrated before cyclization occurs. This means that protonation of the less basic group leads to the product. Satisfy yourself that this is reasonable by writing possible reaction mechanisms that would result from protonation of the hydroxyl and carbonyl hydrogens.

b.

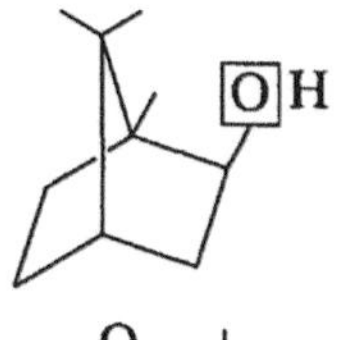

c.

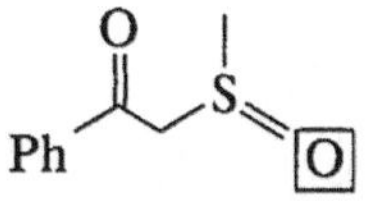

Appendix C is of help here. The pK_a of protonated acetophenone is -4.3, whereas the pK_a of protonated DMSO is -1.5. Thus, DMSO is almost 10^3 times more basic than aceto-phenone. These compounds are excellent models for the two functional groups in the compound given.

If hard data like those in Appendix C are unavailable, it is very difficult to decide which oxygen is more basic, by simply considering the functional groups and the structure of the molecule. On one hand, because sulfur is slightly more electronegative than carbon (see Table 1.2), one might predict that the carbonyl oxygen is more basic than the sulfoxide oxygen.

Furthermore, protonation of the carbonyl group leads to a cation that is stabilized by resonance delocalization onto the aromatic ring:

On the other hand, the π bond between C and O should be stronger than the π overlap between S and O because C and O both use $2p$ orbitals to form π bonds, whereas a π bond between S and O utilizes the less effective $3p–2p$ overlap. This should make the oxygen of the sulfoxide functional group more negative. Additionally, sulfur stabilizes charges more effectively than do carbon and oxygen because of its higher polarizability.

In the absence of hard data, qualitative arguments suggest that both oxygens are fairly basic. In a situation like this, it is usually best to adopt a trial-and-error approach, first writing mechanisms that start with the protonation of one functional group and then writing mechanisms that start with the protonation of the other functional group. Remember, a mechanism does not always proceed through protonation of the most basic atom.

Problem 2.5

a. Start numbering the product at the carboethoxy group, which has not been altered by the reaction. Once you notice that the keto group of the ketoester reactant is also unchanged, the remaining atoms easily fall into place. For the isocyanate, N-6 is obvious, and C-7 and O-8 follow.

b. The first thing to notice is that the position of the C—O bond, relative to the methyl group, has not changed. Thus, an initial attempt to number the product would leave those atoms in the same relative positions as in the starting material. This gives the following numbers:

A strong possibility is that the left-hand ring of the product contains the same carbons that this ring contained in the starting material. By minimizing the changes in bonding, the following results:

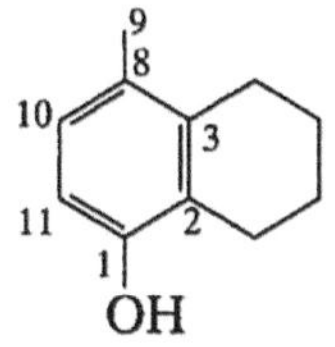

This leaves two possible numbering schemes for the right-hand ring:

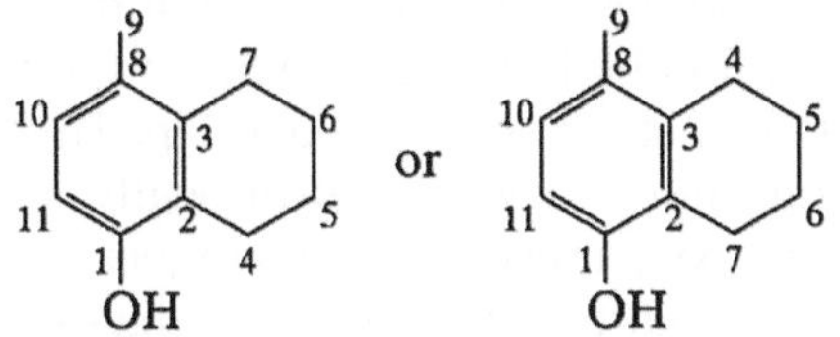

The structure on the right involves less rearrangement; however, if we examine the symmetrical intermediate involved in this reaction (see Example 4.10), we see that, in fact, the two numbering schemes are equivalent in terms of bond reorganization.

Problem 2.6

a. Hint 2.6: This reaction is taking place in a strongly acidic medium. Therefore, strong bases like ⁻OH will be in such low concentration that they cannot be effective reagents in the reaction. A better mechanism would be to protonate the oxygen of one of the hydroxyl groups to convert it into a better (and neutral) leaving group, the water molecule.

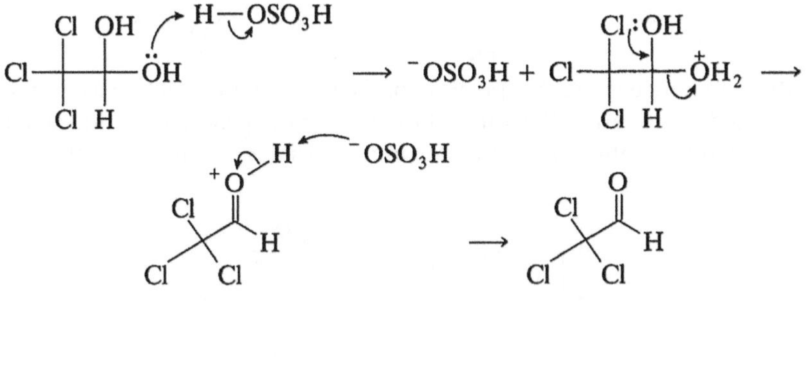

b. Hint 2.9: Break up the trimolecular step. First, protonate the carbonyl group, converting it to a better electrophile. Follow this with nucleophilic reaction with the carbonyl group by the oxygen of another molecule of starting material.

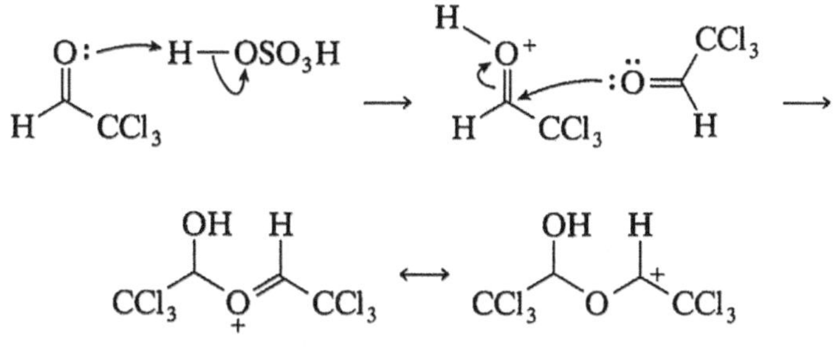

The nucleophilic addition of the carbonyl compound to another protonated molecule can be written either as shown or by using the π bond as the nucleophile instead of a lone pair on oxygen. This second kind of addition leads to the second resonance form shown instead of the first:

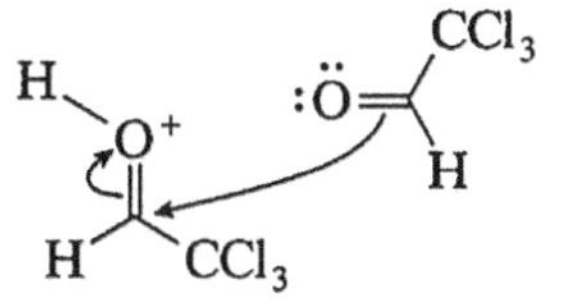

The two representations for this step are equivalent.

The oxygen of one starting aldehyde molecule must be protonated before nucleophilic addition of the carbonyl oxygen of the second. Otherwise, a very basic anion is formed in a very acidic medium (the same problem discussed in part a).

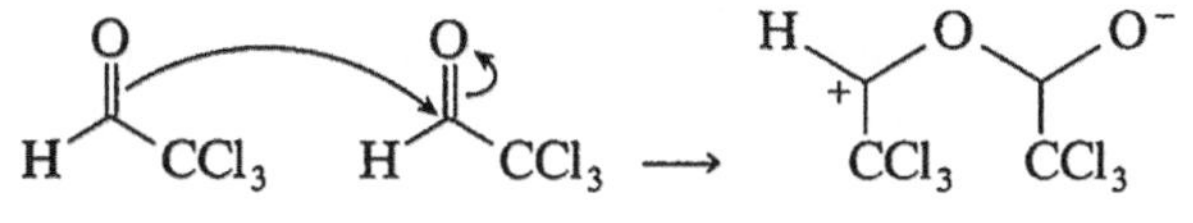

c. There are two major problems with the mechanism shown. Data from Appendix C indicate that the substituted phenol would not be ionized significantly in DMSO. Protonated DMSO has a pK_a of -1.5, whereas m-nitrophenol has a pK_a of 8.3. Also, the intermediate carbanion, resulting from the addition, is very unstable; it is not stabilized by resonance in any way. The following mechanism avoids these problems by forming an intermediate cation, **2-9**, which is stabilized by resonance, and by leaving removal of the proton to the last step.

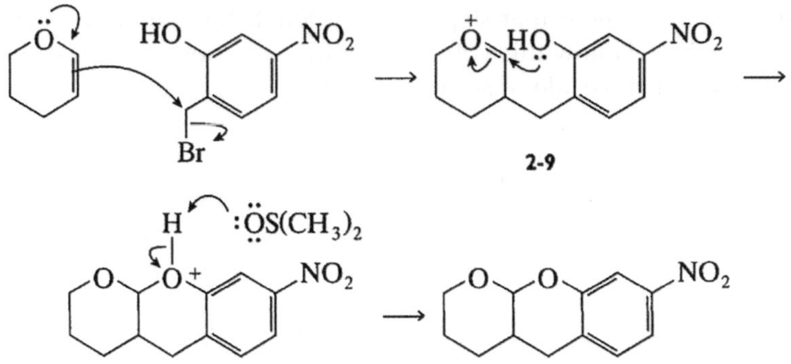

2-9

Intermediate **2-9** could also be written as the other resonance form, **2-10**, which contributes to the stabilization of the positive charge:

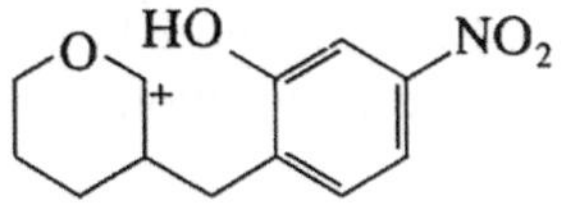

2-10

Often it is useful to draw resonance forms for proposed intermediates because their existence is an indication of stability, which represents a driving force for the reaction. Removal of a proton from the OH group of the resonance hybrid of **2-9** and **2-10** is unlikely because bromide ion is an even weaker base than DMSO (the pK_a HBr is -9). Therefore, it is expected that ring closure, by nucleophilic reaction with the cation, takes place before removal of the proton.

d. Hint 2.6: The given mechanism has two steps that are unlikely in strong acid: loss of hydroxide ion and its subsequent nucleophilic reaction. Water is a much better leaving group than hydroxide, so that in acidic solution, where protonation of the oxygen can occur, water acts as the leaving group. Notice that because hydrochloric acid is completely dissociated in aqueous solution, hydronium ion, not hydrochloric acid, acts as the protonating agent. In acidic solution, the concentration of hydroxide ion is so low that, although hydroxide ion is a better nucleophile than water, it cannot compete with water for the nucleophilic site.

It is always a good idea to keep track of lone pairs of electrons on heteroatoms, such as oxygen and sulfur, by drawing them in. As the mechanism proceeds, it also is a good idea to keep track of formal charges on the atoms of interest, in this case oxygen and sulfur.

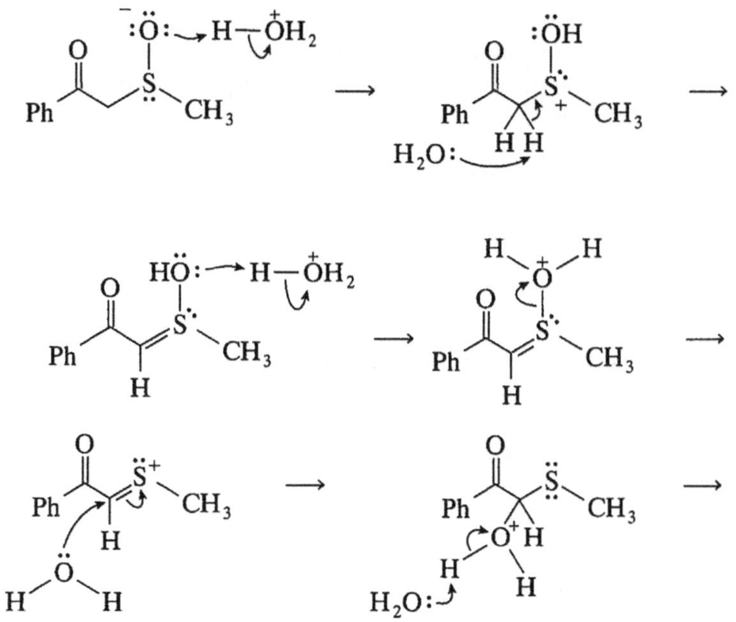

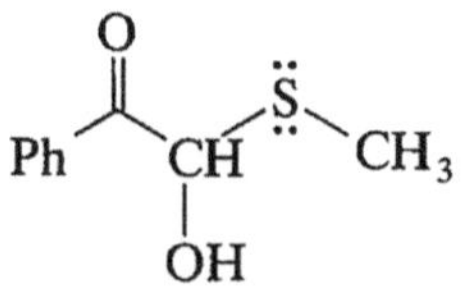

Note that the first two steps of the mechanism represent a tautomerization, in which the overall result is the movement of a proton from carbon to oxygen and movement of a double bond. Notice, also, that there are several intermediates in which sulfur has an expanded octet.

Take a look at some alternative mechanistic steps:

(i) One possibility is formation of 2-12 (by doubly protonating the sulfoxide oxygen of the starting material), from which simultaneous elimination of water and loss of a proton from the methylene group might be written. This would not be as good a step as those shown previously because the development of two adjacent positive centers is destabilizing.

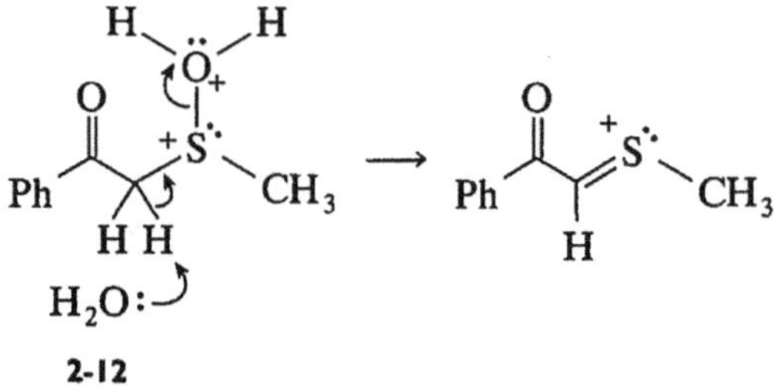

(ii) Removal of a proton always requires reaction with a base, which in this step could be water. Thus, the following representation is not strictly correct because no base is indicated to remove the proton; however, as stated previously, this would be considered acceptable by a number of instructors.

(iii) Showing **2-13** for removal of a proton is not accurate for two reasons. First, the arrow indicates electron movement in the wrong direction; this would produce H^- instead of H^+ (see Chapter three for further discussion of hydride loss). Second, proton removal always requires reaction with a base, even if it is a weak base.

2-13

(iv) The following equation is an example of a [1,3] sigmatropic intramolecular shift of hydrogen. Chapter 6 discusses why this type of tautomeric reaction is unlikely to occur.

References

Alberola, A.; Gonzalez, A. M.; Laguna, M. A.; Pulido, F. J. *J. Org. Chem.* **1984,** *49* (18), 3423–3424.

Bernier, J. L.; Henichart, J. P.; Warin, V.; Trentesaux, C.; Jardillier, J. C. *J. Med. Chem.* **1985,** *28* (4), 497–502.

Cohen, T.; Bhupathy, M.; Matz, J. R. *J. Am. Chem. Soc.* **1983,** *105* (3), 520–525.

Ford-Moore, A. H. *Org. Synth., Coll.* **1964,** Vol. 4, 84.

House, H. O. *Modern Synthetic Reactions*, 2nd ed.; Benjamin: Menlo Park, CA, 1972, 726.

Jacobson, R. M.; Lahm, G. P. *J. Org. Chem.* **1979,** *44* (3), 462–464.

Jaffe, K.; Cornwell, M.; Walker, S.; Lynn, D. G. *Abstracts of Papers, 190th National Meeting of American Chemical Society, Chicago*; American Chemical Society: Washington DC, 1985. ORGN 267.

Jessop, P. G. *Green. Chem.* **2011,** *13* (6), 1391–1398.

Kirby, A. J.; Martin, R. J. *J. Chem. Soc., Perkin Trans. II* **1983,** (9), 1627–1632.

Kirby, G. W.; Mackinnon, J. W. M.; Elliott, S.; Uff, B. C. *J. Chem. Soc. Perkin Trans. 1* **1979,** 1298–1302.

Kirby, G. W.; McGuigan, H.; Mackinnon, J. W. M.; Mallinson, P. R. *J. Chem. Soc. Perkin Trans. 1* **1985,** 405–408.

Kunieda, N.; Fujiwara, Y.; Suzuki, A.; Kinoshita, M. *Phosphorus Sulfur* **1983,** *16*, 223–232.

Warnhoff, E. W.; Martin, D. G.; Johnson, W. S. *Org. Synth. Coll.* **1963,** Vol. 4, 162.

3

Reactions of Nucleophiles and Bases

Many reactions of nucleophiles and bases present an array that often is bewildering. These reactions usually are organized on the basis of the reacting group and the overall reaction (e.g., nucleophilic addition to carbonyl groups), which gives rise to a very large number of categories. Although there are a very large number of reactions, it soon becomes apparent that we can write mechanisms for most of them once we are familiar with a few general patterns and the general principles outlined in Chapter 2. As an example, consider that the hydrolysis of an ester in base is classified as a nucleophilic substitution at an aliphatic sp^2 carbon, whereas the reaction of hydrazine with a ketone is classified as a nucleophilic addition. However, as the following reactions show, both reactions involve nucleophilic addition to the carbonyl group, followed by loss of a leaving group.

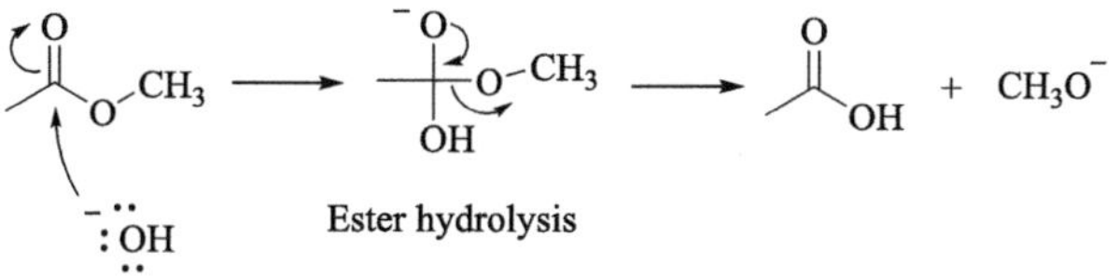

Ester hydrolysis

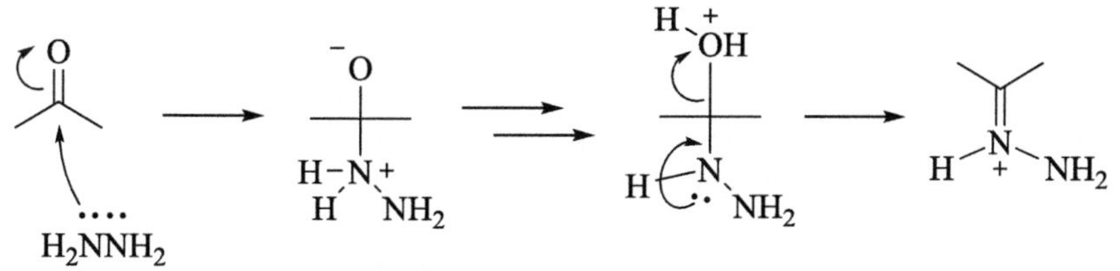

Hydrazone formation

As you work your way through a number of reaction mechanisms, you will find that mechanistic patterns offer a way to organize organic reactions in a way that complements the organization based on functional groups.

This chapter includes examples of aliphatic nucleophilic substitution at both sp^3 and sp^2 centers, aromatic nucleophilic substitution, E2 elimination, nucleophilic addition to carbonyl compounds, 1,4-addition to α, β-unsaturated carbonyl compounds, and rearrangements promoted by base.

Writing Reaction Mechanisms in Organic Chemistry
http://dx.doi.org/10.1016/B978-0-12-411475-3.00003-8

Copyright © 2014 Elsevier Inc. All rights reserved.

1. NUCLEOPHILIC SUBSTITUTION

The classification within this section is based on the structural (rather than the mechanistic) relationship between the starting materials and products. Mechanistically, all the reactions considered in this section involve nucleophilic substitution as the first step, *except for* aromatic substitution via the aryne mechanism, which involves elimination followed by nucleophilic addition.

A. The S_N2 Reaction

The S_N2 reaction is a concerted bimolecular nucleophilic substitution at carbon. It involves an electrophilic carbon, a leaving group, and a nucleophile. The partial positive charge on the electrophilic carbon is due to the electron-withdrawing effect of the electronegative leaving group. This partial positive charge can be augmented by the presence of other electron-withdrawing groups attached to the electrophilic carbon, and the presence of such groups enhances reaction at the electrophilic center. For example, α-halocarbonyl groups react much faster than simple alkyl halides (Examples 3.3 and 3.21).

The S_N2 reaction occurs only at sp^3-hybridized carbons. The relative reactivities of carbons in the S_N2 reaction are $CH_3 > 1° > 2° \gg 3°$, due to steric effects. Methyl, 1° carbons, and 1° and 2° carbons that also are allylic, benzylic, or α to a carbonyl group are especially reactive.

$$\text{Me}-\text{Br} \quad > \quad \overset{\text{Me}}{\diagdown}-\text{Br} \quad > \quad \overset{\text{Me}}{\underset{\text{Me}}{\diagup}}-\text{Br} \quad > \quad \text{Me}-\overset{\text{Me}}{\underset{\text{Me}}{\overset{|}{\underset{|}{\text{C}}}}}-\text{Br}$$

| Relative rates | 145 | 1 | 8x10-3 | 5x10-4 |

The source is *Principles of Organic Synthesis*, R. O. C. Norman, Methuen and Co. Ltd., London, **1968,** pp. 126–131.

EXAMPLE 3.1. THE S_N2 REACTION: A CONCERTED PROCESS

The electrons of the nucleophile interact with carbon at the same time that the leaving group takes both the electrons in the bond between carbon and the leaving group. This particular example involves both a good leaving group and a good nucleophile.

$$^-\text{CN} \quad \text{PhCH}_2\text{—OSO}_2\text{CF}_3 \longrightarrow \text{PhCH}_2\text{CN} + {}^-\text{OSO}_2\text{CF}_3$$

Leaving Groups

Nucleophiles are discussed in Chapter 1, Section 9. Table 3.1 lists the typical leaving groups and gives a qualitative assessment of their effectiveness. Usually, the less basic the substituent, the easier it will act as a leaving group. This is because both basicity and leaving group ability are related to the stability of the anion involved. Frequently, these are both

TABLE 3.1   Leaving Group Abilities

Excellent

N_2, $^-OSO_2CF_3$ (triflate), $^-O-\!\!\overset{\overset{\displaystyle O}{\|}}{\underset{\underset{\displaystyle O}{\|}}{S}}\!\!-\!\!\langle\ \rangle\!\!-NO_2$ (nosylate = Nos),

$^-O-\!\!\overset{\overset{\displaystyle O}{\|}}{\underset{\underset{\displaystyle O}{\|}}{S}}\!\!-\!\!\langle\ \rangle\!\!-Br$ (brosylate = Bs), $^-O-\!\!\overset{\overset{\displaystyle O}{\|}}{\underset{\underset{\displaystyle O}{\|}}{S}}\!\!-\!\!\langle\ \rangle\!\!-CH_3$ (tosylate = Tos or Ts),

$^-OSO_2CH_3$ (mesylate = Ms)

Good	Fair
I^-, Br^-, Cl^-, SR_2	OH_2, NH_3, $^-OCOCH_3$ (acetate = OAc)

Poor	Very Poor
F^-, ^-OH, ^-OR	$^-NH_2$, ^-NHR, $^-NR_2$, R^-, H^-, Ar^-

related to charge dispersal in the anion, with greater charge dispersal being associated with greater stability of the ion. In looking at Table 3.1, we see that the best leaving groups are those for which resonance, inductive effects, or size results in distribution of any negative charge.

Relative leaving group abilities also depend upon the solvent and the nature of the nucleophile. For example, negatively charged leaving groups will be stabilized by interactions with protic solvents, so that protic solvents will increase the rate of bond breaking for these groups. Although these effects are important in modifying reaction conditions and yields, they rarely are large enough to completely change the mechanism by which a reaction proceeds, and we will not consider them here in detail.

HINT 3.1

Hydride (H^-) rarely acts as a leaving group. Exceptions are the Cannizzaro reaction and hydride abstraction by carbocations.

The S_N2 reaction rarely occurs with poor leaving groups. However, in other reactions, such as the nucleophilic substitution of carboxylic acid derivatives (Sections 1.B and 3), reactions with a poor leaving group like OH^- or RNH^- are encountered more frequently. Hydroxide may act as a leaving group, but only when there is considerable driving force for the reaction, as in certain elimination reactions where the double bond formed is stabilized by resonance (Example 3.10).

PROBLEM 3.1

Explain why the following mechanistic step in the equilibrium between a protonated and an unprotonated alcohol is a poor one.

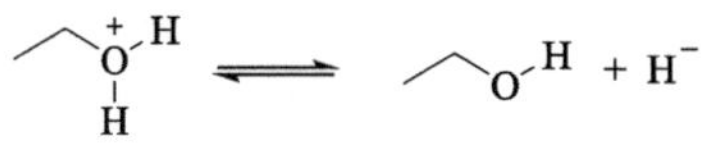

EXAMPLE 3.2. THE S$_N$2 REACTION OF AN ALCOHOL REQUIRES PRIOR PROTONATION

The alcohol oxygen is protonated before substitution takes place. Thus, the leaving group is a water molecule, a fair leaving group, rather than the hydroxide ion, a poor leaving group.

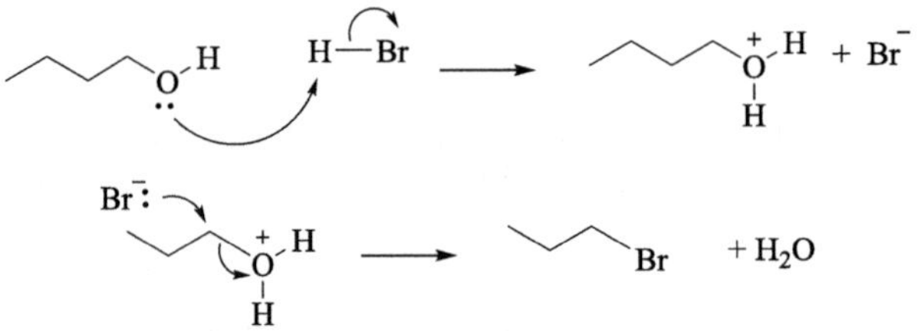

A poorer mechanistic option would show hydroxide as the leaving group:

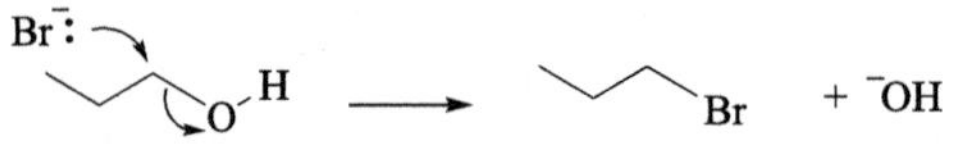

Stereochemistry

The S$_N$2 reaction always produces 100% inversion of configuration at the electrophilic carbon. Thus, as shown in Example 3.3, the nucleophile approaches the electrophilic carbon on the side opposite to the leaving group (there is a 180° angle between the line of approach of the nucleophile and the bond to the leaving group).

EXAMPLE 3.3. STEREOCHEMISTRY OF THE S$_N$2 REACTION

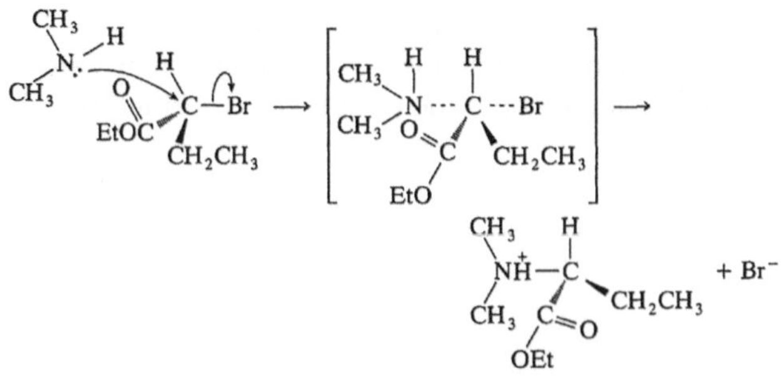

PROBLEM 3.2

Consider the following synthesis, which involves alkylation of the phenolic oxygen (attachment of the benzyl group onto the oxygen).

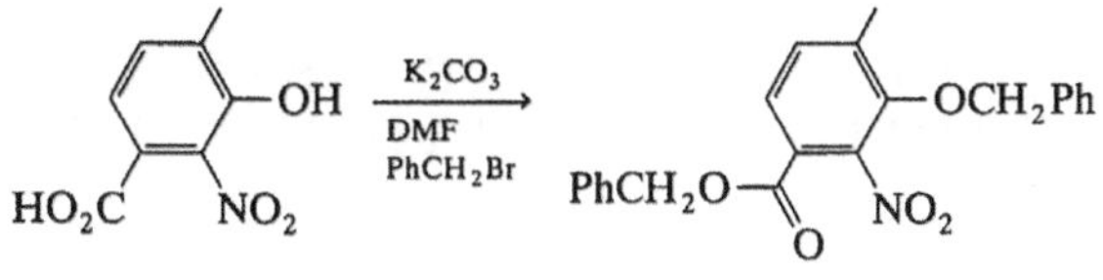

Propose a more reasonable mechanism for the alkylation than that shown in the following step. Formation of the product would involve deprotonation of the positively charged oxygen.

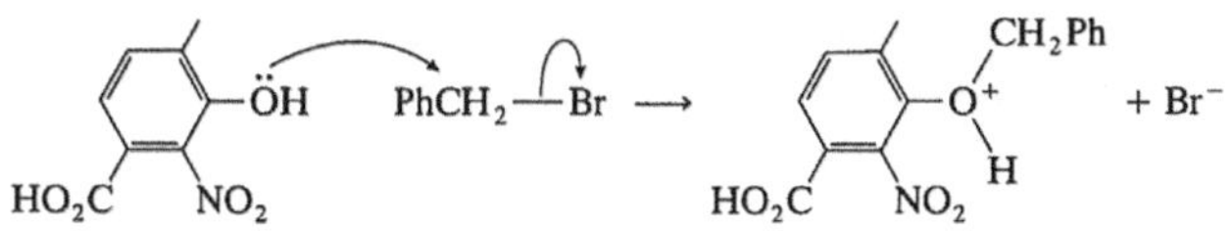

Source: Pena and Stille, (1989).

PROBLEM 3.3

Pick out the electrophile, nucleophile, and leaving group in each of the following reactions and write a mechanism for the formation of products.

a.

b. $H_3\overset{+}{N}CH_2CH_2CH_2CH(CH_3)Cl + CO_3^{2-} \longrightarrow$

c.

d.

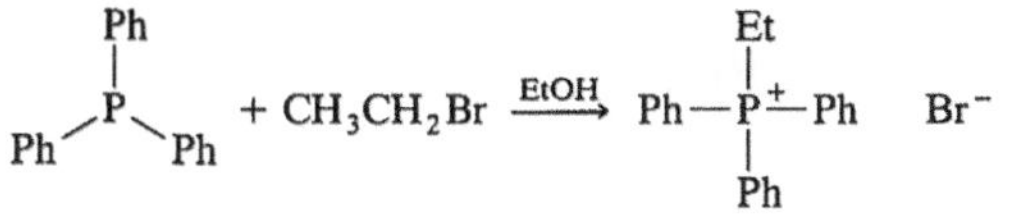

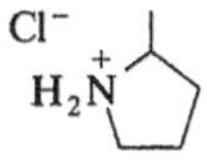

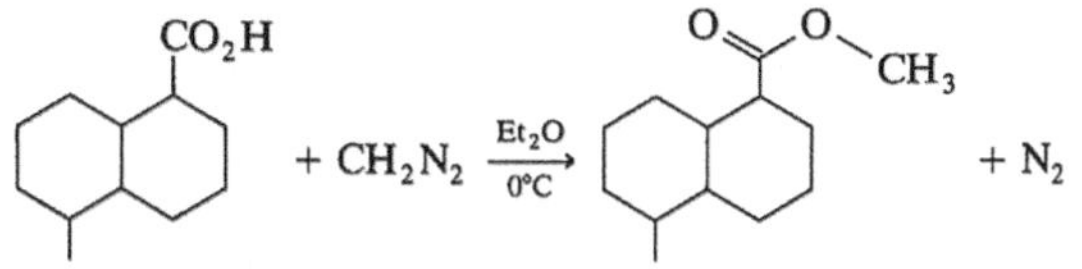

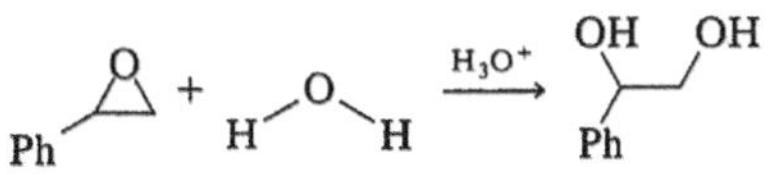

Neighboring Group Participation

On occasion, a molecule undergoing nucleophilic substitution may contain a nucleophilic group that participates in the reaction. This is known as the *neighboring group effect* and usually is revealed by *retention of stereochemistry* in the nucleophilic substitution reaction or by an *increase in the rate* of the reaction.

EXAMPLE 3.4. NEIGHBORING GROUP PARTICIPATION IN THE HYDROLYSIS OF ETHYL-2-CHLOROETHYL SULFIDE

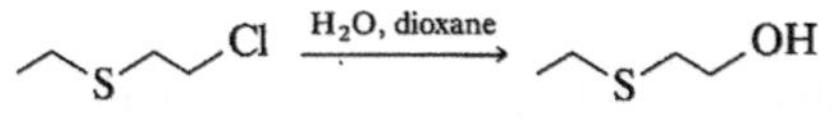

The hydrolysis of the chlorosulfide proceeds to give the expected product. However, the reaction is 10,000 times faster than the reaction of the corresponding ether, $ClCH_2CH_2OCH_2CH_3$. This rate enhancement has been credited to the ready formation of a cyclic sulfonium ion due to intramolecular displacement of chloride by sulfur, followed by rapid nucleophilic reaction of water with the intermediate sulfonium ion.

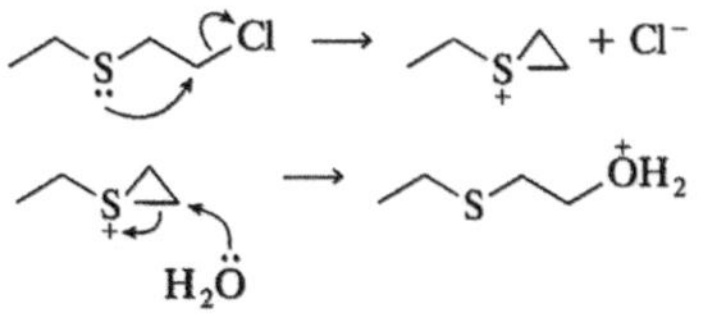

Other groups that exhibit this behavior include thiol, sulfide, alkoxy (RO^-), ester, halogen, and phenyl.

PROBLEM 3.4

Write step-by-step mechanisms for the following transformations:

a.

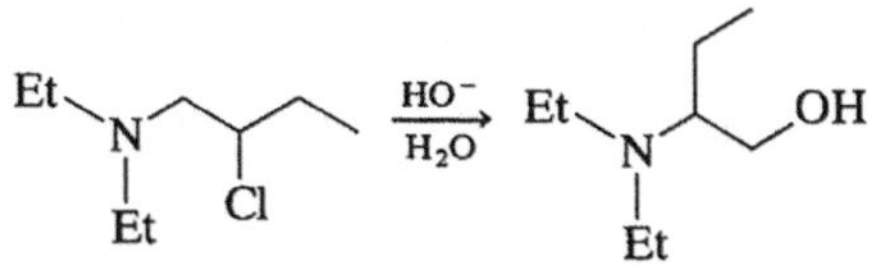

b.

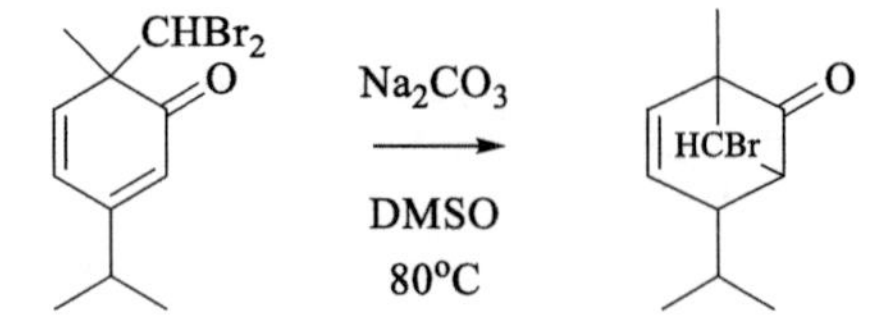

Source: Wenkert et al., (1990).

B. Nucleophilic Substitution at Aliphatic sp^2-Carbon (Carbonyl Groups)

The familiar substitution reactions of derivatives of carboxylic acids with basic reagents illustrate nucleophilic substitution at aliphatic sp^2-carbons. (Substitution reactions of carboxylic acids, and their derivatives, with acidic reagents are covered in Chapter 4.) The mechanisms of these reactions involve two steps: (1) addition of the nucleophile to the carbonyl group and (2) elimination of some other group attached to that carbon. Common examples include the basic hydrolysis and aminolysis of acid chlorides, anhydrides, esters, and amides.

EXAMPLE 3.5. MECHANISM FOR HYDROLYSIS OF AN ESTER IN BASE

Unlike the one-step S_N2 reaction, the hydrolysis of esters in base is a two-step process. The net result is substitution, but the first step is nucleophilic addition to the carbonyl group, during which the carbonyl carbon becomes sp^3-hybridized. The second step is an elimination, in which the carbonyl group is regenerated as the carbon rehybridizes to sp^2.

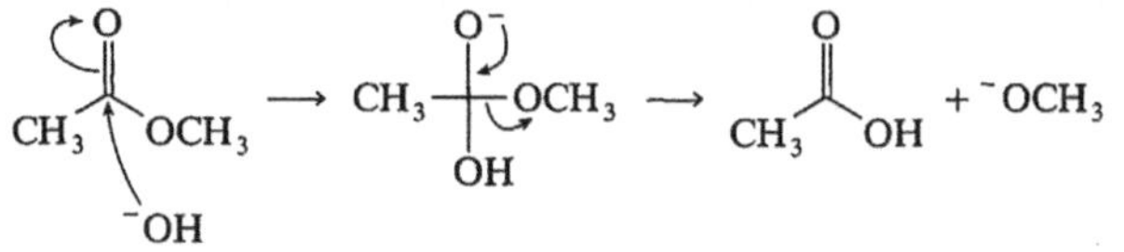

This is followed by removal of a proton from the acid, by the methoxide ion, to yield methanol and the carboxylate ion:

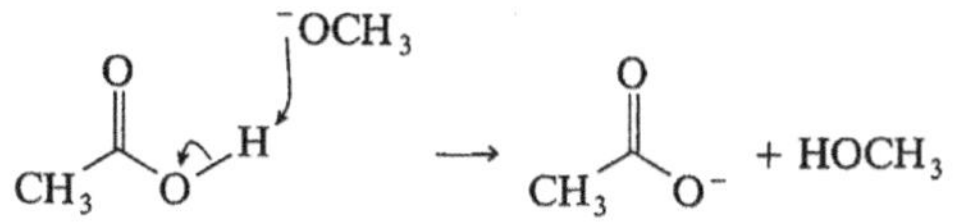

Another possible mechanism for this hydrolysis is an S_N2 reaction at the alkyl carbon of the ester:

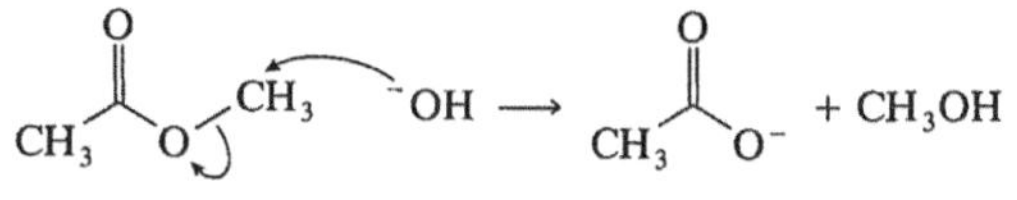

This single-step mechanism appears reasonable because carboxylate is a fair leaving group and hydroxide is a very good nucleophile. However, labeling studies rule out this mechanism under common reaction conditions. The two-step mechanism must be favored because the higher mobility of the π electrons of the carbonyl group makes the carbonyl carbon especially electrophilic.

HINT 3.2

Direct nucleophilic substitution at an sp^2-hybridized center is not likely under common reaction conditions. Thus, nucleophilic substitution reactions at such centers usually are broken into two steps. (For exceptions to this hint, see Dietze and Jencks, (1989), and references cited therein.)

There are several reasons why direct substitutions occur at sp^2-hybridized centers less readily than at sp^3 centers. First, because there is more s character in the bond to the leaving group, this bond is stronger than the corresponding bond to an sp^3-hybridized carbon. Second, the greater mobility of the π electrons at an sp^2 center increases the likelihood that the interaction will cause electron displacement. Third, because of the planar configuration of the substituents around an sp^2 center, there is strong steric interference to the approach of a nucleophile to the side opposite to the leaving group. On the other hand, in the addition of a nucleophile to the carbonyl group, the nucleophile approaches perpendicular to the plane of the sp^2-orbitals so that there is maximum overlap with the π electron system. This means that the relatively unhindered addition step occurs in preference to direct substitution.

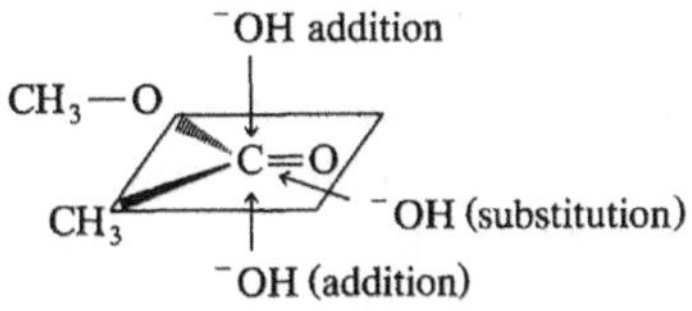

Thus, the mechanism for basic hydrolysis of an ester would *not be written* as follows:

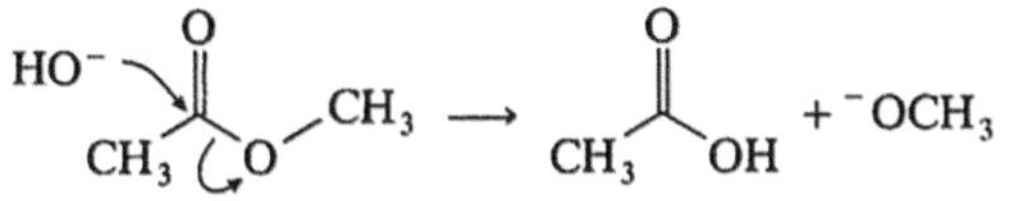

We will cover the influences that drive the direction of attack at a carbonyl in more detail in Chapter 7.

PROBLEM 3.5

Consider the mechanism shown for the following transformation. Propose a more reasonable alternative.

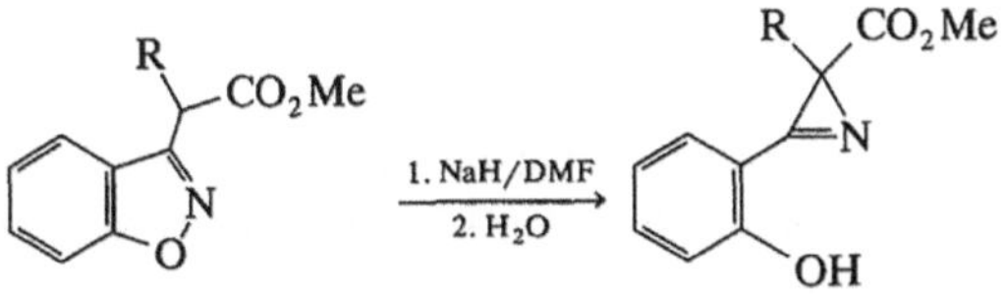

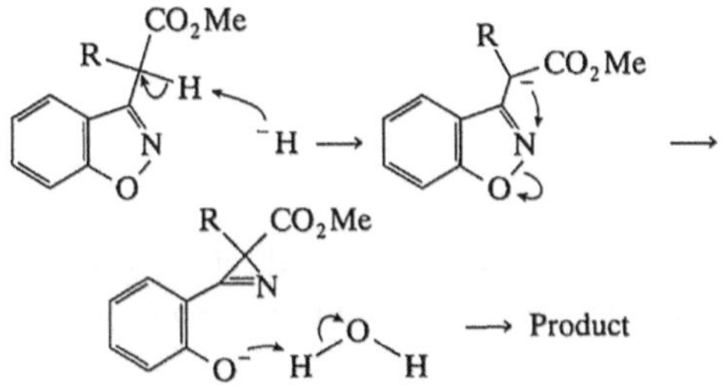

PROBLEM 3.6

Write step-by-step mechanisms for the following transformations:

a.

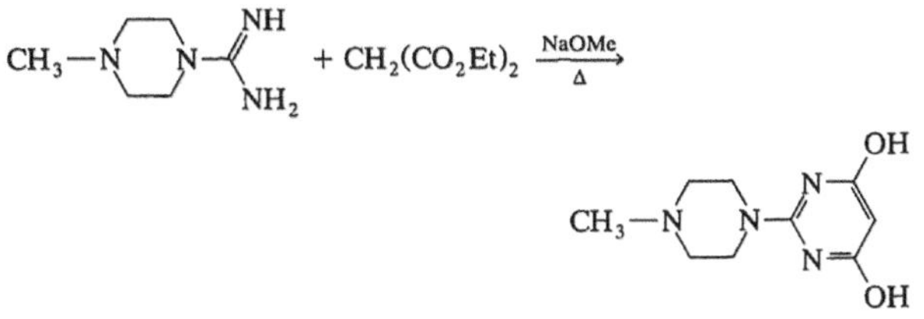

b.

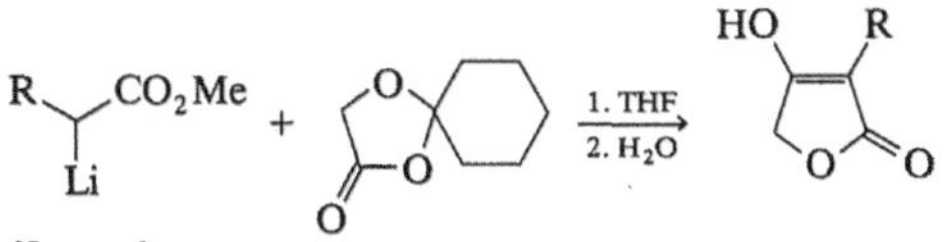

62 mmol   25 mmol

c. Critically evaluate the following partial mechanism for the reaction given in part a:

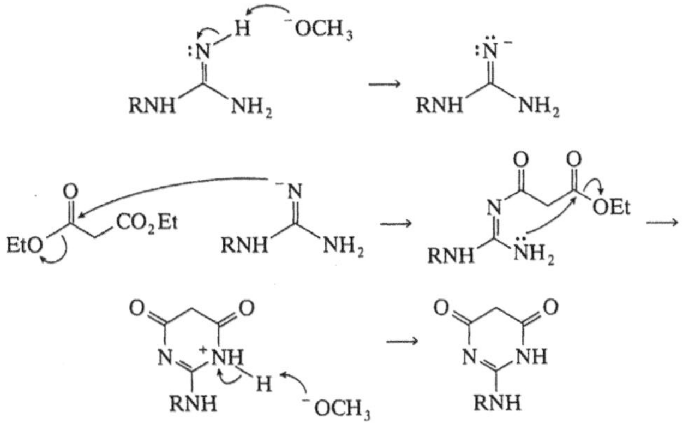

Sources: Gueremy et al., (1986); Ramage et al., (1984).

C. Nucleophilic Substitution at Aromatic Carbons

There are two mechanisms for nucleophilic aromatic substitution. Both occur in two important steps. In one mechanism, an addition is followed by an elimination. In the other mechanism, an elimination is followed by an addition.

Addition–Elimination Mechanism

The addition–elimination mechanism generally requires a ring activated by electron-withdrawing groups. These groups are especially effective at stabilizing the negative charge in the ring when they are located at positions *ortho* and/or *para* to the eventual leaving group.

EXAMPLE 3.6. RELATIVE REACTIVITY IN THE ADDITION–ELIMINATION MECHANISM

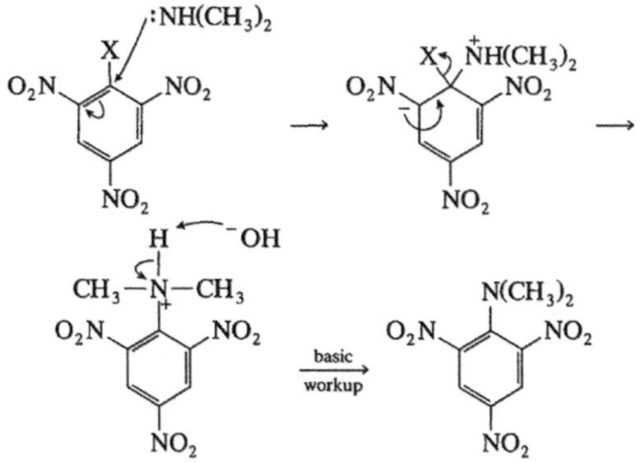

When X = halogen, the observed relative reactivities of the starting materials are F > Cl > Br > I. This indicates that the first step is rate-determining step because the greater the electron-withdrawing power of the halogen (Table 3.2), the more it increases the electrophilicity of the aromatic ring, making it more reactive to nucleophiles. If the second step were rate-determining step, the relative reactivities would be reversed because the relative abilities of the leaving groups are $I^- > Br^- > Cl^- > F^-$.

TABLE 3.2 Addition of Organometallic Reagents to Carbonyl Compounds and Carboxylic Acid Derivatives

	R'C(=O)R"	R'C(=O)OEt	R'C(=O)Cl	R'C(=O)OH	R'C≡N
RMgX	R'—C(OH)(R)—R"	R'—C(OH)(R)—R	R'—C(OH)(R)—R	R'CO$_2$MgX	R'—C(=NH)—R → R'C(=O)R
RLi	R'—C(OH)(R)—R"	R'—C(OH)(R)—R, R'C(=O)R	—	R'—CH(OH)R	—

PROBLEM 3.7

By drawing the appropriate resonance forms, show that the negative charge in the intermediate anion in Example 3.6 is stabilized by extensive electron delocalization.

PROBLEM 3.8

Write a step-by-step mechanism for the following transformation:

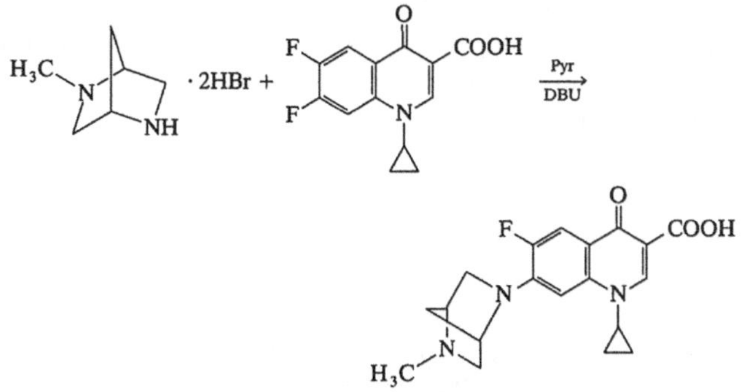

Braish and Fox, (1990). [This is the last step in the synthesis of danofloxacin, an antibacterial. Pyr (or Py) is a common acronym for pyridine; DBU is 1,8-diazabicyclo[5.4.0]undec-7-ene. A good reference for the translation of acronyms is Daub et al., (1984)]

Elimination–Addition (Aryne) Mechanism

In reactions that proceed by the elimination–addition mechanism (often called the aryne mechanism), the bases used commonly are stronger than those used in reactions proceeding by the addition–elimination mechanism. Also, in this reaction, the aromatic ring does not need to be activated by electron-withdrawing substituents, although a reasonable leaving group (usually a halide) must be present.

EXAMPLE 3.7. AN ELIMINATION–ADDITION MECHANISM—ARYNE INTERMEDIATE

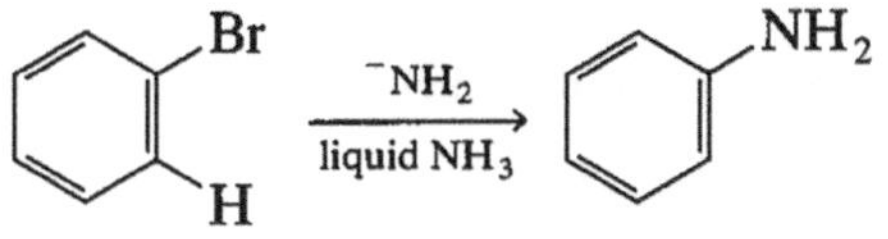

The following mechanism can be written for this reaction:

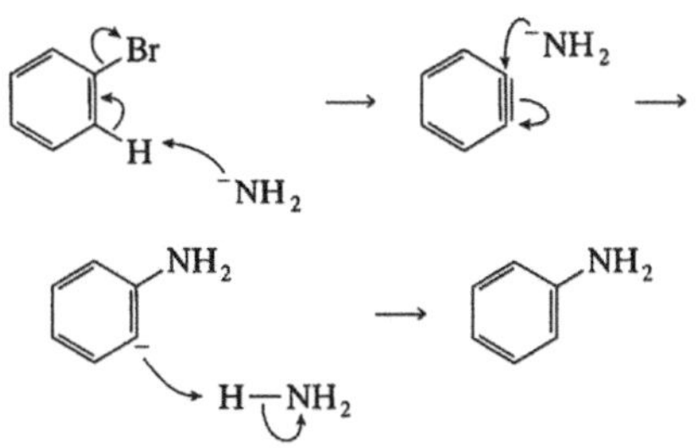

The intermediate with a triple bond is called *benzyne*. For substituted aromatic compounds, this type of intermediate is called an *aryne*. In benzyne, the ends of the triple bond are equivalent, and either can react with a nucleophile.

The triple bond in an aryne is not a normal triple bond. The six-membered ring does not allow the normal linear configuration of two *sp*-hybridized carbon atoms and their substituents. Thus, the carbons remain sp^2-hybridized, and the triple bond contains the σ bond, the π bond, and a third bond formed by overlap of the sp^2-hybridized orbitals that formerly bonded with the bromine and hydrogen atoms. This third bond is in the plane of the benzene ring and contains two electrons.

The rate-determining step can be either proton removal or departure of the leaving group, depending on the acidity of the proton and the ability of the leaving group. In many cases, the relative rates are so close that the reaction cannot be distinguished from a concerted process.

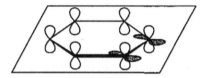

PROBLEM 3.9

Assume that in Example 3.7, the carbon bound to the bromine in bromobenzene is labeled by enrichment with ^{13}C. Where would this label be found in the product aniline?

PROBLEM 3.10

Write a step-by-step mechanism for the following transformation:

Source: Bunnett and Skorcz, (1962).

2. ELIMINATIONS AT SATURATED CARBON

Important eliminations at saturated carbon are the E2 (bimolecular elimination) and Ei (intramolecular elimination) processes.

A. E2 Elimination

The E2 reaction is a concerted process, with a bimolecular rate-determining step. In this case, "concerted" means that bonding of the base with a proton, formation of a double bond, and departure of the leaving group all occur in one step.

Stereochemistry

The stereochemistry is usually *anti*, but in some cases is *syn*. The term *anti* means that the proton and leaving group depart from opposite sides of the bond, which then becomes a double bond. That is, the dihedral angle (measured at this bond) between their planes of departure is 180°. If they depart from the same side (the dihedral angle is 0°), the stereochemistry of the elimination is called *syn*.

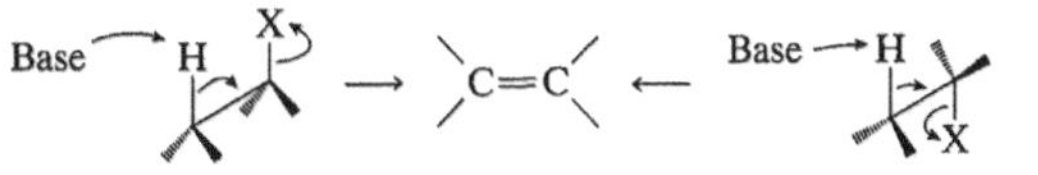

syn elimination
(dihedral angle 0°)

anti elimination
(dihedral angle 180°)

EXAMPLE 3.8. AN ANTI-E2 ELIMINATION

The dihedral angle between the proton and bromide is 180°, so this is an antielimination.

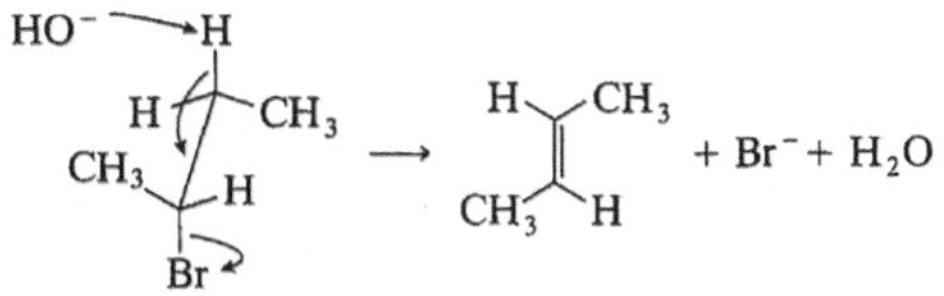

Leaving Groups

The nature of the leaving group influences whether the reaction proceeds by an E2 mechanism. An excellent leaving group like $CH_3SO_3^-$ (mesylate) will favor competing reactions that proceed through a carbocation. Poor leaving groups, due to their failure to react, will allow competing reactions via anionic mechanisms.

EXAMPLE 3.9. IN AN ACID-CATALYZED ELIMINATION OF WATER FROM AN ALCOHOL, WATER IS THE LEAVING GROUP

The mechanism for an elimination step in the acid-catalyzed aldol condensation is written as follows:

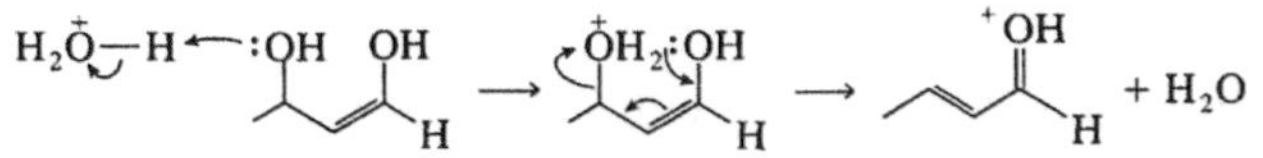

The following step is less likely for the formation of an α, β-unsaturated aldehyde in acid (Hint 2.6):

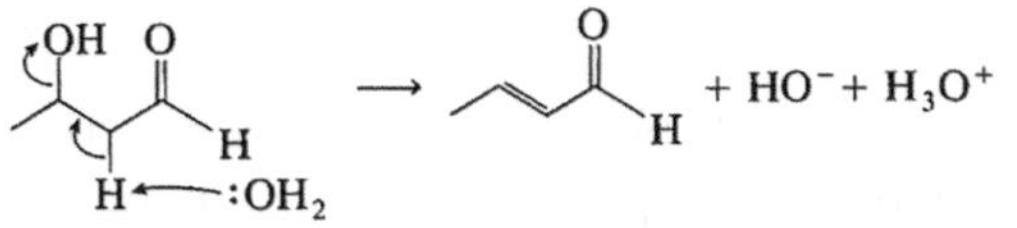

EXAMPLE 3.10. UNDER SOME CONDITIONS, HYDROXIDE CAN ACT AS A LEAVING GROUP

A 3-hydroxyaldehyde (or ketone) will undergo elimination under basic conditions if the double bond being formed is especially stable, e.g., conjugated with an aromatic system. Such eliminations can occur under the reaction conditions of the base-promoted aldol condensation. An example is the formation of 3-phenyl-2-butenal by an E2 elimination from 3-hydroxy-3-phenylbutanal.

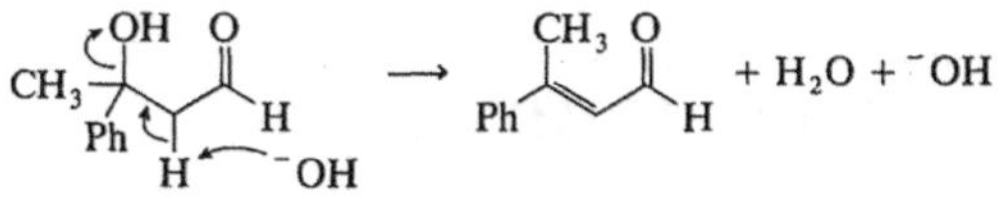

B. Ei Elimination

In another type of elimination reaction, called Ei or intramolecular, the base, which removes the proton, is another part of the same molecule. Such eliminations from amine oxides or sulfoxides have five-membered-ring transition states. These transition states are more stable with *syn* than with *anti* orientations of proton and leaving group, producing very high *syn* stereoselectivity.

EXAMPLE 3.11. AN EI REACTION: PYROLYTIC ELIMINATION FROM A SULFOXIDE

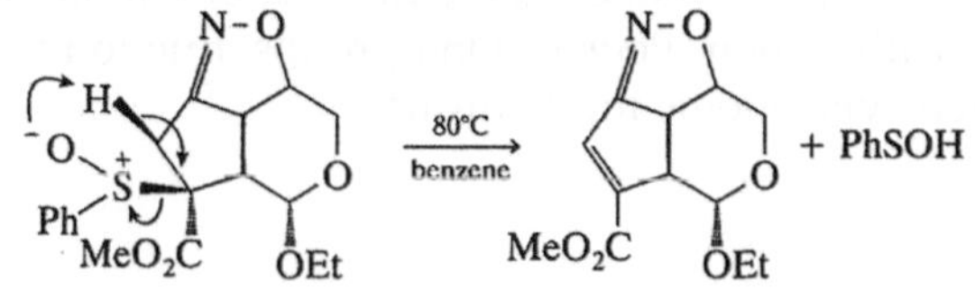

Source: Curran et al., (1987).

PROBLEM 3.11

Write a mechanism for the following reaction. What is the other product?

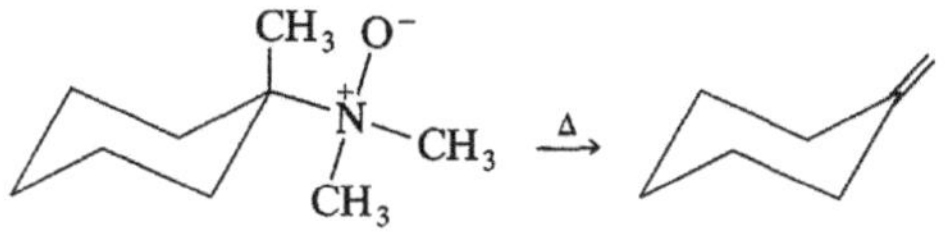

3. NUCLEOPHILIC ADDITION TO CARBONYL COMPOUNDS

Nucleophilic addition to the carbonyl groups of aldehydes and ketones occurs readily, and the carbonyl groups of carboxylic acids and their derivatives (acid chlorides, anhydrides, amides, and esters) also react with nucleophiles. In this section, the numerous nucleophilic addition reactions of carbonyl groups are organized first by the type of nucleophile (e.g., organometallics, nitrogen nucleophiles, and carbon nucleophiles) and then according to the kind of carbonyl group.

Addition reactions with organometallic reagents are usually irreversible, but many other addition reactions are reversible. In these reversible additions, equilibrium may favor the starting materials. When the equilibrium does not favor product formation, the reaction can be made productive if the initial product is removed, either physically (a salt dropping out of solution or a gas coming out of the mixture, for example) or by undergoing further reaction, as in Example 3.18.

A. Additions of Organometallic Reagents

Reactions of either Grignard or organolithium reagents with most aldehydes, ketones, or esters produce alcohols. Reactions of organolithium reagents with carboxylic acids, or of Grignard reagents with nitriles, produce ketones.

The addition reactions of Grignard and organolithium reagents with carbonyl groups and carboxylic acid derivatives are summarized in Table 3.2.

Additions to Aldehydes and Ketones

The reactions of Grignard reagents (RMgX) and alkyl lithium reagents (RLi) with aldehydes and ketones are similar. The mechanism is illustrated by the following example.

EXAMPLE 3.12. THE GRIGNARD REACTION OF PHENYLMAGNESIUM BROMIDE WITH BENZOPHENONE

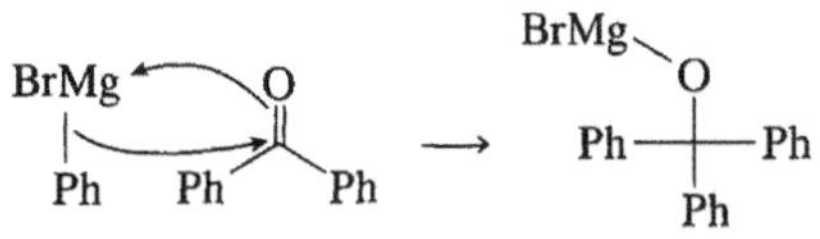

3-1

The electron pair in the carbon–magnesium bond of phenylmagnesium bromide is the nucleophile, and the carbonyl carbon of the ketone is the electrophile. Also, magnesium is an electrophile and the carbonyl oxygen is a nucleophile, so that the salt of an alcohol is the product of the reaction. The alcohol itself is generated by an acidic workup.

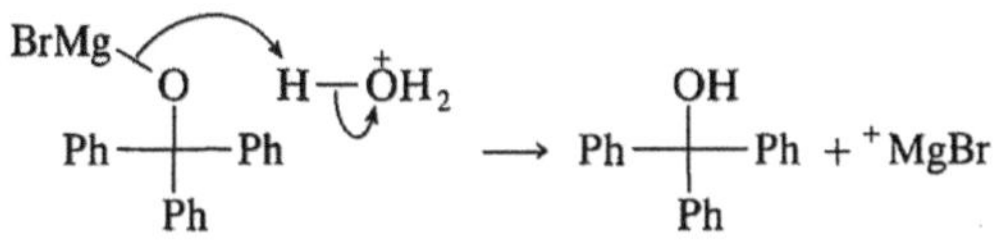

3-1

Additions to Carboxylic Acid Derivatives

HINT 3.3

The stability of the intermediate formed by the addition of an organometallic reagent to a carboxylic acid derivative determines the product produced in the subsequent steps.

With organolithium reagents, stable intermediates are produced by addition to a carboxylic acid. On workup, these intermediates produce ketones. As with Grignard reagents, the reaction of organolithium reagents with esters produces tertiary alcohols because the intermediates decompose to ketones under the reaction conditions. The mechanisms for these processes are illustrated in the examples that follow.

EXAMPLE 3.13. GRIGNARD ADDITION TO A NITRILE

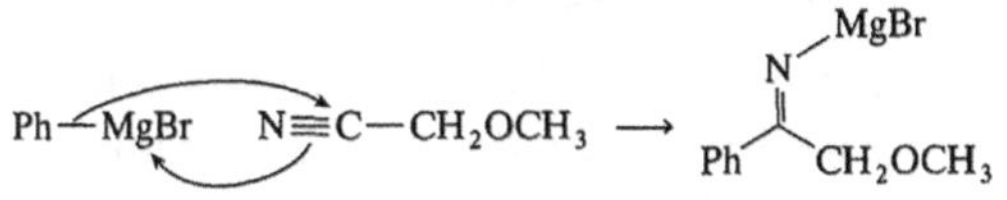

3-2

The protonation of the intermediate, 3-2, to give 3-3 is similar to the protonation of 3-1 in the previous example.

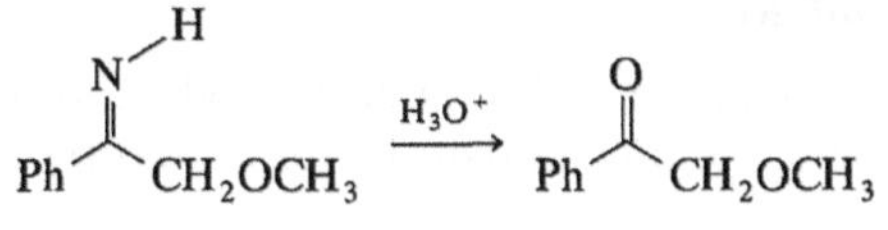

3-3   3-4

Then the initial product 3-3 can be hydrolyzed to the ketone 3-4. (See the answer to Problem 3.14 for the mechanism of this reaction.)

EXAMPLE 3.14. ADDITION OF AN ORGANOLITHIUM REAGENT TO A CARBOXYLIC ACID

The reaction requires 2 mol of organolithium reagent per mole of acid. The first mole of organolithium reagent neutralizes the carboxylic acid, giving a salt.

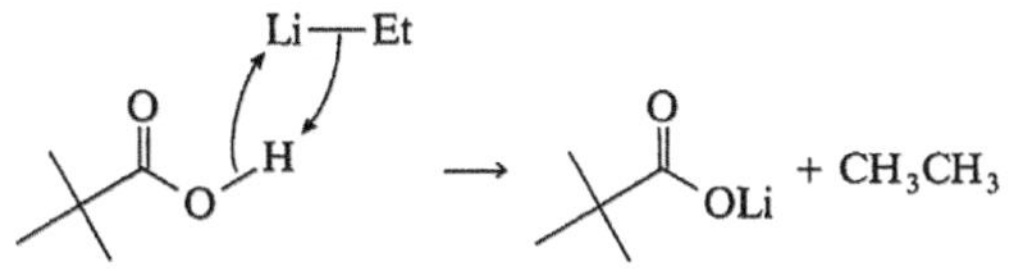

The second mole adds to the carbonyl group to give a dilithium salt, **3-5**, which is stable under the reaction conditions. Sequential hydrolysis of each OLi group in acid, during workup, gives a dihydroxy compound, **3-6**, which is the hydrate of a ketone. A series of protonations and deprotonations transforms the hydrate into a species that can eliminate a molecule of water to form the ketone.

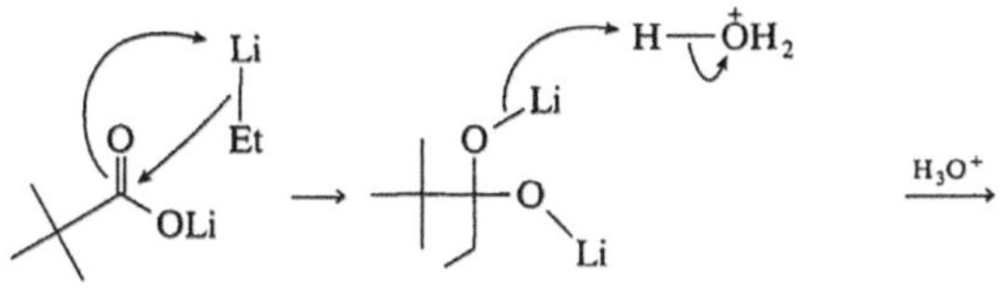

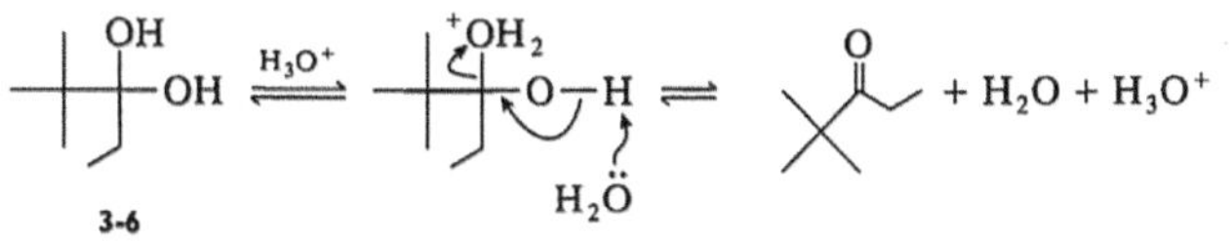

EXAMPLE 3.15. GRIGNARD REACTION OF AN ESTER

Esters react with 2 mol of Grignard reagent to give the salt of an alcohol. For example, reaction of ethyl benzoate with 2 mol of phenylmagnesium bromide gives a salt of triphenylmethanol. The first addition gives an intermediate, **3-7**, which is unstable under the reaction conditions.

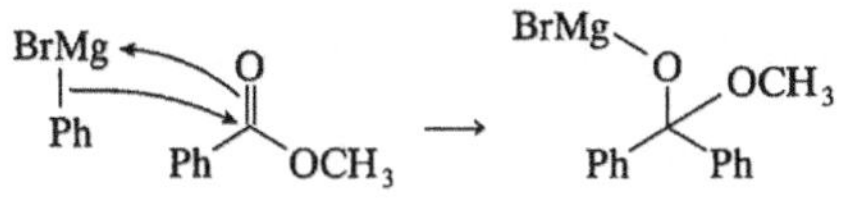

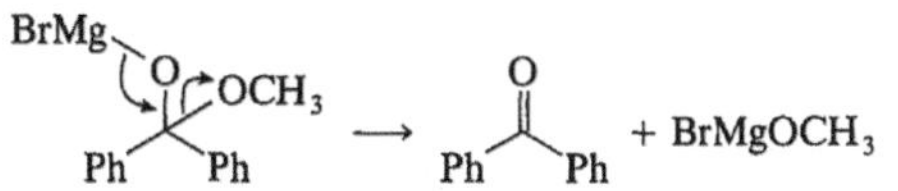

Another molecule of phenylmagnesium bromide now reacts with benzophenone, as shown in Example 3.12.

PROBLEM 3.12

Write step-by-step mechanisms for the following reactions:

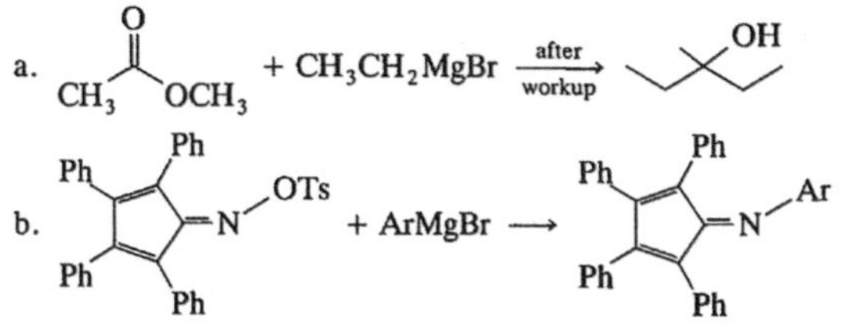

a.

b.

(Ar = Aryl)

Source: Hagopian et al., (1984).

B. Reaction of Nitrogen-Containing Nucleophiles with Aldehydes and Ketones

A number of reactions of nitrogen-containing nucleophiles with aldehydes and ketones involve addition of the nitrogen to the carbon of the carbonyl group, followed by elimination of water to produce a double bond. Common examples are reactions of primary amines to produce substituted imines, reactions of secondary amines to produce enamines, reactions of hydrazine or substituted hydrazines to produce hydrazones, reactions of semicarbazides to give semicarbazones, and reactions of hydroxylamine to produce oximes. Usually these reactions are run with an acid catalyst.

In the synthesis of imines and enamines by this method, the water produced in the reaction must be removed azeotropically to drive the reaction to the right. In aqueous acid, equilibrium conditions favor the ketone rather than the imine. This relationship is the reason why Grignard reaction of a nitrile provides a good route to the synthesis of ketones. The intermediate imine formed is hydrolyzed easily to the corresponding ketone (e.g., the transformation of **3-3** to **3-4** in Example 3.13).

EXAMPLE 3.16. MECHANISM FOR FORMATION OF A HYDRAZONE

The first step in a mechanism for the following synthesis of a phenylhydrazone is an equilibrium protonation of the carbonyl oxygen.

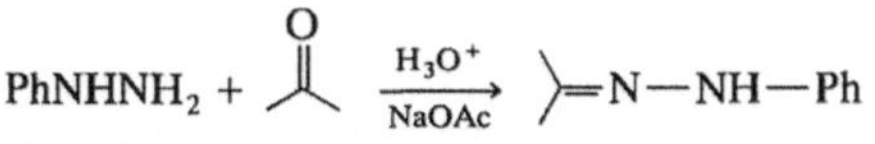

The protonated carbonyl group then is more susceptible to reaction with a nucleophile than the neutral compound (note that the protonated carbonyl group is a resonance hybrid).

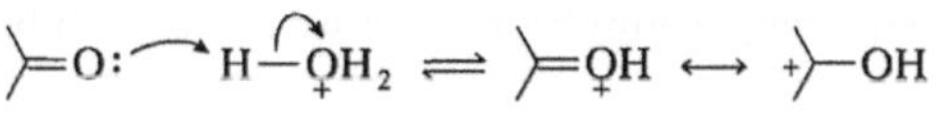

The more nucleophilic nitrogen of the hydrazine reacts at the electrophilic carbon of the carbonyl group. Loss of a proton, facilitated by base ($^-$OAc) is followed by acid-catalyzed elimination of water.

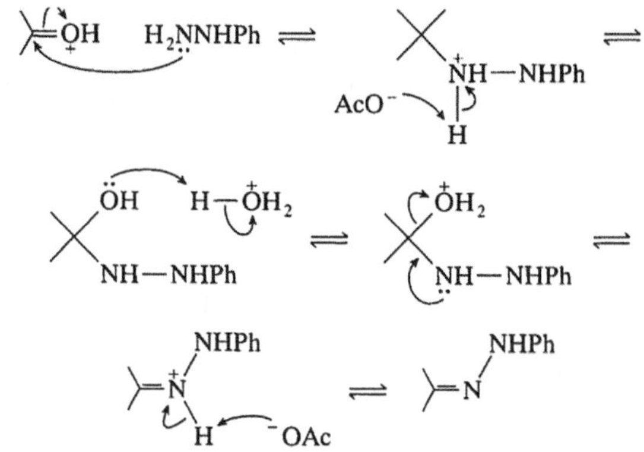

PROBLEM 3.13

Explain why the nitrogen in phenylhydrazine that acts as the nucleophile is the nitrogen without the phenyl substituent.

PROBLEM 3.14

Write step-by-step mechanisms for the following transformations:

a.

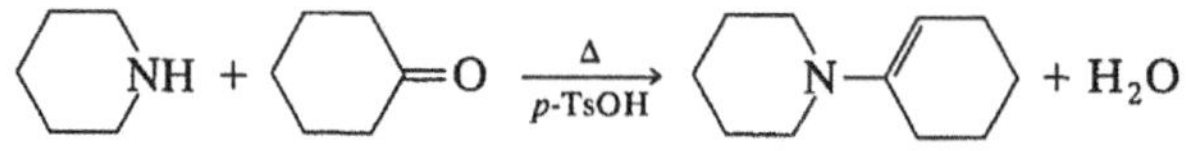

b.

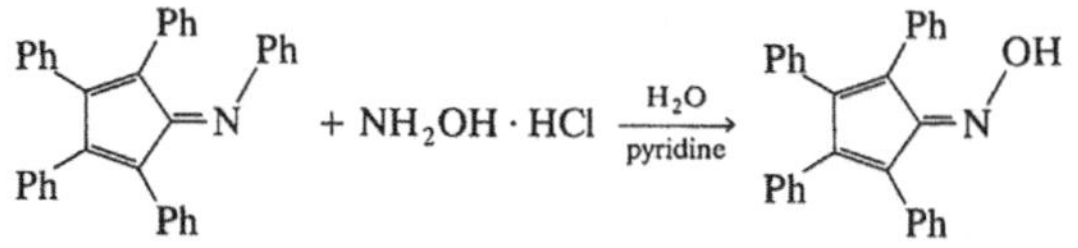

c.

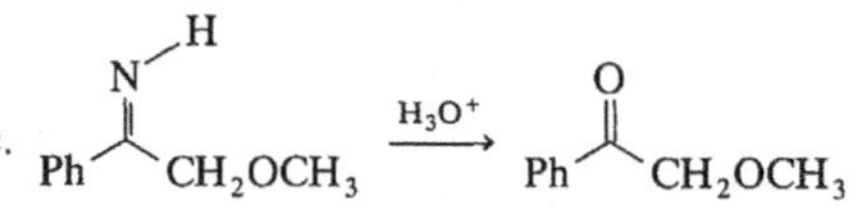

Source: Hagopian et al., (1984).

C. Reactions of Carbon Nucleophiles with Carbonyl Compounds

The Aldol Condensation

The aldol condensation involves the formation of an anion on a carbon α to an aldehyde or ketone carbonyl group, followed by nucleophilic reaction of that anion at the carbonyl group of another molecule. The reaction may involve the self-reaction of an aldehyde or ketone or the formation of the anion of one compound and reaction at the carbonyl of a different compound. The latter is called a mixed aldol condensation.

EXAMPLE 3.17. CONDENSATION OF ACETOPHENONE AND BENZALDEHYDE: NUCLEOPHILIC ADDITION OF AN ANION TO A CARBONYL GROUP FOLLOWED BY AN ELIMINATION

This reaction is a mixed aldol condensation of an aldehyde and a ketone.

Consider a step-by-step mechanism for this process. The first step is removal of a proton from the carbon α to the carbonyl group of the ketone to give a resonance-stabilized anion. (Note that removal of the proton directly attached to the aldehyde carbonyl carbon does not give a resonance-stabilized anion, and there are no hydrogens on the carbon α to the aldehyde carbonyl.)

The equilibrium in this reaction favors starting material; in Problem 1.14.b, the equilibrium constant for this reaction was calculated to be approximately 10^{-9}. Nonetheless, the reaction continues to a stable product because the subsequent step has a much more favorable equilibrium constant. In this next step, the carbonyl group of the aldehyde undergoes nucleophilic addition by the enolate anion to give 3-8:

3-8

Why is there preferential reaction at the aldehyde carbon? In other words, why does the acetophenone anion react with the aldehyde instead of another acetophenone molecule? The ketone carbonyl is less reactive for two reasons. First, the tetrahedral intermediate formed by addition to the carbonyl group of a ketone is less stable than the intermediate formed by addition to an aldehyde because there is more steric interaction with the alkyl group of the ketone than with the corresponding hydrogen of the aldehyde. Second, inductive effects due to the two alkyl groups stabilize the carbonyl bond of a ketone relative to that of an aldehyde.

Anion **3-8** can remove a proton from the solvent, which often is ethanol.

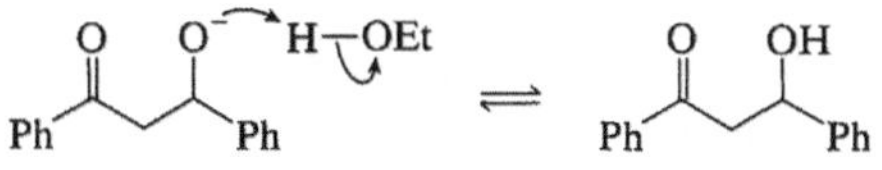

Finally, a base-promoted E2 elimination of water occurs to give the product. This elimination is driven energetically by the formation of a double bond, which is stabilized by conjugation with both a phenyl group and a carbonyl group.

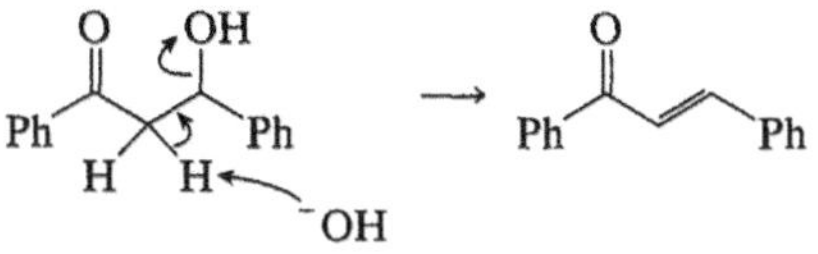

PROBLEM 3.15

Write step-by-step mechanisms for the following reactions:

a.

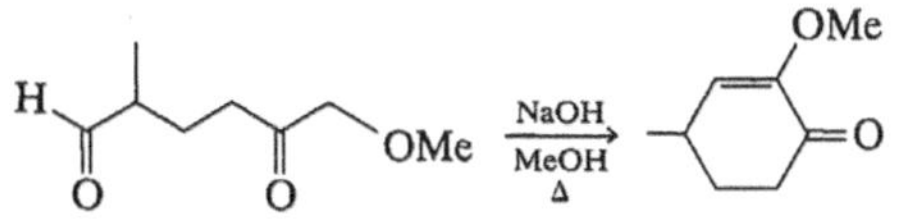

b. PhN=O + NCCH$_2$CO$_2$Et $\xrightarrow[\text{EtOH}]{\text{K}_2\text{CO}_3}$

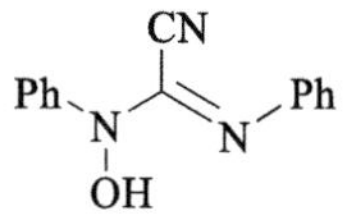

Sources: Gadwood et al., (1984); Bell, (1957).

The Michael Reaction and Other 1,4-Additions

The Michael reaction is addition of a carbon nucleophile to the β position of an α, β-unsaturated carbonyl compound or its equivalent. It also may be called a 1,4-addition reaction (the carbonyl oxygen is counted as 1 and the β-carbon as 4). The conjugation of the π bond with the carbonyl group imparts positive character to the β position, making it susceptible to reaction with a nucleophile. The product of this reaction, an enolate ion, also is stabilized by resonance.

When nucleophiles other than carbon add to α, β-unsaturated carbonyl compounds, the process is called a 1,4-addition.

EXAMPLE 3.18. A TYPICAL MICHAEL REACTION

This example shows the addition of a fairly stable carbanion (stabilized by two adjacent carbonyl groups) to an α, β-unsaturated ketone.

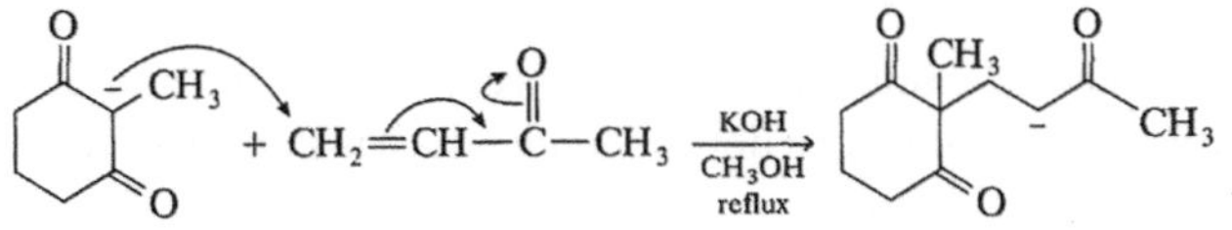

The adduct formed initially is itself an enolate ion stabilized by resonance.

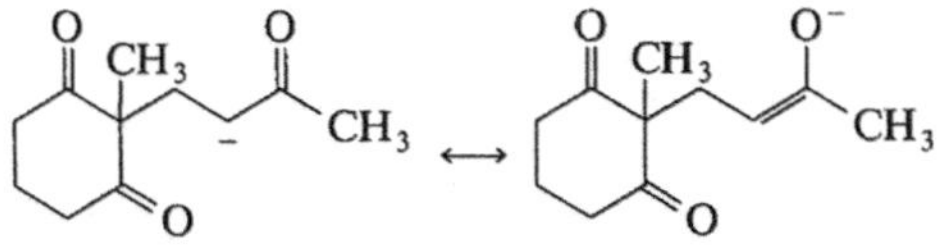

The enolate can remove a proton from the solvent to give the neutral product.

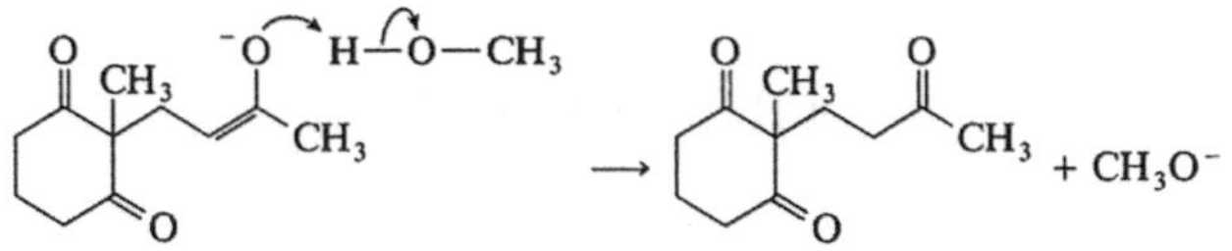

There are many instances in the literature where the Michael reaction is followed by subsequent steps. The following example is one of them.

Source: Ramachadran and Newman, (1961).

EXAMPLE 3.19. 1,4-ADDITION FOLLOWED BY SUBSEQUENT REACTION

The overall reaction is

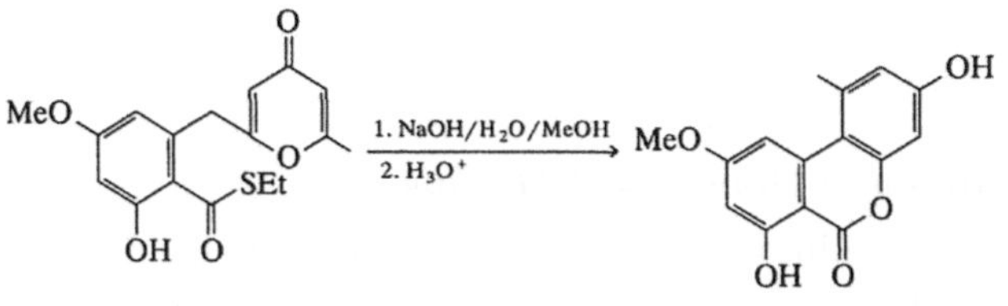

Source: Abell et al., (1986).

Analysis of the starting material indicates an acidic phenolic hydroxyl, a thioester susceptible to base-promoted hydrolysis, and an α, β-doubly unsaturated ketone that could undergo 1,4-addition followed by subsequent reaction. From the structure of the product, it is clear that both the thioester and the α, β-unsaturated ketone undergo reaction. Because hydroxide is the base, a proton will be removed readily from the phenolic hydroxyl group, forming **3-9-1**.

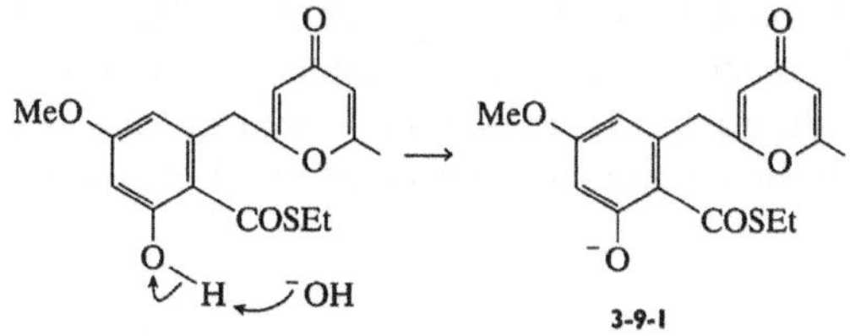

3-9-1

Whether the ester or one of the positions β to the carbonyl group reacts first cannot be ascertained from the data given. Thus, both possibilities are discussed.

Mechanism 1: A drawing of another resonance form of **3-9-1**, **3-9-2**, shows that the oxygen of the thioester has negative character and could act as an intramolecular nucleophile:

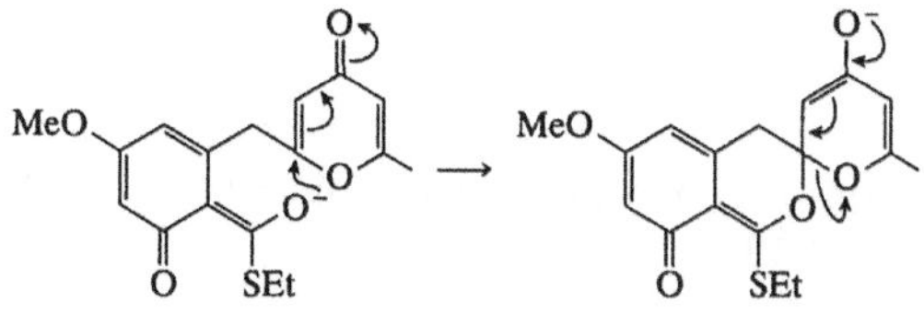

3-9-2   3-10

Comparison of **3-10** with the product reveals a central ring with the same atomic skeleton as the product, but a right-hand ring that does not. Thus, the latter opens to an enolate ion, **3-11**.

The enolate **3-11** can remove a proton from solvent to give **3-12**, which can undergo addition of hydroxide to give a resonance-stabilized anion, **3-13**.

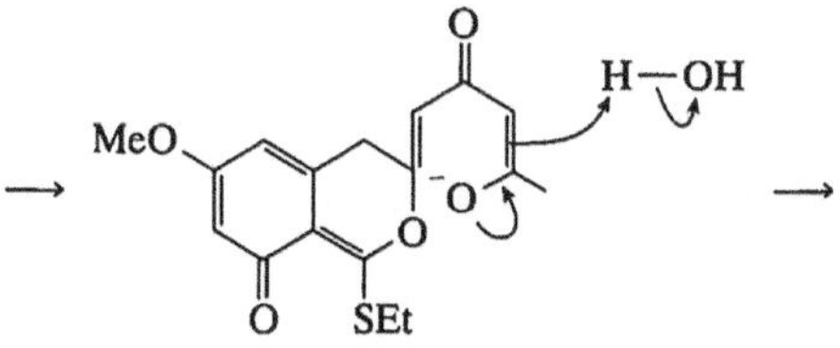

3-11

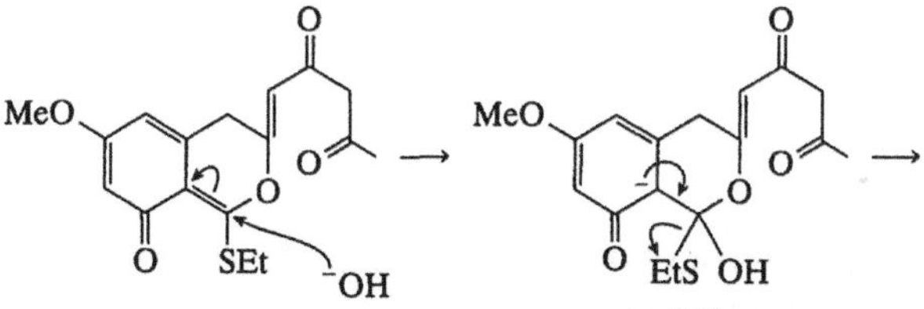

3-12   3-13

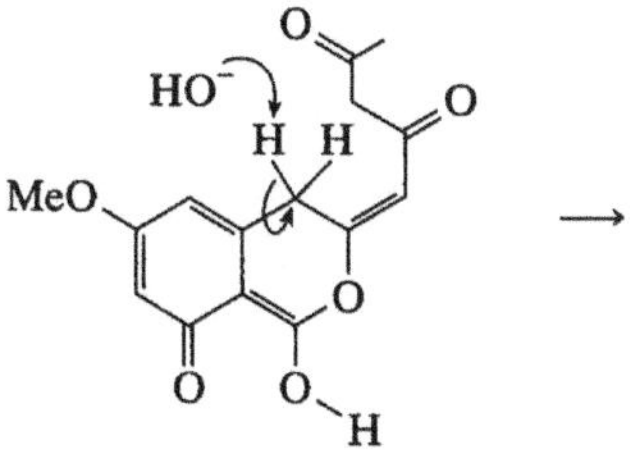

3-14

Ion **3-13** can lose ethyl thiolate to give **3-14**.

Intermediate **3-14** contains a number of acidic protons. Removal of some of these would give anions that probably react to give side products, and removal of others may result in unproductive equilibria. Removal of the proton shown gives a resonance-stabilized anion, **3-15**, which can react with the terminal carbonyl group of the side chain to form the tricyclic structure **3-16**.

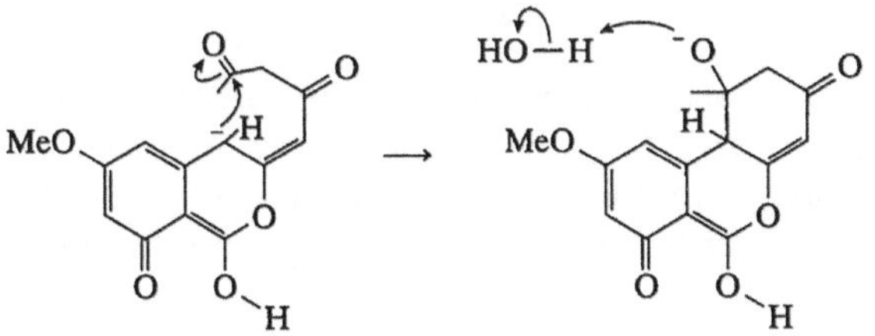

3-15   **3-16**

The intermediate **3-16** removes a proton from solvent to give **3-17**, and **3-17** undergoes elimination of water to give **3-18**.

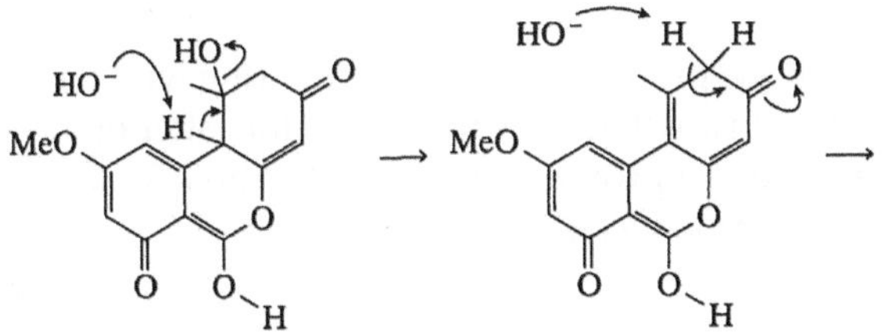

3-17   **3-18**

Removal of a proton from the right-hand ring of **3-18** gives a phenolate ion, **3-19**.

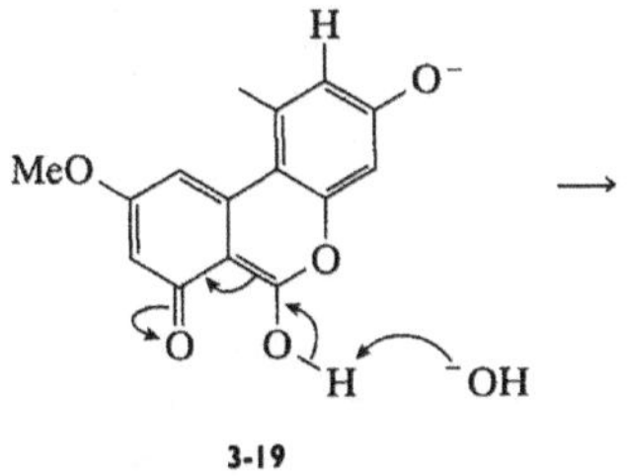

3-19

Removal of another proton from **3-19** gives a diphenolate, **3-20**, which can be protonated to give the product upon workup in aqueous acid.

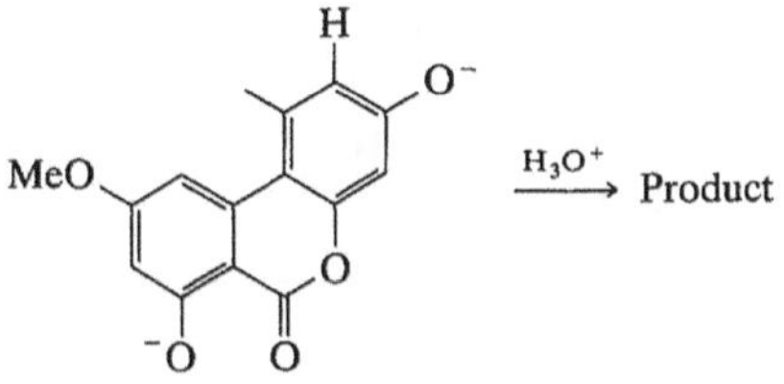

H_3O^+ → Product

3-20

In this reaction, as in many others, the exact timing of steps, especially proton transfers, is difficult to anticipate. For example, the proton removed from **3-19** actually may be removed in an earlier step.

Mechanism 2: In this mechanism, formation of the right-hand ring occurs before formation of the middle ring. After the formation of **3-9**, there is 1,4-addition of hydroxide ion to the α, β-unsaturated ketone.

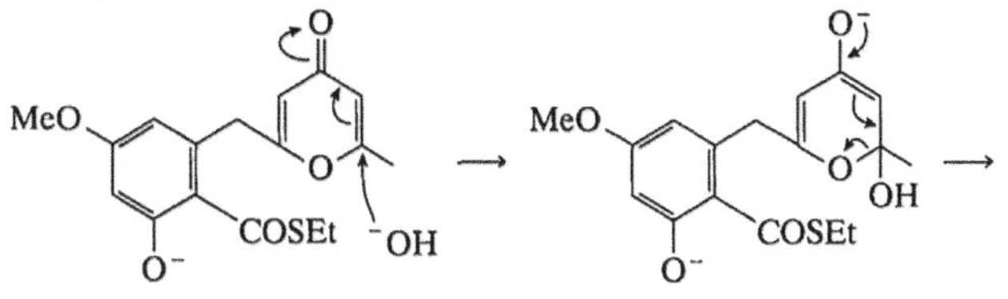

Ring opening of **3-21** follows to give a new enolate, **3-22**, which can be protonated at carbon by water to give **3-23**.

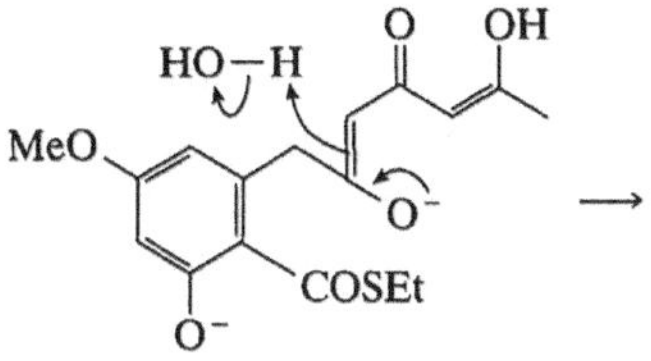

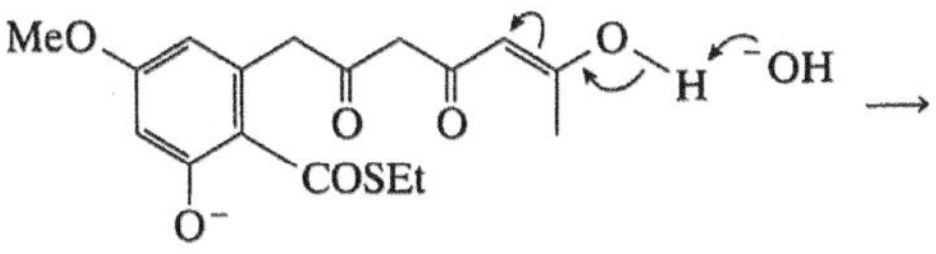

Base-promoted tautomerization of the enol in **3-23** gives **3-24**. An alternative route to the methyl ketone **3-24** starts with reaction of hydroxide on the other carbon β to the ketone in the right-hand ring of **3-9**. These steps follow a course analogous to that depicted (**3-21** to **3-22** to **3-23**).

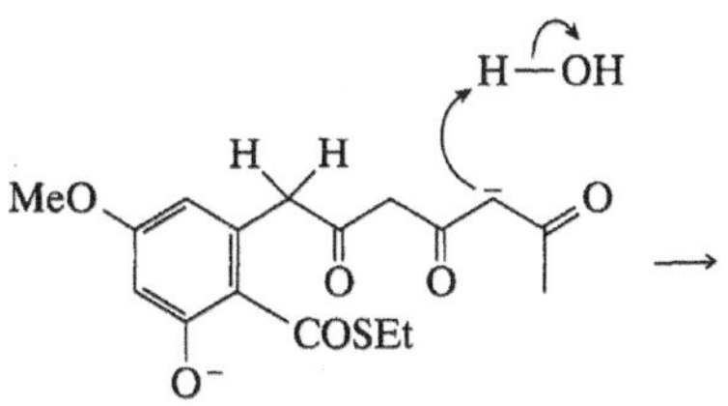

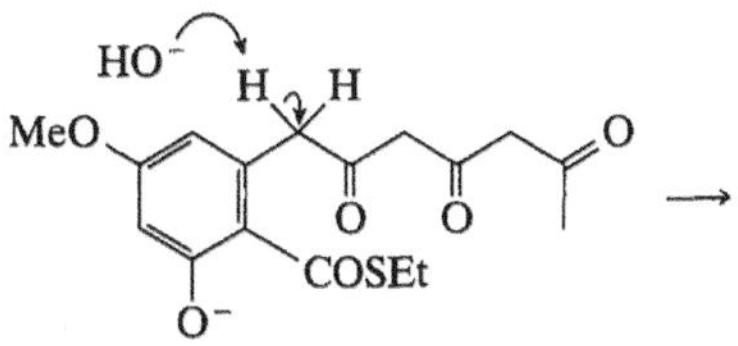

The mechanism then continues with an intramolecular aldol condensation.

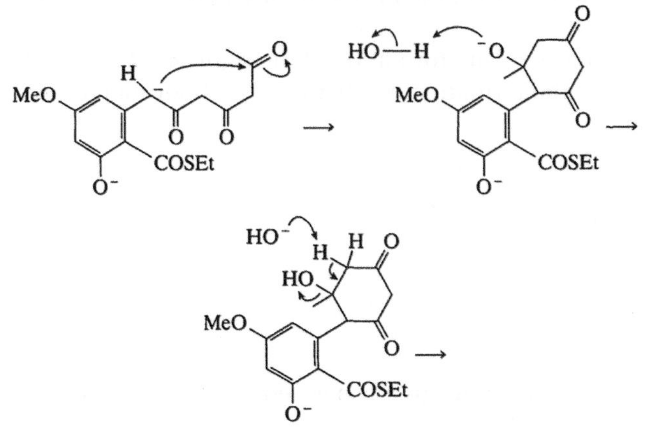

3-25

The resulting β-hydroxy ketone, **3-25**, can eliminate water to give **3-26**. Aromaticity is established as the base-promoted tautomerization of one of the protons in the box in **3-26** gives the enol, and removal of the proton in the circle then gives the phenolate ion, **3-27**.

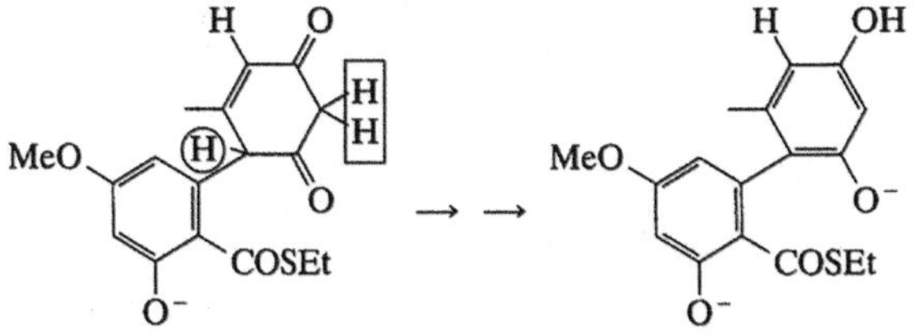

3-26   **3-27**

The nucleophilic phenolate oxygen atom of **3-27** adds to the carbon of the thioester group; then ethylthiolate is eliminated.

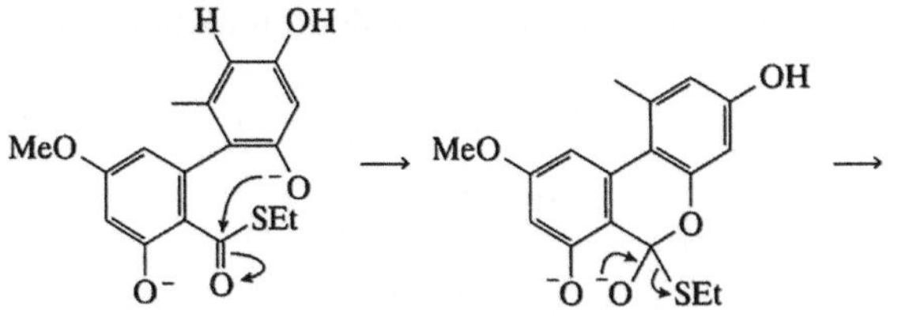

3-28

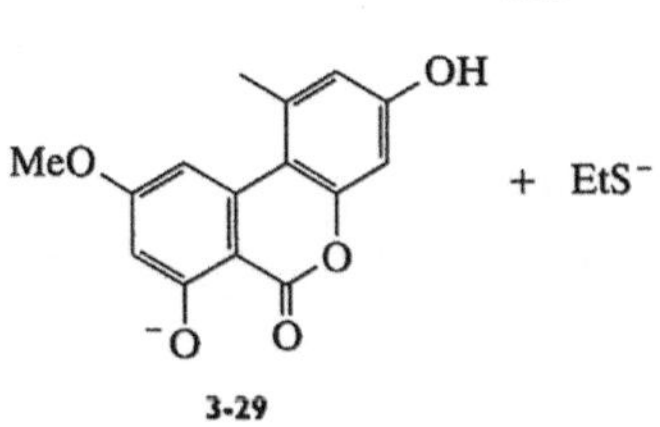

+ EtS⁻

3-29

Once the phenolate ion **3-27** has reacted, the phenol in the right-hand ring of either **3-28** or **3-29** reacts with hydroxide to give a new phenolate in this ring. One of these possibilities is represented next. (The phenolate ion in the right-hand ring of **3-27** reduces the acidity of the other phenolic group in that ring. Thus, we anticipate that the proton of the second phenolic group is removed in a later step.)

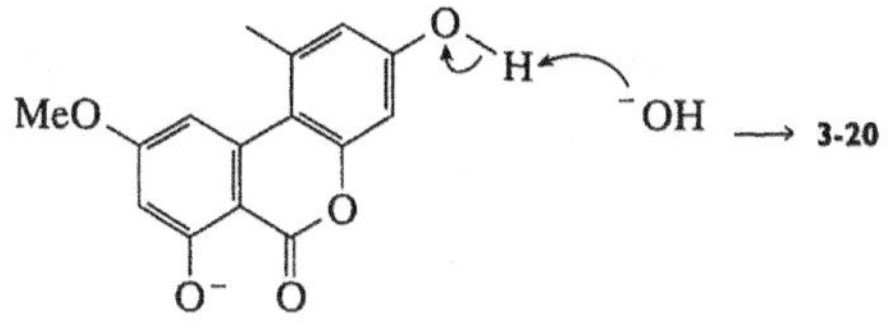

3-29

Both mechanisms for the reaction seem reasonable. The authors of the paper cited (Abell et.al. 1986) showed that **3-30** also cyclizes to the product in excellent yield. Note that **3-30** is the phenol corresponding to the intermediate phenolate **3-24** of mechanism 2. This evidence does not prove that **3-24** is an intermediate in the reaction, but does support it as a viable possibility.

PROBLEM 3.16

Why is the following mechanistic step unlikely? How would you change the mechanism to make it more reasonable?

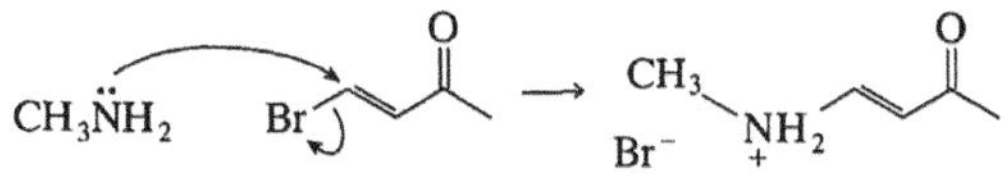

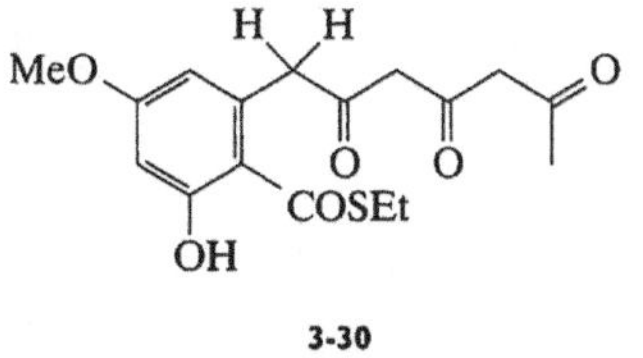

3-30

Other processes similar to the aldol transformation are the Claisen and Diekmann condensations. The Claisen condensation is a base-catalyzed condensation of esters to form a β-keto ester.

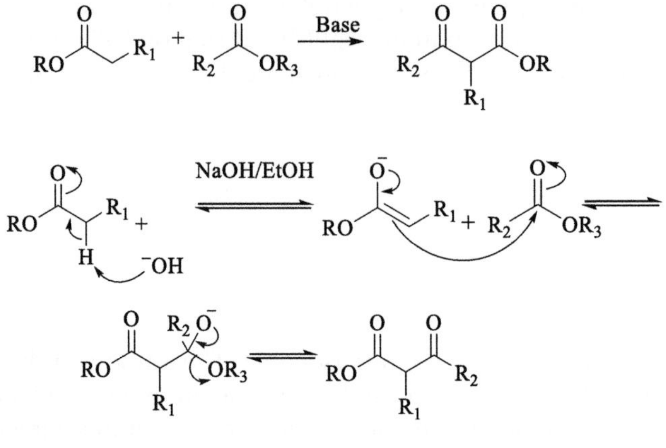

See reference: Mogilaiah and Reddy (2003); Nakatsuji et al., (2009).

The intramolecular (within the molecule) form of the Claisen condensation is called the Diekmann condensation. In this transformation, two esters separated by four or more carbons cyclize when treated with base.

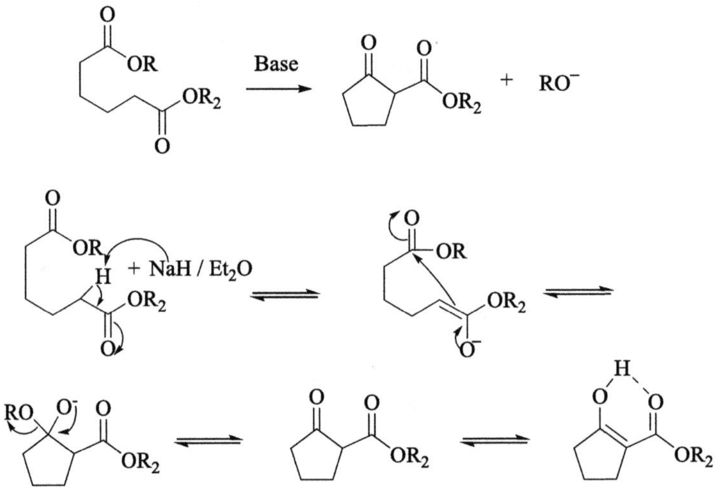

The resulting β-keto ester is very acidic and it is therefore common, under the conditions of the reaction, for the substrate to tautomerize and deliver the unsaturated β-hydroxy enol-ester.

PROBLEM 3.17

Write step-by-step mechanisms for the following transformations.

a.

$$\xrightarrow[\text{2. LDA, CH}_2=\text{CHNO}_2]{\text{1. LDA, PhSSPh}}$$

Lithium diisopropylamide (LDA) is a strong base, but not a good nucleophile because of steric inhibition by the two isopropyl groups directly attached to the nitrogen anion. For this challenging problem, consider numbering the atoms to follow the outcome of the transformation.

b.

$$\xrightarrow[\substack{\text{DMF/H}_2\text{O} \\ \text{room temperature}}]{\text{KCN}}$$

$+ \quad$ $+ \text{ CO}_2$

c. $\text{Ph}\diagdown\text{CN} \; + \; 2 \text{ eq.}$ $\xrightarrow[\text{THF}]{\text{t-BuO-}}$

Sources: Curran et al., (1987); Yogo et al., (1984); DeGraffenreid et al, (2007).

4. BASE-PROMOTED REARRANGEMENTS

A. The Favorskii Rearrangement

A typical Favorskii rearrangement involves reaction of an α-halo ketone with a base to give an ester or carboxylic acid, as in the following example:

$$+ \text{ CH}_3\text{O}^- \longrightarrow$$

Labeling studies have shown that the two α-carbons in the starting ketone become equivalent during the course of the reaction. This means that a symmetrical intermediate must be formed. One possible mechanism, which is consistent with this result, is as follows:

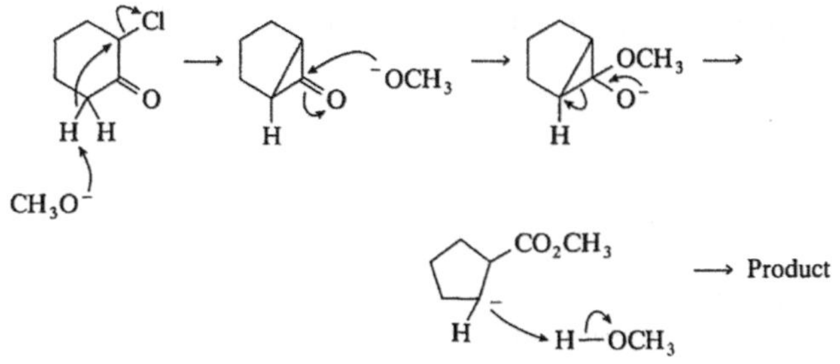

B. The Benzilic Acid Rearrangement

This is a rearrangement of an α-diketone, in base, to give an α-hydroxycarboxylic acid. The reaction gets its name from the reaction of benzil to give benzilic acid:

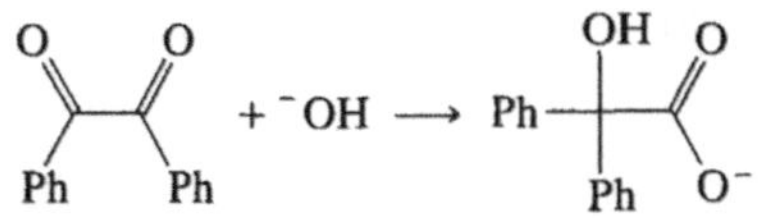

The mechanism involves nucleophilic addition of the base to one carbonyl group, followed by transfer of the substituent on that carbon to the adjacent carbon:

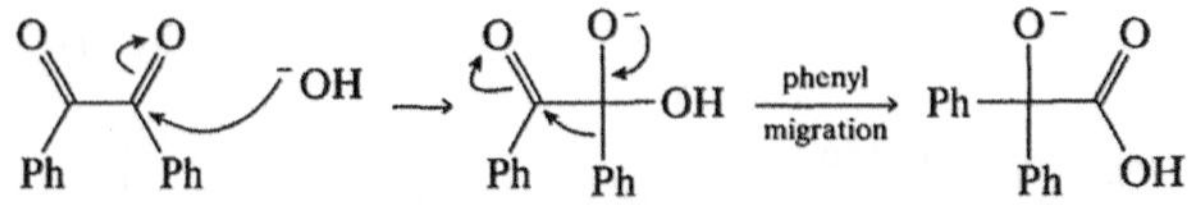

The final steps, under the reaction conditions, are protonation of the alkoxide and deprotonation of the carboxylic acid to give the corresponding carboxylate salt.

PROBLEM 3.18

Write step-by-step mechanisms for the following transformations:

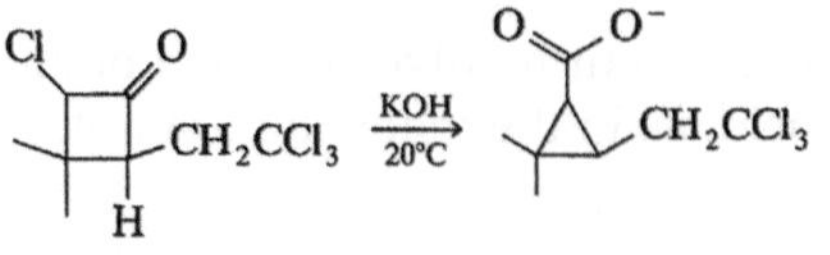

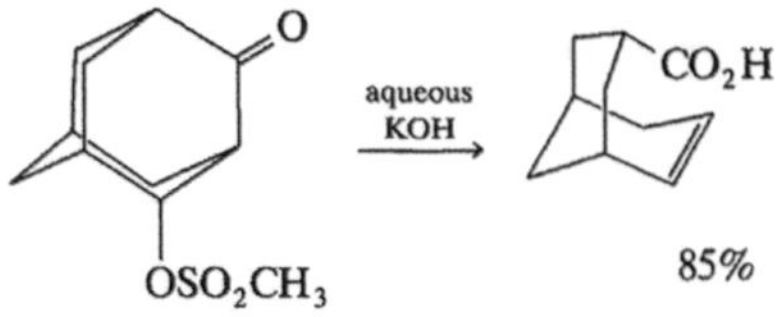

c.

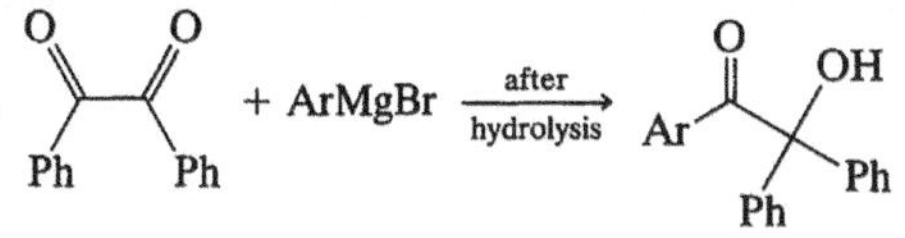

$$\text{Ph} \overset{O}{\underset{}{\|}} \text{Ph} + \text{ArMgBr} \xrightarrow[\text{hydrolysis}]{\text{after}} \text{Ar} \overset{O}{\underset{}{\|}} \overset{OH}{\underset{Ph}{\underset{|}{C}}} \text{Ph}$$

Sources: Martin et al., (1979); Sasaki et al., (1969); March, (1985).

PROBLEM 3.19

Propose a mechanism for the following transformation and offer an explanation for the difference in stereochemistry obtained when the reaction is run in methanol and in the ether dimethoxyethane (DME).

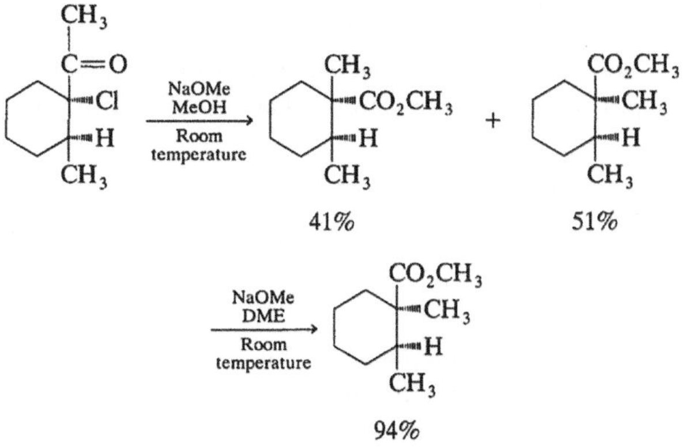

41%   51%

94%

Source: House and Gilmore, (1961).

5. ADDITIONAL MECHANISMS IN BASIC MEDIA

EXAMPLE 3.20. NUCLEOPHILIC ADDITION FOLLOWED BY REARRANGEMENT

Write a mechanism for the following transformation:

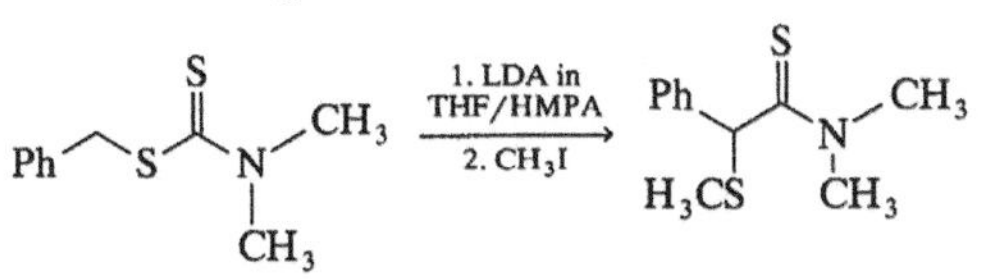

(THF = tetrahydrofuran; HMPA = hexamethylphosphoramide)

First, number the atoms in starting material and product to ascertain how the atoms have been reorganized.

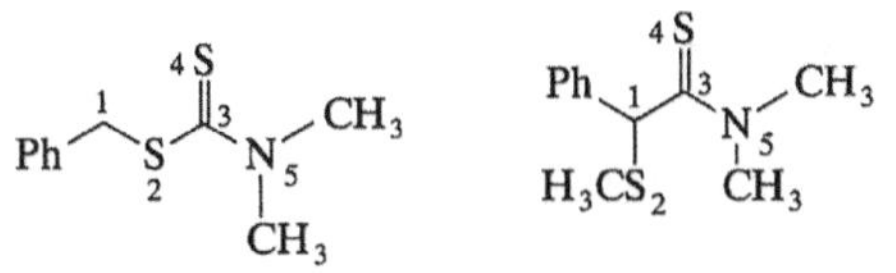

Numbering indicates that C-1 has become attached to C-3 and that the methyl group on S-2 must come from methyl iodide. Focusing attention on positions 1 and 3 of the starting material reveals the following: (1) the protons at position 1 are acidic because they are benzylic, that is, if a proton is removed from this position, the resulting anion is stabilized by resonance; (2) position 3, athiocarbonyl carbon, is an electrophile and should react with nucleophiles. Thus, the first step of the reaction might be as follows:

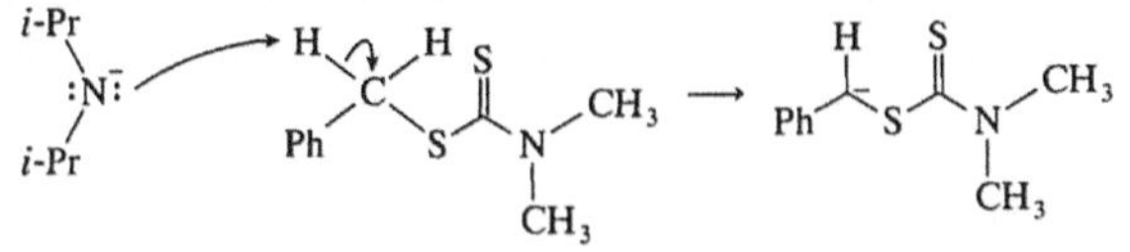

The next step then would be nucleophilic reaction of the carbanion with the electrophilic carbon of the thiocarbonyl group. This reaction joins carbons 1 and 3, as was predicted from the numbering scheme.

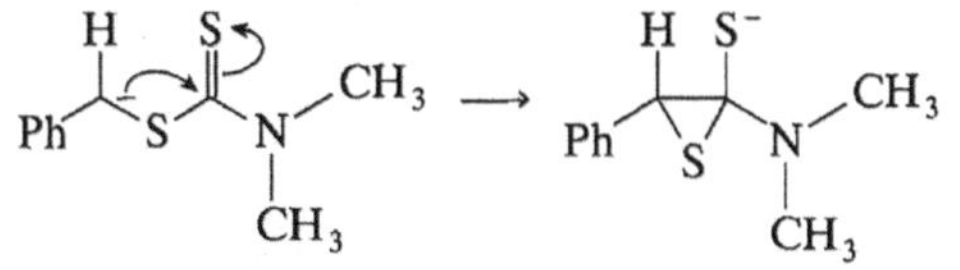

The resulting three-membered-ring intermediate (thiirane) is not stable under the reaction conditions. We know this because there is no three-membered ring in the product! The ring strain in a three-membered ring and the negatively charged sulfur facilitate the ring opening. There are three possible bonds that could be broken when the electron pair on the thiolate makes a π bond with the carbon to which it is attached. Each possibility gives an anion whose stability can be approximated by comparing the relative strengths of the corresponding acids (formed when each anion is protonated). These values can be approximated by choosing compounds from Appendix C with structures as close as possible to the structural features of interest. Breaking the C—N bond would give the dimethylamide ion (the pK_a's of aniline and diisopropylamine are 31 and 36, respectively); breaking the C—C bond would give back the starting material (the pK_a of toluene is 43); and breaking the C—S bond would give a new thiolate ion (the pK_a ethanethiol is 11). Thus, on the basis of the thermodynamic stability of the product, the C—S bond of the ring would break. In fact, breaking the C—S bond leads to an anion that is related structurally to the final product of the reaction. However, keep in mind that although thermodynamics often is helpful, it does not always predict the outcome of a reaction.

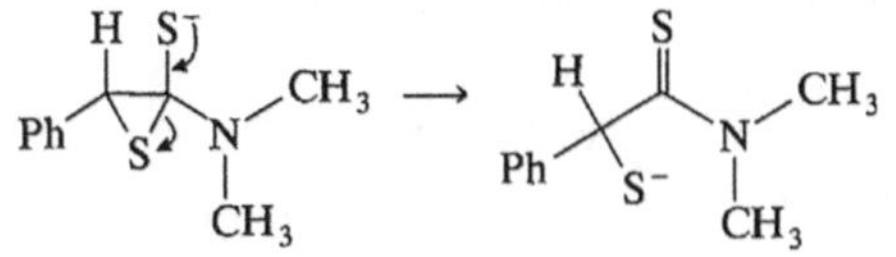

3-31

The original transformation is completed by an S_N2 reaction of the thiolate ion, **3-31**, with methyl iodide.

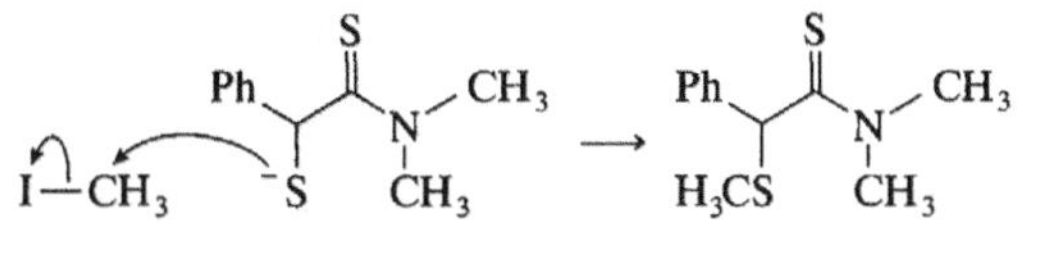

EXAMPLE 3.21. A COMBINATION OF PROTON EXCHANGE, NUCLEOPHILIC ADDITION, AND NUCLEOPHILIC SUBSTITUTION

Write a mechanism for the following transformation:

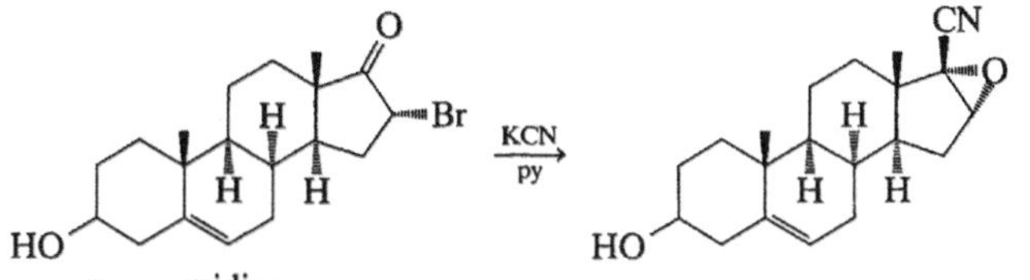

py = pyridine

When a reaction involves only part of a large molecule, such as the steroids in these reactions, it is common to abbreviate the structure. In the structures that follow, the wavy lines indicate the location of the A and B rings that are left out.

First, consider the reaction medium. In pyridine, cyanide is very basic and is also an excellent nucleophile. Because nucleophilic substitution at a position α to a carbonyl is facile, one possible step is nucleophilic substitution of Br by CN. This would be an S_N2 reaction with 100% inversion.

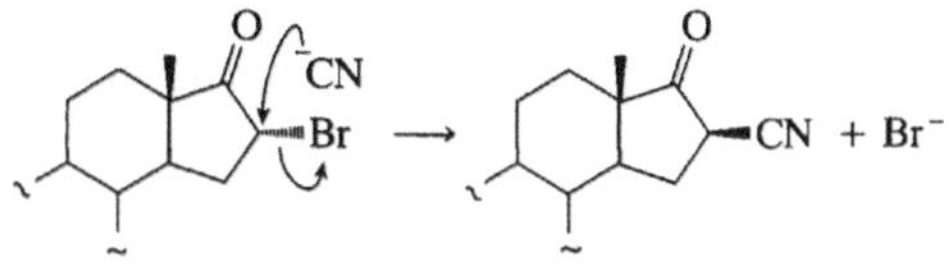

However, there is no reasonable pathway from the product of this reaction to the final product. Another possible step is a nucleophilic reaction of cyanide at the electrophilic carbonyl carbon:

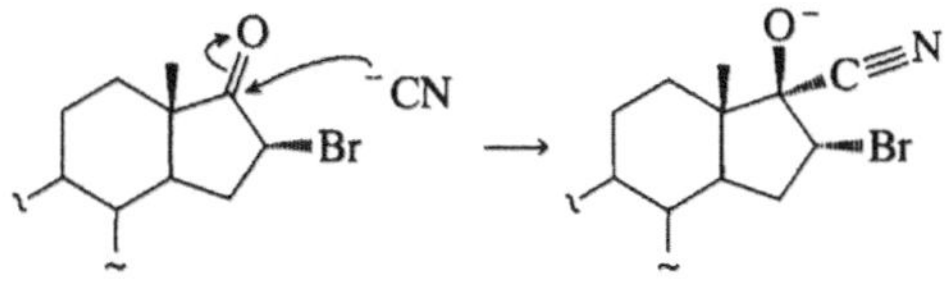

The cyanide approach has been directed so that the alkoxide produced is *anti* to the halide. In this position, the alkoxide is situated in the most favorable orientation for backside nucleophilic reaction at the carbon bearing the bromo group to give an epoxide:

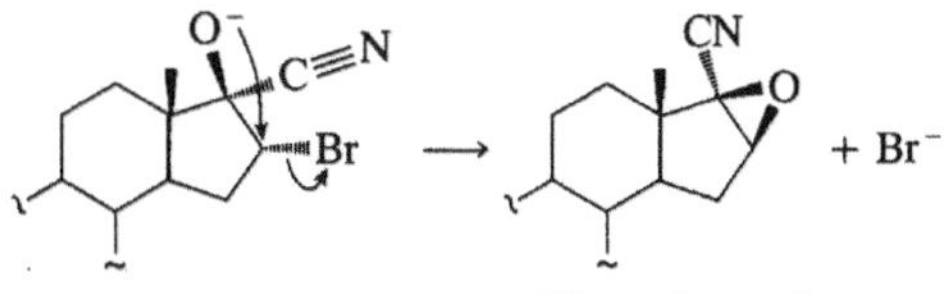

Wrong isomer!

However, this reaction leads to the wrong stereochemistry for the product.

A third possible mechanism can explain the stereochemical result. In this mechanism, a proton is removed from and then returned to the α-carbon, such that the starting material is "epimerized" before cyanide reacts. (Epimerization is a change in stereochemistry at one carbon atom.)

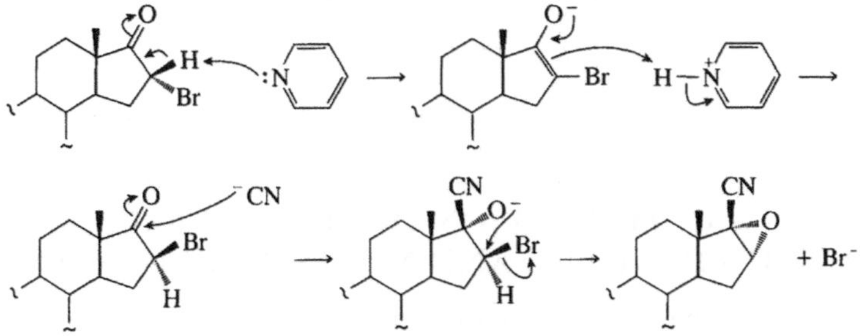

The epimerization mechanism is supported by the following findings: the starting material and the epimeric bromo compound are interconverted under the reaction conditions, and both isomers give a 75% yield of the epoxide product.

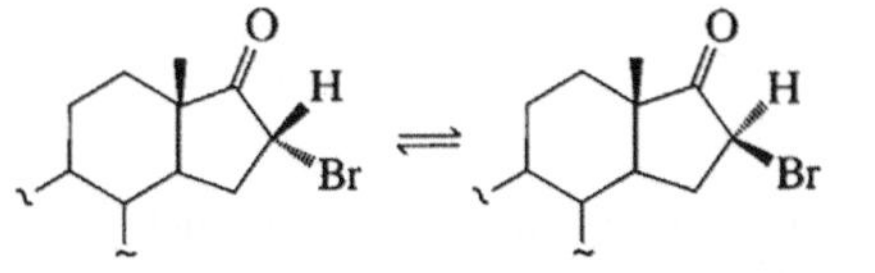

Source: Numazawa et al., (1986).

PROBLEM 3.20

Write reasonable step-by-step mechanisms for the following transformations:

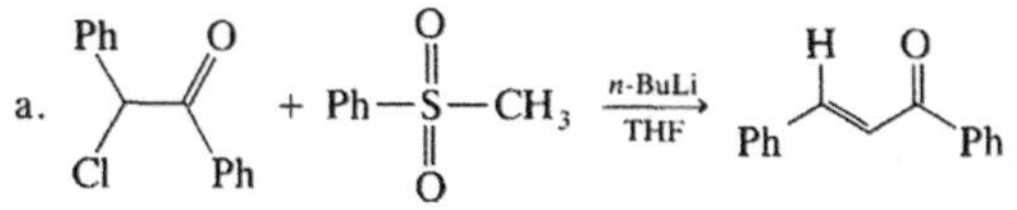

Note: The sulfone is first treated with excess butyllithium to form a dianion and then the α-chlorocarbonyl compound is added.

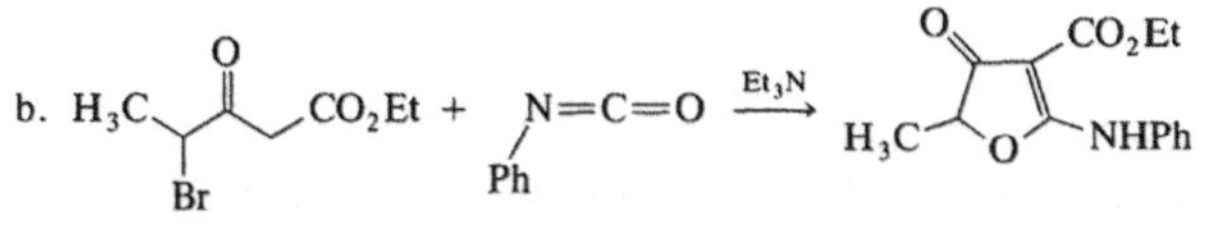

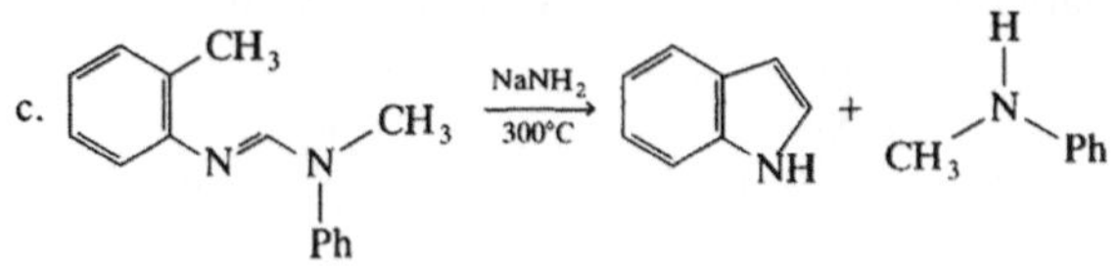

d. 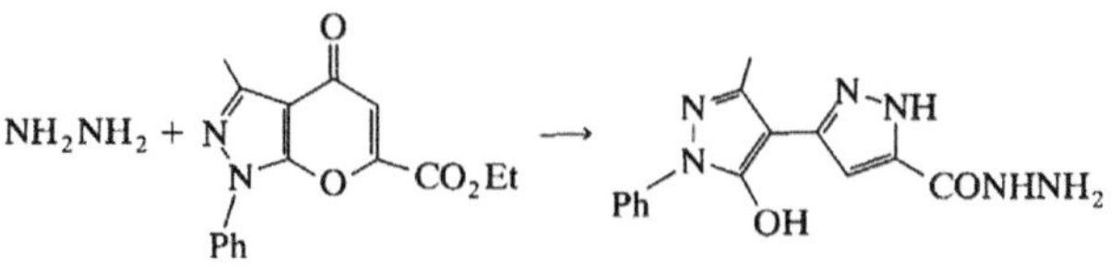 NH_2NH_2 + [structure] $\longrightarrow$ [structure with $CONHNH_2$]

e. 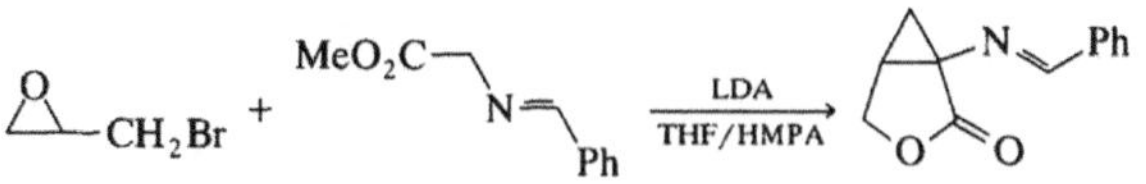 [epoxide]$-CH_2Br$ + MeO_2C-[structure]$N=$[structure]Ph $\xrightarrow[\text{THF/HMPA}]{\text{LDA}}$ [structure with N, Ph]

Sources: Eisch et al., (1985); Mack et al., (1988); Lorenz et al., (1965); Khan and Cosenza, (1988); Bland et al., (1988).

PROBLEM 3.21

Consider the following reaction:

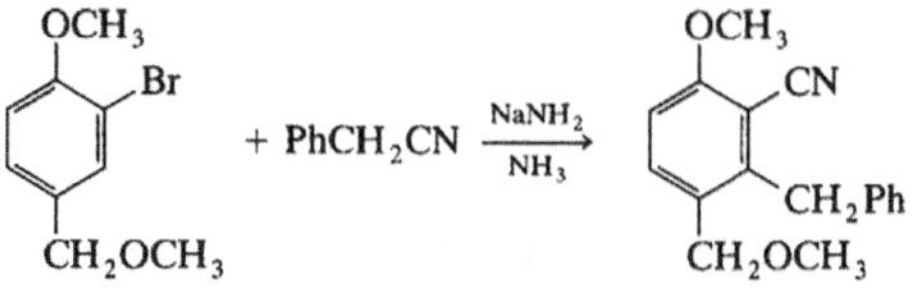

[structure with OCH₃, Br, CH₂OCH₃] + $PhCH_2CN$ $\xrightarrow[\text{NH}_3]{\text{NaNH}_2}$ [structure with OCH₃, CN, CH₂Ph, CH₂OCH₃]

43% yield

Two mechanisms for the reaction are written next. Both proceed through formation of the anion of phenylacetonitrile:

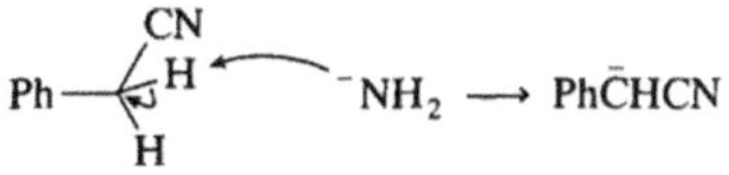

Ph—[C with CN, H, H] $^-NH_2$ $\longrightarrow$ $Ph\bar{C}HCN$

Decide which is better mechanism and discuss the reasons for your choice.

Mechanism 1

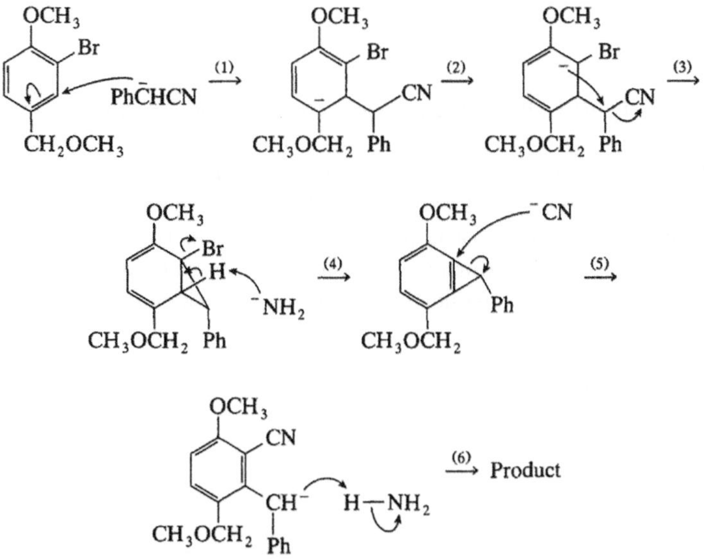

[reaction scheme showing steps (1) through (6)]

(1) → (2) → (3) → (4) → (5) → (6) → Product

Mechanism 2

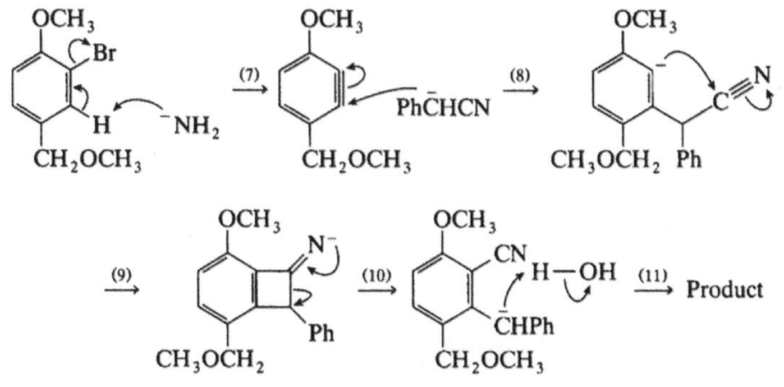

Source: Khanapure et al., (1988).

PROBLEM 3.22

The following esterification is an example of the Mitsunobu reaction. Notice that there is inversion of configuration at the asymmetric carbon bearing the alcohol group in the starting material.

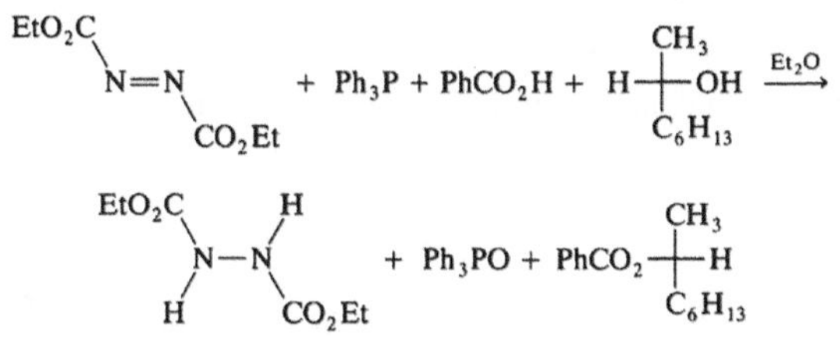

Mitsunobu and Eguchi, (1971). For a review article on the versatility of the reaction, see Mitsunobu, (1981).

The reaction mechanism is believed to proceed according to the following outline: triphenylphosphine reacts with the diethyl azodicarboxylate to give an intermediate, which is then protonated by the carboxylic acid to form a neutral salt. This salt then reacts with the alcohol to form dicarboethoxyhydrazine and a new salt. This salt then reacts further to give a triphenylphosphine oxide and the ester. Using this outline as a guide, write a mechanism for the reaction.

ANSWERS TO PROBLEMS

Problem 3.1

First, although the atoms are balanced on both sides of the equation, the charges are not. Also, there is no way to write a reasonable electronic state for the oxygen species on the right. If the electrons were to flow as written, the oxygen on the right would have a double positive charge (highly unlikely for an electronegative element) and only six electrons lacking two for an octet.

A proton, H^+, should be shown on the right side of the equation, rather than a hydride, H^- (Hint 2.2). Also, as stated in Hint 3.1, hydride is an extremely poor leaving group. The curved arrow, on the left side of the equation, is pointed in the wrong direction (Hint 2.4). This should be apparent from the fact that the positively charged oxygen atom is far more electronegative than hydrogen. Thus, the arrow should point toward oxygen, not away from it.

Loss of a proton is often written in mechanisms simply as $-H^+$. We will not do this because protons are always solvated in solution. Therefore, in this book, with the exception of Chapter 7, the loss of a proton will always be shown as assisted by a base. However, the base need not be a strong base. Thus, for the transformation indicated in the problem, the following could be written:

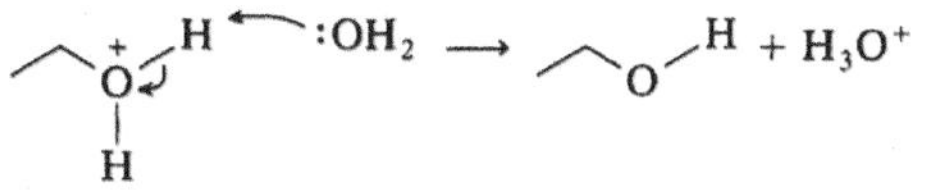

Problem 3.2

According to Appendix C, the pK_a of bicarbonate is 10.2, whereas the pK_a of phenol is 10.0 and the pK_a's of m-nitrophenol and p-nitrophenol are 8.3 and 7.2 (i.e., an increase of 1.8 and 2.8 pH units), respectively. The combined effect of the substituents would make this phenol sufficiently acidic to be converted almost entirely to the corresponding phenoxide ion, which will act as a nucleophile. (The negatively charged phenoxide ion is a much better nucleophile than neutral phenol.)

The carboxylic acid group in the starting material will also be converted to a salt in carbonate solution. Although phenoxide is a better nucleophile than carboxylate ion, this phenoxide may not react as rapidly as the carboxylate with benzyl bromide because it is much more sterically hindered (by the *ortho* methyl and nitro groups). S_N2 reactions are slowed considerably by steric hindrance.

Problem 3.3

a. The methylene carbon of the ethyl bromide is the electrophile, the lone pair of electrons on the phosphorus of triphenylphosphine is the nucleophile, and the bromide ion is the leaving group.

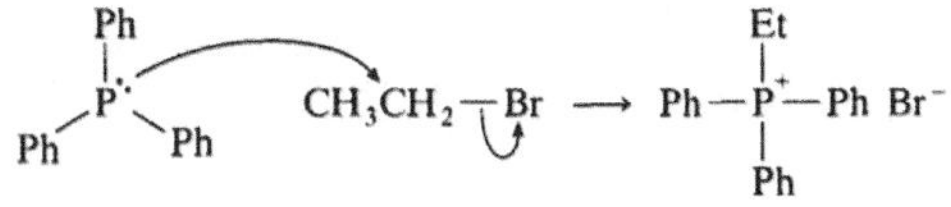

The product salt is stable, and no further reaction takes place. If you wrote that ethoxide was formed and acted as a nucleophile to give further reaction, you neglected to consider the relative acidities of HBr and EtOH (see Table 3.3 and Appendix C). The relative pK_a's indicate that the following reaction does not occur:

$$Br^- + EtOH \rightleftharpoons HBr + EtO^-$$

HBr has a pK_a of -9.0 and EtOH has a pK_a 16. Thus, the K for this reaction is 10^{-25}! Prior ionization of ethyl bromide to the carbocation is unlikely, because the primary carbocation is very unstable.

b. Base is present to neutralize the amine salt, giving free amine. The free amine is the nucleophile, the carbon bearing the chlorine is the electrophile, and chloride ion is the leaving group. The driving force for this intramolecular substitution reaction is greater than that of an intermolecular reaction, because of entropic considerations.

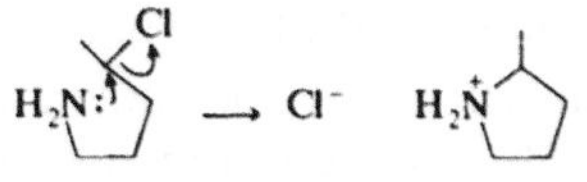

Under the basic reaction conditions, the salt shown and the neutral product would be in equilibrium.

c. The overall reaction can be separated into two sequential substitution reactions. In the first step, the proton of the carboxylic acid is the electrophile, the carbon of diazomethane is the nucleophile, and the carboxylate anion is the leaving group. In the second step, the carboxylate ion is the nucleophile, the methyl group of the methyldiazonium ion is the electrophile, and nitrogen is the leaving group. Loss of the small stable nitrogen molecule provides a lot of driving force to the reaction (Hint 2.14).

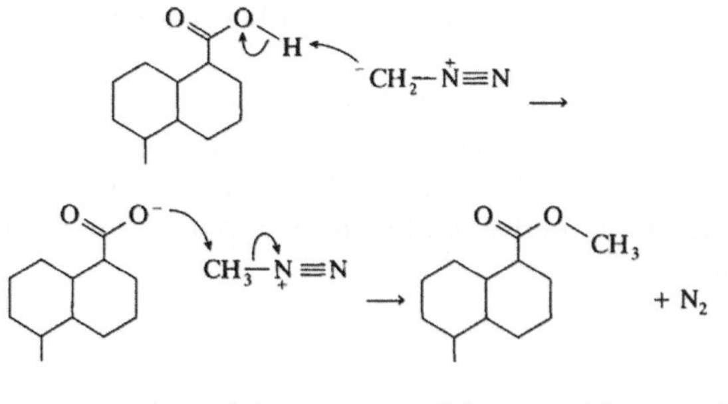

d. In acid, the first step is protonation of the oxygen of the epoxide in order to convert it into a better leaving group. This is followed by ring opening to the more stable carbocation (the one stabilized by conjugation with the phenyl group), followed by nucleophilic reaction of water at the positive carbon to give the product. Thus, in this reaction, the oxygen of water is the nucleophile, the more highly substituted carbon is the electrophile, and the protonated epoxide oxygen is the leaving group.

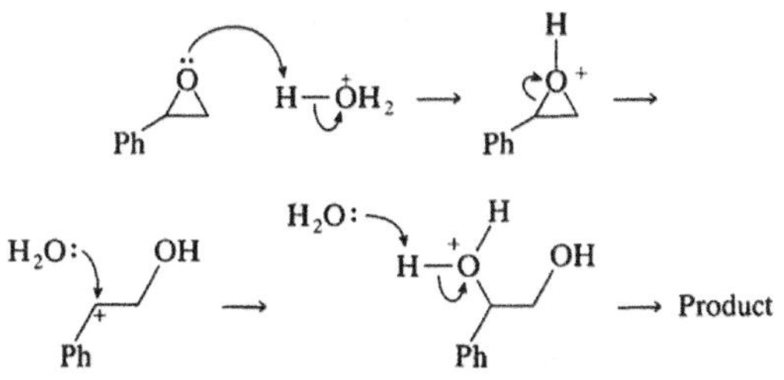

If this reaction were run in base, the following mechanism would apply:

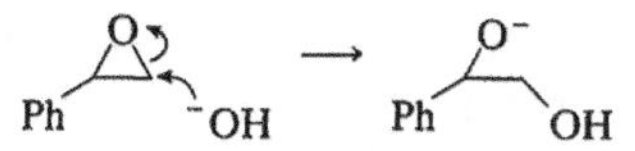

This is one of the few examples where RO⁻ acts as a leaving group. The reason that this reaction takes place is that it opens a highly strained three-membered ring. Note that in base, the nucleophile reacts at the less-substituted carbon. This is because the S_N2 reaction is sensitive to steric effects.

Water is not a strong-enough nucleophile to open an epoxide in the absence of acid, so that in neutral water, the following mechanistic step is invalid:

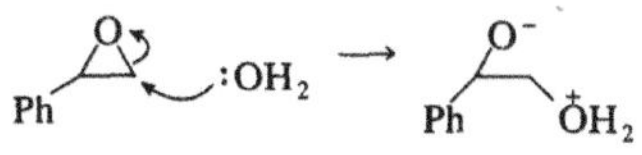

Problem 3.4

a. The carbon skeleton has rearranged in this transformation. Moreover, numbering of corresponding atoms in product and starting material indicates that it is not the ethyl group, but the nitrogen, that moves. In other words, carbon-3 of the ethyl group is attached to carbon-2 in both the starting material and the product. This focuses attention on the nitrogen, which is a nucleophile. Because chlorine is not present in the product, an intramolecular nucleophilic substitution is a likely possibility. This intramolecular S_N2 reaction is an example of neighboring group participation. This reaction gives a three-membered-ring intermediate, which can open in a new direction to give the product. In this second S_N2 reaction, hydroxide is the nucleophile and the CH_2 group of the three-membered ring is the electrophile. Notice that as in the previous problem, S_N2 reaction of the nucleophile on the three-membered ring occurs at the less-hindered carbon.

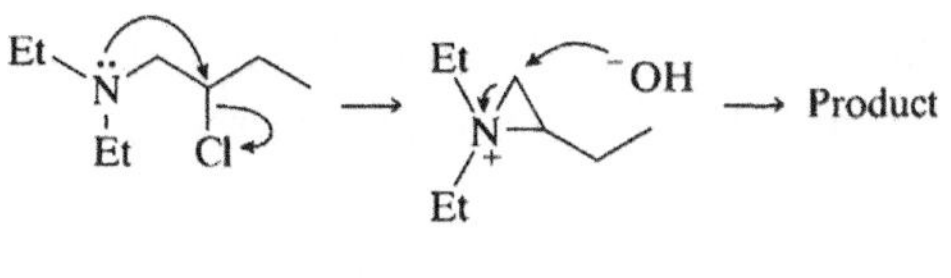

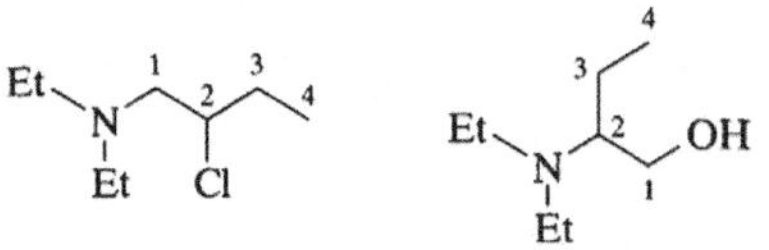

Some steps in an alternative mechanism, written by a student, are shown here:

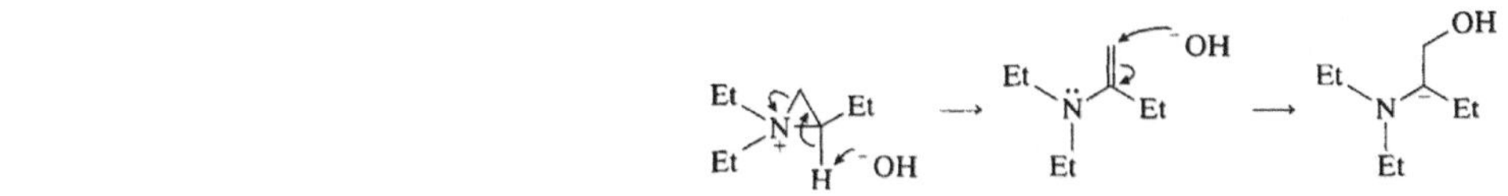

An unlikely step is the addition of hydroxide to the double bond. Double bonds of enamines, like this one, tend to be nucleophilic rather than electrophilic. That is, the resonance interaction of the lone pair of electrons on nitrogen with the double bond is more important than the inductive withdrawal of electrons by the nitrogen. Another way of looking at this is to realize that the final carbanion is not stabilized by resonance and, thus, is not likely to be formed in this manner.

b. The first step is removal of the most acidic proton, the one on the central carbon of the isopropyl group. The anion produced is stabilized by conjugation with both the carbonyl group and the new double bond to the isopropyl group. Removal of no other proton would produce a resonance-stabilized anion. The second step is nucleophilic reaction of the anion with the electrophilic carbon, the one activated by two bromines. Bromide ion acts as the leaving group.

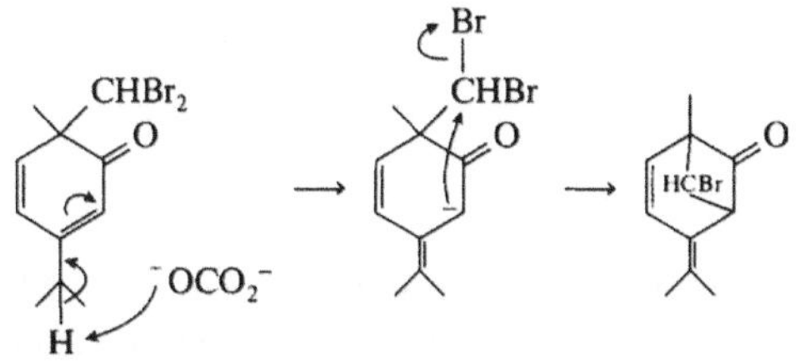

Problem 3.5

The mechanism shown violates Hint 3.2 because the second step shows nucleophilic substitution at an sp^2-hybridized nitrogen occurring as a single concerted process. Change it to two steps, namely, addition followed by elimination:

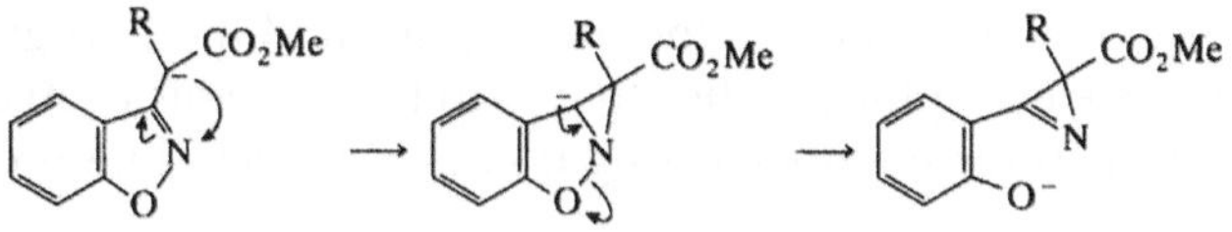

An alternative possibility is a sequence leading through the resonance structure, then onto the ring opening and three-membered ring-forming process.

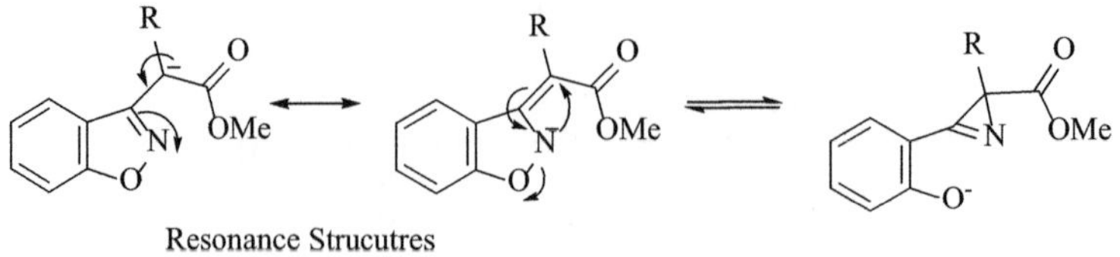

Resonance Strucutres

Problem 3.6

a. Because a strong base is present, the mechanism is not written with one of the neutral guanidino nitrogens acting as the nucleophile. Instead, the first step is removal of a proton from the guanidino NH$_2$ group. This gives a resonance-stabilized anion, **3-32**. Removal of the proton from the imino nitrogen would produce a less-stable anion because it is not resonance stabilized (Problem 1.13.d). In the following, R = the N-methylpyrazine ring.

Another reasonable step that can take place under the reaction conditions is removal of one of the protons on the carbon α to the carbonyl groups in diethyl malonate. However, this reaction does not lead to the product and is an example of an unproductive step. Another unproductive reaction is ester interchange in diethyl malonate. In this case, sodium methoxide would react with the ethyl ester to produce a methyl ester. However, this reaction is inconsequential because methyl and ethyl esters have similar reactivity and the alkyl oxygens with their substituents are lost in the course of the reaction.

The nucleophilic nitrogen of **3-32** adds to the carbonyl group, and then ethoxide ion is lost.

Intermediate **3-33** reacts by a route completely analogous to the previous steps to give **3-35**, a tautomer of the product. That is, the proton removed from the amino nitrogen of **3-33** leads to a resonance-stabilized anion, **3-34**. (The anion formed by removal of a proton from the imide nitrogen would not be resonance stabilized.) The nucleophilic anion, **3-34**, adds to the remaining ester carbonyl. Elimination of ethoxide then gives **3-35**.

Intermediate **3-35** undergoes two tautomerizations to give the product. The first tautomerization involves removal of the proton on nitrogen because this gives an anion that is considerably more delocalized than the anion that would be produced by removal of a proton from carbon. Either enolate oxygen of **3-36** can pick up a proton from either ethanol or methanol. Finally, removal of a second acidic proton forms the product phenolate, **3-37**.

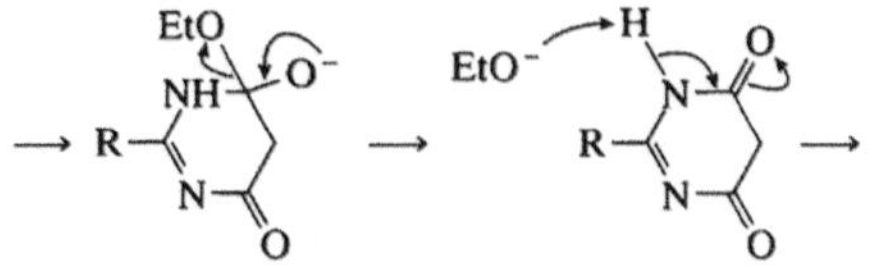

3-35

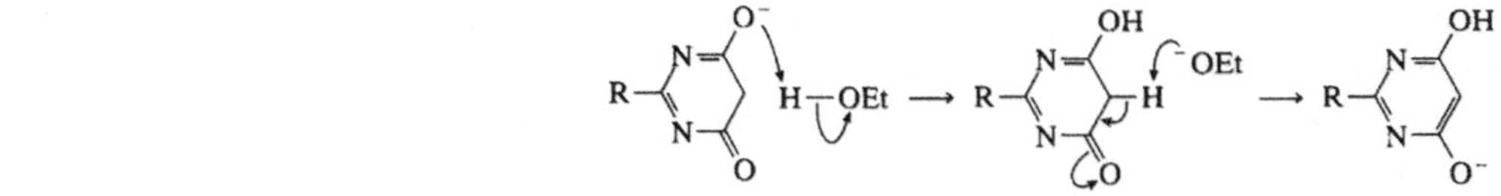

3-36   3-37

In fact, the reactions shown for this mechanism are reversible, and an important driving force for the reaction is production of the stable phenolate salt. The neutral product would be obtained by acidification of the reaction mixture upon workup.

Note: In a tautomerism like that shown for **3-36**, transfer of a proton generally is written as an intermolecular process, not as an intramolecular process as pictured here.

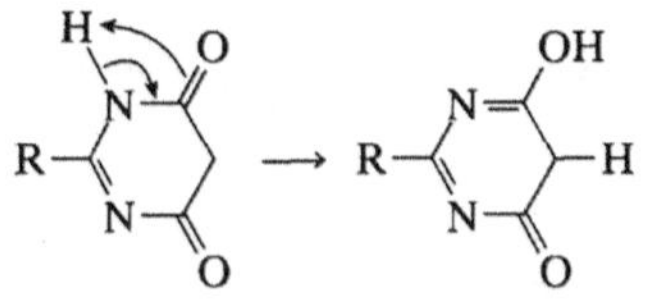

3-35

b. Because the product contains no saturated carbon chain longer than one carbon, it is unlikely that the carbons of the six-membered ring are part of the product. Furthermore, because the functionality attached to the six-membered ring is an acetal and this functional group is easily cleaved (e.g., during aqueous workup), it is probable that cyclohexanone is the other product. Also, the methoxy group of the ester is not present in the product, which suggests a nucleophilic substitution at the carbonyl of the ester.

Numbering of starting material and product also aids in analysis. The relationship between the starting materials and the product is seen more clearly if **3-39-1**, a tautomer of the product, is depicted.

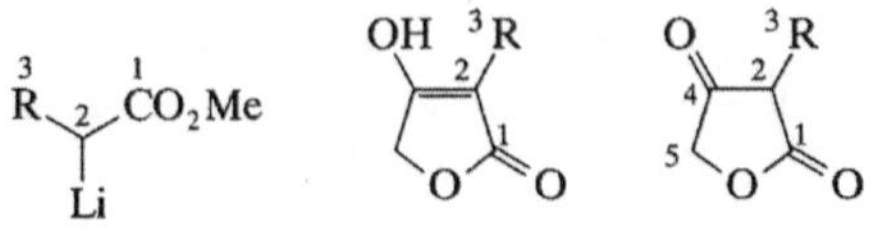

3-38   3-39-1

It is reasonable to assume that three carbons of the product **3-39-1** arise from the lithio reagent and that the other two carbons are introduced from **3-40**. The simplest assumption is that carbons 1, 2, and 3 in **3-39-1** are derived from the lithium reagent and carbons 4 and 5 and oxygen-6 are derived from **3-40**. If, instead, the carbons of the lithio reagent become carbons 1, 2, and 3 of structure **3-39-2**, carbons 4 and 5 from the reagent **3-40** would no longer be bonded to one another. It is difficult to write a mechanism that would account for this bonding change. If the lithio reagent is incorporated as in **3-39-2**, the ring oxygen (O-6) could be derived from either of the ring oxygens in **3-40**, as illustrated in the following structures.

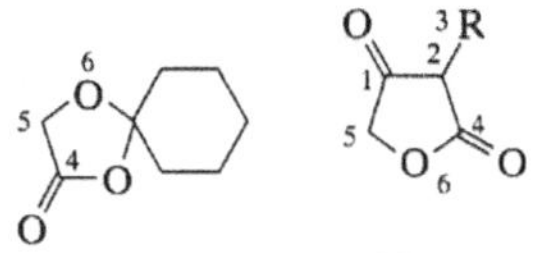

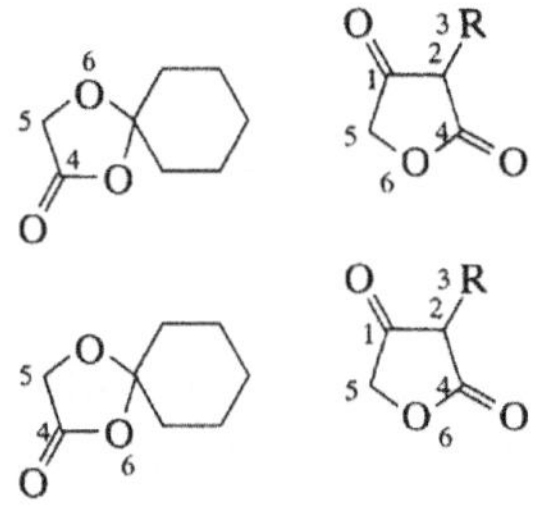

Occam's razor then leads us to assume that the source of the carbons is as shown in structure **3-39-1**.

Analysis by numbering the atoms, the presence of a nucleophile (the lithium reagent) and an electrophile (the ester carbonyl group of **3-40**), all suggest nucleophilic substitution of the ester as a likely first step.

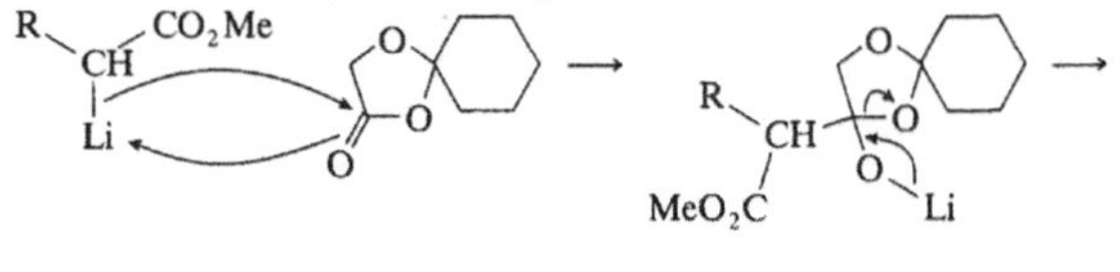

The elimination part of this initial nucleophilic substitution might be concerted with the loss of cyclohexanone, or it might occur in a separate step, as follows:

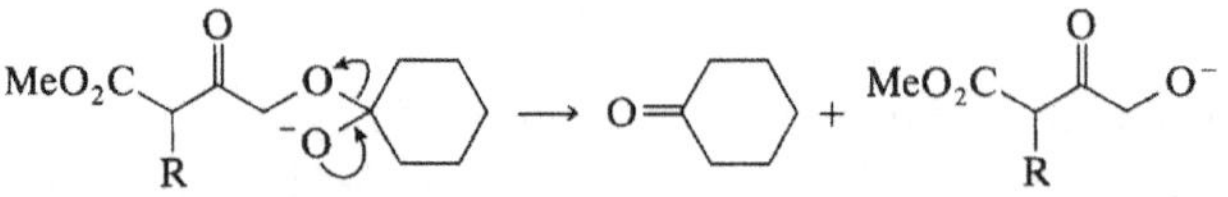

An alternative elimination with breaking of a carbon–carbon bond instead of a carbon–oxygen bond is less likely because the carbanion formed would be much less stable than the alkoxide ion.

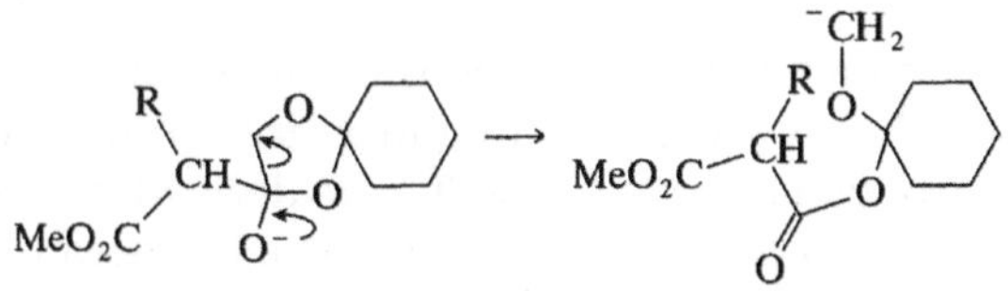

The alkoxide ion, **3-41**, reacts intramolecularly with the ester to produce a lactone.

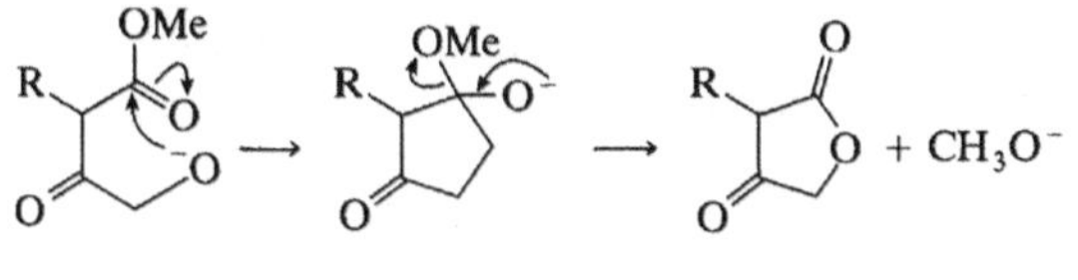

3-41

The remaining steps show tautomerization of the ketone.

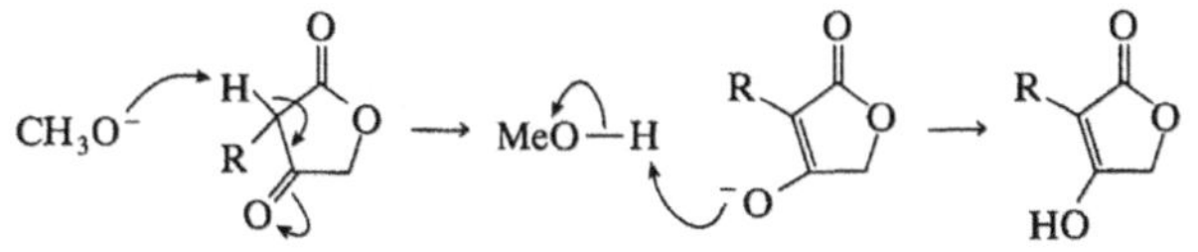

Because there is a considerable excess of the lithium reagent, it is possible that the acidic proton α to both the ester and the ketone is removed to form the dianion **3-42** and that this is the species that undergoes cyclization.

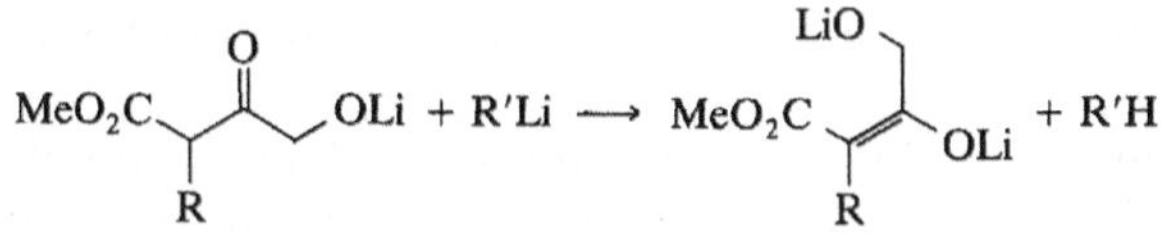

3-42

However, this reaction would be slower than the reaction of the alkoxide **3-41** because of the delocalization of charge shown in **3-43**.

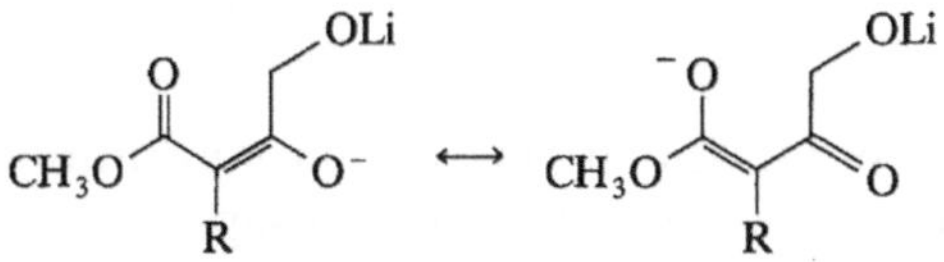

3-43

Because of repulsion between the two centers of negative charge, it may be that the cyclization reaction proceeds through the alkoxide **3-41** even if the dianion **3-42** is the predominant species in the reaction mixture.

c. The first step should be removal of a proton from an amino group because this gives a resonance-stabilized anion. (Use the amino group that becomes substituted during the course of the reaction.)

Substitution of the ethoxy group by the guanidino group should be a two-step process: addition followed by elimination.

In the next step, because of the basicity of the medium and acidic protons that are present, the amino proton is removed prior to nucleophilic substitution of the second ester carbonyl. Because of the resonance stabilization possible for the resulting anion, the amino proton, not the imino proton, reacts. Removal of a proton prior to cyclization also eliminates the need for the last step shown in the problem.

Problem 3.7

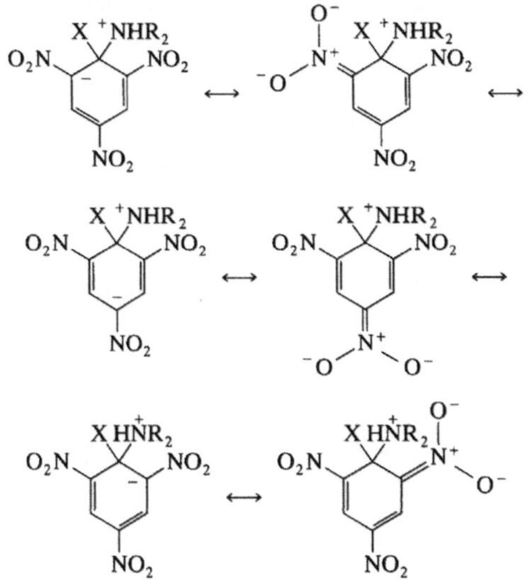

Resonance forms, in which the nitro group is shown in its alternative forms, do not add to the stability of the anion relative to the stability of the starting material because both the starting material and the intermediate have such resonance forms. Thus, these forms often are omitted from answers to questions like this one.

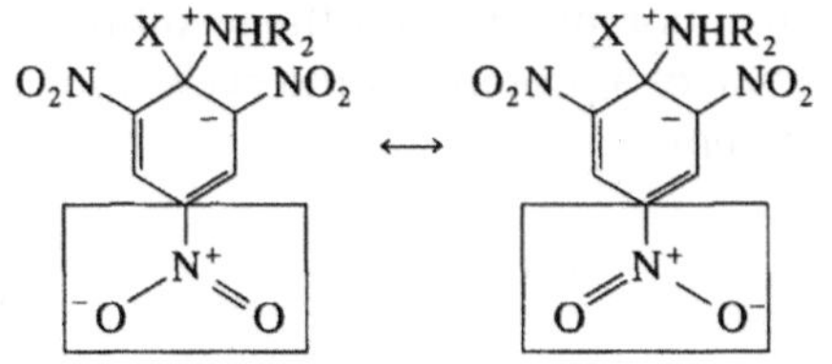

Problem 3.8

The first steps, which are not shown, are the removal of the protons from the protonated amine starting material. Those steps look similar to the final step of the mechanism shown.

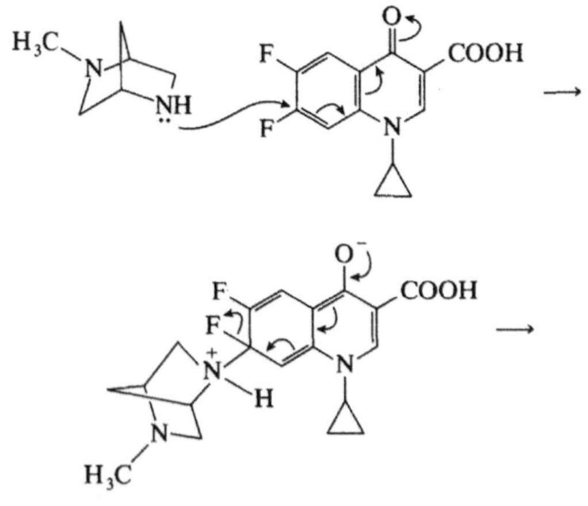

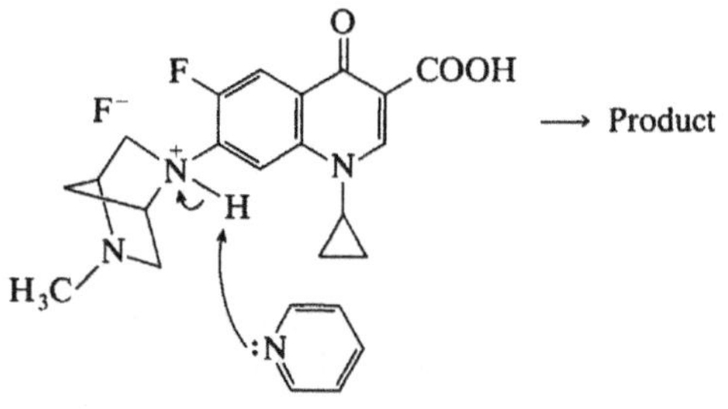

Substitution of the particular fluorine shown is favored because the intermediate anion formed is stabilized by delocalization of the charge on the keto oxygen (resonance). Reaction at the carbon bearing the other fluorine would result in an intermediate in which the negative charge could not be delocalized onto this oxygen.

Initial formation of an aryne, followed by nucleophilic attack, is not a likely mechanism. The most important factors that detract from such a mechanism are the following: (1) the base is not a strong base, and (2) the carbon–fluorine bond is very strong.

Problem 3.9

Because the aryne intermediate is symmetrical, half of the ^{13}C label will be on the carbon bearing the amino group and the other half will be on the carbon *ortho* to the amino group.

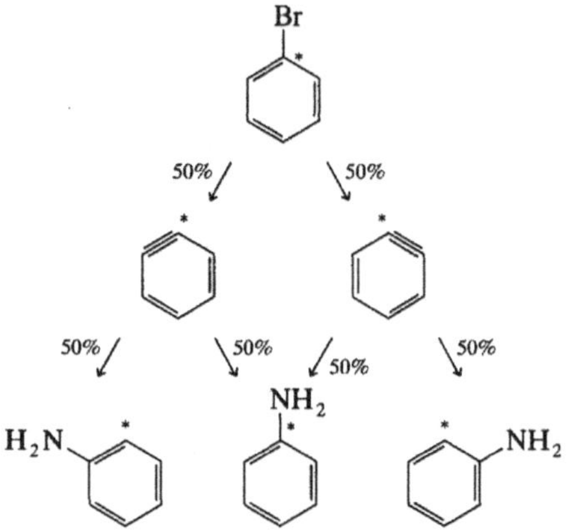

The two outer products at the base of the pyramid are the same. Both are aniline with the label at the *ortho* position.

Problem 3.10

The most acidic proton in the molecule is on the carbon α to the nitrile. This proton is removed first. The second step is elimination of HCl to give an aryne. (As indicated in

Section 1.C, this might also be a two-step process in which removal of the proton is followed by loss of the chloride ion.)

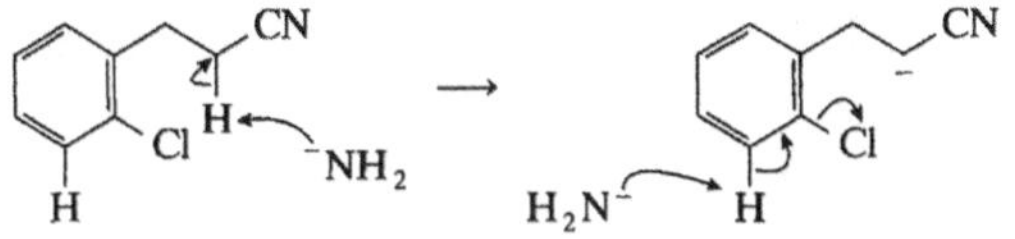

The aryne intermediate is usually written with a triple bond and a delocalized aromatic system, as shown in **3-44**. The anion in the side chain reacts as a nucleophile with the electrophilic aryne. The resulting anion, **3-45**, can remove a proton from ammonia to give **3-46**. Because the product has been reached, we usually stop writing the reaction mechanism at this point. However, in the reaction mixture, amide will remove a proton from the carbon α to the cyano group of **3-46**. Only during workup will the anion be protonated to give back **3-46**.

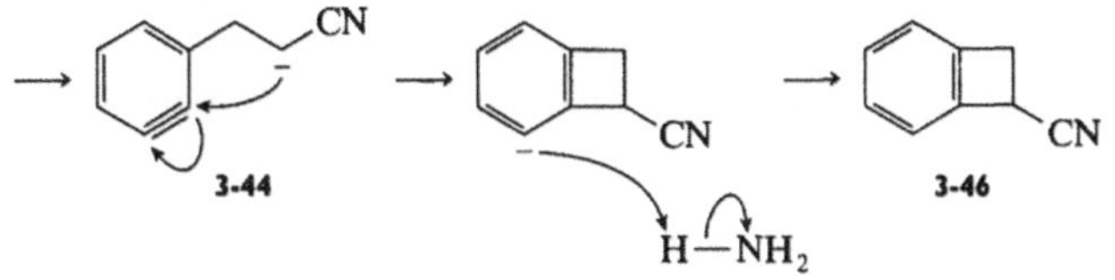

Some students, when answering this question, have used NH_4^+ as the reagent for protonation of **3-45**. However, because amide ion in ammonia is a very strongly basic medium, the concentration of ammonium ion would be essentially zero.

The representation, **3-47**, shown here is equivalent to the structure usually written, **3-44**; however, keep in mind that the extra π bond drawn, the one highlighted, is not an ordinary π bond. It is formed by the overlap of two sp^2-orbitals, one on each carbon. This bond is perpendicular to the other π bonds shown in **3-47**.

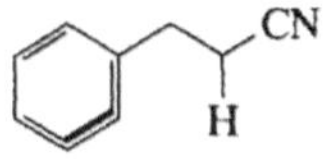

Problem 3.11

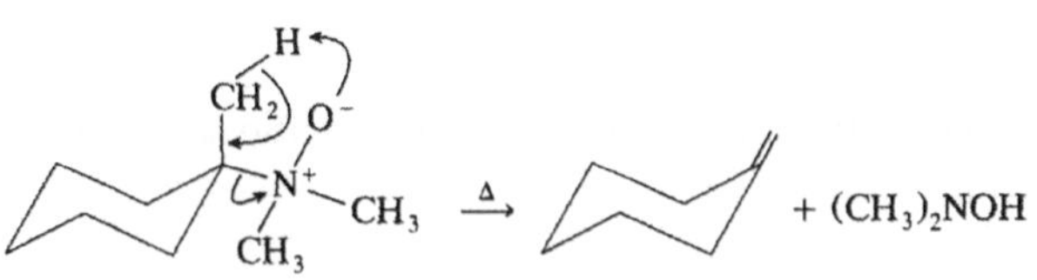

An interesting aspect of this elimination reaction is that it gives only the isomer with the exocyclic double bond. This is a result of the strict stereochemical requirements of the five-membered-ring transition state: all the atoms must lie in the same plane. This rules out the alternate reaction, removal of a proton from a ring carbon, because too much distortion of the cyclohexane ring would be required.

Problem 3.12

a. This reaction is analogous to that in Example 3.15. However, we can also apply our knowledge that esters undergo nucleophilic substitution. Therefore, the initial reaction of the Grignard reagent will give a ketone:

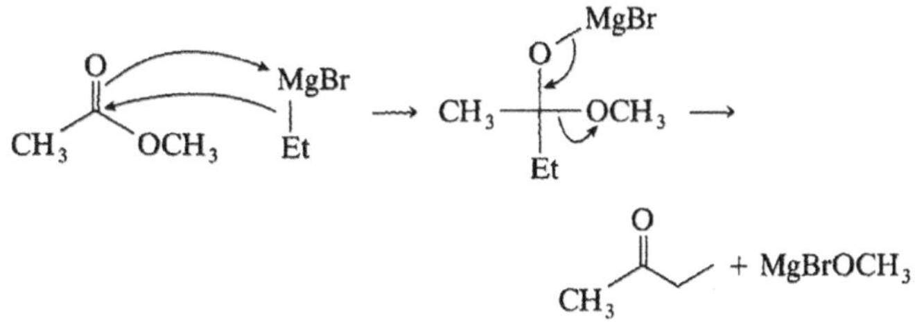

The ketone then reacts with another mole of Grignard reagent to give the salt of an alcohol, which will be converted to the alcohol during acidic workup.

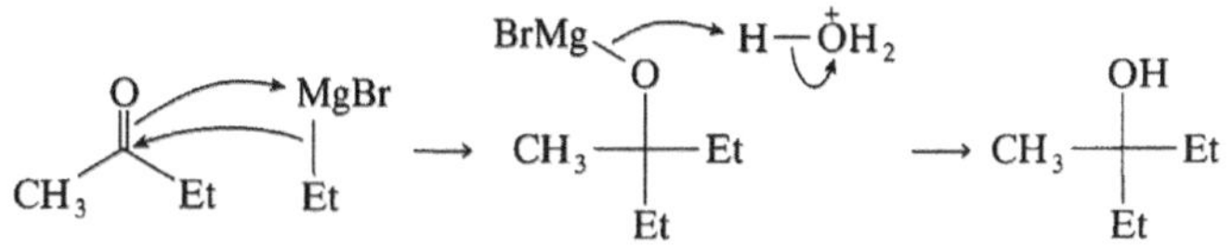

b. This reaction is an unusual addition to a carbonyl derivative. Normally, the nucleophile would react at the carbon atom of the C=N group, but because the resonance form with a positive charge on this carbon would impart antiaromatic character to the ring, the electrophilic character of this carbon is decreased and the aryl group reacts at the unsaturated nitrogen instead. The direction of addition is also influenced by stabilization of the intermediate **3-48**. To the extent that the carbon of the C–Mg bond in **3-48** is negative, it can increase the aromatic character, and thus the stability, of the five-membered ring system.

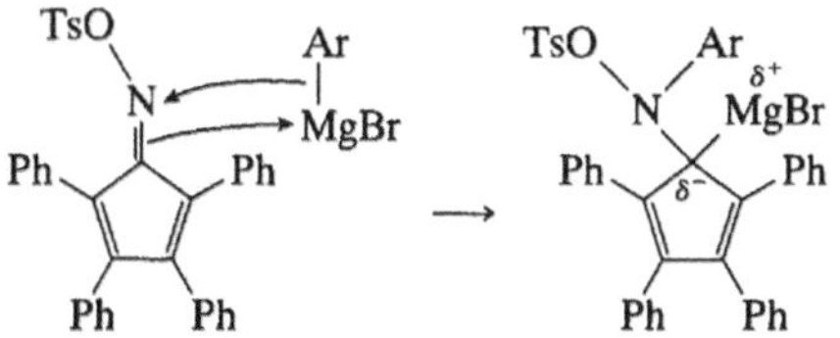

Subsequent elimination of the tosylate group gives the product:

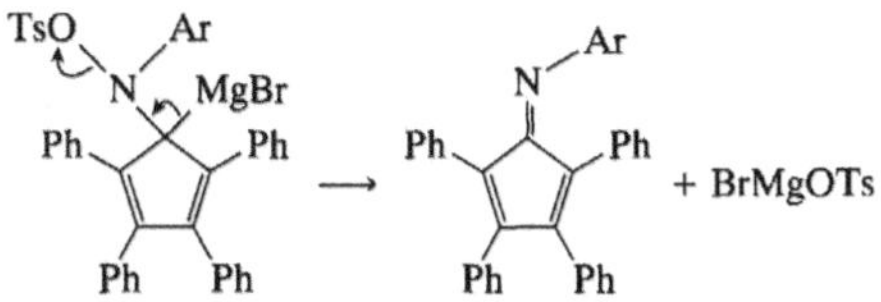

A less likely mechanism would be the direct displacement of the tosylate anion by the Grignard reagent because this would be an S_N2 reaction at an sp^2 atom (Hint 3.2).

Problem 3.13

It can be argued that sterics play some role in the lower reactivity of the proximal nitrogen. But, the main reason for the decreased nucleophilicity at the nitrogen attached to the phenyl ring is because the lone pair of electrons on that nitrogen is delocalized onto the aromatic ring.

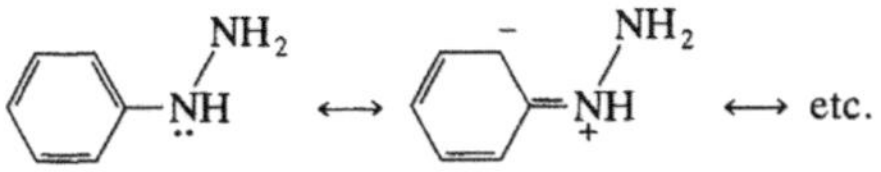

Problem 3.14

a. This is another example of the reaction of an amine with a carbonyl compound in the presence of an acid catalyst. The initial steps are protonation of the carbonyl group, nucleophilic addition of the amine, and deprotonation of the nitrogen to give intermediate **3-49**.

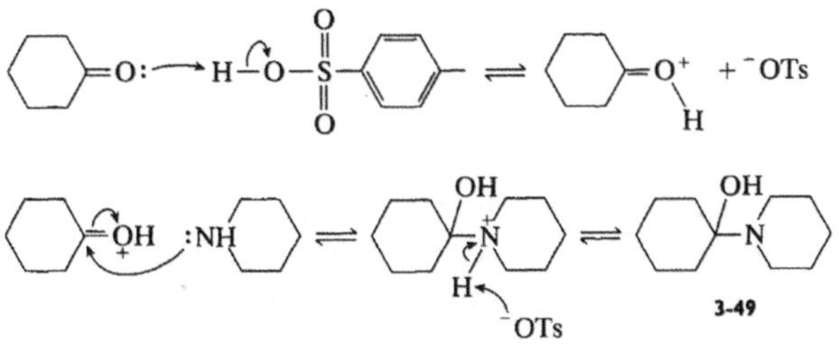

Unlike the intermediate in Example 3.16, intermediate **3-49** has no proton on the nitrogen. Thus, it is not possible to form an imine via loss of a water molecule. Instead, dehydration occurs via loss of a proton from carbon to give the product, which is called an enamine.

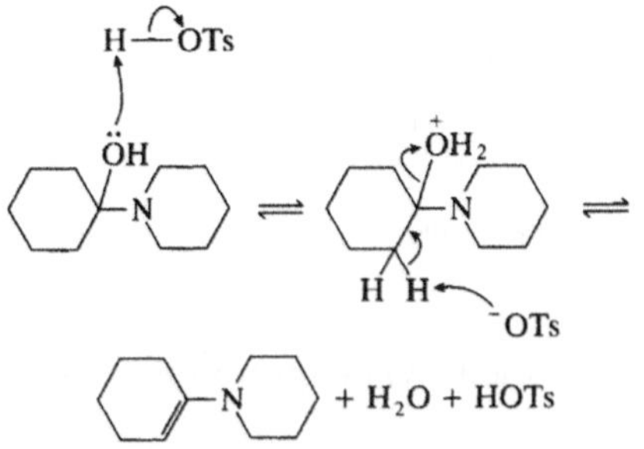

Notice that in the last step, the catalyst is regenerated and water is produced. The equilibria in this problem favor toward starting material. However, the enamine can be obtained in significant amounts if the product water is removed from the reaction mixture as it is produced.

b. Analysis of the starting materials and products reveals that the =NPh group has been replaced by =NOH. An outline of steps can be formulated on the basis of this observation: (1) reaction of hydroxylamine at the carbon of the C=N and (2) elimination of the NPh group. Details to be worked out include identifying the actual nucleophile and electrophile in (1) and the actual species eliminated in (2). These can be ascertained by considering the relative acidities and basicities of the species involved.

Because the nitrogen in hydroxylamine hydrochloride has no lone pairs of electrons, the salt cannot be the nucleophile. The pyridine in the reaction mixture can remove a proton from the nitrogen of hydroxylamine hydrochloride. This equilibrium favors starting material (note the relative pK_a's in the following), but excess pyridine will release some hydroxylamine.

$$pK_a = 8.03 \qquad pK_a = 5.2$$

Moreover, because the salts of pyridine and hydroxylamine are sources of protons, it is unlikely that the anion derived from hydroxylamine, $^-$NHOH, or an anion like **3-50** will be formed as an intermediate. The anion in **3-50** can be compared to the anion formed when aniline ($pK_a = 30$) acts as an acid. Such a strong base will not be produced in significant amounts in a medium in which there is protonated amine (see pK_a's mentioned previously).

3-50

Thus, the nitrogen of the imine will be protonated prior to nucleophilic reaction at the carbon.

The electrophilic protonated imine reacts with nucleophilic hydroxylamine.

Before the phenyl-substituted nitrogen acts as a leaving group, it too is protonated to avoid the poor leaving group PhNH⁻.

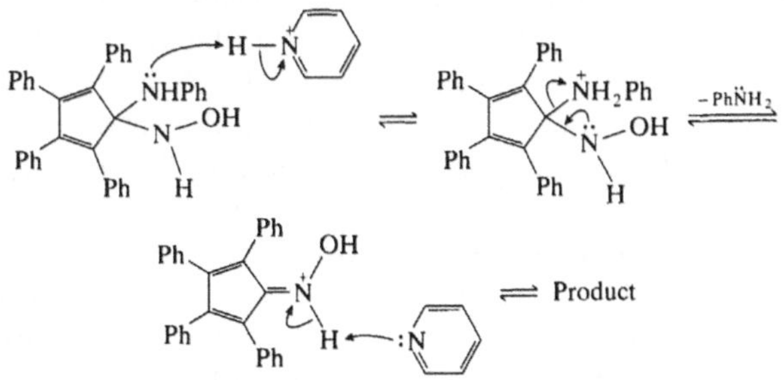

c. In the first step, the basic nitrogen of the imine is protonated. This converts the molecule into a better electrophile, and water, acting as a nucleophile, adds at the positive carbon.

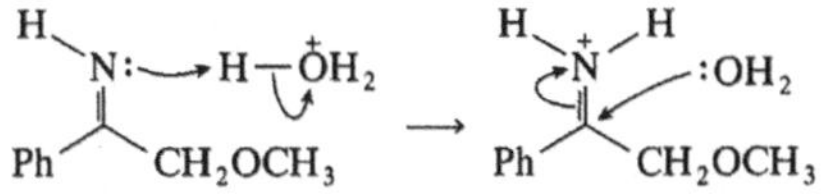

The resulting intermediate can be deprotonated at oxygen and protonated at nitrogen.

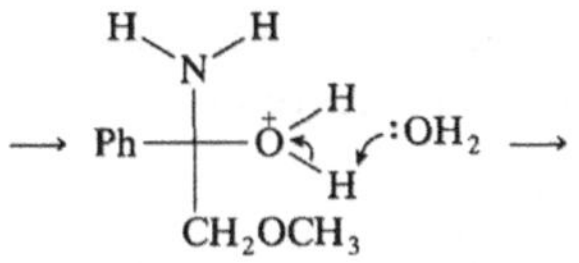

The result is to convert the nitrogen into a better leaving group.

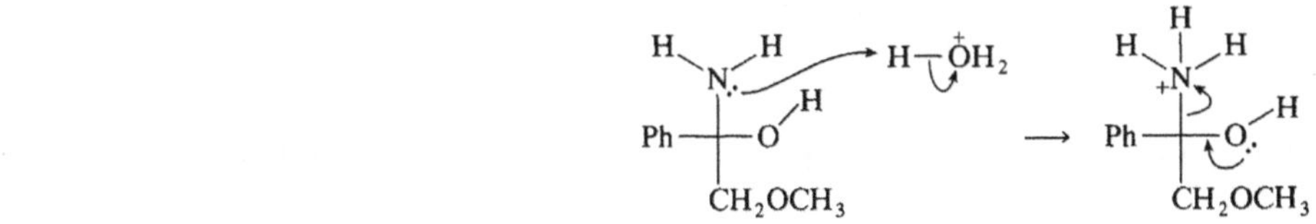

The oxygen in the molecule provides the driving force for the reaction by stabilizing the positive charge. Deprotonation gives the product.

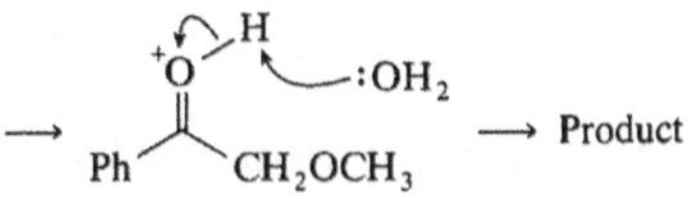

Problem 3.15

a. Although there are three carbons from which a proton could be removed to produce an enolate ion, only one of the possibilities leads to the product shown.

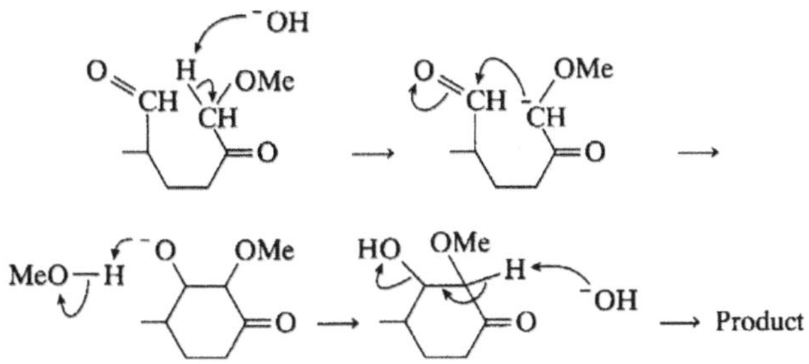

A common shortcut that students take is to write the following mechanistic step for loss of water:

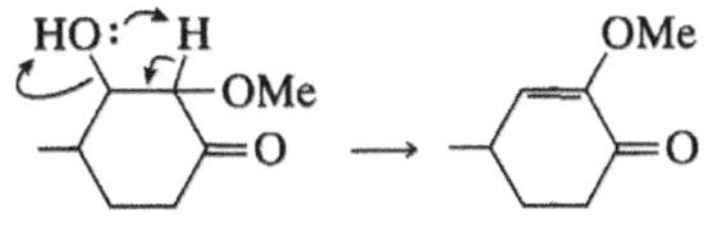

Because hydroxide ion is a much stronger base than the alcohol used as a base in this step, the elimination using hydroxide ion is a better step.

b. The first step of this reaction is removal of the very acidic proton α to both a cyano and a carbethoxy group. There are hydroxide ions present in 95% ethanolic solutions of carbonate, so that either hydroxide ion or carbonate ion can be used as the base.

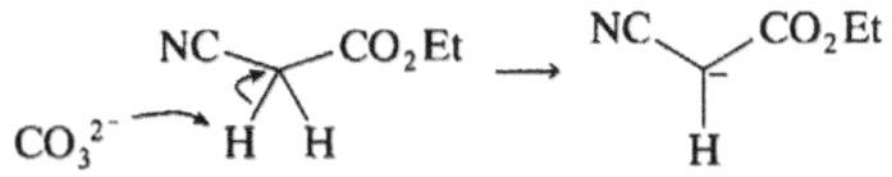

The resulting anion can then carry out a nucleophilic reaction with the electrophilic nitrogen of the nitroso group of nitrosobenzene.

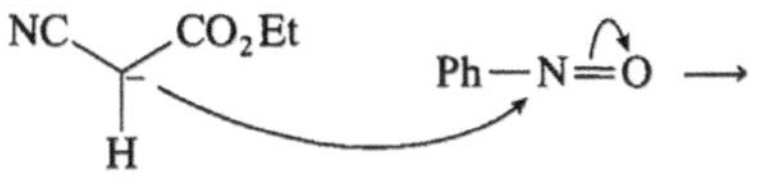

The resulting oxyanion can remove a proton from solvent.

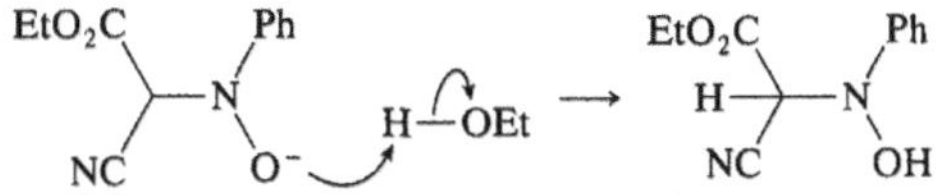

3-51

After the first addition, the reaction repeats itself with removal of the second α proton in **3-51** and addition to a second molecule of nitrosobenzene.

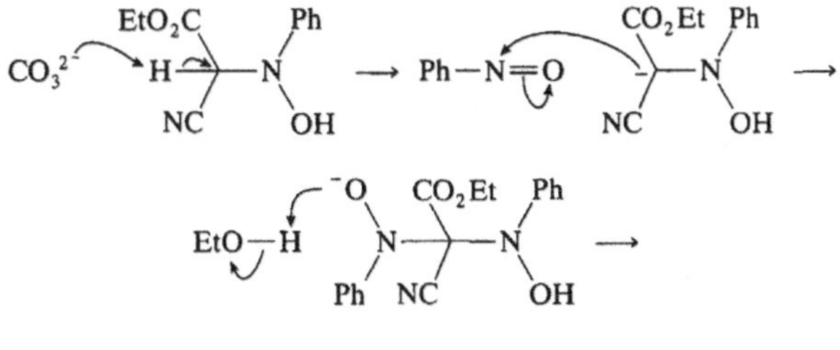

Finally, hydroxide can add to the carbethoxy group, and the intermediate undergoes elimination to give the products.

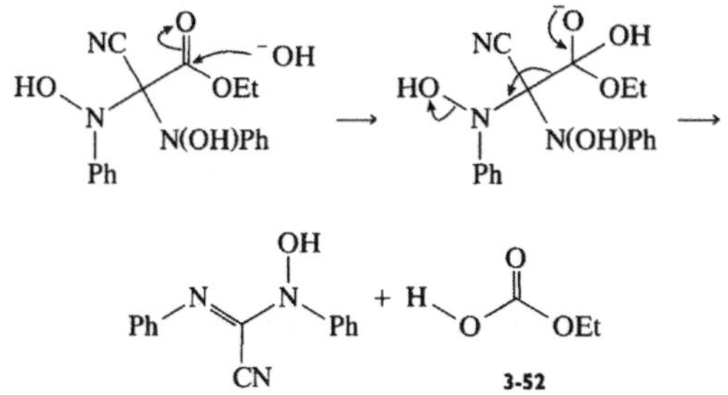

Product **3-52**, a half ester of carbonic acid, is unstable and would decompose to CO_2 and HOEt under the reaction conditions.

Another possible mechanism for the final stages of the reaction involves an intramolecular nucleophilic reaction:

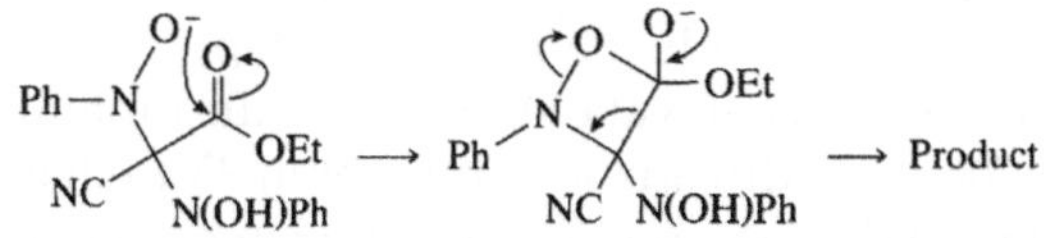

The fact that **3-53** is not necessary does not eliminate it as a possible intermediate in the reaction.

A student wrote the following as a mechanism for the final elimination step:

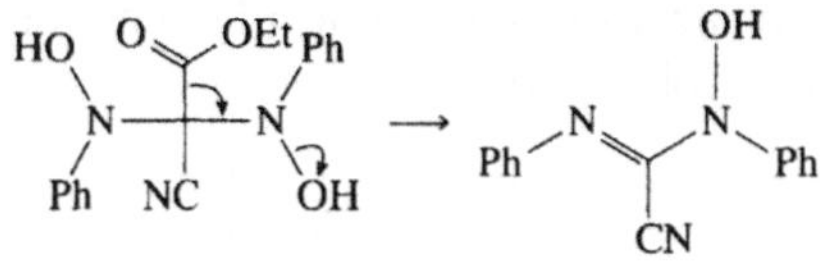

The student's mechanism involves the elimination of hydroxide ion and the formation of the following cation:

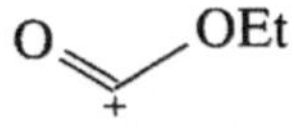

Formation of this cation, a very strong acid, would not be expected in a basic medium (Hint 2.6.). If the structures of the eliminated fragments had been drawn, this unlikely step might not have been suggested.

Problem 3.16

This mechanism shows a direct nucleophilic substitution at an sp^2-hybridized carbon, which is unlikely. An alternative is addition of the amine to the α,β-unsaturated system, followed by elimination of bromide.

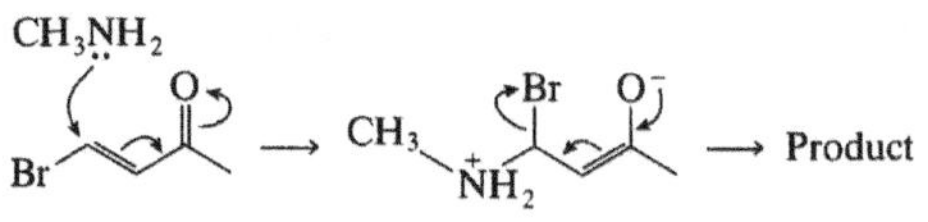

This addition follows the usual course for a 1,4-addition reaction, but the subsequent addition of a proton at the 1 position (oxygen of the carbonyl group) is replaced by elimination of bromide ion.

Problem 3.17

a. Notice that both new groups in the molecule are attached to the carbon next to the carbonyl group. Thus, the first step is removal of a proton from the carbon α to the carbonyl group to give an anion, followed by nucleophilic substitution effected by that anion.

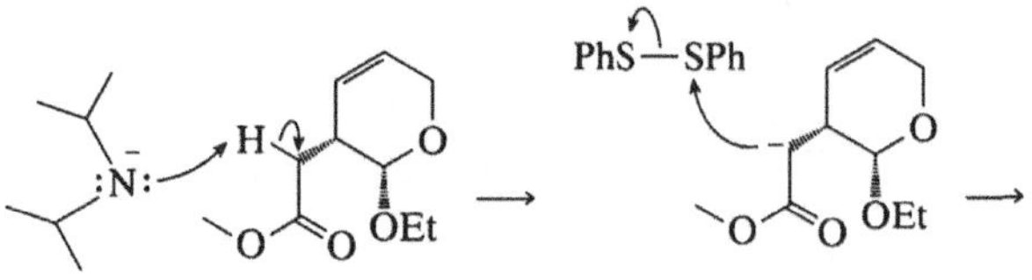

Then a second anion is formed, which adds to the β carbon of the α,β-unsaturated nitro compound. The nitro group can stabilize the intermediate anion by resonance, analogous to a carbonyl group.

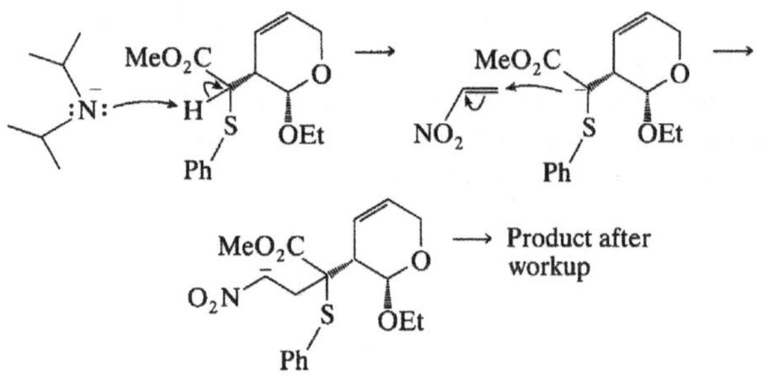

3-54

The diisopropylamine formed when LDA acts as a base is much less acidic than the proton α to the nitro group (Appendix C). Thus, protonation of the anion **3-54** must take place during workup.

b. Cyanide is a good nucleophile. A 1,4-addition to the α,β-unsaturated carbonyl puts the cyano group in the right place for subsequent reaction.

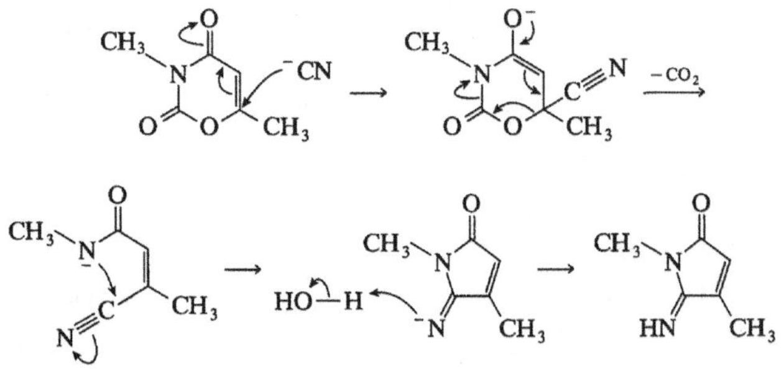

Formation of the other product involves addition to a carbonyl group as the first step.

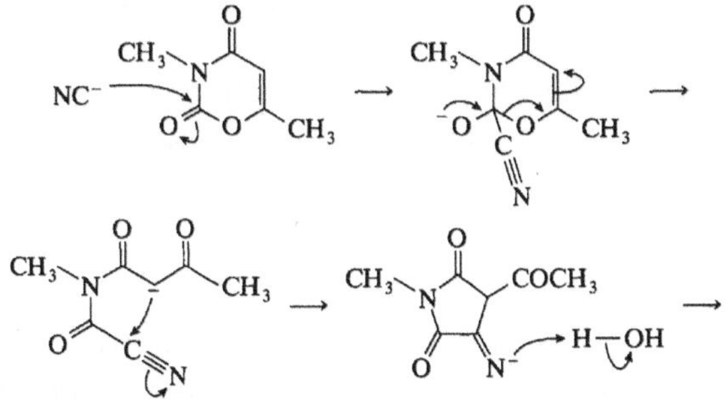

The final steps are tautomerization.

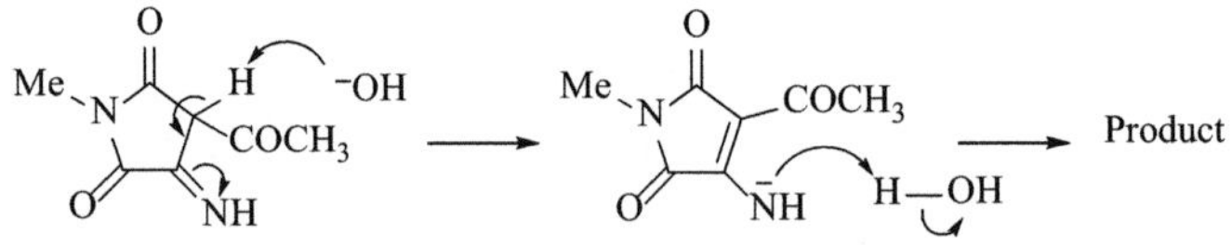

The final steps proceed through tautomerization of the b-keto ester to produce the enol ester **3-58**.

c. Benzyl cyanide **3-55** is relatively acidic (protons both adjacent to a cyano group and benzylic) and the anion adds into the methyl acrylate **3-56** in a Michael fashion (1,4 addition). The addition product of this first series is again deprotonated and adds into a second methyl acrylate. The ester enolizes and then cyclizes via a Diekmann condensation (intramolecular condensation of esters) **3-57**. The final steps proceed through tautomerization of the β-keto to produce the enol ester **3-58**.

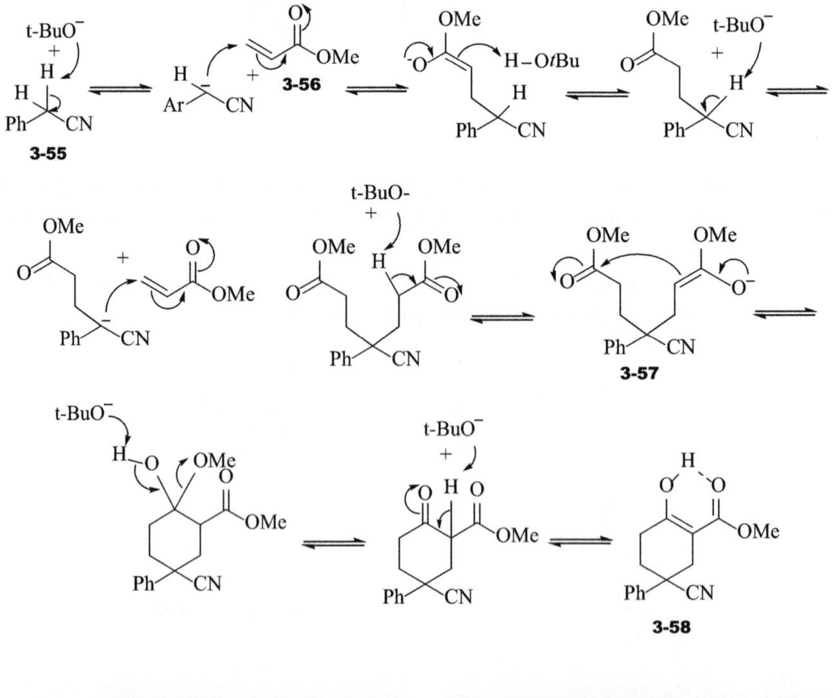

Problem 3.18

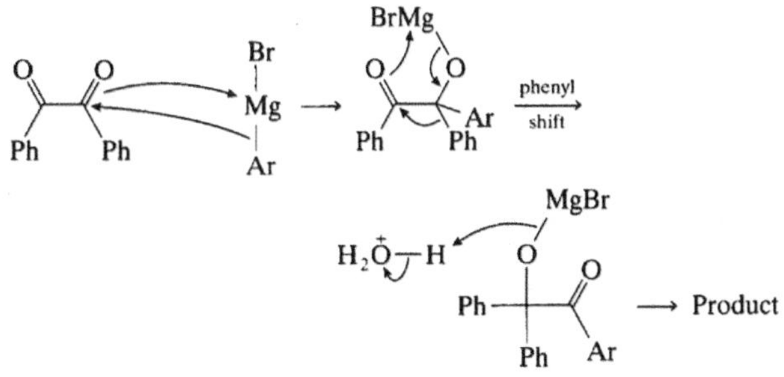

The last step shows the intramolecular transfer of a proton because five-membered cyclic transition states are readily achieved. The process could also be represented by two intermolecular steps.

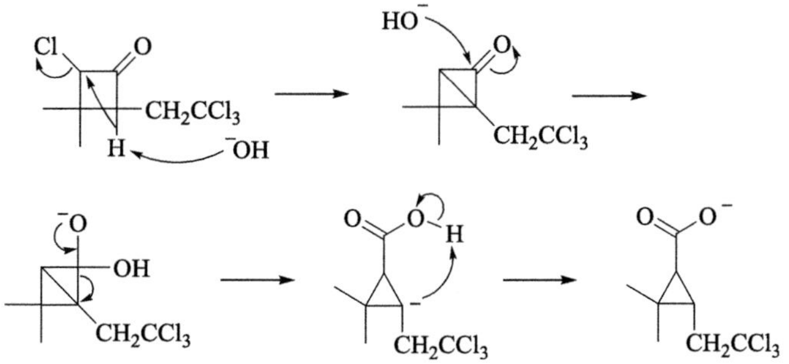

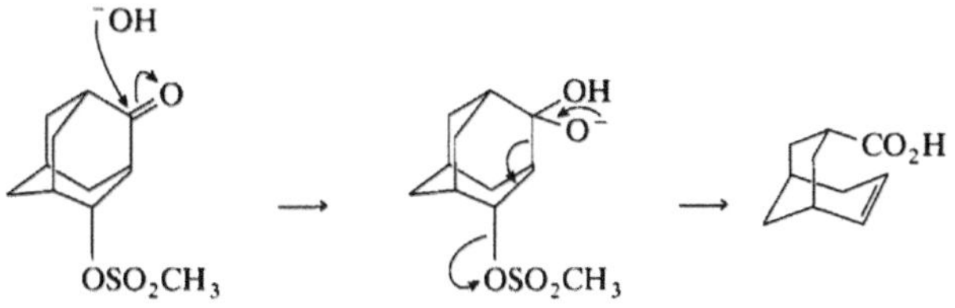

Problem 3.19

The stereospecific reaction occurring in DME could arise by a concerted mechanism such as the following:

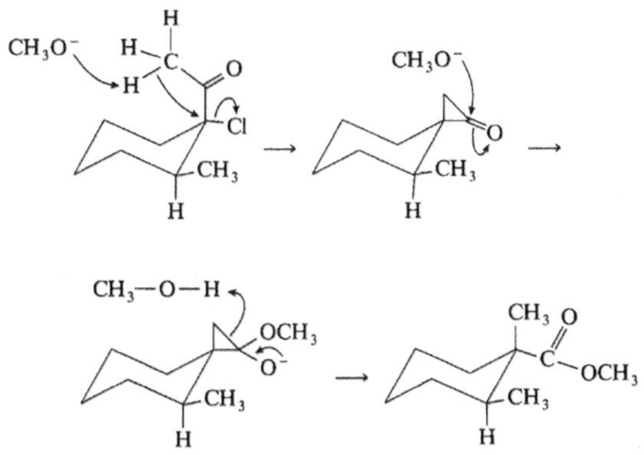

The loss of stereochemistry when the reaction is carried out in methanol may be the result of a stepwise mechanism:

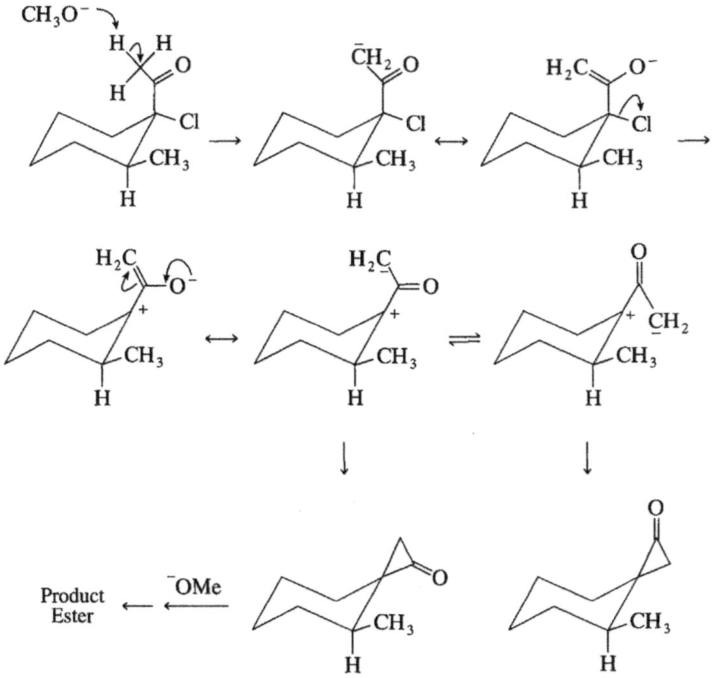

Development of a carbocation center due to loss of chloride ion can account for the loss of stereochemistry at this carbon. It seems reasonable to write this step after removal of a proton to give the anion because, if chloride loss occurs before anion formation, the carbocation would be expected to react with solvent before the anion could be formed. In that case, one might expect to isolate some of the compound in which chloride had been replaced by methoxyl. The effect of the solvent on the stereochemistry of the reaction is due to its ability to solvate the intermediate charged species involved in a stepwise mechanism.

Problem 3.20

a. The reaction mechanism requires two separate nucleophilic steps, i.e., the mechanism requires two nucleophiles rather than one. The reaction cannot be run with the α-chloro compound in the presence of n-butyllithium because reactions between these reagents would give several important side products. Thus, 3-59, the dilithium derivative of the starting sulfone, is formed first, and then the chloro compound is added.

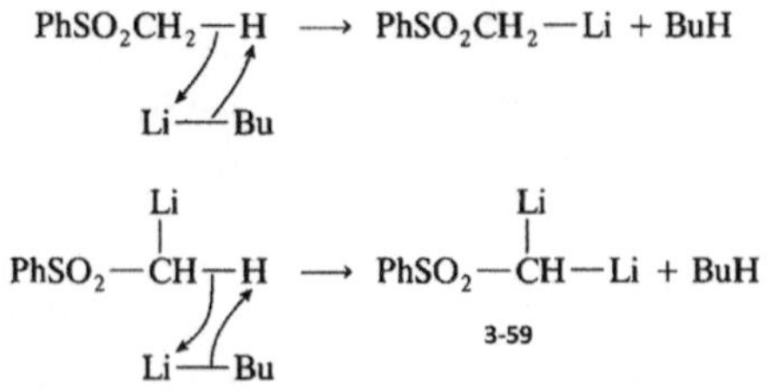

There are two electrophilic positions in the chloro compound; the carbonyl carbon and the carbon α to it, which bears chloride as a leaving group. Nucleophilic reaction at either position by **3-59** is a reasonable reaction, and we will illustrate both possibilities. Reaction at the carbonyl gives **3-60**, which can be close to a cyclopropane.

3-60

The cyclopropane ring then opens to give the product.

3-61

The other mechanism involves **3-59** as the nucleophile in the S$_N$2 displacement at the highly reactive chloro-substituted carbon α to the carbonyl. The remaining anion, **3-62**, reacts with the carbonyl group to give **3-61**.

The authors of the studies cited favor the first mechanism by analogy to the reaction of **3-59** with 1-chloro-2,3-epoxypropane.

Notice how this reaction resembles Example 3.20. In Example 3.20, the benzylic carbon is inserted between the S and C=S. In this reaction, the carbon introduced by the phenylsulfonyl anion is inserted between the benzoyl group and the α carbon of the starting carbonyl compound. In both cases, the "insertion" is affected by forming one bond to close a three-membered ring and then breaking a different bond to open the three-membered ring.

b. This reaction is run in the presence of base. The most acidic hydrogen in the starting materials is on the carbon between the ester and ketone functional groups. If that proton is removed, the resulting anion can act as a nucleophile and add to the carbonyl group of the isocyanate. The oxyanion formed (stabilized by resonance with the nitrogen) can undergo an intramolecular nucleophilic substitution to produce the five-membered ring. Base-catalyzed tautomerization gives the final product.

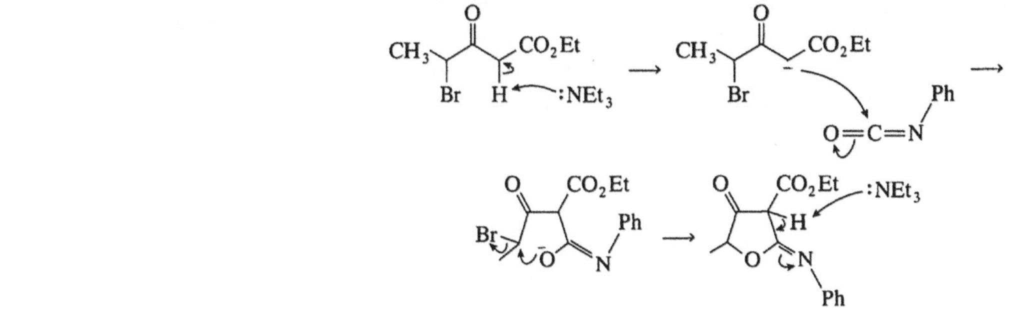

The anionic intermediate through which the tautomers **3-63** and **3-64** interconvert is a resonance hybrid:

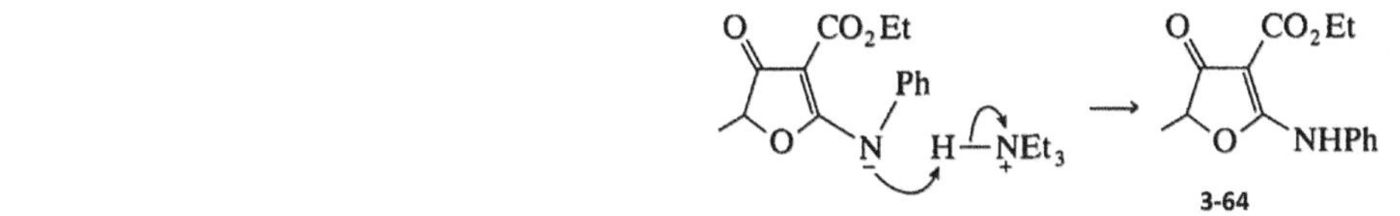

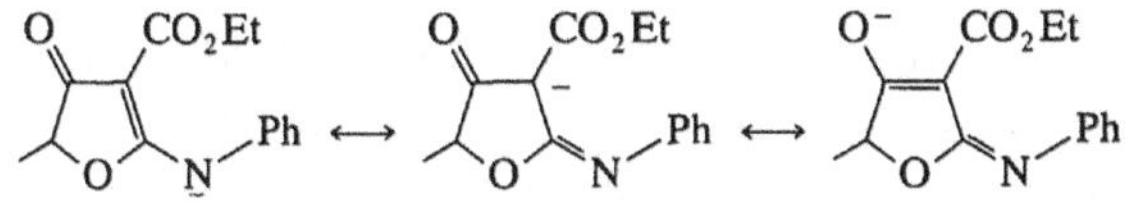

A much less likely first step is nucleophilic reaction of the carbonyl oxygen of the isocyanate with the carbon attached to bromine:

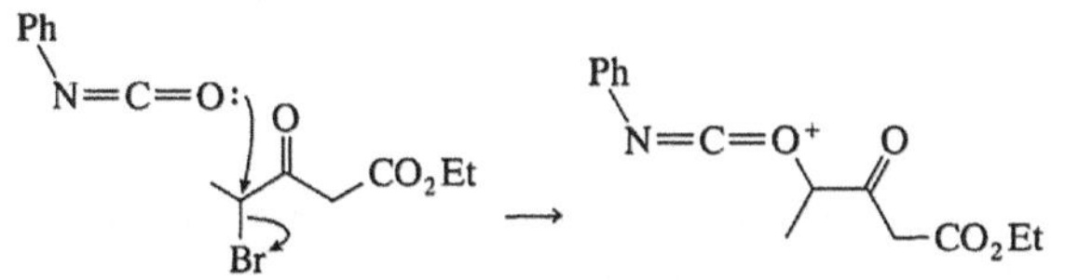

Although carbonyl groups will act as bases with strong acids, their nucleophilicity generally is quite low. We can get a rough idea of the basicity of the carbonyl group, relative to the acidity of the proton actually removed, from Appendix C. The basicity of acetone is estimated from the pK_a of its conjugate acid (-2.85). The pK_a of the proton should be similar to that of ethyl acetoacetate (11). Thus, the ketoester is much more acidic than the carbonyl group is basic, and it is much more likely that the proton would be removed. Nonetheless, a positive feature of this mechanism is that there is a viable mechanism leading to product. (The next step would be removal of a proton on the carbon between the ketone and ester.)

c. In this reaction, the amide anion, a very strong base, removes a benzylic proton. Cyclization, followed by loss of methylphenylamide, and tautomerization lead to the product. Notice that nucleophilic reaction of the benzylic anion and loss of methylphenylamide are separate steps, in accord with Hint 3.2.

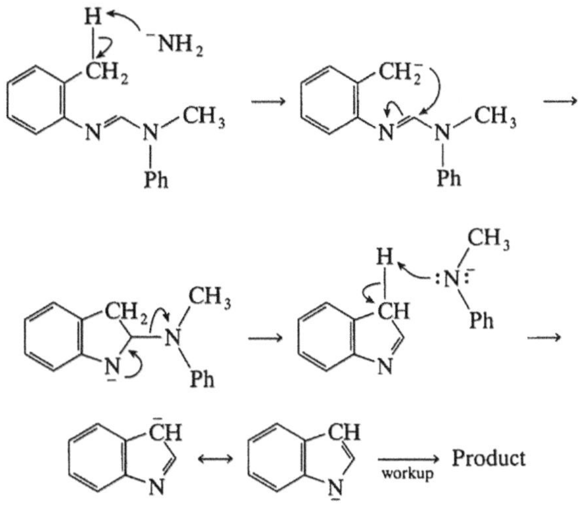

The anion corresponding to the product is considerably more stable than either the amide ion or the phenylamide ion, so protonation of the final anion will take place during workup.

As in Problem 3.19.b, it is necessary to be able to distinguish between tautomers and resonance forms. The following two structures are tautomers, so a mechanism needs to be written for their interconversion.

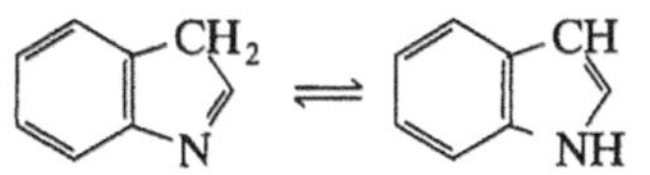

d. Note that 2 mol of hydrazine have reacted to give the product. It also appears that the introduction of each mole is independent of the other and, thus, each mechanism can be shown separately. The hydrazinolysis of the ester is shown first.

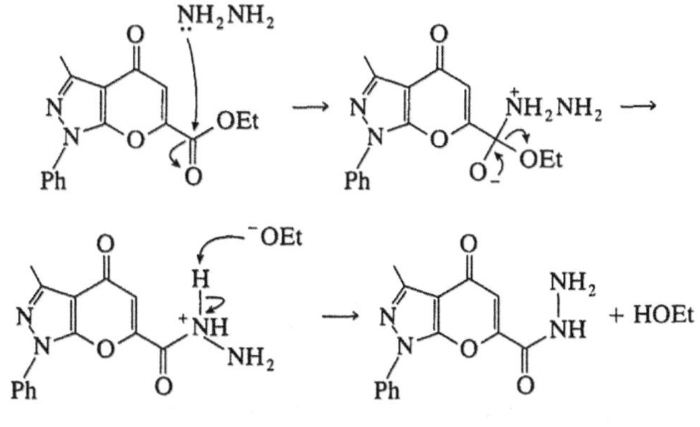

A mechanism for reaction with the other mole of hydrazine follows. The reaction sequence shown is preferred because the anion produced by the first step, addition at the β-position, is resonance-stabilized. The anion produced by reaction of hydrazine at the keto carbonyl carbon would not be resonance stabilized.

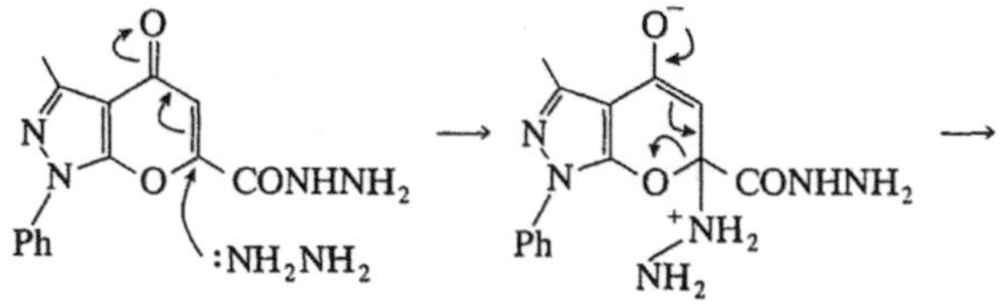

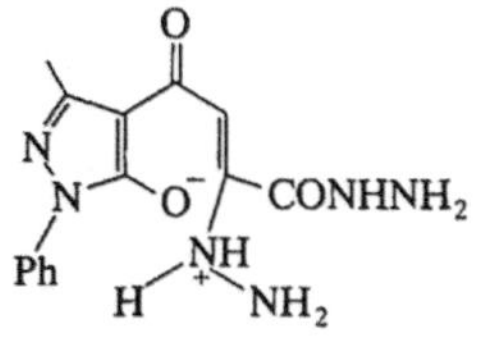

3-65

The enolate ion, **3-65**, will pick up a proton from solvent, and the hydrazinium ion will lose a proton to a base, such as a molecule of hydrazine. The distance between the groups in the molecule makes intramolecular transfer of a proton quite unlikely. The nucleophilic hydrazine group in the neutral intermediate, **3-66**, can react intramolecularly with the electrophilic carbon of the carbonyl group.

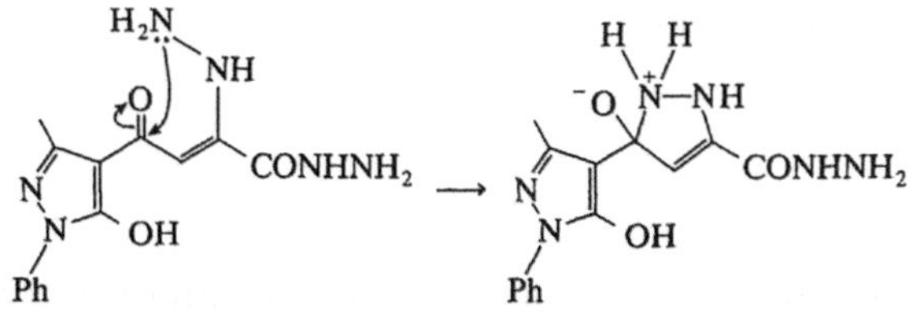

3-66

Again, the positive nitrogen loses a proton and the negative oxygen picks up a proton. In the final step, base-promoted elimination of water occurs. The driving force for loss of water is formation of an aromatic ring.

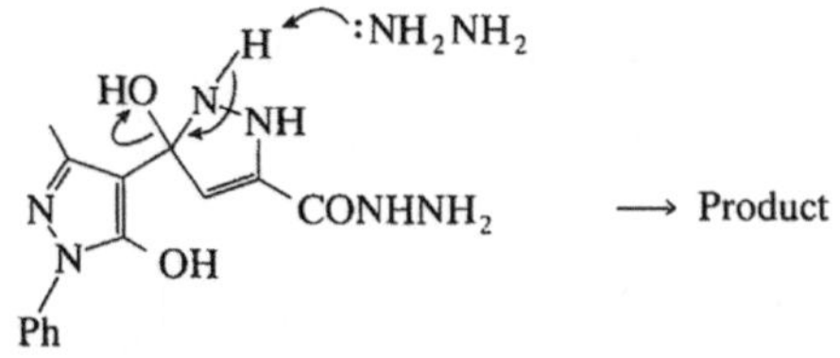

Do not write intramolecular loss of water in the last step. The mechanism shown is better for two reasons. First, the external base, hydrazine, is a much better base than the hydroxyl group. Second, most eliminations go best when the proton being removed and the leaving group are *anti* to one another.

The following would not be a good step:

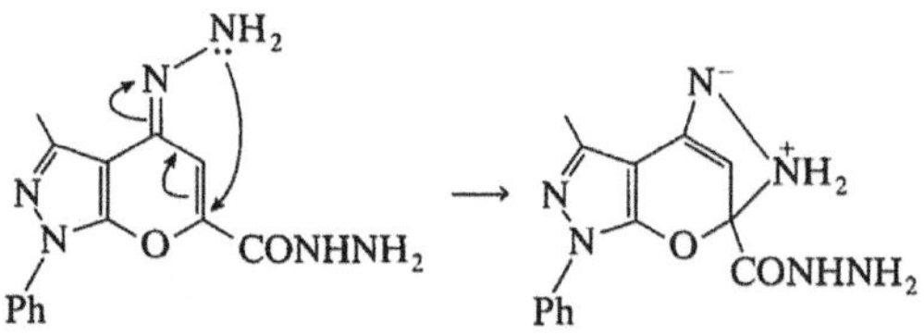

All the atoms involved in this step lie in a plane (the C=N nitrogen is sp^2-hybridized), but because the conjugated double-bond system is locked into the transoid arrangement due to the position of the C=C double bond in the six-membered ring, the amino group cannot reach close enough to cyclize in the manner shown. A further point to consider in this type of reaction is that the most favorable approach of the nucleophile is in line with and at a slight angle into the conjugated system. With two double bonds holding the nitrogen out and away, it is not possible for the intrarmolecular nitrogen nucleophile to attain the right incoming trajectory. The proper trajectory for attack on a sp2 hybridized center will be discussed in chapter 7.

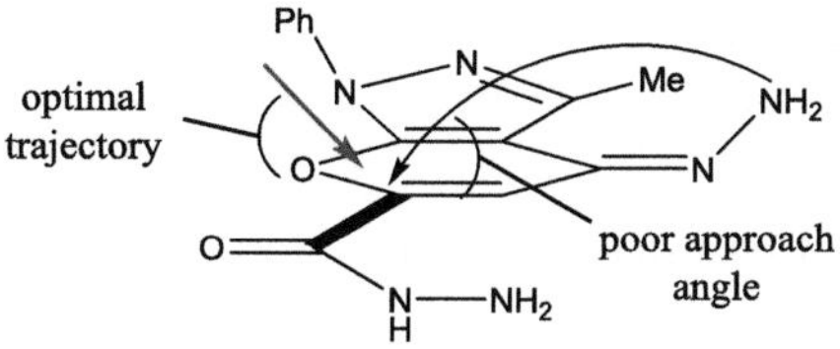

e. Bromine is not present in the product, so one step in the process probably involves nucleophilic substitution at the carbon bearing the bromine. A simple analysis (by numbering or by inspection) also reveals that the methylene group in the imine ester is substituted twice in the reaction. Thus, removal of a proton from the methylene group is a good first step, followed by a nucleophilic substitution reaction.

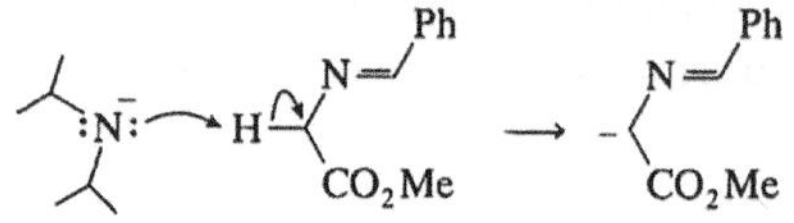

3-67

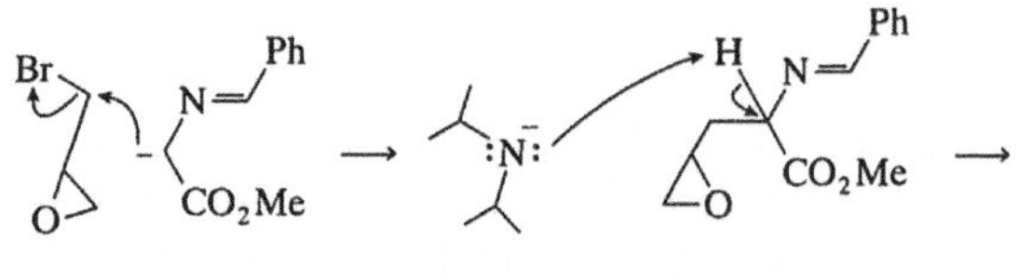

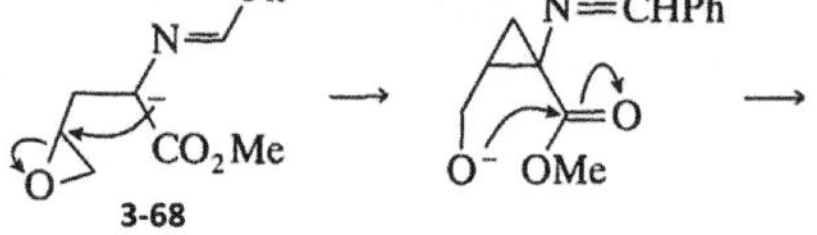

3-68

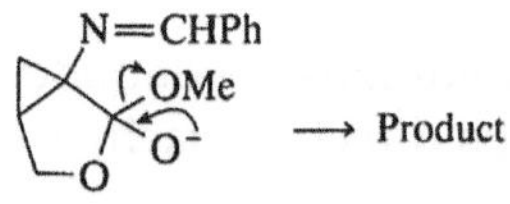

Another possible nucleophilic reaction of anion **3-68** would be at the other electrophilic carbon of the epoxide:

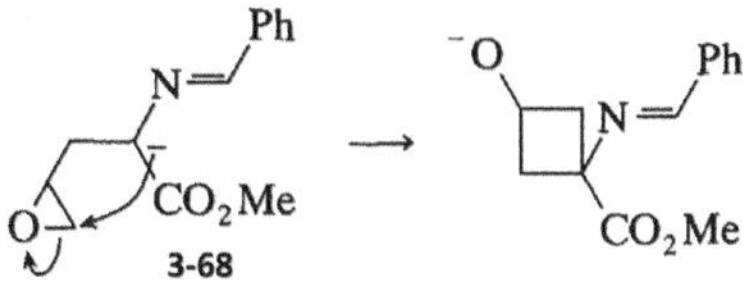

This reaction might have been favored for two reasons. First, reaction with the epoxide, an S_N2 reaction, should occur best at the least hindered carbon. Second, formation of a four-membered ring would be favored by enthalpy because it would have less strain energy than the three-membered ring. The fact that the three-membered ring is actually formed must mean that the reaction is directed by entropy.

Another possible mechanism starts with nucleophilic reaction by **3-67** at the epoxide. The alkoxide ion then displaces bromide to produce a new epoxide.

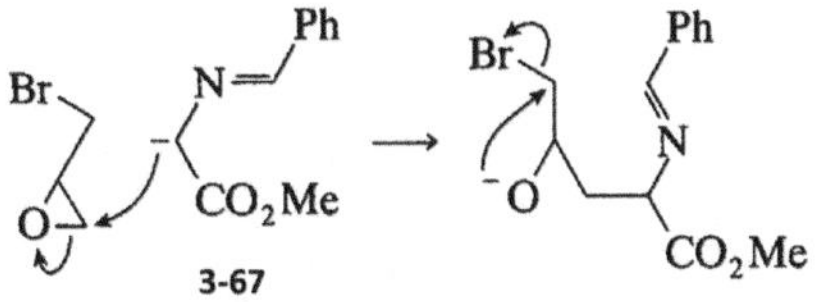

Proton removal followed by nucleophilic reaction gives a cyclopropane.

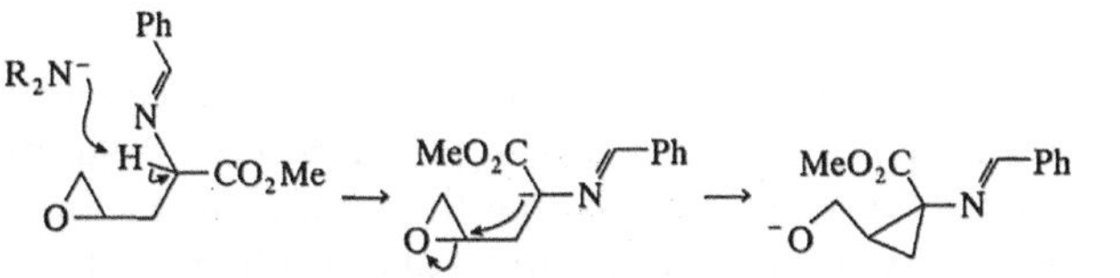

The nucleophilic alkoxide, **3-69**, reacts intramolecularly with the carbon of the ester functional group. The resulting intermediate loses methoxide to give the product.

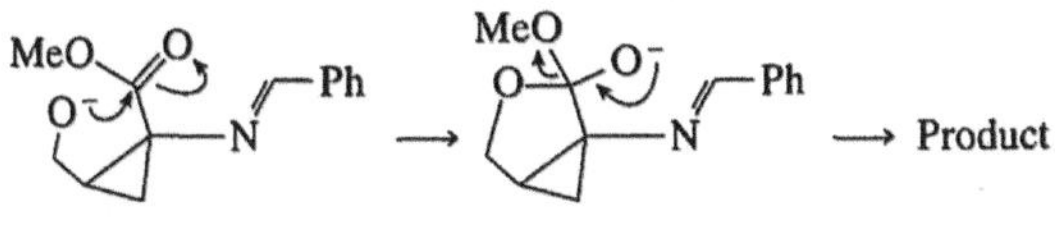

Initial reaction of **3-68** at the CH of the epoxide (rather than the CH_2) is at the more hindered carbon, which is not preferred for an S_N2 reaction. It has been observed that 3-membered rings are formed preferentially over 4-membered rings in many cases. Although the reasons for this preference are not fully understood, it seems to be, at least in part, due to a kinetic advantage for forming 3-membered systems.

Problem 3.21

The reaction conditions, sodamide in ammonia, as well as the lack of electron-withdrawing groups directly attached to the aromatic ring, suggest the aryne mechanism

(mechanism 2) rather than the nucleophilic aromatic substitution (mechanism 1). There are several additional problems with mechanism 1.

Step 1, the nucleophilic addition of the anion to the aromatic ring, is less likely than the aryne formation (step 7) because the intermediate anion is not very stable. The methoxy and bromine substituents can remove electron character from the ring only by inductive effects. Nucleophilic aromatic substitution (except at elevated temperatures) ordinarily requires substituents that withdraw electrons by resonance. The usual position for nucleophilic reaction in this mechanism is at the carbon bearing the leaving group, in this case, bromide. However, in some other mechanism, involving several steps, addition of the nucleophile at other positions might be acceptable.

For step 2, the arrow between the two structures should be replaced with a double-headed arrow, indicating that these are resonance structures.

In step 3, cyanide can act as a leaving group, even though it is not a very good one. Furthermore, the product contains a cyano group. Thus, another problem with this step is that if cyanide leaves the molecule, it will be diluted by the solvent to such a low concentration that subsequent addition will occur very slowly. (However, that does not mean that it cannot happen.)

In step 4, the elimination of HBr is reasonable because it gives an aromatic system and because *syn*-E2 elimination is a common reaction.

In step 5, direct substitution by cyanide ion at an sp^2 center, as shown, is an unlikely process.

In step 6, removal of a proton from ammonia, by an anion that is much more stable than the amide ion, is very unlikely; this step would have a very unfavorable equilibrium. In other words, the product-forming step, like step 11, would occur on workup.

Other comments:

1. Writing the aryne formation (step 7) as a two-step process would be acceptable.
2. Direct substitution of cyanide for bromine on the ring is not a good mechanistic step. Initially, nucleophilic substitution at an sp^2-hybridized carbon is unlikely.
3. An aryne intermediate is unlikely to react as a nucleophile. Because of the high s character in the orbitals forming the third bond, arynes tend to be electrophilic, not nucleophilic.
4. Other mechanisms, which involve formation of a carbanion in the ring, are also unlikely because this carbanion is not stabilized by strongly electron-withdrawing substituents on the ring.

Problem 3.22

There are two apparent ways that triphenylphosphine can react with the diethyl azocarboxylate. One is nucleophilic reaction at the electrophilic carbonyl carbon, and the other is 1,4-addition. Because the ester groups are intact in the hydrazine product, the 1,4-addition is more likely. The intermediate anion can be protonated by the carboxylic acid to give a salt.

The electrophilic phosphorus atom then undergoes nucleophilic reaction with the alcohol. An intriguing aspect of this reaction is the fate of the proton on the alcohol. There is no strong base present in the reaction mixture. A good possibility is that the carboxylate anion removes the proton from the alcohol as it is reacting with the phosphorus.

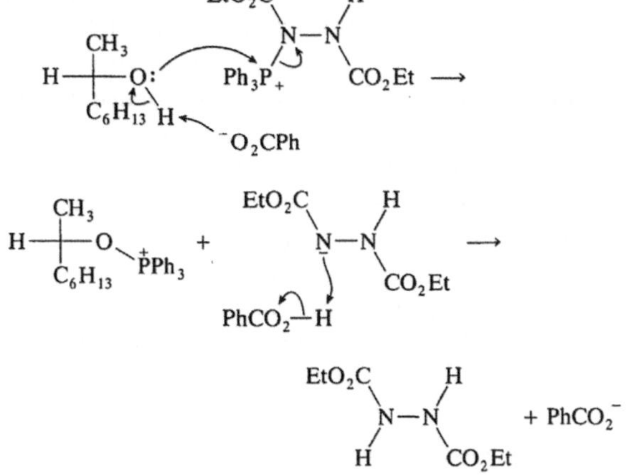

How do we rationalize what appears to be a trimolecular reaction? Because the solvent is a nonpolar aprotic solvent, the phosphonium carboxylate must be present as an ion pair and can be considered as a single entity. The carboxylate ion also may be properly situated to remove the proton. It can also be postulated that if there is an existing tight connection between the alcohol and the activated phosphorous containing species, the system is a bond rotation away from being able to abstract the proton as the oxygen adds into the phosphorous.

Finally, the carboxylate anion acts as a nucleophile to displace the very stable (a key driving force for the process) triphenylphosphine oxide and give the inverted product.

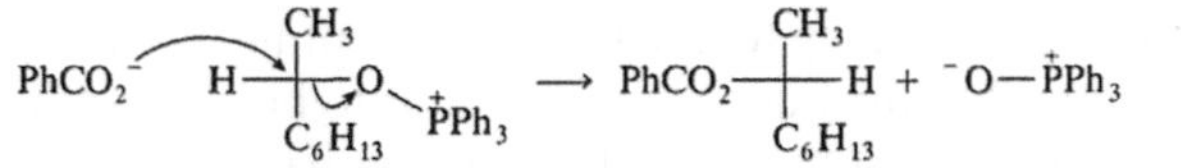

An alternate mechanism, in which the alcohol reacts with the phosphonium salt without assistance from the carboxylate anion, is less likely because the intermediate produced has two positive centers adjacent to each other.

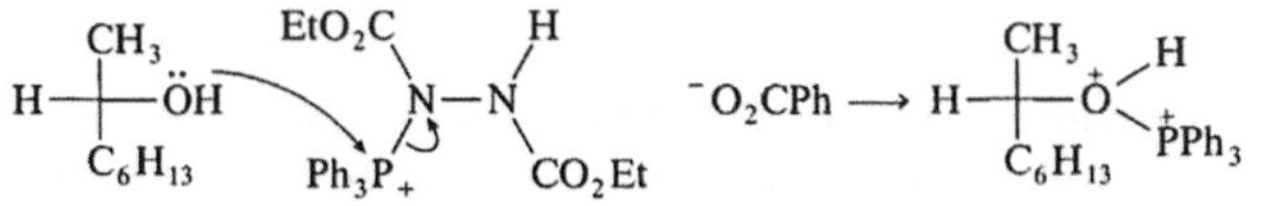

References

Abell, C.; Bush, B. D.; Staunton, J. *J. Chem. Soc., Chem. Commun.* **1986**, 15–17.
Bell, F. *J. Chem. Soc.* **1957**, 516–518.
Bland, J.; Shah, A.; Bortolussi, A.; Stammer, C. H. *J. Org. Chem.* **1988**, *53*, 992–995.

Braish, T. F.; Fox, D. E. *J. Org. Chem.* **1990,** *55,* 1684–1687.

Bunnett, J. F.; Skorcz, J. A. *J. Org. Chem.* **1962,** *27,* 3836–3843.

Curran, D. P.; Jacobs, P. B.; Elliott, R. L.; Kim, B. H. *J. Am. Chem. Soc.* **1987,** *109,* 5280–5282.

Daub, G. H.; Leon, A. A.; Silverman, I. R.; Daub, G. W.; Walker, S. B. *Aldrichim. Acta* **1984,** *17,* 13–23.

DeGraffenreid, M. R.; Bennett, S.; Caille, S.; Gonzalez-Lopez de Turiso, F.; Hungate, R. W.; Julian, L. D.; Kaizerman, J.; McMinn, D. L.; Sun, D.; Yan, X.; Powers, J. P. *J. Org. Chem.* **2007,** *72,* 7455–7458.

Dietze, P.; Jencks, W. P. *J. Am. Chem. Soc.* **1989,** *111,* 5880–5886.

Eisch, J. J.; Dua, S. K.; Behrooz, M. *J. Org. Chem.* **1985,** *50,* 3674–3676.

Gadwood, R. C.; Lett, R. M.; Wissinger, J. E. *J. Am. Chem. Soc.* **1984,** *106,* 3869–3870.

Gueremy, C.; Audiau, F.; Renault, C.; Benavides, J.; Uzan, A.; Le Fur, G. *J. Med. Chem.* **1986,** *29,* 1394–1398.

Hagopian, R. A.; Therian, J. J.; Murdoch, J. R. *J. Am. Chem. Soc.* **1984,** *106,* 5753–5754.

House, H. O.; Gilmore, W. F. *J. Am. Chem. Soc.* **1961,** *83,* 3980–3985.

Khan, M. A.; Cosenza, A. G. *Afinidad* **1988,** *45,* 173–174; Chem. Abstr. **1988,** *109,* 128893.

Khanapure, S. P.; Crenshaw, L.; Reddy, R. T.; Biehl, E. R. *J. Org. Chem.* **1988,** *53,* 4915–4919.

Lorenz, R. R.; Tullar, B. F.; Koelsch, C. F.; Archer, S. *J. Org. Chem.* **1965,** *30,* 2531–2533.

Mack, R. A.; Zazulak, W. I.; Radov, L. A.; Baer, J. E.; Stewart, J. D.; Elzer, P. H.; Kinsolving, C. R.; Georgiev, V. S. *J. Med. Chem.* **1988,** *31,* 1910–1918.

March, J. *Advanced Organic Chemistry,* 3rd ed; Wiley: New York, 1985, 970.

Martin, P.; Greuter, H.; Bellus, D. *J. Am. Chem. Soc.* **1979,** *101,* 5853–5854.

Mitsunobu, O.; Eguchi, M. *Bull. Chem. Soc. Jpn.* **1971,** *44,* 3427–3430.

Mitsunobu, O. *Synthesis* **1981,** 1–28.

Mogilaiah, K.; Reddy, N. V. *Synth. Commun.* **2003,** *33,* 73.

Nakatsuji, H.; Nishikado, H.; Ueno, K.; Tanabe, Y. *Org. Lett.* **2009,** *11,* 4258–4261.

Norman, R. O. C. *Principles of Organic Synthesis;* Methuen and Co. LTD.: London, 1968, 126–131.

Numazawa, M.; Satoh, M.; Satoh, S.; Nagaoka, M.; Osawa, Y. *J. Org. Chem.* **1986,** *51,* 1360–1362.

Pena, M. R.; Stille, J. K. *J. Am. Chem. Soc.* **1989,** *111,* 5417–5424.

Ramachadran, S.; Newman, M. S. *Org. Synth.* **1961,** *41,* 38–41.

Ramage, R.; Griffiths, G. J.; Shutt, F. E.; Sweeney, J. N. A. *J. Chem. Soc., Perkin Trans. I* **1984,** 1539–1545.

Sasaki, T.; Eguchi, S.; Toru, T. *J. Am. Chem. Soc.* **1969,** *91,* 3390–3391.

Wenkert, E.; Arrhenius, T. S.; Bookser, B.; Guo, M.; Mancini, P. *J. Org. Chem.* **1990,** *55,* 1185–1193.

Yogo, M.; Hirota, K.; Maki, Y. *J. Chem. Soc., Perkin Trans. I* **1984,** 2097–2102.

4

Reactions Involving Acids and Other Electrophiles

Acids and electrophiles are electron-deficient species. According to the Lewis concept, all electrophiles (e.g., cations, carbenes, and metal ions) are acids by definition. However, from long usage, the term *acid* is frequently used to refer to a proton donor, whereas the term *Lewis acid* usually refers to charged electrophiles in general.

1. STABILITY OF CARBOCATIONS

Reactions in acid often involve the formation of carbocations—trivalent, positively charged carbon atoms—as intermediates. The order of stability of carbocations containing only alkyl substituents is $3° > 2° > 1° > CH_3$. Cation stability is influenced by several factors:

1. *Hyperconjugation.* An increase in the number of alkyl substituents increases the stability of the carbocation due to orbital overlap between the adjacent σ bonds and the unoccupied p orbital of the carbocation. The resulting delocalization of charge, which can be represented by resonance structures, stabilizes the cation.
2. *Inductive effects.* Neighboring alkyl groups stabilize a cation because electrons from an alkyl group, which is relatively large and polarizable compared to hydrogen, can shift toward a neighboring positive charge more easily than can electrons from an attached hydrogen.
3. *Resonance effects.* Conjugation with a double bond increases the stability of a carbocation. Thus, allylic and benzylic cations are more stable than their saturated counterparts. (For example, see Problem 1.4.c.) Heteroatoms with unshared electron pairs, e.g., oxygen, nitrogen, or halogen, can also provide resonance stabilization for cationic centers as shown in the following examples:

Copyright © 2014 Elsevier Inc. All rights reserved.

Cations at sp^2- or sp-hybridized carbons are especially unstable. In general, the more s character in the orbitals, the less stable the cation. An approximate order of carbocation stability is CH_3CO^+ (acetyl cation) $\sim (CH_3)_3C^+ \gg PhCH_2^+ > (CH_3)_2CH^+ > H_2C= CH-CH_2^+ \gg CH_3CH_2^+ > H_2C=CH^+ > Ph^+ > CH_3^+$. The stabilities of various carbocations can be determined by reference to the order of stability for alkyl carbocations, $3° > 2° > 1° > CH_3$. The acetyl cation has stability similar to that of the t-butyl cation. Secondary carbocations, primary benzylic cations, and primary allylic cations are all more stable than primary alkyl cations. Vinyl, phenyl, and methyl carbocations are less stable than primary alkyl cations.

2. FORMATION OF CARBOCATIONS

A. Ionization

A compound can undergo unimolecular ionization to a carbocation and a leaving group. If the final product formed is due to substitution, the process is called S_N1. If it is due to elimination, the process is called E1. In both cases, the rate-determining step is the ionization, not the product-forming step.

EXAMPLE 4.1. ACID-CATALYZED LOSS OF WATER FROM A PROTONATED ALCOHOL

In this process, protonation of the alcohol group is the first step. This occurs much faster than the rate-determining step, loss of water from the protonated alcohol.

EXAMPLE 4.2. SPONTANEOUS IONIZATION OF A TRIFLATE

In this example, ionization is favored by several factors. First, the benzyl ion formed is resonance-stabilized and bears a methoxy group in the *para* position that can further stabilize the cation by resonance. In addition, the leaving group is triflate, an exceptionally good leaving group. Finally, the ionization takes place in water, a polar solvent that can stabilize the two charged species formed.

B. Addition of an Electrophile to a π Bond

Intermediate cations are often produced by addition of a proton or a Lewis acid to a π bond.

EXAMPLE 4.3. PROTONATION OF AN OLEFIN

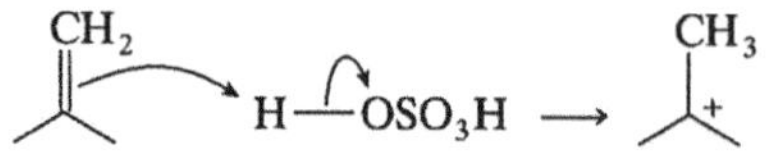

EXAMPLE 4.4. PROTONATION OF A CARBONYL GROUP

In acid, carbonyl compounds are in equilibrium with their protonated counterparts. Protonation is often the first step in nucleophilic addition or substitution of carbonyl groups. For aldehydes and ketones, the protonated carbonyl group is a resonance hybrid of two forms: one with positive charge on the carbonyl oxygen and the other with positive charge on the carbonyl carbon.

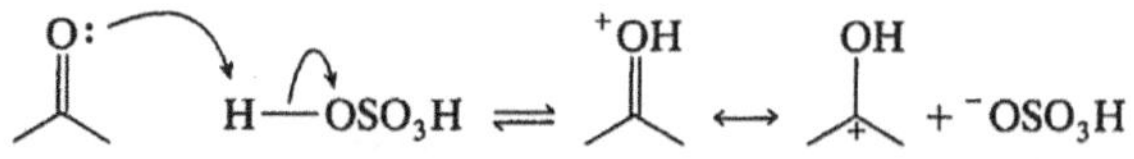

When esters are protonated at the carbonyl group, there are three resonance forms: two corresponding to the ones that form with aldehydes and ketones and the third with positive charge on the alkylated oxygen.

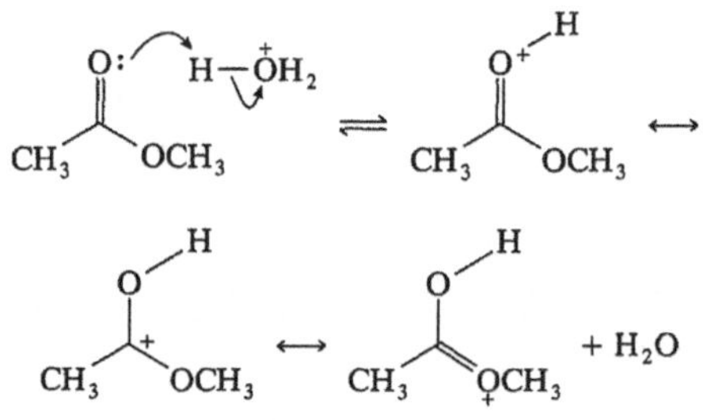

EXAMPLE 4.5. REACTION OF A CARBONYL COMPOUND WITH A LEWIS ACID

Carbonyl groups form complexes or intermediates with Lewis acids like $AlCl_3$, BF_3, and $SnCl_4$. For example, in the Friedel–Crafts acylation reaction in nonpolar solvents, an aluminum chloride complex of an acid chloride is often the acylating agent. Because of the basicity of ketones, the products of the acylation reaction are also complexes. For more detail on electrophilic aromatic substitution, see Section 7.

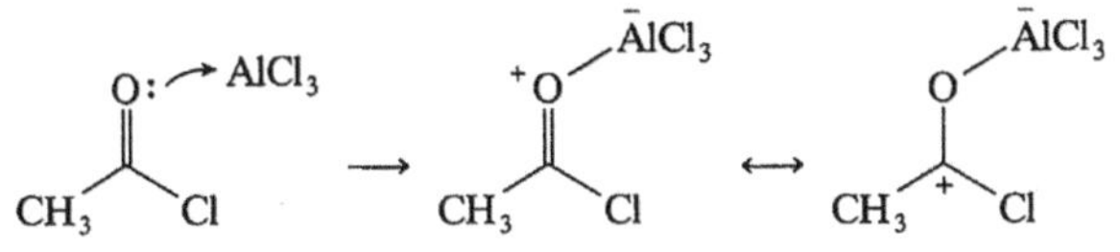

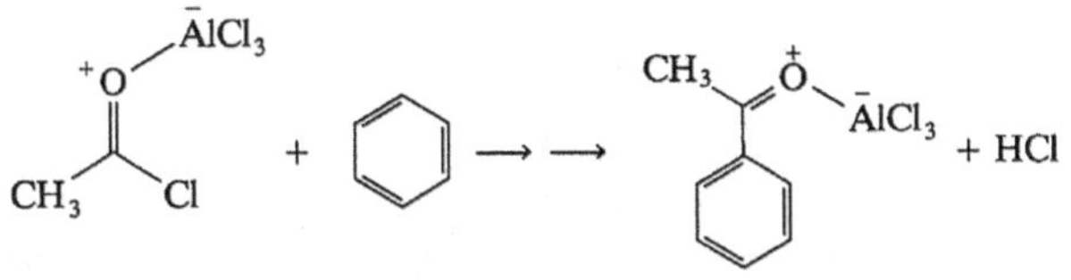

C. Reaction of an Alkyl Halide with a Lewis Acid

EXAMPLE 4.6. REACTION OF 2-CHLOROBUTANE AND ALUMINUM TRICHLORIDE

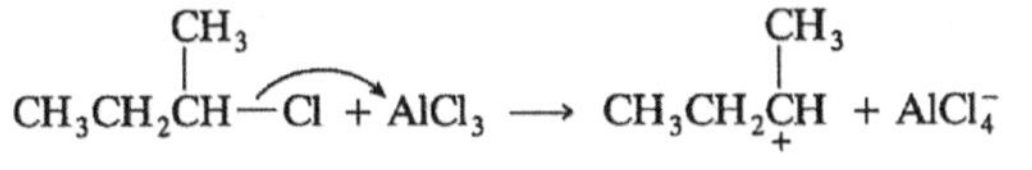

The Lewis acid removes the halide ion to give a carbocation.

3. THE FATE OF CARBOCATIONS

Once formed, carbocations have several options for further reaction. Among these are substitution, elimination, addition, and rearrangement.

1. Substitution (S_N1) occurs when the carbocation reacts with a nucleophile.
2. Elimination (E1) usually occurs with loss of a proton, as in the formation of 2-methyl-1-propene from the *t*-butyl cation.
3. The carbocation can react with an electron-rich reagent.

EXAMPLE 4.7. CATIONIC OLEFINIC POLYMERIZATION

$$CH_3CH=CH_2 \xrightarrow{BF_3} CH_3CHCH_2\overset{\ominus}{BF_3}$$

$$CH_3CH=CH_2 \longrightarrow CH_3CHCH_2\overset{\ominus}{BF_3} \longrightarrow \longrightarrow$$

$$\underset{+}{CH_3CHCH_2}$$

4. The carbocation can undergo rearrangement (see the following section). (The negative sign on the boron indicates charge only; there is not an unshared pair of electrons on boron.)

4. REARRANGEMENT OF CARBOCATIONS

Carbocations tend to rearrange much more easily than carbanions. Under common reaction conditions, a carbocation rearranges to another carbocation of equal or greater stability. For example, a secondary carbocation will rearrange to a tertiary carbocation or a different secondary carbocation, but ordinarily it will not rearrange to a less stable primary carbocation. This generalization is not absolute, and because there is no high-energy barrier to the rearrangement of carbocations, rearrangement to a less-stable cation can occur if it offers the chance to form a more stable product.

HINT 4.1

Rearrangement of a carbocation frequently involves an alkyl, phenyl, or hydride shift to the carbocation from an adjacent carbon (a 1,2-shift).

In many cases, there are several different pathways by which rearrangement may take place. In these situations, the question of which group will migrate (migratory aptitude) is a complex one. In general, aryl and branched alkyl chains migrate in preference to unbranched chains, but the selectivity is not high. Similarly, the tendency of hydrogen to migrate is unpredictable: sometimes hydrogen moves in preference to an aryl group, at other times it migrates less readily than an alkyl. Very often, other factors such as stereochemistry, relief of strain, and reaction conditions are as important as the structure of the individual migrating group. *Frequently it is difficult to predict the product of a reaction in which a carbocation is formed; it is much easier to identify a reasonable pathway by which an experimentally obtained product is derived from starting material.*

EXAMPLE 4.8. A HYDRIDE SHIFT IN THE REARRANGEMENT OF A CARBOCATION

Treatment of isobutyl alcohol with HBr and H_2SO_4 at elevated temperatures leads to *t*-butyl bromide. In the first step, the hydroxyl group is protonated by the sulfuric acid to convert it into a better leaving group, the water molecule. Water then leaves, giving the primary isobutyl carbocation, **4-1**.

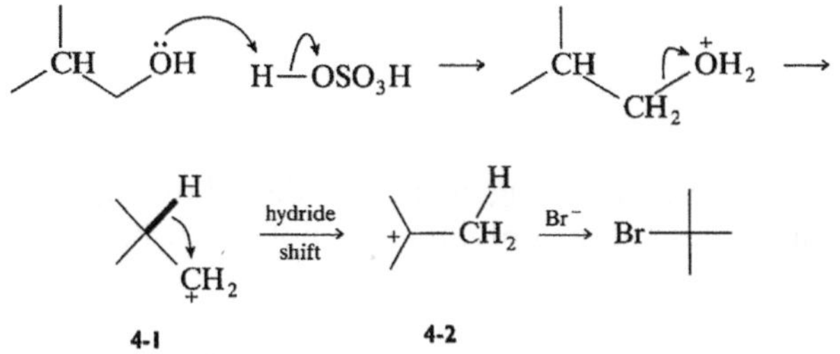

The hydrogen and the electrons in its bond to carbon, highlighted in **4-1**, move to the adjacent carbon. Now the carbon from which the hydride left is deficient by one electron and is, thus, a carbocation. Because the new carbocation, **4-2**, is tertiary, the molecule has gone from a relatively unstable primary carbocation to the much more stable tertiary carbocation. In the final step, the nucleophilic bromide ion reacts with the positive electrophilic tertiary carbocation to give the alkyl halide product.

HINT 4.2

In rearrangement reactions, the method of numbering both the starting material and the product, introduced in Chapter 2, can be very helpful.

EXAMPLE 4.9. AN ALKYL SHIFT IN THE REARRANGEMENT OF A CARBOCATION

Consider the following reaction:

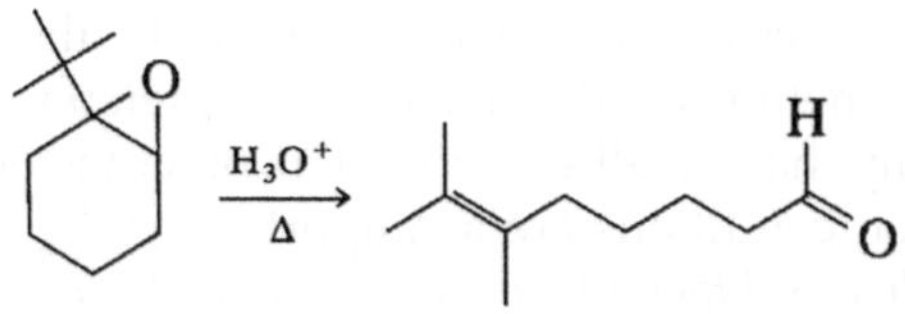

(This example is derived from Corona et al., (1985).)

Because the product has neither a *t*-butyl group nor a six-membered ring, a rearrangement must have taken place. Numbering of both the starting material and the product helps to visualize what takes place during the course of the reaction.

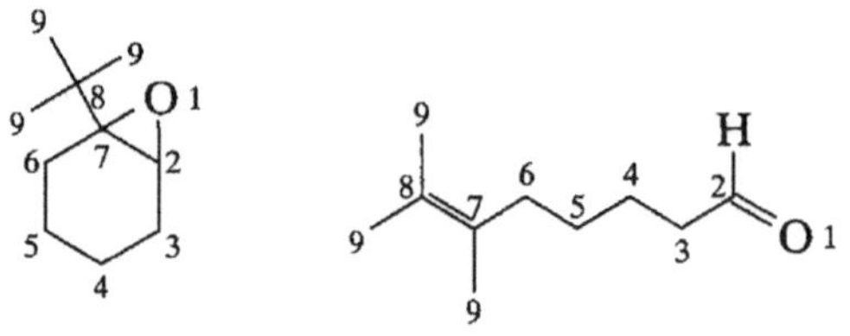

4-3   **4-4**

The three methyls in **4-3** are given the same number because they are chemically equivalent. The product **4-4** has been numbered so that the system conforms as closely as possible to that of the starting material, i.e., with the least possible rearrangement. Comparison of the numbering in **4-3** to that in **4-4** shows that one of the methyl groups shifts from carbon-8 to carbon-7 and that the bond between carbon-2 and carbon-7 is broken. Because a rearrangement to carbon-7 takes place, that carbon must have a positive charge during the course of the reaction.

The oxygen is the only basic atom in the molecule, so protonation of the oxygen must be the first step.

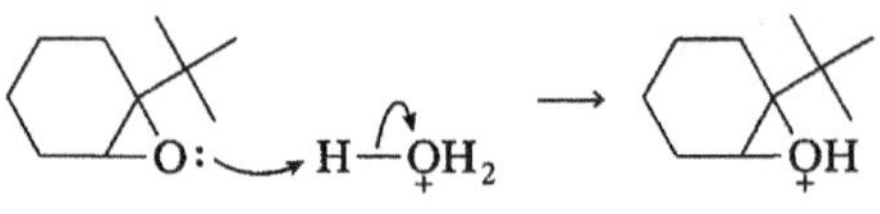

The protonated epoxide is unstable because of the high strain energy of the three-membered ring and opens readily. There are two possible modes of ring opening:

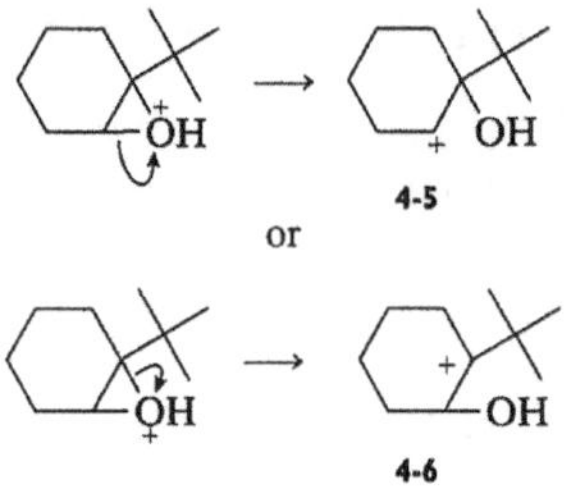

4-5

or

4-6

Because **4-6**, a tertiary carbocation, is more stable than **4-5**, a secondary cation, **4-6**, would be expected to be formed preferentially. (However, if the tertiary carbocation did not lead to the product, we would go back to consider the secondary cation.) In addition, the formation of **4-6** appears to lead toward the product because the carbon bearing the positive charge is number 7 in **4-3**. The tertiary carbocation can undergo rearrangement by a methyl shift to give another tertiary carbocation.

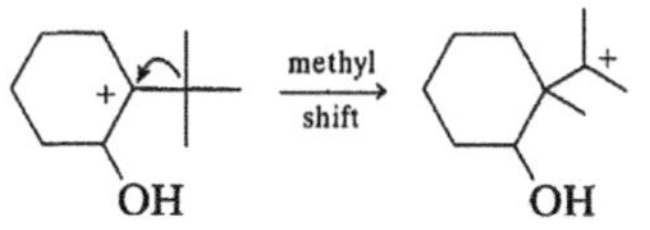

4-6

Now, one of the lone pairs of electrons on oxygen facilitates breaking the bond between C-2 and C-7. The formation of two new π bonds compensates for the energy required to break the bond between C-2 and C-7.

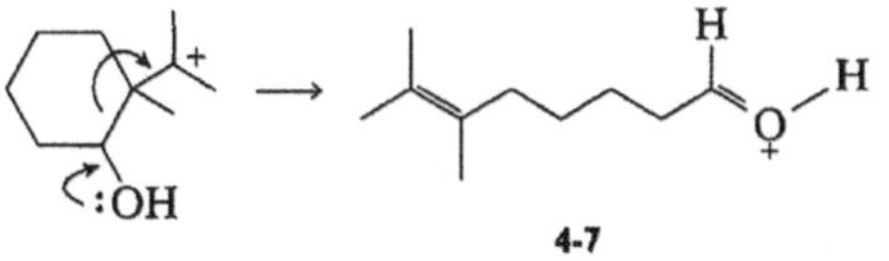

4-7

The only remaining step is deprotonation of the protonated aldehyde, **4-7**, to give the neutral product. This would occur during the workup, probably in a mild base like sodium bicarbonate.

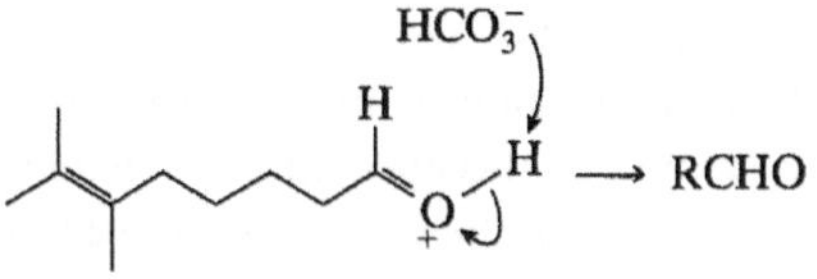

A. The Dienone–Phenol Rearrangement

The dienone–phenol rearrangement is so named because the starting material is a dienone and the product is a phenol.

EXAMPLE 4.10. REARRANGEMENT OF A BICYCLIC DIENONE TO A TETRAHYDRONAPHTHOL SYSTEM

Inspection shows that a skeletal change occurs in the following transformation, that is, rearrangement occurs. (See the answer to Problem 2.4.b for an application of Hint 2.14 to analyzing the bonding changes involved.)

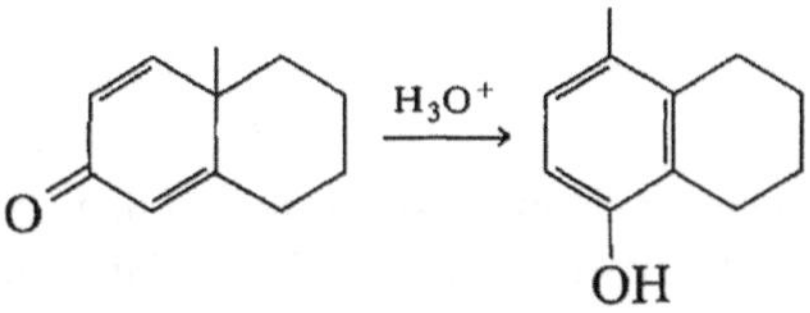

The first step in the mechanism of this reaction is protonation of the most basic atom in the molecule, the oxygen of the carbonyl.

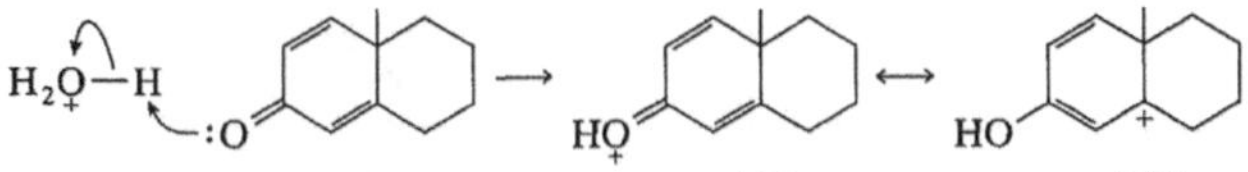

4-8-1   **4-8-2**

The intermediate carbocation, **4-8**, undergoes an alkyl shift to give another resonance-stabilized carbocation, **4-9**.

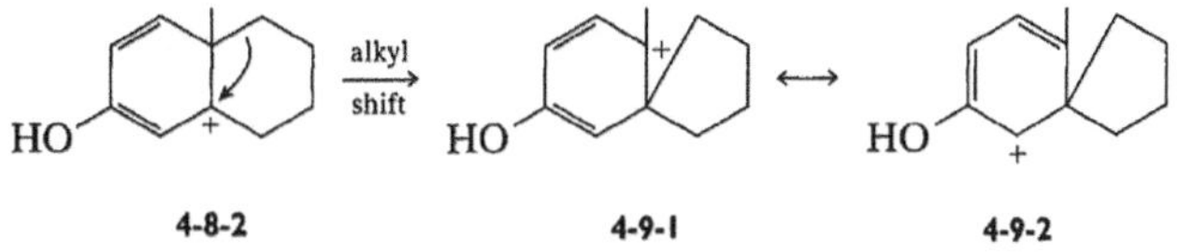

Finally, **4-9** undergoes another alkyl shift, followed by loss of a proton, to give the product. Looking at the resonance structures that can be drawn for all the cations involved in the mechanism suggests that their energy should be comparable to that of the initial cation, **4-8-2**. The driving force for the reaction comes from the formation of the aromatic ring.

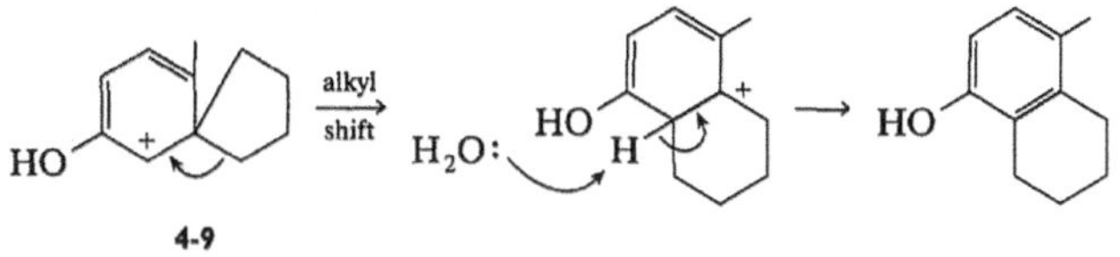

What initially appears to be a complicated reaction is the result of a series of simple steps. For other examples of this reaction, see Miller, (1975).

PROBLEM 4.1

Write step-by-step mechanisms for the following transformations.

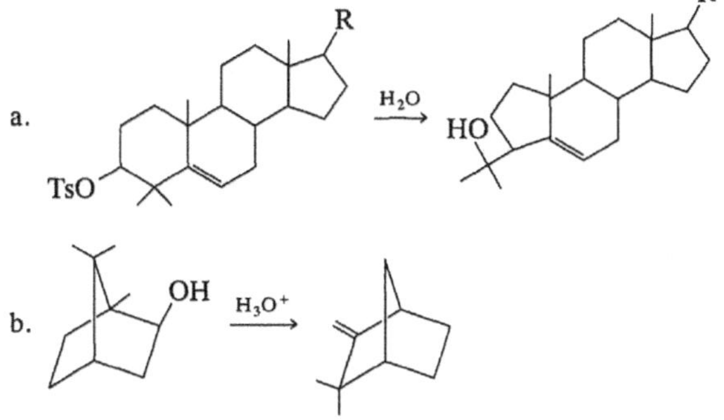

B. The Pinacol Rearrangement

Many common rearrangement reactions are related to the rearrangement of 1,2-dihydroxy compounds to carbonyl compounds. Often these reactions are called *pinacol rearrangements* because one of the first examples was the transformation of pinacol to pinacolone:

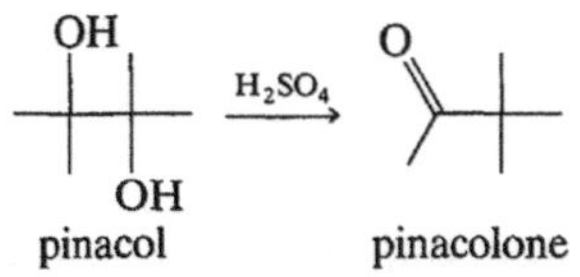

pinacol   pinacolone

EXAMPLE 4.11. THE REARRANGEMENT OF 1,2-DIPHENYL-1,2-ETHANEDIOL TO 2,2-DIPHENYL-ETHANAL

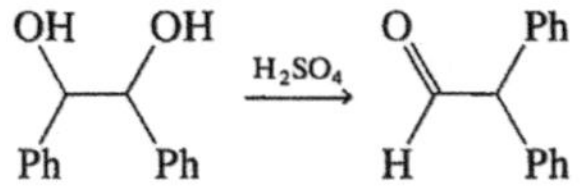

The mechanism of this reaction involves formation of an intermediate carbocation, a 1,2-phenyl shift, and loss of a proton to form the product:

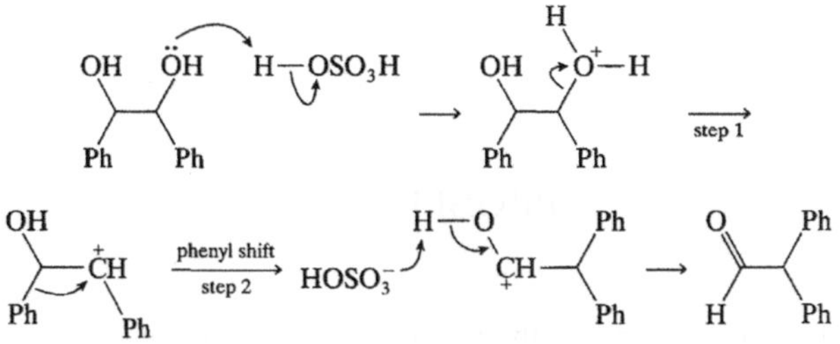

4-10

In reactions of this type, it is possible to form more than one initial carbocation if the starting material is not symmetrical. In this situation, the more stable carbocation is usually formed in step 1. Once this initial carbocation has been formed, the course of step 2 is more difficult to predict because it depends on the propensity of one group to migrate in preference to another (*migratory aptitude*), which often depends on reaction conditions.

It has been suggested that in pinacols containing an aryl group, the initial carbocation formed by loss of hydroxyl can be stabilized by neighboring group participation of the phenyl group to give a bridged phenonium ion:

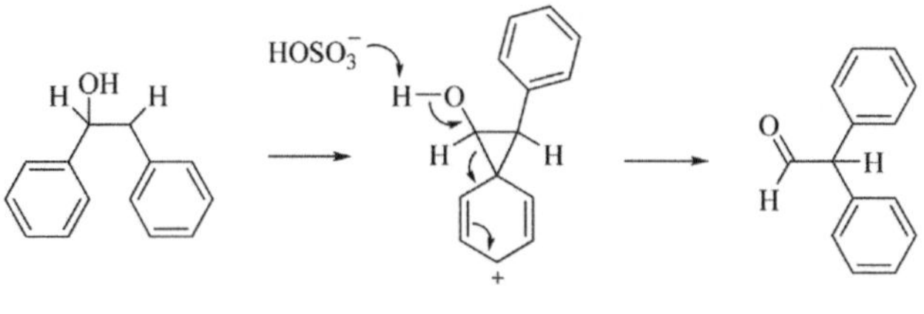

4-11

Although these types of bridged phenonium ions are accepted intermediates in a number of reactions, they do not appear to be involved in the pinacol rearrangement (see Schubert and LeFevre, (1972)). Stabilization of the carbocation **4-10** by resonance with the oxygen substituent may be a factor in determining the preference for phenyl migration over phenonium ion formation in the pinacol rearrangement.

PROBLEM 4.2

Write a step-by-step mechanism for the transformation of pinacol to pinacolone in the presence of sulfuric acid.

PROBLEM 4.3

For the following reaction, write a step-by-step mechanism that accounts for the observed stereospecificity.

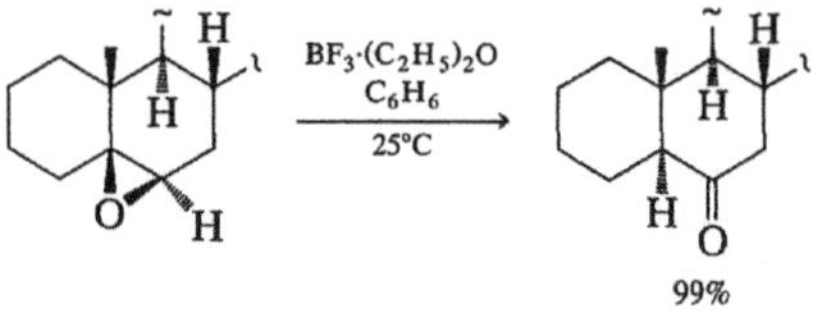

Source: Heubest and Wrigley, (1957).

PROBLEM 4.4

Write a step-by-step mechanism for the following reaction.

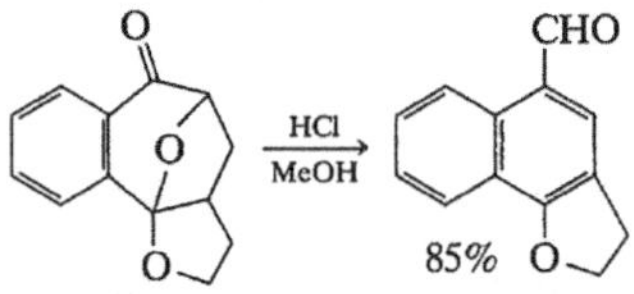

Source: Padwa et al., (1988).

5. ELECTROPHILIC ADDITION

Addition of electrophiles is a reaction typical of aliphatic π bonds (Example 4.3). Such additions involve two major steps: (1) addition of the nucleophilic π bond to the electrophile to give an intermediate carbocation, and (2) reaction of the carbocation with a nucleophile. Typical electrophiles are bromine, chlorine, a proton supplied by HCl, HBr, HI, H_2SO_4, or H_3PO_4, Lewis acids, and carbocations. The nucleophile in step 2 is often the anion associated with the electrophile, e.g., bromide, chloride, iodide, etc., or a nucleophilic solvent like water or acetic acid.

A. Regiospecificity

Because the more stable of the two possible carbocations is formed predominantly as the intermediate in addition of the electrophile (step 1), electrophilic additions are often regiospecific.

EXAMPLE 4.12. REGIOSPECIFICITY IN ELECTROPHILIC ADDITIONS

When HI adds to a double bond, the proton acts as an electrophile, giving an intermediate carbocation that then reacts with the nucleophilic iodide ion to give the product. In the reaction of HI with 1-methylcyclohexene, there is only one product, 1-iodo-1-methylcyclohexane; no 1-iodo-2-methyl-cyclohexane is formed.

This reaction is said to be regiospecific because the iodide might occupy the ring position at either end of the original double bond, but only one of these products is actually formed. The reaction is regiospecific because the proton adds to form the more stable tertiary carbocation, **4-12**, and not the secondary carbocation, **4-13**. The regiochemical outcome that is derived from the more stable cationic intermediate (more substituted carbocation), is often referred to as "Markovnikov product".

4-12   4-13

B. Stereochemistry

Anti *Addition*

In some electrophilic additions, an unusual three-membered-ring intermediate is formed. When this intermediate is stable under the reaction conditions, an *anti* addition of electrophile and nucleophile takes place.

EXAMPLE 4.13. STEREOSPECIFIC *ANTI* ADDITION OF BROMINE TO *CIS-* AND *TRANS*-2-BUTENE

The bromination of *cis-* and *trans*-2-butene occurs stereospecifically with each isomer.

4-14   4-15

4-14   4-16   4-14   4-17

From *cis*-2-butene, the products are the enantiomeric dibromides **4-16** and **4-17**, which are formed in equal amounts. The enantiomers are formed in equal amounts because bromine adds to the top and bottom faces of the alkene to give the intermediate bromonium ions, **4-14** and **4-15**, in equal amounts. Either carbon of each of these bromonium ions can then react with the nucleophile on the side opposite to the bromine to give the product dibromides.

Note that **4-16** and **4-17** are mirror images and are nonsuperimposable. In the same manner, the reaction of **4-15** with bromide ion also gives **4-16** and **4-17**.

From *trans*-2-butene, *anti* reaction of bromide at either carbon of the bromonium ion, **4-18**, gives only **4-19-1**, which is a *meso* compound because it has a mirror plane. This is most easily recognized in the eclipsed conformation, **4-19-2**, rather than the staggered conformation, **4-19-1**.

4-18   4-19-1   4-19-2

In the addition of bromine to *cis-* and *trans*-2-butene, the stereochemistry of each product occurs because the two bromines were introduced into the molecule on opposite faces of the original double bond.

Under some experimental conditions, electrophilic addition of either Cl^+ or a proton may form stable three-membered-ring intermediates. Thus, when a double bond undergoes stereospecific *anti* addition, formation of a three-membered-ring intermediate analogous to the bromonium ion is often part of the mechanism.

Syn *Addition*

Sometimes *syn* addition to a double bond may occur. These reactions usually occur in very nonpolar media.

EXAMPLE 4.14. THE CHLORINATION OF INDENE TO GIVE *CIS*-1,2-DICHLOROINDANE

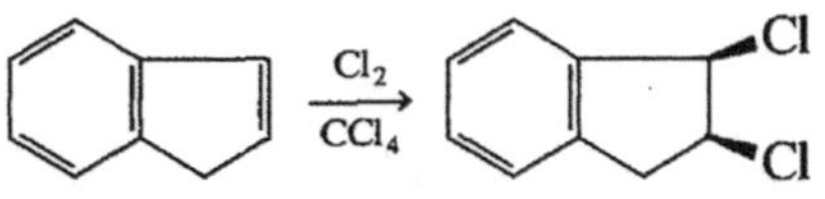

There are several factors that influence the course of this halogen addition: (1) the reaction takes place with chlorine rather than bromine; (2) the double bond of the starting material is conjugated with an aromatic ring; and (3) the reaction takes place in a nonpolar solvent.

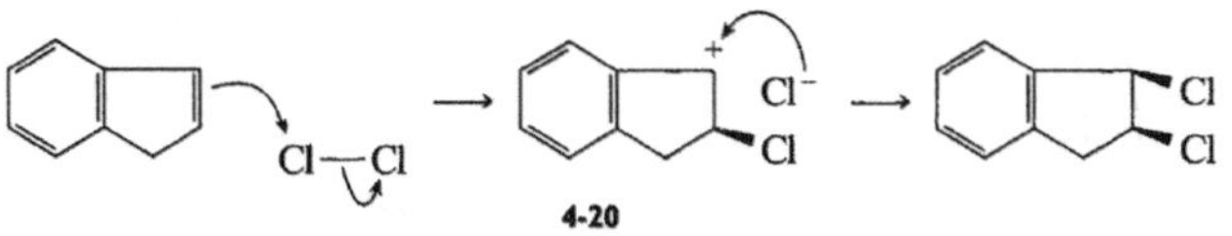

4-20

Chlorine is smaller and less polarizable than bromine and so has less of a tendency to form a bridged halonium ion than bromine. Also, the position of the double bond means that electrophilic addition of chlorine gives a stabilized benzylic cation, which is expected to be planar. In the nonpolar solvent, the planar cation is strongly attracted to the negative chloride ion to form the ion pair, **4-20**, because the carbocation and the chloride ion are not as strongly stabilized by solvation as they are in more polar solvents. Thus, the chloride ion remains in the position at which it was originally formed. Whereas a bridged chloronium ion would react most easily by backside reaction with the planar benzylic cation, the chloride ion can recombine without having to move to the other face of the cation. This step gives *syn* addition.

Nonstereospecific Addition

Electrophilic addition is not always stereospecific. Some substrates and reaction conditions lead to products from both *syn* and *anti* additions.

EXAMPLE 4.15. THE NONSTEREOSPECIFIC BROMINATION OF *CIS*-STILBENE IN ACETIC ACID

4-21 *dl* : **4-22** *meso* ?

(*anti* addition) (*syn* addition)

$\xrightarrow[Br_2]{CCl_4}$ **4-21** (sole product).

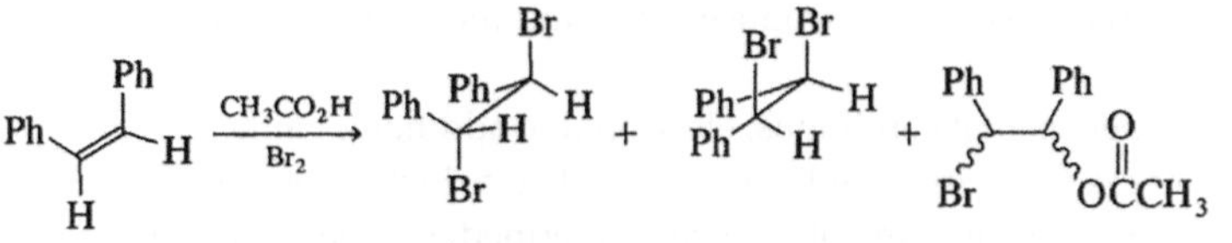

In the highly polar solvent, acetic acid, the reaction is completely nonstereospecific. The product distribution is consistent with reaction of *cis*-stilbene with bromine to form an intermediate carbocation, followed by reaction of bromide on either face of the planar intermediate to give **4-21** and **4-22**. In the nonpolar solvent, carbon tetrachloride, the product is exclusively the *dl* product **4-21** that would result from *anti* addition of bromide to a bridged bromonium ion. In the polar solvent, the localized charge on an intermediate planar carbocation would be more stabilized than the bromonium ion by solvent interactions because the charge on the bridged bromonium ion is more dispersed.

The side product of the reaction is most likely a mixture of bromoacetoxy compounds (unspecified stereochemistry is indicated by the wavy bond lines). Electrophilic additions in nucleophilic solvents often give a mixture of products because the nucleophile derived from the electrophilic reagent (e.g., Br$^-$) and the solvent compete for the intermediate carbocation.

Source: Buckles et al., (1962).

PROBLEM 4.5

Write step-by-step mechanisms for the following transformations.

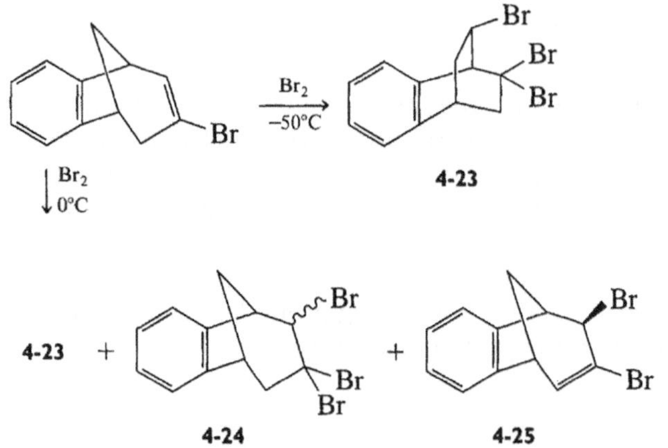

At −50 °C, **4-23** is the only product; at 0 °C, **4-23** is still the major product, but **4-24** and **4-25** are also produced. Note that, as the wavy bond lines indicate, both the *exo* and *endo* isomers of the 2-bromo compound **4-24** are produced.

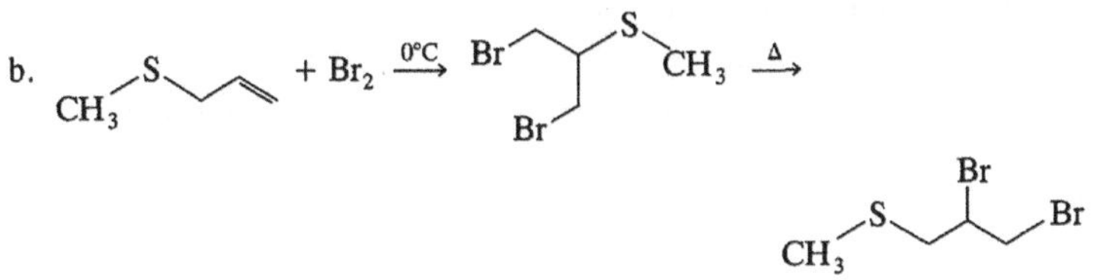

Sources: *Harmandar and Balci, (1985); Bland and Stammer, (1983).*

6. ACID-CATALYZED REACTIONS OF CARBONYL COMPOUNDS

Several examples of the importance of acid catalysis have already been given: Example 2.5 gives one of the steps in the acid-catalyzed formation of an ester, and Example 3.16 shows the acid-catalyzed mechanism for the formation of a hydrazone.

A. Hydrolysis of Carboxylic Acid Derivatives

Acidic hydrolysis of all derivatives of carboxylic acids (e.g., esters, amides, acid anhydrides, and acid chlorides) gives the corresponding carboxylic acid as the product. These hydrolyses can be broken down into the following steps.

(1) Protonation of the oxygen of the carbonyl group. This enhances the electrophilicity of the carbonyl carbon, increasing its reactivity with nucleophiles.
(2) The oxygen of water acts as a nucleophile and adds to the carbonyl carbon.
(3) The oxygen of the water, which has added and which is positively charged, loses a proton.
(4) A leaving group leaves. In the case of acid halides, the leaving group leaves directly; in the case of esters or amides, the leaving group leaves after prior protonation.
(5) A proton is lost from the protonated carboxylic acid.

EXAMPLE 4.16. HYDROLYSIS OF AN AMIDE

The overall reaction is as follows:

$$CH_3CONH_2 + H_3O^+ \rightarrow CH_3CO_2H + {}^+NH_4$$

The mechanism of this reaction is as follows:

(1) The initial protonation of the carbonyl oxygen gives a cation that is a resonance hybrid with positive character on carbon and nitrogen, as well as an oxygen.

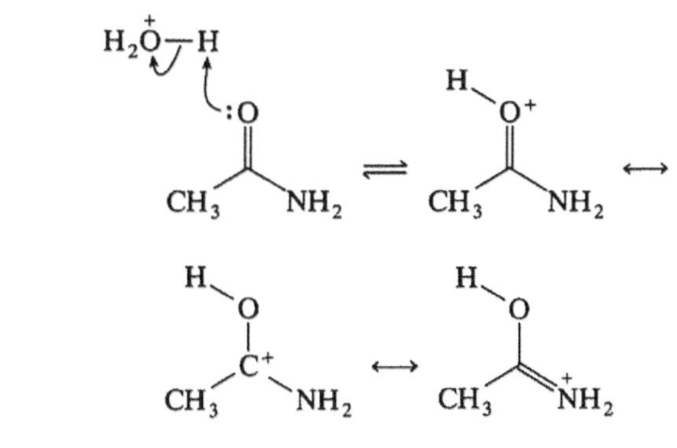

(2) The electrophilic cation reacts with the nucleophilic oxygen of water.

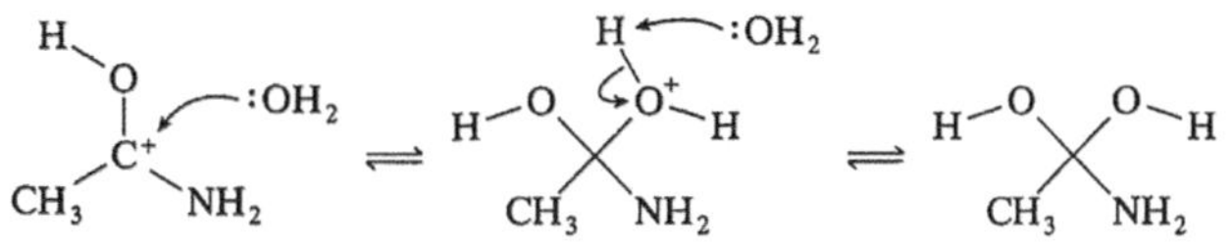

4-26

(3) Loss of a proton gives **4-26**.

(4) The neutral intermediate, **4-26**, can be protonated on either oxygen or nitrogen, but only protonation on nitrogen leads to product formation. Notice that the NH_2 group is now an amine and is much more basic than the NH_2 group of the starting amide.

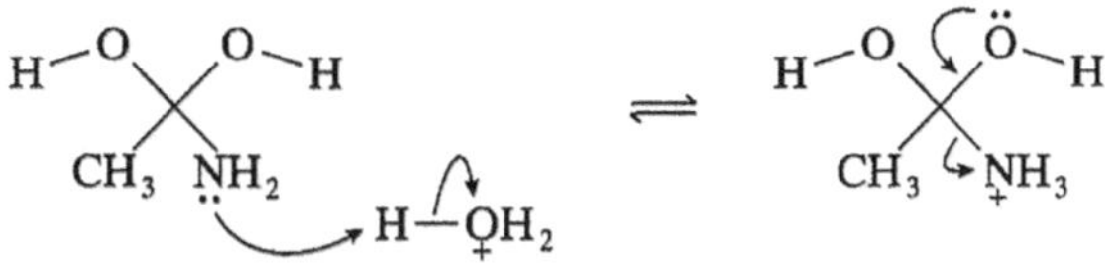

4-26

Note that the leaving group is ammonia rather than $-NH_2$, which would be a very poor leaving group. Whereas the loss of ammonia is a potentially reversible process, the protonation of ammonia to give the ammonium ion occurs much more rapidly in the acidic medium. Thus, the loss of ammonia is irreversible not because the addition of ammonia in the reverse process is energetically unfavorable, but because there is no ammonia present.

(5) Finally, a proton is removed from the protonated carboxylic acid to give the carboxylic acid product.

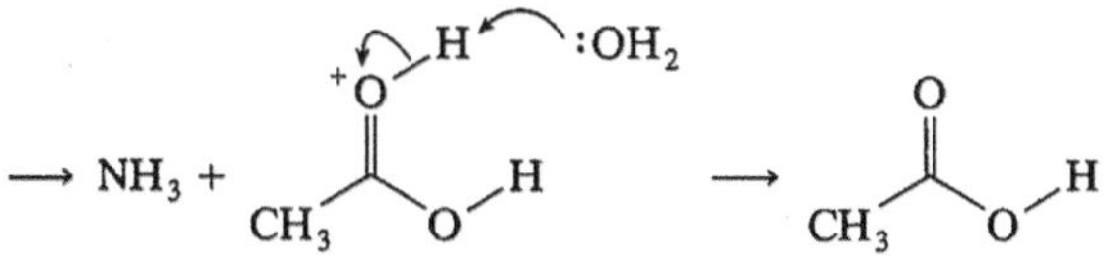

EXAMPLE 4.17. HYDROLYSIS OF AN ESTER

$$CH_3\overset{O}{\overset{\|}{C}}OCH_3 \underset{}{\overset{H^+, H_2O}{\rightleftharpoons}} CH_3\overset{O}{\overset{\|}{C}}OH + CH_3OH$$

For esters, all the steps of the hydrolysis reaction are reversible, and the mechanism of ester formation is the reverse of ester hydrolysis. The course of the reaction is controlled by adjusting the reaction conditions, chiefly the choice of solvent and the concentration of water, to drive the equilibrium in the desired direction. For hydrolysis, the reaction is carried out in an excess of water; for ester formation, the reaction is carried out with an excess of the alcohol component under anhydrous conditions. Frequently, an experimental set-up is designed to remove water as it is formed in order to favor ester formation.

PROBLEM 4.6

Write step-by-step mechanisms for the following transformations.

a.

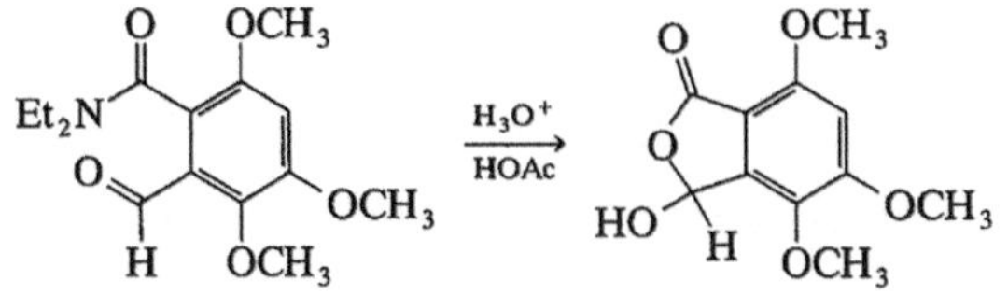

b.

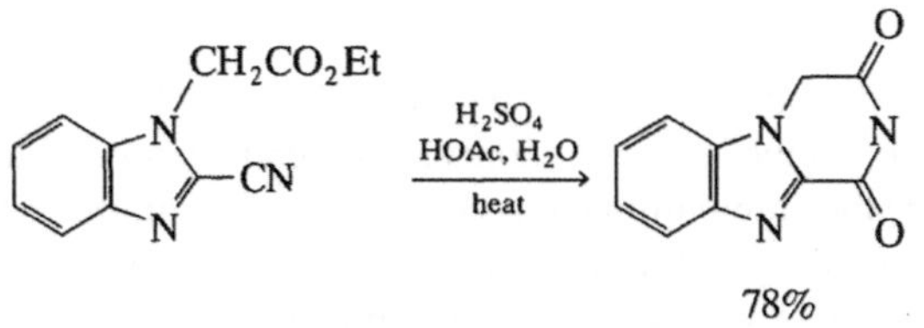

78%

Sources: Hauser et al., (1988); Serafin and Konopski, (1978).

B. Hydrolysis and Formation of Acetals and Orthoesters

Acetals, ketals, and orthoesters are polyethers, represented by the following structural formulas:

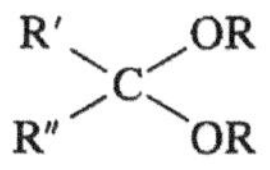

ketal

If R′ = H, the compound
is an acetal.

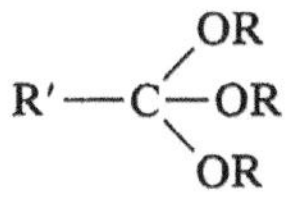

orthoester

The formation and hydrolysis of these groups are acid-catalyzed processes. Other derivatives of carbonyl groups, e.g., enamines, are also formed and hydrolyzed under acidic conditions by very similar mechanisms.

EXAMPLE 4.18. HYDROLYSIS OF ETHYL ORTHOFORMATE TO ETHYL FORMATE

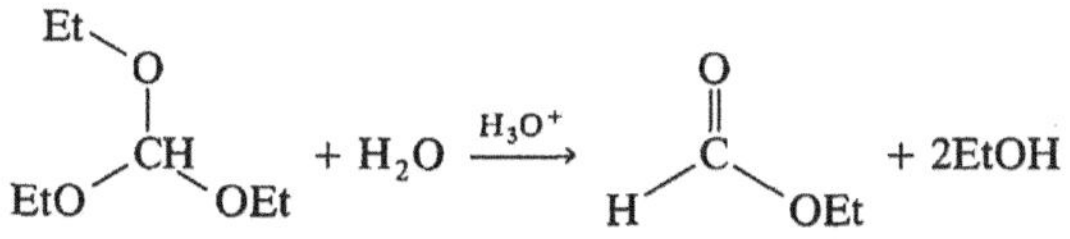

As in the case of other acid-catalyzed hydrolyses, the first step involves protonation of the most basic atom in the molecule. (In the case of ethyl orthoformate, all three oxygen atoms are equally basic.)

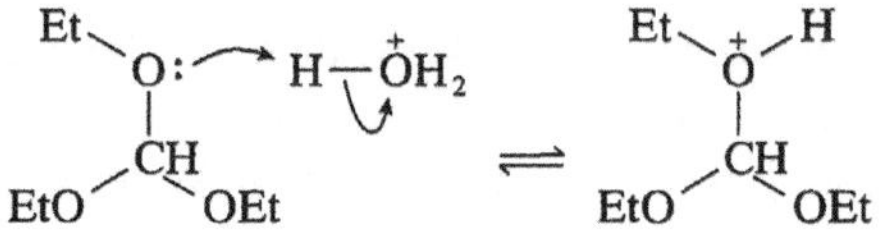

Protonation creates a better leaving group, that is, ethanol is a better leaving group than ethoxide ion.

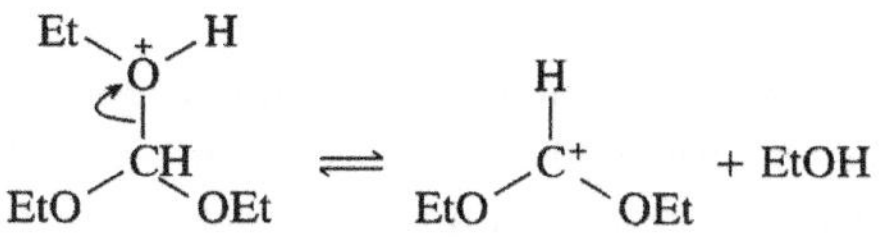

4-27

The electrophilic carbocation, **4-27**, reacts with nucleophilic water. Because water is present in large excess over ethanol, this reaction occurs preferentially and shifts the equilibrium toward the hydrolysis product. The protonated intermediate loses a proton to give **4-28**.

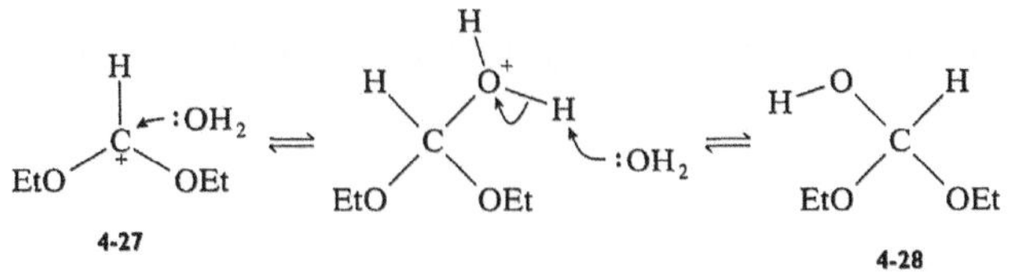

4-27   4-28

The neutral intermediate, **4-28**, can be protonated on either a hydroxyl or an ethoxy oxygen. Protonation on the hydroxyl oxygen is simply the reverse of the deprotonation step. Although this is a reasonable step, it leads to starting material. However, protonation on the oxygen of the ethoxy group leads to product.

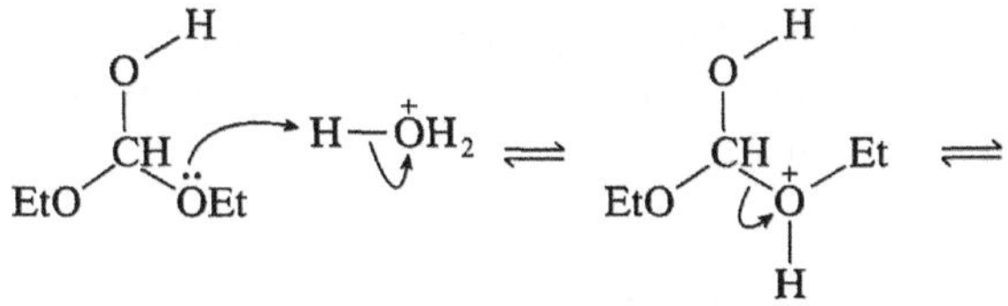

Loss of ethanol, followed by removal of a proton by water, gives the product ester.

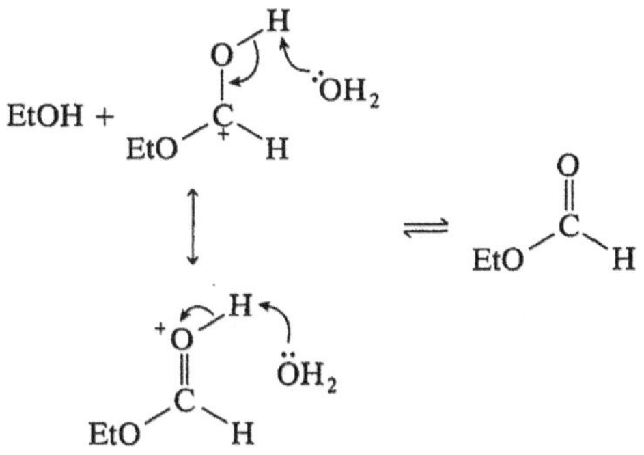

All the steps in this reaction are reversible. Why, then, do the hydrolysis conditions yield the formate ester and not the starting material? The key, as we saw in the preceding section with ester formation and hydrolysis, lies in the overall reaction. Water is present on the left-hand side of the equation and ethanol on the right-hand side. Thus, an excess of water would shift the equilibrium to the right, and an excess of ethanol would shift the equilibrium to the left. In fact, in order to get the reaction to go to the left, water must be removed as it is produced. Depending on the reaction conditions, the hydrolysis may proceed further to the corresponding carboxylic acid.

In Problem 3.14.a, we saw that the reaction of an amine with a carbonyl compound in the presence of an acid catalyst can be driven toward the enamine product by removing water from the reaction mixture as it is formed. The reverse of this reaction is an example of the acid hydrolysis of an enamine, a mechanism that is very similar to that of the orthoacetate hydrolysis shown in Example 4.18.

C. 1,4-Addition

Electrophilic addition to α,β-unsaturated carbonyl compounds is analogous to electrophilic addition to isolated double bonds, except that the electrophile adds to the carbonyl oxygen, the most basic atom in the molecule. After that, the nucleophile adds to the β carbon, and the resulting intermediate enol tautomerizes to the more stable carbonyl compound. These reactions may also be considered as the electrophilic counterparts of the nucleophilic Michael and 1,4-addition reactions discussed in Chapter 3, Section 3.C.

EXAMPLE 4.19. ELECTROPHILIC ADDITION OF HCl TO ACROLEIN

The overall reaction is as follows:

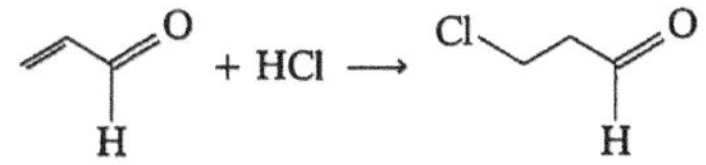

The first step in a mechanism is reaction of the nucleophilic oxygen of the carbonyl group with the positive end of the HCl molecule.

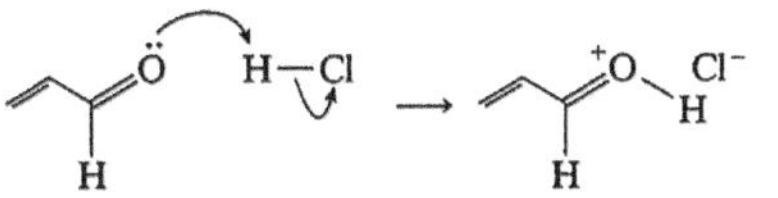

The resulting cation is a resonance hybrid with a partial positive charge on carbon as well as on oxygen:

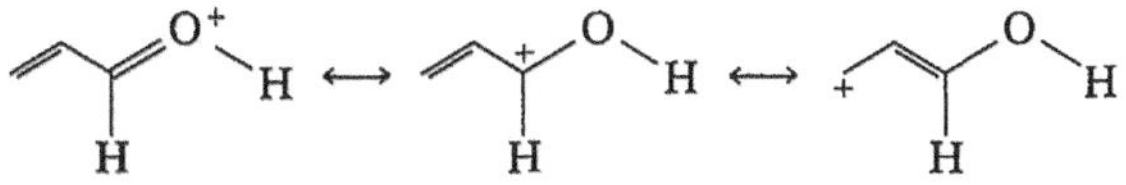

The electrophilic β position now reacts with the nucleophilic chloride ion to give an enol, which then tautomerizes to the keto form.

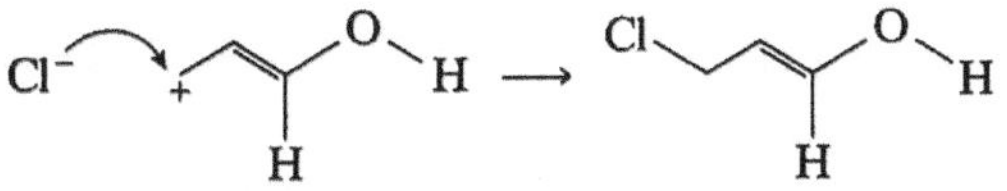

This reaction is the acid-catalyzed counterpart of a 1,4-addition reaction to an α, β-unsaturated carbonyl compound. Chloride ion, without an acid present, will not add to acrolein. That is, chloride ion is not a strong-enough nucleophile to drive the reaction to the right. However, if the carbonyl is protonated, the intermediate cation is a stronger electrophile and will react with the chloride ion.

PROBLEM 4.7

Rationalize the regiochemistry of the protonation shown in Example 4.19 by comparing it to protonation at other sites in the molecule.

PROBLEM 4.8

Write a mechanism for the following tautomerization in the presence of anhydrous HCl.

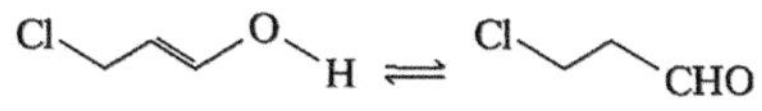

PROBLEM 4.9

Write a step-by-step mechanism for the following transformation.

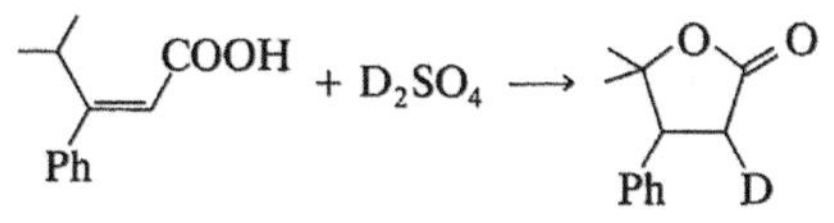

7. ELECTROPHILIC AROMATIC SUBSTITUTION

The interaction of certain electrophiles with an aromatic ring leads to substitution. These electrophilic reactions involve a carbocation intermediate that gives up a stable, positively charged species (usually a proton) to a base to regenerate the aromatic ring. Typical electrophiles include chlorine and bromine (activated by interaction with a Lewis acid for all but highly reactive aromatic compounds), nitronium ion, SO_3, the complexes of acid halides and anhydrides with Lewis acids (see Example 4.5) or the cations formed when such complexes decompose ($R-\overset{+}{C} = O$ or $Ar\overset{+}{C} = O$), and carbocations.

EXAMPLE 4.20. ELECTROPHILIC SUBSTITUTION OF TOLUENE BY SULFUR TRIOXIDE

In this reaction, the aromatic ring is a nucleophile and the sulfur of sulfur trioxide is an electrophile.

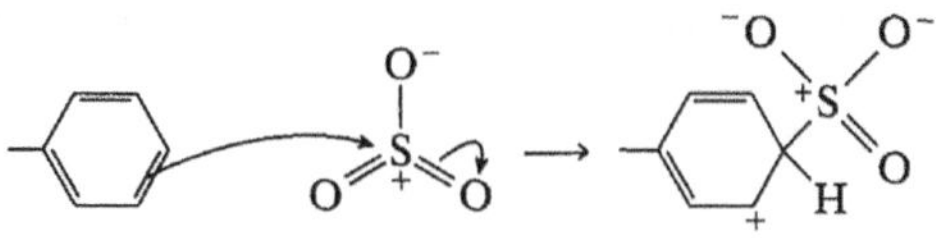

The positive charge on the ring is stabilized by resonance.

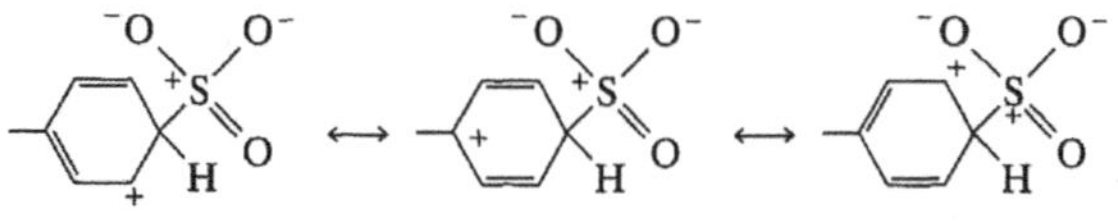

4-29-1   4-29-2   4-29-3

A sulfonate anion, acting as a base, can remove a proton from the intermediate to give the product:

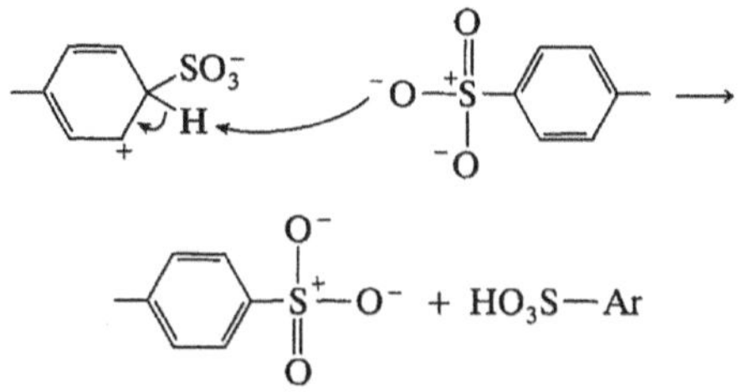

Because the aromatic ring acts as a nucleophile, the reaction rate will be enhanced by electron-donating substituents and slowed by electron-withdrawing substituents. Furthermore, the intermediate cation is especially stabilized by an adjacent electron-donating group, as in resonance structure 4-29-2. Because of this, electrophiles react at positions *ortho* or *para* to electron-donating groups. Such groups are said to be *ortho-* and *para*-directing substituents. Conversely, because electron-withdrawing groups destabilize a directly adjacent positive charge, the electrophile will react at the *meta* position in order to avoid this stabilization. A review of directing and activating–deactivating effects of various substituents is given in Table 4.1.

TABLE 4.1   Influence of Substituents in Electrophilic Aromatic Substitution

Strongly activating and *ortho-* or *para*-directing $-NR_2$, $-NRH$, $-NH_2$, $-O^-$, and $-OH$

Moderately activating and *ortho* or *para*-directing $-OR$ and $-NHCOR$

Weakly activating and *ortho-* or *para*-directing $-R$ and $-Ph$

Weakly deactivating and *ortho-* or *para*-directing $-F$, $-Cl$, and $-Br$

Strongly deactivating and *meta*-directing $-\overset{+}{S}R_2$, $-\overset{+}{N}R_3$, $-NO_2$, SO_3H, $-CO_2H$, $-CO_2R$, $-CHO$, $-COR$, $-CONH_2$, $-CONHR$, $-CONR_2$, and $-CN$

The effect of fluorine, chlorine, or bromine as a substituent is unique in that the ring is deactivated, but the entering electrophile is directed to the *ortho* and *para* positions. This can be explained by an unusual competition between resonance and inductive effects. In the starting material, halogen-substituted benzenes are deactivated more strongly by the inductive effect than they are activated by the resonance effect. However, in the intermediate carbocation, halogens stabilize the positive charge by resonance more than they destabilize it by the inductive effect.

PROBLEM 4.10

For the following reactions, explain the orientations in the product by drawing resonance forms for possible intermediate carbocations and rationalize their relative stabilities.

a. The reaction in Example 4.20.

b.

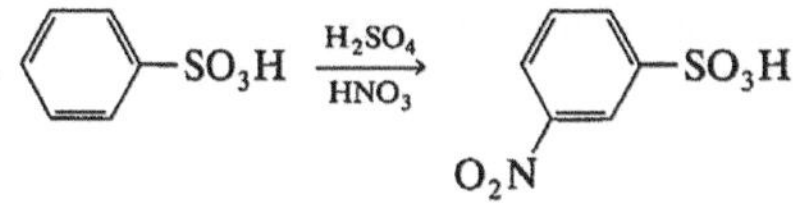

c.

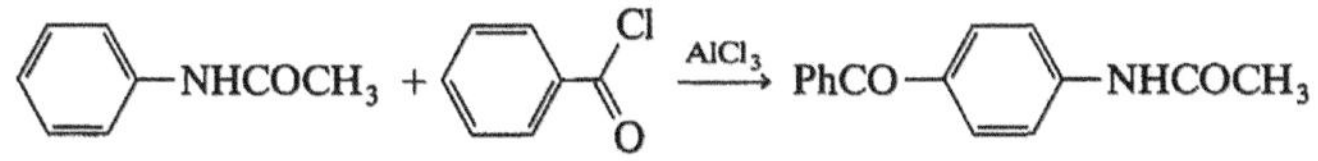

EXAMPLE 4.21. A METAL-CATALYZED, INTRAMOLECULAR, ELECTROPHILIC AROMATIC SUBSTITUTION

Write a mechanism for the following transformation.

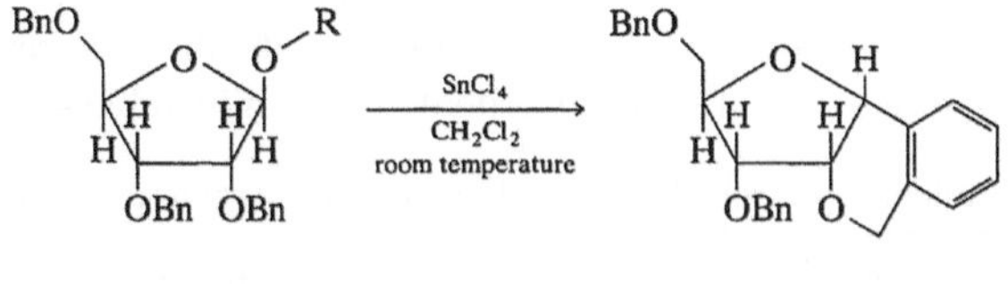

Bn = benzyl (PhCH$_2$−)

Tin(IV) chloride can undergo nucleophilic substitution, which converts the acetal OR group into a better leaving group. (The positively charged oxygen of the OR group of an acetal is a much better leaving group than the oxygen of an ether because the resulting cation is stabilized by resonance interaction with the remaining oxygen of the original acetal.) After the leaving group leaves, the aromatic ring of a benzyl group at position 2 in **4-30** (suitably situated geometrically for this interaction) acts as a nucleophile toward the positive center. The resulting carbocation then loses a proton to give the product:

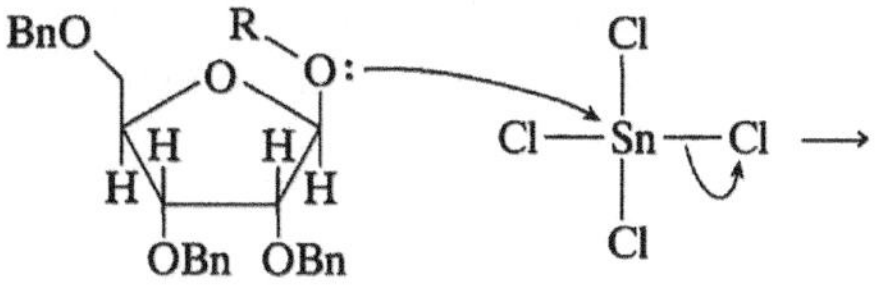

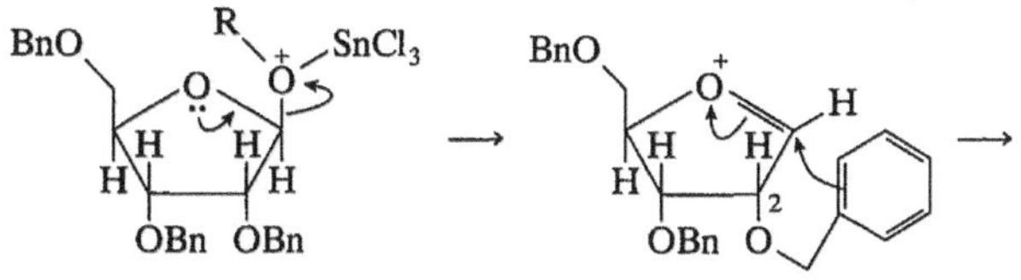

4-30

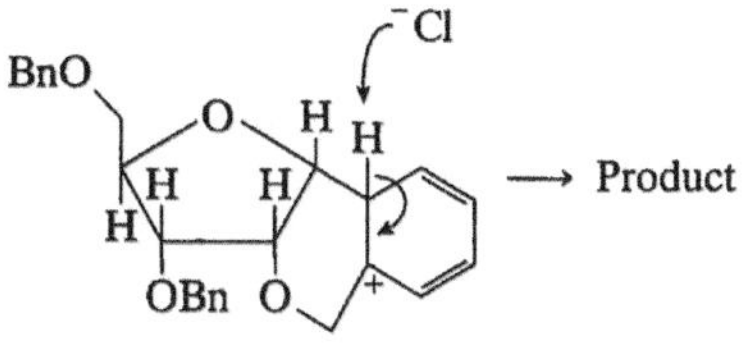

Similar intramolecular electrophilic aromatic substitution reactions are common, especially when five- or six-membered rings are formed.

Source: Martin, (1985).

PROBLEM 4.11

Write step-by-step mechanisms for the following transformations.

a.

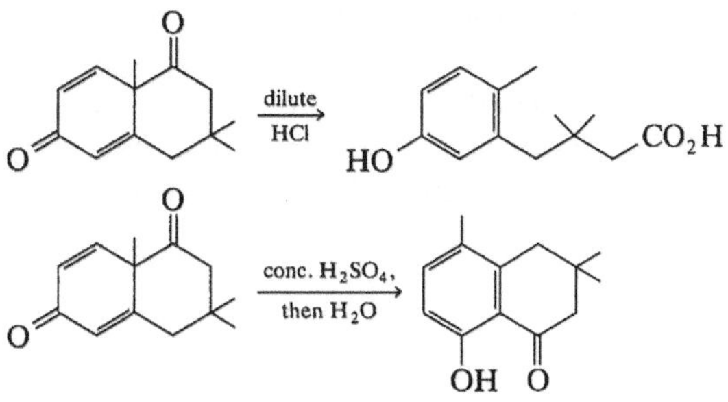

(conc. = concentrated)

The reaction in dilute HCl was discussed in another context in Example 2.11. Modify the mechanism given there to account for the result shown here for concentrated sulfuric acid.

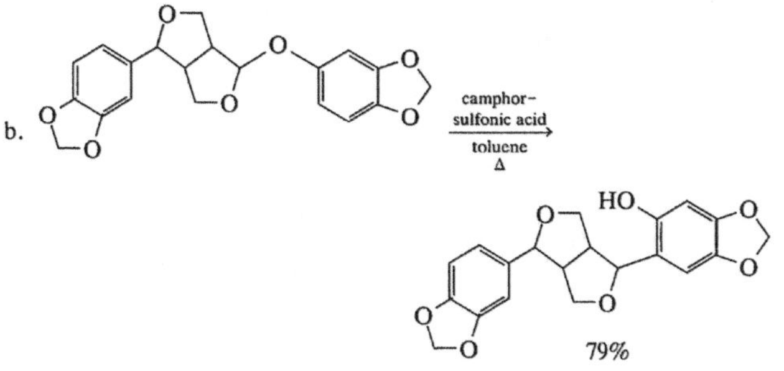

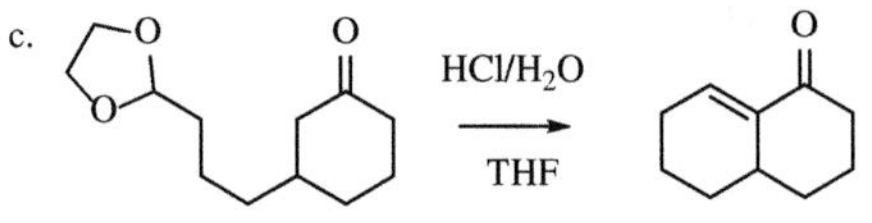

b.

camphor–
sulfonic acid
toluene
Δ

79%

c.

HCl/H₂O
THF

Sources: Waring and Zaidi, (1985); Fukuda et al., (1986); Abbot and Spencer, (1980).

8. CARBENES

Carbenes are very reactive intermediates. Some have been isolated in matrices at low temperatures, but with few exceptions (see Dagani, (1991)), they are very short-lived at ambient temperatures. Although carbenes are often generated in basic media, they usually act as electrophiles.

A. Singlet and Triplet Carbenes

A carbene is a neutral divalent carbon species containing two electrons that are not shared with other atoms. When these two electrons have *opposite* spins, the carbene is designated a *singlet* carbene; when they have *parallel* spins, the carbene is a *triplet*. In the ground state, a singlet carbene has a pair of electrons in a single orbital, whereas the triplet has two unpaired electrons, each occupying a separate orbital. The designations singlet and triplet originate in spectroscopy.

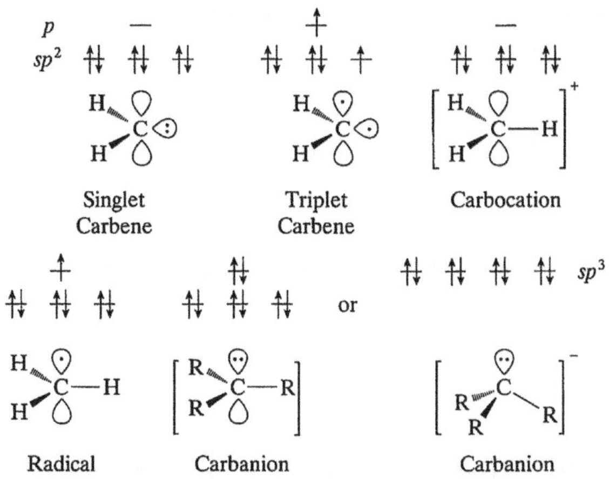

Note that the formal charge on the carbon atom in either a singlet or a triplet carbene is zero; the singlet carbene is *not* a carbanion.

The term *carbenoid* is used to refer to a carbene when the exact nature of the carbene species is uncertain and especially when referring to neutral electron-deficient species that are coordinated with a metal.

B. Formation of Carbenes

EXAMPLE 4.22. GENERATION OF CARBENES FROM ALKYL HALIDES AND BASE

(1) Dichlorocarbene can be formed from aqueous KOH and chloroform:

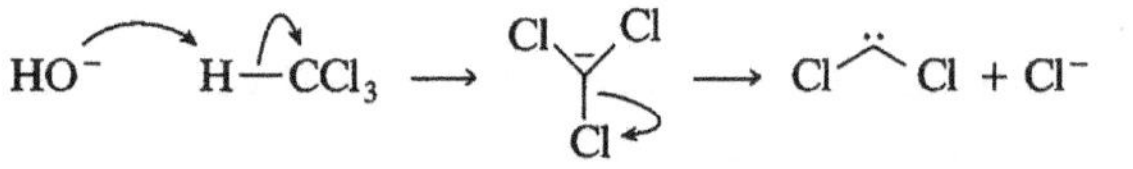

The base removes the acidic proton from chloroform. The resulting anion then loses chloride ion, giving a divalent carbon with two unshared electrons. In this case, the unshared electrons are paired (occupy the same orbital), i.e., dichlorocarbene is a singlet. The carbon of the carbene has no charge.

(2) A carbenoid is formed when potassium *t*-butoxide reacts with benzal bromide in benzene. The reactivity of the carbenoid, **4-31**, is similar to that of phenylbromocarbene. The exact nature of the interaction between the carbenoid carbon and the metal halide, in this case potassium bromide, is not known.

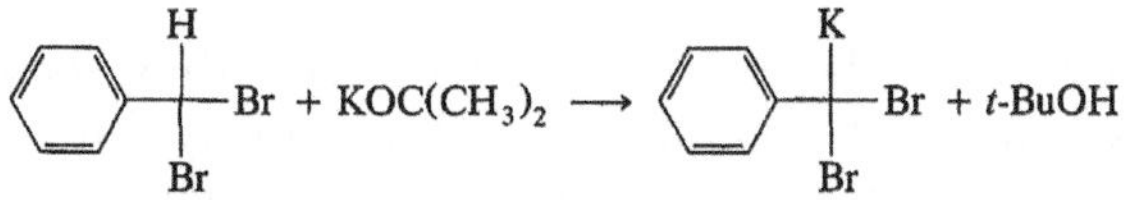

4-31

(3) Some alkyl halides react with alkyl lithium reagents under aprotic conditions to give carbenoids by a process called "halogen–metal exchange."

4-32

In some cases, proton removal from the carbon may predominate over exchange with the halogen. For details, see Kobrich, (1972).

EXAMPLE 4.23. GENERATION OF ICH$_2$ZNI (THE SIMMONS–SMITH REAGENT) FROM METHYLENE IODIDE AND ZINC–COPPER COUPLE

When diiodomethane is treated with zinc–copper couple, a carbenoid is formed.

$$CH_2I_2 \ + \ ZnCu \ \longrightarrow \ ICH_2ZnI$$
4-33

The Simmons–Smith reagent (**4-33**), other carbenoids, and carbenes are very useful in the synthesis of cyclopropanes (Example 4.25).

EXAMPLE 4.24. GENERATION OF CARBENES FROM DIAZO COMPOUNDS

Loss of nitrogen from a diazo compound can be affected by heat, light, or a copper catalyst. This gives either a carbene (heat or light) or a carbenoid (copper).

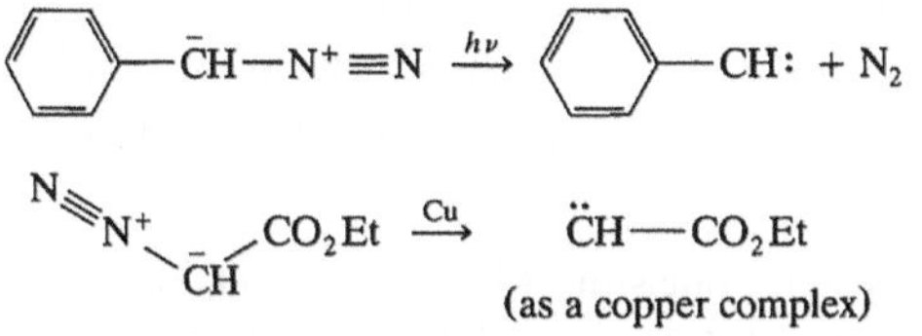

(as a copper complex)

C. Reactions of Carbenes

Once generated by the methods outlined in the preceding section, the highly reactive carbene intermediates can react in a number of different ways, including addition, substitution, insertion, rearrangement, and hydrogen or halogen abstraction. Two important reactions involving carbenes are addition to carbon—carbon double and triple bonds to generate cyclopropanes and cyclopropenes, respectively, and electrophilic aromatic substitution (the Reimer—Tiemann reaction), in which electrophilic addition of a carbene is the first step.

Addition

Electrophilic addition of carbenes to carbon—carbon double and triple bonds has been extremely useful synthetically. In many cases, the reaction goes with 100% stereospecificity, so that the stereochemistry about a double bond in the starting material is maintained in the product. Cases in which addition is not 100% stereospecific are rationalized on the basis of a triplet or diradical intermediate. If the triplet carbene is relatively unreactive, the formation of the two new carbon—carbon bonds may be a stepwise process that allows for rotation and, therefore, loss of stereochemistry in the intermediate.

EXAMPLE 4.25. ADDITION OF SINGLET DICHLOROCARBENE TO CIS-2-BUTENE AND TRANS-2-BUTENE

The addition of dichlorocarbene to the double bond of *cis*-2-butene goes with 100% stereospecificity, that is, the only product is *cis*-1,2-dimethyl-3,3-dichlorocyclopropane. The addition of dichlorocarbene to *trans*-2-butene gives only the corresponding *trans*-isomer. The stereospecificity of the reaction has been interpreted to mean that dichlorocarbene is a singlet and that both ends of the double bond react simultaneously, or nearly so, with the carbene.

EXAMPLE 4.26. A NONSTEREOSPECIFIC ADDITION OF A CARBENE

Because the reaction is nonstereospecific, the mechanism would be written as a stepwise addition to the double bond by a diradical carbene.

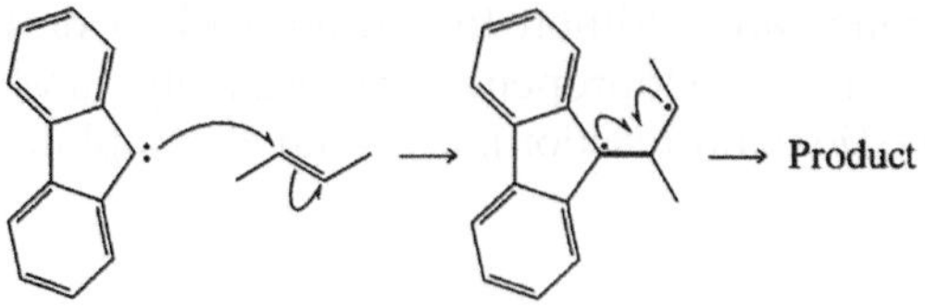

Note that the arrows used to show the flow of unpaired electrons (radicals) have only a half head. Also, the intermediate in the reaction is a diradical. (Radicals are discussed in more detail in Chapter 5.) Rotation about the highlighted single bond takes place fast enough that the stereochemistry of the starting olefin is lost.

Substitution

Carbenes also add to other nucleophiles, such as hydroxide, thiolate, and phenoxide. The reaction with phenoxide is the classic Reimer–Tiemann reaction.

EXAMPLE 4.27. THE REIMER–TIEMANN REACTION

Phenols react with chloroform in the presence of hydroxide ion in water to give *o*- and *p*-hydroxy benzaldehydes. The steps of the reaction are (1) the formation of dichlorocarbene, as shown in Example 4.22; (2) nucleophilic reaction of the phenoxide with the electrophilic carbene to form intermediate **4-34**; and (3) hydrolysis to form the aldehyde **4-35**.

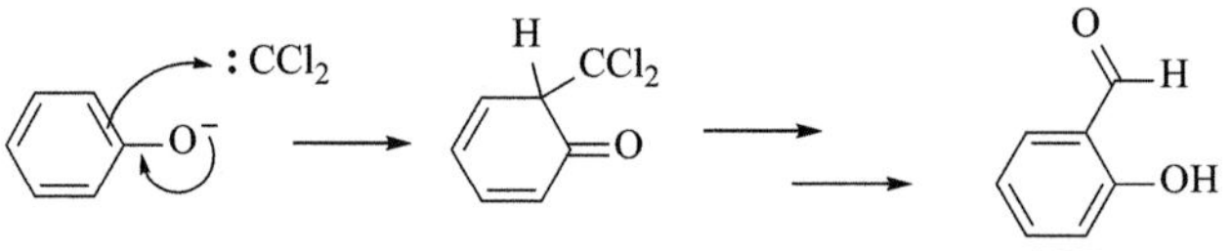

PROBLEM 4.12

Complete the mechanism for Example 4.27 showing the steps leading from the intermediate carbanion to the product *o*-salicylaldehyde.

PROBLEM 4.13

Write a step-by-step mechanism for the following transformation.

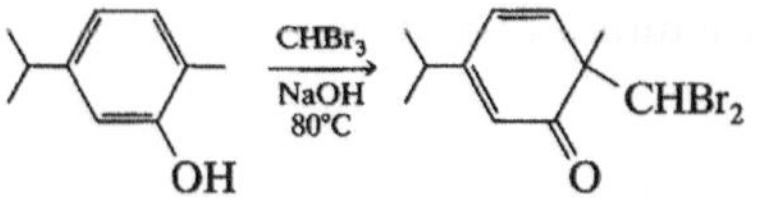

Source: Wenkert et al., (1990).

EXAMPLE 4.28.

Although the reactivity of carbenes can lead to a variety of products from a single reaction, their utility is the result of interesting transformations that would otherwise be very difficult to effect. In addition to the "normal" results (path B shown below, a process leading to the aldehyde products) that we have described for the Reimer–Tiemann reaction, the same conditions can also produce a so-called "abnormal" product (path A, the ring expanded product). In the course of the reaction shown below, in addition to the 3-chloro pyridine **4-36**, the 3-pyrolle aldehyde **4-37** (path B) is produced as well.

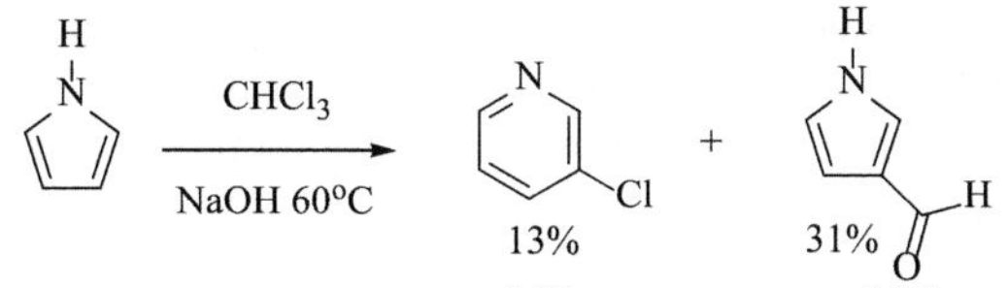

In the first step, the carbene is generated through the action of hydroxide on chloroform as we saw above.

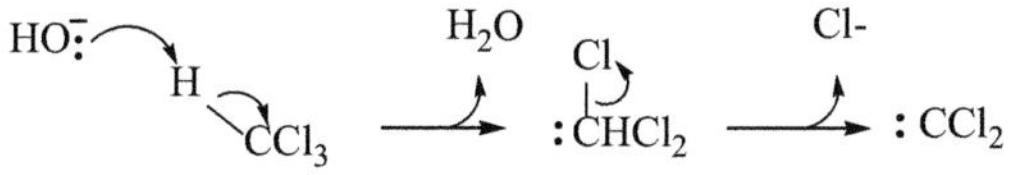

The carbene reacts with the pyrrole to form the dichloro carbanion **4-38** which can then move through one of two manifolds. Closure of the anion through route A to form the cyclopropane **4-39** and then ring expansion leads to the 3-chloro pyridine **4-36** with the expulsion of HCl (trapped out as NaCl). The second manifold (path B) moves through steps where a chlorine ion is lost **4-41**, hydroxide adds into the cationic pyrrole **4-42** and then the second chlorine ion is lost **4-43** resulting in the product aldehyde **4-37**. Although the pyrrole is shown protonated and positively charged for the sake of simplicity, it is reasonable to show the pyrrole nitrogen at step **4-42** unprotonated and neutral and pulling a proton from water as the hydroxide adds in.

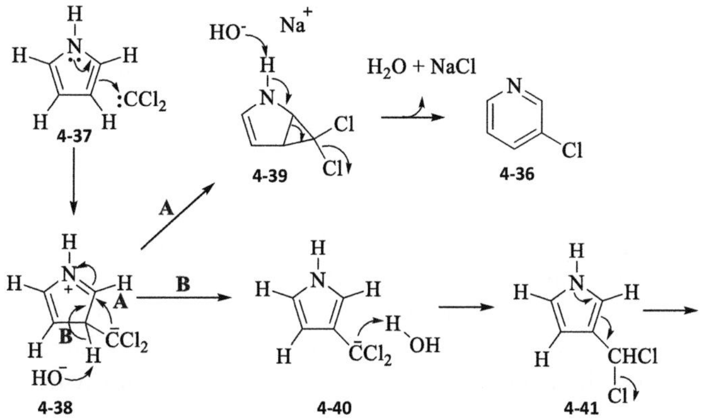

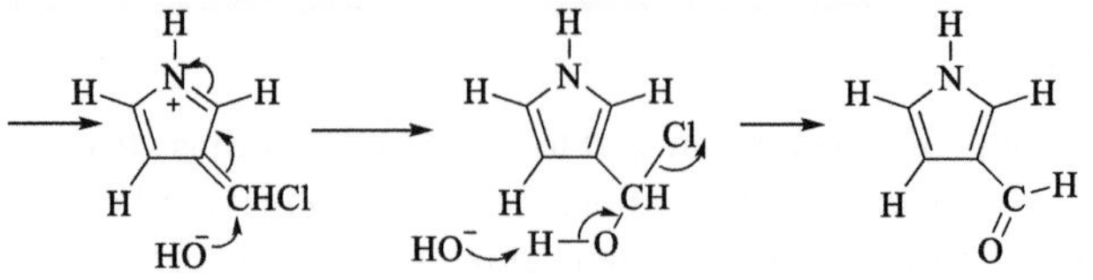

4-42   **4-43**   **4-37**

Insertion

In insertion, as the name suggests, the carbene inserts itself between two atoms. Insertions have been observed into C–H, C–C, C–X, N–H, O–H, S–S, S–H, and M–C bonds, among others. The mechanism of the process is often concerted. A three-centered transition state is usually written for the concerted mechanism:

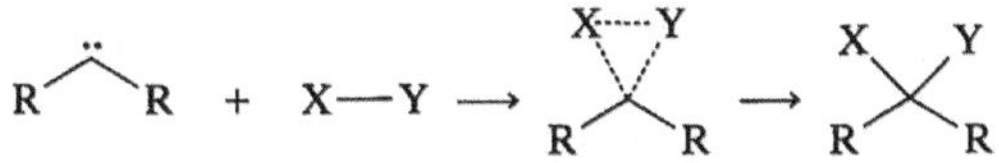

Source: Wynberg and Meijer, (1982).

EXAMPLE 4.29. A SYNTHETICALLY USEFUL INSERTION REACTION

Carbenoid **4-32** can be generated as shown in Example 4.22. The insertion of **4-32** into the C1–C2 bond leads to the formation of cycloheptanone in 70% yield. Other homologs give even higher yields of ring-expanded products.

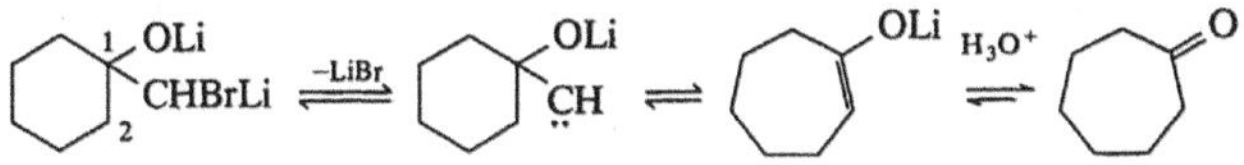

4-32   **4-44**

The intermediate enolate, **4-44**, forms the corresponding ketone in acid. From Taguchi et al., (1974).

Rearrangement

Because of their reactivity, it is frequently difficult to decide whether a free carbene has been generated or whether an electrophilic carbon undergoes a synchronous reaction. It is generally accepted that a free carbene is an intermediate in the Wolff rearrangement, in which a diazoketone rearranges to a highly reactive ketene. The versatile ketene intermediate then reacts to give the carboxylic acid, ester, or amide by reaction with water, alcohol, or amine, respectively.

EXAMPLE 4.30. THE WOLFF REARRANGEMENT OF DIAZOACETOPHENONE TO METHYL PHENYL-ACETATE

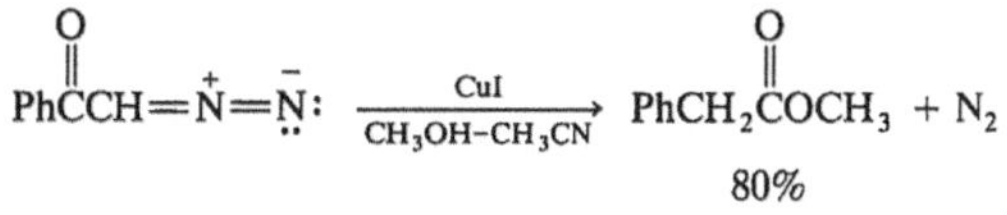

$$PhCCH = \overset{+}{N} = \overset{-}{N}: \xrightarrow[CH_3OH-CH_3CN]{CuI} PhCH_2COCH_3 + N_2$$

80%

The mechanism involves decomposition of the diazoketone **4-45** to a carbene **4-46**. The decomposition can be brought about by activation with metal salts, particularly copper and silver, as well as by heat or light.

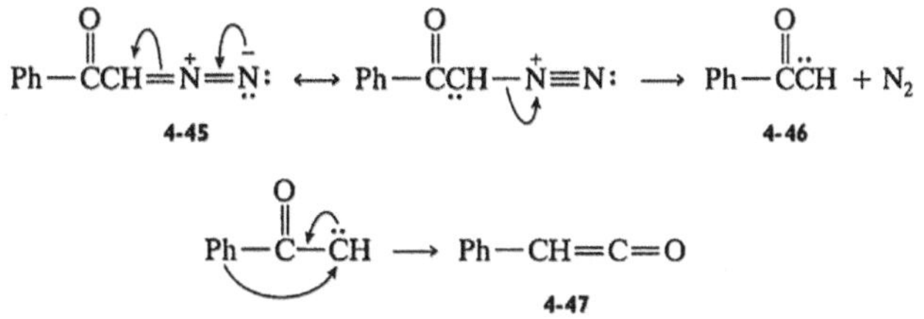

$$Ph-CCH = \overset{+}{N} = \overset{-}{N}: \longleftrightarrow Ph-CCH-\overset{+}{N} \equiv N: \longrightarrow Ph-CCH + N_2$$

4-45   **4-46**

$$Ph-C-CH \longrightarrow Ph-CH = C = O$$

4-47

The carbene then rearranges to the ketene **4-47**. An important feature of the rearrangement is that the migrating group moves with *retention* of configuration. In the example shown, the ester is formed by reaction of the ketene with methanol in the solvent.

9. ELECTROPHILIC HETEROATOMS

A number of reactions involve electrophilic heteroatom species that are analogous to their carbon counterparts. In this section, we will consider electrophilic nitrogen and oxygen species.

A. Electron-Deficient Nitrogen

Nitrenes

A nitrene is the nitrogen analog of a carbene. In other words, it is a neutral, univalent nitrogen that contains two lone pairs of electrons not shared with other atoms.

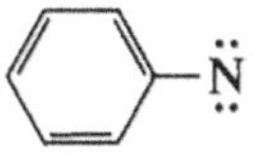

phenyl nitrene

Nitrenes are generated and react very similarly to carbenes but because of their greater reactivity, the presence of free nitrenes is difficult to demonstrate experimentally.

EXAMPLE 4.31. GENERATION OF NITRENES

Common methods for generating nitrene intermediates are the photolysis and thermolysis of azides. Generation of nitrenes from acyl azides can only be effected photochemically; thermolysis of an acyl azide gives the corresponding isocyanate.

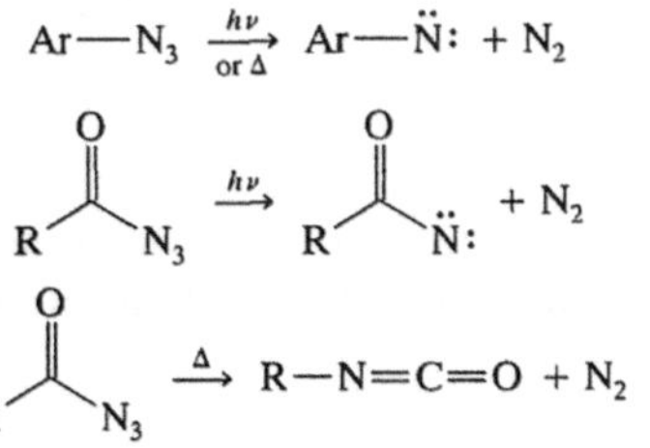

$$Ar\!-\!N_3 \xrightarrow[\text{or }\Delta]{h\nu} Ar\!-\!\ddot{\ddot{N}}\!: \; + N_2$$

The reactivity of nitrenes is similar to that of carbenes. They readily add to double bonds to give the corresponding three-membered heterocycles, aziridines. Insertion reactions are also common.

Nitrenium Ions

A nitrenium ion is a positively charged divalent nitrogen with one lone pair of electrons. Thus, it is a cation that is isoelectronic with a carbene, a neutral species.

EXAMPLE 4.32. GENERATION AND REACTION OF A NITRENIUM ION

Silver-ion-assisted loss of chloride ion from the starting material gives a nitrenium ion, which acts as an electrophile toward the aromatic ring to give **4-48**.

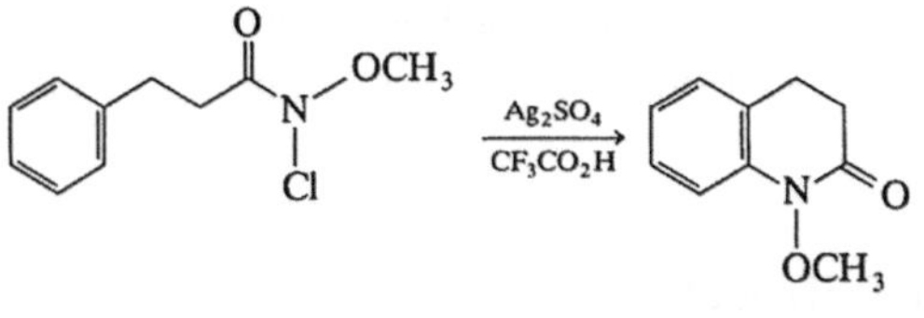

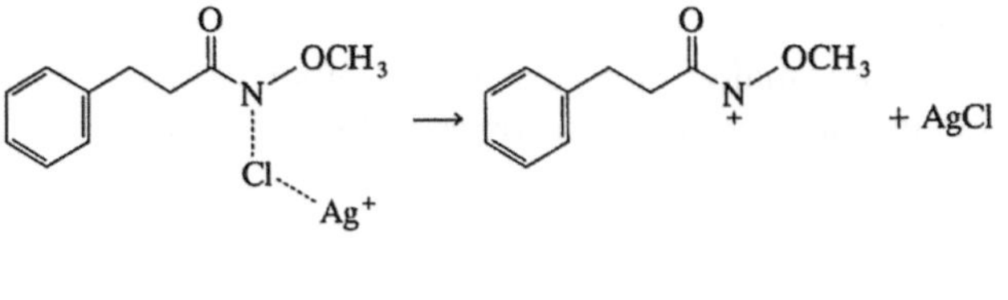

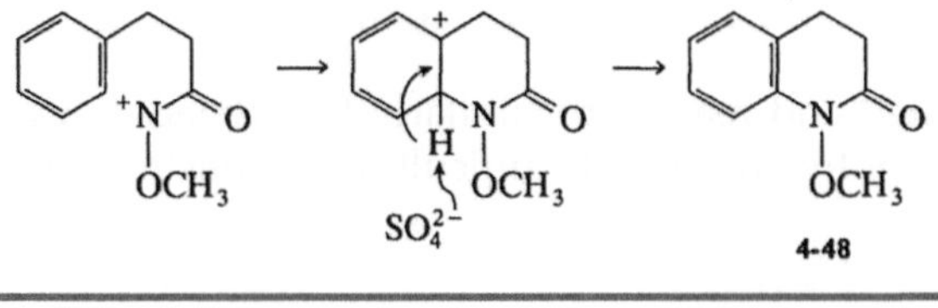

B. Rearrangements Involving Electrophilic Nitrogen

The Beckmann Rearrangement

In strong acids, or when treated with reagents such as thionyl chloride or phosphorus pentachloride, an oxime will react to give a rearranged amide. This is known as the Beckmann rearrangement. When the reaction gives products other than amides, these products are referred to as abnormal products. One such abnormal pathway is illustrated in Problem 4.14.
The source is *Yambe et al. (2005)*.

EXAMPLE 4.33. BECKMANN REARRANGEMENT OF BENZOPHENONE OXIME

The overall reaction is as follows:

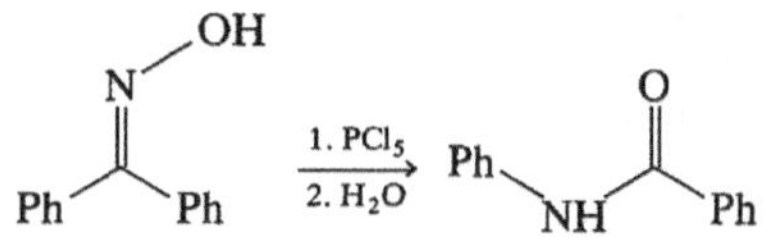

The first two steps of the reaction mechanism convert the original oxime to a derivative, **4-49**. This process converts the hydroxyl group of the oxime into a better leaving group.

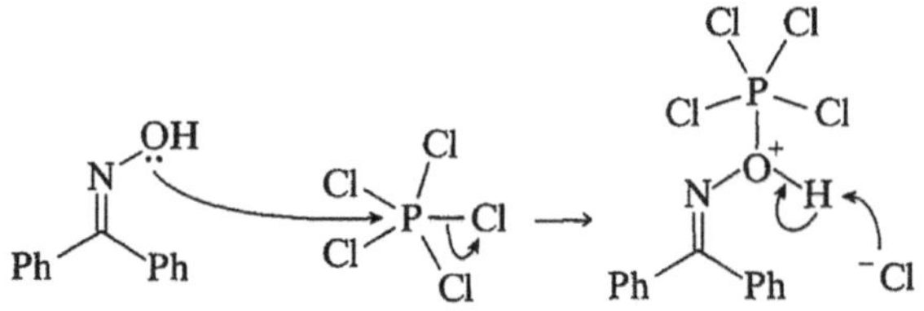

The strong electron-withdrawing leaving group creates a substantial partial positive charge on nitrogen. The phenyl group, on the side opposite to the leaving group, moves with its pair of electrons to the electron-deficient nitrogen as the leaving group leaves. The resulting ions then collapse to form **4-50**. During workup, **4-50** will undergo hydrolysis and tautomerization to the final product, the amide.

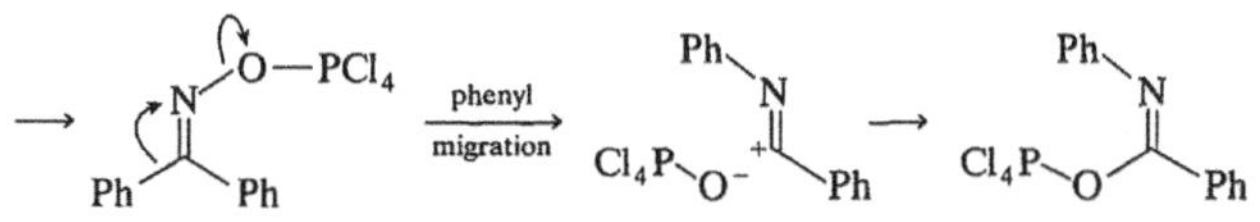

PROBLEM 4.14

Write a step-by-step mechanism for the following reaction.

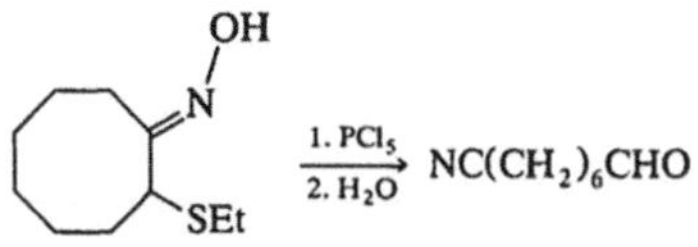

Source: Ohno et al., (1966).

Nitrogen Analogs of the Wolff Rearrangement

Important rearrangements involving electrophilic nitrogen are the Hofmann, Curtius, and Schmidt rearrangements, which are nitrogen analogs of the Wolff rearrangement discussed in Section 8. In all these rearrangements, it is possible to write a discrete nitrene intermediate; however, it is generally considered more likely that the reactions proceed through a less reactive intermediate in which rearrangement accompanies loss of the leaving group. The electron deficiency of the nitrogen is due to withdrawal of electron density by a good leaving group. As with the Wolff rearrangement, all these reactions proceed with *retention* of configuration by the group migrating to nitrogen. Table 4.2 compares the rearrangements involving various electron-deficient heteroatom species.

EXAMPLE 4.34. THE HOFMANN REARRANGEMENT OF HEXANAMIDE TO PENTYLAMINE

In the Hofmann rearrangement, a halide leaving group confers electrophilic character on the nitrogen atom.

$$CH_3(CH_2)_4\overset{\overset{\displaystyle O}{\|}}{C}NH_2 \xrightarrow[\text{2. HCl}]{\text{1. NaOCl, H}_2\text{O}} CH_3(CH_2)_4NH_2$$

95%

Sources: Shioiri, (1991); Magnien and Baltzly, (1958).

TABLE 4.2   Parallels in Electrophilic Rearrangements

Name reaction	Starting material	Reaction conditions	Reactive intermediate	Intermediate product	Final product
Wolff	$R-\overset{\overset{O}{\|}}{C}-CH-\overset{+}{N}\equiv N:$	$Ag^+, Cu^+, \Delta, h\nu$		$N_2 + O=C=CHR \xrightarrow{H_2O}$	RCH_2CO_2H
Hofmann	$R-\overset{\overset{O}{\|}}{C}-NH-Br$	OH^-, H_2O	$R-\overset{\overset{O}{\|}}{C}-\ddot{N}-Br$	$Br^- + O=C=N-R \xrightarrow{H_2O}$	$RNH_2 + CO_2$
Curtius	$R-\overset{\overset{O}{\|}}{C}-\ddot{N}-\overset{+}{N}\equiv N:$	Δ		$N_2 + O=C=N-R \xrightarrow{H_2O}$	$RNH_2 + CO_2$
Schmidt	$R-\overset{\overset{O}{\|}}{C}R'$	$H^+ + NaN_3$	$R-C=\ddot{N}-\overset{+}{N}\equiv N:$ $\underset{R'}{\|}$	$N_2 + R-\overset{+}{C}=N-R' \xrightarrow{H_2}$	$R\overset{\overset{O}{\|}}{C}NHR'$
Beckman	$\underset{R}{\overset{R}{>}}C=\underset{\ddot{}}{N}\overset{OH}{\diagup}$	H^+	$\underset{R}{\overset{R}{>}}C=\underset{\ddot{}}{N}\overset{\overset{+}{O}H_2}{\diagup}$	$H_2O + R-\overset{+}{C}=N-R \xrightarrow{H_2O}$	$R-\overset{\overset{O}{\|}}{C}NHR$
Baeyer–Villager	$\underset{R'}{\overset{R}{>}}C=O$	$H^+, R''CO_3H$	$\underset{R}{\overset{R'}{\diagdown}}\overset{OH}{\underset{O-\overset{\overset{O}{\|}}{C}R''}{C}}$	$R-\overset{OH}{\underset{+}{\overset{\|}{C}}}OR' \xrightarrow{H_2O}$	$R-\overset{\overset{O}{\|}}{C}OR'$
				$R'-\overset{OH}{\underset{+}{\overset{\|}{C}}}OR \longrightarrow$	$R'-\overset{\overset{O}{\|}}{C}OR$

The first step in the mechanism is formation of the *N*-chloroamide.

$$CH_3(CH_2)_4C\overset{O}{\overset{\|}{-}}N\overset{H}{\underset{H}{\big\langle}} \quad :\ddot{O}H \longrightarrow CH_3(CH_2)_4C\overset{O}{\overset{\|}{-}}\ddot{N}^- - H \longrightarrow$$

$$CH_3(CH_2)_4C\overset{O}{\overset{\|}{-}}\ddot{N} - H + OH^-$$
$$\underset{Cl}{|}$$

$$CH_3(CH_2)_4C\overset{O}{\overset{\|}{-}}\underset{Cl}{\overset{|}{\ddot{N}}} H \quad :\ddot{O}H \longrightarrow CH_3(CH_2)_4\overset{O}{\overset{\|}{C}} \ddot{N}: \longrightarrow$$

$$CH_3(CH_2)_4 - N{=}C{=}O + Cl^-$$

4-51

Because of the electron-withdrawing effect of the chloro group, the remaining hydrogen attached to the amide nitrogen is acidic and easily removed by aqueous base. The tendency of the chloro group to leave with the N—Cl bonding electrons confers electrophilic character on the amide nitrogen, so that as the chloro group departs, the alkyl group moves to compensate for the developing electron deficit on nitrogen, yielding the isocyanate **4-51**.

$$CH_3(CH_2)_4 - N{=}C{=}O \longrightarrow CH_3(CH_2)_4 - N{=}C - O^-$$
$$\underset{:\ddot{O}H}{} \qquad\qquad \underset{OH}{|}$$

4-51

$$CH_3(CH_2)_4\ddot{N}H - C{=}O \longrightarrow CH_3(CH_2)_4\ddot{N}H + CO_2 \longrightarrow CH_3(CH_2)_4NH_2$$
$$\underset{:\ddot{O}:^-}{|} \qquad\qquad \underset{H-OH}{}$$

4-52

Basic hydrolysis of the reactive isocyanate **4-51** leads to an intermediate that tautomerizes under the reaction conditions to give **4-52**, which spontaneously decarboxylates. The irreversible decarboxylation yields the amine, which contains one less carbon (lost as CO_2) than the starting material.

PROBLEM 4.15

Propose a reasonable mechanism for the transformation shown.

46%

Source: Greco, (1970).

C. Rearrangement Involving Electron-Deficient Oxygen

The Baeyer–Villager reaction involves rearrangement of an oxygen-deficient species. Because of the high electronegativity of oxygen, we would not expect a positively charged electron-deficient oxygen, and the reaction is considered to proceed by a concerted mechanism. As in the rearrangement involving electron-deficient nitrogen, the configuration at the migrating carbon is maintained.

EXAMPLE 4.35. BAEYER–VILLAGER OXIDATION OF CYCLOPENTANONE TO δ-VALEROLACTONE

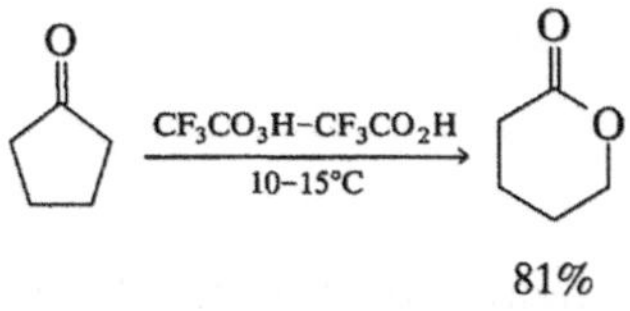

81%

Protonation of the ketone carbonyl is the first step. The generally accepted mechanism is as follows:

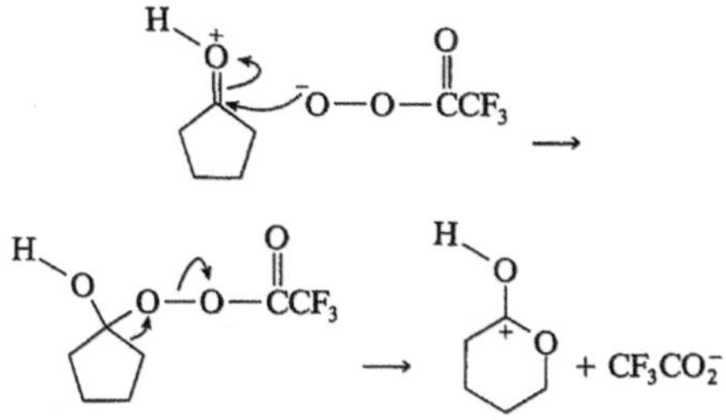

Deprotonation yields the lactone product.

PROBLEM 4.16

Baeyer–Villager oxidation of the heavily functionalized cyclopentanone yields a mixture of products. Use a step-by-step mechanism to decide whether this result is unexpected.

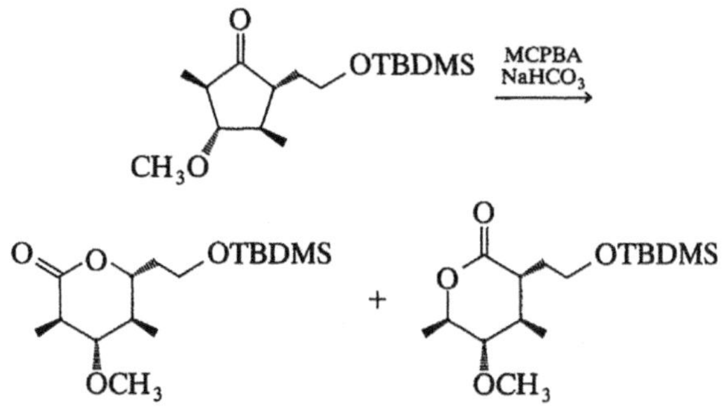

TBDMS = *t*-butyldimethylsilyl.
MCPBA = *m*-chloroperbenzoic acid.

Source: Clissoid et al., (1997).

PROBLEM 4.17

Write step-by-step mechanisms for the following transformations:

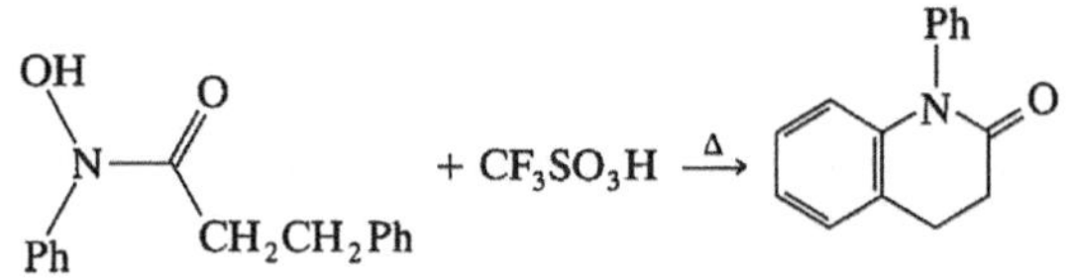

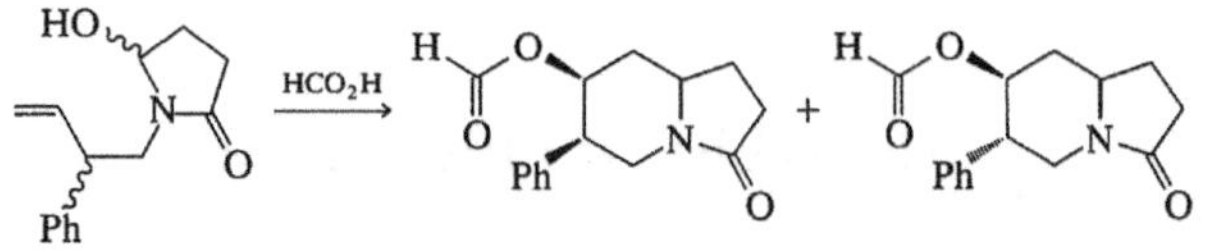

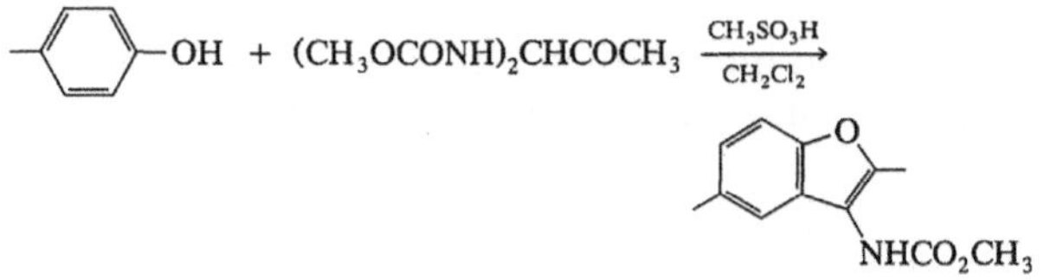

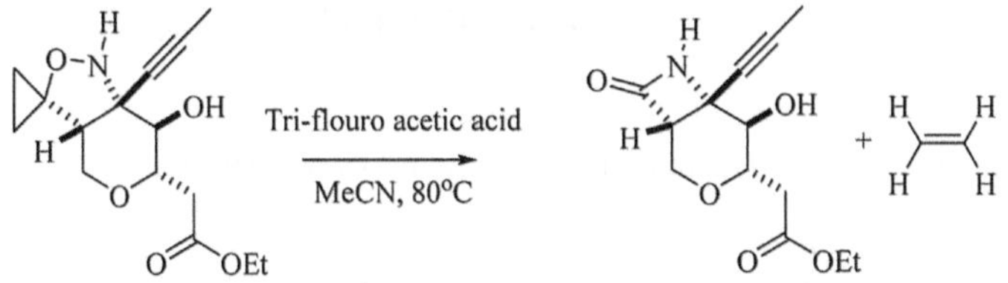

Sources: *Capozzi et al., (1984); Ent et al., (1986); Ben-Ishai and Denenmark, (1985); Hantawong et al., (1985); Kikugawa and Kawase, (1984); White et al., (1985); Diethelm and Carreira, (2013); Cordero et al., (2011); Cordero et al., (2000).*

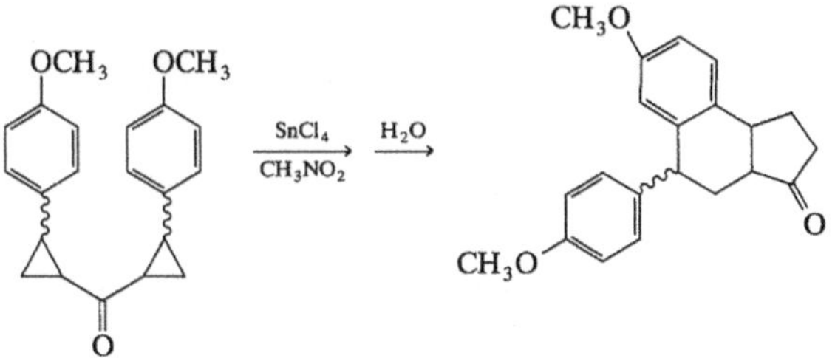

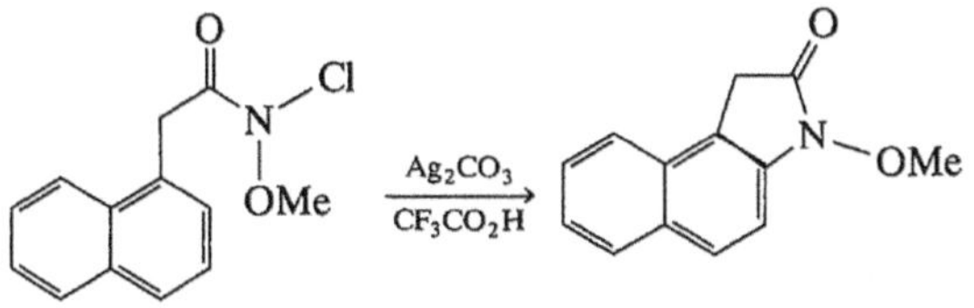

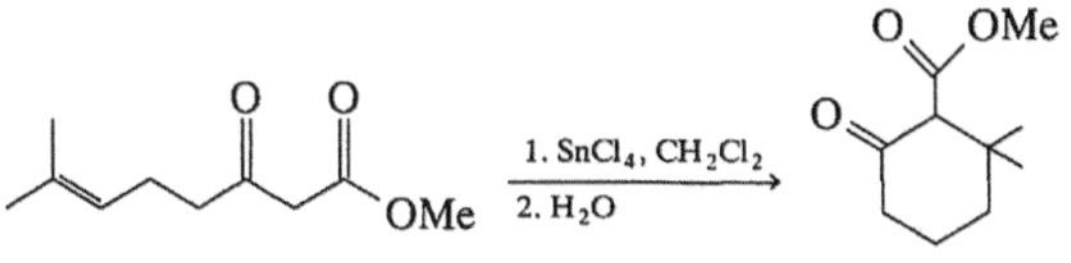

ANSWERS TO PROBLEMS

Problem 4.1

a. This is an ionization, followed by an alkyl shift and nucleophilic reaction of solvent with the electrophilic carbocation intermediate. In other words, it is an example of an S$_N$1 reaction with rearrangement. As noted in Table 3.1, the tosylate anion is an excellent leaving group.

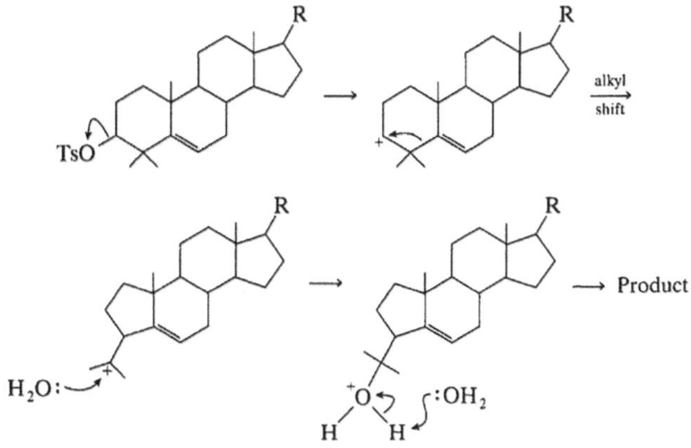

b. Because the hydroxyl group is lost and an alkene is formed, the reaction appears to be a dehydration involving a carbocation rearrangement. Numbering of the carbons in starting material and product is helpful. Numbering of the first five carbons in the product is straightforward because of the methyl carbons.

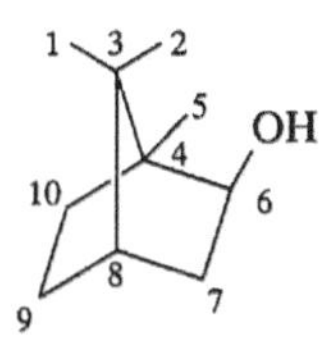

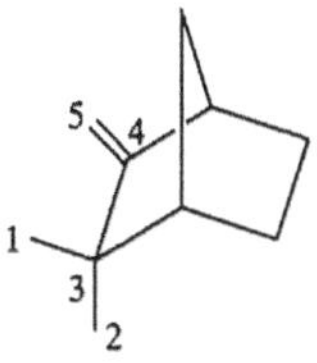

If we minimize bond reorganization, there are only two ways to number the remaining carbons of the product:

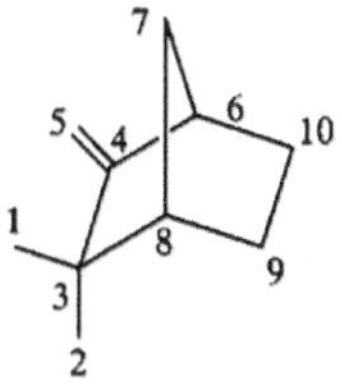

4-53 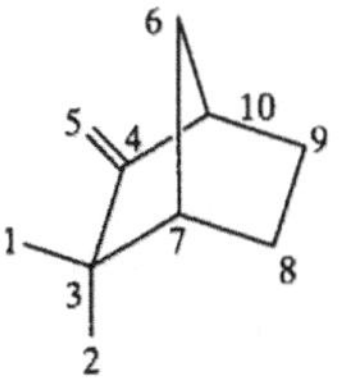   **4-54**

A major difference between **4-53** and the starting material is that, at carbon-10 (C-10), a bond to C-4 has been replaced by a bond to C-6. This is equivalent to an alkyl shift of C-10 from C-4 to C-6. In **4-54** at C-6, the bond to C-4 has been replaced by a bond to C-10. This is equivalent to an alkyl shift of C-6 from C-4 to C-10. However, the carbocation will be formed at C-6, and thus this carbon cannot be the one that shifts. Thus, **4-53** must be the product.

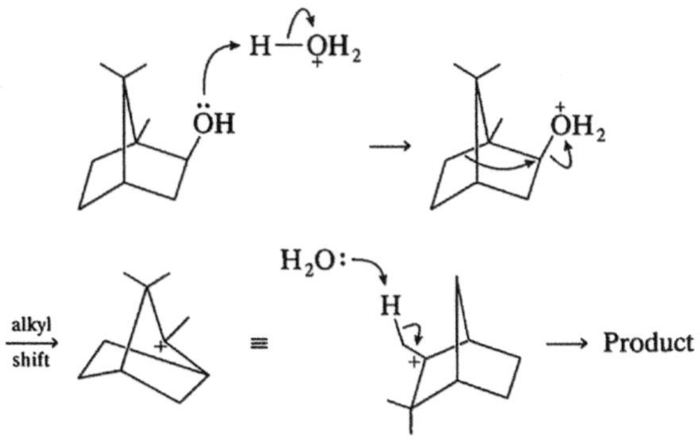

An alternative approach to solving this problem is to start by forming the secondary carbocation at C-6 and then to assess possible 1,2-alkyl shifts that could occur. If we recognize that the relationship of atoms C-1 through C-5 is unchanged in the product, two 1,2-alkyl shifts are possible.

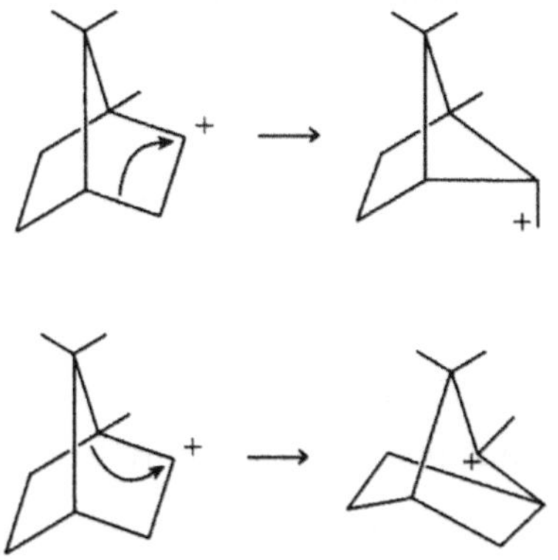

The first leads to formation of a primary carbocation. The instability of the primary carbocation is not so great that it is grounds for eliminating this step from consideration; however, the intermediate ion also contains a four-membered ring, a structural feature not found in the product. The second alkyl shift leads to the carbocation encountered in the mechanism shown previously.

For some reactions, it will be necessary to consider both alkyl and hydride shifts in order to account for the products formed.

Problem 4.2

The mechanism of this reaction involves the protonation of oxygen and loss of water to form a carbocation. Because the molecule is symmetrical, the hydroxyls are equivalent; thus, protonation of either oxygen leads to the product.

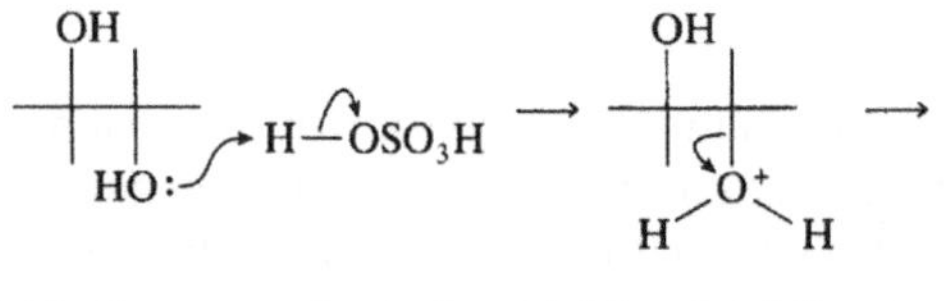

The subsequent alkyl shift gives a carbocation, **4-55**, which is resonance-stabilized by inter-action with the adjacent oxygen. Loss of a proton then leads to the product.

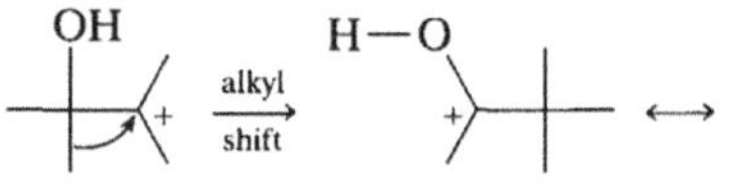

4-55-1

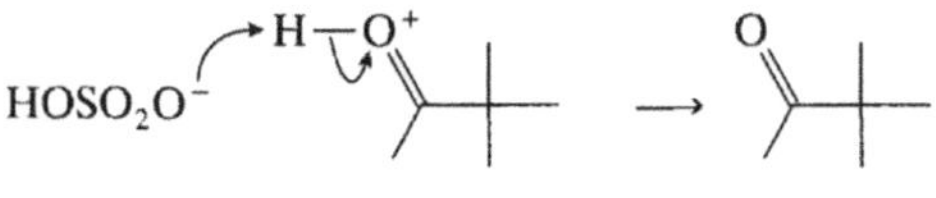

4-55-2

Problem 4.3

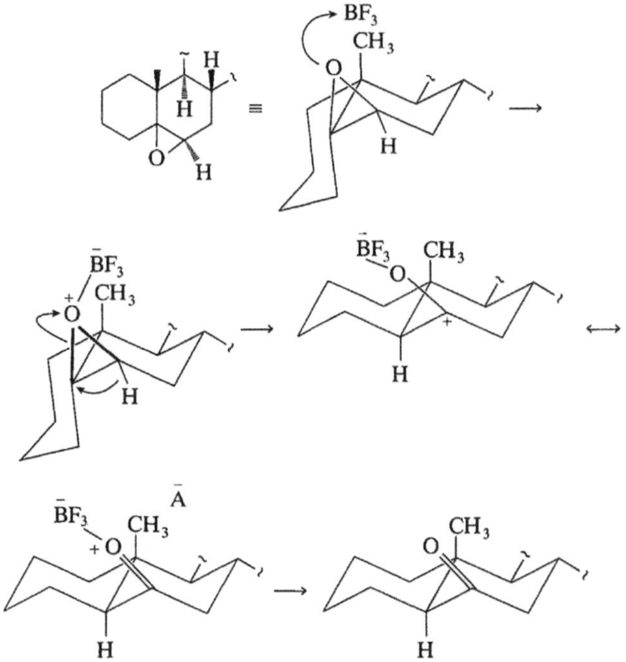

The Lewis acid BF_3 can complex with the epoxide oxygen to induce ring opening.

If migration of the hydride and ring opening of the epoxide are concerted, stereospecificity of the reaction is assured. Nucleophilic displacement by ethyl ether (derived from boron tri-fluoride etherate) removes boron trifluoride to generate the product ketone.

Problem 4.4

There are several possible mechanistic variations for this reaction. Because the starting material contains several basic oxygens and acid is one of the reagents, protonation is a reasonable first step. Dimethyl ether is more basic than acetophenone by only 0.5 pK_a units (Appendix C). An acetal would be expected to be less basic than an ether because of the electron-withdrawing effect of the second oxygen. Thus, protonation at both the acetal oxygen of the center ring and the carbonyl oxygen will be examined. (Protonation of the other acetal oxygen is unlikely to lead to product because the ring containing this oxygen is not rearranged in the final product.)

Mechanism 1

In this sequence, initial protonation of the acetal oxygen in the center ring is considered.

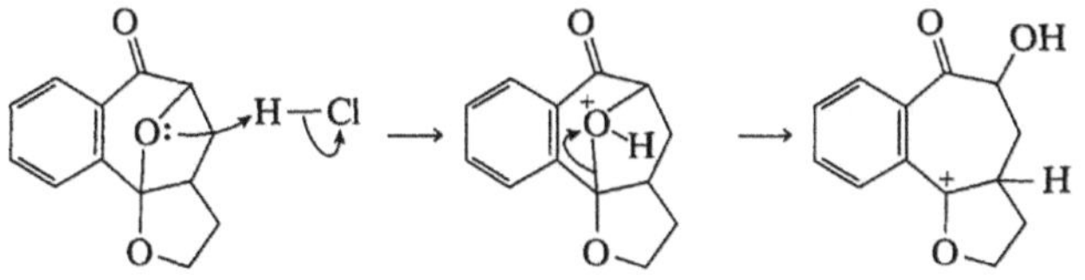

4-56

Ring opening of the protonated intermediate could produce either cation **4-56** or cation **4-57** depending upon which bond is broken. Carbocation **4-56** is stabilized by resonance with both the aromatic ring and the adjacent ether oxygen. Carbocation **4-57** is a hybrid of only two resonance forms, and **4-57-2** is quite unstable because the positive oxygen does not have an octet of electrons. Thus, the ring-opening reaction to give **4-56** is preferred.

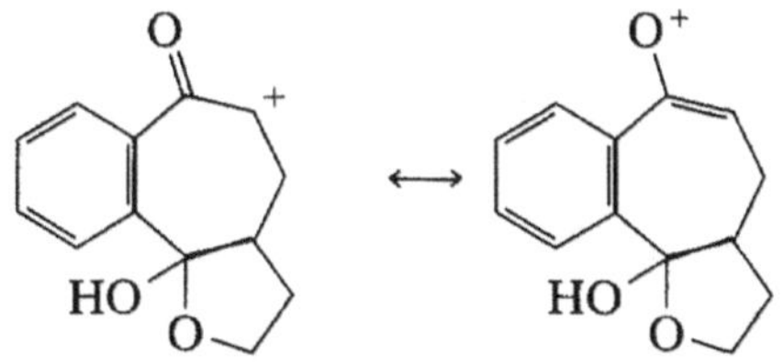

4-57-1   **4-57-2**

Reaction of **4-56** continues with elimination of a proton to form a double bond, followed by protonation of the carbonyl group.

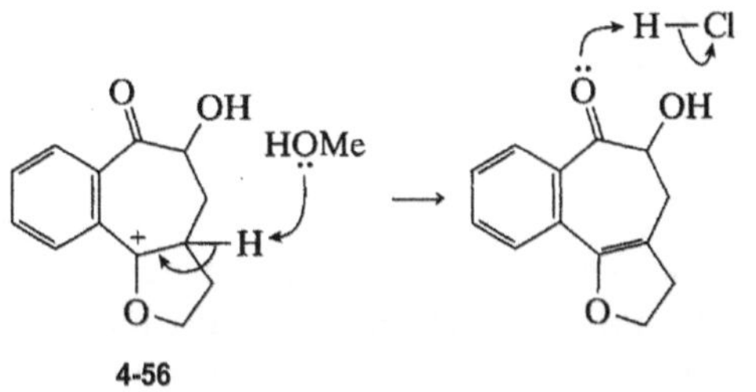

4-56

Then an alkyl shift occurs, followed by a deprotonation.

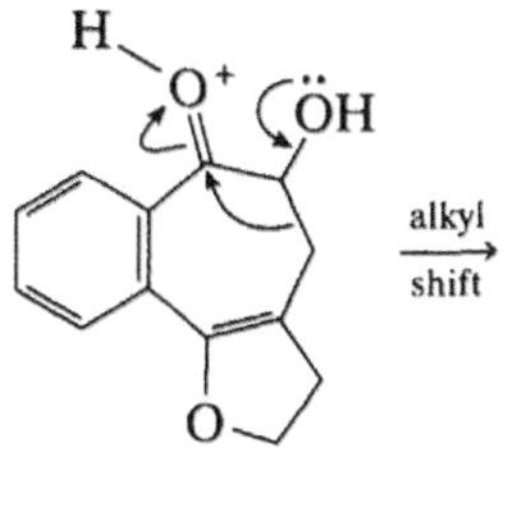

Finally, a dehydration takes place, forming the product.

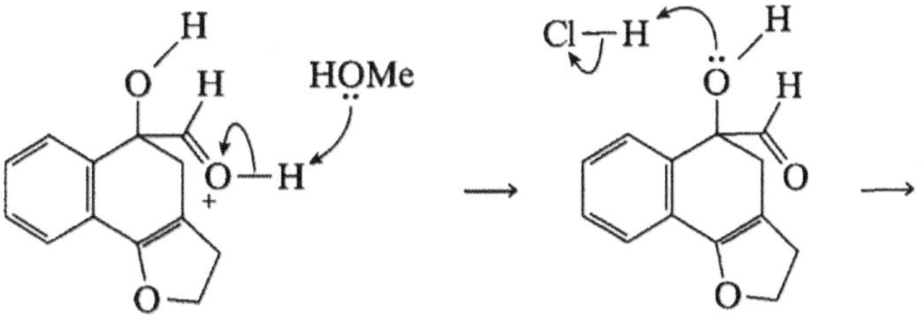

4-58

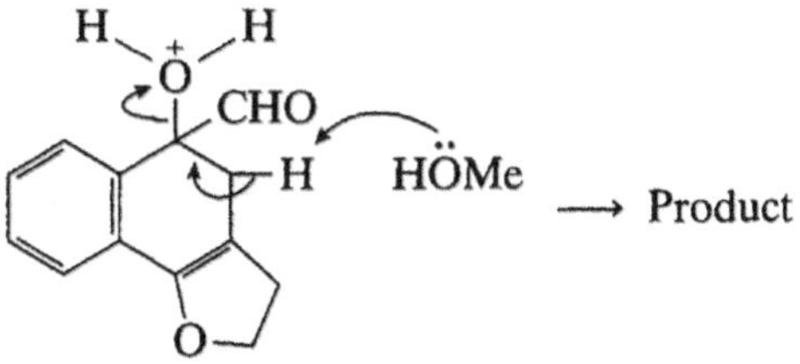

In this sequence (mechanism 1), methanol was consistently used as the base, because data in Appendix C show that it is a stronger base than chloride ion.

Mechanism 2

Mechanism 2 starts with protonation of the carbonyl oxygen to produce carbocation **4-59**.

An alkyl shift gives carbocation **4-60**, which is resonance-stabilized. This cation then undergoes bond cleavage to form **4-61**, which is stabilized in the same manner as **4-56** in mechanism 1. Loss of a proton gives **4-58**, which is also an intermediate in mechanism 1.

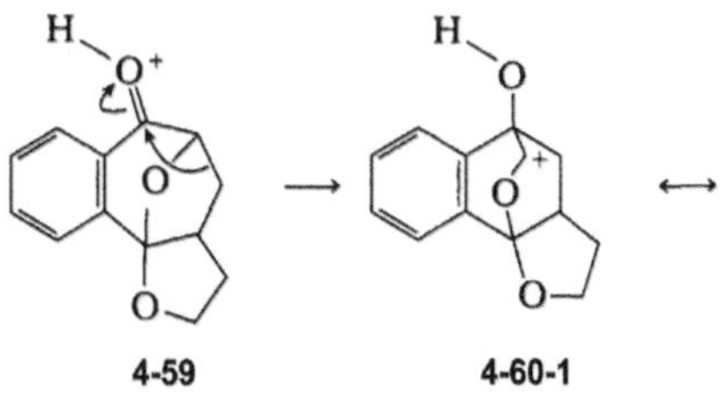

4-59   **4-60-1**

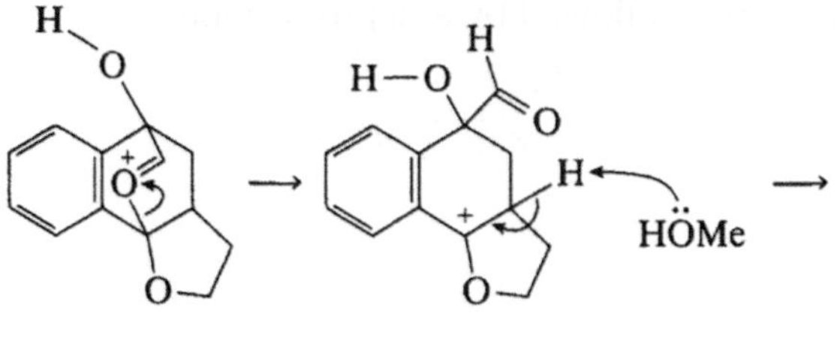

4-60-2   4-61

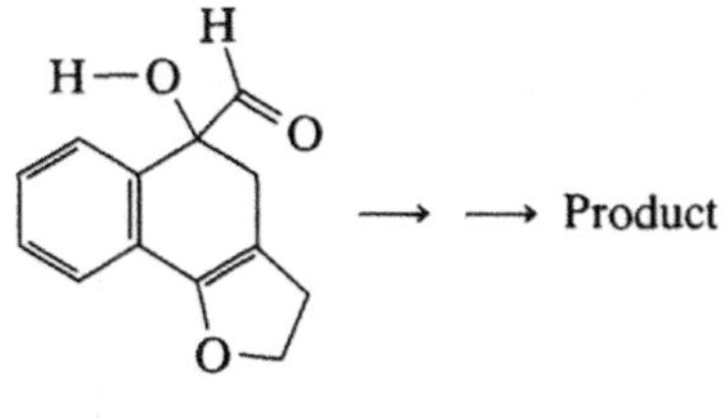

4-58

Mechanism 3

Intermediate **4-59** of mechanism 2 undergoes a hydride shift, instead of an alkyl shift, to give the following:

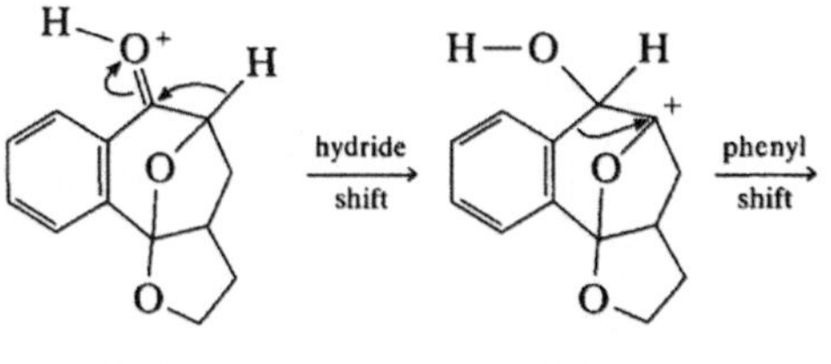

4-59   4-62

A phenyl shift gives the cation, **4-63**, which can lose a proton to give the aldehyde functional group. The acetal oxygen in the center ring can be protonated and open to give carbocation, **4-61**, which behaves as shown previously in mechanism 2.

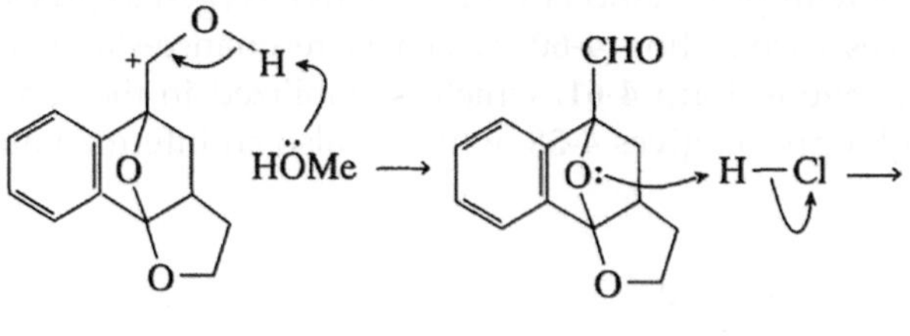

4-63

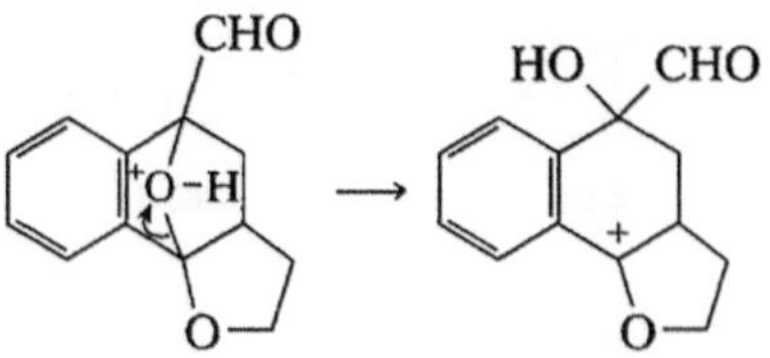

4-61

The problem with this mechanism is that the likelihood of forming a carbocation at the position shown in intermediate **4-62** is vanishingly small. Several factors combine to make this cation very unstable. First, although it is tertiary, the carbocation is at a bridgehead position and is constrained from assuming the planar geometry of an sp^2-hybridized carbocation. The bridgehead location also precludes resonance stabilization by the adjacent bridging oxygen, which nevertheless has a destabilizing inductive effect. Finally, there is a hydroxyl group on the carbon adjacent to the cationic center, and its inductive effect would further destabilize the carbocation in **4-62**. Because of these factors, this would not be considered a viable mechanism.

Problem 4.5

a. The generation of **4-23** could involve formation of a bromonium ion, **4-64-1**, similar to **4-14**, **4-15**, and **4-18**. This ion undergoes an aryl shift and then reacts with bromide to give the product. The bromonium ion must form on the *exo* side for the rearrangement to take place because the migrating aryl group must enter from the side opposite to the leaving bromo group.

4-64-1

4-23

4-23

In the last step of the preceding sequence, the bromide ion enters on the side opposite to the phenonium ion, in analogy to the stereochemistry of reaction of bromide ion with a bromonium ion.

There are several possible explanations for the different results at higher temperatures. One is that the bromonium ion forms on both the *exo* and *endo* sites. Reaction of the *exo* bromonium ion gives **4-23**, as before, and **4-25** is formed in an elimination reaction.

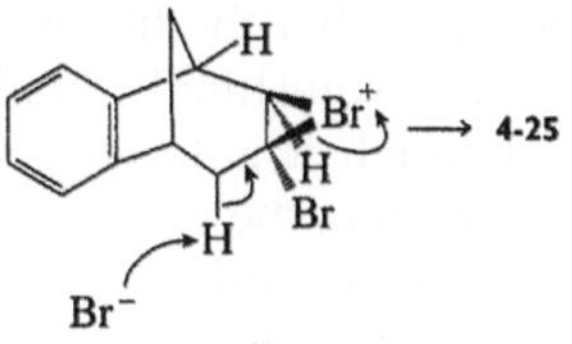

4-64-1

Reaction of the *endo* bromonium ion gives the isomers, **4-24**.

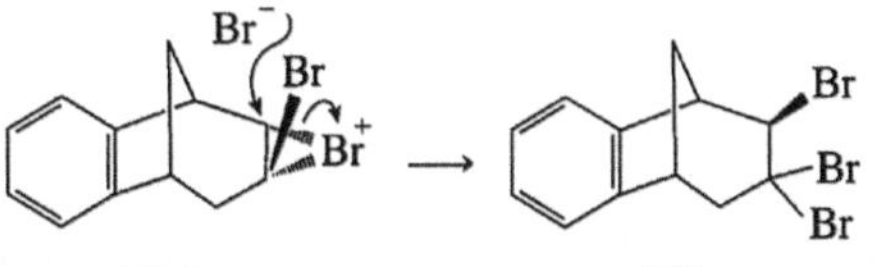

4-64-2   4-24

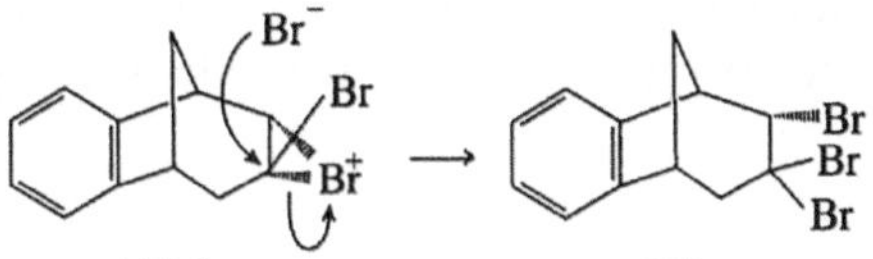

4-64-2   4-24

Another explanation is that at the higher temperature, only the *exo* bromonium ion **4-64-1** forms, but that under these conditions, **4-64-1** can rearrange to give **4-23**, undergo elimination to form **4-25**, or react with bromide ion approaching from the *endo* side to give the two isomers **4-24**.

b. The displacement of Br by sulfur in the intermediate bromonium ion, **4-65**, is similar to the ring closure in Problem 3.4a.

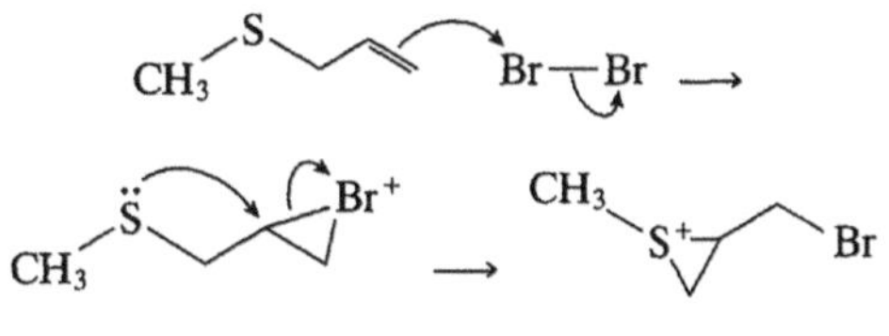

4-65   4-66

Representing the reaction as follows would be incorrect. The doubly charged structure **4-67** is not a resonance form of the starting material because the positions of the atoms and the bond angles of the sigma bonds are different. Also, because it is not stabilized by resonance, **4-67** would be very unstable.

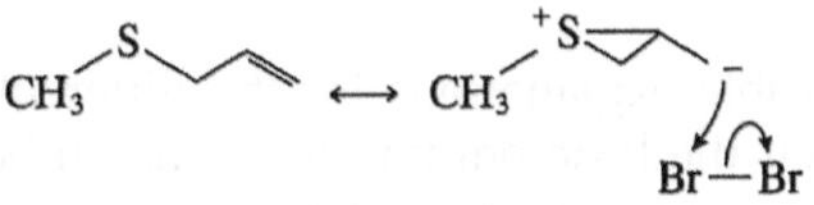

4-67

At low temperatures, the three-membered-ring episulfonium ion, **4-66**, which resembles a bromonium ion, reacts with bromide only at the primary carbon. The product of this S_N2 reaction is determined by the reaction rate, which is faster at the primary carbon than at the secondary carbon. This is a reaction in which product formation is rate-controlled.

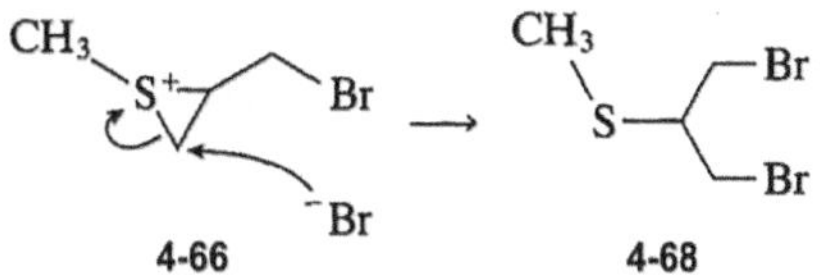

At elevated temperatures, **4-68** is in equilibrium with **4-66**. Under these conditions, **4-66** also reacts at the secondary carbon to give **4-69**. Because **4-69** is more stable than **4-68**, its reverse reaction to **4-66** is slower, and **4-69** accumulates. Thus, at higher temperatures, product formation is equilibrium-controlled.

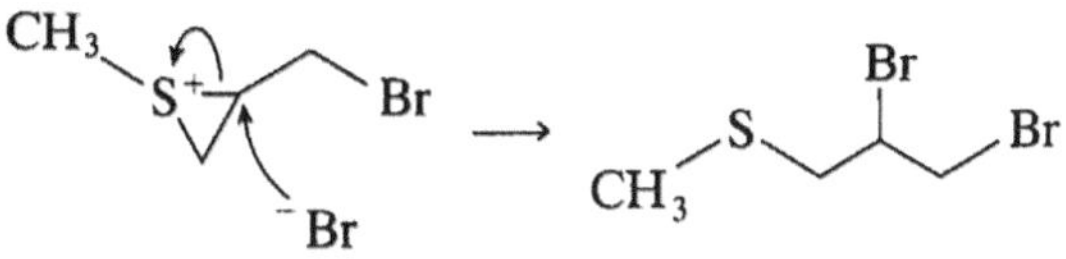

Problem 4.6

a. This reaction involves loss of the diethylamino group of the amide and ring closure with the aldehyde to form the hydroxylactone product. Because the timing of the ring closure is open to question, two possible mechanistic sequences are given.

Mechanism 1

In this mechanism, the second ring is formed early. The carbonyl oxygen of the amide is protonated, and then the oxygen of the aldehyde adds to this protonated functional group to form a new ring.

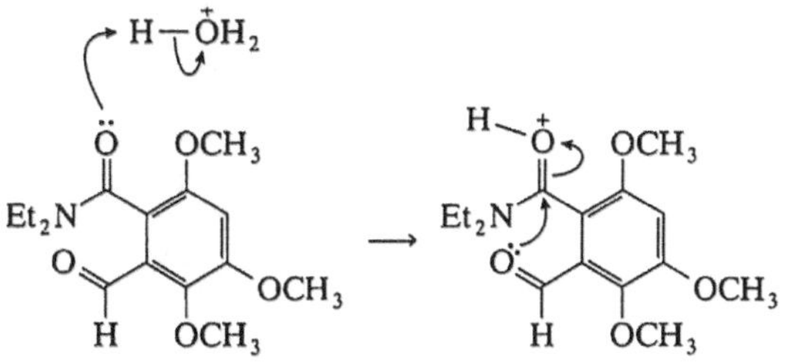

The four steps, after participation of the aldehyde oxygen, are (1) nucleophilic reaction of water with the most electrophilic carbon; (2) loss of a proton; (3) protonation of the nitrogen; and (4) loss of diethylamine. Writing the steps in this order avoids formation of more than one positive charge in any intermediate.

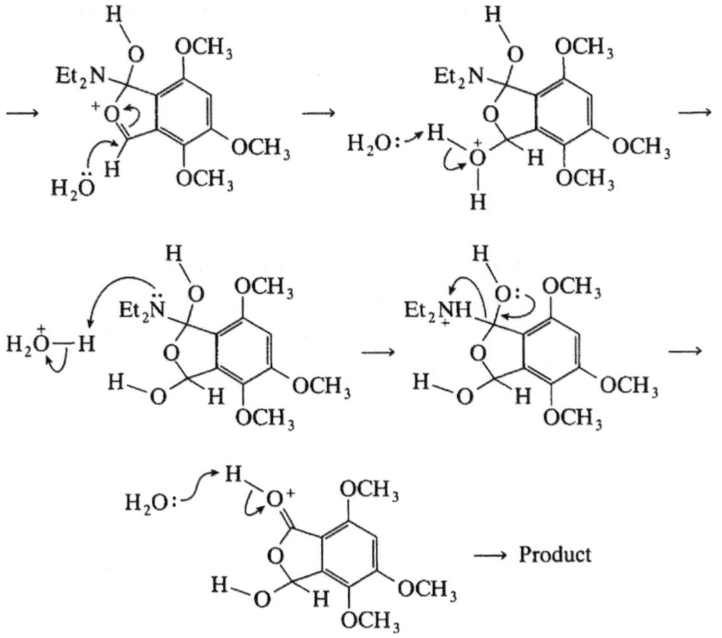

4-70

Mechanism 2

Another possibility is hydrolysis of the amide to a carboxylic acid followed by closure with the aldehyde. The hydrolysis of the amide mimics the steps in Example 4.16.

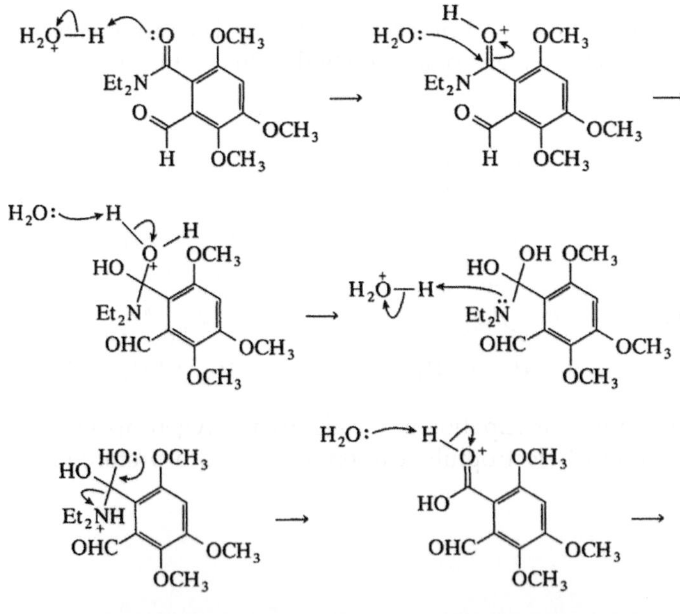

Because it is easier to protonate an aldehyde than a carboxylic acid (compare ben-zoic acid and benzaldehyde in Appendix C), the ring closure would be written best as protonation of the aldehyde oxygen followed by nucleophilic reaction of the carbox-ylic acid carbonyl oxygen. The carbonyl oxygen acts as the nucleophile because a resonance-stabilized cation is produced. If the hydroxyl oxygen acts as a nucleophile, the cation is not resonance-stabilized.

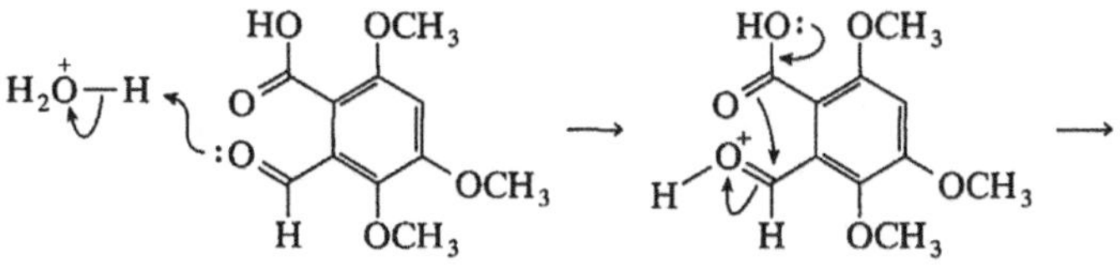

The last step is the same as that for mechanism 1. Mechanism 1 seems better than mech-anism 2 because no tetrahedral intermediate is formed at the amide carbon prior to cycli-zation. The amide position is sterically congested because it is flanked by *ortho* substituents. A tetrahedral intermediate (sp^3-hybridized carbon) adds to the congestion and might be of such high energy that its formation would be unlikely.

b. Both the ester and nitrile undergo hydrolysis. (There are many acids present. The strongest acids, H_3O^+ and H_2SO_4, can be used interchangeably.) Normally, esters are hydrolyzed more rapidly than nitriles, so that the ester will be hydrolyzed first.

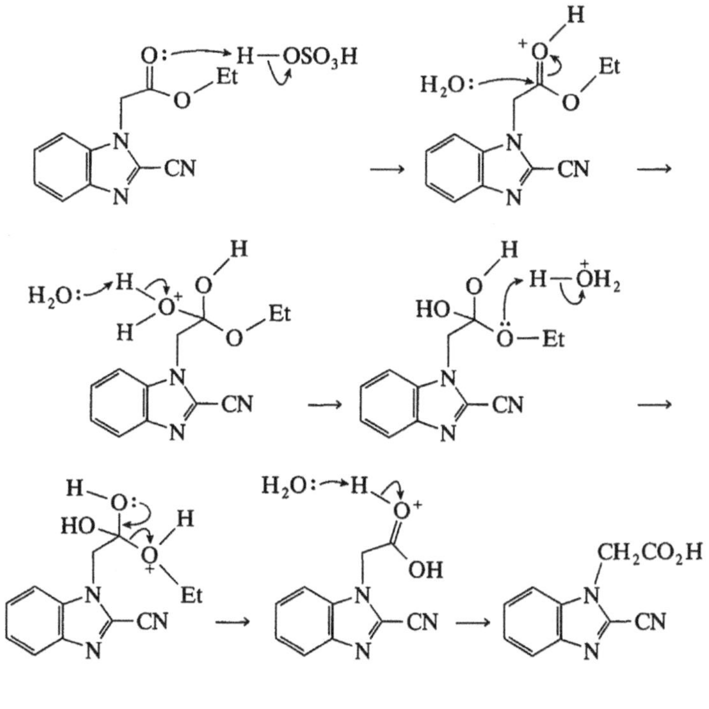

Hydrolysis of the nitrile produces the imino form of the amide, which readily tautomerizes by the usual mechanism to form the amide.

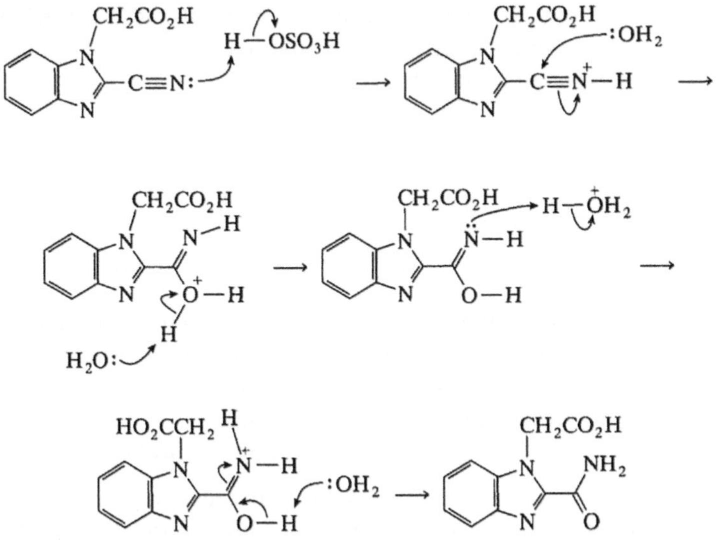

If reaction conditions are controlled, the hydrolysis of nitriles can be stopped at the amide stage. In this case, once the amide is produced, the nitrogen will react as a nucleophile with the protonated carbonyl of the acid. Because a six-membered-ring transition state is involved, this intramolecular reaction is very favorable.

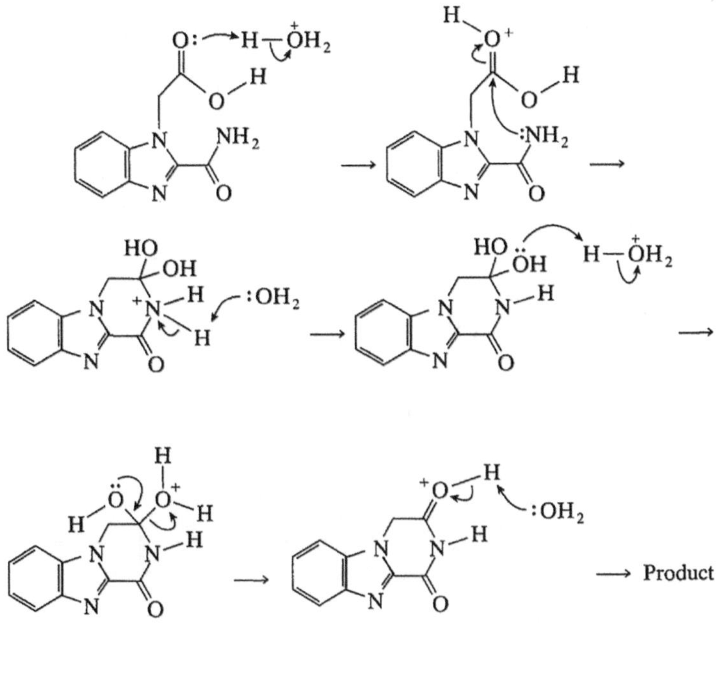

Reaction of the nitrile, acting as a nucleophile, with the electrophilic protonated ester is unlikely as a ring-forming reaction. Because of the linearity of the nitrile group, it is difficult for the lone pair of electrons on nitrogen to reach close enough to overlap the p-orbital on the carbon of the carbonyl group. (See the boxed groups on the last structure in the sequence.) Also, because the lone pair of electrons on the nitrile occupies an sp-orbital, these electrons would be much less nucleophilic than the lone pair occupying a p-orbital on the amide nitrogen.

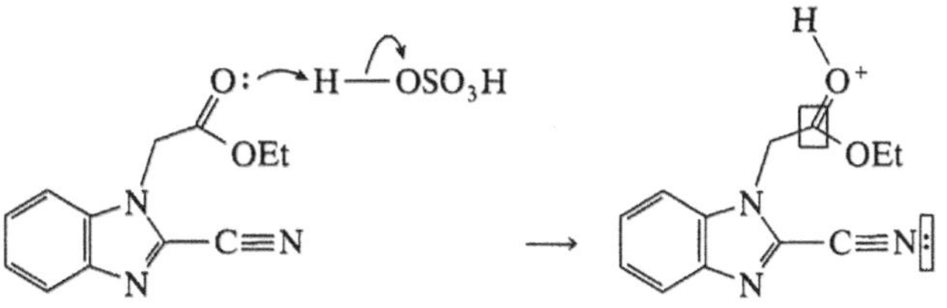

Problem 4.7

None of the cations produced by protonation at the carbons of acrolein is as stable as the cation produced by protonation at oxygen. We will consider the possibilities.

Protonation of the aldehyde carbonyl carbon gives cation **4-71**. This is a very unstable intermediate. It is not stabilized by resonance, and the positive oxygen lacks an octet of electrons. Notice that the double bond and the positive oxygen are not conjugated because they are separated by an sp^3-hybridized carbon.

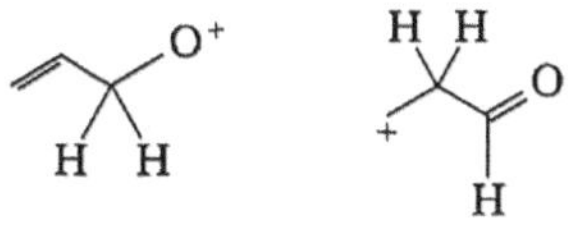

The intermediate produced by protonation at the α carbon also gives a very unstable primary cation, **4-72**, which is not stabilized by resonance. Protonation at the β carbon gives a delocalized cation, **4-73**, but the resonance form (**4-73-2**) is especially unstable because the oxygen has a positive charge and does not have an octet of electrons.

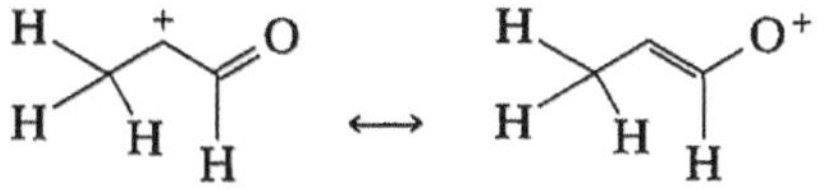

Problem 4.8

HCl is a catalyst for the transformation, so it must be regenerated at some point in the mechanism.

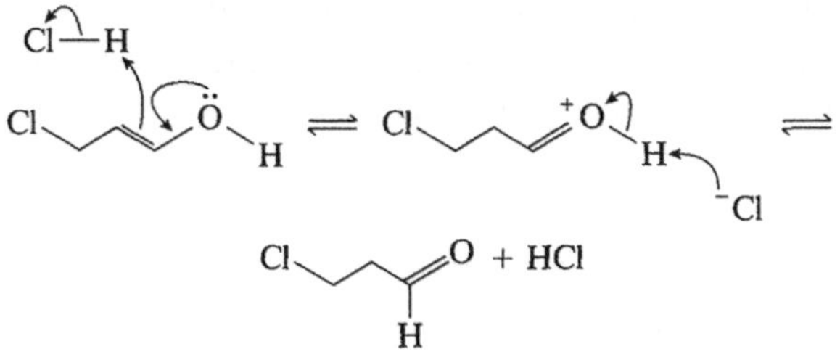

Problem 4.9

By analogy to Example 4.19 and Problem 4.7, protonation at the carbonyl group, rather than at the double bond, of the starting material will give a more stable cation. (That is not to say that all reactions occur via a mechanism involving formation of the most stable carbocation. But if the most stable cation leads to product, that is the one to use.)

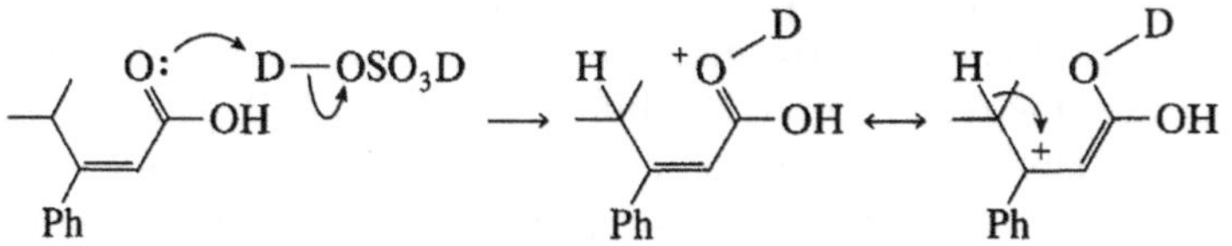

The carbocation formed initially, **4-74**, can undergo a hydride shift to give a tertiary carbocation. The oxygen of the OH or OD group reacts as a nucleophile with the electrophilic carbocation. In order for the ring to form, the oxygen that reacts must be the one on the same side of the double bond as the carbocation. In **4-75**, this happens to be the OD oxygen, but because these two oxygens can equilibrate rapidly under the reaction conditions, the oxygen *cis* to the alkyl group could just as easily be protonated as deuterated.

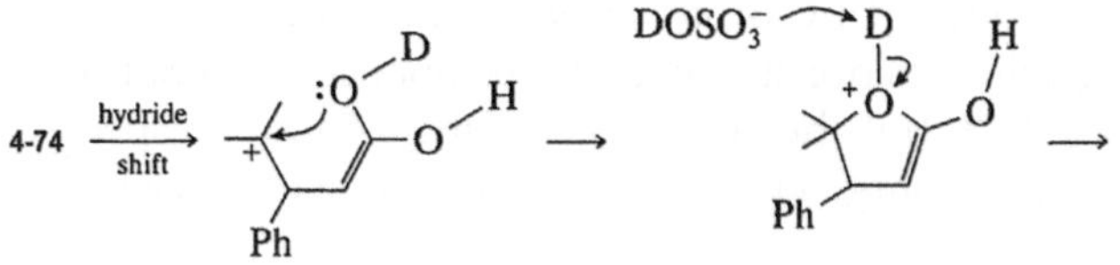

Intermediate **4-76** undergoes acid-catalyzed tautomerization to give the product.

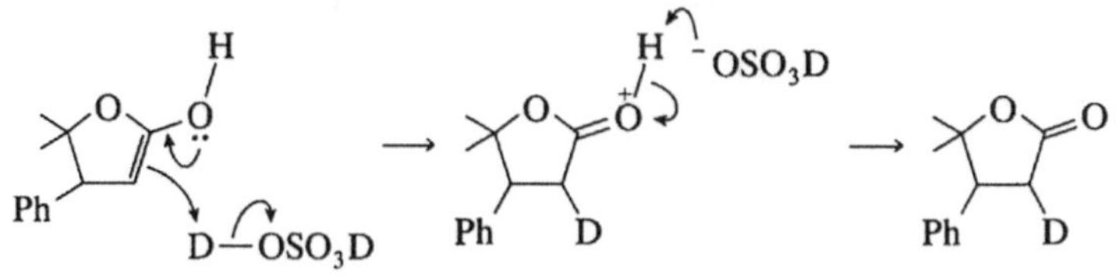

There is another cation that could undergo the same rearrangement as this one. This cation could be formed by direct protonation of the double bond with Markovnikov regiospecificity:

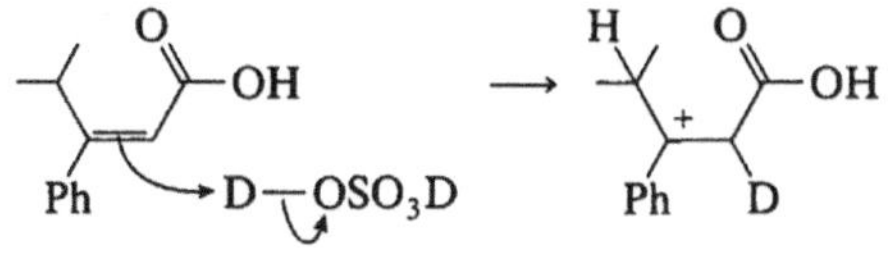

The remaining steps in this mechanism would be very similar to those of the first. However, the first mechanism is better because the carbocation formed initially is more highly stabilized by resonance and, thus, should form faster.

Problem 4.10

a. The intermediate, formed by reaction of the electrophile at the *para* position, has three resonance forms. Form **4-77** is especially stable because the positive charge is next to an alkyl group, which stabilizes positive charge by the inductive effect and by its polarizability.

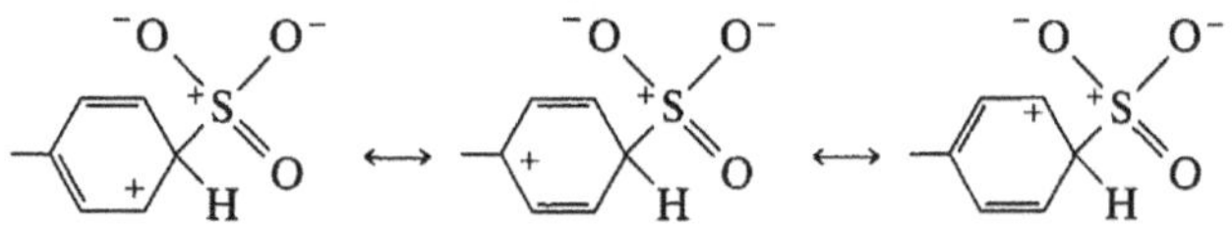

4-77

Reaction of the electrophile at the *ortho* position would give an intermediate of essentially the same stability. However, reaction of the electrophile at the *meta* position gives an intermediate, **4-78**, in which the positive character cannot be located at the alkyl group position. Thus, this intermediate is not as stable and is not formed as rapidly as the intermediates from reaction at either the *ortho* or *para* position.

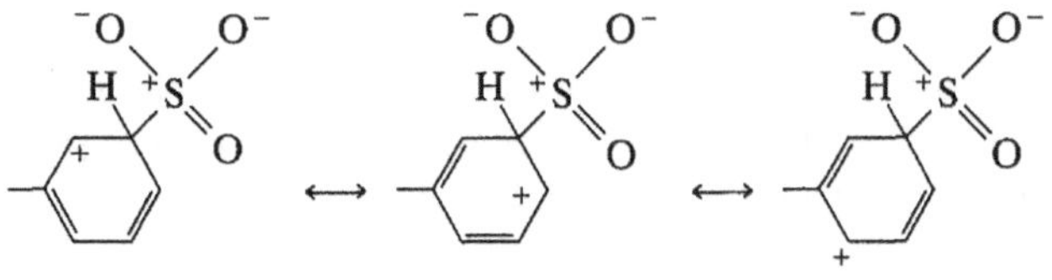

4-78

b. The intermediate, formed by reaction of a nitronium ion at the *meta* position, does not have positive charge on the carbon bearing the positively charged sulfur of the sulfonic acid group.

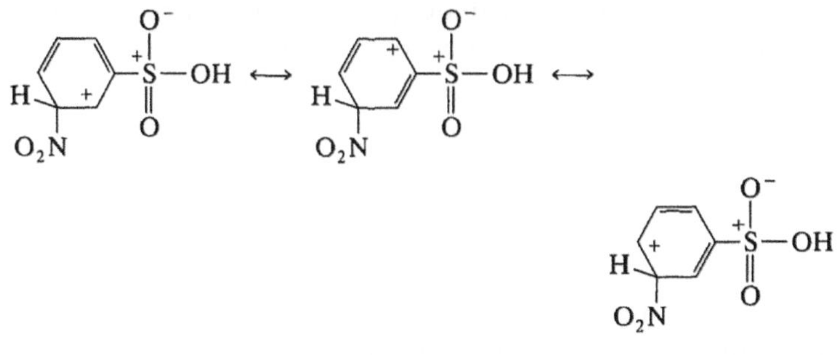

The intermediates, formed by electrophilic reaction at the *ortho* or *para* positions, are not as stable as the intermediate formed by reaction at the *meta* position because, for *ortho* and *para* attack, one of the three resonance forms has positive charge on the carbon bearing the positively charged sulfur of the sulfonic acid group. For example, for the intermediate resulting from reaction at the *para* position, we can draw three resonance forms. Of these, **4-79-2** is particularly unstable because of the destabilizing effect of two centers of positive charge in close proximity. Because the intermediate involved in *meta* substitution lacks this unfavorable interaction, it is more stable and *meta* substitution is favored.

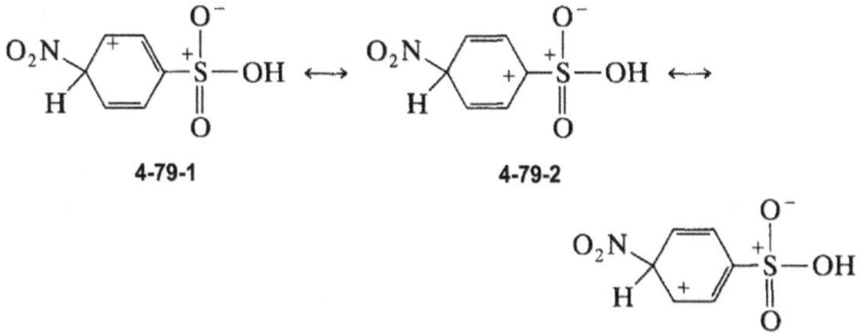

4-79-1   **4-79-2**

4-79-3

For reaction at the *ortho* position, the resonance forms of the resulting intermediate are destabilized in the same way as those that result from reaction at the *para* position.

c. The intermediate, formed by reaction of the electrophile at the *para* position, has a resonance form, **4-80-2**, with positive character at the substituent position. This means that the unshared pair of electrons on the nitrogen of the acetamido group can overlap with the adjacent positive charge, as shown in **4-80-4**.

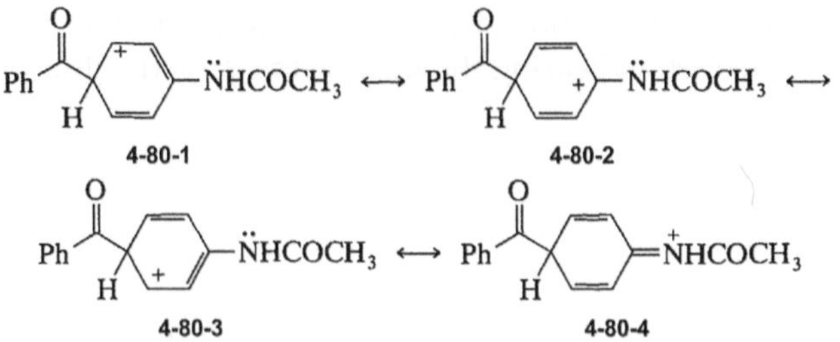

4-80-1   **4-80-2**

4-80-3   **4-80-4**

This extra delocalization of the positive charge adds to the stability of the intermediate. On the other hand, if the electrophile reacts at the *meta* position, the positive charge cannot be placed on the nitrogen.

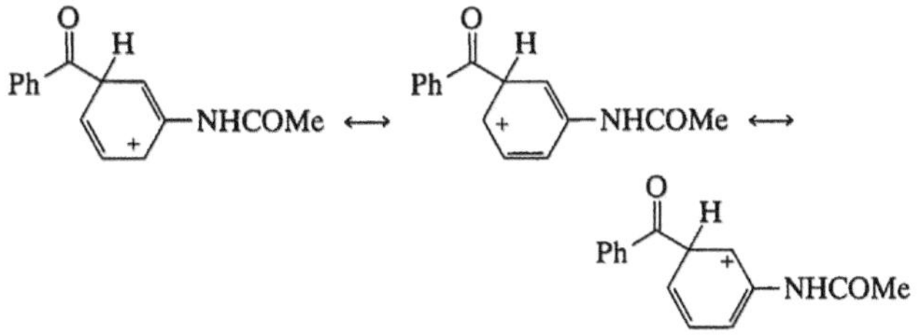

Problem 4.11

a. The mechanism in sulfuric acid might occur by the steps typical for a dienone–phenol rearrangement. Protonation of the A-ring carbonyl gives a more highly resonance-stabilized intermediate than protonation of the B-ring carbonyl.

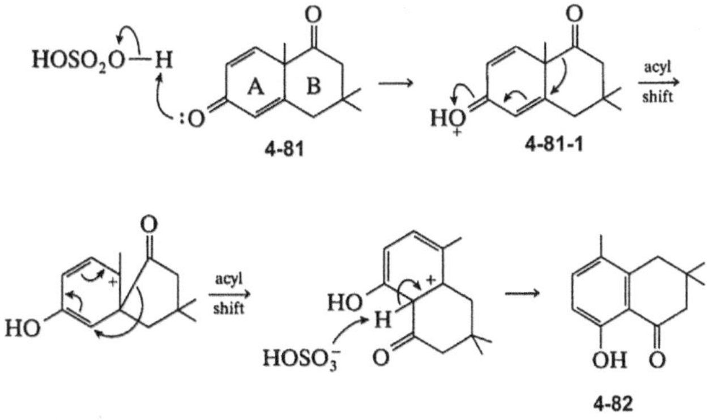

Formation of the product, **4-83**, formed in dilute HCl requires cleavage of the B ring. A mechanism involving the formation of the hydrate of the keto group in the B ring, followed by ring opening, is shown next. (Example 2.11 shows a slightly different possibility.)

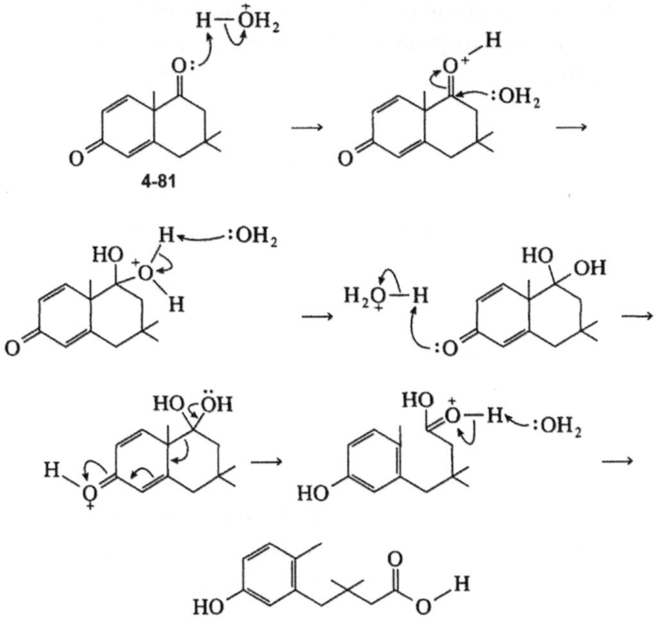

4-81

4-83

The mechanism in dilute acid suggests an alternative route to **4-82** in the stronger acid, sulfuric acid, in which initial cleavage of the B ring occurs. In this pathway, the ring of the intermediate cation **4-81-1** can open to give the acylium ion, **4-84**.

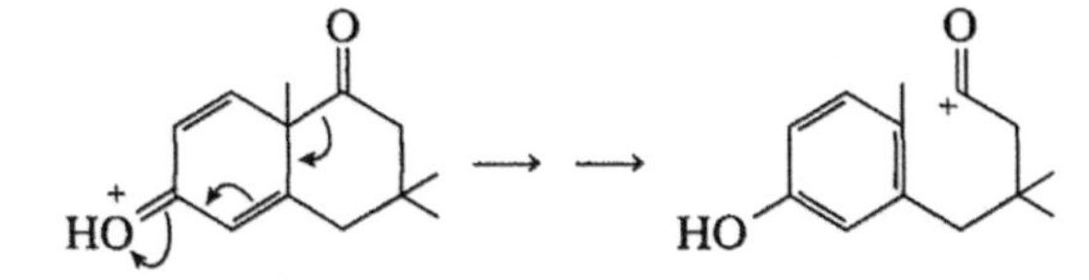

4-81-1   **4-84**

Rewriting **4-84**, with the acylium ion in proximity to the position *ortho* to the hydroxyl group clarifies the intramolecular acylation reaction that forms a six-membered ring.

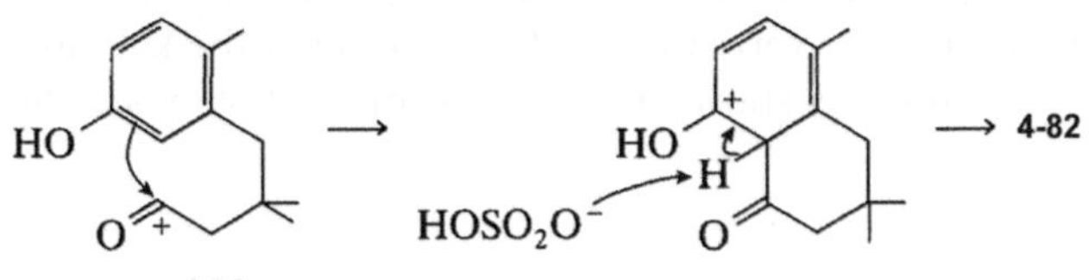

4-84

The acylium ion would have a longer lifetime in concentrated sulfuric acid than in aqueous hydrochloric acid because the concentration of water in the former acid is extremely low. In aqueous hydrochloric acid, the starting ketones are more likely to be in equilibrium with their hydrates, and the intermediate acylium ion might never form. If it does form, it will react so rapidly with the nucleophilic oxygen of water that the electrophilic aromatic substitution cannot occur.

Experimental data in the cited paper support the opening of the B ring in concentrated as well as dilute acid. It was found that on treatment with sulfuric acid, **4-83** and **4-81** both react to give **4-82** at the same rate. However, the alternative mechanism involving successive acyl shifts appears to occur in trifluoroacetic acid. In this medium, **4-81** rearranges to **4-82**, but **4-83** does not give **4-82**. This means that **4-83** cannot be an intermediate in the reaction. Consequently, the mechanism involving opening of ring B is ruled out, and the acyl shift mechanism is a reasonable alternative.

b. The reaction proceeds by a cleavage–recombination reaction.

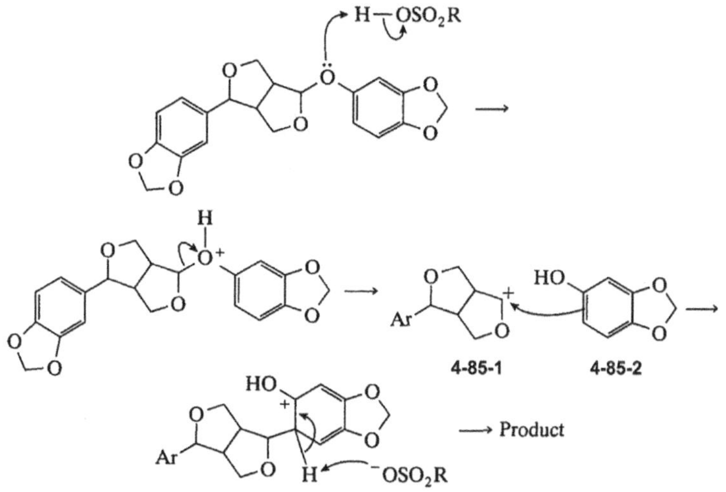

Support for the formation of **4-85-1** and **4-85-2** comes from a cross-over experiment reported in the paper. In the presence of *m*-cresol, **4-86** was obtained.

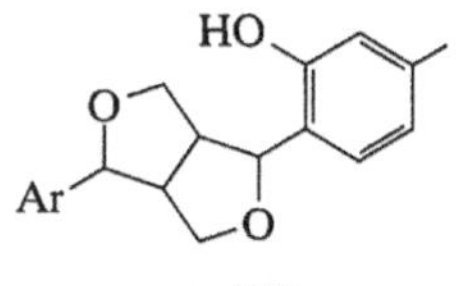

4-86

This product is formed by capture of the intermediate carbocation **4-85-1** by the added *m*-cresol. Loss of a proton gives **4-86**. Furthermore, as the concentration of *m*-cresol increases, the amount of **4-86** increases. This supports the idea that the *m*-cresol is competing with **4-85-2** for **4-85-1**. The term cross-over experiment refers to the fact that **4-85-1** reacts with an external reagent rather than **4-85-2**, its co-cleavage product.

The occurrence of cross-over rules out the following mechanism:

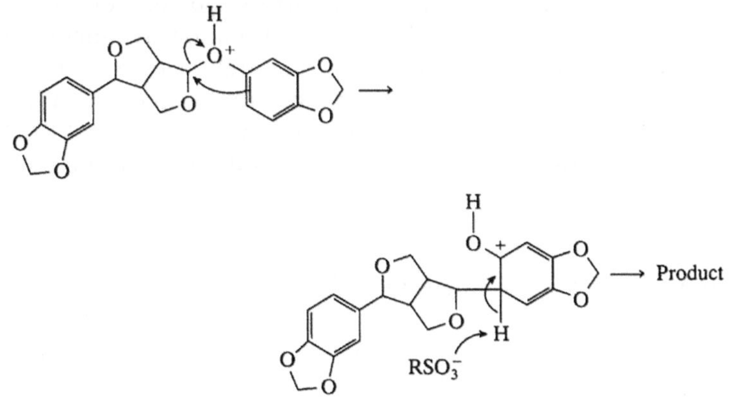

c. The transformation can be described as occurring in two parts. In the first part, the acetal is hydrolyzed to the aldehyde. In this sequence, HCl or H_3O^+ acts as a catalytic proton source. The acetal is protonated, and then one acetal oxygen kicks out the adjacent oxygen 4-87 and the hydroxyl hemi hemiacetal 4-87–1 is formed leading to the aldehyde 4-88.

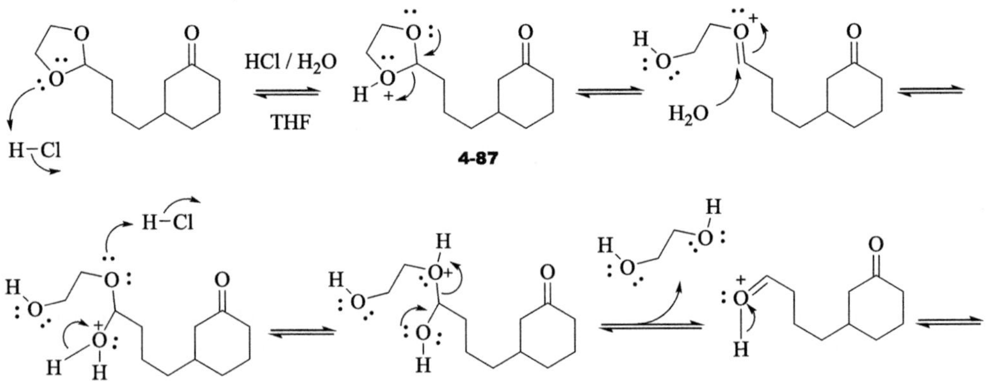

In the second part of the transformation, the ketone enolizes 4-88-1 and the aldehyde is activated 4-89 by another proton leading to the aldol condensation (cyclization). The acidic proton alpha to the ketone eliminates with the activated hydroxyl in a net dehydration step 4-90 leading to the final bicyclo product 4-91.

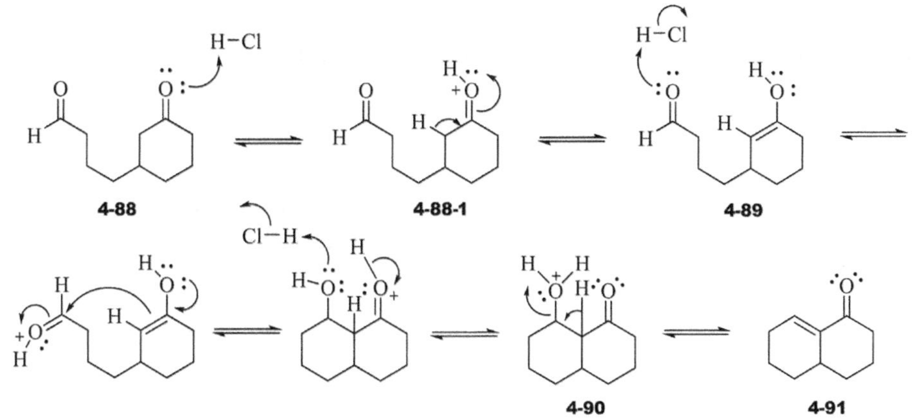

Problem 4.12

The intermediate anion picks up a proton: removal of the proton on the sp^3-hybridized carbon α to the carbonyl group gives the phenolate ion. The phenolate loses chloride ion to give **4-92**, which undergoes addition of hydroxide at the carbon at the *ortho* position. Loss of the remaining chloride and removal of a proton gives the product phenolate.

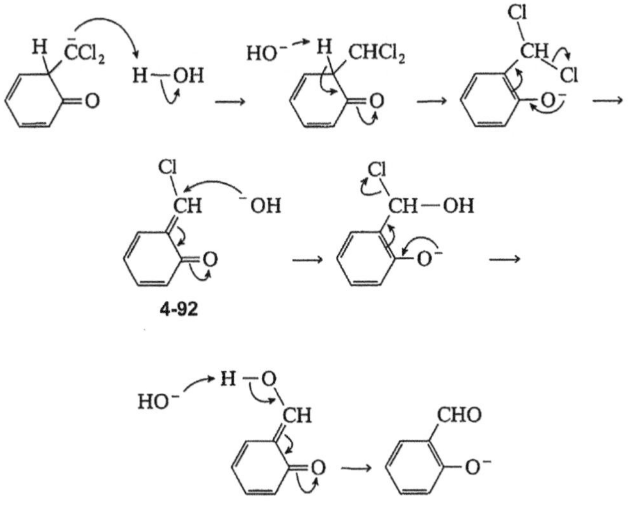

Acidic workup will give the phenol itself.

Problem 4.13

Dibromocarbene can be produced from bromoform and base:

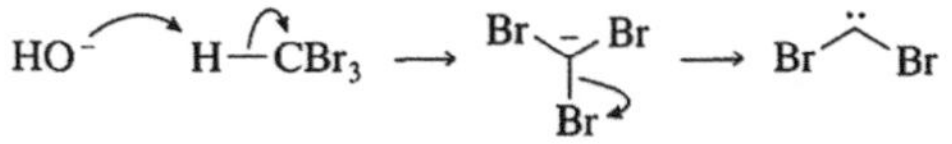

Phenolate, formed in the basic medium, reacts as a nucleophile with the electrophilic carbene. The resulting anion is protonated by water to give the product.

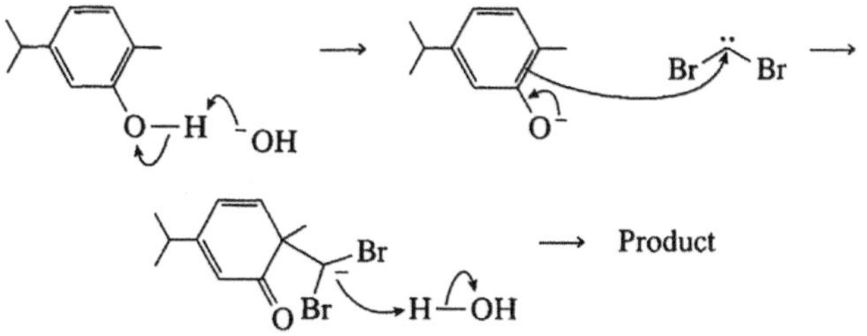

Problem 3.4b gives a subsequent reaction of this product and its literature reference.

An alternative mechanism with the ring acting as an electrophile and the carbene as a nucleophile would be incorrect. The dihalocarbenes are quite electrophilic. Also, because the phenol readily forms a salt with sodium hydroxide, the aromatic ring would not be electrophilic.

Problem 4.14

This reaction is a fragmentation, which often accompanies a Beckmann rearrangement. Reaction of the hydroxyl group with phosphorus pentachloride converts it into a better leaving group. In **4-93**, instead of an alkyl group migration, the ring bond on the side opposite to the leaving group cleaves, producing a ring-opened cation, **4-94**. This mimics the stereochemical requirement of the Beckmann rearrangement.

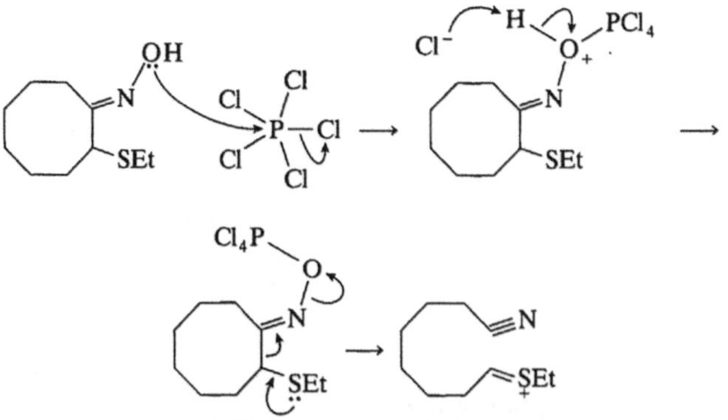

4-93   4-94

The electrophilic cation, **4-94**, reacts with a nucleophile at carbon to give the neutral derivative, **4-95**.

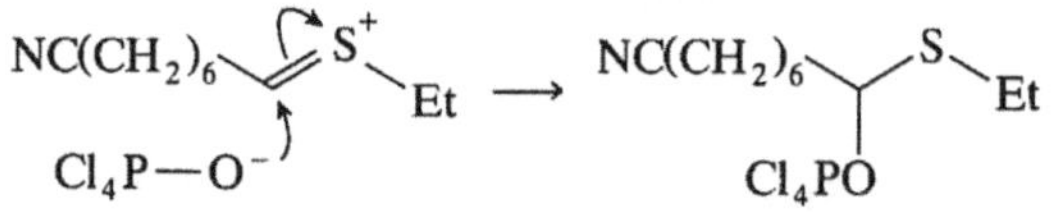

4-94   **4-95**

The product, **4-95**, will undergo hydrolysis readily when the reaction is worked up in water. This hydrolysis is acid-catalyzed because the phosphorus compounds in the reaction mixture are acidic.

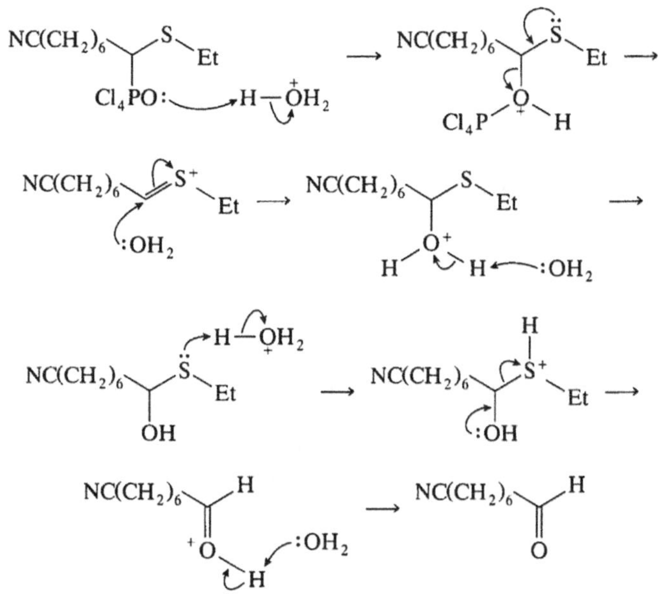

One student wrote an alternative mechanism that involved ring expansion of **4-93** rather than fragmentation.

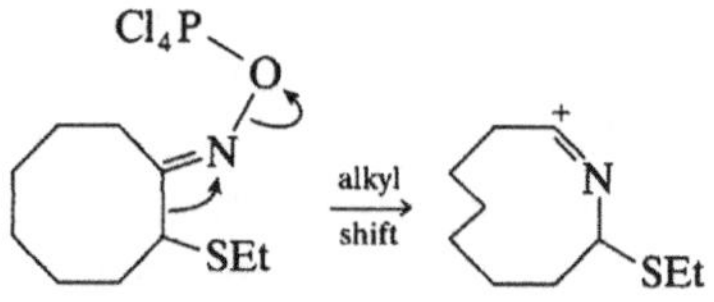

4-93

This is a good step because it is the normal Beckmann rearrangement. However, to get from this intermediate to the open-chain product took a fairly large number of "creative" steps. Application of Occam's razor suggests that the initial fragmentation occurs early in the mechanism.

Problem 4.15

The initial step is a protonation, followed by nucleophilic reaction of azide ion.

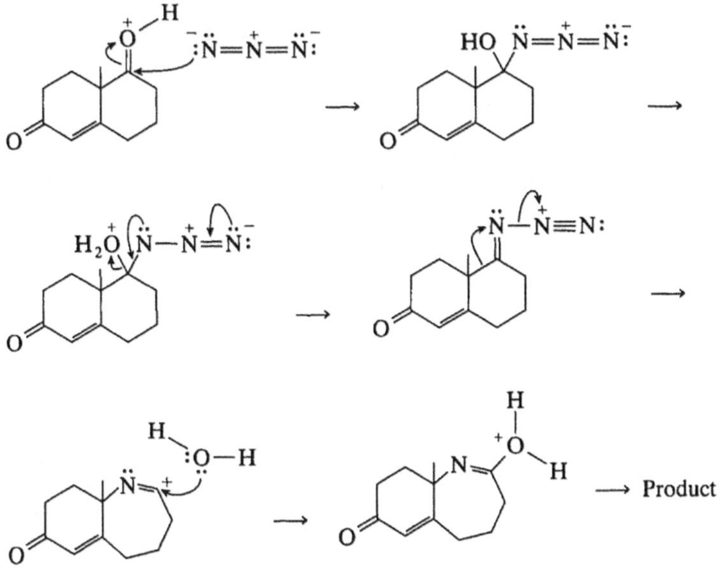

The reaction is a Schmidt reaction and its mechanism closely resembles that of the Beckman and Wolff rearrangements. The final step is deprotonation, followed by tautomerization to the lactam. Note that the alternate lactam formed by migration of the secondary carbon was not found.

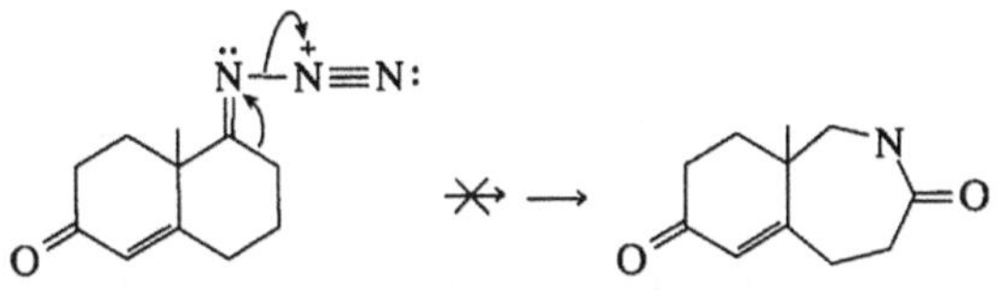

One might expect that reaction could also occur at the α, β-unsaturated keto group, and, depending on the reaction conditions, several different products were formed by reaction at this site. The course of reaction when there are two similar functional groups present in a molecule frequently is difficult to predict. The authors of the article cited offer rationalizations for the formation of the various products of the reaction.

Problem 4.16

In the presence of NaHCO$_3$, deprotonation of the peracid is the first step, followed by nucleophilic attack on the carbonyl group:

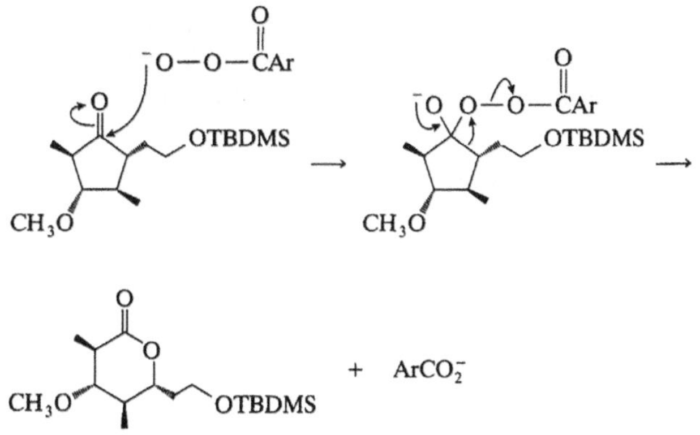

The minor product is formed by migration of the other alkyl substituent. Note that both products are formed with retention of configuration at the migrating carbon.

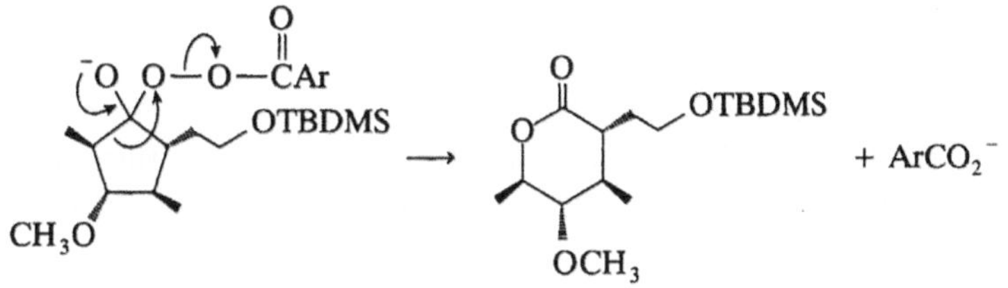

There is no readily apparent reason why one group should migrate in preference to the other. The rearrangement of a closely related structure that differs only in the side-chain substitution is regiospecific:

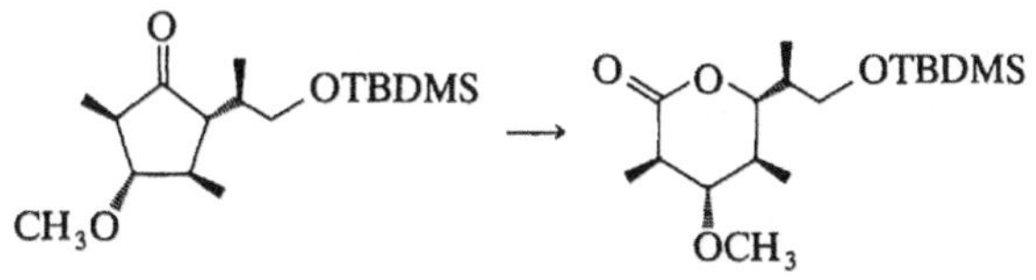

sole product

The reaction is buffered by NaHCO$_3$ to prevent hydrolysis of the TBDMS-protecting group by the acid produced in the course of the reaction.

Problem 4.17

a. Trifluoromethanesulfonic acid is a very strong acid, and the most basic atom in the amide is the carbonyl oxygen. Protonation of the carbonyl group is a likely process, but it is not one that leads to the product. That is, protonation of the hydroxyl oxygen is on the pathway to product, whereas protonation of the carbonyl oxygen is not. The nitrogen bearing the protonated hydroxyl group can act as an electrophile, and the phenyl ring, situated to form a favorable six-membered ring, can act as a nucleophile.

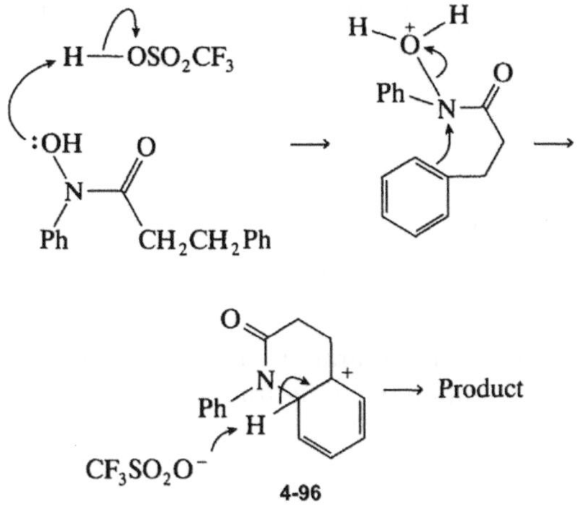

4-96

In an alternate mechanism, water leaves the protonated starting material prior to involvement of the nucleophile (the aromatic ring). This gives a nitrenium ion, which acts as the electrophile. Cyclization leads to the same intermediate carbocation, **4-96**, as before.

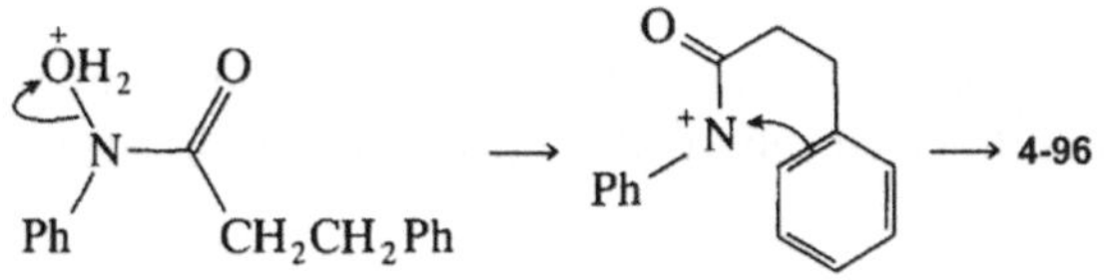

The following reaction in trifluoromethanesulfonic acid provides support for initial formation of positive nitrogen (Endo et al., (1977)).

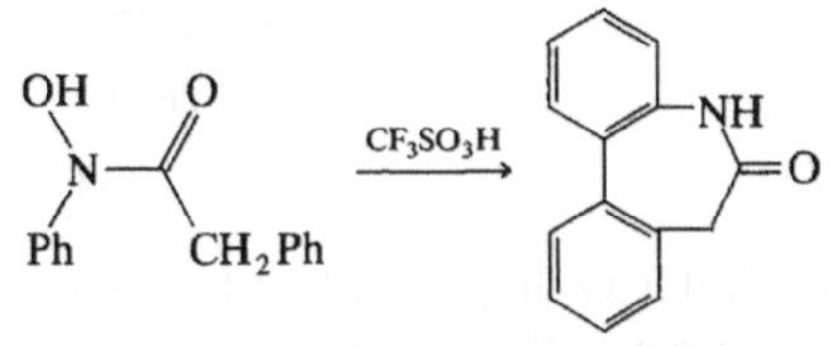

76%

This reaction is an electrophilic aromatic substitution in which one benzene ring acts as the electrophile and the other benzene ring acts as the nucleophile. In order to develop substantial positive charge in the electrophilic ring (a driving force for the reaction), resonance must occur. This requires prior loss of water. The intermediate nitrenium ion and the direction of the substitution reaction are shown:

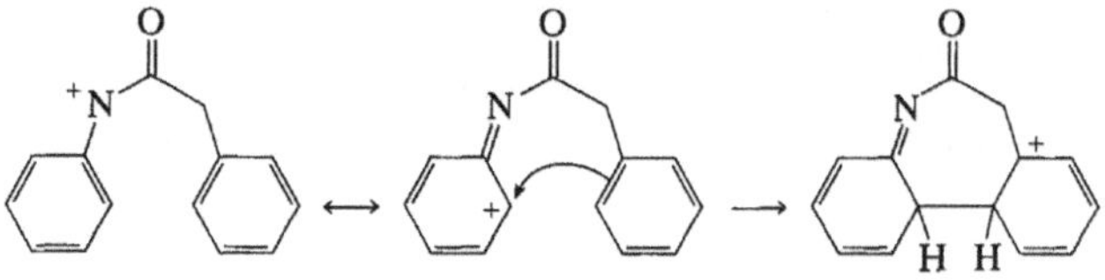

Formation of the product requires loss of a proton to regenerate a neutral compound and tautomerization to regenerate the other benzene ring.

b.

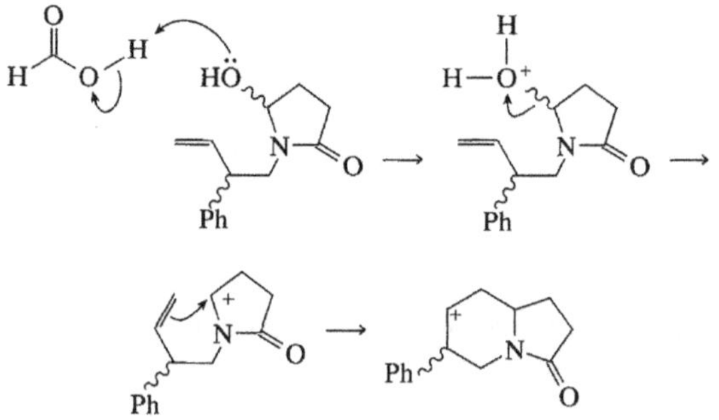

Nucleophilic reaction of formic acid at the electrophilic carbocation can be either *cis* or *trans* to the phenyl group in the bicyclic intermediate and leads to the two products.

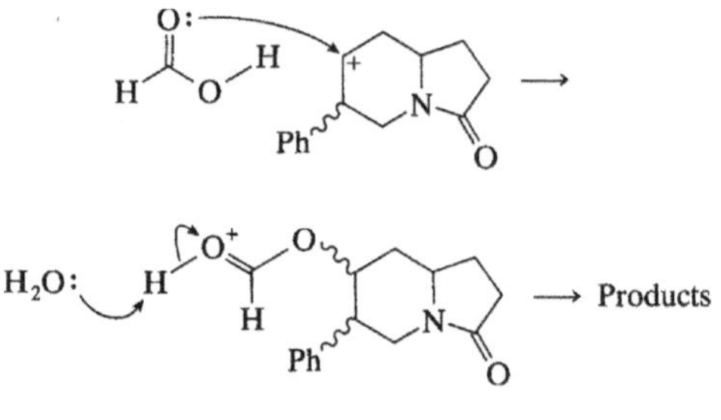

c. There are several different mechanistic sequences that might lead to the product.

Mechanism 1

The two equivalent amide carbonyl oxygens are the most basic atoms in the starting materials. Protonation of one of these oxygens gives a carbocation, **4-97**, which is stabilized by delocalization onto oxygen and nitrogen as well as onto carbon.

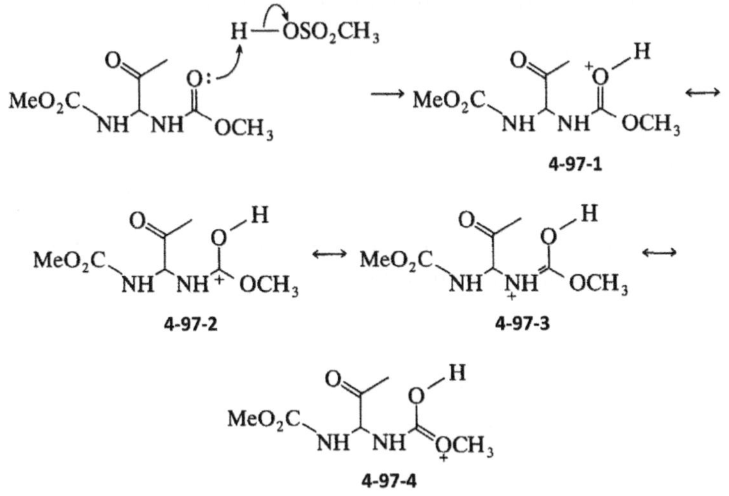

4-97-1

4-97-2   **4-97-3**

4-97-4

The intermediate, **4-97**, can decompose to give a new electrophile, with which the nucleophilic phenol then reacts.

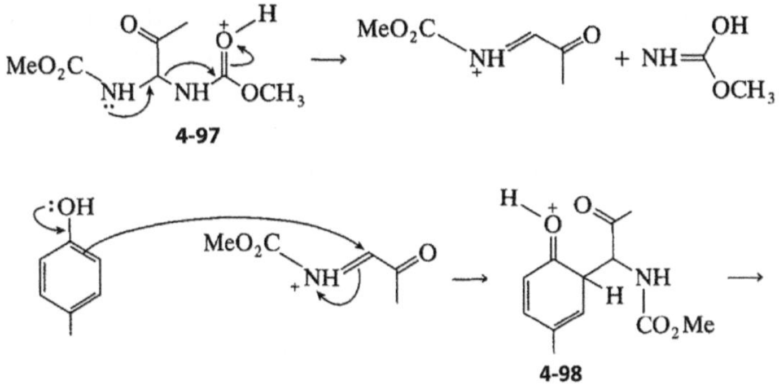

4-97

4-98

An alternative mechanism to **4-98** might be protonation of one of the nitrogens in the starting material, followed by nucleophilic reaction of the phenol with the carbon bearing the protonated leaving group (S$_N$2 reaction). However, this carbon is quite sterically congested, so that this is not an attractive option.

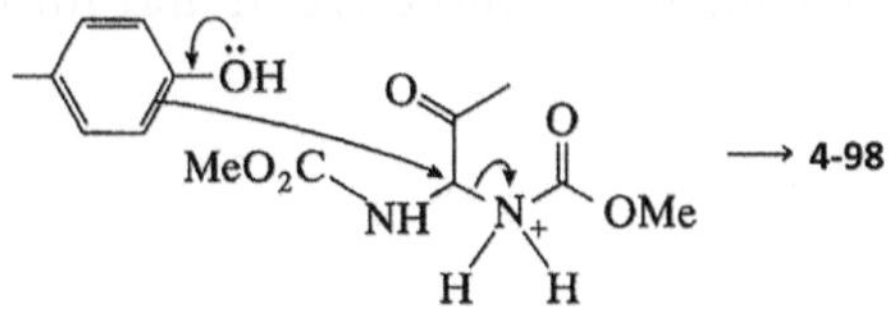

After aromatization of **4-98** and protonation of the keto carbonyl, cyclization can occur:

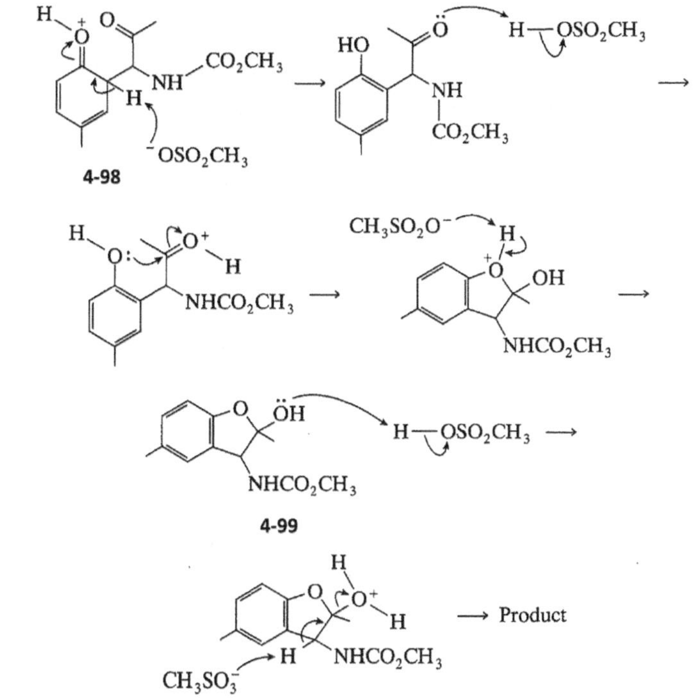

4-98

4-99

→ Product

Mechanism 2

If the keto group in the starting material were protonated, the phenolic oxygen could react as a nucleophile at the carbonyl carbon.

Following proton loss from the positive oxygen and protonation of one of the amide nitrogens, intermediate **4-100** is formed. This intermediate can lose methyl carbamate to form **4-101**, in which the ring acts as a nucleophile.

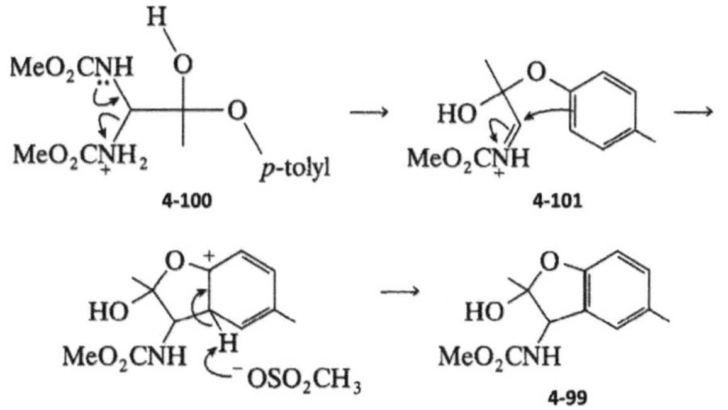

4-100

4-101

4-99

Loss of a proton gives **4-99**, which undergoes dehydration as in mechanism 1.

d. The tin intermediate can be written as either the product of a nucleophilic displacement on tin, or as an expanded orbital on tin, as in **4-102**.

Mechanism 1

Ring opening occurs readily in the cyclopropyl-substituted cation, **4-102**, to give **4-103**. The driving forces are release of the strain energy of the three-membered ring and stabilization of the resulting cation by the *p*-methoxyphenyl group.

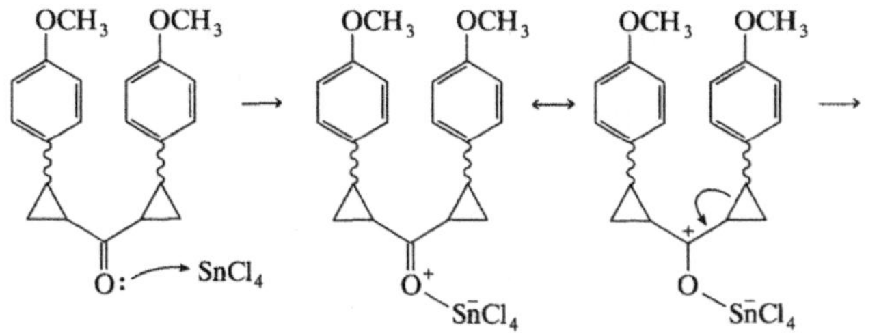

4-102

Intermediate **4-103** must equilibrate with the *cis* isomer, **4-104**, which undergoes further ring opening and cyclization to **4-105**, with the same driving forces as the reaction of **4-102** to give **4-103** (Ar = aryl).

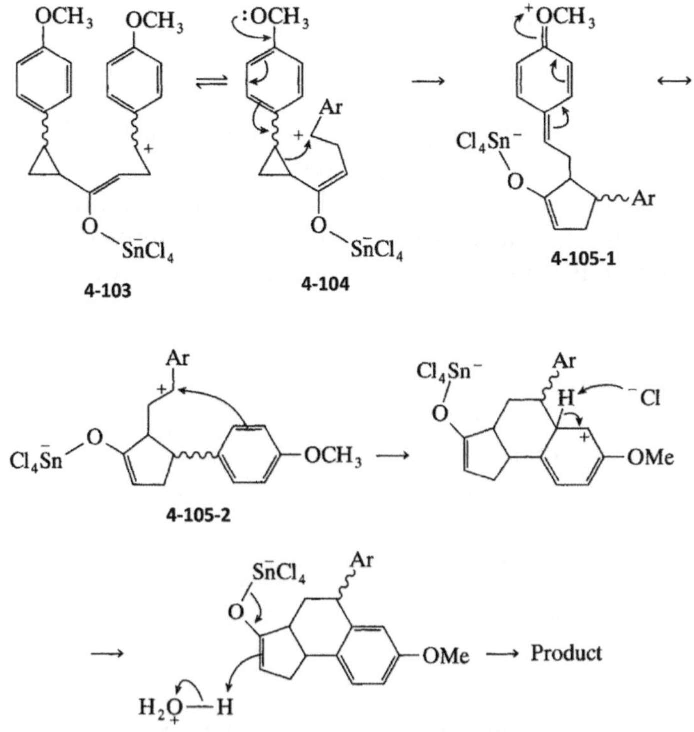

4-103   **4-104**   **4-105-1**

4-105-2

Mechanism 2

An alternative mechanism involves formation of an eight-membered-ring intermediate from **4-104**:

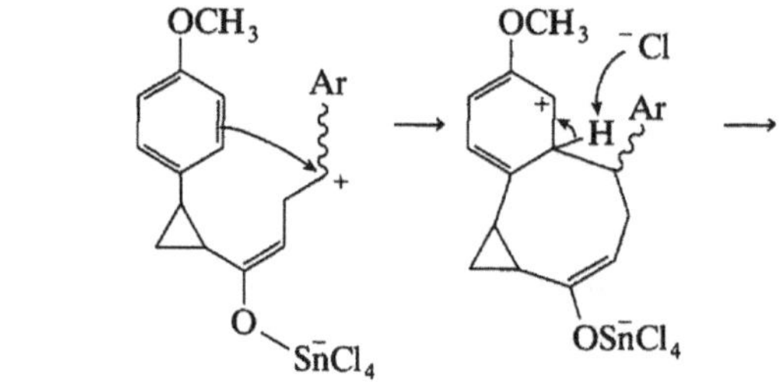

4-104

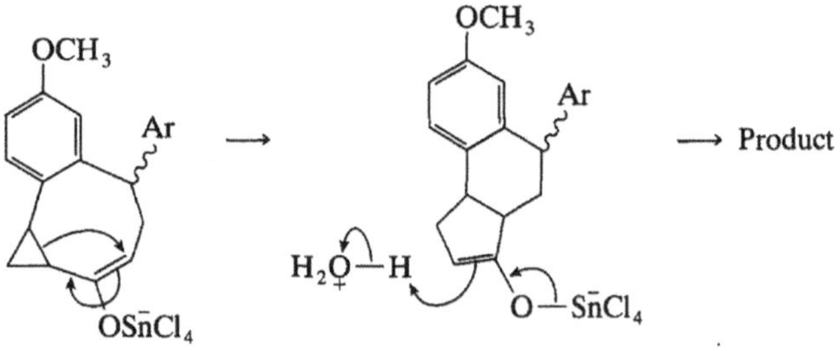

Formation of the eight-membered ring is entropically less favorable than mechanism 1. Another advantage of mechanism 1 is that the methoxy group participates in resonance stabilization of the positive charge produced when the five-membered ring is formed. Direct participation of the methoxy group is not possible in the intermediates formed in mechanism 2.

The source of chloride ion, used as a base in these mechanisms, would be the following equilibrium:

$$ROS\bar{n}Cl_4 \rightleftharpoons ROSnCl_3 + Cl^-$$

Nitromethane is the sole solvent for the reaction.

e. Silver ion coordinates with the chlorine and increases its ability to act as a leaving group. After the positive center (a nitrenium ion) has been created, an intramolecular electrophilic aromatic substitution takes place.

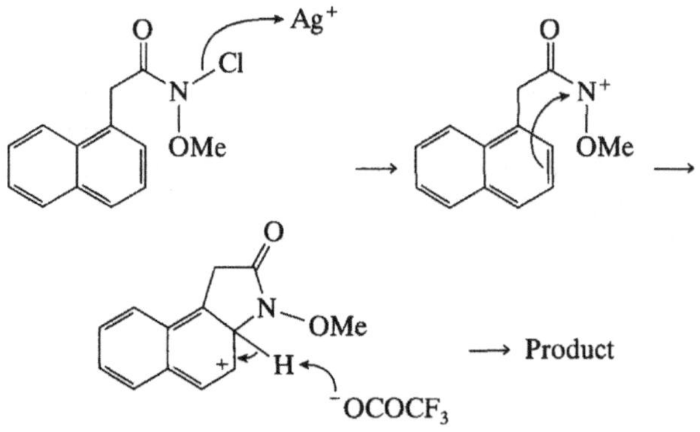

f. The nucleophilic ester carbonyl oxygen reacts with the electrophilic tin(IV) chloride. Loss of a proton from the intermediate formed initially gives **4-106**.

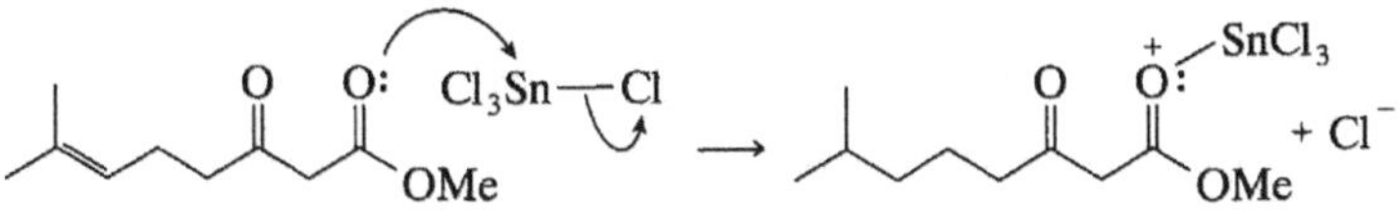

The HCl that is produced can add in a regiospecific manner to the isolated double bond **(4-105)**, giving a tertiary carbocation. An intramolecular nucleophilic reaction with this cation generates the product.

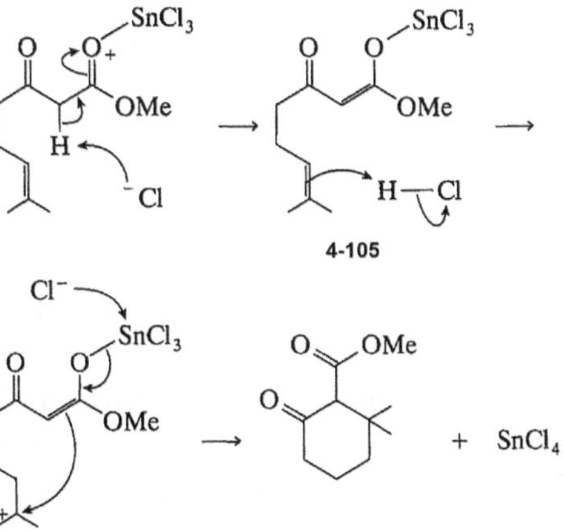

4-105

The following would *not* be a good mechanistic step because the anion that is generated is so unstable. Also, we would not expect such a strong base to be formed in a strongly acidic medium.

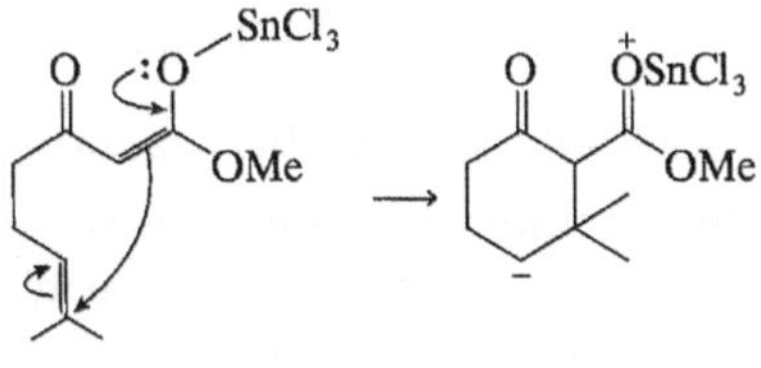

4-106

g. This trifluoroacetic acid-promoted rearrangement moves through a contraction leading to a lactam. The transformation results in the ejection of the alkene (ethylene) as a byproduct. The initial step is activation of the nitrogen by protonation. It is possible that the cyclopropane ring opens and the electrons flow into the carbon–oxygen bond to form the new carbonyl **4-107** and at the same time the oxygen–nitrogen bond is cleaved.

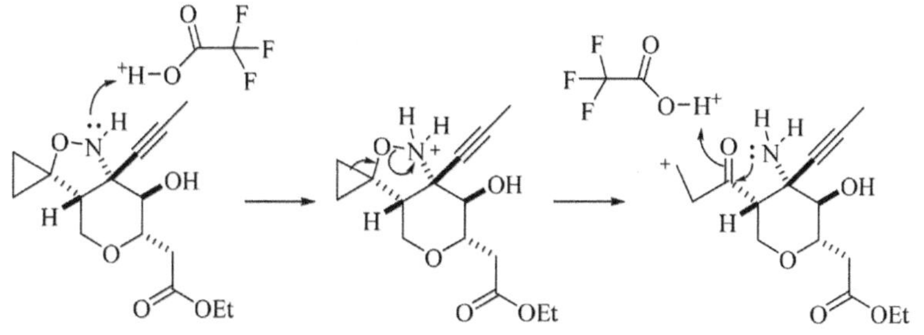

4-107 **4-108**

The nitrogen is then in position to add back into the carbonyl carbon **4-108** leading to the ejection of the molecule of ethylene. The ethylene molecule is small and leaves the reaction mixture as a gas. This mechanism proceeds through a primary carbocation, which is relatively unfavorable, but not unreasonable. The ring strain of a cyclopropane ring is approximately 28 kcal/mol, and relieving this strain could offset the energy of associated with the generation of a primary carbocation (about 25 kcal/mol less stable than a secondary carbocation), but it is always good to probe a little further and look for alternatives when a situation like this arises.

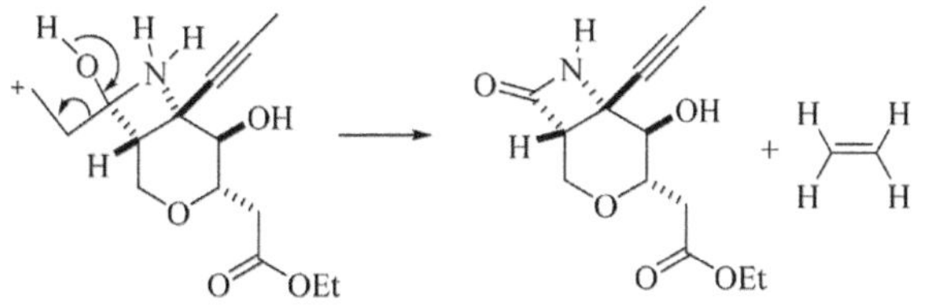

Another plausible mechanism starts with protonation and the shift of the cyclopropane ring carbon **4-109** to facilitate the cleavage of the nitrogen—oxygen bond and formation of the tertiary carbocation **4-110**. This trisubstituted cation (about 35 kcal/mol more stable than the primary carbocation and therefore more likely to be generated than the primary carbocation) then acts as the electrophile in the formation of the new nitrogen—carbon bond.

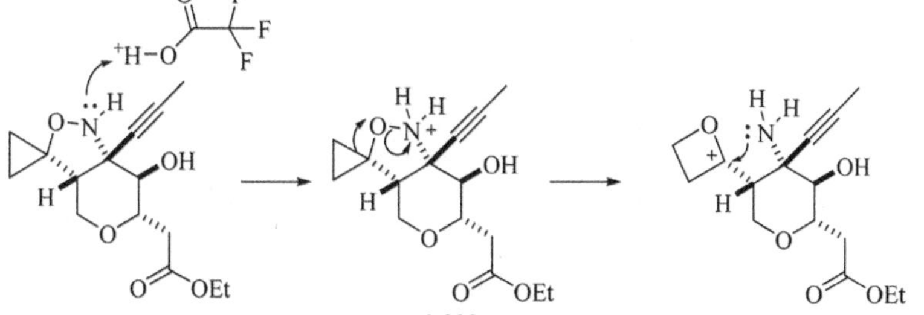

4-109 **4-110**

The spirocycle **4-111** then opens to form the ethylene and the carbonyl of the lactam **4-112**.

4-111   **4-112**

This ring contraction to form the lactam has shown generality in its application as seen by the many examples reference in the paper. It may be considered that the mechanism is not strictly a sequence of defined steps as shown, but rather the result of the bonds "slipping" in a single smooth sequence that has the cyclopropane bonds breaking as the nitrogen shifts over, forming the lactam as the ethylene is ejected.

We are not suggesting that this is a "concerted" process in the strict sense of the word, rather, that the bonds shift in a quick sequence in which no single distinct intermediate plays a significant role.

It would be interesting to consider how one might be able to tell which mechanism is operating or more accurately, what model best reflects what is occurring? The series of papers and references in the papers associated with this problem discuss options and other data that may help direct further exploration.

References

Abbot, R. E.; Spencer, T. A. *J. Org Chem.* **1980**, *45*, 5398–5399.
Ben-Ishai, D.; Denenmark, D. *Heterocycles* **1985**, *23*, 1353–1356.
Bland, J. M.; Stammer, C. H. *J. Org. Chem.* **1983**, *48*, 4393–4394.
Buckles, R. E.; Bader, J. M.; Thurmaier, R. J. *J. Org. Chem.* **1962**, *27*, 4523–4527.
Capozzi, G.; Chimirri, A.; Grasso, S.; Romeo, G. *Heterocycles* **1984**, *22*, 1759–1762.
Clissoid, C.; Kelly, C. L.; Lawrie, K. W. M.; Willis, C. L. *Tetrahedron Lett.* **1997**, *38*, 8105–8108.
Cordero, F. M.; Pisaneschi, F.; Goti, A.; Ollivier, J.; Salaun, J.; Brandi, A. *J. Am. Chem. Soc.* **2000**, *122*, 8075–8076.
Cordero, F. M.; Vurchio, C.; Brandi, A.; Gandolfi, R. *Eur. J. Org. Chem.* **2011**, *28*, 5608–5616.
Corona, T.; Crotti, P.; Ferretti, M.; Macchia, F. *J. Chem. Soc., Perkin Trans.* **1985**, *1*, 1607–1616.
Dagani, R. *Chem. Eng. News* **1991**, *69* (4), 19–20.

Diethelm, S.; Carreira, E. M. *J. Am. Chem. Soc.* **2013,** *135,* 8500−8503.

Endo, Y.; Ohta, T.; Shudo, K.; Okamoto, T. *Heterocycles* **1977,** *8,* 367−370.

Ent, H.; de Koning, H.; Speckamp, W. N. *J. Org. Chem.* **1986,** *51,* 1687−1691.

Fukuda, Y.; Isobe, M.; Nagata, M.; Osawa, T.; Namiki, M. *Heterocycles* **1986,** *24,* 923−926.

Greco, C. V.; Gray, R. P. *Tetrahedron* **1970,** *26,* 4329−4337.

Hantawong, K.; Murphy, W. S.; Boyd, D. R.; Ferguson, G.; Parvex, M. *J. Chem. Soc., Perkin Trans.* **1985,** *2,* 1577−1582.

Harmandar, M.; Balci, M. *Tetrahedron Lett.* **1985,** *26,* 5465−5468.

Hauser, F. M.; Hewawasam, P.; Baghdanov, V. M. *J. Org. Chem.* **1988,** *53,* 223−224.

Heubest, H. B.; Wrigley, T. I. *J. Chem. Soc.* **1957,** 4596−4765.

Kikugawa, Y.; Kawase, M. *J. Am. Chem. Soc.* **1984,** *106,* 5728−5729.

Kobrich, G. *Angew. Chem., Int. Ed. Engl.* **1972,** *11,* 473−485.

Magnien, E.; Baltzly, R. *J. Org. Chem.* **1958,** *23,* 2029−2032.

Martin, O. R. *Tetrahedron Lett.* **1985,** *26,* 2055−2058.

Miller, B. *Acc. Chem. Res.* **1975,** *8,* 245−256.

Ohno, M.; Naruse, N.; Torimitsu, S.; Teresawa, I. *J. Am. Chem. Soc.* **1966,** *88,* 3168−3169.

Padwa, A.; Carter, S. P.; Nimmesgern, H.; Stull, P. D. *J. Am. Chem. Soc.* **1988,** *110,* 2894−2900.

Schubert, W. M.; LeFevre, P. H. *J. Am. Chem. Soc.* **1972,** *94,* 1639−1645.

Serafin, B.; Konopski, L. *Pol J. Chem.* **1978,** *52,* 51−62.

Shioiri, T. *Comp. Org. Syn.* **1991,** *6,* 800−806.

Taguchi, H.; Yamamoto, H.; Nozaki, H. *J. Am. Chem. Soc.* **1974,** *96,* 6510−6511.

Waring, A. J.; Zaidi, J. H. *J. Chem. Soc., Perkin Trans.* **1985,** *1,* 631−639.

Wenkert, E.; Arrhenius, T. S.; Bookser, B.; Guo, M.; Mancini, P. *J. Org. Chem.* **1990,** *55,* 1185−1193.

White, J. D.; Skeean, R. W.; Trammell, G. L. *J. Org. Chem.* **1985,** *50,* 1939−1948.

Wynberg, H.; Meijer, E. W. *Organic Reactions, The Reimer−Tiemann* **1982,** *Vol. 28. Chapter 1, pp. 1−36.*

Yamabe, S.; Tsuchida, N.; Yamazaki, S. *J. Org. Chem.* **2005,** *70,* 10638−10644.

5

Radicals and Radical Anions

I. INTRODUCTION

Radicals are species that contain one or more unpaired electrons. They are encountered in many reactions used in chemical industry (e.g., the production of polyethylene), in the processes responsible for spoiling of foods (e.g., autoxidation by molecular oxygen), and in many biological systems (e.g., signaling by nitrogen oxide, NO). Radical anions, as their name implies, are radical species that have an unpaired electron and a negative charge.

In chemical structures, the unpaired electron of a radical is represented by a dot. Radical mechanisms are depicted in one of two ways. Most commonly, each individual step of the mechanism is written without the use of arrows to show electron movement. The resulting series of equations shows the order of events, and it is assumed that one-electron transfers are taking place throughout. A second method uses curved, half-headed arrows ($\rightharpoonup$) to show electron movement. The half-headed arrow is used to denote movement of a single electron, whereas the normal arrowhead is used to denote movement of a pair of electrons.

Various studies indicate that although radicals tend to be pyramidal, the pyramids are shallow and the barriers to inversion are low. This means that the stereochemical results of radical reactions are very similar to those of carbocations; in other words, stereochemistry is usually lost at a reactive center.

Whereas most carbon free radicals are highly reactive species, there are notable exceptions.

2. FORMATION OF RADICALS

Many radicals are produced by homolytic cleavage of bonds. The energy for this kind of bond breaking comes from thermal or photochemical energy or from electron-transfer reactions effected by either inorganic compounds or electrochemistry. These kinds of processes initiate reactions that proceed by a radical mechanism. Compounds that readily produce radicals are called *initiators* or *free-radical initiators*.

A. Homolytic Bond Cleavage

Radicals produced from chlorine and bromine can be generated photochemically and/or thermally. Because bromine atoms are less reactive than chlorine atoms, brominations are often done in the presence of both heat (Δ) and light ($h\nu$).

Writing Reaction Mechanisms in Organic Chemistry
http://dx.doi.org/10.1016/B978-0-12-411475-3.00005-1

Copyright © 2014 Elsevier Inc. All rights reserved.

$$Cl_2 \xrightarrow{h\nu} 2Cl^{\cdot}$$

$$Br_2 \xrightarrow[\Delta]{h\nu} 2Br^{\cdot}$$

Many peroxides and azo compounds can be heated to generate radicals. Peroxides decompose readily because of the weak O—O bond, whereas azo compounds cleave readily because of the driving force provided by the formation of the stable nitrogen molecule. Common examples of these decompositions follow. (The $t_{1/2}$ is the time it takes for one-half of the material to decompose.)

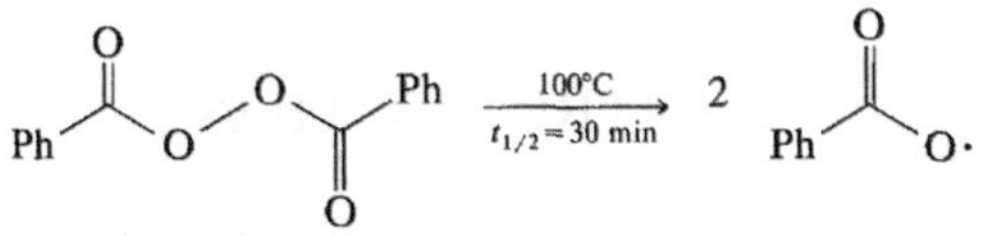

benzoyl peroxide

$$(CH_3)_3C—O—O—C(CH_3)_3 \xrightarrow[t_{1/2}=1\,h]{150°C} 2(CH_3)_3C—O\cdot$$

di-*t*-butyl peroxide

$$(CH_3)_2CN=NC(CH_3)_2 \xrightarrow[t_{1/2}=1\,h]{85°C} 2(CH_3)_2\dot{C}—CN + N_2$$

$$\underset{CN}{|} \qquad \underset{CN}{|}$$

5-1

EXAMPLE 5.1. DECOMPOSITION OF AIBN [AIBN = AZOBISISOBUTYRONITRILE]

Half-headed arrows show the movement of a single electron:

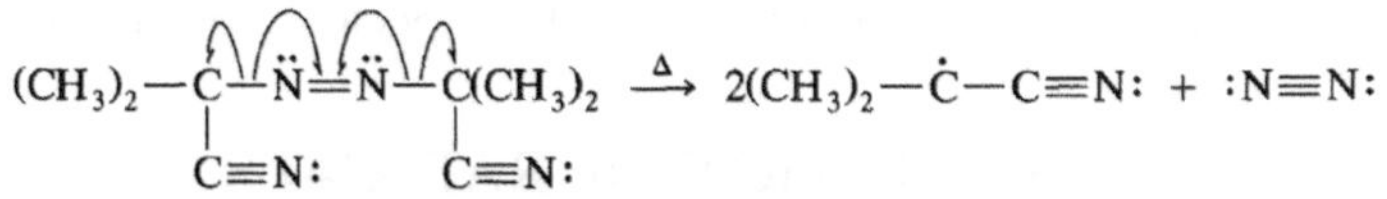

$$(CH_3)_2—\underset{C\equiv N:}{\overset{}{C}}—\ddot{N}=\ddot{N}—\underset{C\equiv N:}{\overset{}{C}}(CH_3)_2 \xrightarrow{\Delta} 2(CH_3)_2—\dot{C}—C\equiv N: + :N\equiv N:$$

PROBLEM 5.1

Why is the reaction shown in Example 5.1 a highly favorable process?

B. Hydrogen Abstraction from Organic Molecules

The carbon−hydrogen bond is relatively stable, so that direct formation of an organic radical by homolytic cleavage of a C−H bond is rarely observed. However, many radicals can remove hydrogen atoms from organic molecules to form carbon radicals. This process is known as *hydrogen abstraction*. For example.

$$F\cdot \ + \ CH_4 \longrightarrow HF \ + \cdot CH_3$$

Some radicals, like fluorine, are so reactive that they form radicals with almost any organic compound, whereas others are so stable that they can abstract hydrogen from very few organic compounds. An example of such a stable radical is **5-1**, formed from AIBN. When selectivity is desired in a radical initiator, a relatively stable radical like the one formed from AIBN is a good choice.

Rates of Hydrogen Abstraction and Relative Stability of Radicals

Because most radicals are electrophilic, the effects of structure on their rate of formation are very similar to those for the formation of carbocations. Thus, the rates of hydrogen abstraction are $1° < 2° < 3°$, corresponding to the order of stability of the resulting radicals: $1° < 2° < 3°$. Also, it is relatively easy to abstract a hydrogen from an allylic or a benzylic position because the resulting radicals are stabilized by delocalization. In contrast, it is quite difficult to abstract vinyl and aromatic hydrogens because the electrophilicity of the resulting radicals is higher due to the higher s character of an sp^2-hybridized orbital relative to an sp^3-hybridized orbital. In these cases, the sp^2-hybridized orbital is perpendicular (orthogonal) to the π system; thus, the radical *cannot* be stabilized by resonance. Finally, it is difficult to abstract a hydrogen from the alcohol functional group. The resulting alkoxy radical is quite unstable due to the high electronegativity of oxygen.

The approximate relative stabilities of organic radicals are as follows:

$$Ph\cdot \ < \ CH_2 = CH\cdot \ < \ RO\cdot \ < \ RCH_2\cdot \ < \ R_2CH\cdot \ < \ R_3C\cdot \ < \ PhCH_2\cdot \ < \ RCH-CH_2\cdot$$

C. Organic Radicals Derived from Functional Groups

Reactive radicals are often produced by abstraction of a halogen atom from a substrate. A commonly used halogen-abstracting reagent is tri-*n*-butyltin radical, formed from tri-*n*-butyltin hydride using AIBN as an initiator. AIBN generates **5-1**, which abstracts a hydrogen atom from the tri-*n*-butyltin hydride, generating the tri-*n*-butyltin radical. This tin radical can abstract a halogen from a variety of substrates (e.g., alkyl, olefinic, or aryl chlorides, bromides, or iodides) to generate the corresponding radical and tri-*n*-butyltin halide.

$$(CH_3)_2\dot{C}-CN \ + \ (n\text{-}Bu)_3SnH \longrightarrow (CH_3)_2CH-CN \ + \ (n\text{-}Bu)_3Sn\cdot$$

   5-1

$$(n\text{-}Bu)_3Sn\cdot \ + \ RBr \longrightarrow (n\text{-}Bu)_3SnBr \ + \ R\cdot$$

Tri-*n*-butyltin radicals can also be used to generate radicals from selenium compounds. An example is the formation of acyl radicals from seleno esters.

The source is *Boger and Mathvink (1988)*.

Radicals can also be synthesized by the reduction of alkyl mercury salts. For example, in the presence of sodium borohydride, compound **5-2** reacts to form the radical, **5-3**.

The source is *Giese et al. (1982)*.

Despite their toxicity, alkyl mercury salts have been used widely in research due to their versatility as synthetic intermediates.

3. RADICAL CHAIN PROCESSES

Most synthetically useful radical reactions occur as chain processes. A radical chain process is one in which many moles of product are formed for every mole of radicals produced. These chain processes include the following steps:

1. *Initiation.* One or more steps produce a radical from starting material.
2. *Propagation.* The radical enters a series of steps that results in the formation of product and a new radical that can start the series of propagation steps over again.
3. *Termination.* Removal of radicals from the propagation steps, thus ending a chain.

Initiation has been discussed in the previous section. In order for propagation to occur, many product molecules need to be formed from just one radical produced in the initiation step, and this occurs *only if the propagation steps are exothermic*. Termination steps include coupling, disproportionation, and abstraction. Abstraction by a chain-transfer agent removes a radical from a propagation step, and a new radical is generated in its place. If this newly generated radical is sufficiently stable, the chain transfer results in termination.

HINT 5.1

The propagation steps of a chain process must add up to the overall equation for the reaction. The overall equation will not contain any radicals. This means that the radical produced in the last propagation step must be the same as a reactant in the first propagation step.

HINT 5.2

In radical reactions, the products are rarely generated by radical coupling.

Radical coupling reactions are those in which two radicals react to form a covalent bond. They are rarely a significant source of product because both radicals are reactive intermediates and are present in extremely low concentrations. This means that the probability that they will react together is very small, and thus the rate of their reaction, which depends on their concentrations, is very low and other processes will compete effectively.

These various aspects of radical chain processes are illustrated in the following example.

EXAMPLE 5.2. RADICAL CHAIN HALOGENATION BY T-BUTYL HYPOCHLORITE

The overall process is as follows:

$$t\text{-BuOCl} + \underset{\substack{\Delta \\ \text{or } h\nu}}{\overset{H}{\diagup\!\!\!\diagdown}} \quad \overset{Cl}{\diagup\!\!\!\diagdown} + t\text{-BuOH}$$

This can be broken down into initiation, propagation, and termination steps.

Initiation Step

$$t\text{-BuOCl} \xrightarrow[\text{or } h\nu]{\Delta} t\text{-BuO}\cdot + \cdot\text{Cl}$$

In continuing with the mechanism, consider t-BuO to be the chain-carrying radical. This means that t-BuO· will be used up in the first propagation step but regenerated in the last propagation step. In this case, the last propagation step is the second step.

Propagation Steps

$$(1) \quad \overset{H}{\diagup\!\!\!\diagdown} + t\text{-BuO}\cdot \longrightarrow t\text{-BuOH} + \diagup\!\!\!\diagdown\cdot$$

$$(2) \quad \diagup\!\!\!\diagdown\cdot + t\text{-BuOCl} \longrightarrow \overset{Cl}{\diagup\!\!\!\diagdown} + t\text{-BuO}\cdot$$

The radical formed in step 2 can start another reaction (step 1); that is why these processes are called *chain processes*. In this way, just one t-BuO· radical can initiate many sets of propagation steps. Addition of steps 1 and 2 gives the equation for the overall reaction; the radicals cancel when the addition is performed.

Is there another possible reaction pathway? Could the chlorine radical be the chain-carrying radical? If so, Equation (3), an exothermic reaction, could represent one of the propagation steps.

$$(3) \quad \overset{H}{\diagup\!\!\!\diagdown} + \cdot\text{Cl} \longrightarrow \diagup\!\!\!\diagdown\cdot + \text{HCl}$$

Now a step must be found that forms the *t*-butyl chloride product and regenerates chlorine radical. This rules out Equation (2) as a product-forming step because that reaction does not generate the same radical that is used in Equation (3). Addition of Equations (2) and (3) does not yield the overall reaction.

If we couple Equation (3) with Equation (4), we regenerate the chlorine radical but arrive at a different product for the reaction.

$$(4) \quad \text{(CH}_3)_3\text{C} \cdot \ + \ t\text{-BuOCl} \ \longrightarrow \ (\text{CH}_3)_3\text{C—O-}t\text{-Bu} \ + \ \text{Cl} \cdot$$

Another possible product-forming step would be reaction of the *t*-butyl radical with a chlorine atom. However, because this reaction removes radicals without forming new ones, it could not be one of the propagation steps of a chain process.

$$(\text{CH}_3)_3\text{C} \cdot \ + \ \cdot\text{Cl} \ \longrightarrow \ (\text{CH}_3)_3\text{C—Cl}$$

Termination Steps
(1) Disproportionate

$$(\text{CH}_3)_3\text{C} \cdot \ + \ (\text{CH}_3)_3\text{C} \cdot \ \longrightarrow \ (\text{CH}_3)_3\text{CH} \ + \ \text{(CH}_3)_2\text{C=CH}_2$$

The disproportionation process involves the abstraction of hydrogen by one radical from another radical. Abstraction of a hydrogen atom from the carbon adjacent to the radical produces a double bond:

$$\longrightarrow$$

(2) Radical Coupling
Some possible coupling reactions follow:

$$(\text{CH}_3)_3\text{C} \cdot \ + \ \cdot\text{Cl} \ \longrightarrow \ (\text{CH}_3)_3\text{C—Cl}$$

$$t\text{-BuO} \cdot \ + \ t\text{-BuO} \cdot \ \longrightarrow \ t\text{-BuOO-}t\text{-Bu}$$

$$(\text{CH}_3)_3\text{C} \cdot \ + \ (\text{CH}_3)_3\text{C} \cdot \ \longrightarrow \ (\text{CH}_3)_3\text{C—C(CH}_3)_3$$

The following radical coupling does not affect the chain process because chlorine atoms are not involved in chain propagation.

$$\cdot\text{Cl} \ + \ \cdot\text{Cl} \ \longrightarrow \ \text{Cl—Cl}$$

PROBLEM 5.2

What other radical coupling reactions are possible for the reaction of *t*-butyl hypochlorite in Example 5.2?

---

4. RADICAL INHIBITORS

Radical reactions can be slowed or stopped by the presence of substances called *inhibitors*.

HINT 5.3

A reaction that is slowed by the addition of a free-radical inhibitor can be assumed to proceed by a radical mechanism.

Common radical inhibitors include 2,6-di-*t*-butyl-4-methylphenol (butylated hydroxytoluene, BHT), 2-nitroso-2-propane, and oxygen.

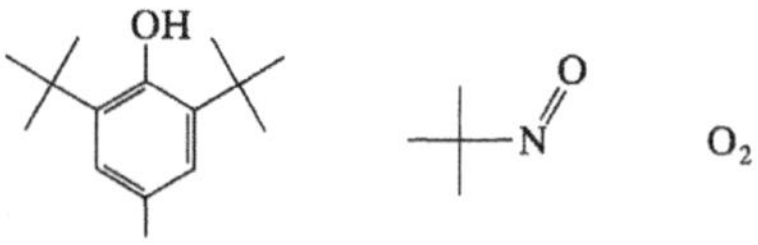

BHT is the common name for 2,6-di-*t*-butyl-4-methylphenol. It is used widely as an antioxidant in foodstuffs and food packaging. The reaction of oxygen with unsaturated fats gives, after several steps, alkylperoxy radicals (ROO·) that, upon further reaction, give smaller odiferous molecules that can ruin the palatability of foods. BHT acts as a scavenger for alkylperoxy radicals, ROO·, because hydrogen abstraction by ROO· gives the stable free radical, **5-4**. The hydroperoxide ROOH, which also is formed in the reaction, is much less reactive than ROO· and consequently causes much less oxidation. The new radical, **5-4**, which is formed from BHT, is relatively unreactive for two reasons: (i) it is stabilized by resonance, and (ii) the groups attached to the ring sterically hinder further reaction.

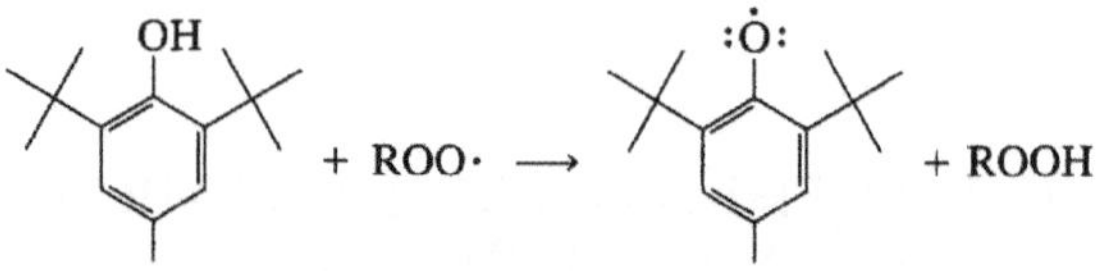

5-4

Radicals also can be trapped by addition to the nitroso group to form nitroxide radicals. In the following equation, diacetyl peroxide is a source of methyl radicals, which are trapped by the nitroso compound. (For further discussion of the fragmentation of radicals, see Section 7.)

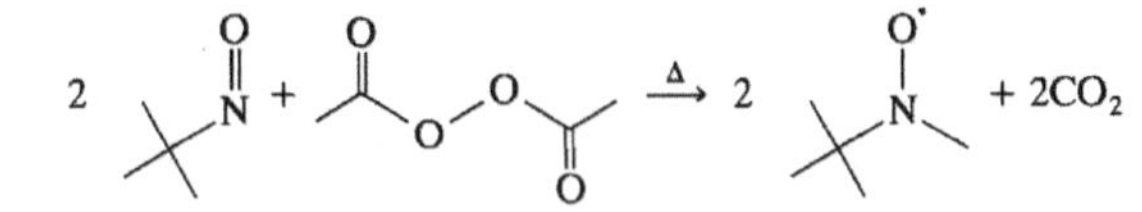

Oxygen can act as an inhibitor by reacting with radicals to produce less reactive hydroperoxyl radicals.

Two compounds that effectively inhibit electron-transfer reactions are di-*t*-butyl nitroxide and 1,4-dinitrobenzene:

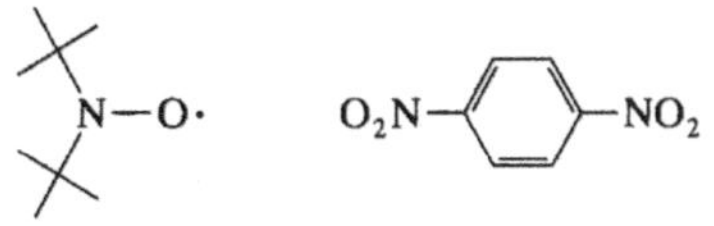

Transfer of an electron to 1,4-dinitrobenzene gives a radical anion so stable that it is unlikely to transfer an electron to anything else. The reaction of benzene radical anion with 1,4-dinitrobenzene proceeds to the right because the radical anion of the 1,4-dinitrobenzene is much more stable than the radical anion of benzene itself.

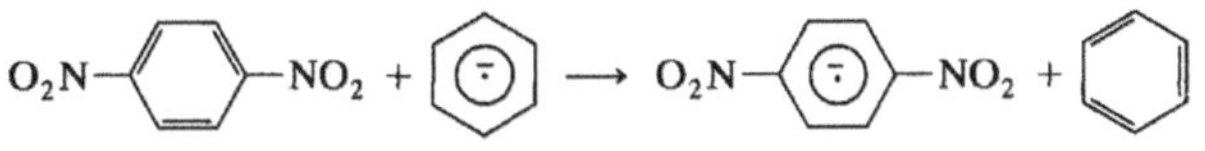

Although 1,3-dinitrobenzene sometimes is used for the inhibition of electron-transfer reactions, it is not as effective as the 1,4-isomer because its radical anion is not as stable (Problem 1.5).

Nitroxides are also used to remove radical anions from a reaction sequence. For example, di-*t*-butyl nitroxide has often been used to inhibit the $S_{RN}1$ reaction (Section 10). Apparently, nitroxides remove radicals via coupling reactions (Hoffmann et al., (1964)).

HINT 5.4

A radical reaction that is initiated only by light can be prevented by omitting light. In this case, any reaction that takes place in the dark must proceed by a nonradical pathway.

5. DETERMINING THE THERMODYNAMIC FEASIBILITY OF RADICAL REACTIONS

The bond dissociation energies (BDEs) in Table 5.1 can be used to determine the feasibility of radical reactions. In general, a radical process will give reasonable synthetic yields only if all propagation steps are exothermic, so that many propagation steps can occur before

TABLE 5.1   Bond Dissociation Energies (BDEs, kcal/mol) at 298 K[a]

		$Cl-Cl^b$	58	$Br-Br^b$	46	$I-I^b$	36	$F-F^b$	37
		$H-Cl^b$	103	$H-Br^b$	87	$H-I^b$	36	$H-F^b$	135
CH_3-H	104	CH_3-Cl	73	CH_3-Br	70				
C_2H_5-H	98	C_2H_5-Cl	81	C_2H_5-Br	69	C_2H_5-I	53	C_2H_5-F	106
$CH_3CH_2CH_2-H$	98	$n\text{-}C_3H_7-Cl$	82	$n\text{-}C_3H_7-Br$	69				
$(CH_3)_2CH-H$	94.5	$(CH_3)_2CH-Cl$	81	$(CH_3)_2CH-Br$	68				
$(CH_3)_3CH-H$	91.0	$(CH_3)_3C-Cl$	79	$(CH_3)_3C-Br$	63				
$PhCH_2-H$	85	$PhCH_2-Cl$	68	$PhCH_2-Br$	51				
$CH_2{=}CHCH_2-H$	85								
CCl_3-H	95.7	CCl_3-Cl	73	CCl_3-Br	54				
$Ph-H$	104			$Ph-Br$	71				
$CH_2{=}CH-H$	104								
$HOCH_2-H$	92								
CH_3O-H	102								
C_2H_5O-H	102								
$i\text{-}C_3H_7O-H$	103								
$t\text{-}C_4H_9O-H$	103	$t\text{-}C_4H_9O-Cl$	44^c						
CH_3COO-H	112								
CH_3S-H	88								
$PhS-H$	75								
CH_3-CH_3	88								
$PhCO-H$	74	$PhCO-Cl$	74						
$HCO-H$	88								
CH_3CO-H	88								
CH_3COCH_2-H	92								

[a]All values from Kerr, (1966), unless otherwise noted.
[b]Handbook of Chemistry and Physics, 63rd ed.; Weast and Astle.
[c]Walling and Jacknow, (1960).

termination. The number of propagation steps per initiation step is the *chain length* of the reaction. Highly exothermic propagation steps involve highly reactive radicals, like alkyl radicals, and have a long chain length. More stable radicals, like aryl radicals, may fail to react until they encounter another radical. In this case, there is no chain reaction.

In Example 5.3, calculations show that the radical chlorination of 2-methylpropane by *t*-butyl hypochlorite is an energetically feasible process for the synthesis of *t*-butyl chloride.

EXAMPLE 5.3. DETERMINATION OF THE ENTHALPY CHANGE IN CHLORINATION BY *T*-BUTYL HYPOCHLORITE

Consider the reactions shown in Example 5.2. If both propagation steps are exothermic, the chain length will be long enough that the thermochemistry of the initiation step(s) is unimportant relative to that of the propagation steps.

We will calculate the enthalpy change for the reactions in the propagation steps. In Equation (1), the C−H bond broken has a BDE of 91.0 kcal/mol, and the O−H bond formed has a BDE of 103 kcal/mol. Therefore, this first reaction is exothermic by 91−103 or −12 kcal/mol.

$$(1) \quad \text{\Large λ}^H + t\text{-BuO}\cdot \longrightarrow t\text{-BuOH} + \text{\Large λ}\cdot \qquad \Delta H = -12 \text{ kcal/mol}$$

For Eqn (2), the O−Cl bond broken has a BDE of 44 kcal/mol and the C−Cl bond formed has a BDE of 79 kcal/mol. Thus, the second reaction is exothermic by 44−79 kcal/mol or −35 kcal/mol.

$$(2) \quad \text{\Large λ}\cdot + t\text{-BuOCl} \longrightarrow \text{\Large λ}^{Cl} + t\text{-BuO}\cdot \qquad \Delta H = -35 \text{ kcal/mol}$$

Because both reactions are substantially exothermic, this should be a highly favorable process with a long chain length. The enthalpy change for the overall reaction is $[(-12)+(-35)]$ or −47 kcal/mol.

PROBLEM 5.3

What are the defects in the following mechanism for the chlorination of 2-methylpropane by *t*-butyl hypochlorite? You will need to consider the thermochemistry of the processes, as well as the fact that a chain process is not involved.

$$t\text{-BuOCl} \xrightarrow{\Delta} t\text{-BuO}\cdot + \cdot\text{Cl}$$

$$\text{\Large λ}^H + t\text{-BuO}\cdot \longrightarrow t\text{-BuOH} + \text{\Large λ}\cdot$$

$$\text{\Large λ}\cdot + \text{Cl}\cdot \longrightarrow \text{\Large λ^{Cl}

6. ADDITION OF RADICALS

Radical addition reactions are commonly used in organic synthesis. These additions range from the simple addition of halocarbons to π bonds to cyclization reactions with demanding stereoelectronic requirements.

A. Intermolecular Radical Addition

HINT 5.5

Common types of radicals that add to π bonds are those that can be generated from alkyl halides, mercaptans, thiophenols, thioacids, aldehydes, and ketones. Like the corresponding electrophilic additions to double bonds, many radical additions are either regiospecific or highly regioselective.

EXAMPLE 5.4. PHOTOCHEMICAL ADDITION OF TRIFLUOROIODOMETHANE TO ALLYL ALCOHOL

In the initiation step of this reaction, light induces homolytic cleavage of the weak C—I bond.

$$CF_3I \xrightarrow{h\nu} \cdot CF_3 + \cdot I$$

The trifluoromethyl radical adds to the double bond, giving the most stable intermediate radical:

$$F_3C \cdot \quad CH_2 = CHCH_2OH \longrightarrow F_3CCH_2\dot{C}HCH_2OH$$

This radical then abstracts an iodide atom from trifluoroiodomethane to generate product and the chain-propagating radical, the trifluoromethyl radical.

$$F_3CCH_2\dot{C}HCH_2OH + ICF_3 \longrightarrow F_3CCH_2CHICH_2OH + \cdot CF_3$$

The product shown is the major regioisomer produced. For further details on this reaction, see Park et al. (1961).

PROBLEM 5.4

Examine the mechanistic steps for the reaction shown.

a. For each step, use half-headed arrows to show the movement of electrons, and dots to indicate all the unpaired electrons.

$$CH_2=CHC_6H_{13} + CBr_4 \xrightarrow[\Delta]{(PhCO_2)_2} Br_3CCH_2CHBrC_6H_{13}$$

Step 1

$$Ph\overset{\overset{\displaystyle O}{\|}}{C}-O-O-\overset{\overset{\displaystyle O}{\|}}{C}Ph \longrightarrow 2Ph\overset{\overset{\displaystyle O}{\|}}{C}-O$$

Step 2

$$Ph\overset{\overset{\displaystyle O}{\|}}{C}-O + Br-CBr_3 \longrightarrow Ph\overset{\overset{\displaystyle O}{\|}}{C}-O-Br + CBr_3$$

Step 3

$$CBr_3 + CH_2=CHC_6H_{13} \longrightarrow Br_3CCH_2CHC_6H_{13}$$

Step 4

$$Br_3CCH_2CHC_6H_{13} + Br-CBr_3 \longrightarrow Br_3CCH_2CHBrC_6H_{13} + CBr_3$$

b. Identify the initiation and propagation steps.
c. Why is the addition in step 3 regiospecific?

EXAMPLE 5.5. ADDITION OF A RADICAL FORMED BY REDUCTION OF A C–HG BOND

Addition of radicals formed from mercury compounds to alkenes often produce good-to-excellent yields. The following mechanism illustrates a reaction with a yield of 64%.

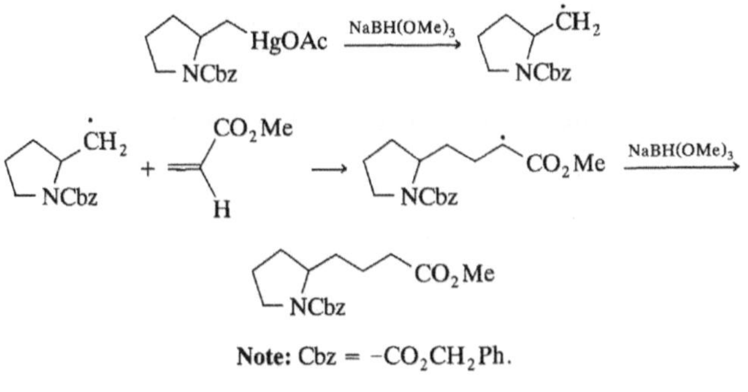

Note: Cbz = $-CO_2CH_2Ph$.

Source: Danishefsky et al., (1983).

B. Intramolecular Radical Addition: Radical Cyclization Reactions

Generation of a radical in a molecule that contains a site of unsaturation presents an opportunity for cyclization. These types of radical cyclizations have been developed into very useful synthetic reactions. The synthetic utility of these cyclizations is enhanced by the ability to predict the regiochemistry of cyclization by applying the Baldwin rules. The Baldwin rules cover the formation of three- to seven-membered rings by various reactions and are based on consideration of both kinetic and thermodynamic factors. (See Beckwith et al., (1980), and Baldwin, (1976).)

The Baldwin rules distinguish two types of ring closure, *exo* and *endo*.

exo adduct   *endo* adduct

With regard to radical cyclization, the most important guidelines are as follows:

1. For unsubstituted ω-alkenyl radicals containing up to eight carbons, with the general structure previously shown, the preferred mode of cyclization is *exo*. (The symbol ω means that the alkene is at the terminus distant from the radical.) The reaction is controlled kinetically; the major product is the one that forms faster. The rates of formation of the two possible adducts are a result of the stereochemical requirements of their transition states. Often the *endo* product is more stable than the *exo* product. When this is the case, however, the more stable product is not the major product formed.
2. If the alkene is substituted at the nonterminal position, reaction to form the *exo* product is sterically hindered and the percentage of *endo* product increases. With substituted alkenes, the more stable product may predominate.

EXAMPLE 5.6. EXO AND ENDO RING CLOSURE IN RADICAL CYCLIZATION

exo adduct

exo adduct   *endo* adduct

Source: Julia et al., (1975).

EXAMPLE 5.7. INTRAMOLECULAR CYCLIZATION OF A VINYL RADICAL

The overall reaction is as follows:

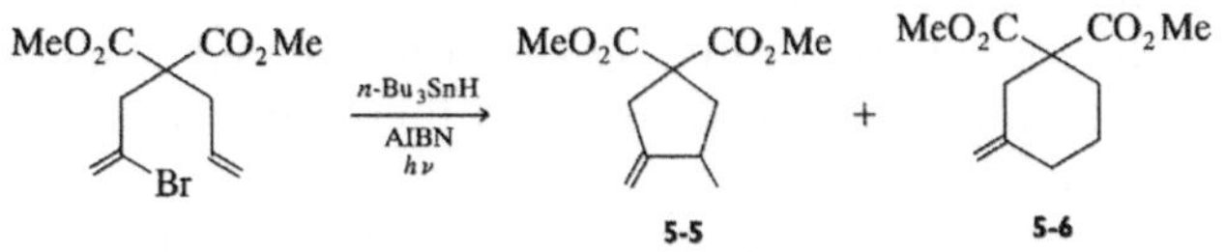

5-5   **5-6**

The vinyl radical, **5-7**, is formed by the process shown in Section 2. Cyclization of **5-7** can give an *exo* adduct or an *endo* adduct. The resulting radical abstracts hydrogen from n-Bu₃SnH to form products, **5-5** and **5-6**, as well as tri-n-butyltin radical, which continues the chain. As is predicted by the guidelines, the *exo* product (**5-5**) predominates, in this case by a factor of 2.

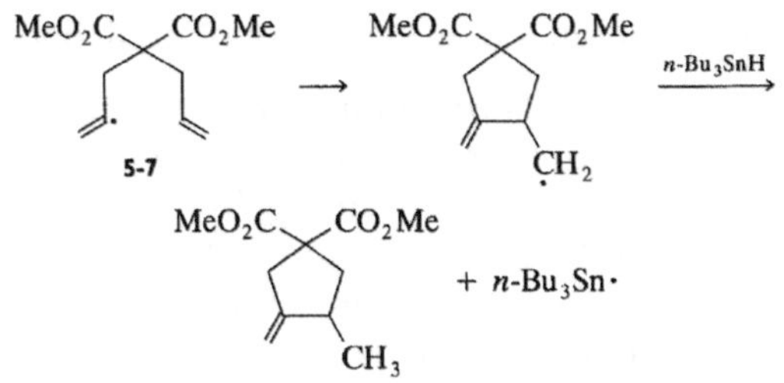

5-7

Source: Stork et al., (1982).

PROBLEM 5.5

Consider the following reaction:

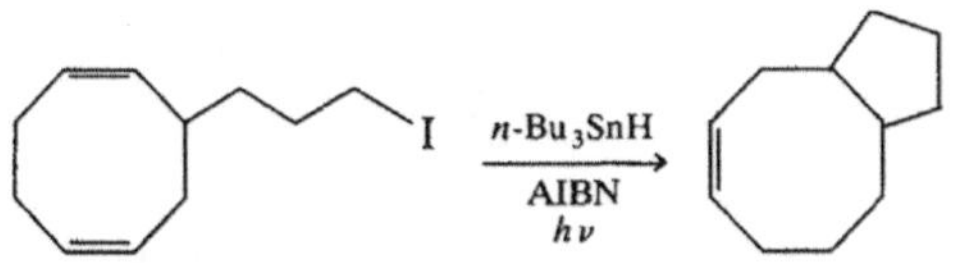

a. Write the initiation and propagation steps.
b. The following structure represents a minor product isolated from the reaction mixture. Show how it might have been formed.

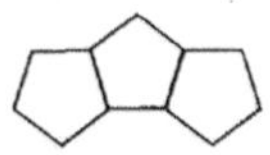

Source: Winkler and Sridar, (1986).

PROBLEM 5.6

Write the initiation and propagation steps for the following reaction:

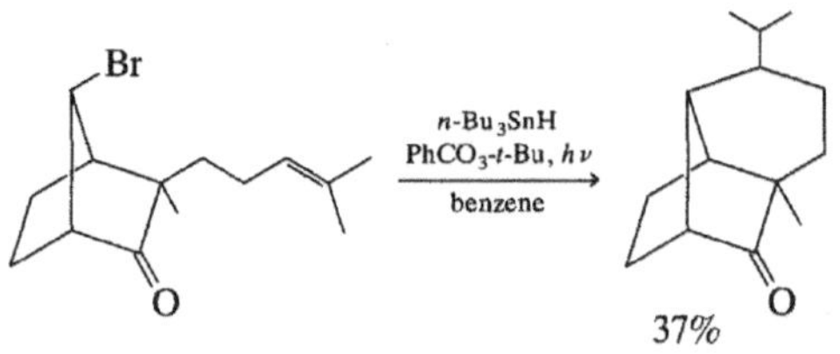

37%

Source: Hart, (1984).

PROBLEM 5.7

Propose a mechanism for the following transformation.

21%

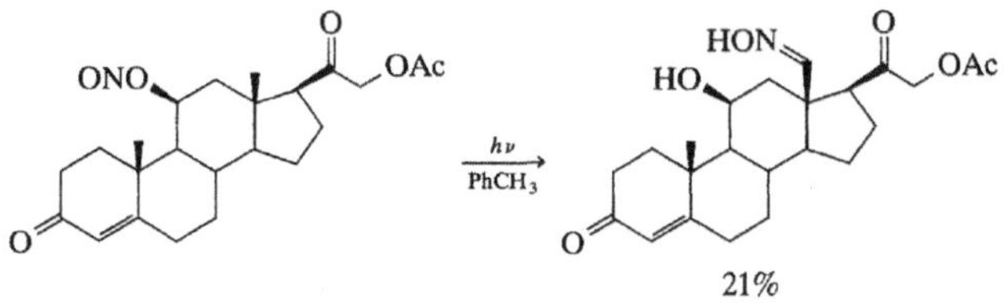

Source: Barton et al., (1960).

PROBLEM 5.8

a. Propose a mechanism for the following transformation. Show the chain initiation step and propagation steps.
b. Explain the resulting stereochemical relationship.

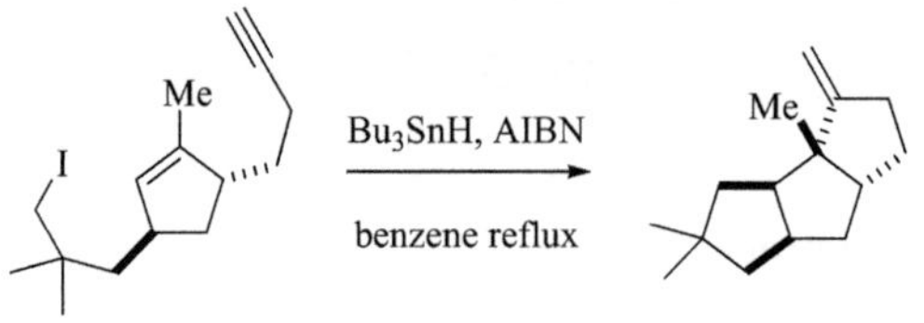

Source: Curran and Rakiewicz, (1985).

7. FRAGMENTATION REACTIONS

HINT 5.6

Many radical processes involve the loss of small, stable molecules, such as carbon dioxide, nitrogen, or carbon monoxide. These kinds of reactions are called *fragmentations*.

A. Loss of CO_2

The radical initiator, diacetyl peroxide, homolyzes at the O—O bond to form carboxy radicals, which then readily lose CO_2 to give methyl radicals. In fact, between 60 and 100 °C, acetyl peroxide can be a convenient source of methyl radicals.

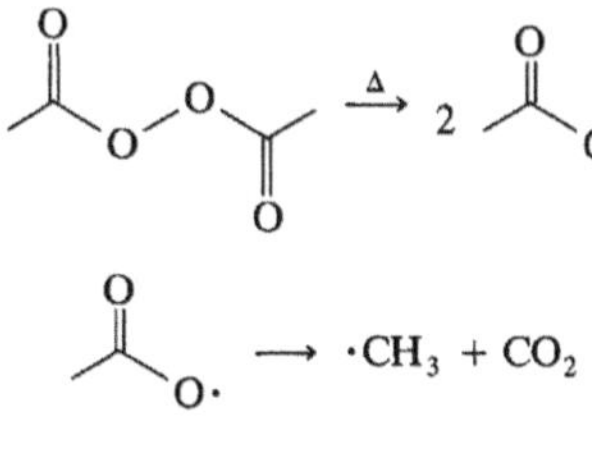

(See Walling, (1957).) Aryl-substituted carboxy radicals also lose CO_2, but they do so much less readily (e.g., the initial radical formed from benzoyl peroxide, Section 2).

Fragmentation with loss of CO_2 also occurs in the Hunsdiecker reaction, in which a silver salt of a carboxylic acid reacts with bromine to produce an alkyl halide. The reaction results in shortening of the carbon chain by one carbon. The overall reaction is as follows:

$$RCO_2Ag + Br_2 \xrightarrow{\Delta} RX + CO_2 + AgBr$$

The mechanism is

$$RCO_2Ag + Br_2 \rightarrow RCO_2Br + AgBr$$

Initiation:

$$RCO_2Br \rightarrow RCO_2 \cdot + Br \cdot$$

Propagation 1:

$$RCO_2 \cdot \rightarrow R \cdot + CO_2$$

Propagation 2:

$$R \cdot + RCO_2Br \longrightarrow RBr + RCO_2 \cdot$$

B. Loss of a Ketone

The radical initially produced by homolytic decomposition of a dialkyl peroxide can undergo further scission. The rate of scission depends on the temperature and the stability of the resulting radical(s). For example, t-butoxy radicals decompose on heating to methyl radicals and acetone.

$$t\text{-BuO} \cdot \longrightarrow \cdot CH_3 + CH_3COCH_3$$

C. Loss of N_2

Azo compounds often decompose with loss of nitrogen. The decomposition of the initiator, AIBN, is an example (Section 2).

D. Loss of CO

At elevated temperatures, the radical, **5-8**, generated by hydrogen abstraction from the aldehyde functional group fragments to give a new carbon radical and carbon monoxide. The fragmentation is temperature-dependent; the higher the temperature, the more favorable the loss of CO from the initial radical.

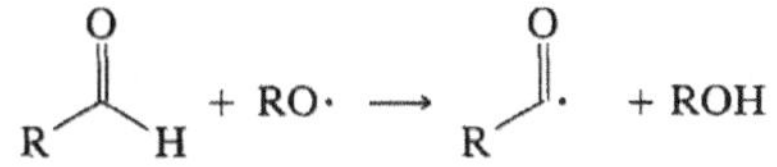

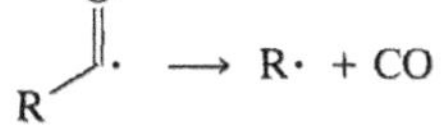

In the presence of a suitable compound, hydrogen abstraction may compete with loss of CO, as in the following example.

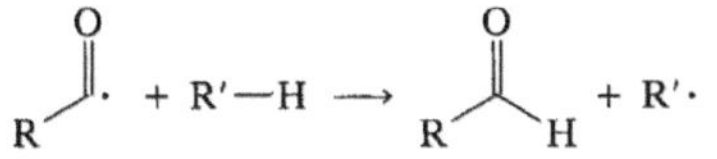

EXAMPLE 5.8. ADDITION FOLLOWED BY FRAGMENTATION

The radical reaction of carbon tetrachloride with aliphatic double bonds involves addition of the trichloromethyl radical to the double bond, followed by chlorine atom abstraction from carbon tetrachloride by the intermediate radical to give the product. After the addition of the trichloromethyl radical to β-pinene, a fragmentation occurs prior to formation of the product.

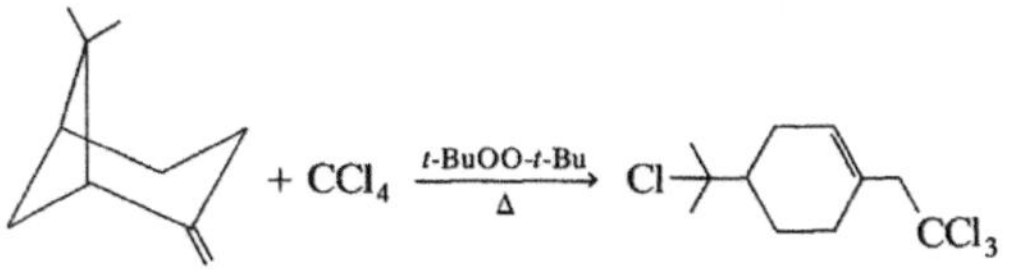

The mechanism for this reaction starts with the generation of the trichloromethyl radical:

$$t\text{-BuOO-}t\text{-Bu} \xrightarrow{\Delta} 2t\text{-BuO·}$$

The trichloromethyl radical now adds regiospecifically to the double bond to form a new carbon radical, **5-9**:

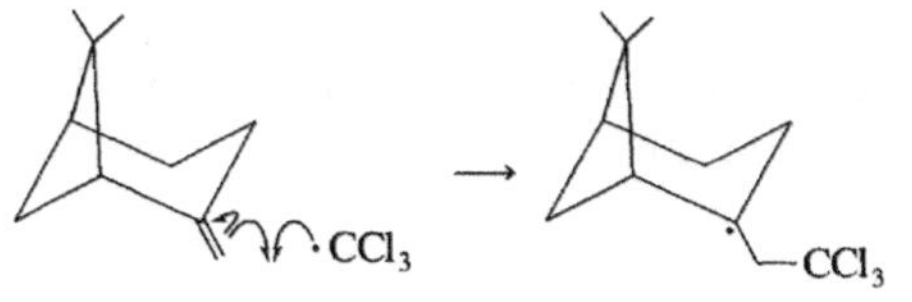

5-9

This radical can now fragment to give another radical, **5-10**, which then abstracts a chlorine atom from another molecule of CCl₄ to give the product:

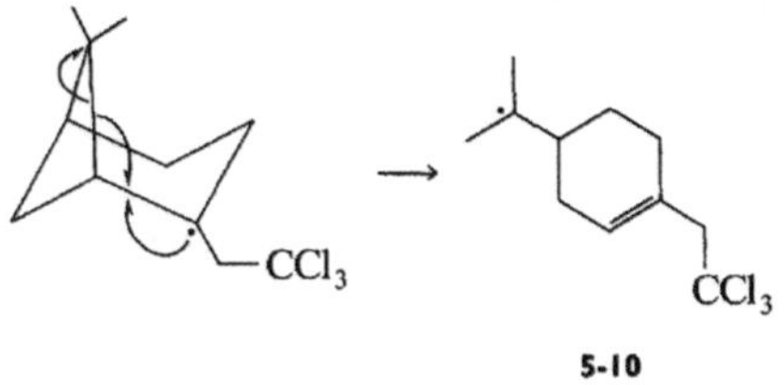

5-10

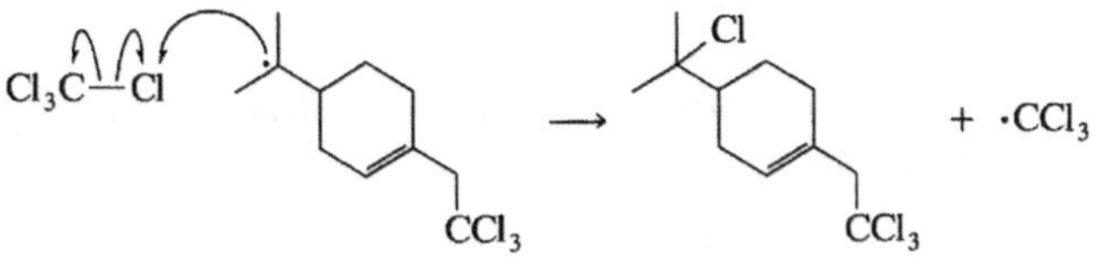

The radical addition of thiolacetic acid to β-pinene gives unrearranged product. This result is evidence for the discrete existence of radical **5-9**. That is, the rate of abstraction of a hydrogen atom from thiolacetic acid by **5-11** is faster than its rate of fragmentation. Note that the product is a mixture of stereoisomers, as indicated by the wavy bond lines.

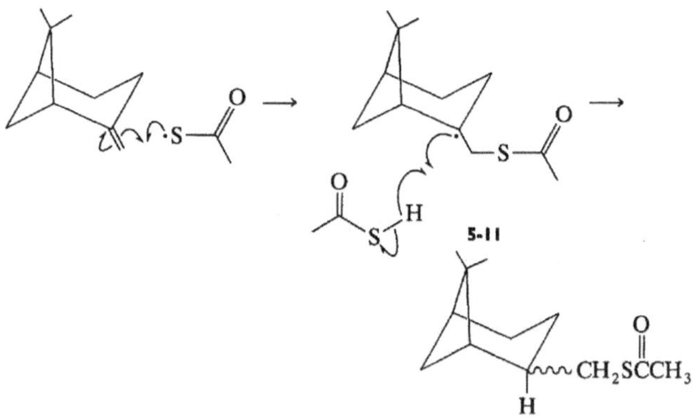
5-11
CH2SCCH3

Source: (See Claisse et al., (1970)).

PROBLEM 5.9

Write step-by-step mechanisms for the following reactions.

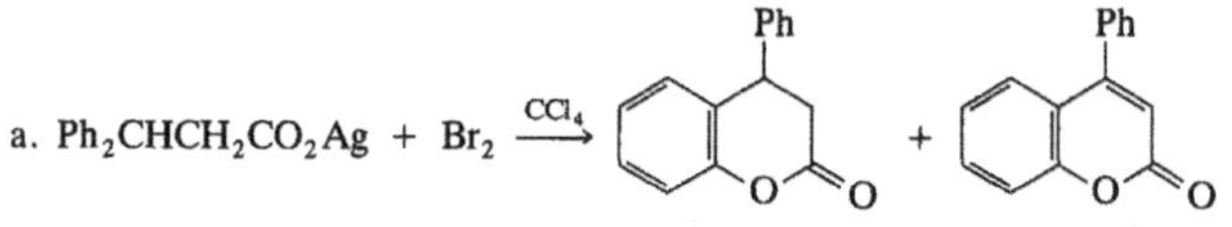
a. Ph2CHCH2CO2Ag + Br2
CCl4
Ph
Ph

43% 25%

5-12 5-13

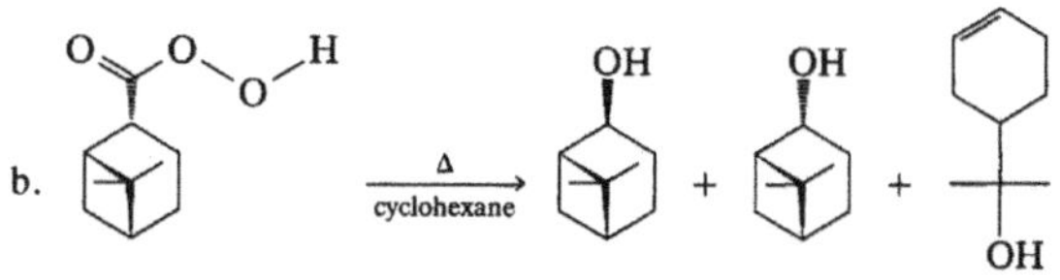
b.
Δ
cyclohexane
OH
OH
OH

5-14 5-15 5-16 5-17

In writing your mechanism, take into account the following experimental observations: (i) the other isomer (with the peracid group up) of the starting material reacts to give roughly the same ratio of **5-15** and **5-16**, and (ii) the ratio of bicyclic to ring-opened products increases with increasing peracid concentration.

Sources: Pandet and Dirk, (1963); Fossey et al., (1986).

8. REARRANGEMENT OF RADICALS

Rearrangements of radicals are much less common than rearrangements of carbocations.

HINT 5.7

In radical rearrangements, the migrating groups are those that can accommodate electrons in a π system (vinyl, aryl, and carbonyl) or atoms that can expand their valence shell, i.e., all halogens but fluorine. *Hydrogen and alkyl do not migrate* to radicals. However, an addition–elimination pathway could give the appearance of alkyl migration (Example 5.11). For a discussion of the reasons why alkyl and hydrogen do not migrate, see Carey and Sundberg, (1990).

EXAMPLE 5.9. ARYL MIGRATION

In the following reaction, **5-19** is a product resulting from migration of an aryl group, and **5-18** is a nonrearranged product. Thus, there is competition between rearrangement and hydrogen abstraction by a radical intermediate.

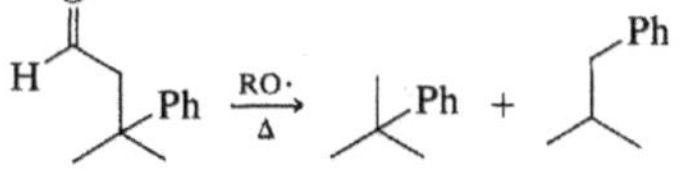

The mechanism for formation of **5-18** and **5-19** involves formation of a hydrocarbon radical from the starting aldehyde. The radical from the initiator abstracts the aldehyde proton to give a carbonyl radical. This loses carbon monoxide to give **5-20**.

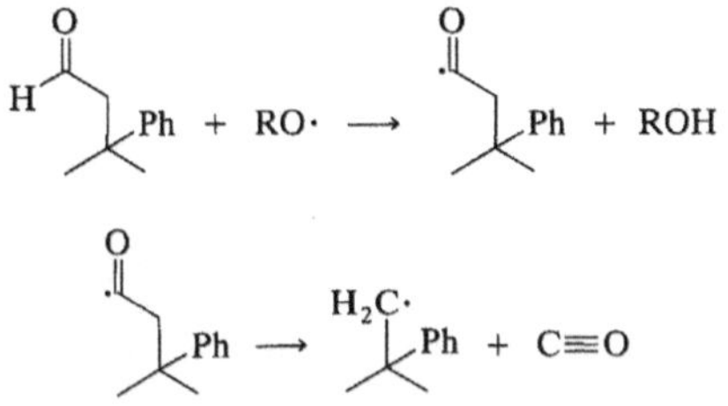

Radical **5-20** can either abstract a hydrogen atom from the starting aldehyde to give nonrearranged product, **5-18**, or rearrange via phenyl migration to **5-21**, which then abstracts a hydrogen atom to give **5-19**.

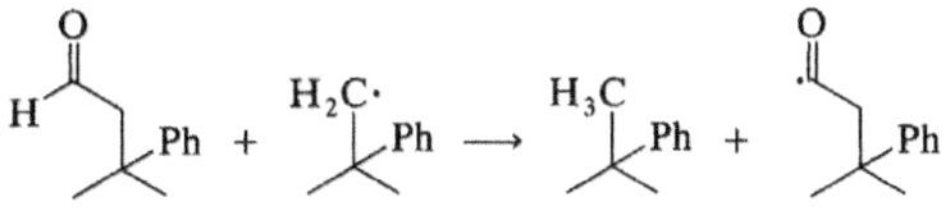

5-20   5-18

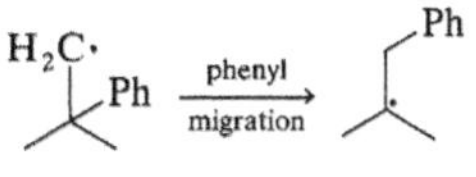

5-20   5-21

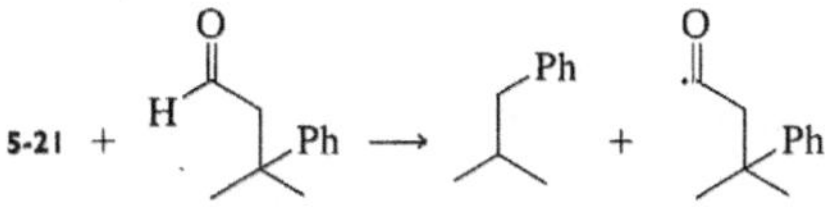

5-21 +   5-19

EXAMPLE 5.10. HALOGEN MIGRATION

The radical addition of HBr to 3,3,3-trichloropropene involves migration of a chlorine atom in an intermediate step.

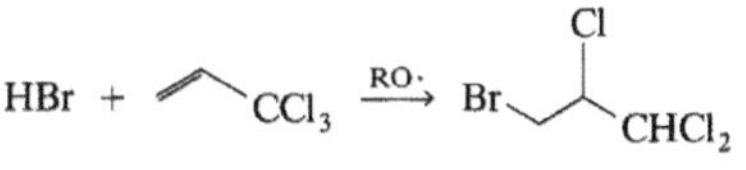

The mechanism for the reaction involves a regiospecific addition of bromine radical to the double bond.

$$RO\cdot + HBr \longrightarrow ROH + Br\cdot$$

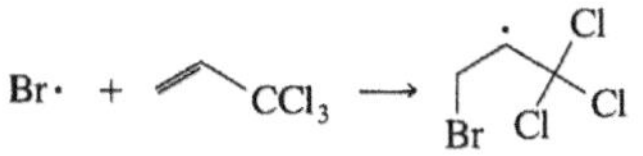

A chlorine atom then migrates to the adjacent radical, giving **5-22**. Finally, **5-22** abstracts a hydrogen from HBr to form the product plus another bromine radical to continue the chain.

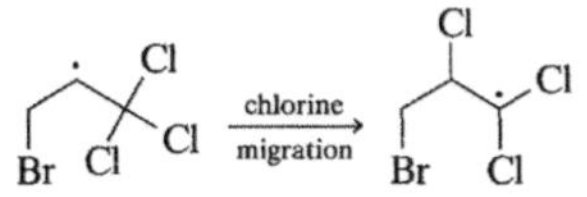

5-22

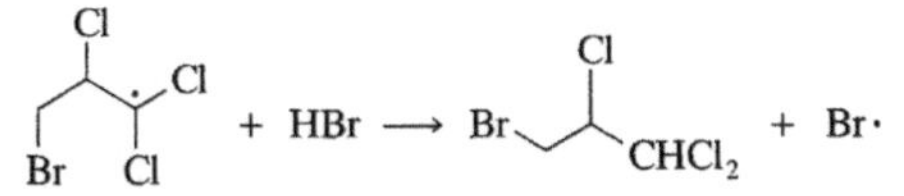

5-22

EXAMPLE 5.11. AN APPARENT ACYL MIGRATION MEDIATED BY RADICAL ADDITION TO A CARBONYL GROUP

The overall reaction is as follows:

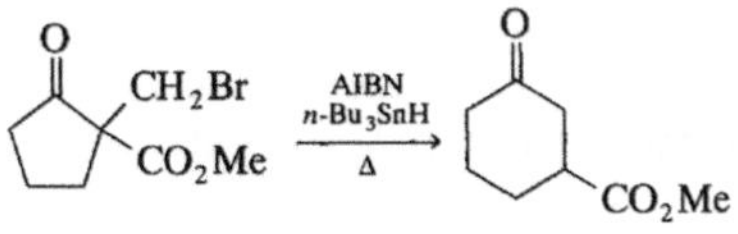

In this reaction, the tri-*n*-butyltin radical, formed by the usual initiation steps, removes the bromine atom from the starting material to give **5-23**. This radical then adds to the carbonyl carbon, forming a three-membered ring that then opens at a different bond to give **5-24**. On the basis of Hint 5.7, we would not expect direct formation of **5-24** by migration of carbon C-1 in **5-23** to the radical center.

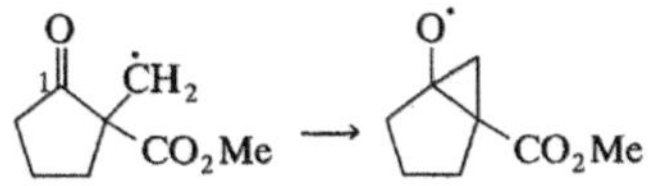

5-23

Radical **5-24** can abstract a hydrogen atom from tri-*n*-butyltin hydride to give the product and a new radical to propagate the chain:

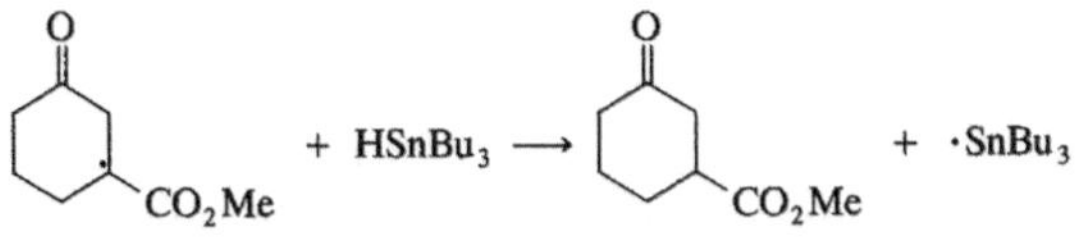

5-24

Source: Dowd and Choi, (1987).

PROBLEM 5.10

Compare the two reactions shown and explain why one process gives rearrangement and the other does not. Write step-by-step mechanisms for both processes.

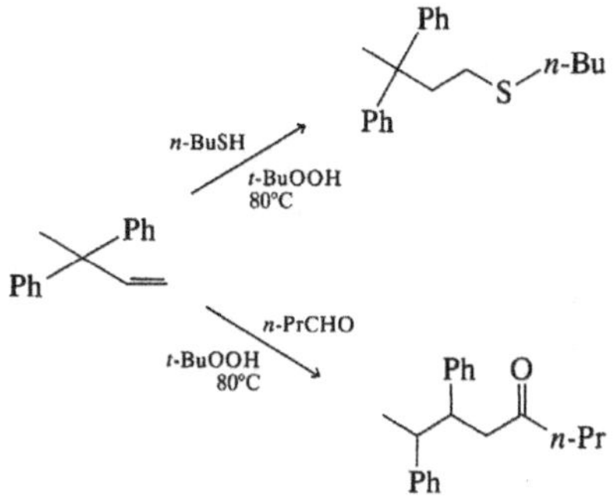

Source: Weinstock and Lewis, (1957).

PROBLEM 5.11

Write a complete mechanism for the following process.

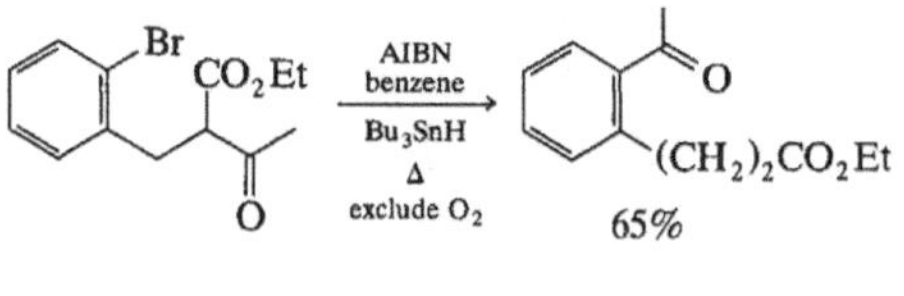

Source: Beckwith et al., (1988).

9. THE S_RN1 REACTION

The $S_{RN}1$ reaction, a versatile synthetic tool, is initiated by generation of a radical anion. The designation $S_{RN}1$ indicates that the reaction is a nucleophilic substitution proceeding through a radical intermediate and that the rate-limiting step is unimolecular decay of the radical anion intermediate formed from the substrate (see a review by Bunnett, (1978)). $S_{RN}1$ reactions occur with both aliphatic and aromatic compounds.

For the general case, the propagation steps for the $S_{RN}1$ reaction can be represented as follows:

$$[RX]^{\bar{\cdot}} \longrightarrow R\cdot + X^-$$
$$R\cdot + Y^- \longrightarrow [RY]^{\bar{\cdot}}$$
$$[RY]^{\bar{\cdot}} + RX \longrightarrow RY + [RX]^{\bar{\cdot}}$$

Note that the radical anion consumed in the first propagation step is regenerated in the third propagation step.

$S_{RN}1$ reactions can be initiated by photochemical excitation, electrochemical reduction, and solvated electrons (alkali metal in ammonia). In some cases, spontaneous thermal initiation can also take place. The leaving group, X^-, is often a halide—frequently bromide or iodide, never fluoride. The nucleophile, Y^-, is commonly a nitroalkane anion (5-25) or another anion such as thiolate (RS^-), phenolate (PhO^-), or various enolates.

5-25

Because these are free-radical reactions, they are inhibited by the addition of free-radical inhibitors such as di-t-butyl nitroxide and 1,4-dinitrobenzene.

Addition of the propagation steps of the general mechanism for an $S_{RN}1$ reaction gives the following equation for the overall reaction:

$$RX + Y^- \longrightarrow RY + X^-$$

This equation also describes the overall reaction of either an S_N2 or a nucleophilic aromatic substitution process. In some cases, the only way to distinguish an $S_{RN}1$ reaction from these processes is that an $S_{RN}1$ is inhibited by radical inhibitors. Another distinguishing feature is that the order of the relative leaving group abilities of halides are opposite to that found for nucleophilic aromatic substitution by the addition—elimination mechanism (Chapter 3).

EXAMPLE 5.12. REACTION OF AN ENOLATE WITH AN AROMATIC IODIDE

The overall reaction is as follows:

70%

The mechanism involves photochemical excitation of the enolate. (An excited state is indicated by an asterisk, *.) The excited enolate transfers an electron to the aromatic π system. Because a single electron has been added, there must be an unpaired electron present; thus, the new intermediate, **5-26**, is a radical. Because an electron has been added to a neutral system, there must be a negative charge. Thus, **5-26** is both an anion and a radical or what is usually called a radical anion or an anion radical.

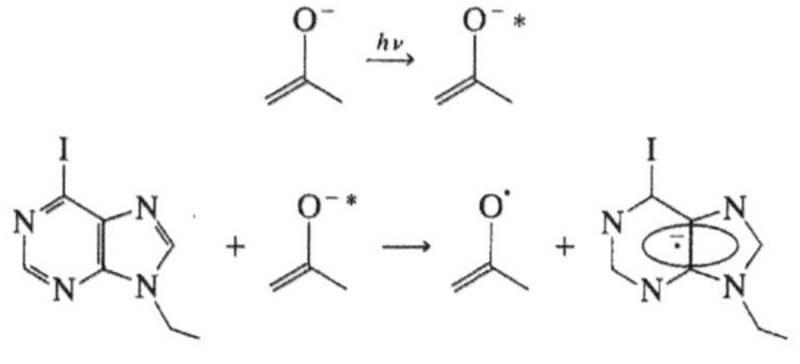

5-26

The radical anion now loses iodide ion to give a radical.

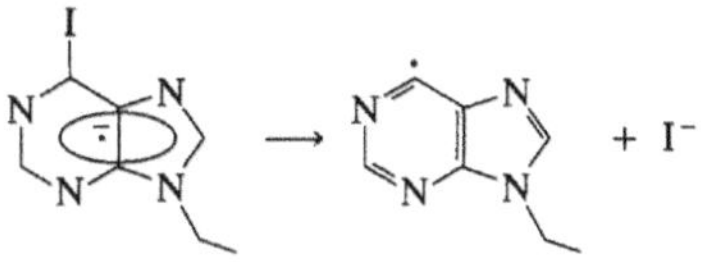

This radical couples with the enolate to give a new radical anion, **5-27**.

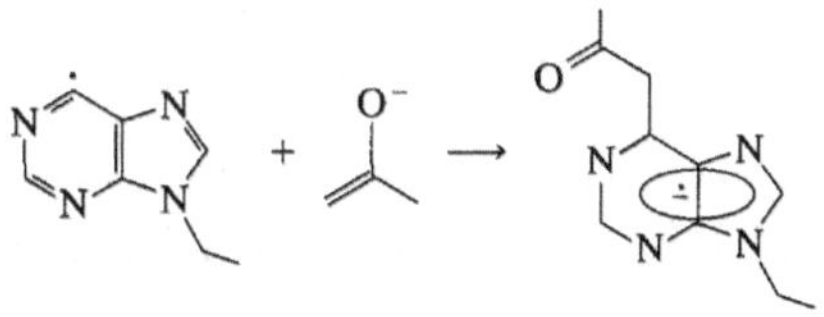

5-27

Intermediate **5-27** transfers an electron to starting material to give a molecule of product and a new molecule of **5-26** to propagate the chain.

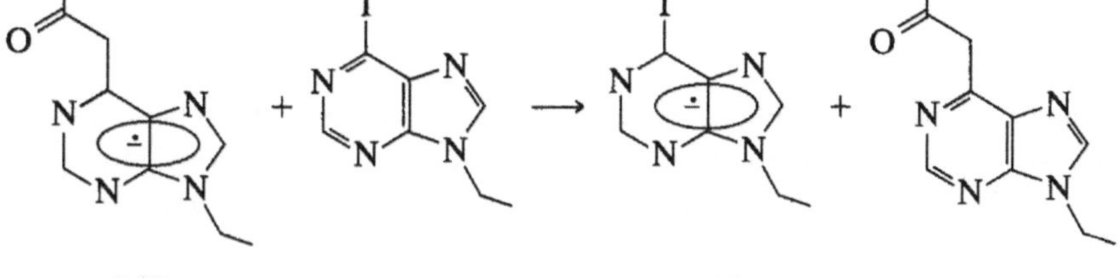

5-27   **5-26**

Source: Nair and Chamberlain, (1985).

PROBLEM 5.12

Write step-by-step mechanisms for the following reactions:

a.

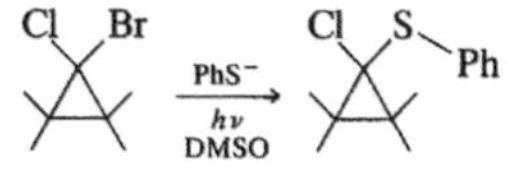

b.

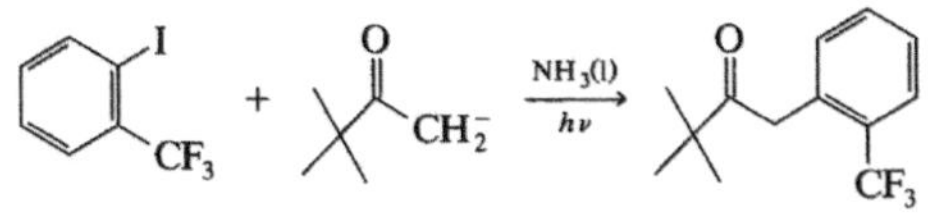

Sources: Meijs, (1986); Bunnett and Galli, (1985).

10. THE BIRCH REDUCTION

The typical conditions for the Birch reduction are sodium in liquid ammonia that contains a small amount of ethanol. Workup generally is in acid.

EXAMPLE 5.13. BIRCH REDUCTION OF BENZOIC ACID

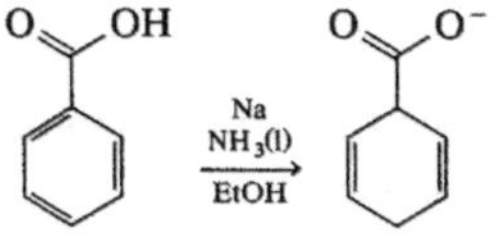

90%

Under the basic reaction conditions, the carboxylate ion is formed from benzoic acid. Sodium then transfers a single electron to the aromatic system to produce a pentadienyl radical anion, **5-28**. Protonation of **5-28** gives the radical, **5-29**, which will be reduced to the corresponding anion, **5-30**, by another sodium atom. Finally, **5-30** is protonated, giving the product.

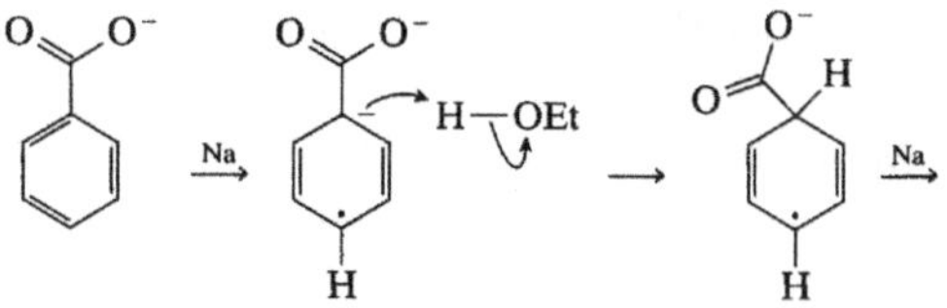

5-28   **5-29**

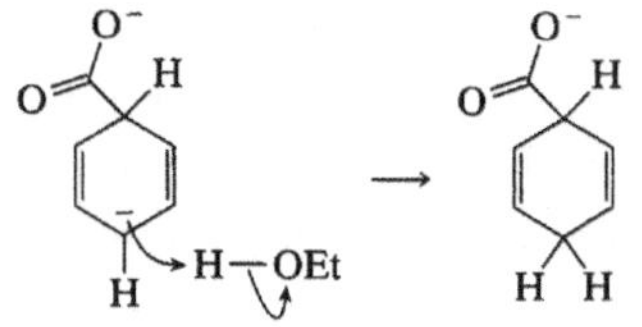

5-30

Molecular orbital calculations indicate that the intermediate radical anions have more anionic character at the positions *ortho* or *meta* to a π-donating substituent and *para* or *ipso* (*ipso* means at the substituent-containing carbon) to a π-accepting substituent. (See Birch et al., (1980)). The penta-dienyl anions formed after transfer of the second electron have more anionic character on the central carbon, and that is where the second protonation occurs.

The result of the various factors is that in the product 1,4-cyclohexadienes, π-withdrawing groups (e.g., $-CO_2H$) are found on the saturated carbons, whereas π-donating groups (e.g., $-OCH_3$) are found on the unsaturated carbons.

PROBLEM 5.13

Write a mechanism for the following reaction that is consistent with the regiochemistry observed.

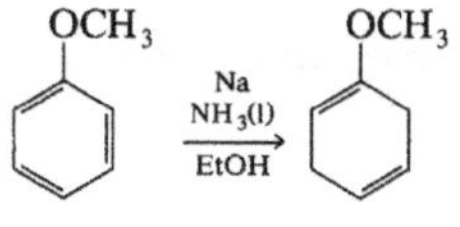

11. A RADICAL MECHANISM FOR THE REARRANGEMENT OF SOME ANIONS

In the Wittig rearrangement, an anion derived from an ether rearranges to the salt of an alcohol.

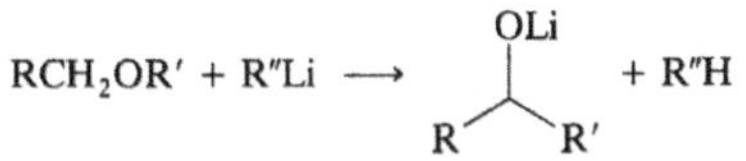

$$RCH_2OR' + R''Li \longrightarrow \underset{R \quad R'}{\overset{OLi}{|}} + R''H$$

In some cases, the mechanism for the formation of at least some of the product(s) of reaction appears to be radical scission of the anion produced initially.

The radicals can now recombine to form the product:

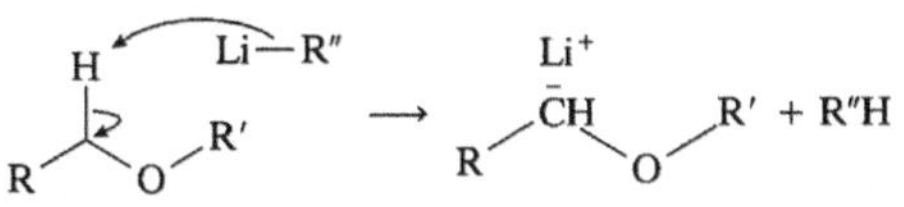

5-31

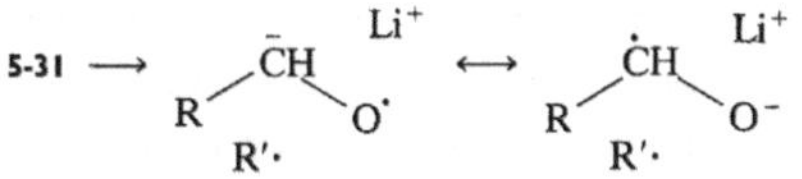

There is evidence to suggest that some of these reactions go by an ionic mechanism. The purpose in introducing a radical mechanism here is to indicate that when there are several possible mechanisms for a reaction, some may be radical in character.

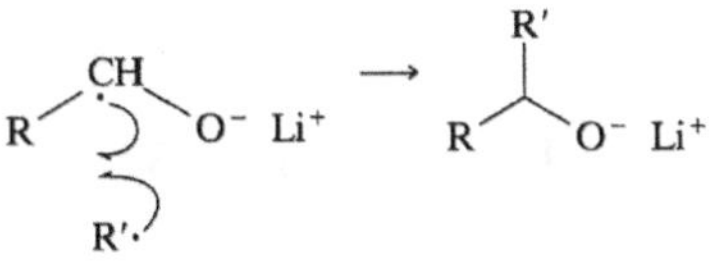

PROBLEM 5.14

Write reasonable mechanisms for the formation of each of the products in the following reaction.

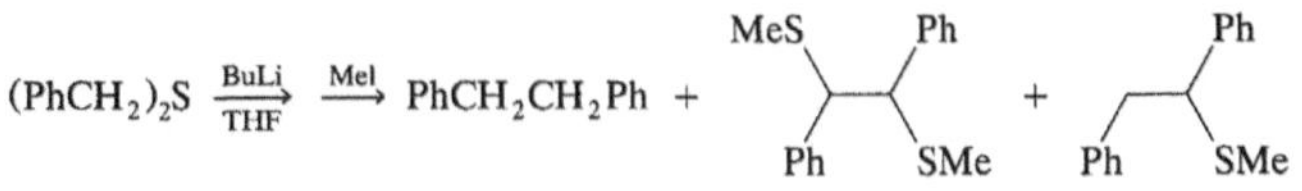

$$(PhCH_2)_2S \xrightarrow[\text{THF}]{\text{BuLi}} \xrightarrow{\text{MeI}} PhCH_2CH_2Ph + \underset{Ph \quad SMe}{\overset{MeS \quad Ph}{\diagup}} + \underset{Ph \quad SMe}{\overset{Ph}{\diagup}}$$

Source: Biellmann and Schmitt, (1973).

Other reactions that may involve radical scission of anion intermediates include the rearrangement of anions adjacent to either trivalent or tetravalent nitrogen.

EXAMPLE 5.14. UNUSUAL REARRANGEMENT OF AN ANION α TO A NITROGEN

In this example, an anion formed initially undergoes radical scission, recombination, and then anionic rearrangement to the final product.

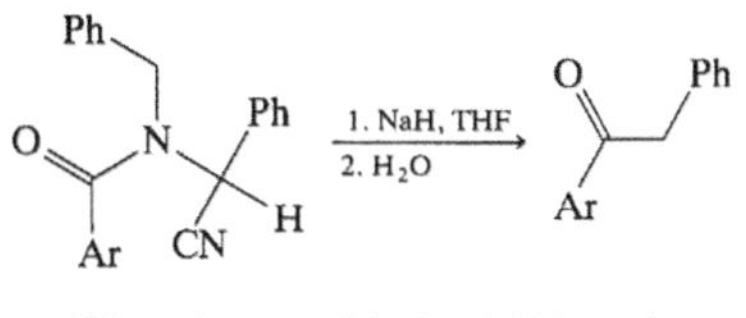

When Ar = p-tolyl, the yield is 57%.

In the paper cited, the following mechanism was proposed. The hydride removes the proton on the carbon α to the nitrile, leaving a carbanion. Then a benzyl radical cleaves, leaving a resonance-stabilized radical anion, **5-32**.

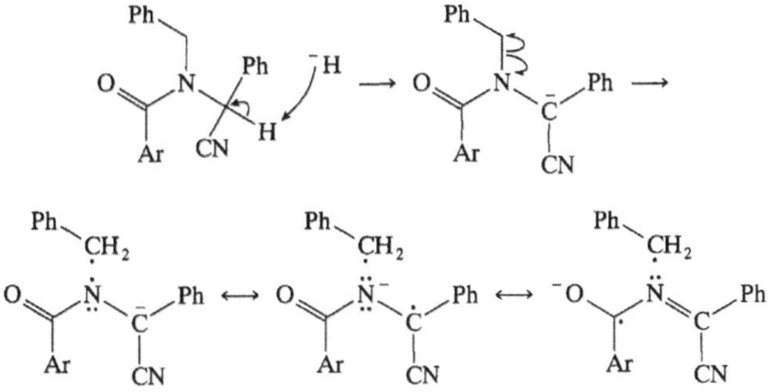

The benzyl radical then recombines at the carbon of the carbonyl group in **5-32**, and the subsequent elimination of benzonitrile and cyanide ion gives the product.

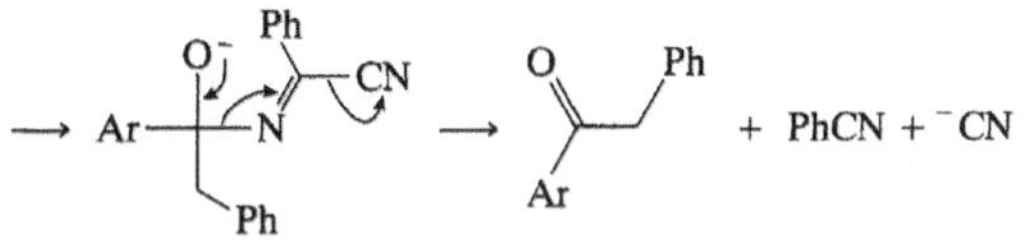

Source: Stamegna and McEwen, (1981).

PROBLEM 5.15

Consider the following data and then write reasonable mechanisms for the formation of **5-33** and **5-34**.

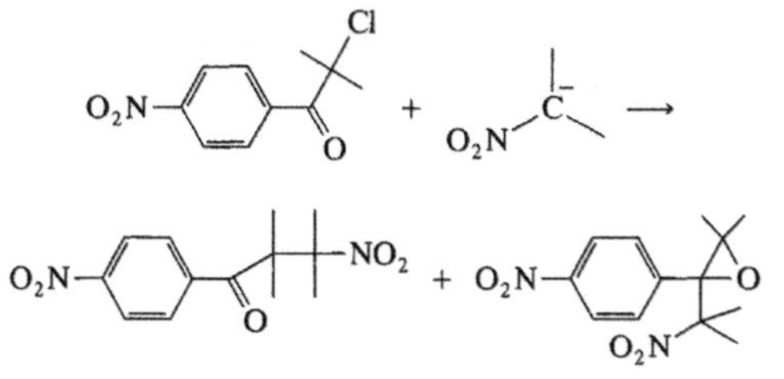

5-33   **5-34**

(a) 10 mol% di-*t*-butyl nitroxide completely suppressed formation of **5-33**.
(b) The yield of **5-34** was higher in the dark or in the presence of nitroxide.
(c) The yield of **5-33** was lower in the dark than with sunlamp irradiation.

Source: Russell and Ros, (1985).

PROBLEM 5.16

Write step-by-step mechanisms for the following reactions.

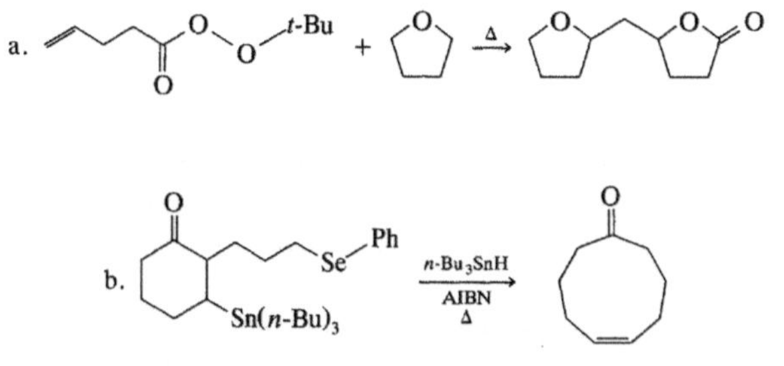

a.

b.

Only 10 mol% tin hydride is used.

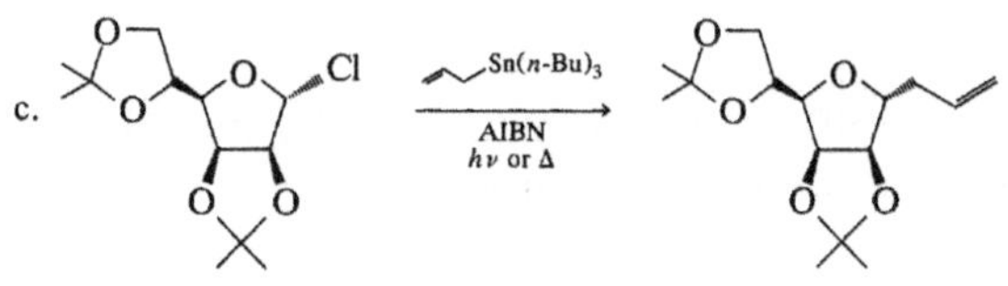

c.

79%

d.

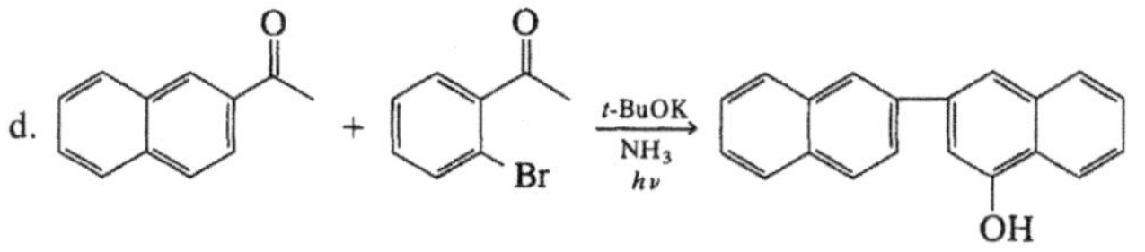

5-35

The yield of **5-35**, when the reaction mixture is irradiated, is 72%; in the dark, none of this product is formed.

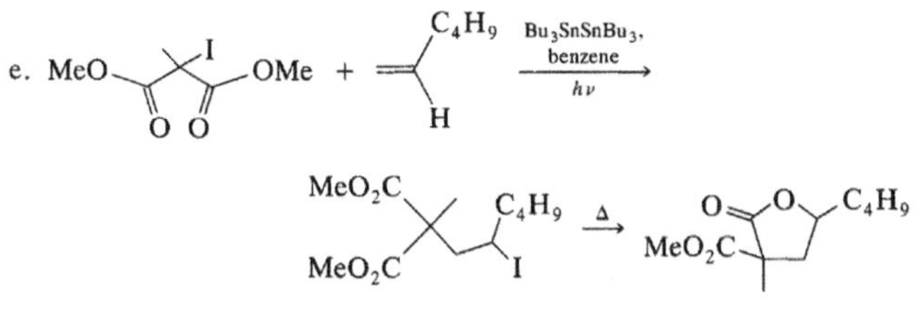

e.

5-36   **5-37**

Hexabutylditin was present in 10 mol%. The yield of **5-37** from 1-hexene and iodomalonate was 69%.

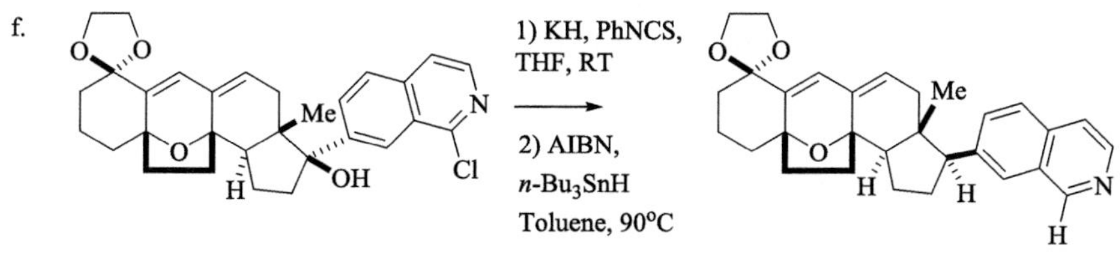

f.

1) KH, PhNCS, THF, RT

2) AIBN, n-Bu₃SnH
Toluene, 90°C

Sources: Hart, (1984); Kharrat et al., (1984); Baldwin et al., (1988); Beugelmans et al., (1987); Curran et al., (1989); Yamashita and Iso, (2011).

ANSWERS TO PROBLEMS

Problem 5.1

The radical formed is stabilized by resonance and by two alkyl groups. In addition, the small, stable molecule N_2 is formed in the process (Hints 2.14 and 2.15).

$$CH_3-\overset{\underset{\displaystyle CH_3}{|}}{\dot{C}}-C\equiv N: \longleftrightarrow CH_3-\overset{\underset{\displaystyle CH_3}{|}}{C}=C=\dot{N}:$$

Problem 5.2

$$t\text{-BuO}\cdot \; + \; \diagup\!\!\!\diagdown \longrightarrow \; \diagup\!\!\!\!\diagup\!\!\!\diagdown\!\!-O\text{-}t\text{-Bu}$$

$$t\text{-BuO}\cdot \; + \; Cl\cdot \longrightarrow t\text{-BuOCl}$$

The last coupling is the reverse of an initiation process. It removes t-butoxy radical from the chain propagation steps and, thus, is a termination process.

Problem 5.3

The three equations shown add up to the same overall reaction as Eqns (1) and (2) in Example 5.3. Thus, the total enthalpy change, −47 kcal/mol, is the same as the total enthalpy change calculated in Example 5.3. Why then do these equations not represent a mechanism for the reaction? There are two reasons. First, the equations in the problem do not represent a chain process, so that the thermochemistry of each step, *including the first one*, must be considered. Because this first reaction is highly endothermic, the overall process would be very slow. Second, the third step involves the coupling of two radicals. Whereas the enthalpy for this reaction is very favorable, both radicals will be present in very low concentrations. Thus, the probability that they will collide and react is very low. In contrast, both propagation steps in Example 5.3 have a radical colliding with a stable molecule whose concentration is much higher than that of any radical intermediate.

Problem 5.4

Step 1

$$\underset{Ph}{}\overset{O}{\|}\!\!-\!O\!-\!O\!\!\overset{O}{\|}\underset{Ph}{} \longrightarrow 2\; \underset{Ph}{}\overset{O}{\|}\!\!-\!O\cdot$$

Step 2

$$\underset{Ph}{}\overset{O}{\|}\!\!-\!O\cdot \; + \; Br\!\!-\!CBr_3 \longrightarrow \underset{Ph}{}\overset{O}{\|}\!\!-\!O\!-\!Br \; + \; \cdot CBr_3$$

Step 3

$$\cdot CBr_3 \; + \; \diagup\!\!\!\!\diagdown\!\!\overset{H}{\underset{C_6H_{13}}{}} \longrightarrow Br_3C\!\!\diagdown\!\!\overset{H}{\underset{C_6H_{13}}{\cdot}}$$

5-38

Step 4

$$Br_3C\!\!\diagdown\!\!\overset{H}{\underset{C_6H_{13}}{}} \; + \; Br\!\!-\!CBr_3 \longrightarrow Br_3C\!\!\diagdown\!\!\overset{H\;\;Br}{\underset{C_6H_{13}}{}} \; + \; \cdot CBr_3$$

a. The initiation steps are steps 1 and 2. The propagation steps are steps 3 and 4.
b. Addition of the tribromomethyl radical to the double bond could give either a primary or a secondary radical. Regiospecific addition leads to the more stable secondary radical, **5-38**.

Problem 5.5

a. *Initiation Steps*

Thermal decomposition of AIBN and then hydrogen abstraction from tri-*n*-butyltin hydride.

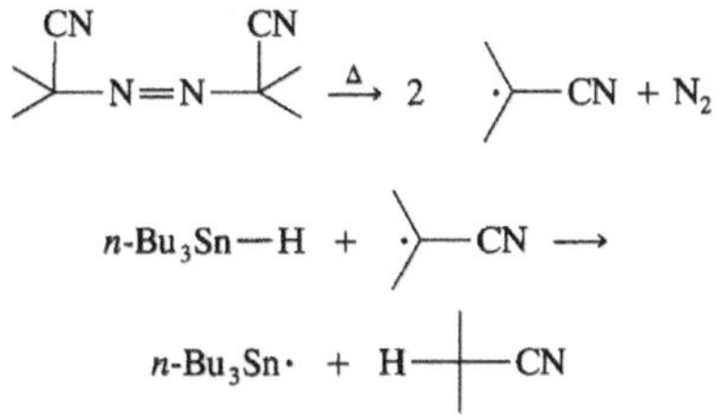

Propagation Steps

(1) The tin radical abstracts an iodine atom.

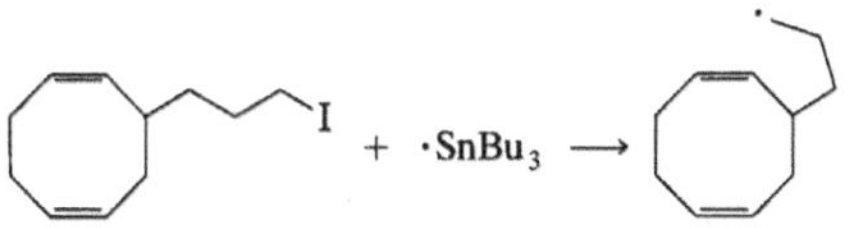

(2) The resulting radical cyclizes.

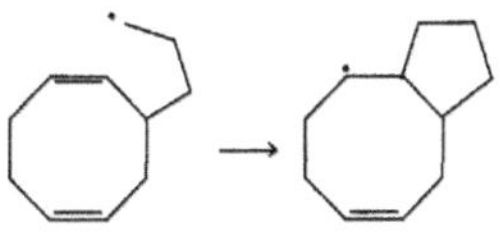

5-39

(3) The cyclized radical abstracts a hydrogen atom from tri-*n*-butyltin hydride to give the product and a new tri-*n*-butyltin radical to continue the chain.

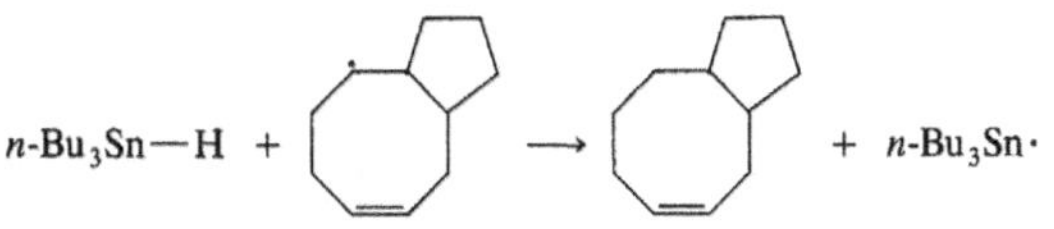

5-39

b. Radical **5-39** can undergo further cyclization across the ring. Such transannular reactions are common in medium-sized rings.

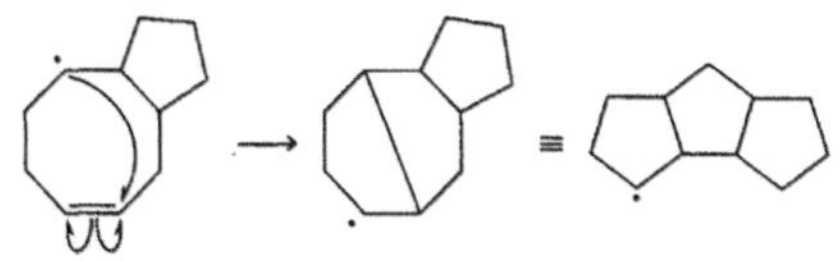

The resulting radical also can abstract a hydrogen from tri-*n*-butyltin hydride to produce the hydrocarbon and a tin radical to continue the chain.

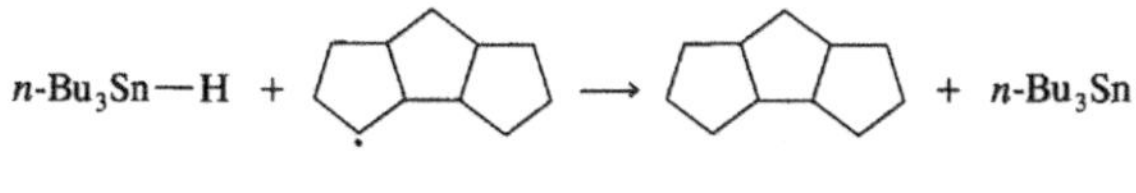

The relative ease with which transannular reactions occur in medium-sized rings becomes apparent when we look at the actual three-dimensional structure of these rings. In their most stable conformation, the hydrogen atoms attached to the ring are staggered, and this means that the carbon atoms across the ring are sufficiently close to form bonds when the appropriate functionality is present. Although medium-sized rings are often represented as regular polygons, alternative representations give a much more realistic picture of the molecular configuration.

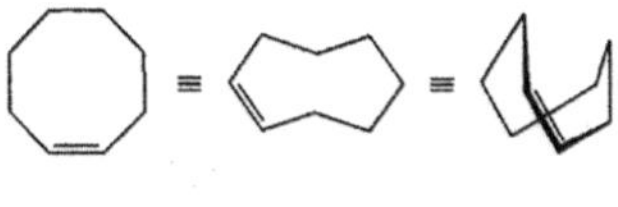

These kinds of pictures make it easier to rationalize transannular reactions.

Problem 5.6

Initiation Steps

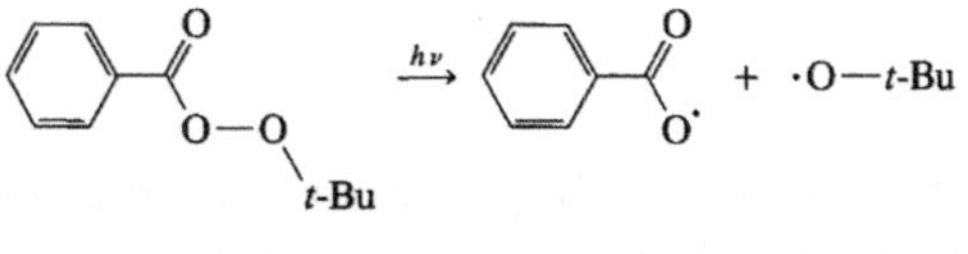

Either of the resulting radicals can remove a hydrogen from tri-*n*-butyltin hydride.

$$t\text{-BuO}\cdot \ + \ \text{H}-\text{SnBu}_3 \ \longrightarrow \ t\text{-BuOH} + \cdot\text{SnBu}_3$$
$$\text{PhCO}_2\cdot \ + \ \text{H}-\text{SnBu}_3 \ \longrightarrow \ \text{PhCO}_2\text{H} + \cdot\text{SnBu}_3$$

Propagation Steps
The tributyltin radical abstracts a bromine atom from the starting material.

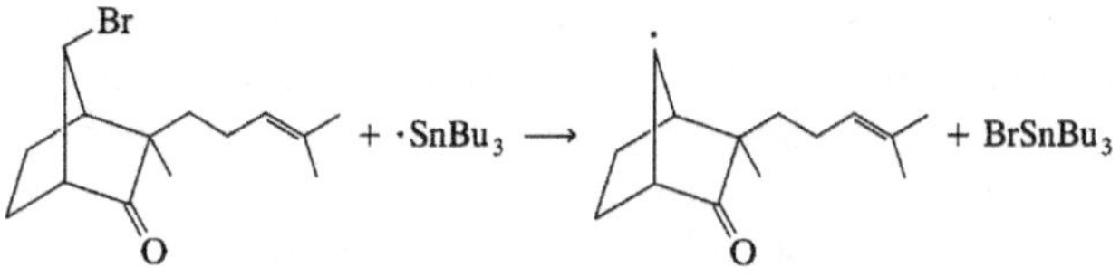

The radical undergoes intramolecular cyclization. This cyclization is regiospecific to form a new radical on the exocyclic carbon. This is expected on the basis of the Baldwin–Beckwith guidelines (Section 6.B).

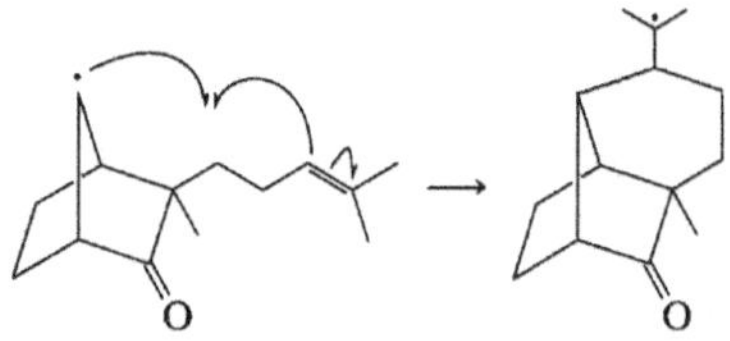

The cyclized radical abstracts a hydrogen from another molecule of tributyltin hydride to generate the product and another tributyltin radical.

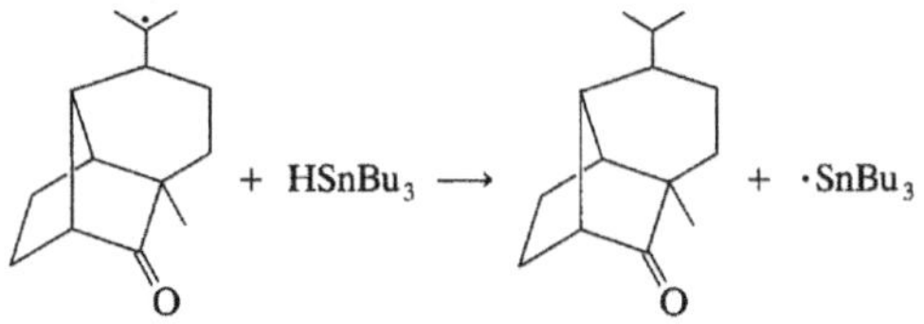

Problem 5.7

Note that this is not a chain process. The reaction, known as the Barton nitrite photolysis, was developed to functionalize steroids in positions that are difficult to activate by other routes. The first step is photolysis of the nitrite ester, cleaving the relatively weak N–O bond.

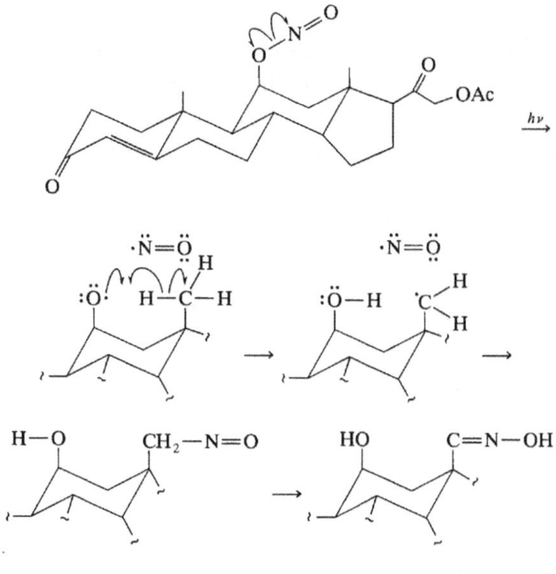

Because of the conformational rigidity of the steroid ring system, the nitrite ester and the methyl group that becomes functionalized are in close proximity, and the oxyradical formed by photolysis of the nitrite ester is well-positioned to abstract a hydrogen from the methyl group via a six-membered cyclic transition state. (The formation of a five- or six-membered ring is a common feature of the transition state for many reactions because bond lengths, bond angles, and entropy all combine to stabilize these ring sizes.) Presumably, the intramolecular hydrogen abstraction is so rapid that the nitroso radical is still close enough to recombine with the primary carbon radical. The resulting nitroso compound tautomerizes to the oxime on workup.

Problem 5.8

This reaction is initiated by heating the mixture to benzene reflux which causes the AIBN decomposition. The AIBN breaks down into nitrogen gas and two equivalents of the cyano-prop-2-yl radicals **5-40**. The cyanopropyl radicals abstract hydrogen from the tin hydride generating tri-*n*-butyltin radicals **5-41**.

Initiation

The tin radical abstracts the iodine generating an alkyl radical **5-42** that adds into the olefin to form a second five-membered ring. In this first step, the terminal acetylene, which is more reactive than the olefin, is held out away from the newly formed radical and cannot be easily reached to react. This first cyclization occurs to form a *cis* ring fusion (generating the five-membered ring on one side of the original center ring) **5-43**. The new radical then cyclizes into the easily accessed alkyne to form the third ring via a 5-exo cyclization **5-44**. This occurs from the other side of the original core ring system.

The terminal olefin radical then abstracts a hydrogen **5-45** from tri-*n*-butyltin to generate the product and more tin radical in the chain-propagating step **5-46**.

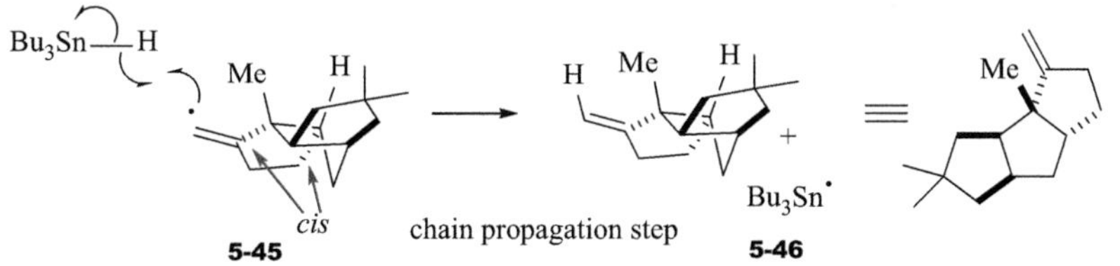

5-45 **5-46**

This example illustrates an important trend for radical reactions leading to the bond fusion of small rings. Systems that generate ring fusions for small rings (five-membered rings or smaller) will form *cis*-bond fusions. The corresponding *trans* relationship is unfavored geometrically (*trans* five-membered ring junctions are not possible).

Problem 5.9

Bromine reacts with the silver salt to form silver bromide and the acyl hypobromite. The latter undergoes homolytic scission to give the carboxy radical, **5-47**.

$$Ph_2CHCH_2CO_2Ag + Br_2 \longrightarrow Ph_2CHCH_2CO_2Br + AgBr$$

$$Ph_2CHCH_2CO_2Br \xrightarrow{\Delta} \underset{\textbf{5-47}}{Ph_2CHCH_2CO_2\cdot} + Br\cdot$$

Hydrogen abstraction by the bromine radical gives the radical **5-48**, which cyclizes, regenerating a bromine radical.

$$Ph_2CHCH_2CO_2Br + Br\cdot \longrightarrow \underset{\textbf{5-48}}{Ph_2\dot{C}HCH_2CO_2Br} + HBr$$

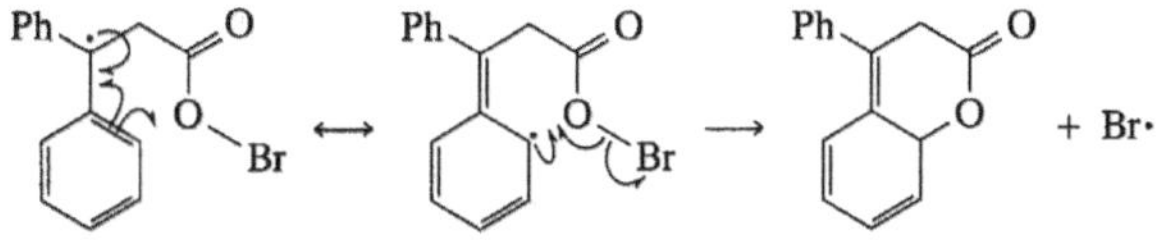

5-48-1 **5-48-2** **5-49**

The intermediate **5-49** can isomerize to product **5-12** by a protonation–deprotonation mediated by the strong acid HBr formed in the reaction medium.

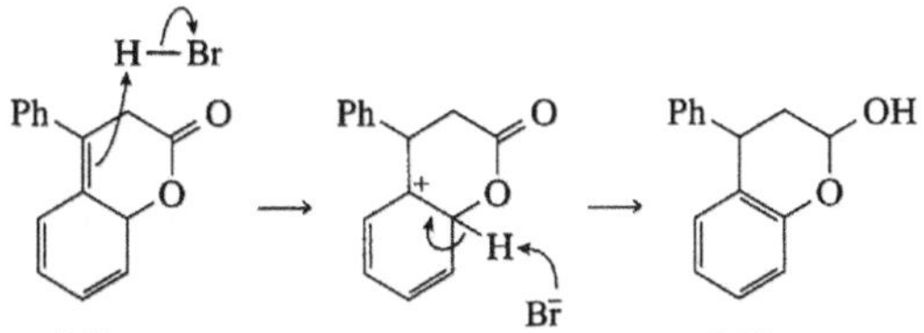

5-49 **5-12**

An alternative mechanism can be written in which **5-47** cyclizes to **5-50**, which is then oxidized by bromine to the corresponding carbocation. The product is then obtained by abstraction of a proton by bromide ion.

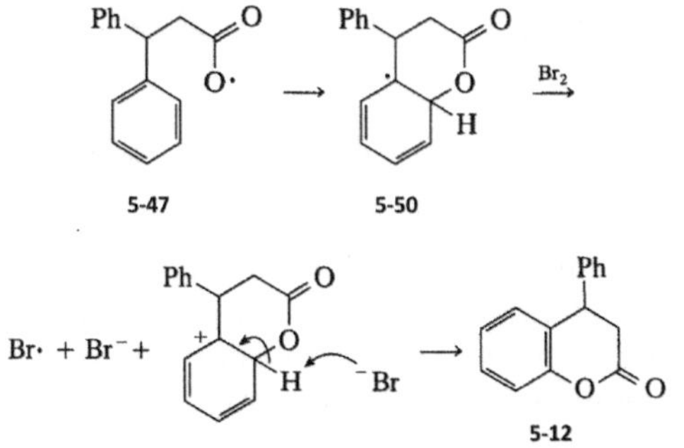

This second mechanism is less satisfactory because cyclization of the carboxy radical **5-47** is unlikely. Alkyl carboxy radicals are highly unstable and they decarboxylate so rapidly that fragmentation would be expected to occur faster than cyclization. The previous mechanism has at least two other advantages. First, it proceeds through the radical **5-48**, which is stabilized by resonance with two phenyl substituents, so that hydrogen abstraction to form **5-48** should compete effectively with homolytic scission of the acyl bromide to form **5-47**. Second, it will occur rapidly because it is a radical chain process; the bromine radical formed in the cyclization of **5-48** regenerates **5-48** from the acyl hypobromite.

Product **5-13** can be produced via hydrogen atom abstraction from **5-12** to form radical **5-51**, followed by mechanistic steps analogous to those for the transformation of **5-50** to **5-12**.

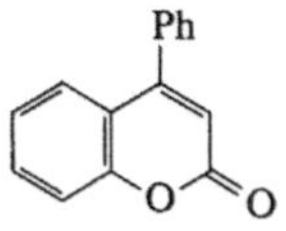

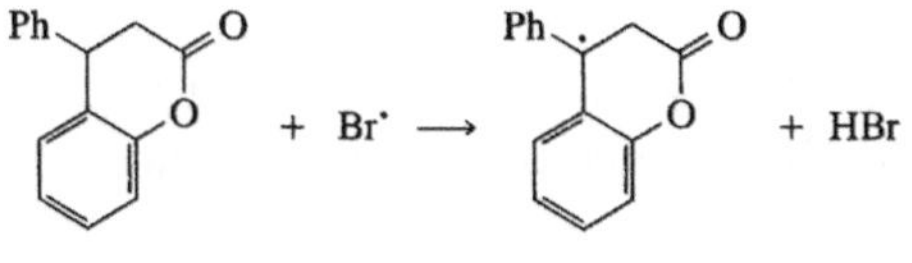

b. Note that the bicyclic structure of the starting material is the same as the bicyclic structure in Example 5.8. It has been written in a different (more old-fashioned) way. Once this is recognized, a mechanism can be written for the reaction, which is analogous to Example 5.8.

The first step is homolysis of the weakest bond in **5-14**, the O—O bond. This is followed by loss of CO_2.

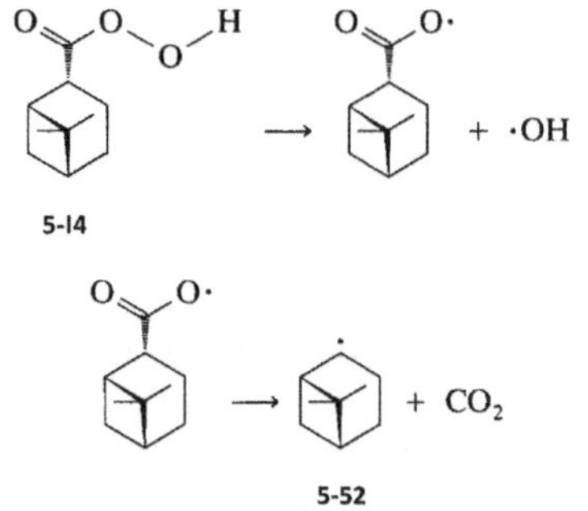

5-14

5-52

Because the ratio of bicyclic to ring-opened products increases with increasing peracid concentration, the bicyclic products (**5-15** and **5-16**) must be formed by reaction of the intermediate **5-52** with **5-14**. This bimolecular process competes with the unimolecular fragmentation of **5-52** that leads to the ring-opened product **5-17**.

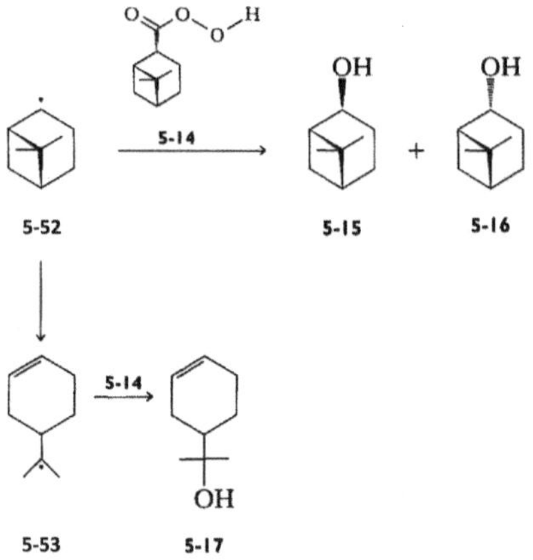

The intermediate radicals **5-52** and **5-53**, symbolized by •R′ in the following equation, can abstract OH from **5-14** to produce the product alcohols.

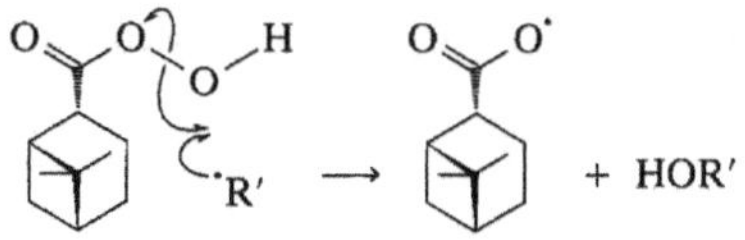

Because roughly the same amounts of **5-15** and **5-16** are produced from either isomer of starting material, **5-14**, the mechanism of the reaction cannot be concerted:

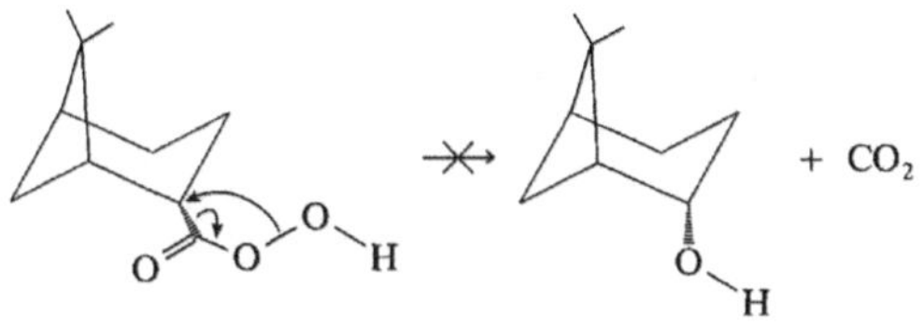

Problem 5.10

n-Butyl mercaptan is a better chain-transfer agent than the aldehyde because hydrogen abstraction from the mercaptan is much easier. Thus, the radical, formed by addition of *n*-butylthio radical to the aliphatic double bond, is captured by mercaptan before it can rearrange. This is not the case for reaction with aldehyde.

$$t\text{-BuOOH} \longrightarrow t\text{-BuO}\cdot + \cdot\text{OH}$$

$$t\text{-BuO}\cdot + n\text{-BuSH} \longrightarrow t\text{-BuOH} + n\text{-BuS}\cdot$$

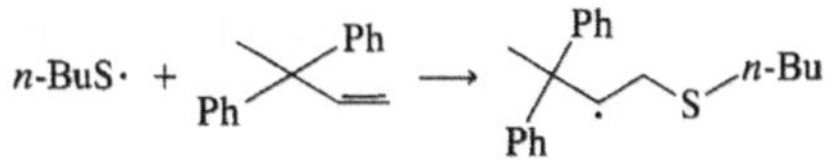

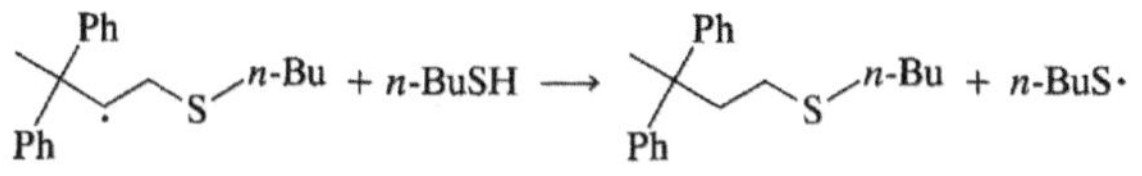

The reaction with aldehyde is as follows:

$$t\text{-BuO}\cdot + n\text{-PrCHO} \longrightarrow t\text{-BuOH} + n\text{-PrC}\dot{=}\text{O}$$

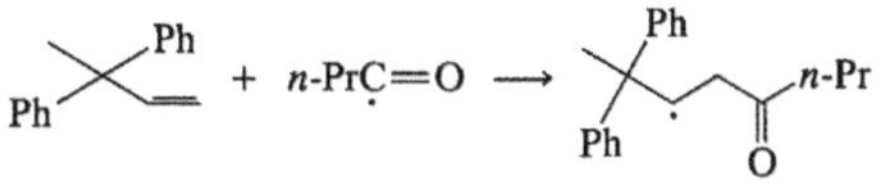

5-54

In this case, rearrangement of **5-54** competes effectively with its abstraction of hydrogen from starting material.

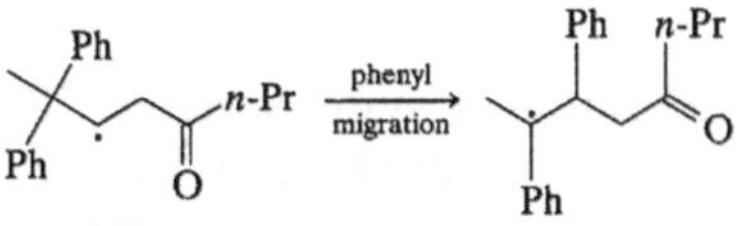

5-54

The rearranged radical abstracts a hydrogen from aldehyde to continue the chain.

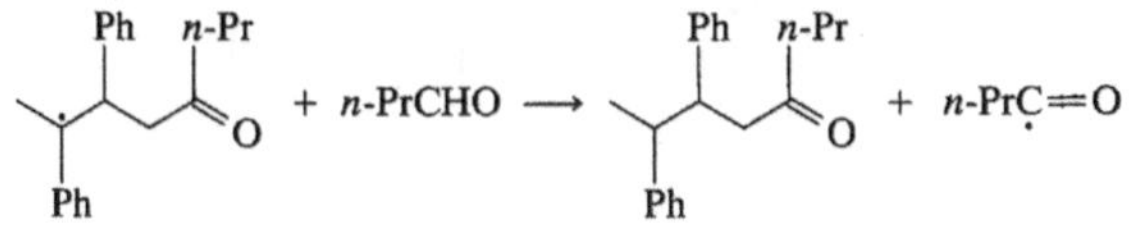

Problem 5.11

Numbering of the critical carbon atoms in the starting material and product gives a good indication of the course of the reaction.

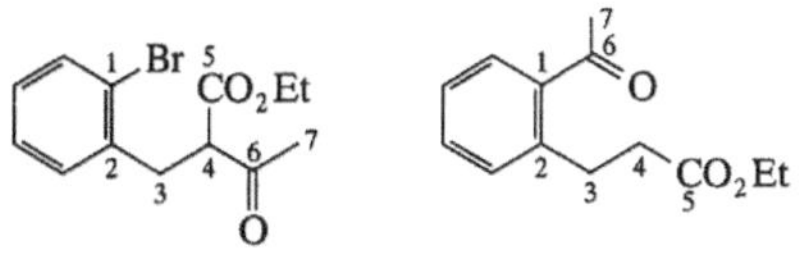

The acetyl group has "migrated" from C-4 in the starting material to C-1 in the product. Thus, the following sequence of events can be anticipated: (i) removal of the bromine atom from C-1, (ii) addition of the resulting radical to C-6, (iii) cleavage of the C-6 to C-4 bond, and (iv) reaction with tri-n-butyltin hydride to give the product and a radical to continue the chain.

The initiation steps, to form the tri-n-butyltin radical, are the same as those in Problem 5.5. This radical then abstracts bromine from the starting material.

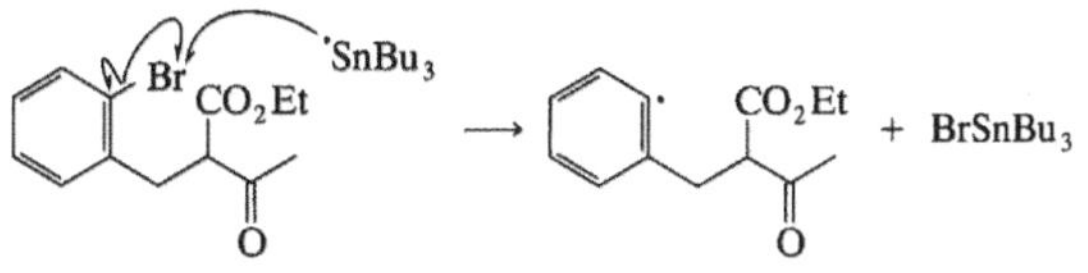

The aryl radical adds to the carbonyl group, a bond cleaves, and the resulting radical abstracts hydrogen from tri-n-butyltin hydride to form the product and a new chain-carrying radical.

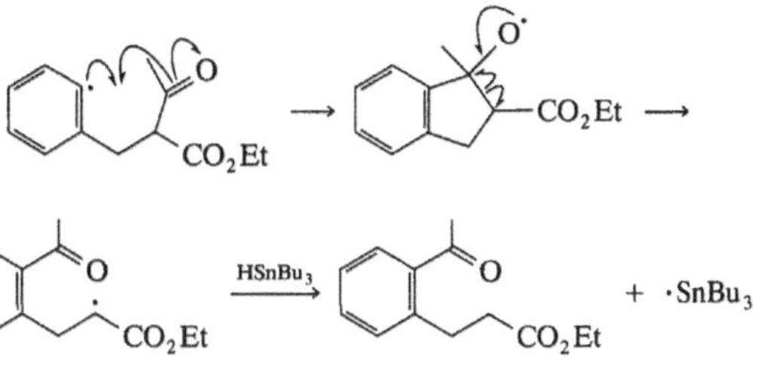

Problem 5.12

The fact that reactions in both (a) and (b) require light suggests that a radical mechanism is involved for each of them. This rules out a simple S_N2 substitution in part a or a nucleophilic aromatic substitution by an addition–elimination reaction in part b. When substitution occurs under basic conditions in the presence of light, the most likely mechanism is $S_{RN}1$.

a. The thiolate anion is excited by the light.

$$PhS^- \xrightarrow{h\nu} PhS^{*-}$$

The excited thiolate anion transfers an electron to cyclopropane.

$$\text{PhS*}^- \ + \ \underset{}{\overset{\text{Cl} \quad \text{Br}}{\diagup\!\!\triangle\!\!\diagdown}} \ \longrightarrow \ \text{PhS·} \ + \ \left[\underset{}{\overset{\text{Cl} \quad \text{Br}}{\diagup\!\!\triangle\!\!\diagdown}}\right]^{\overline{\cdot}}$$

The better leaving group, bromide, leaves.

$$\left[\underset{}{\overset{\text{Cl} \quad \text{Br}}{\diagup\!\!\triangle\!\!\diagdown}}\right]^{\overline{\cdot}} \ \longrightarrow \ \underset{}{\overset{\text{Cl}}{\diagup\!\!\triangle\!\!\diagdown}}· \ + \ \text{Br}^-$$

Thiolate reacts, as a nucleophile, with the electrophilic radical to produce the radical anion of the product.

$$\underset{}{\overset{\text{Cl}}{\diagup\!\!\triangle\!\!\diagdown}}· \ + \ \text{PhS}^- \ \longrightarrow \ \left[\underset{}{\overset{\text{Cl} \quad \overset{\text{Ph}}{|}\,\text{S}}{\diagup\!\!\triangle\!\!\diagdown}}\right]^{\overline{\cdot}}$$

Electron transfer to another molecule of starting cyclopropane gives the product and another cyclopropane radical anion to continue the chain.

$$\left[\underset{}{\overset{\text{Cl} \quad \overset{\text{Ph}}{|}\,\text{S}}{\diagup\!\!\triangle\!\!\diagdown}}\right]^{\overline{\cdot}} \ + \ \underset{}{\overset{\text{Cl} \quad \text{Br}}{\diagup\!\!\triangle\!\!\diagdown}} \ \longrightarrow \ \left[\underset{}{\overset{\text{Cl} \quad \text{Br}}{\diagup\!\!\triangle\!\!\diagdown}}\right]^{\overline{\cdot}} \ + \ \underset{}{\overset{\text{Cl} \quad \overset{\text{Ph}}{|}\,\text{S}}{\diagup\!\!\triangle\!\!\diagdown}}$$

The following is not a proper product-forming step because it does not also produce a radical to continue the chain. It actually is a termination step, involving the coupling of two radicals. Formation of a major product by the coupling of two radicals is unlikely because the concentration of radical intermediates usually is quite low.

$$\underset{}{\overset{\text{Cl}}{\diagup\!\!\triangle\!\!\diagdown}}· \ + \ \text{PhS·} \ \longrightarrow \ \underset{}{\overset{\text{Cl} \quad \overset{\text{Ph}}{|}\,\text{S}}{\diagup\!\!\triangle\!\!\diagdown}}$$

b. Each step in the mechanism of this reaction has a direct counterpart in the mechanism for part a.

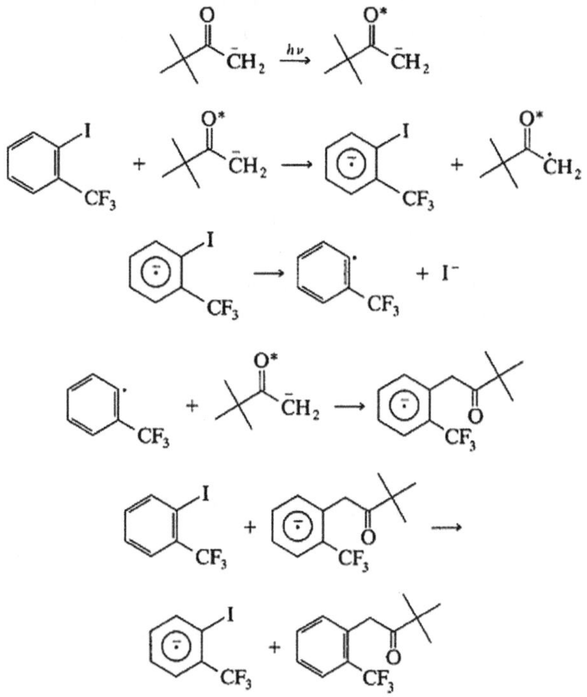

Problem 5.13

Sodium transfers a single electron to the aromatic system to produce a pentadienyl radical anion, **5-55**. This anion is protonated at the position either *ortho* or *meta* to the π-donating methoxy group to give an intermediate radical. When a second electron is added to this radical, the pentadienyl anion is again protonated at its central carbon, *ortho* or *meta* to the methoxy substituent.

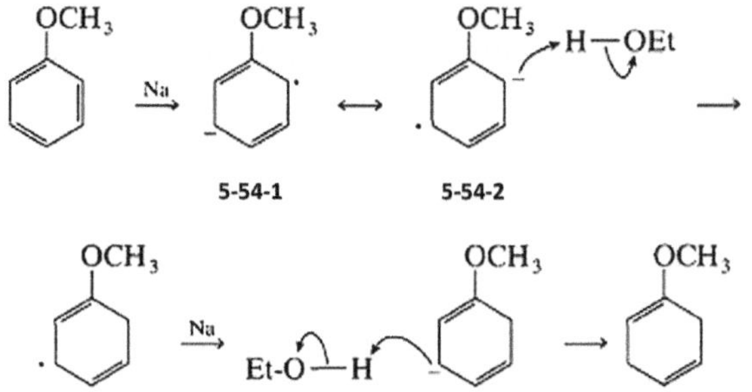

Problem 5.14

Comparison of the starting material with products shows that methyl iodide is used to methylate sulfur. The reaction takes place in base, so that the most likely mechanism for the methylation is an S_N2 reaction with thiolate ion:

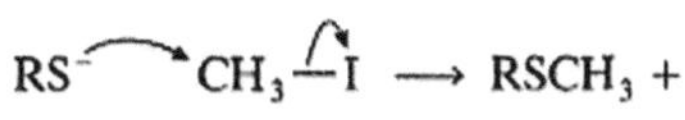

$$RS^- \frown CH_3 - I \longrightarrow RSCH_3 + I^-$$

Despite the fact that radical coupling is rarely a significant source of reaction products, the structural symmetry of the first two products suggests that they are formed by this route. Butyl lithium can remove a proton from the carbon α to the sulfur. The resulting anion could then fragment into a radical and a radical anion by scission of the S—C bond.

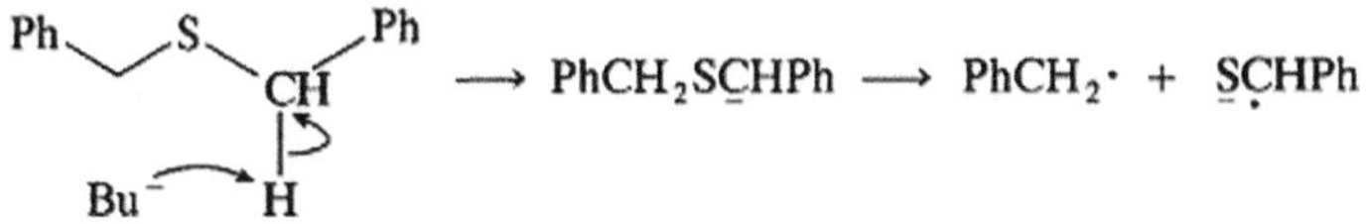

$$\longrightarrow PhCH_2SCHPh \longrightarrow PhCH_2\cdot + \underset{\cdot}{S}CHPh$$

Two benzyl radicals can couple to give the first product.

$$PhCH_2\cdot + \cdot CH_2Ph \longrightarrow PhCH_2CH_2Ph$$

Two thiolate radical anions can couple to give the dithiolate precursor, **5-56**, of the second product. This is analogous to the well-documented coupling of ketyls (radical anions derived from one-electron reduction of carbonyl compounds), which results in the formation of pinacols (1,2-diols).

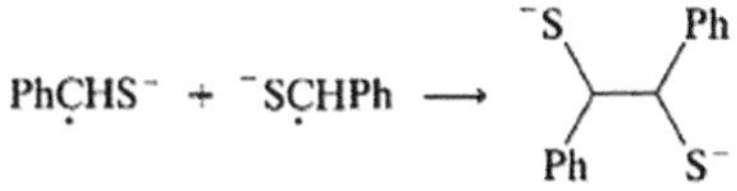

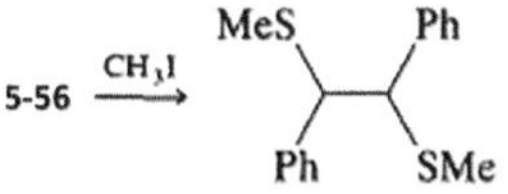

Finally, the two radicals formed initially can couple, at carbon, to give the thiolate precursor, **5-57**, of the third product.

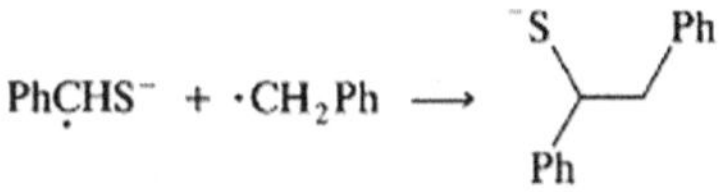

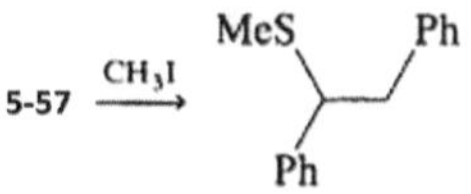

This is one of those relatively few reactions that proceed through radical coupling. The reaction occurs in part because the radicals involved are sufficiently stable to remain in existence until they can couple with another radical.

Problem 5.15

The evidence suggests that **5-33** and **5-34** are formed by different types of mechanisms: structure **5-33** by a radical process and structure **5-34** by a nonradical process. Formation of **5-33** is suppressed completely by addition of a free-radical inhibitor, but is stimulated by light. These observations strongly support a radical pathway to **5-33**. Other factors also make ionic S_N pathways unlikely. An S_N2 reaction is ruled out because **5-33** is formed by substitution at a tertiary carbon. An S_N1 ionization also is unlikely because formation of a carbocation next to the partially positive carbonyl would be required. Therefore, we are left with the strong possibility that **5-33** is formed by an $S_{RN}1$ mechanism. On the other hand, **5-34** forms in better yield in the dark or in the presence of a radical inhibitor, data which support a nonradical reaction. Note the regiochemistry of addition is different in the formation of **5-33** and **5-34**.

An $S_{RN}1$ mechanism for the formation of **5-33** can be written as follows:

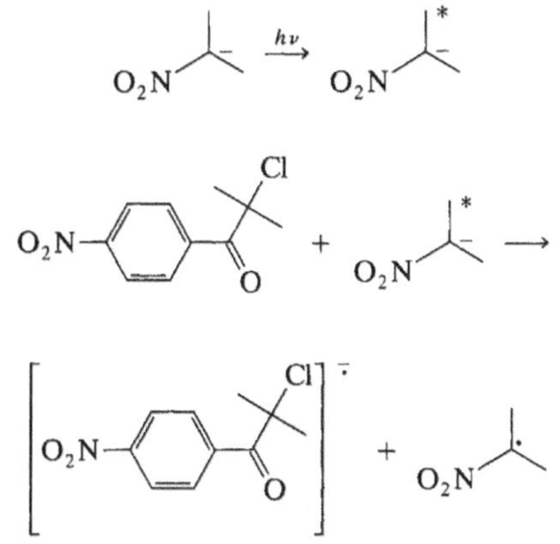

5-58

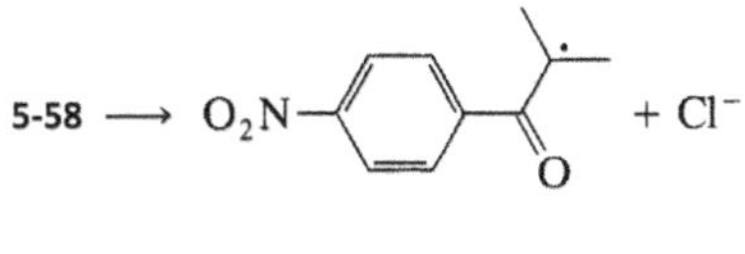

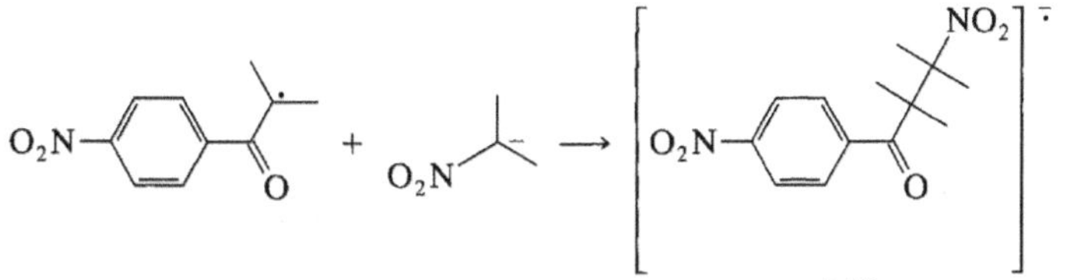

5-58 $\longrightarrow$

5-59

5-59 +

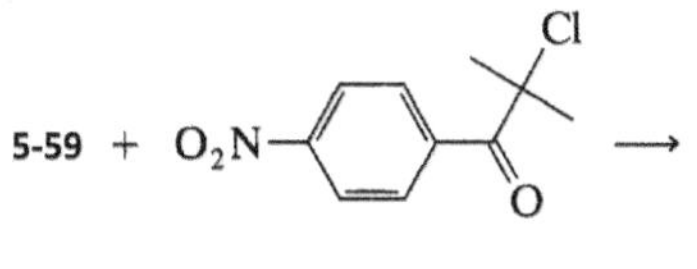

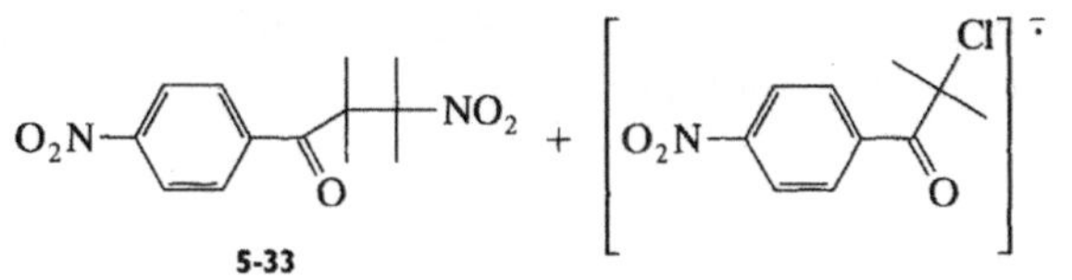

5-33

It is tempting to write a radical coupling to form product **5-33** because it involves fewer steps. However, this mechanism would be energetically inefficient because it does not continue the chain process. Thus, the following *is not* the major product-forming step:

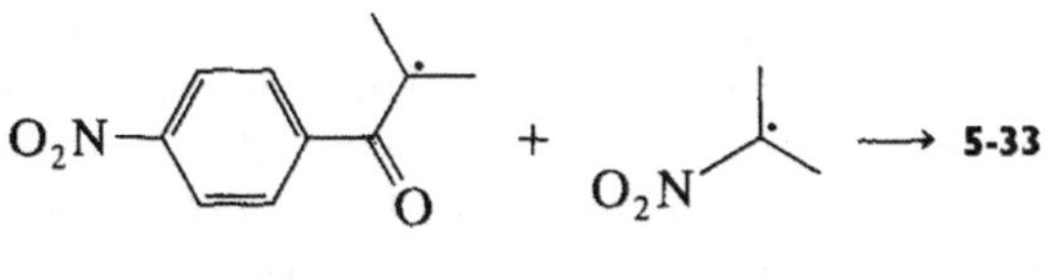

Noting the formation of the epoxide group and numbering the atoms in the starting materials and in **5-34** help to indicate how **5-34** must be formed.

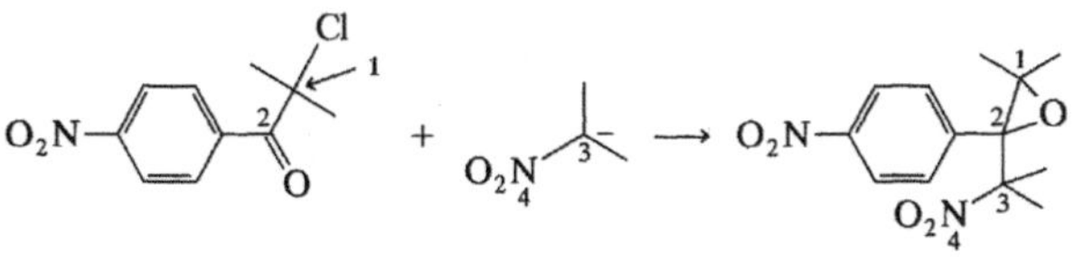

5-34

Because C-3 becomes attached to C-2, a nucleophilic reaction of the nitranion at the electrophilic carbonyl carbon must take place. Sterically this carbon is the most accessible. This step is followed by an intramolecular nucleophilic reaction of the alcoholate with the adjacent carbon (numbered 1 in the previous equation).

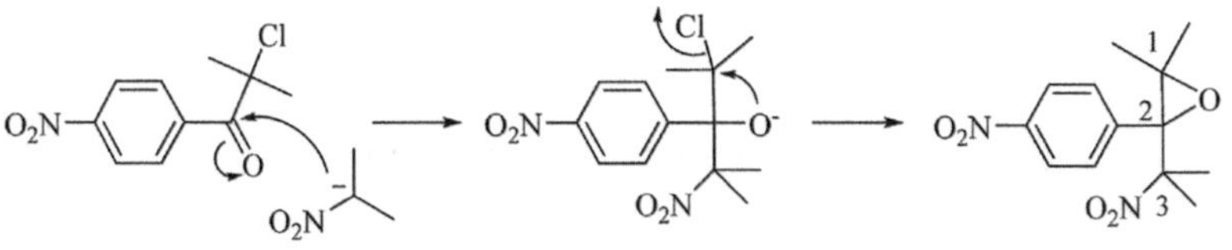

5-34

Problem 5.16

a. The perester could serve as a free-radical initiator. At first glance, it appears that two tetrahydrofuran molecules form the two rings of the product, but the product contains one additional carbon atom. Therefore, the perester must be the source of the atoms of one ring, as well as a source of radicals. Because the perester contains a carbonyl group, the simplest explanation is that the lactone ring is derived from the perester. These considerations suggest the following chain mechanism:

Initiation

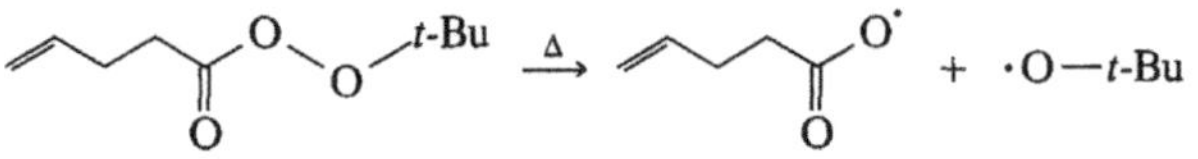

Propagation

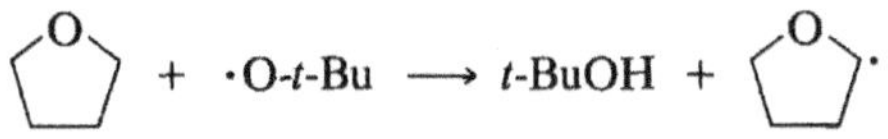

5-60

The tetrahydrofuranyl radical, **5-60**, adds to the carbon–carbon double bond of the perester.

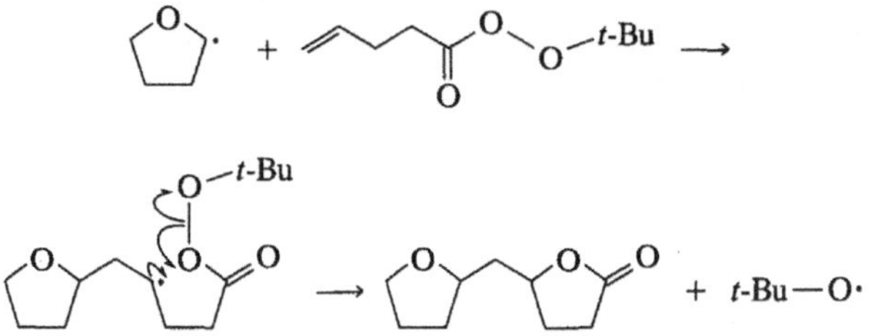

The following mechanism is not as satisfactory.

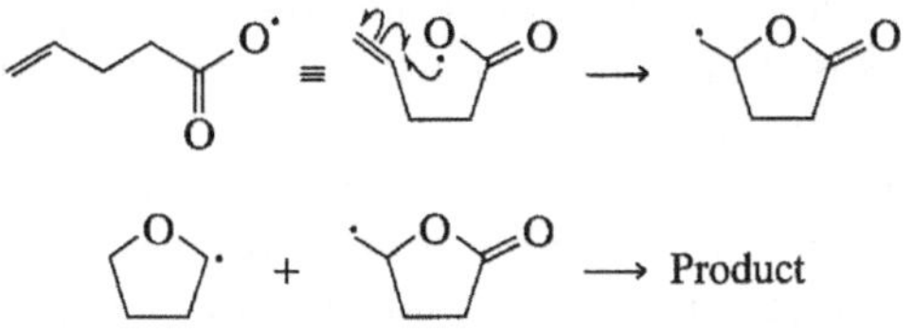

Rapid decarboxylation of the alkyl carboxyl radical to the corresponding alkyl radical would be expected to compete with intramolecular cyclization. Moreover, the reaction scheme shown would not be efficient. Not only would the reaction be slow because two radicals must collide in order to form the product, but the step fails to regenerate a radical to continue the reaction. (Note: Aryl carboxyl radicals lose carbon dioxide more slowly than alkyl carboxyl radicals.)

b. From the reaction conditions, we can make the following speculations and/or conclusions: (i) the tri-*n*-butyltin radical will be the chain carrier; (ii) tri-*n*-butyltin will be lost from the starting material; and (iii) a bond in the six-membered ring breaks at some point in the mechanism.

In Section 2, we learned that tri-*n*-butyltin radicals react with selenium compounds to produce radical intermediates. Therefore, this might be a logical first step, after the initiation reaction between AIBN and tri-*n*-butyltin hydride has produced tri-*n*-butyltin radicals.

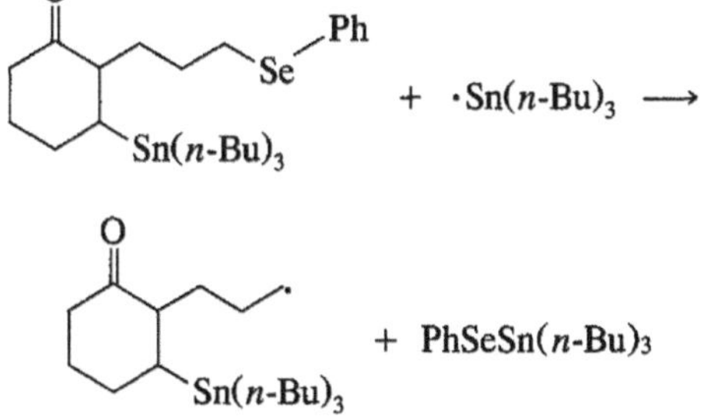

O
Ph
Se
Sn(n-Bu)3
+ ·Sn(n-Bu)3
O
Sn(n-Bu)3
+ PhSeSn(n-Bu)3
5-61

The carbonyl group of radical **5-61** can undergo an intramolecular cyclization reaction with the alkyl radical. The shared bond between the five- and six-membered rings then breaks to form the product and a new tri-*n*-butyltin radical, which continues the chain.

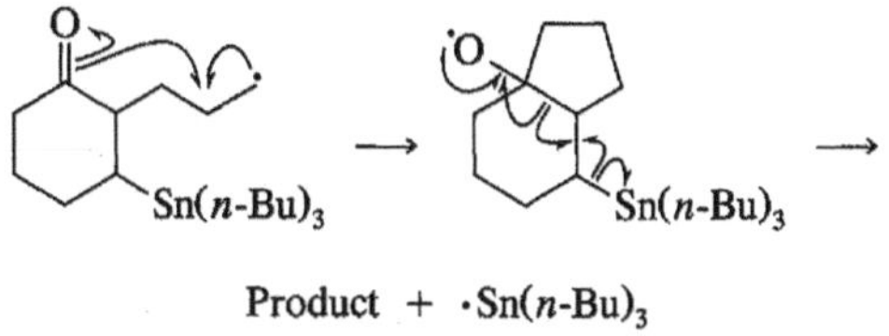

O
Sn(n-Bu)3
O
Sn(n-Bu)3
Product + ·Sn(n-Bu)3

The major difference between this reaction and that of Example 5.11 is that only 10 mol% tin hydride is present. This is enough to initiate the reaction, but not enough to reduce intermediate radicals significantly.

c. Clues to the reaction mechanism are (i) chlorine is missing from the product; (ii) AIBN is present as an initiator; and (iii) the tributyltin group is present. Thus, it appears that tributyltin radicals abstract chlorine from the starting material. In the initiation steps, AIBN forms 2-cyano-2-propyl radicals in the usual way, and these radicals react with allyltri-*n*-butyltin to produce tri-*n*-butyltin radicals by an addition–elimination mechanism.

Initiation

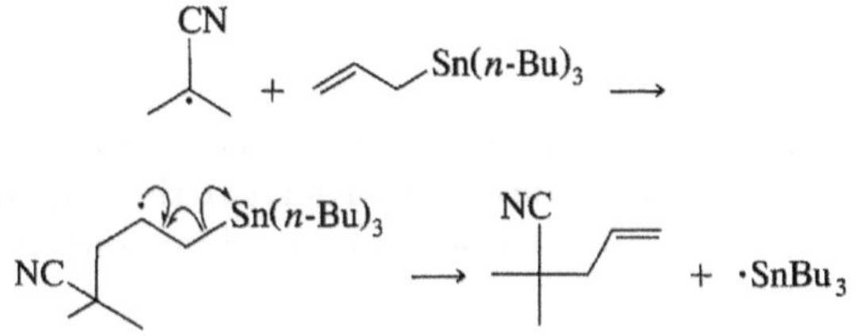

CN
+
Sn(n-Bu)3
Sn(n-Bu)3
NC
NC
+ ·SnBu3

Propagation

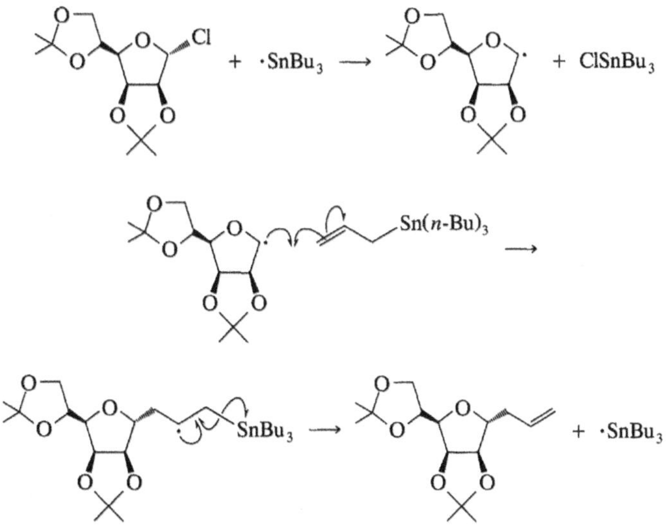

+ ·SnBu₃ → + ClSnBu₃

Sn(n-Bu)₃ →

SnBu₃ → + ·SnBu₃

d. Reaction in the presence of light and strong base suggests the involvement of a radical anion intermediate and the S_RN1 mechanism for at least part of the reaction pathway. In the initiation steps, the anion of the naphthyl ketone, excited by light, donates an electron to the bromoketone to form a radical anion. The product of an S_RN1 reaction would be **5-62**.

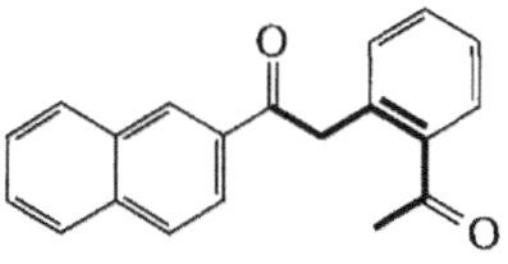

5-62

If the six highlighted carbon atoms in **5-62** form a six-membered ring, the correct carbon skeleton for the final product is obtained. In fact, an intramolecular aldol condensation (discussed in Chapter 3), followed by elimination of water, gives the final product.

S_RN1 Reaction
(1) *Initiation Steps*

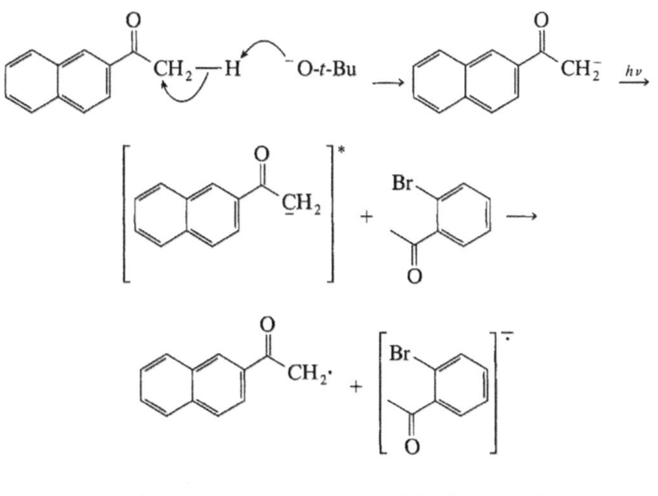

CH₂—H ⁻O-t-Bu → CH₂⁻ hν →

+ Br → →

CH₂· + Br

(2) Propagation Steps

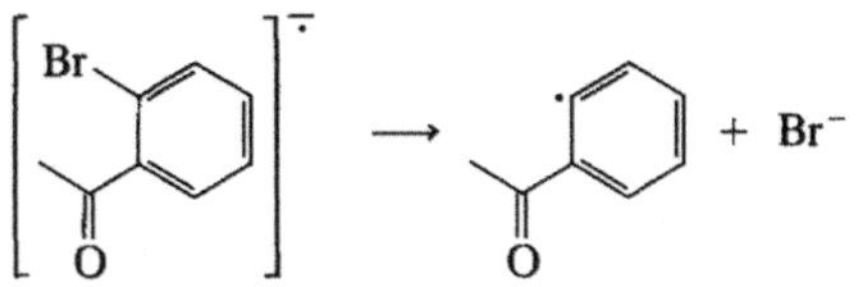

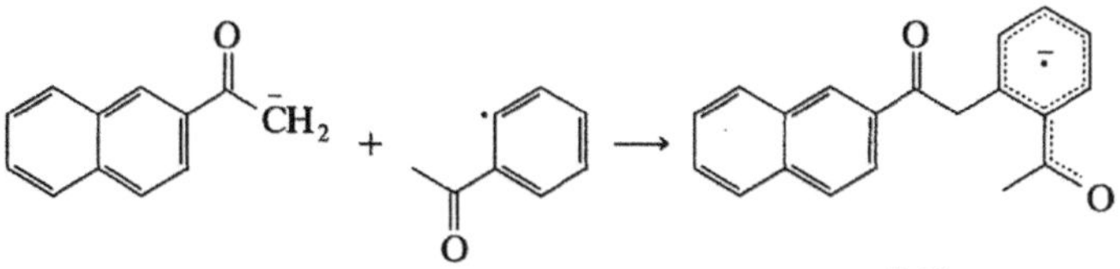

5-63

In the last step, a new radical anion of the starting bromo compound is formed, which continues the chain.

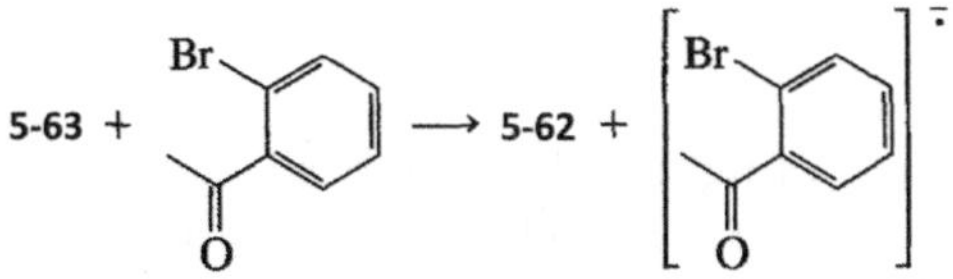

Aldol Condensation

The aldol condensation of **5-62** involves several steps. First, a proton is removed from the methyl group by the *t*-butoxide ion. (If the proton were removed from the methylene group α to the other carbonyl, condensation would be at the carbonyl attached to the methyl. This would produce a five-membered ring.) The resulting enolate adds to the other carbonyl group.

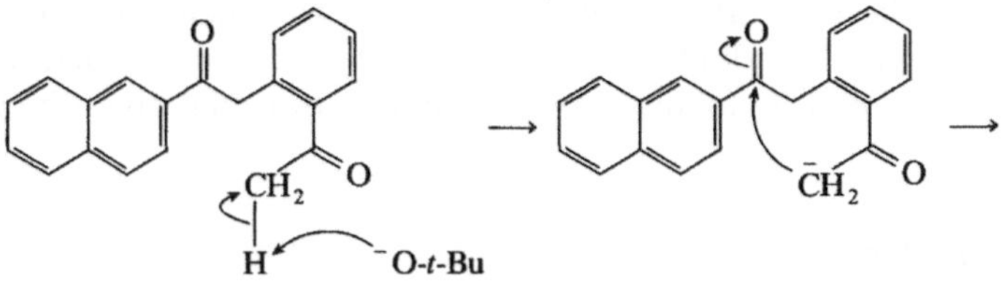

Protonation of the resulting alkoxide ion gives alcohol **5-64**.

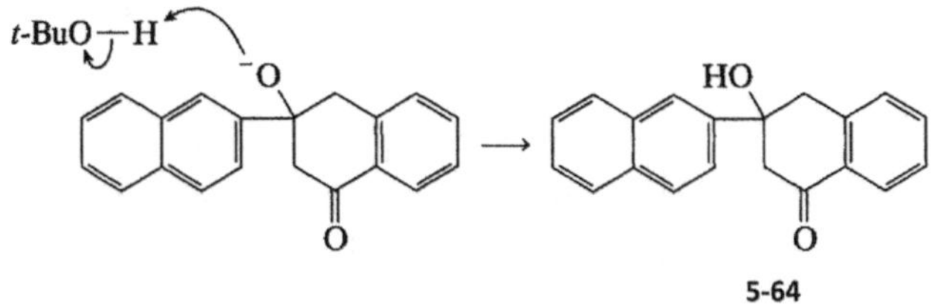

5-64

The alcohol can undergo base-promoted elimination to give two possible products, **5-65** and **5-66**. The new double bond in each of these products is stabilized by conjugation with the carbonyl group and at least one of the aromatic rings.

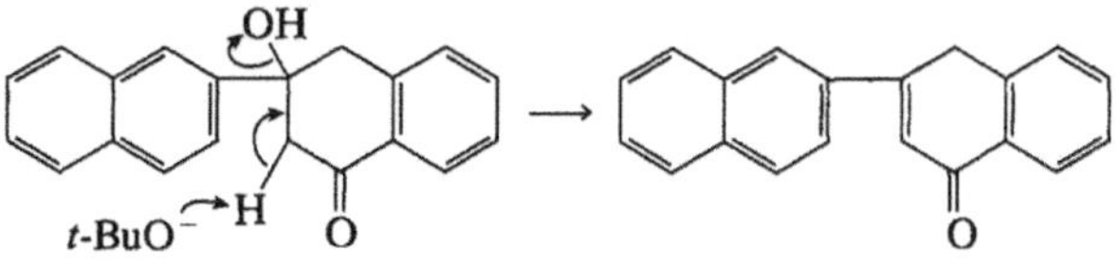

5-65

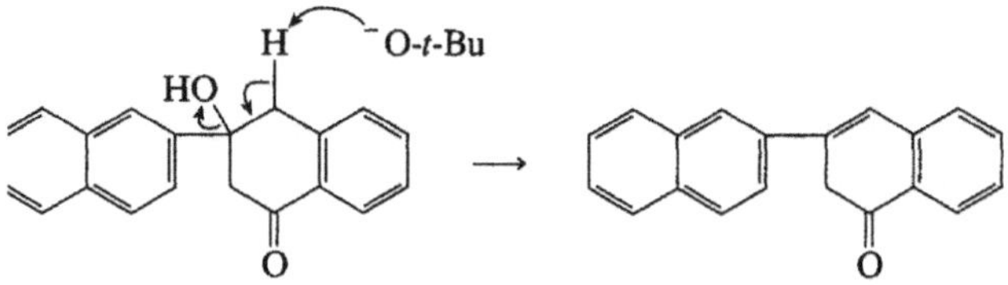

5-64   **5-66**

Either **5-65** or **5-66** can readily tautomerize to give the final product. Both mechanisms are very similar, so only one is shown.

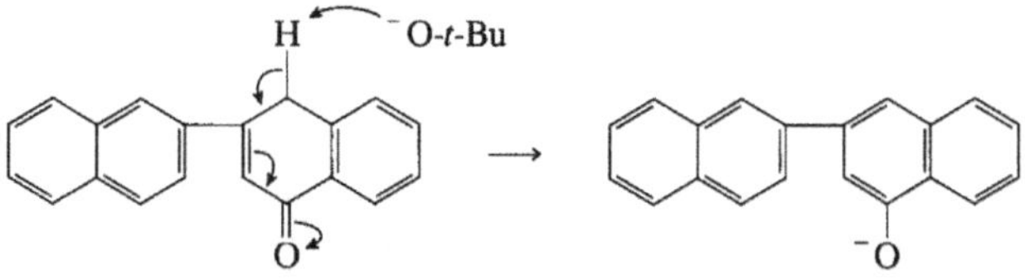

5-65   **5-67**

Upon acidic workup, the naphthoxide ion **5-67** is protonated to give **5-35**.

Another plausible mechanism for the final stages of the reaction is tautomerization of **5-64** to **5-68**, followed by elimination directly to the phenol. (Structure **5-69** would not be produced because the aromaticity of the right-hand ring is interrupted.) However, the mechanism previously discussed is preferable because alcohol **5-68** would not be as stable as **5-65** or **5-66**.

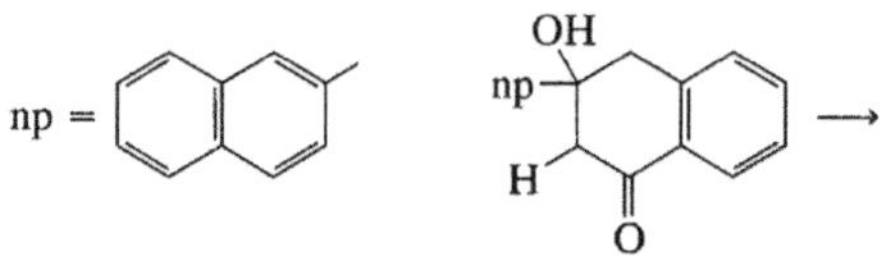

5-64

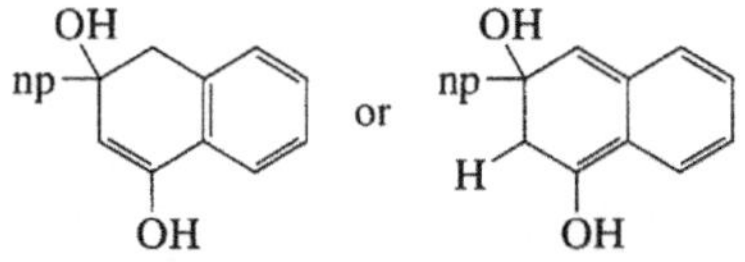

5-68   or   **5-69**

e. The overall reaction yield is 69%, with only 10 mol% hexabutylditin present, which indicates that this must be a chain process. Because iodide is present, probable initiation steps would be photochemical decomposition of the ditin compound to tri-n-butyltin radicals, followed by abstraction of iodine from the starting material.

$$(n\text{-Bu})_3\text{SnSn}(n\text{-Bu})_3 \xrightarrow{h\nu} 2n\text{-Bu}_3\text{Sn}\cdot$$

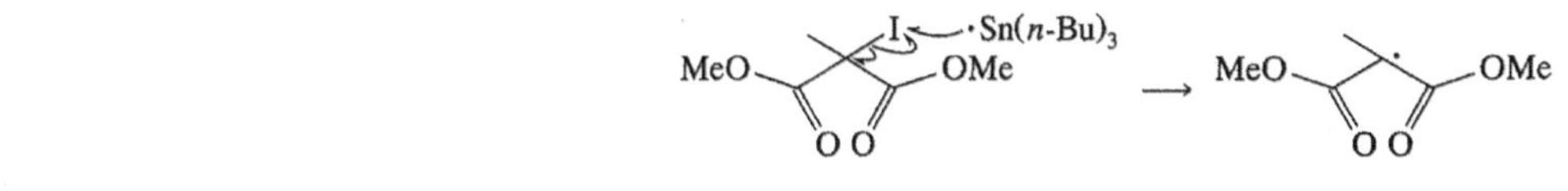

5-70

Radical 5-70 can add regiospecifically to the 1-alkene to produce an intermediate radical, which can then abstract iodine from the starting ester to produce 5-36 and a new radical to continue the chain.

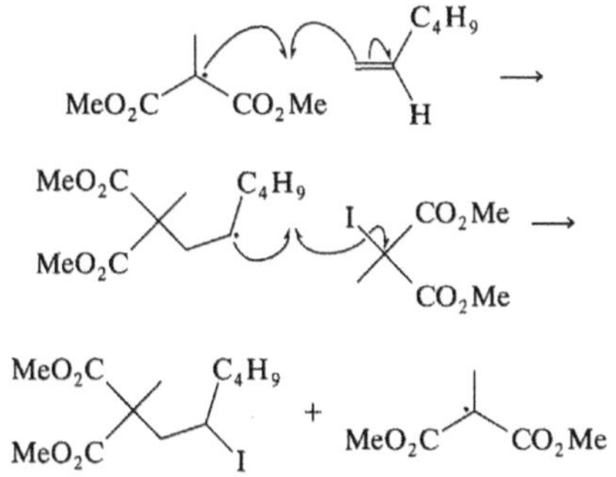

5-36

Because both homolytic and heterolytic cleavages readily occurs with C—I bonds, plausible reaction mechanisms can be written with either radical or ionic intermediates. However, in the nonpolar solvent benzene, the radical mechanism is more likely because the ionic mechanism requires the carbocation and anion to be separated, and these species cannot be stabilized by interaction with the nonpolar solvent.

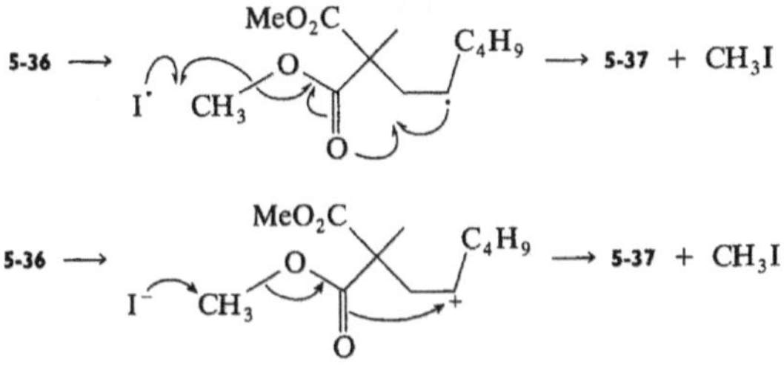

f. In the first step of this two-step process, the proton of the tertiary alcohol is abstracted using potassium hydride to form the alkoxide. The alkoxide then adds into the phenyl isothiocyanate to form the thiocarbamate 5-71.

The chlorines and the thiocarbamate are removed simultaneously through the action of AIBN and tri-*n*-butyltin hydride. The AIBN is heated in the presence of *n*-Bu₃SnH to form the tin radical. Tin radical then effects the removal of chloride and then the thiocarbamate through the cascade process **5-72**.

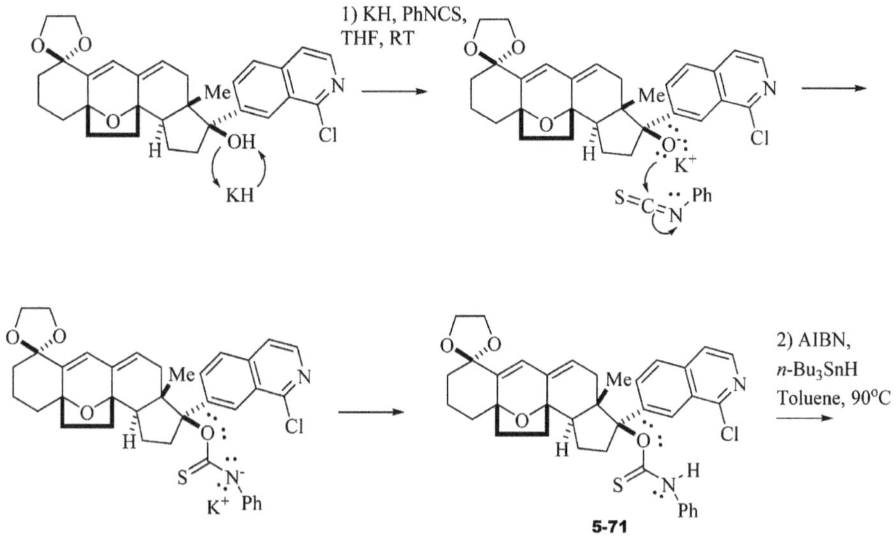

5-71

Initiation Steps

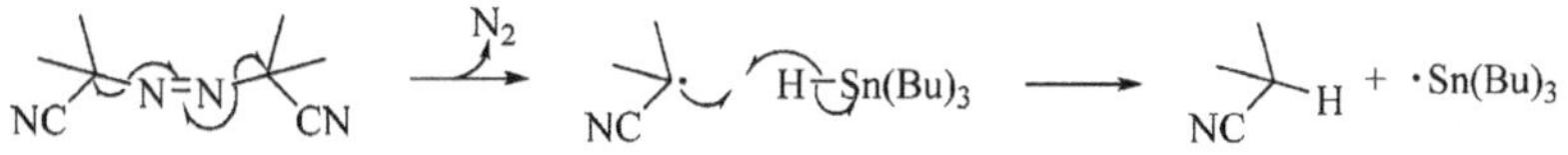

The transformation occurs through the interaction of the tin radical with the chloride. We show the process leading through the radical dechlorination (homolytic cleavage: each atom leaves with one of the bond electrons) to form the aryl radical and the tri-*n*-butyltin chloride. The aryl radical abstraction of the hydrogen propagates the chain reaction and provides the dehalogenated product **5-74**. The tin radical then causes the cascade decomposition of the thiocarbamate leading to the trisubstituted radical **5-76**. Another tri-*n*-butyltin hydride is converted to the tin radical as the product **5-77** is formed.

Transformation Steps

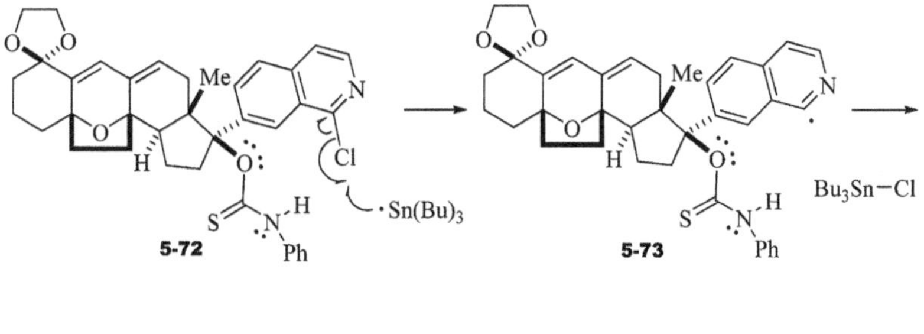

5-72   **5-73**

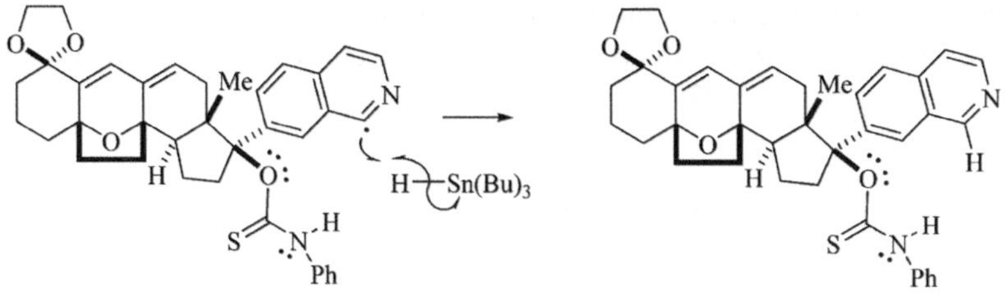

5-74

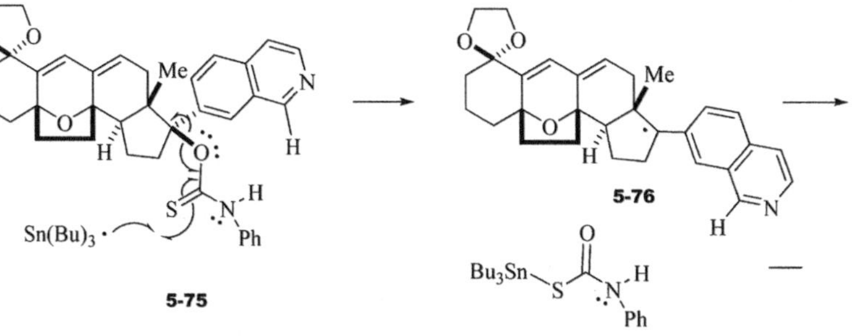

5-75

5-76

The product is delivered as a single isomer at the C-17 position. The resulting stereo-chemistry of the C-17 position is most likely due to the delivery of the hydride to the more accessible face of the molecule rather than the radical favoring one side of the mole-cule. The tertiary radical of the intermediate **5-76** is most likely in the orbital on both sides of the shallow pyramidal carbon center. The tri-*n*-butyltin hydride is relatively large and sterically demanding and the top-face of the molecule is blocked by the methyl group and the attached ring systems. This leads to the hydride being delivered to the back face of the molecule (or bottom face as depicted in **5-76**, where the five-membered ring is substituted with a hydrogen and ring carbon). This will be the approach regardless of the stereochem-ical configuration of the C-17 carbon in the original tertiary thiocarbamate since the start-ing configuration is irrelevant once the radical is formed and the C-17 carbon center becomes planar.

The geometry of alkyl radicals can be described
as "inverting shallow pyramidal" strucutres

Propagation and Product Formation

5-76

(Bu)₃Sn—H

5-77

·Sn(Bu)₃

less accessible approach

more accessible approach

5-76

References

Baldwin, J. E.; Adlington, R. M.; Robertson, J. *J. Chem. Soc., Chem. Commun.* **1988**, 1404–1406.

Baldwin, J. E. *J. Chem. Soc., Chem. Commun.* **1976**, 734–736.

Barton, D. H. R.; Beaton, J. M.; Geller, L. E.; Pechet, M. M. *J. Am. Chem. Soc.* **1960**, *82*, 2640–2641.

Beckwith, A. L. J.; Easton, C. J.; Serelis, A. K. *J. Chem. Soc., Chem. Commun.* **1980**, 482–483.

Beckwith, A. L. J.; O'Shea, D. M.; Westwood, S. W. *J. Am. Chem. Soc.* **1988**, *110*, 2565–2575.

Beugelmans, R.; Bois-Choussy, M.; Tang, Q. *J. Org. Chem.* **1987**, *52*, 3880–3883.

Biellmann, J. F.; Schmitt, J. L. *Tetrahedron Lett.* **1973**, 4615–4618.

Birch, A. J.; Hinde, A. L.; Radom, L. *J. Am. Chem. Soc.* **1980**, *102*, 3370–3376.

Boger, D. L.; Mathvink, R. J. *J. Org. Chem.* **1988**, *53*, 3377–3379.

Bunnett, J. F.; Galli, C. *J. Chem. Soc., Perkin Trans. I* **1985**, 2515–2519.

Bunnett, J. F. *Acc. Chem. Res.* **1978**, *11*, 413–420.

Carey, F. A.; Sundberg, R. J. *Advanced Organic Chemistry*, 3rd ed; Plenum Press: New York, 1990, 704.

Claisse, J. A.; Davies, D. I.; Parfitt, L. T. *J. Chem. Soc. C* **1970**, 258–262.

Curran, D. P.; Rakiewicz, D. M. *J. Am. Chem. Soc.* **1985**, *107*, 1448–1449.

Curran, D. P.; Chen, M. H.; Spletzer, E.; Seong, C. M.; Chang, C. T. *J. Am. Chem. Soc.* **1989**, *111*, 8872–8878.

Danishefsky, S.; Taniyama, E.; Webb, R. R., II *Tetrahedron Lett.* **1983**, *24*, 11–14.

Dowd, P.; Choi, S.-C. *J. Am. Chem. Soc.* **1987**, *109*, 3493–3494.

Fossey, J.; Lefort, D.; Sorba, J. *J. Org. Chem.* **1986**, *51*, 3584–3587.

Giese, B.; Horler, H.; Zwick, W. *Tetrahedron Lett.* **1982**, *23*, 931–934.

Hart, D. J. *Science* **1984**, *223*, 883–887.

Hart, D. J. *Science* **1984**, *223*, 883–887.

Hoffmann, A. K.; Feldman, A. M.; Gelblum, E.; Hodgson, W. G. *J. Am. Chem. Soc.* **1964**, *86*, 639–646.

Julia, M.; Descoins, C.; Baillarge, M.; Jacquet, B.; Uguen, D.; Groeger, F. A. *Tetrahedron* **1975**, *31*, 1737–1744.

Kerr, J. A. *Chem. Rev.* **1966**, *66*, 465–500.

Kharrat, A.; Gardrat, C.; Maillard, B. *Can. J. Chem.* **1984**, *62*, 2385–2390.

Meijs, G. F. *J. Org. Chem.* **1986**, *51*, 606–611.

Nair, V.; Chamberlain, S. D. *J. Am. Chem. Soc.* **1985**, *107*, 2183–2185.

Pandit, U. K.; Dirk, I. P. *Tetrahedron Lett.* **1963**, 891–895.

Park, J. D.; Rogers, F. E.; Lacher, J. R. *J. Org. Chem.* **1961**, *26*, 2089–2095.

Russell, G. A.; Ros, F. *J. Am. Chem. Soc.* **1985**, *107*, 2506–2511.

Stamegna, A. P.; McEwen, W. E. *J. Org. Chem.* **1981,** *46,* 1653−1655.

Stork, G.; Baine, N. H. *J. Am. Chem. Soc.* **1982,** *104,* 2321−2323.

Walling, C.; Jacknow, B. B. *J. Am. Chem. Soc.* **1960,** *82,* 6108−6112.

Walling, C. *Free Radicals in Solution;* John Wiley: New York, 1957, 493.

Handbook of Chemistry and Physics, 63rd ed; Weast, R. C.; Astle, M. J., eds.; CRC Press: Boca Raton, FL, pp. 1982−1983.

Weinstock, J.; Lewis, S. N. *J. Am, Chem. Soc.* **1957,** *79,* 6243−6247.

Winkler, J. D.; Sridar, V. *J. Am. Chem. Soc.* **1986,** *108,* 1708−1709.

Yamashita, S.; Iso, K.; Kitajima, K.; Himuro, M.; Hirama, M. *J. Org. Chem.* **2011,** *76,* 2408−2425.

Pericyclic Reactions

1. INTRODUCTION

Many reactions involve a cyclic transition state. Of these, some involve radical or ionic intermediates and proceed by stepwise mechanisms. *Pericyclic* reactions are *concerted*, and in the transition state, the redistribution of electrons occurs in a single continuous process. In this chapter, we will consider several different types of pericyclic reactions, including electrocyclic transformations, cycloadditions, sigmatropic rearrangements, and the ene reaction.

A. Types of Pericyclic Reactions

- *Electrocyclic transformations* involve intramolecular formation of a ring by bond formation at the ends of a conjugated π system. The product has one more σ bond and one less π bond than the starting material. The reverse reaction, ring opening of a cyclic polyene, is also an electrocyclic process.

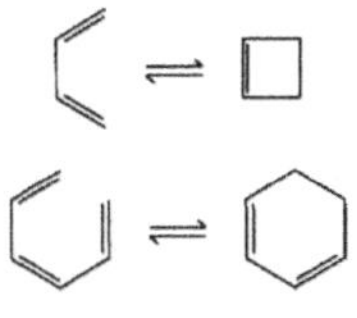

- *Cycloadditions* involve bonding between the termini of two π systems to produce a new ring. The product has two more σ bonds and two less π bonds than the reactants. Common examples are the Diels–Alder reaction and many 1,3-dipolar cycloadditions.

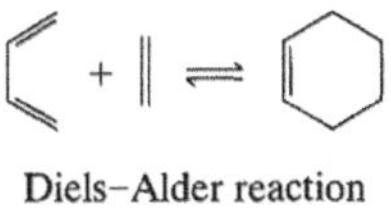

Diels–Alder reaction

- *Sigmatropic Reactions.* In these, an allylic σ bond at one end of a π system appears to migrate to the other end of the π system. The π bonds change position, but the total number of σ and π bonds is the same, as in the Cope and Claisen rearrangements.

Copyright © 2014 Elsevier Inc. All rights reserved.

Cope rearrangement

- *Ene reactions.* These combine aspects of cycloadditions and sigmatropic reactions. They may be inter- or intramolecular.

B. Theories of Pericyclic Reactions

Pericyclic reactions can be initiated either thermally or photochemically, but in either case, they show great stereospecificity. The conditions under which pericyclic reactions occur and the stereochemistry of the products formed are dependent on the symmetry characteristics of the *molecular orbitals* (MOs) involved. On this basis, pericyclic processes are classified as either *symmetry-allowed* or *symmetry-forbidden*. There are several theoretical approaches for deriving the *selection rules* governing concerted pericyclic reactions, and each of these theories represents the work of many contributors. All these approaches assume that a cyclic transition state is formed by π orbital overlap.

- *Conservation of Orbital Symmetry.* This approach relies on a detailed analysis of the symmetry properties of the MOs of starting materials and products. Orbital correlation diagrams link the orbital characteristics of starting materials and products.
- *Frontier Orbital Theory.* This view focuses on the symmetry characteristics of the highest occupied and lowest unoccupied orbitals, particularly the symmetry at the termini of the systems.
- Moebius–Huckel Theory. The MO array of the transition state is analyzed in terms of aromaticity, which is determined by the number of π electrons.

Each of these theoretical approaches leads to the same predictions regarding reaction conditions and stereochemistry. For a wide range of reactions, *the selection rules can be used empirically, based on a simple method of electron counting, without regard to their theoretical basis.* The selection rules for pericyclic reactions relate three features:

1. The number of π electrons involved.
2. The method of activation.
3. The stereochemistry.

If any two of these features are specified, the third feature is determined by the selection rules.

Note: Reactions involving a cyclic transition state are not always concerted, and the selection rules and their stereochemical consequences apply only to *concerted* pericyclic processes. Indeed, *failure to conform to the selection rules is usually taken as proof that a reaction does not proceed by a concerted mechanism.* On the other hand, failure to react may simply mean that the reaction is symmetry-allowed but does not occur because the thermodynamics is unfavorable. This point will be discussed further in the next section.

The following sections present an empirical approach to applying the selection rules. This chapter continues with a basic introduction to the analysis of symmetry properties of orbitals and the application of orbital correlation diagrams to the relatively simply cyclobutene–butadiene interconversion; it concludes with some examples of the frontier orbital approach to pericyclic reactions.

2. ELECTROCYCLIC REACTIONS

Electrocyclic reactions are the intramolecular ring openings or ring closures. The interconversion of substituted cyclobutenes and butadienes under different conditions illustrates which modes of reaction are symmetry-allowed and the stereochemical consequences of the selection rules. (In the reactions that follow, because there are no polar substituents, the directions of electron flow are arbitrary.) By the principle of microscopic reversibility, the bonding changes involved in going from starting material to product are the exact reverse of the changes involved in going from product to starting material. Thus, for electrocyclic reactions, any analysis we make for cyclization also applies for the ring-opening reaction and vice versa.

Whereas, in theory, symmetry-allowed electrocyclic reactions may proceed in either direction, in practice, one side of the equation is usually favored over the other. Because of the large strain energy of the four-membered ring, the equilibrium for the thermal opening of cyclobutene rings is usually favored over the reverse reaction (ring closure). On the other hand, equilibrium usually favors the six-membered ring rather than the ring-opened compound.

A. Selection Rules for Electrocyclic Reactions

Like other pericyclic reactions, electrocyclic reactions may be initiated either thermally or photochemically. The selection rules enable us to correlate the stereochemical relationship of the starting materials and products with the method of activation required for the reaction and the number of π electrons in the reacting system.

HINT 6.1

To apply the selection rules for electrocyclic reactions, count the number of π electrons in the open-chain polyene.

EXAMPLE 6.1. RING OPENING OF CIS-3,4-DIMETHYLCYCLOBUTENE

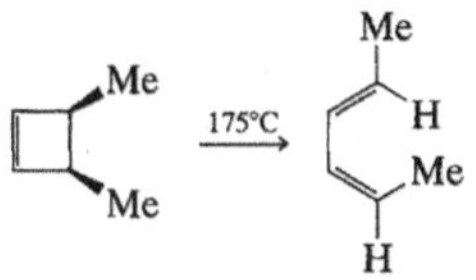

The cyclobutene–butadiene interconversion involves four π electrons and is designated a π^4 process. Note that by the principle of microscopic reversibility, the number of π electrons involved in the transformation is the same for ring opening as for ring closing. Once we know the number of π electrons involved in an electrocyclic reaction and the method of activation, the stereochemistry of the process is fixed according to the rules outlined in Table 6.1.

TABLE 6.1 Selection Rules for Electrocyclic Reactions

Number of Electrons	Mode of Activation	Allowed Stereochemistry[a]
$4n$	Thermal	Conrotatory
	Photochemical	Disrotatory
$4n + 2$	Thermal	Disrotatory
	Photochemical	Conrotatory

[a]*The terms* conrotatory *and* disrotatory *are explained in Section 6.B.*

B. Stereochemistry of Electrocyclic Reactions (Conrotatory and Disrotatory Processes)

EXAMPLE 6.2. THE THERMAL RING OPENING OF CIS-3-CHLORO-4-METHYL-CYELOBUTENE, A CONROTATORY PROCESS

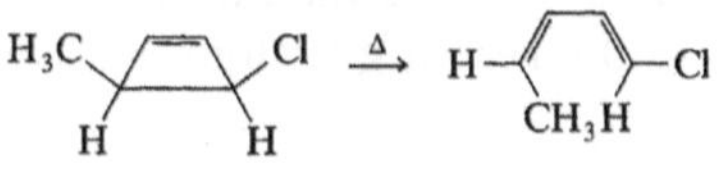

By applying Hint 6.1, we see that the number of π electrons involved in the transformation is four. The reaction takes place thermally, so that the selection rules as outlined in Table 6.1 indicate that a conrotatory process should occur. The following discussion will show what this means.

When the cyclobutene ring is transformed into a butadiene, the C-3—C-4 bond breaks. As it does so, there is rotation about the C-1—C-4 bond and the C-2—C-3 bond, so that the substituents on the breaking C-3—C-4 σ bond rotate into the plane of the conjugated diene system of the product. In the starting cyclobutene, the methyl and chloro groups lie above the plane of the four ring carbons; in the product, these groups lie in the same plane as the four carbon atoms of the butadiene. If all possible rotations about the C-3—C-4 σ bond were allowed, four different products would be formed, two by conrotatory processes and two by disrotatory processes. We will take a closer look at this process.

Conrotatory Process

In a conrotatory process, the substituents on C-3 and C-4 rotate *in the same direction.*

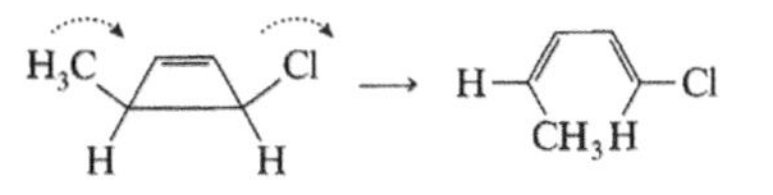

6-2

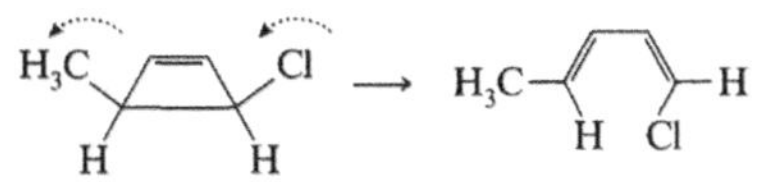

6-3

Clockwise rotation of both substituents gives product **6-2**, whereas *counterclockwise* rotation gives product **6-3**. (The arrows in the figure show the direction of rotation, not the movement of electrons.)

The selection rules predict that the thermal ring opening should be conrotatory for *cis-*3-chloro-4-methylcyclobutene, but they do not distinguish between products **6-2** and **6-3**. Both products are allowed by the selection rules, but experimentally only product **6-2** is formed. An explanation for this preference of one allowed process over the other is beyond the scope of this book; however, possible explanations can be found in the following references: Dolbier et al., (1984); Rondan and Houk, (1985); Krimse et al., (1984).

Disrotatory Process

In a disrotatory process, the substituents on C-3 and C-4 rotate *in opposite directions.*

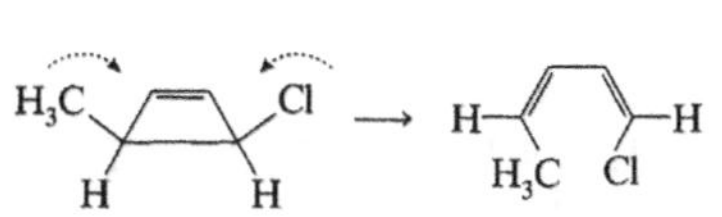

6-4

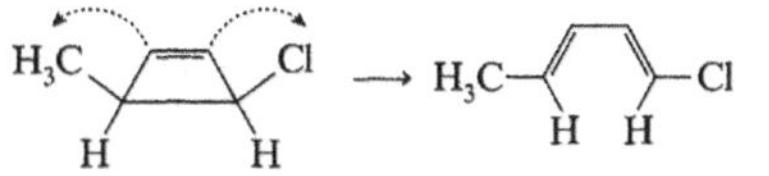

6-5

Rotation of the substituents *toward one another* would give the product **6-4**, whereas rotation of the substituents *away from one another* would give the product **6-5**. For the thermal reaction, neither of these rotations is allowed by the selection rules.

Summary of the Results in the Preceding Example: Each of the four rotational modes of ring opening leads to a particular stereoisomer of the product. For the concerted ring opening of a π^4 system, only the two conrotatory modes are allowed. Of these, only one is observed experimentally.

PROBLEM 6.1

By applying Hint 6.1 and the rules in Table 6.1, decide whether the following *thermal* reactions are symmetry-allowed or symmetry-forbidden.

a.

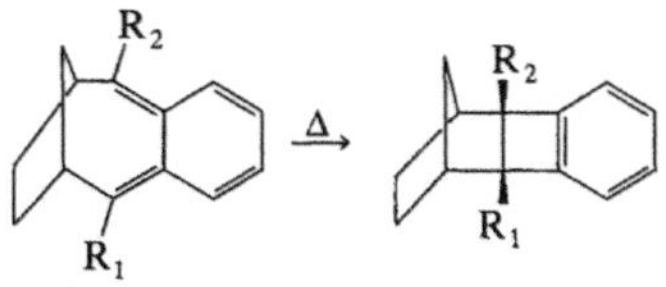

b.

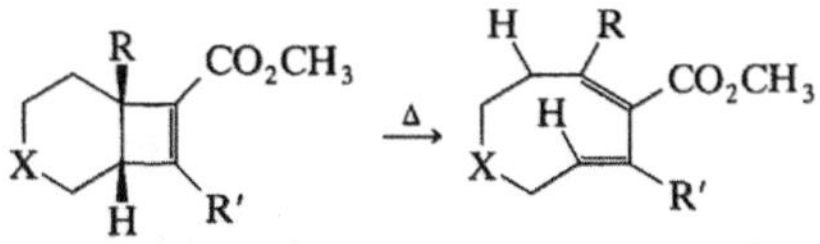

$$X = (CH_2)_3$$

Sources: Spellmeyer et al., (1989); Vos et al., (1985).

HINT 6.2

When applying the selection rules, only those π electrons taking part in the cyclization are counted.

EXAMPLE 6.3. THE THERMAL CYCLIZATION OF CYCLOOCTATETRAENE TO BICYCLO[4.2.0]-OCTATRIENE

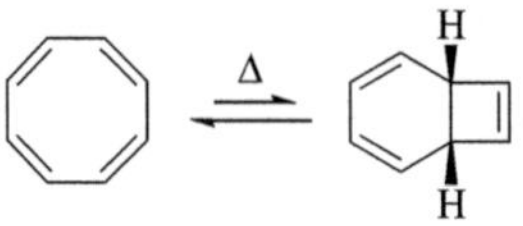

Because the cyclic transition state of the electrocyclic reaction involves joining the ends of the π system, this reaction may be regarded as a π^4 ring closure to a cyclobutene or a π^6 ring closure to a cyclohexadiene. Under thermal conditions, the selection rules require a conrotatory ring closure for the π^4 system, leading to a *trans* ring junction, which is unlikely for such a fusion because of ring strain. On the other hand, the π^6 system should close by a disrotatory process to give a *cis* ring junction. The product observed experimentally is consistent with a disrotatory closure of the π^6 system. Cyclooctatetraene is tub-shaped rather than planar, and the disrotatory process can be visualized as shown in the following drawing:

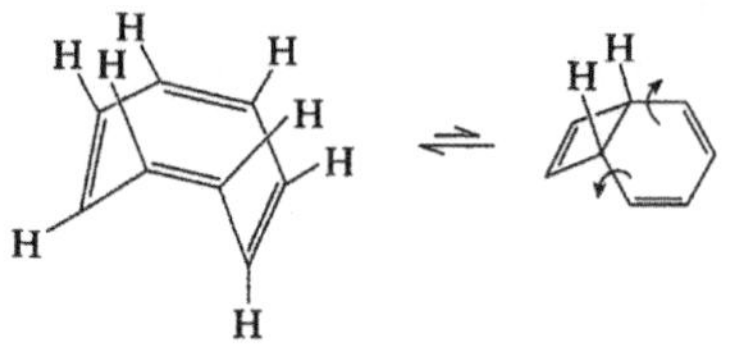

In practice, cyclooctatetraene is the more stable isomer, and the presence of the bicyclic compound has been demonstrated only through trapping experiments.

PROBLEM 6.2

What is the relative stereochemistry of the ambiguous groups in the following concerted processes?

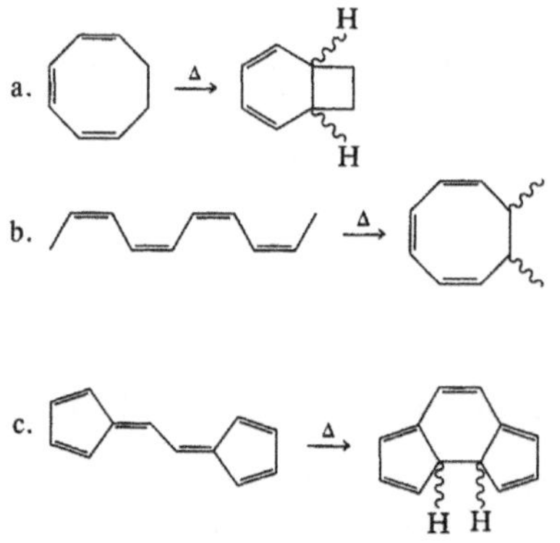

Sources: Huisgen et al., (1967); Sauter et al., (1977).

PROBLEM 6.3

Is the following reaction symmetry-allowed or symmetry-forbidden? Explain.

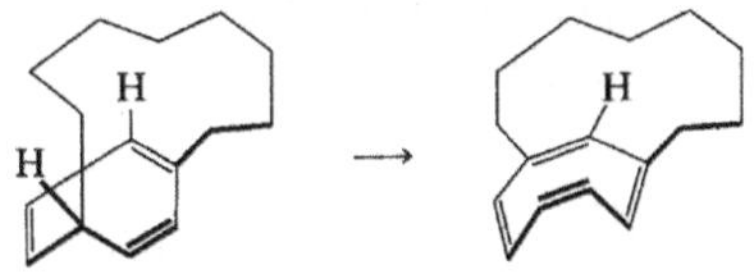

Source: Paquette and Wang, (1988).

PROBLEM 6.4

Consider the following concerted process:

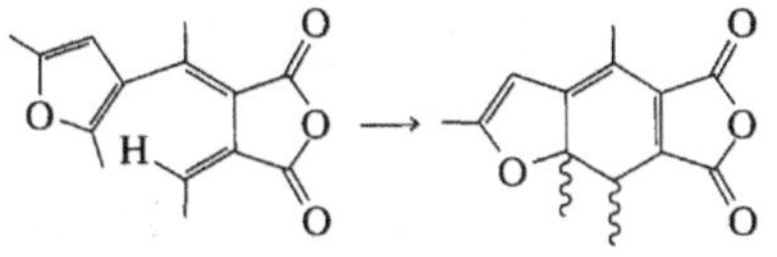

a. What is the relative stereochemistry of the product formed when the reaction is initiated by heat?
b. What is the relative stereochemistry of the product formed when the reaction is initiated by light?
c. What is the relative stereochemistry of the product of the photochemical ring opening of the product formed in part a?

Source: Darcy et al., (1981).

C. Electrocyclic Reactions of Charged Species (Cyclopropyl Cations)

The selection rules can be applied to charged species as well as to neutral molecules. The only requirement is that the reaction be a concerted process involving electrons in overlapping p orbitals. For example, the conversion of a cyclopropyl cation to the allyl cation can be considered as a π^2-electrocyclic process. For this process, the selection rules predict a disrotatory process.

EXAMPLE 6.4. THE SOLVOLYSIS OF CYCLOPROPYL TOSYLATES

The relative rates of concerted ring opening for substituted cyclopropyl tosylates can be explained on the basis of the selection rules and the principle of maximum orbital overlap. The relative rates of solvolysis for two different dimethyl cyclopropyl tosylates are shown in the accompanying diagram:

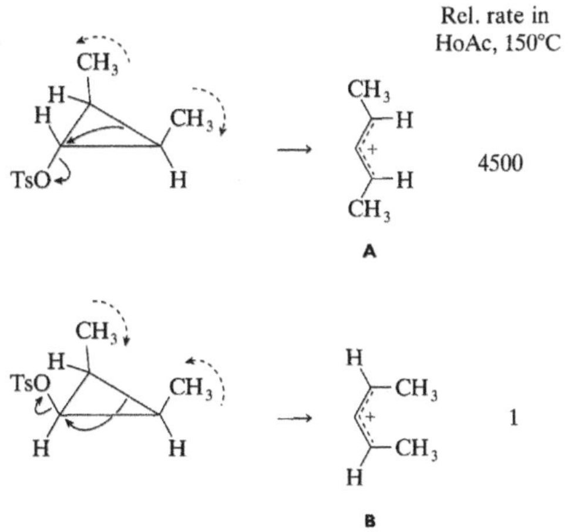

The large difference in relative rates suggests that the rate-limiting step does not involve unimolecular ionization by loss of the tosyl group (an S_N1 process). If it did, we would expect that the all-*cis* isomer would react faster due to relief of steric interaction between the tosyl and methyl groups. Instead, the results can be explained by assuming that loss of the tosyl group occurs with concerted ring opening to the allyl cation and that loss of the tosyl group is assisted by backside reaction of the electrons in the breaking C—C bond of the cyclopropyl group.

As shown in the diagram, opening of a cyclopropyl cation by a disrotatory process can occur in two ways (compare the discussion of the disrotatory and conrotatory openings of the cyclobutene system). The methyl groups can rotate outward (away from one another) or inward (toward one another). Disrotatory opening in which the methyl groups rotate away from each other gives the allyl cation A, whereas if the methyl groups rotate toward each other, the result is the allyl cation B. The opening in which the methyl groups rotate outward to give allyl cation A is expected to be more favorable due to lack of steric interaction. (Note that because of the delocalized π system in the allyl cation, there is restricted rotation around the C—C bonds and A and B are not interconvertible.) When the tosyl group is *trans* to the methyl groups, the electrons move so that they can assist the loss of the tosyl group by a backside reaction reminiscent of the neighboring group effect in the S_N2 reaction. For the isomer in which the tosyl group is *cis* to the methyl groups, outward rotation of the methyl groups would result in an increase in electron density in the vicinity of the tosyl group and a consequent increase in repulsion between centers of increased electron density.

Source: von et al., (1966).

EXAMPLE 6.5. GEOMETRIC CONSTRAINT TO DISROTATORY RING OPENING OF A BICYCLO[3.1.0]HEXANE SYSTEM

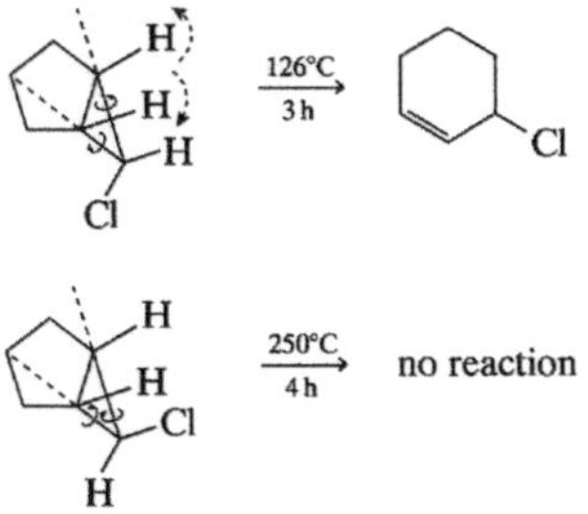

In small, fused cyclopropyl systems, only the disrotatory mode that moves the bridgehead hydrogens outward is geometrically feasible. This means that the electrons of the breaking cyclopropyl C–C bond can assist in the loss of the chloride group only when it is *trans* to the bridgehead hydrogens.

Source: Baird et al., (1969).

PROBLEM 6.5

Explain why the reactivity is so different in the following two reactions:

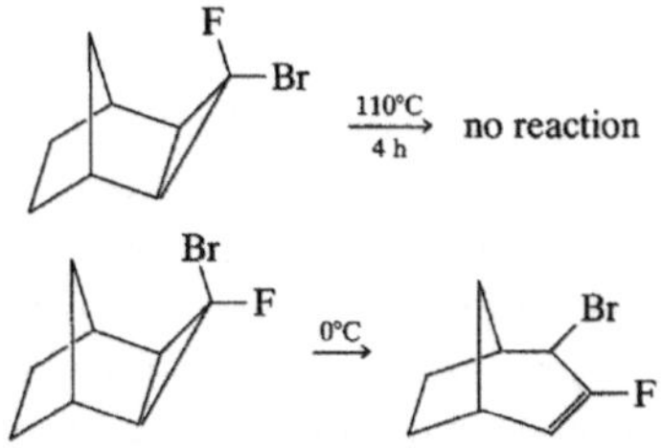

Source: Jefford and Hill, (1969).

---

3. CYCLOADDITIONS

A cycloaddition is the reaction of two (occasionally more) separate π systems, in which the termini are joined to produce a ring. Cycloadditions may be intermolecular or intramolecular. One way of describing a cycloaddition is to record separately the number of electrons in each component involved in the reaction.

A. Terminology of Cycloadditions

Number of Electrons

Cycloadditions can be described on the basis of the number of electrons of each of the components. Additional symbols are used to designate the type of orbital and the type of process involved.

EXAMPLE 6.6. THE DIELS–ALDER REACTION OF (2E, 3E)-2,4-HEXADIENE AND DIMETHYL MALEATE

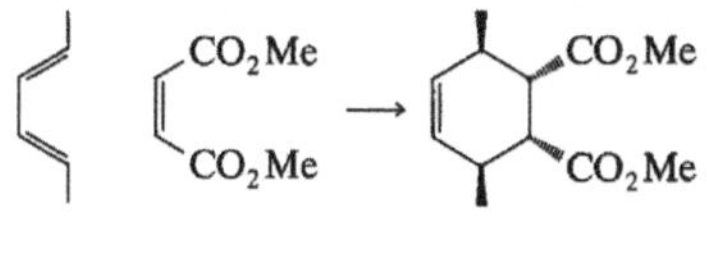

This is a six-electron process because there are four π electrons from the diene and two π electrons from the dienophile. It is also referred to as a [4 + 2] cycloaddition. Note that the carbonyl groups in the dimethyl maleate starting material are conjugated with the π bond undergoing reaction. However, because the π electrons of these carbonyl groups are not forming new bonds in the course of the reaction, they are not counted.

Stereochemistry (Suprafacial and Antarafacial Processes)

A cycloaddition reaction can be classified not only by the number of electrons in the individual components but also by the stereochemistry of the reaction with regard to the plane of the π system of each reactant. For each component of the reaction, there are two possibilities: the reaction can take place on only one side of the plane or across opposite faces of the plane. If the reaction takes place across only one face, the process is called *suprafacial*; if across both faces, *antarafacial*. The four possibilities are shown in the following diagram:

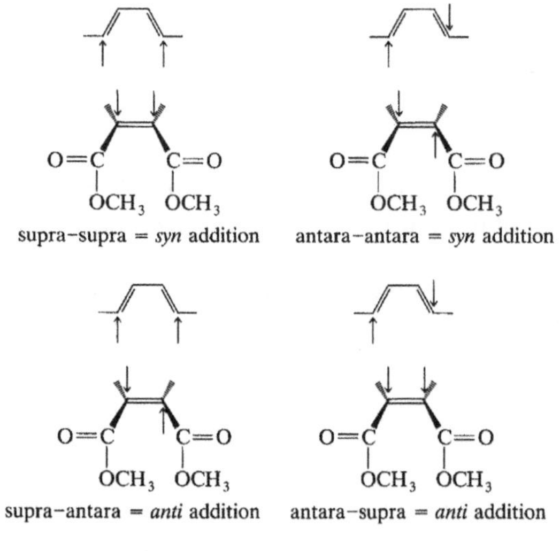

supra–supra = *syn* addition   antara–antara = *syn* addition

supra–antara = *anti* addition   antara–supra = *anti* addition

Essentially, a *suprafacial–suprafacial* or an *antarafacial–antarafacial* cycloaddition is equivalent to a concerted *syn* addition. A *suprafacial–antarafacial* or an *antarafacial–suprafacial* process is equivalent to a concerted *anti*-addition. The Diels–Alder reaction is suprafacial for both components, so that the stereochemical relationships among the substituents are maintained in the product. In Example 6.6, suprafacial addition to the dienophile component means that the two carbomethoxy groups that are *cis* in the starting material also are *cis* in the product. Suprafacial reaction at the diene component leads to a *cis* orientation of the two methyl groups in the product.

When classifying cyclizations, the subscripts "s" and "a" are used to designate suprafacial and antarafacial, respectively. Thus, a more complete designation of the Diels–Alder reaction is $[\pi_s^4 + \pi_s^2]$. This type of designation can be applied to other concerted processes, such as electrocyclic reactions, sigmatropic rearrangements, and the ene reaction; however, we will use these designations only for cycloadditions.

EXAMPLE 6.7. A CYCLOADDITION REACTION WITH THREE π COMPONENTS

78%

Because the two π bonds in the bicycloheptadiene are not conjugated, each is designated separately in the description of the reaction. Only the carbons, at each end of the C=C bond of maleic anhydride, are forming new bonds; thus, only the two π electrons of this bond are counted and this is a $[\pi_s^2 + \pi_s^2 + \pi_s^2]$ cycloaddition.

Source: Cookson et al., (1964).

Number of Atoms

Some confusion has been occasioned by the introduction of another kind of terminology based on counting the number of atoms of the component systems taking part in the cycloaddition. In this system, the total number of atoms in each component is counted. This number includes both termini and all of the atoms in between. The description of a cycloaddition based on the number of atoms does not give the same designation as that based on the number of electrons.

EXAMPLE 6.8. COUNTING THE ATOMS INVOLVED IN A CYCLOADDITION

Consider the following reaction:

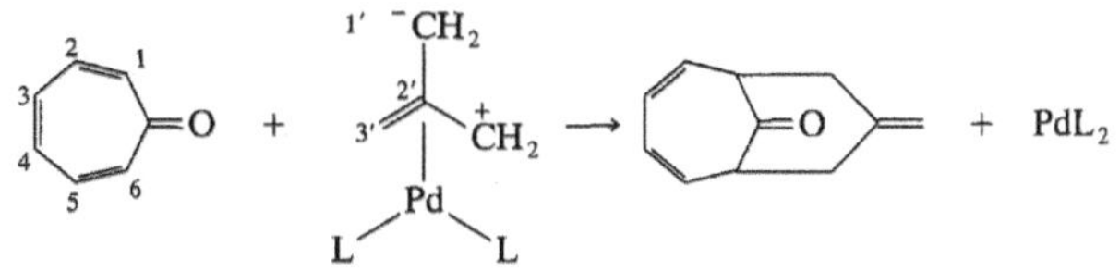

The authors call this a [6 + 3] cycloaddition because the reacting portion of tropone (the ketone) contains six atoms and the alkene component contains three atoms. The mechanism could be shown with arrows as follows:

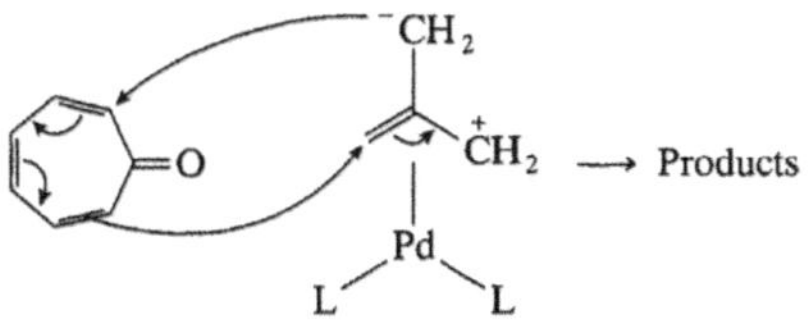

In tropone, 6π electrons are reacting, and in the alkene, 4π electrons are reacting. Therefore, if this is a concerted reaction, it also could be called a [6 + 4] cycloaddition. If the distinction is made that [6 + 3] refers to the cycloadduct and [6 + 4] refers to the cycloaddition, the two methods of nomenclature are compatible. However, this distinction is not always made in the literature, and unless one thinks about the details of the particular reaction, the descriptions may be confusing.

Source: Trost and Seoane, (1987).

B. Selection Rules for Cycloadditions

HINT 6.3

To apply the selection rules for cycloadditions, add the number of π electrons from each component undergoing reaction and then apply the rules outlined in Table 6.2.

Many reactions encountered are of the supra–supra variety. For these systems, a good rule of thumb is that $4n$ systems are activated photochemically and $4n + 2$ systems are activated thermally.

TABLE 6.2   Stereochemical Rules for Cycloaddition Reactions

Number of Electrons	Mode of Activation	Allowed Stereochemistry
$4n$	Thermal	Supra–antara
		Antara–supra
	Photochemical	Supra–supra
		Antara–antara
$4n + 2$	Thermal	Supra–supra
		Antara–antara
	Photochemical	Supra–antara
		Antara–supra

PROBLEM 6.6

Designate the following cycloadditions according to the number of electrons contributed by each component. Is the stereochemistry shown in accordance with that predicted by the selection rules?

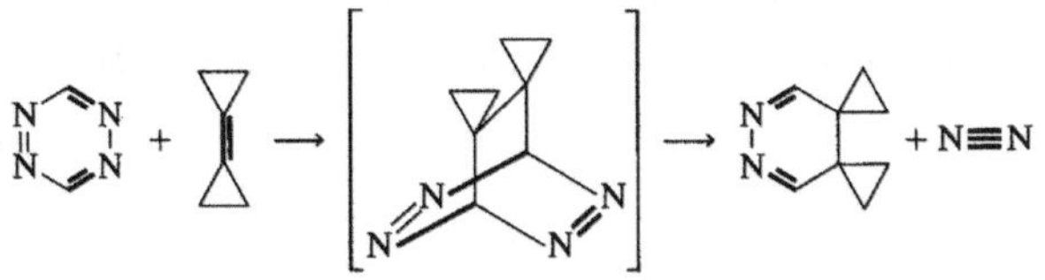

The reaction of the unstable intermediate is called a cycloreversion or retro-Diels–Alder reaction. Ordinarily it is designated by considering the reverse cycloaddition of the products of the reaction.

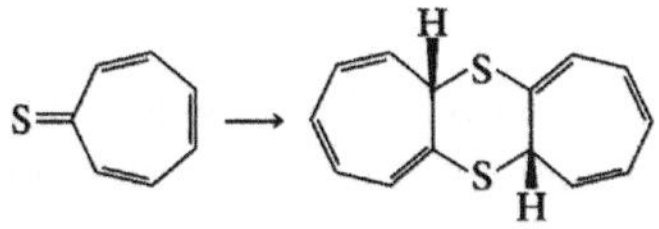

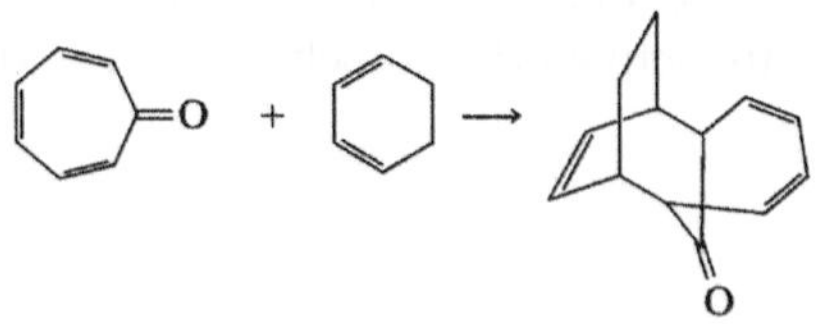

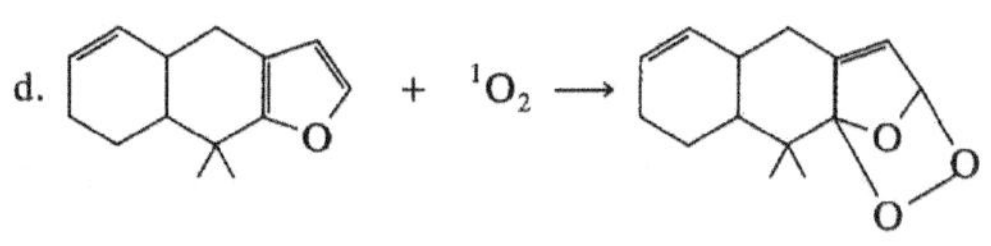

d.   + $^1O_2$ ⟶

$^1O_2$ = singlet oxygen

Sources: de Meijere et al., (1988); Machiguchi et al., (1989); Takeshita et al., (1986); Carte et al., (1986).

PROBLEM 6.7

Give the complete formulation (π_s^2, etc.) for each of the following cycloadditions.

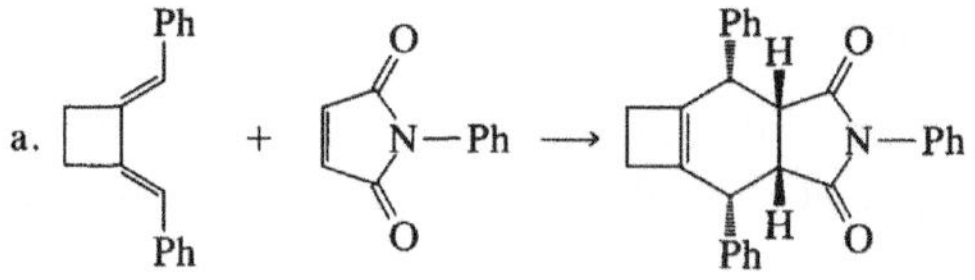

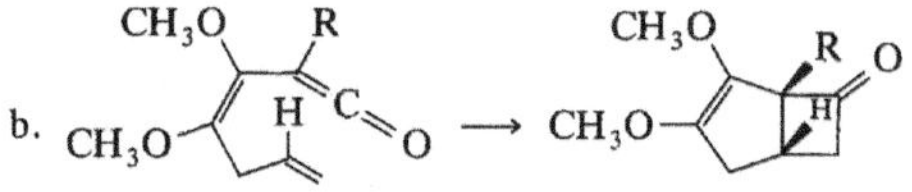

a.

b.

Sources: Minami et al., (1986); Xu and Moore, (1989).

PROBLEM 6.8

If it were concerted, how would the following reaction be described by both systems of nomenclature?

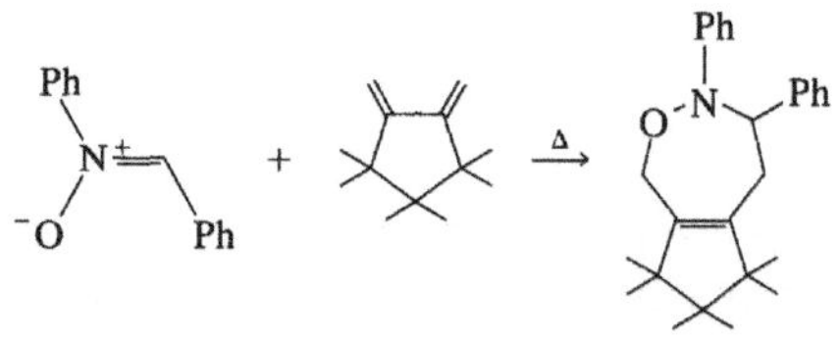

Source: Baran and Mayr, (1987).

PROBLEM 6.9

What is the product from an intramolecular $[4+2]$ cycloaddition of the following molecule?

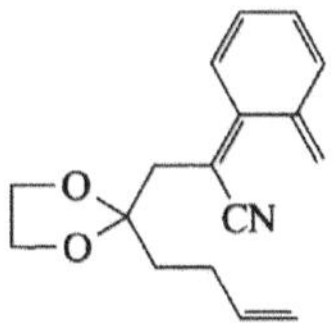

Source: Kametani et al., (1986).

C. Secondary Interactions

In the Diels–Alder reaction of cyclopentadiene with dimethyl maleate, both *exo* and *endo* products are theoretically possible. Only the *endo* product **6-6** is found.

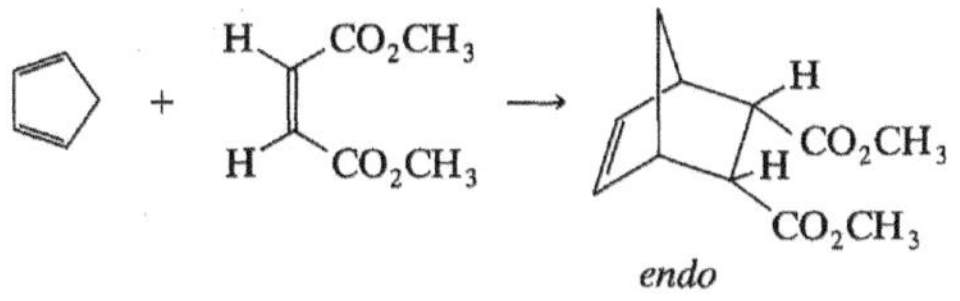

Because the carbonyl groups are part of the π system of the dienophile, there is an opportunity for secondary interactions between orbitals that are not involved in the bonding changes taking part in the cycloaddition. MO calculations have indicated that stabilizing interactions can take place when the reactants are oriented in such a way as to produce the *endo* isomer.

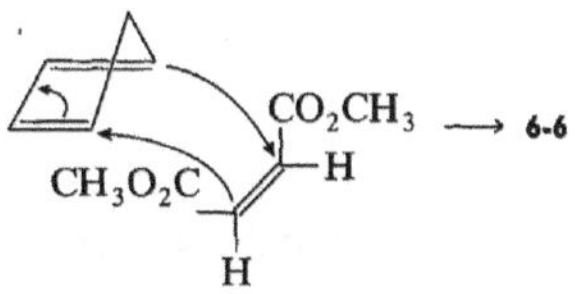

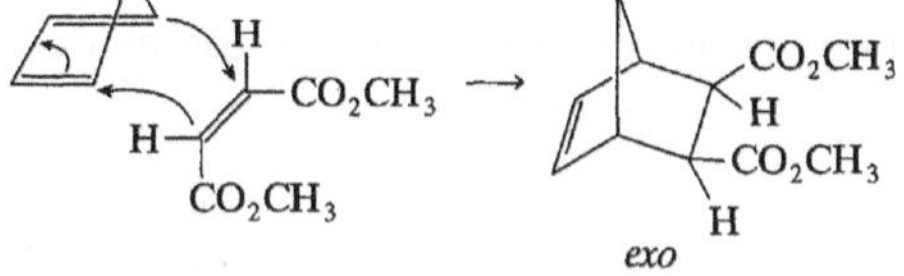

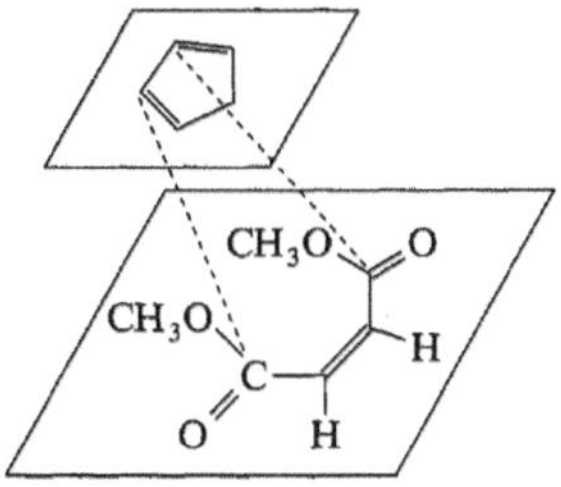

These secondary attractive interactions are represented by the dashed lines between the carbonyl groups and the diene system. Visual inspection of the orientation required for *endo* addition shows that there is greater overlap of the MOs of the two components than for the orientation that leads to *exo* addition.

MO interactions are only one of the factors that can influence the *exo:endo* ratio. Solvent interactions and the structure of the starting materials also are important, and these factors can result in predominantly *exo* addition products.

D. Cycloadditions of Charged Species

Allyl Cations

EXAMPLE 6.9. THE ALLYL CATION AS DIENOPHILE

A mechanism with several steps can be envisioned for the following reaction:

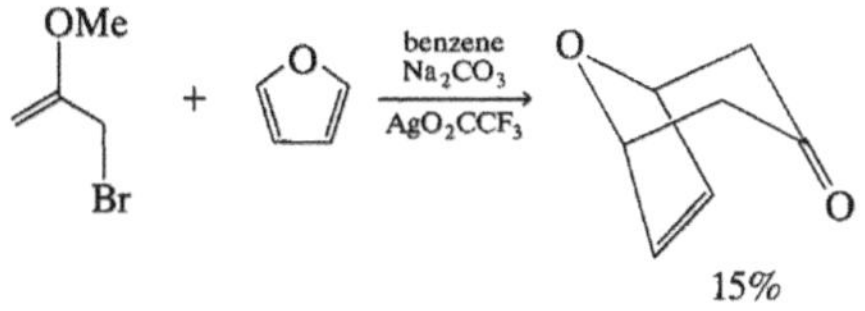

The first step, ionization of the bromide to the 2-methoxyallyl cation, **6-8**, is assisted by the silver ion:

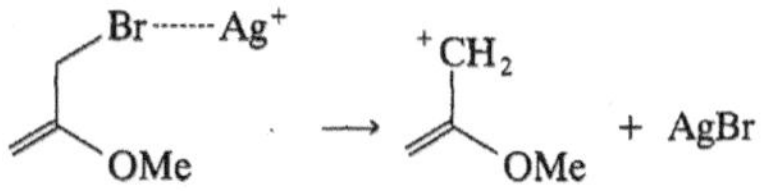

Cycloaddition occurs between cation **6-8**, which contains 2π electrons, and furan, which has 4π electrons, to give the cyclized cation, **6-9**:

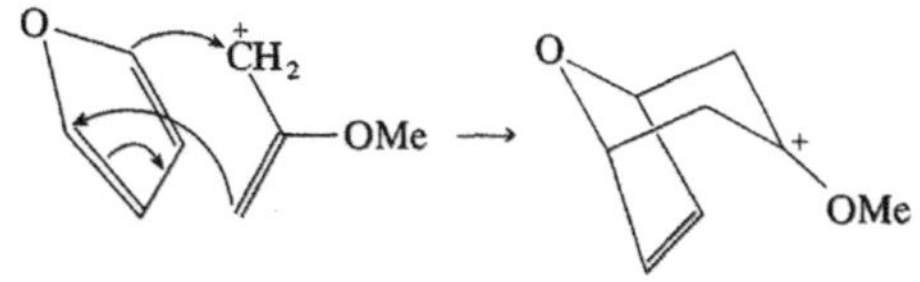

Under the workup conditions (dilute nitric acid), **6-9** hydrolyzes to the product ketone.

Source: Hill et al., (1973).

1,3-Dipoles

From a synthetic standpoint, the 1,3-dipolar cycloaddition is a very important reaction. In this $[4+2]$ cycloaddition, the four-electron component is dipolar in nature, and the two-electron component usually is referred to as the dipolarophile. When the thermal reaction is concerted, these reactions are suprafacial in both components, i.e. they are $[\pi_s^4 + \pi_s^2]$ cycloadditions.

EXAMPLE 6.10. 1,3-DIPOLAR CYCLOADDITION OF A NITRILE OXIDE AND AN ALKYNE

In the following reaction, the 4π-electron system of the nitrile oxide reacts with the 2π system of one of the acetylenic bonds.

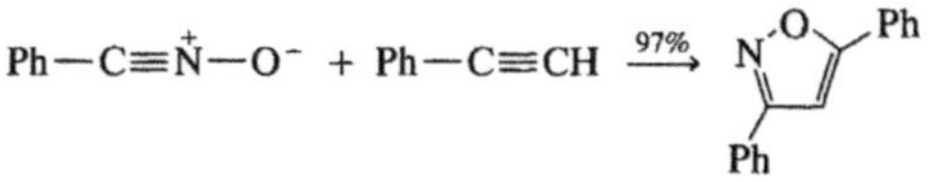

According to the newer terminology, they react to form what can be called a $[3+2]$ adduct. If the mechanism is written with arrows, the usual rules apply. That is, the flow of electrons is away from negative charge and toward positive charge.

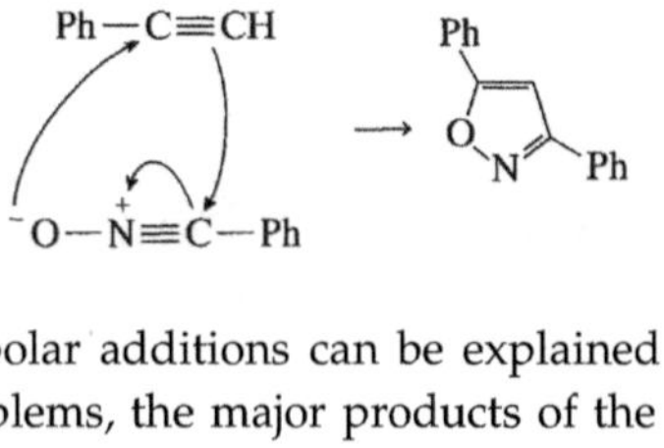

The regiochemistry of 1,3-dipolar additions can be explained on the basis of frontier orbital theory. In the examples and problems, the major products of the reactions are given. For further information, see Houk, (1975).

Source: Huisgen, (1963).

EXAMPLE 6.11. THE INTRAMOLECULAR 1,3-DIPOLAR CYCLOADDITION OF AN AZIDE

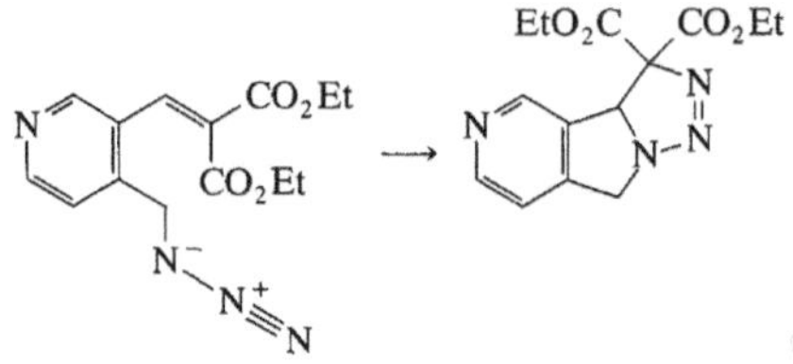

This is a concerted reaction in which four of the π electrons of the azide group undergo cycloaddition with the 2π electrons of the carbon–carbon double bond of the α-β-unsaturated ester:

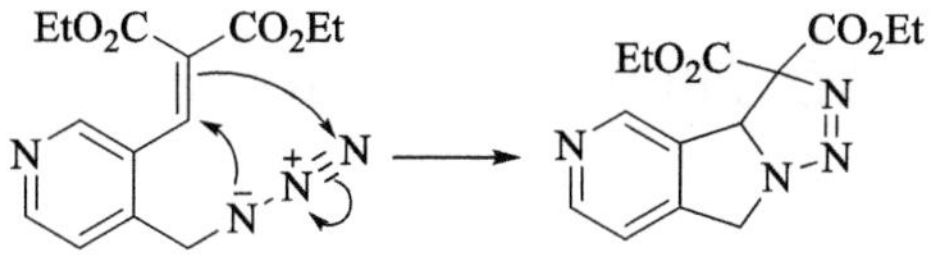

These kinds of intramolecular cycloadditions are very powerful synthetic tools for the synthesis of fused ring systems.

Source: Tsai and Sha, (1987).

PROBLEM 6.10

Write the possible products of the 1,3-dipolar cycloaddition of **6-10** with acrylonitrile. Many 1,3-dipoles, like the nitrile ylide **6-10**, are unstable and therefore are formed in situ when needed, as shown in the first step of the following reaction sequence.

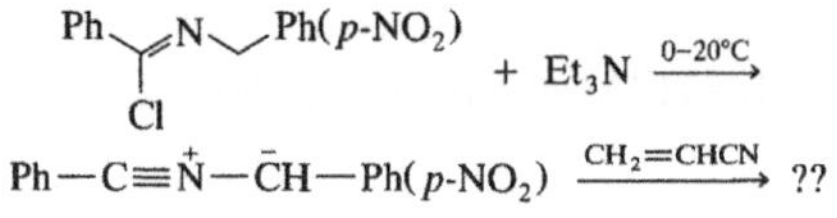

Source: Huisgen, (1963).

PROBLEM 6.11

Show how the following reaction might occur.

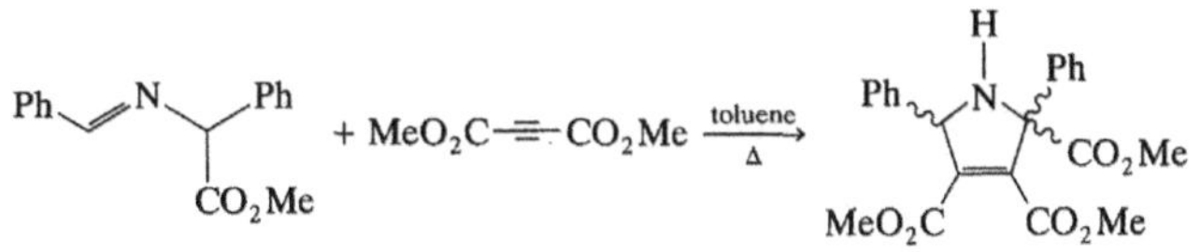

Source: Grigg et al., (1978).

---

4. SIGMATROPIC REARRANGEMENTS

A. Terminology

In a sigmatropic reaction, movement of a σ bond takes place, producing rearrangement (tropic is from the Greek word "tropos," to turn). Common types of sigmatropic reactions are the familiar 1,2-hydride or alkyl shifts in carbocations and the Cope rearrangement.

Sigmatropic reactions can be designated by the same scheme we have used for π electron reactions, i.e., $[\pi_s^4 + \sigma_s^2]$, etc. However, sigmatropic reactions are more often designated in a different way.

HINT 6.4

To determine the order of a sigmatropic reaction, first label both atoms of the original (breaking) σ bond as I. Then count the atoms along the chains on both sides until you reach the atoms that form the new σ bond. The numbers assigned to these atoms are given in brackets, separated by a comma, e.g., [1,5] or [3,3]. This nomenclature is illustrated in Examples 6.12 through 6.14.

EXAMPLE 6.12. A [1,2]-SIGMATROPIC SHIFT IN A CARBOCATION

The σ bond that is broken in the starting material and the σ bond formed in the product are highlighted. Hydrogen (1′) has moved from carbon 1 to carbon 2, so that the reaction is designated as a [1,2]-sigmatropic shift. The 1 of this designation does not refer to the number on carbon, but to the number on hydrogen. This indicates that the same atom (hydrogen) is at one end of the σ bond in both starting material and product. The 2 of the designation [1,2] is the number of the carbon where the new σ bond is formed, relative to the number 1 for the carbon where the old σ bond was broken.

This reaction can also be designated as $[\pi_s^0 + \sigma_s^2]$. This is more appropriate than labeling the reaction as $[\pi_s^2 + \sigma_s^0]$ because this is considered to be a hydride shift, not a proton shift.

EXAMPLE 6.13. A [3,3]-SIGMATROPIC SHIFT

Consider the classical Cope rearrangement:

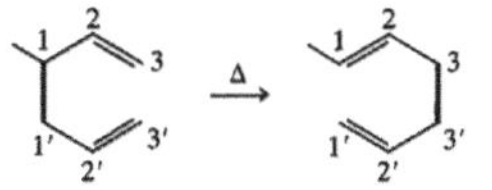

The atoms at the ends of the σ bond being broken are numbered 1 and 1'. Atoms are then numbered along each chain until the atoms at the ends of the new σ bond are reached. The atoms of the new σ bond are numbered 3 and 3'. Formally, the σ bond, originally at the 1,1' position, has moved to the 3,3' position. Thus, this is called a [3,3]-sigmatropic shift.

EXAMPLE 6.14. A [1,5]-SIGMATROPIC SHIFT

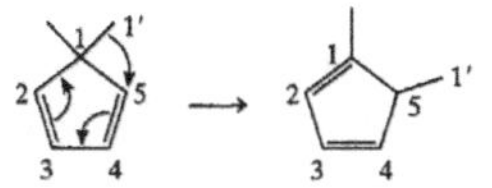

All the carbons in the ring must be counted as part of the process. That is, this reaction is *not* a simple [1,2] shift because the π electrons in the ring also must rearrange. Thus, this is a [1,5]-shift.

PROBLEM 6.12

Designate the type of sigmatropic shift that occurs in each of the following reactions (e.g., [1,3]). Also give the appropriate reaction designation.

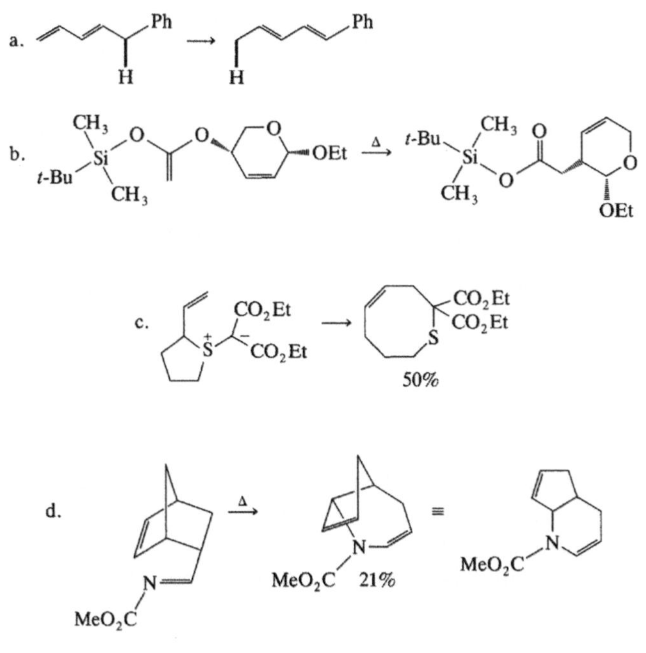

Sources: Curran et al., (1987); Vedejs, (1984); Wu et al., (1988).

B. Selection Rules for Sigmatropic Rearrangements

For sigmatropic reactions, as for electrocyclic reactions and cycloadditions, the course of reaction can be predicted by counting the number of electrons involved and applying the selection rules. A comprehensive rationalization of all the stereochemical aspects of these reactions requires application of the frontier orbital or orbital symmetry approaches, but, at this point, we will focus on the salient features of the more common reactions of this class (Table 6.3).

TABLE 6.3   Selection Rules for Sigmatropic Hydrogen Shifts

Order	Number of Electrons	Mode of Activation	Allowed Stereochemistry
[1,3]	$4n$	Thermal	Antarafacial
		Photochemical	Suprafacial
[1,5]	$4n + 2$	Thermal	Suprafacial
		Photochemical	Antarafacial
[1,7]	$4n$	Thermal	Antarafacial
		Photochemical	Suprafacial

HINT 6.5

For a sigmatropic reaction, the number of electrons involved is the number of π electrons plus the pair of electrons in the migrating σ bond.

Hydrogen Shifts

Thermal [1,3]-hydrogen shifts are unknown. The migrating hydrogen must maintain overlap with both ends of the π system but the geometry required by antarafacial migration makes this impossible. A few photochemical [1,3]-hydrogen shifts are known. Because the hydrogen moves suprafacially with photochemical activation, the reaction is geometrically feasible, as well as being symmetry-allowed.

PROBLEM 6.13

The following compound has been prepared and is stable at dry ice temperatures ($-78\,^{\circ}$C), even though its isomer, toluene, an aromatic compound, is much more stable. Why is it possible to isolate the nonaromatic triene?

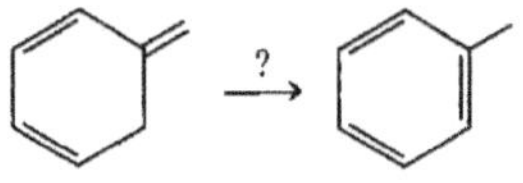

Source: Bailey and Baylouny, (1962).

EXAMPLE 6.15. THE STEREOCHEMICAL CONSEQUENCES OF A CONCERTED, THERMALLY ALLOWED [1,5]-SIGMATROPIC SHIFT

An elegant demonstration of the stereochemistry of a thermally allowed [1,5]-sigmatropic shift was reported by Roth et al. in 1970. They studied the stereochemistry of the reaction of the optically active starting material, **6-11**. There are four possible products: two arise from suprafacial reaction and two more from antarafacial reaction.

Suprafacial rearrangement of hydrogen across the *top* face of the π system gives **6-12**.

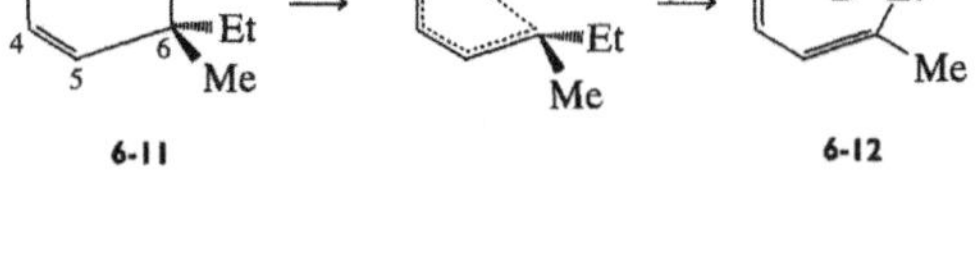

Rotation about the C-5–C-6 bond axis gives a new conformation of **6-11**. The suprafacial movement of hydrogen across the *bottom* face of the π system in this conformation produces **6-13**, a geometric isomer of **6-12**.

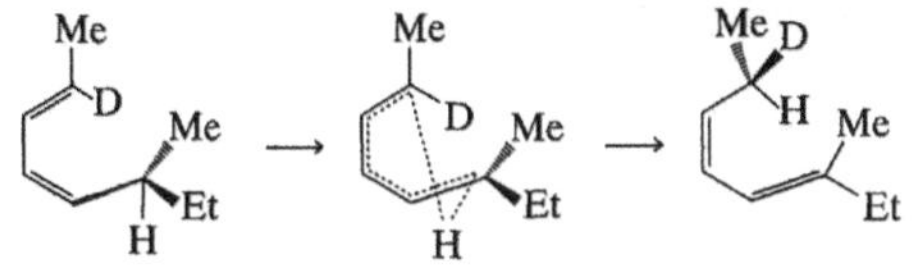

Antarafacial movement of hydrogen also gives two possible products, **6-14** and **6-15**. In the reaction to give **6-14**, hydrogen moves from the top face to the bottom face; in the reaction to give **6-15**, hydrogen moves from the bottom face to the top face.

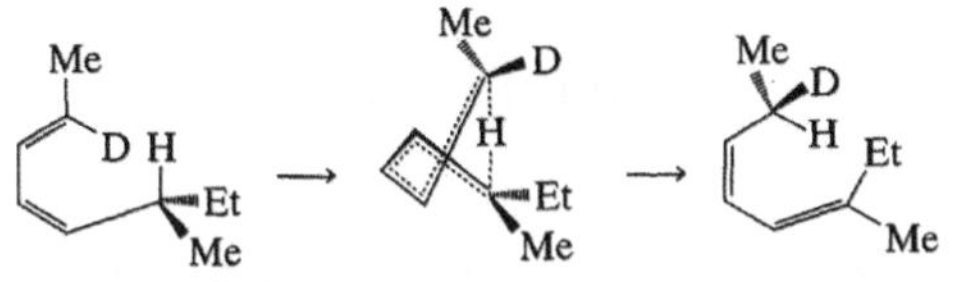

In agreement with the theoretical prediction of suprafacial reaction, compounds **6-12** and **6-13** were produced rather than **6-14** and **6-15**. Because of the favorable geometry for suprafacial migration, there are many examples of thermal [1,5]-sigmatropic shifts, which occur with ease.

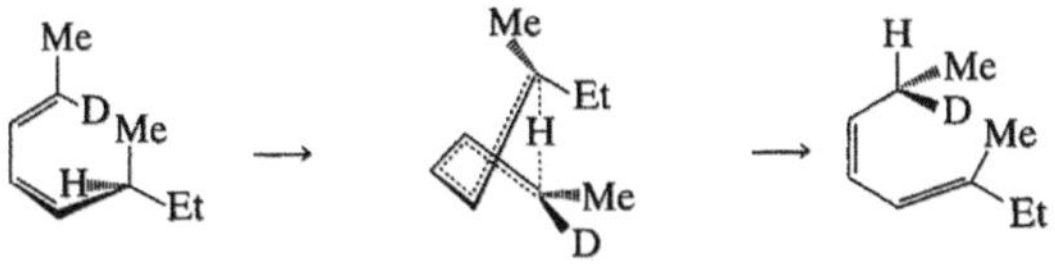

These elegant experiments are reported by Roth et al., (1970).

PROBLEM 6.14

The industrial synthesis of vitamin D_2 involves two pericyclic processes. Identify these processes in the following transformations and comment on the stereochemistry observed.

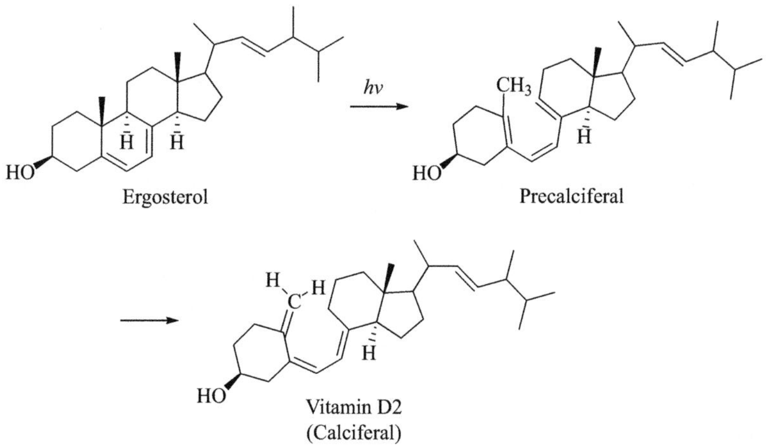

Source: Schlatmann et al., (1964).

Alkyl Shifts

When an alkyl group migrates, there is an additional stereochemical aspect to consider, namely, the carbon atom can migrate with inversion or retention of configuration. Commonly encountered processes include suprafacial [1,3]-shifts with inversion and suprafacial [1,5]-shifts with retention. (Note that this contrasts with the case for hydrogen migration, where the [1,3]-shift is antarafacial and the [1,5]-shift is suprafacial.) Inversion at carbon is an antarafacial process because the bond formed is on the opposite side of the carbon atom to the bond broken; retention at carbon is a suprafacial process (Table 6.4).

TABLE 6.4   Selection Rules for Sigmatropic Alkyl Shifts

Number of Electrons	Mode of Activation	Allowed Stereochemistry
$4n$	Thermal	Suprafacial with inversion
	Photochemical	Antarafacial with retention
$4n + 2$	Thermal	Suprafacial with retention
	Photochemical	Antarafacial with inversion

EXAMPLE 6.16. SIGMATROPIC SHIFTS IN THE EXO-6-METHYLBICYCLO[3.1.0]HEXENYL CATION

This is an interesting example of the stereochemical consequences of orbital symmetry. In this cation, the *exo*-6-methyl group remains *exo* as the migration of carbon-6 proceeds around the ring.

(Note that the numbers shown in the equation are not those used to name the compound. Thus, the 1' carbon is numbered 6 for nomenclature purposes, and the methyl on this carbon is referred to as the *exo*-6-methyl.)

Inspection of the structures involved might suggest that the electrons of the C-1—C-1' bond can "slide over" to form the C-4—C-1' bond, as represented by the arrows in the structure. If this is the reaction pathway, the configuration at C-1' is maintained because both bond breakage and bond formation occur at the same location relative to C-1', and we would expect to see the methyl group change from an *exo*- to an *endo*-orientation because of the rotation required about the C-5—C-1' bond. Consideration of the selection rules leads to a different prediction.

By applying Hint 6.4, we can classify the reaction as a [1,4]-sigmatropic shift. Counting the electrons in the bonds undergoing rearrangement shows that there are four electrons involved; consequently, a thermal process should proceed suprafacially with inversion at the migrating carbon. This means that in the bicyclo[3.1.0]hexenyl cation, bond breaking and bond formation occur on opposite sides of C-1', so that the methyl group remains in the *exo* position. Because these transformations frequently are difficult to visualize, molecular models are very useful for studying these kinds of reactions.

For the actual structures studied, see Hart et al., (1969).

PROBLEM 6.15

Show how the following concerted processes can be explained on the basis of one or more sigmatropic shifts. What are the designations for the shift(s) involved?

a.

b.

Show a mechanism explaining the following process and propose a numerical designation for the shift that occurred?

c.

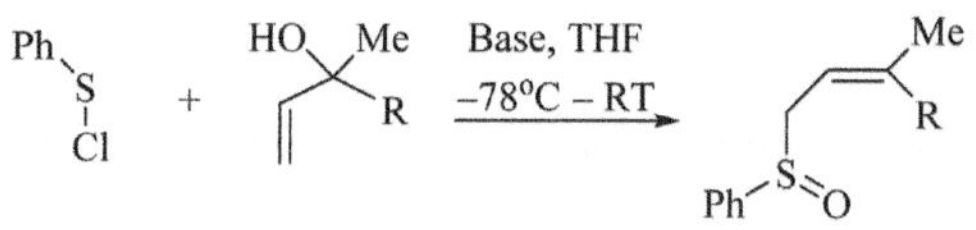

Sources: Miller and Baghdadchi, (1987); Barrack and Okamura, (1986); Evans and Andrews, (1974); Nakai and Mikami, (1986).

5. THE ENE REACTION

The ene reaction appears to combine the characteristics of cycloaddition and sigmatropic reactions.

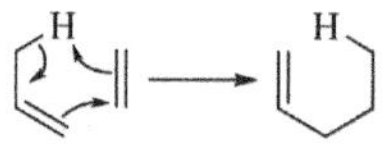

This looks similar to the Diels–Alder reaction, in which a C–H bond replaces the double bond of the diene component. In some cases, the ene reaction actually competes with the Diels–Alder reaction. Because of the similarities, the allyl component often is called the enophile, and the other component is called the ene.

The ene reaction also can take place intramolecularly and, thus, lead to new rings.

EXAMPLE 6.17. AN INTRAMOLECULAR ENE REACTION

The unconjugated diene, **6-16**, reacts to form a cyclic molecule.

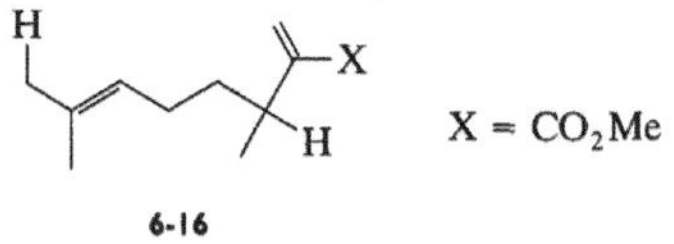

$X = CO_2Me$

6-16

EtAlCl$_2$

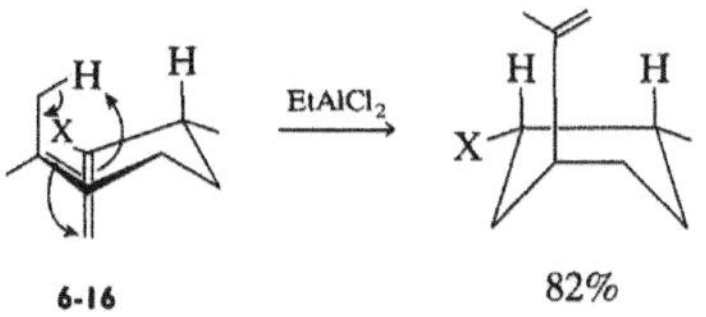

6-16   82%

Because the proton is transferred to the top of the double bond, the carbomethoxy group, X, is forced down. Molecular models are extremely useful for visualizing the conformation of **6-16** needed for cyclization to take place. *Cautionary note:* there is no evidence that this is a concerted reaction.

Source: Snider and Phillips, (1984).

PROBLEM 6.16

Draw the product for the ene reaction of the following components:

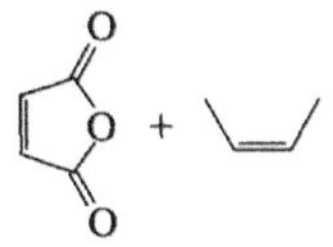

PROBLEM 6.17

Propose a mechanism for the following cyclization.

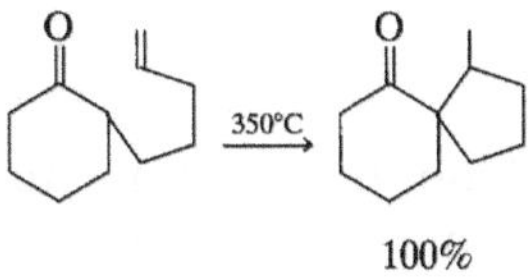

100%

Source: Conia and Robson, (1975).

PROBLEM 6.18

Write appropriate mechanisms for each of the following reactions.

a.

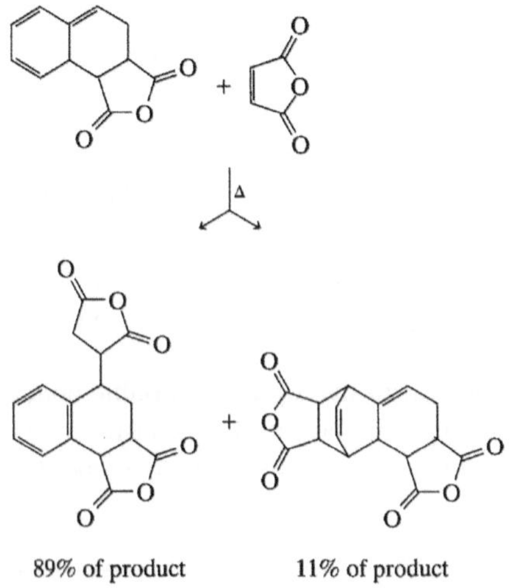

89% of product   11% of product

b.

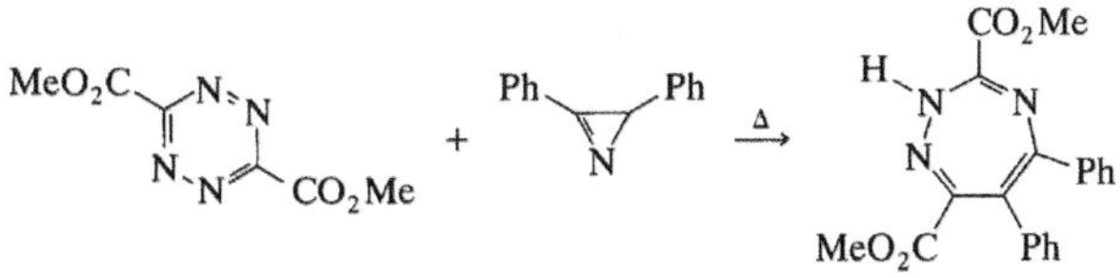

c.

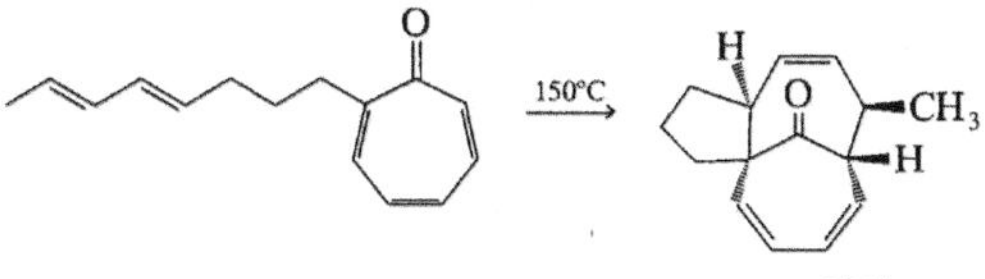

88%

d.

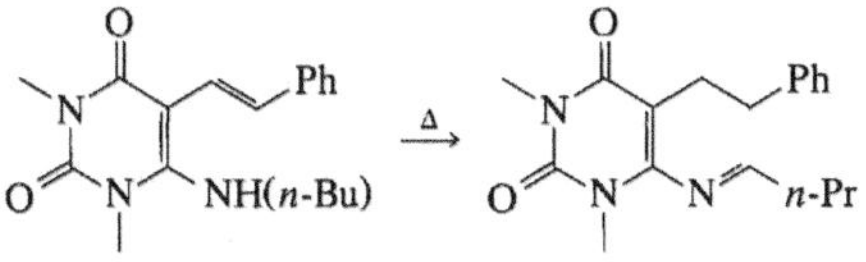

e.

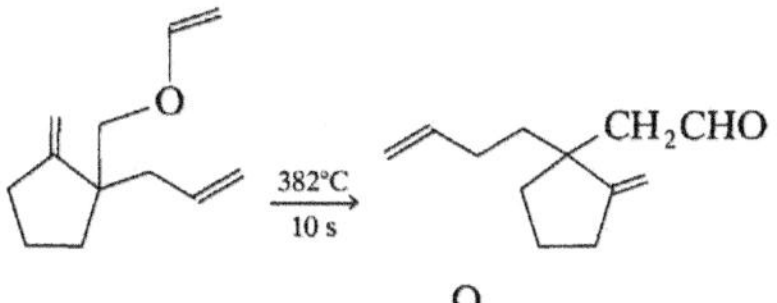

f.

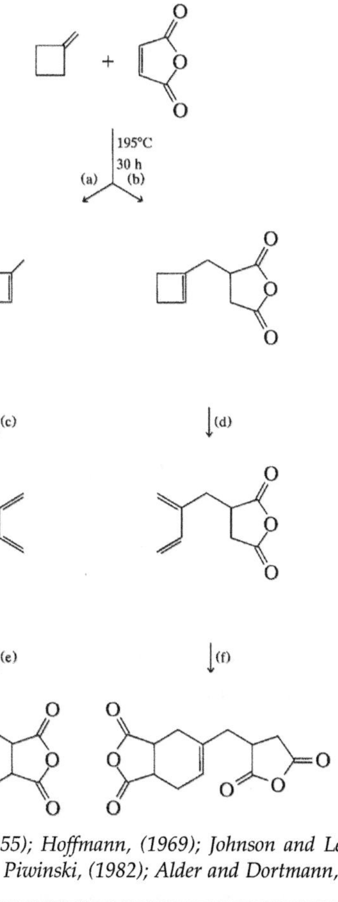

Sources: Alder and Schmitz-Johnson, (1955); Hoffmann, (1969); Johnson and Levin, (1974); Funk and Bolton, (1986); Matyus et al., (1986); Ziegler and Piwinski, (1982); Alder and Dortmann, (1952).

PROBLEM 6.19

The following thermal transformation involves three pericyclic changes. The first two are electrocyclic and the third is a sigmatropic rearrangement. Give structures for the two intermediates in the reaction.

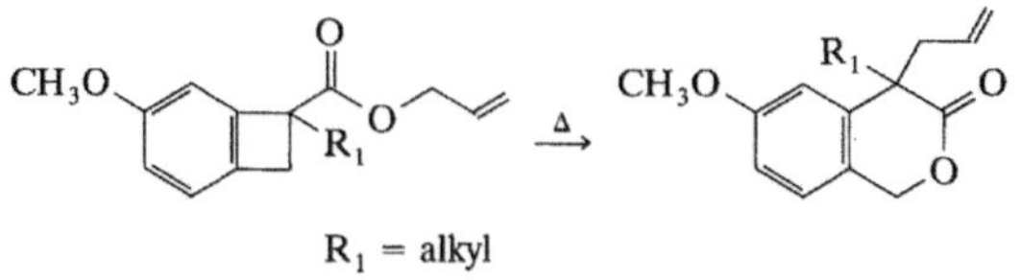

R_1 = alkyl

Source: Shishido et al., (1985).

PROBLEM 6.20

The following equation represents a thermal [4 + 4] cycloaddition, which occurs readily. Discuss orbital symmetry considerations in detail and propose a mechanism for the reaction.

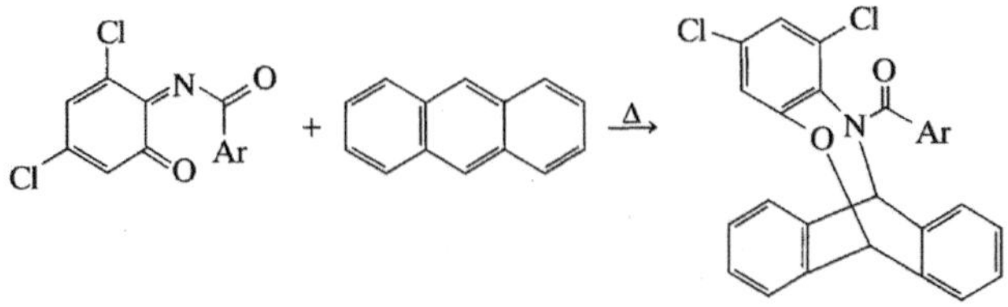

Source: Heine et al., (1989).

PROBLEM 6.21

Transformation of starting material to **6-17** involves a concerted electrocyclic transformation, followed by anion-promoted ring opening of the epoxide. The other transformations are one-step concerted processes. Explain the stereochemical preferences of each reaction. Why is the electrocyclic transformation of **6-17** to **6-19** favored when such reactions usually occur more readily in the opposite direction?

Source: Coates and Last, (1983).

6. A MOLECULAR ORBITAL VIEW OF PERICYCLIC PROCESSES

Although electron counting and application of the selection rules provide a practical method of analyzing and predicting pericyclic reactions, a greater understanding of these reactions can be gained by analyzing the MOs involved. A quantitative mathematical approach is of interest for a detailed analysis of specific reactions, but qualitative analysis can be applied much more rapidly and with very good effect to a much larger number of systems. The theory and methods of the MO approach to pericyclic reactions are beyond the scope of this book. We simply provide some introductory material on how to derive MOs for the π systems involved in pericyclic reactions, along with a few illustrative examples of how MO theory looks at selected pericyclic reactions.

A. Orbitals

An orbital, whether atomic or molecular, represents the region of space where a particular electron is most likely to be found. Orbitals are derived by solving wave equations, which are based on the experimental finding that electrons, like photons, can behave like waves as well as like particles. The wave equations that can be written for the electrons associated with individual atoms are analogous to the equations that can be written to describe a standing wave in a vibrating string. This is a useful analogy because a wavefunction has amplitude and nodes, just like the standing waves in a vibrating string. For a given electron, the probability of finding an electron at a particular location can be obtained by squaring the amplitude given by the wavefunction. Plotting of these probabilities gives a visual representation of the orbital shape and electron density.

HINT 6.6

A node, where the amplitude and electron density are zero, separates an orbital into lobes that have different algebraic signs. An atomic s orbital has no nodes, whereas in an atomic p orbital, a nodal plane divides the orbital into two lobes of differing sign.

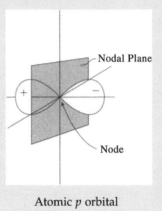

Atomic p orbital

In the diagram used in this chapter, the sign of the amplitude is represented by $+$ and $-$ signs within the lobes of the orbital. (An alternative representation uses shaded or unshaded lobes.)

B. Molecular Orbitals

MOs are derived mathematically by a linear combination of the wavefunctions for the atomic orbitals (AOs) of the individual atoms in a molecule. Usually, only the AOs of the valence electrons are considered because these are the electrons involved in bonding. We can visualize the formation of MOs as proceeding from overlap of the AOs of the valence electrons.

Just as we can consider the *formation of molecules* from atoms in terms of the *interaction of AOs* to form MOs, we can consider *reactions of molecules*, both intramolecular and intermolecular, in terms of the *interaction of the MOs*. In the case of pericyclic reactions, a very useful simplification can be made on the basis of Huckel MO theory. We assume that because the p orbitals that overlap to form π bonds are orthogonal (lie at right angles) to the σ bonds, the π bonding system can be considered independently of the single (σ) bonds in the reacting molecules and that the π bonding system is the most important factor in determining the chemical reactivity of conjugated polyenes and aromatic compounds.

HINT 6.7

MOs can be classified with regard to three elements:

1. The number of nodes
2. Symmetry with respect to a mirror plane (σ)
3. Symmetry with respect to rotation about a twofold axis (C_2).

For each MO in a molecule, the classification according to these three characteristics is unique. In other words, for any molecule, these three descriptors cannot be identical for two different MOs.

In MOs, as in AOs, *nodes* are regions of zero electron density that divide an orbital into lobes with amplitudes of opposite sign. When a node coincides with a nuclear position, there are no lobes depicted on that atom. In the following diagram, we see that the bonding π MO for ethylene has no nodes perpendicular to the bond axis, whereas the antibonding π^* orbital has one node perpendicular to the bond axis.

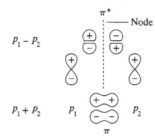

Electron distribution of atomic p orbitals and MOs (π and π^*).

The orbital representation is an approximation of the electron distribution.

The symmetry of a π system can be examined with respect to a *mirror plane* bisecting the π system and perpendicular to the σ bond molecular framework. (A mirror plane similarly bisects a fork or a spoon.) An orbital is symmetric (S) with respect to a mirror plane (usually called a σ plane) if it has the same sign on both sides, and antisymmetric (A) if the two sides have opposite signs. For example, in ethylene, the bonding π orbital is symmetric with respect to the mirror plane, whereas the antibonding orbital is antisymmetric, as shown in Table 6.5. Note that antisymmetric does *not* mean the same as asymmetric.

TABLE 6.5   Symmetry Characteristics of Ethylene π Molecular Orbitals

Molecular Orbital	Reflection through mirrror plane (σ)	Rotation about the C_2 axis
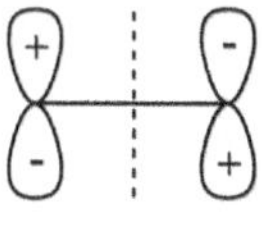 π^* Nodes = 1	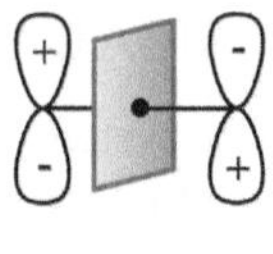 Antisymmetric (A)	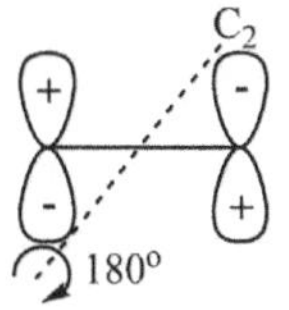 Symmetric (S) 180°
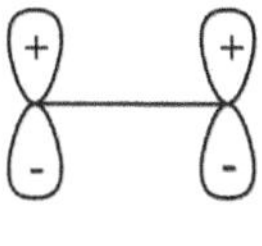 π Nodes = 0	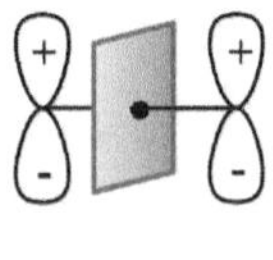 Symmetric (A)	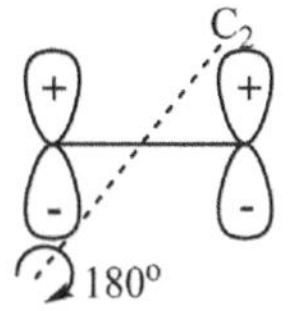 Antisymmetric (A) 180°

The symmetry of a π system also can be classified with respect to *rotation about a C_2-axis bisecting the π system and lying in the plane defined by its σ bond framework.* A C_2-axis is defined operationally: rotation of 180° about a C_2-axis gives an arrangement indistinguishable from the original. (A twin-bladed fan or propeller is an everyday object with a C_2-axis. A flathead screwdriver has both a mirror plane and a C_2-axis.) The following diagram shows the C_2-axis for the antibonding π^* orbital in ethylene. When the orbitals are rotated 180° with respect to this axis, the + and − orbitals coincide with the + and − orbitals of the original arrangement.

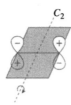

C. Generating and Analyzing π Molecular Orbitals

In analyzing the π MOs of reacting molecules, we need to determine the number of orbitals, their relative energy, and the number and placement of nodes. Once we have determined the number and relative energy of the orbitals, we can determine the electron configurations of the reactants and analyze the amplitudes of the orbitals that overlap in the course of the reaction.

HINT 6.8

The number of MOs is equal to the number of AOs that was used to construct them.

EXAMPLE 6.18. THE *II* MOLECULAR ORBITALS OF ETHYLENE

For ethylene ($H_2C{=}CH_2$), the simplest π system, there are two ways that we can combine the two p AOs from the two carbon atoms so that we get two π MOs.

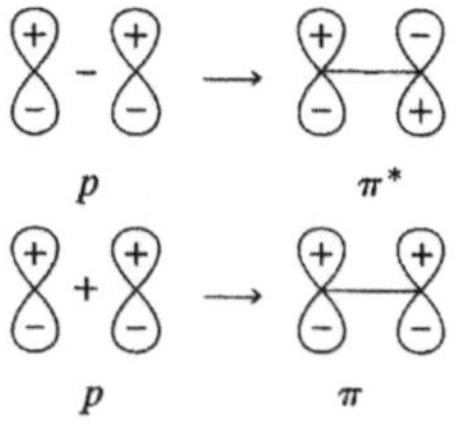

Basis sets and MOs for ethylene.

If the wavefunctions for the two p orbitals are added, we form a bonding orbital (π); if the wavefunctions are subtracted, we form an antibonding orbital (π^*). Molecular π orbitals are commonly represented by the AOs that combine to form them. In the accompanying diagram, the orbitals on the left are the AOs. These are used to form the MOs of ethylene, which appear on the right. The AOs that are combined to form the MOs are called the *basis set*.

HINT 6.9

When AOs are combined to give MOs, the MOs can be bonding, antibonding, or nonbonding. The number of bonding and antibonding orbitals is equal. If the number of AOs used is odd, there is a nonbonding orbital in addition to the bonding and antibonding orbitals.

Electrons in the bonding (π) orbital are lower in energy than electrons in the isolated p orbitals, and electrons in the antibonding (π^*) orbital are higher in energy than electrons in the isolated p orbitals, as illustrated here for ethylene.

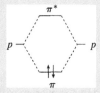

Energy levels of π MOs for ethylene.
The energy of a nonbonding orbital is comparable to that of the p orbitals of the isolated atoms.

HINT 6.10

For linear, conjugated polyenes, each MO has a unique energy.

The conjugated polyene systems taking part in pericyclic reactions are linear, so all the π MOs of the starting materials have a different energy. The distribution of energy levels is important because it affects the electronic configuration of the starting materials (Example 6.19) and consequently the course of the reaction.

In symmetrical *cyclic conjugated* systems, such as benzene, there are sets of MOs with the same energy (degenerate orbitals). (The pattern of energy levels for these aromatic systems can be determined by Frost diagrams, which we will not consider; see Carey and Sundberg, (1990)). The symbol Ψ is commonly used to designate the various MOs, which are distinguished by subscripts (i.e. in order of increasing energy, Ψ_1, Ψ_2, Ψ_3, ...).

HINT 6.11

We determine the electron configuration for molecular π electrons in the same way that we do for atomic electron configurations: we count the total number of electrons and then use these to fill each orbital with a maximum of two electrons, starting with the lowest energy orbital and filling the orbitals in the order of increasing energy.

EXAMPLE 6.19. ENERGY LEVELS FOR THE ALLYL MOLECULAR ORBITALS AND ELECTRON CONFIGURATIONS OF THE ALLYL CATION, RADICAL, AND ANION

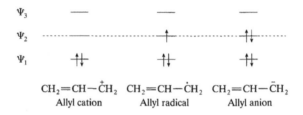

The MOs of the allyl system are formed by the overlap of three atomic p orbitals. Because there is an odd number of AOs, one of the MOs is a nonbonding orbital, whose energy is comparable to that of the isolated p orbitals from which it was derived. Note that if there were degenerate MOs in the allyl system, the electronic configurations of various allyl species would be different. For example, if Ψ_2 and Ψ_3 for the allyl system had identical energy levels, the allyl anion would have two unpaired electrons.

HINT 6.12

The MO of lowest energy has no nodes, the next has one node, the next two nodes, and so on. The nodes are arranged symmetrically with respect to the center of a linear π system.

Nodes in the π MOs bisect the bonds forming the σ framework; for π systems derived from an odd number of AOs, nodes may also coincide with nuclei (Example 6.20).

HINT 6.13

If the MOs are examined in the order of increasing energy, the orbitals alternate in symmetry with regard to the mirror plane and C_2-axis of symmetry. The first orbital is symmetric (S) with respect to a mirror plane bisecting the linear system, the second orbital is antisymmetric (A) with respect to the same plane, and so on. Similarly, the first orbital is antisymmetric with respect to the C_2-axis, the second is symmetric with respect to C_2, and so on.

EXAMPLE 6.20. MOLECULAR ORBITALS OF THE ALLYL SYSTEM

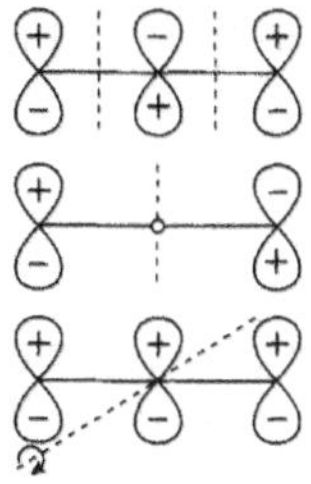

| | | Nodes | Symmetry | |
			Mirror (σ)	C_2
Ψ_3		2	S	A
Ψ_2		1	A	S
Ψ_1		0	S	A

PROBLEM 6.22

Why would the following MO diagram for the allyl system be incorrect?

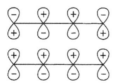

PROBLEM 6.23

Explain why the following would not be the two lowest energy MOs for 1,3-butadiene.

PROBLEM 6.24

Write all the MOs for the four- and five-carbon π systems. Indicate the number of nodes and designate the symmetry of each orbital with respect to the mirror plane (σ) and the C_2-axis of rotation.

D. HOMOs and LUMOs

In analyzing pericyclic reactions, two MOs are of particular interest: the π MO of highest energy that contains one or two electrons (the highest occupied molecular orbital, HOMO) and the MO of lowest energy that contains no electrons (the lowest unoccupied molecular orbital, LUMO). For electrocyclic reactions, where there is only one π system, the important orbital is the HOMO. When more than one π system is involved, as in cycloaddition, reactions are considered to occur through a transition state in which the HOMO of one component overlaps the LUMO of the other.

In photochemically activated reactions, the HOMO and LUMO are not the same as those for thermal reactions because absorption of light promotes an electron from the HOMO to the LUMO of the ground state. Thus, the LUMO of the ground state becomes the HOMO of the photochemically excited state.

EXAMPLE 6.21. HOMOs and LUMOs for the Butadiene System

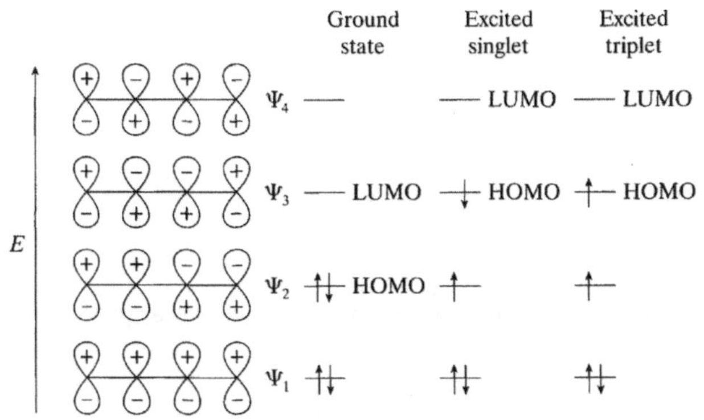

PROBLEM 6.25

By using the orbitals drawn for Problem 6.24, indicate the electronic configuration for the pentadienyl cation, radical, and anion. Label the HOMO and LUMO for each case.

HINT 6.14

For bond formation to occur, the overlapping (i.e. interacting) MOs must have amplitudes of like sign.

Bond formation during reactions is entirely analogous to the process whereby we form MOs from AOs in individual molecules. Interaction between orbitals of like sign results in constructive interference (i.e. bonding), whereas interaction between orbitals of opposing sign results in destructive interference.

E. Correlation Diagrams

In a pericyclic reaction, the pathway predicted by the selection rules is the one that allows maximum orbital overlap along the reaction pathway, including the transition state. Maximum orbital overlap corresponds to the path of minimum energy and is achieved if the orbitals involved are similar in energy and if the symmetry of the orbitals is maintained throughout the reaction path.

When pericyclic reactions are analyzed in terms of correlation diagrams, all the π and σ MOs taking part in the reaction are analyzed in terms of their symmetry properties with respect to reflection in a mirror plane and rotation about a C_2-axis. The symmetry elements of importance are those that are found in both reactants and products *and* are preserved during the course of the reaction. If the symmetry properties of all the orbitals remain unchanged (are conserved) throughout the reaction, then maximum orbital interaction is maintained throughout the course of the reaction and the reaction is *symmetry-allowed*. If the symmetry properties change, then the reaction is *symmetry-forbidden*.

EXAMPLE 6.22. ORBITAL CORRELATION DIAGRAMS FOR THE INTERCONVERSION OF BUTADIENE AND CYCLOBUTENE

Classifying the Relevant Orbitals

As in the electron-counting approach, we consider only the electrons and bonds that change in the reaction.

This is a 4π electron process. The bonds involved are the π bonds of the conjugated diene system and the π bond and the ring-forming σ bond of the cyclobutene. The electronic configurations and symmetry characteristics of the orbitals involved are as follows:

Butadiene			Cyclobutene		
	σ	C_2		σ	C_2
Ψ_4	A	S	σ^*	A	A
Ψ_3	S	A	π^*	A	S
Ψ_2 ⇅	A	S	π ⇅	S	A
Ψ_1 ⇅	S	A	σ ⇅	S	S

Symmetry Correlations between Bonding Orbitals of Starting Materials and Products

If we examine the classification of orbitals of starting material and product with respect to each of the two symmetry elements in turn, we see that the bonding orbitals, Ψ_1 and Ψ_2 starting material and product share symmetry only with respect to the C_2-axis. (With respect to the σ plane, both bonding orbitals of the cyclobutene system are symmetric, but in the butadiene system, only Ψ_1 is symmetric with respect to the σ plane.)

C_2				σ			
Ψ_4 — S	A — σ^*			Ψ_4 — A	A — σ^*		
Ψ_3 — A	S — π^*			Ψ_3 — S	A — π^*		
Ψ_2 — S	A — π			Ψ_2 — A	S — π		
Ψ_1 — A	S — σ			Ψ_1 — S	S — σ		
Butadiene	Cyclobutene			Butadiene	Cyclobutene		

This means that during the course of a reaction, maximum bonding overlap can be maintained only if all the intermediates along the reaction pathway also have C_2 symmetry; in other words, the whole process must be symmetrical with respect to C_2. If we look at symmetry correlations with respect to the σ plane, we see that to maintain symmetry with respect to the σ plane, the butadiene orbital Ψ_2 has to correlate with the π^* orbital. This means that in order to maintain symmetry with respect to the σ plane, the reaction would have to form the product in an excited state. The energy requirement for such a conversion would be very large, so that the conversion requiring the correlation of Ψ_2 with π^* would be *symmetry-forbidden*.

Symmetry Characteristics of the Reaction

As we saw in Section 2.B, the interconversion of butadiene and cyclobutene can occur through either a disrotatory or a conrotatory process. As the following diagram shows, the disrotatory opening is symmetrical with respect to a σ plane, whereas conrotatory opening is symmetrical with respect to the C_2-axis.

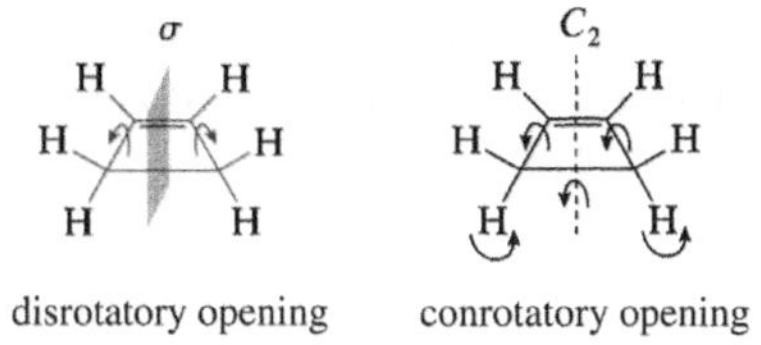

disrotatory opening   conrotatory opening

Because conrotatory ring opening proceeds so that intermediate structures have C_2 symmetry at all points along the reaction pathway linking starting material and product, it is logical that thermal ring opening of cyclobutene occurs by a conrotatory process. Furthermore, because the disrotatory process does not proceed along a pathway that maintains symmetry, it is not expected to occur.

Orbital Phase Correlations

An alternative way of analyzing the thermal ring-opening reaction is to look at the *phases* of the MOs involved. These can be visualized as follows:

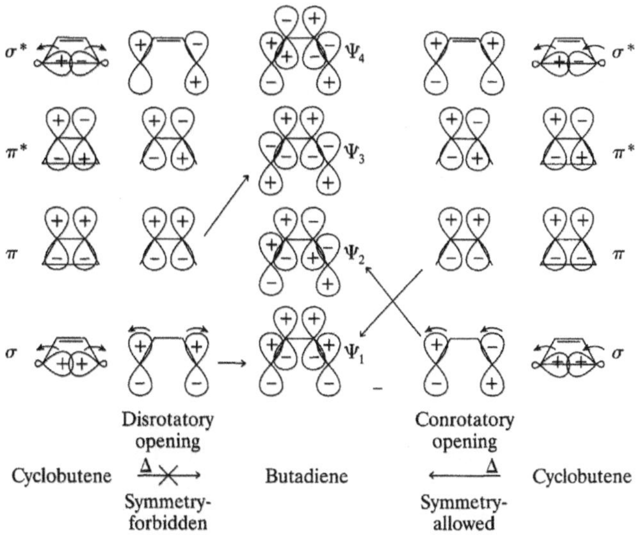

Cyclobutene   Disrotatory opening   Butadiene   Conrotatory opening   Cyclobutene
Symmetry-forbidden   Symmetry-allowed

In conrotatory ring opening, the reoriented σ orbitals derived from cyclobutene look like part of the butadiene MOs Ψ_2 and Ψ_4. The orbitals derived from the double bond of cyclobutene look like part of the butadiene MOs Ψ_1 and Ψ_3. Because the signs of the cyclobutene orbitals can be correlated with bonding orbitals of butadiene by conrotatory opening, this mode of ring opening is allowed, whereas disrotatory opening, in which the signs of the cyclobutene orbitals must be correlated with an antibonding orbital, is forbidden.

F. Frontier Orbitals

The frontier orbital approach leads to the same predictions as those based on an analysis of correlation diagrams, but instead of considering all the orbitals of the π system, only the frontier orbitals, the HOMOs and LUMOs are considered. A general discussion of frontier orbital theory is given in Fukui (1971). The choice of which HOMOs and LUMOs to combine requires an analysis that we will not undertake here. We will simply show that when these orbitals are combined, we can rationalize reactions observed experimentally in terms of the orbitals derived from this type of analysis.

Electrocyclic Reactions

The frontier orbital analysis of electrocyclic reactions focuses on the HOMO of the open-chain polyene.

HINT 6.15

In the frontier orbital approach to electrocyclic transformations, the course of reaction is determined by the amplitude of the lobes at the termini of the HOMO of the polyene. Rotation of bonds during reaction will occur in such a way that orbitals of like sign can overlap.

EXAMPLE 6.23. THE FRONTIER ORBITAL APPROACH TO THE THERMAL RING OPENING OF CYCLOBUTENES

The HOMO for butadiene is Ψ_2. In the conrotatory cyclizations shown here, rotation of the lobes at the ends of the HOMO leads to overlap of orbitals with like signs and production of a σ bond.

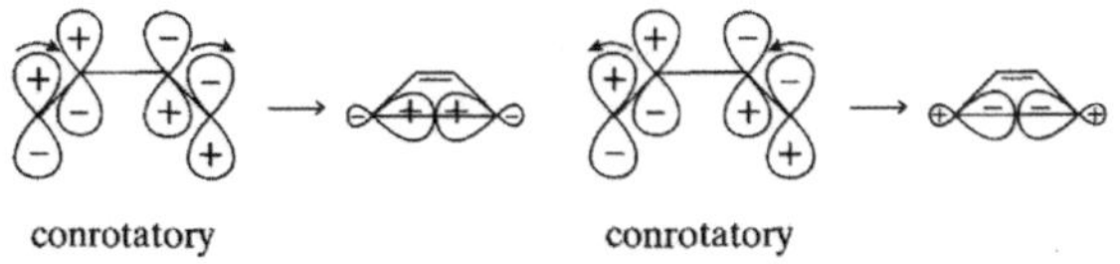

conrotatory   conrotatory

A disrotatory process, on the other hand, would lead to an antibonding interaction between the ends of the chain and would not lead to formation of the σ bond needed for cyclization.

The π-bonding orbital of cyclobutene arises from reorganization of the butadiene π system. This process is considered in Woodward and Hoffmann, (1970). and references cited therein.

Cycloaddition

In cycloaddition reactions, frontier orbital analysis considers the interaction of the HOMO of one component and the LUMO of the other.

EXAMPLE 6.24. FRONTIER ORBITAL OVERLAP IN A [4 + 2] CYCLOADDITION

In a [4 + 2] cycloaddition, we can consider overlap between the HOMO of the diene and the LUMO of the olefin.

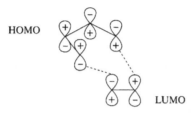

Overlap of lobes with like signs can occur by suprafacial–suprafacial addition or by an antarafacial–antarafacial process. The latter would not be expected for a [4 + 2] reaction because of geometric constraints.

Sigmatropic Rearrangements

EXAMPLE 6.25. FRONTIER ORBITAL ANALYSIS OF SIGMATROPIC [1,3]- AND [1,5]-HYDROGEN SHIFTS

We can think of a hydrogen shift as resulting from cleavage of the C−H sigma bond, followed by movement of a hydrogen atom across a π radical system to form a new sigma bond. (Alternative schemes in which a proton moves across a π anion system or a hydride moves across a π cation system would require more energy because these would require separating charges as well as breaking a carbon−hydrogen bond.) For a [1,3]-shift, the interacting orbitals that are closest in energy are the 1s orbital of hydrogen and the Ψ_2 orbital of the allyl system. Each of these orbitals contains a single electron, unlike the HOMOs involved in electrocyclic reactions and cycloadditions, each of which contains a pair of electrons. In the case of the [1,5]-hydrogen shift, the orbitals involved are the 1s orbital of hydrogen and the Ψ_3 orbital of the pentadienyl system. We can represent the orbital interactions as follows:

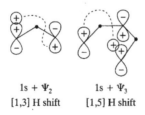

1s + Ψ_2 1s + Ψ_3
[1,3] H shift   [1,5] H shift

In the [1,3]-shift, a total of 4π electrons are involved. A bonding interaction can be maintained only if the hydrogen moves to the opposite site of the π system (i.e. the rearrangement is antarafacial). Geometry is against this move. Either the hydrogen has to enter a region outside the

π system in order to get from one side to the other, or the bond angles of the allyl system have to be strained severely in order for the hydrogen to move from one face to the other while maintaining bonding overlap. Both of these are very high-energy options, so the reaction is not observed.

In the [1,5]-shift, 6π electrons are involved. In this case, bonding can be maintained throughout a suprafacial process, and this type of reaction occurs with ease.

EXAMPLE 6.26. ORBITALS INVOLVED IN SIGMATROPIC SHIFTS OF CARBON

Frontier orbitals can help to explain the results seen when a sigmatropic shift involves migration of carbon rather than hydrogen. A [1,3] migration of carbon involves 4π electrons. We can think of the reaction as migration of a carbon radical across the π system of an allylic radical. In this case, overlap of the p orbital of the carbon radical with the allylic π system can be visualized as shown in the following diagram.

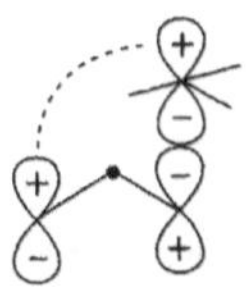

Because the p orbital has a node, the carbon atom can move across the π system. As bonding interaction between the $+$ lobes of the p orbital and the π system decreases, interaction develops between the $-$ lobes. In the process, the migrating carbon undergoes inversion analogous to that seen in an S_N2 process. In both cases, the bonds breaking and forming are at 180°.

EXAMPLE 6.27. ORBITALS INVOLVED IN SIGMATROPIC SHIFTS IN THE EXO-6-METHYL BICYCLO[3.1.0]HEXENYL CATION

In this system, as we saw previously, the methyl group remains *exo* as the migration of carbon-6 proceeds around the ring.

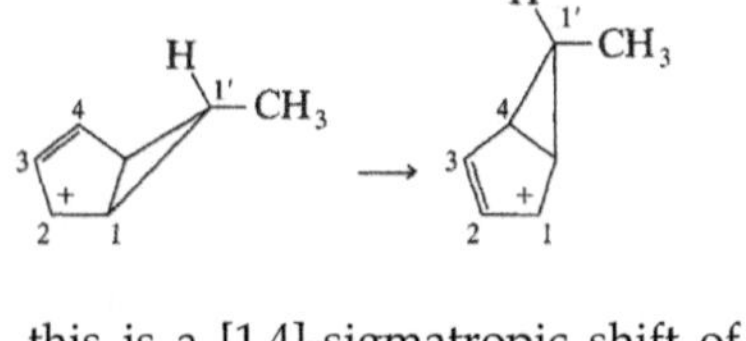

As we saw in Example 6.16, this is a [1,4]-sigmatropic shift of the carbon labeled 1′ in the equation. (Note that the numbers shown in the equation are not those used to name the compound. Thus, the carbon labeled 1′ for classification of the sigmatropic shift is numbered 6 for the purpose of nomenclature.)

One way to consider the orbital interactions is to think of the process as migration of a carbo-cation across a butadiene π system. This is a reasonable choice because, in contrast to Example 6.25, the reacting species is already charged. If we consider the reaction as interaction between the LUMO of the carbocation (atomic p orbital) and the HOMO of the butadiene π system (Ψ_2), we can represent the interaction visually as follows:

In this scheme, inversion of configuration occurs at C-6, which means that the methyl group remains *exo*. If it is difficult to see that such a transformation leads to the methyl group remaining *exo*, molecular models can be of assistance.

For the actual structures studied, see Hart et al., (1969).

ANSWERS TO PROBLEMS

Problem 6.1

a. For the stereochemistry shown, the process must be disrotatory. This is a four-electron process and, thus, is symmetry-forbidden thermally. The term symmetry-forbidden does not mean the reaction is impossible, merely that so much energy is required that other processes usually occur instead.

This is an example of a symmetry-forbidden reaction that, nonetheless, actually takes place. Failure to conform to the selection rules usually is taken to mean that the reaction proceeds by a nonconcerted mechanism. The electronic effects of the substituents, R_1 and R_2, on the course of the reaction are complex.

b. In order to rotate the substituents into their positions in the product, a conrotatory mode is required. This is a 4π electron process and is thermally allowed. Note that the diene, and not the cyclobutene, is considered in giving the designation and stereochemistry.

Note the unusual *trans* orientation of the ring carbons across the lower double bond in the product.

Problem 6.2

a. According to Table 6.1, a concerted thermal reaction of a 6π electron system requires a disrotatory mode of ring closing, which leads to a *cis* orientation of the hydrogens.

b. To determine the stereochemical consequences of the reaction, first rotate about the single bonds of the starting material to get a conformation, in which the ends of the tetraene system are close enough to react. These rotations change the conformation of the molecule, but not the stereochemistry. Note, however, that you may *not* rotate about any of the double bonds to get the ends into position because that would change the stereochemistry of the tetraene. All the double bonds have *cis* stereochemistry. Because this is an 8π electron system, the thermally allowed process will be conrotatory. Conrotatory rotation gives the product with the two methyl groups *trans*.

c. As in the previous example, the starting material must be placed in a conformation in which the ends of the π system are close enough to react. If it is not clear to you which electrons are necessary for the transformation, drawing arrows for the redistribution of electrons can be helpful.

    In order to get the product, all 12π electrons must be involved. Thus, the thermal reaction should be conrotatory, and the hydrogens will be *trans*.

Problem 6.3

    To obtain the stereochemistry observed in the product, the long chain (carbon A) must move behind the plane of the page and hydrogen B (H_B) must move in front of the plane of the page. This is a disrotatory process.

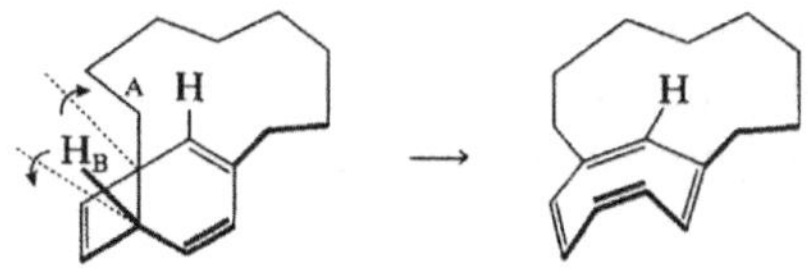

We can consider the change from starting material to product as involving either 4π electrons or 6π electrons, depending on whether we consider the process to be the ring opening of a four-membered ring or a six-membered ring. If it is a 4π-electron process, disrotatory opening can occur only under photochemical conditions. If the reaction is a 6π-electron process, it is symmetry allowed under thermal conditions.

Problem 6.4

a. The thermal reaction will be disrotatory, leading to *trans*-methyl groups:

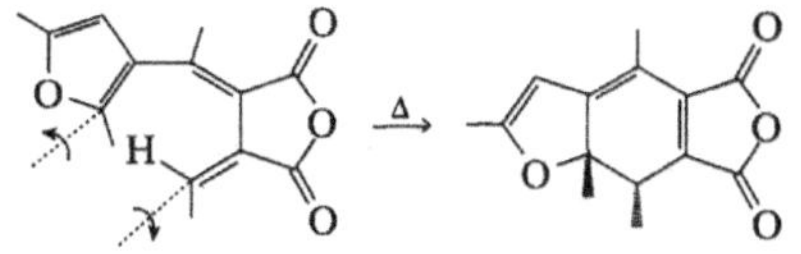

b. This is a 6π electron process. (Only one of the π bonds in the furan ring is involved in the reaction.) The photochemical reaction will be conrotatory (antarafacial), leading to *cis*-methyl groups:

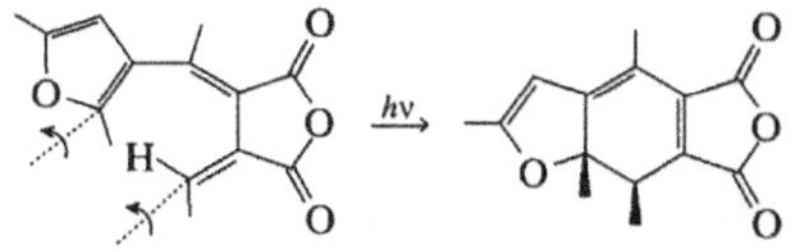

c. The photochemical ring opening of the thermal product will be conrotatory. (Remember, the total number of electrons counted in a process is that of the open-chain compound.) Thus, the product will be an isomer of the starting material for the original process.

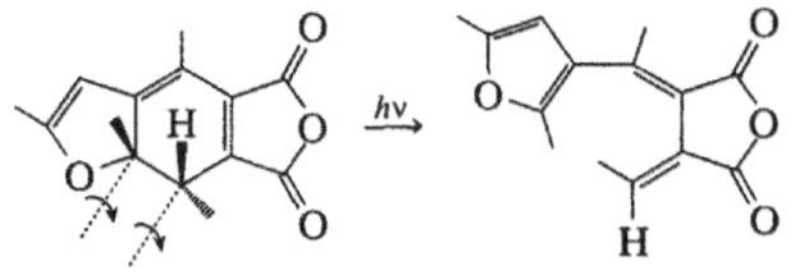

Problem 6.5

Ionization occurs to a 2π electron allylic cation. To be thermally allowed, therefore, the motion must be disrotatory. Only the disrotatory motion, which moves the hydrogens outward, will occur. This places the developing p orbitals opposite the bromide leaving group only for the second compound. The bromide ion then reacts with the cation to produce the product. In the first compound, the developing p orbitals will be opposite to the fluoride group, which is such a poor leaving group that no reaction occurs.

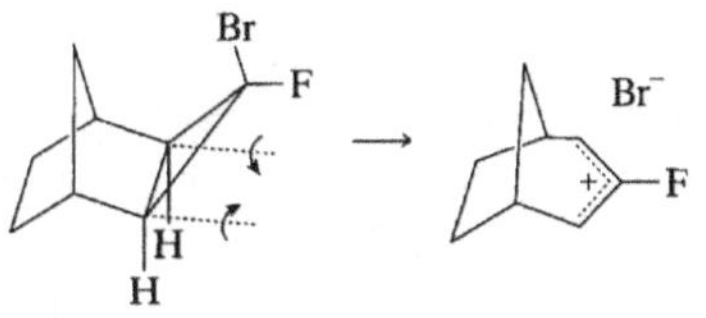

Problem 6.6

a. Both the cycloaddition to give the intermediate and the cycloreversion of the intermediate to give the product are [4 + 2] reactions. The electrons involved in each process are highlighted in the following structures:

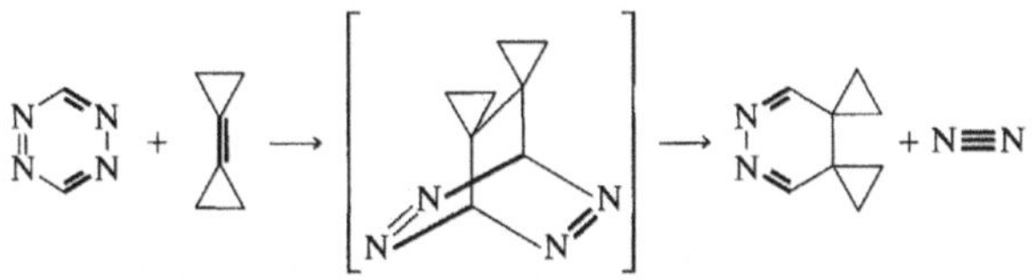

In the retrocycloaddition, only two of the π electrons of the N_2 product are used. The other two π electrons are perpendicular (orthogonal) to the first two and cannot take place in the reaction. Thus, by considering the products of the reaction, this is a $[\pi_s^4 + \pi_s^2]$ process. On the other hand, the reaction of the intermediate would be designated as a $[\sigma_s^2 + \sigma_s^2 + \pi_s^2]$ process.

b. This is an [8 + 8] cycloaddition as becomes evident when arrows are used to show the flow of electrons leading to product. Because this is a $4n$ cycloaddition, the selection rules dictate a supra–antara cyclization, which is equivalent to a concerted *trans* addition. This is difficult to visualize. It is easier to visualize the supra–supra cyclization, which would lead to the isomer in which the hydrogen atoms are *trans*. From this, we can conclude that because the product has the two hydrogens *cis*, the reaction must be supra–antara or antara–supra.

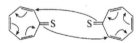

c. This is a [6 + 4] addition. The π electrons of the carbonyl are not involved. The selection rules predict a supra–supra or antara–antara process, which results in the observed *cis* stereochemistry.

d. The singlet oxygen has all electrons paired and so reacts as the two-electron component of a [4 + 2] addition. In the more commonly encountered triplet state, oxygen has two unpaired electrons and reacts as a diradical.

Problem 6.7

a. $[\pi_s^4 + \pi_s^2]$. The phenyls of the diene and the carbonyls of the heterocycle are not involved in the ring-forming process, so the π electrons in these substituents are not counted. The two hydrogens, which are shown up (*cis*) in the product, also are *cis* in the starting material. In addition, the two phenyl groups of the diene end up on the same side of the six-membered ring. Both these stereochemical consequences are the result of suprafacial processes. This can be demonstrated readily by using models.

b. $[\pi_s^2 + \pi_s^2]$. The best way to understand the stereochemistry of the process is to look at models. Placing the appropriate termini together in a model of the starting material shows that the stereochemistry of the product is produced when both components react suprafacially. The π electrons of the C=O group and the dimethoxy-substituted π bond are not counted because they are not involved in the reaction. This is a thermal reaction, so that a nonconcerted pathway is expected.

Problem 6.8

$[\pi^4 + \pi^4]$. The nitrone component contains four electrons: the two π electrons of the double bond and the two π electrons on the oxygen that overlap with them. The diene component also contains four electrons. The nitrone component contains three atoms and the diene component contains four atoms, so that a $[4 + 3]$ adduct is formed. The authors cited propose that this adduct is formed by a diradical mechanism. For more on this reaction, see Problem 7.2.b.

Problem 6.9

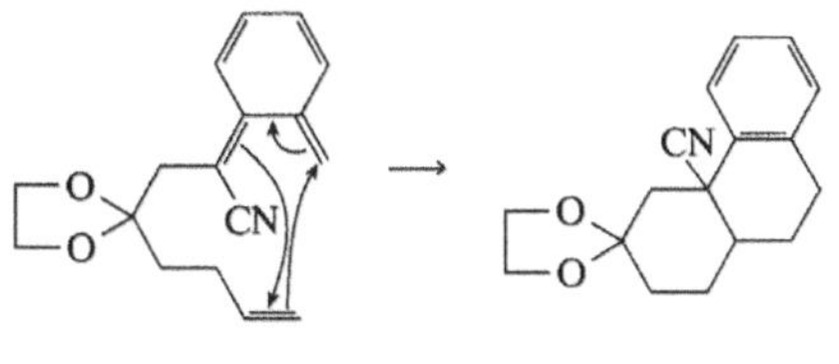

Problem 6.10

The following reaction shows the actual regiochemistry observed:

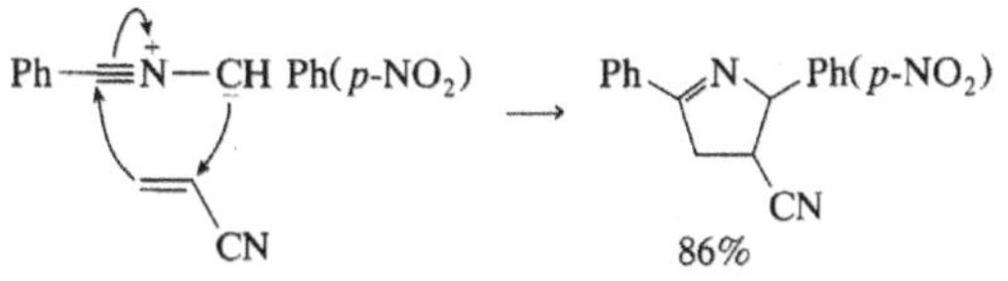

Because the double bond is polarized by the nitrile group, we might have expected the following reaction to predominate:

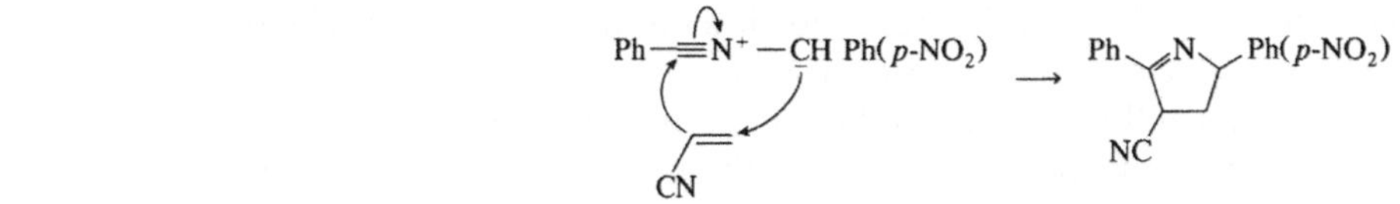

As indicated in the text, many of these reactions show high regioselectivity. However, it is not possible to predict this regiochemistry without MO calculations.

Problem 6.11

A possible mechanism is intermolecular proton transfer from one molecule of the starting ester–imine to another to give a 1,3-dipole, which then cycloadds to the acetylene.

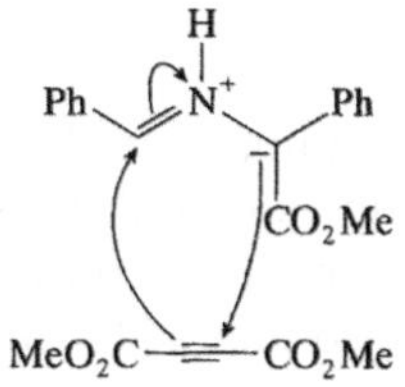

Problem 6.12

a. This is a [1,5]-sigmatropic shift. The highlighted σ bond has been broken, the π system has shifted, and a new σ bond has been formed at the other end of the π system. Numbering of atoms from each end of the original σ bond through the terminal atoms of the new σ bond gives the following:

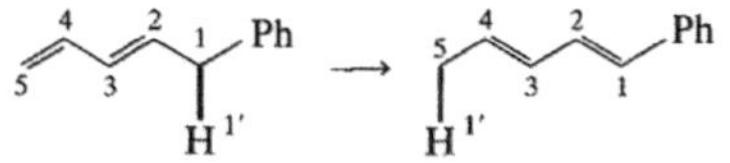

Thus, the hydrogen has moved from the 1 position to the 5 position on the chain. The 1 in the designation [1,5] indicates that at one end of the new σ bond is an atom (the hydrogen) that also was at one end of the old σ bond. The 5 indicates that the other end of the new σ bond is formed at the 5 position along the carbon chain, atom number 5.

b. This is a [3,3]-sigmatropic shift. It is an example of the Claisen rearrangement, a [3,3]-sigmatropic rearrangement of a vinyl ether.

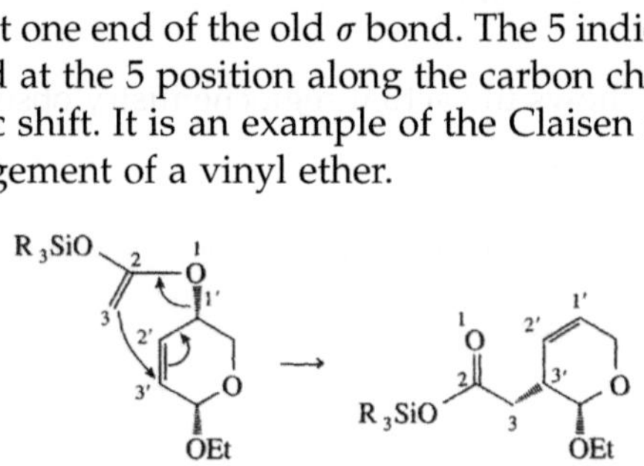

c. [2,3]

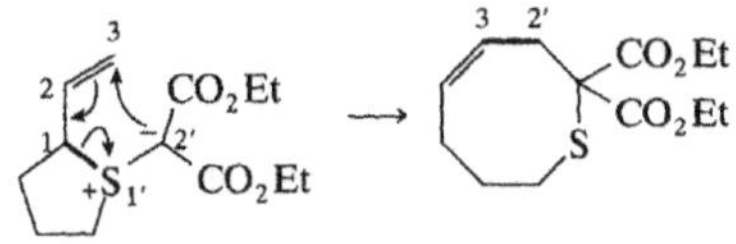

d. This is a [3,3]-sigmatropic shift. It is an example of the Cope rearrangement, in which one of the carbons has been replaced by nitrogen. The σ bond broken and the new σ bond formed are highlighted in the structures.

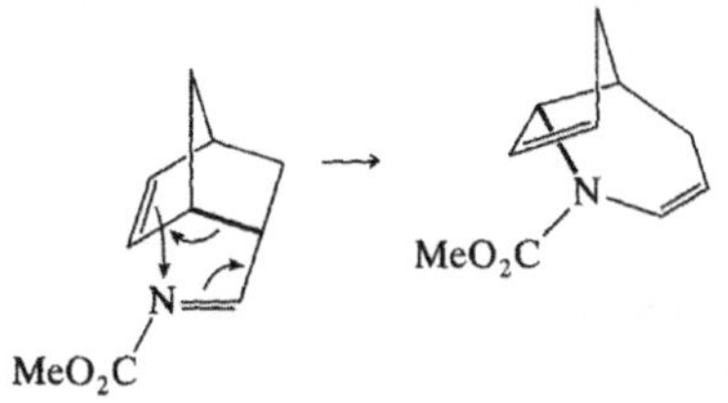

Problem 6.13

Rearrangement to the aromatic toluene requires a [1,3]-sigmatropic shift of hydrogen.

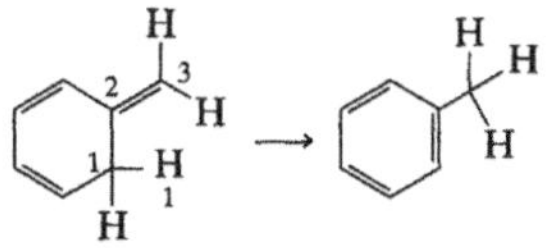

A suprafacial shift does not obey the selection rules and the geometry required makes an antarafacial shift very difficult, so that the compound is kinetically stable. The activation energy for rearrangement is large enough that the compound can be isolated even though it is thermodynamically unstable, just like diamond, which is kinetically stable even though it is thermodynamically unstable compared to graphite.

Problem 6.14

The photochemical transformation is a 6π electron $(4n + 2)$ electrocyclic ring opening. The selection rules predict a conrotatory process as illustrated:

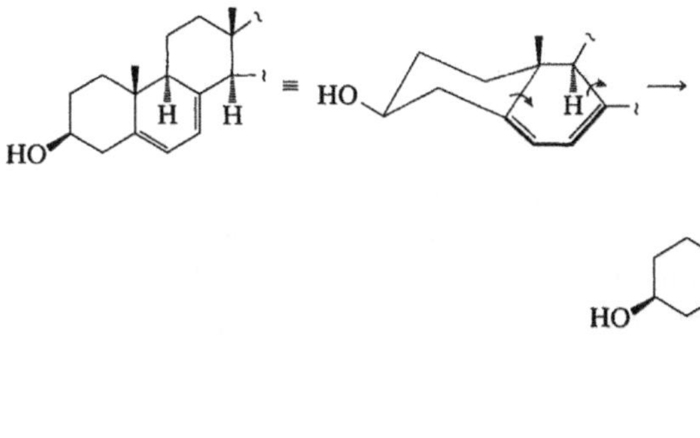

The alternative conrotatory opening would not be possible because it would result in the formation of *trans* double bonds in both cyclohexene rings. The second reaction is a thermal [1,7]-sigmatropic hydrogen shift.

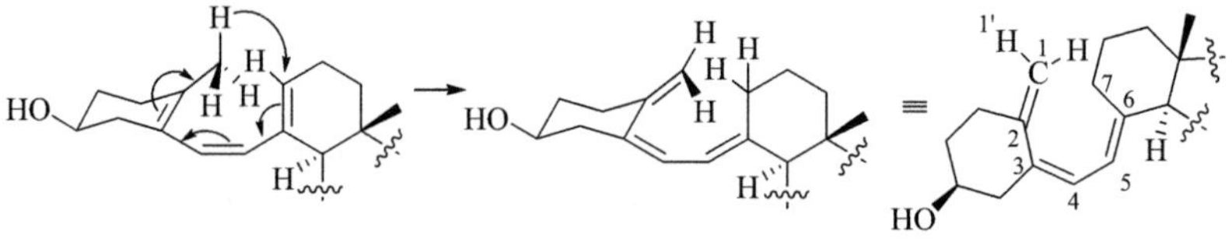

The selection rules predict an antarafacial process. In this case, suprafacial and antarafacial processes would lead to the same product.

Problem 6.15

a. This rearrangement could take place by two different mechanisms. One is a [1,5] sigmatropic shift to give the product directly:

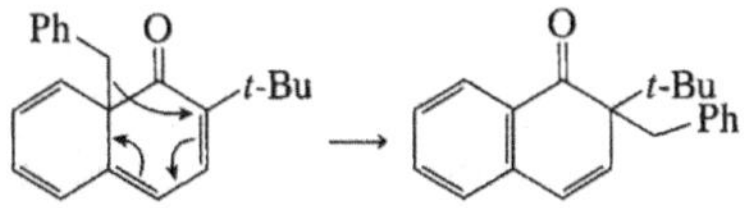

The other is two sequential [3,3] sigmatropic shifts:

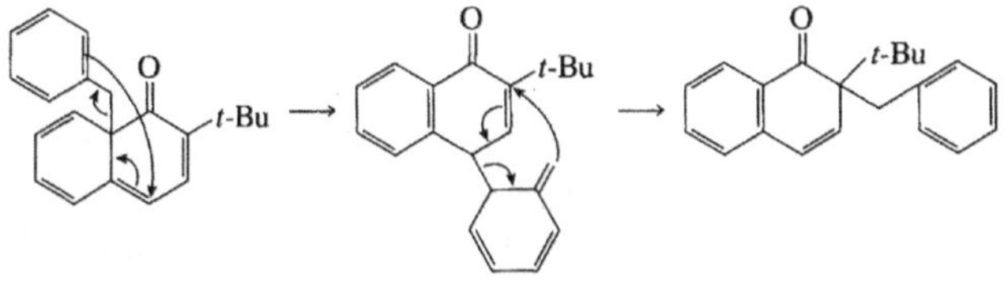

The authors of the paper cited favor the [1,5]-rearrangement because direct formation of the product is thermodynamically more favorable than proceeding through the intermediate from the first [3,3]-sigmatropic shift.

b. This transformation was described by the authors cited as a [1,5]-shift, followed by a "spontaneous" [1,7]-shift.

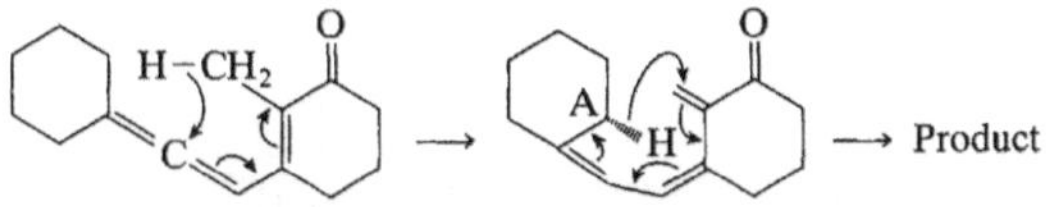

6-20

The [1,5]-shift would be a suprafacial reaction. The [1,7]-shift must be antarafacial in the π component to be thermally allowed. The proton, located at carbon **A** in **6-20**, moves from the bottom side of the six-membered ring to the top side of the π system at the other end. In general, the antarafacial process, necessary for a thermal [1,7]-sigmatropic shift, occurs with ease, especially when compared to the antarafacial process that would be required for a thermal [1,3]-sigmatropic shift.

c. This is an example of a [2,3]-sigmatropic rearrangement. In the first step, the alcohol is added into the phenyl sulfenyl chloride to form the sulfenate and the base soaks up the HCl by-product. The sulfenate undergoes the [2,3] rearrangement as the mixture warms to room temperature. Although the process is reversible, generally the temperature needs to be raised to about 55 °C for the true equilibrium to be attained. The reaction equilibrium lies far to the right as the newly generated sulfoxide bond is very strong and energetically favorable pulling the reaction in that direction. The energy gain going to the sulfoxide is approximately 1.5 kcal/mol and the equilibrium is better than 99:1 in favor of the sulfoxide (Reich et al., (1983)).

The sulfenate–sulfoxide (also known as the Evans–Mislow) rearrangement may initially form a mixture of the *cis* and *trans* olefins, which can be coalesced to a majority of the more stable *trans* olefin upon heating. Another interesting aspect of this transformation is the stereocenter at the sulfoxide sulfur, whose configuration can change through the process of heating and reaching equilibrium as well. In general, it requires very high temperatures to epimerize a sulfoxide stereocenter, but through the [2,3] rearrangement, the epimerization can occur at lower temperatures ($\sim$55 °C) as the molecules move back and forth through rearrangement.

Problem 6.16

Problem 6.17

The mechanism can be written as an ene cyclization if we write the enol form of the keto group.

Problem 6.18

a. The major pathway, a $[\pi_s^2 + \pi_s^2 + \sigma_s^2]$ reaction, is an ene reaction.

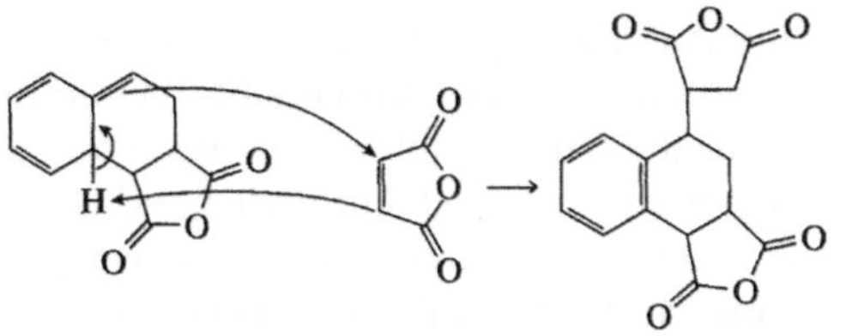

A stepwise mechanism also could be written for this reaction:

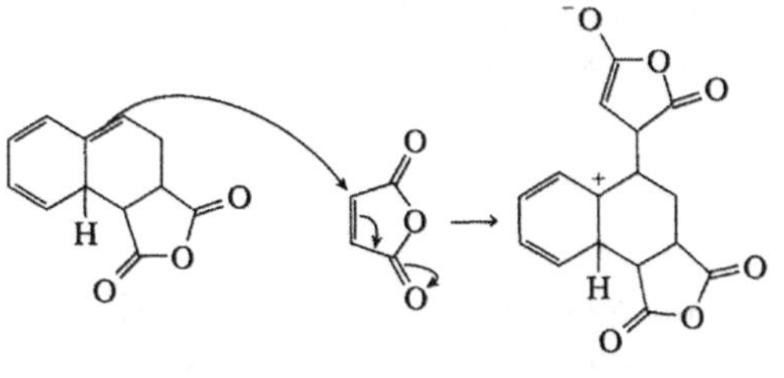

6-21

The intermediate formed, initially, **6-21**, is both an acid and a base. Appropriate intermolecular acid–base reactions, followed by tautomerization, can generate the product readily.

The minor product, formed in 11% yield, results from a Diels–Alder reaction, a $[\pi_s^4 + \pi_s^2]$ cycloaddition.

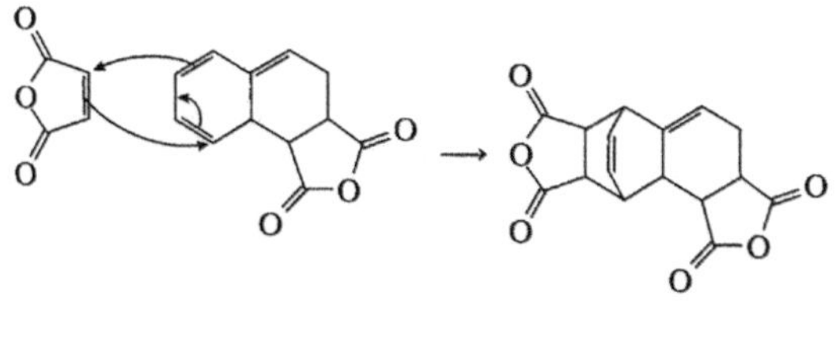

b. The first step is a $[\pi_s^4 + \pi_s^2]$ cycloaddition to give **6-22**.

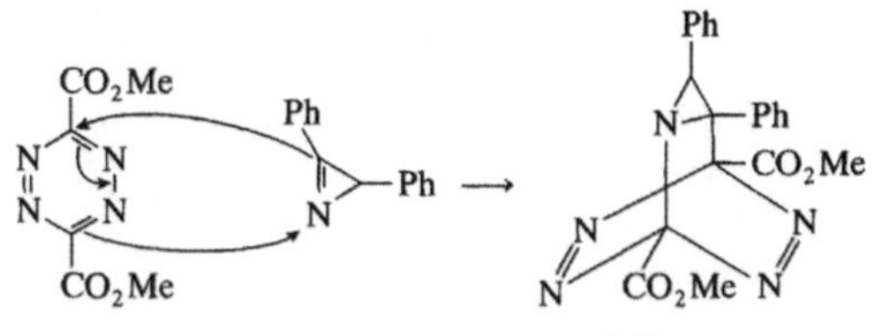

6-22

The second step is retrocycloaddition to give nitrogen and **6-23**, a $[\pi_s^4 + \pi_s^2]$ process. Finally, **6-23** undergoes a [1,5]-sigmatropic shift to form the final product. In **6-23**, the bonds involved in the sigmatropic shift are highlighted.

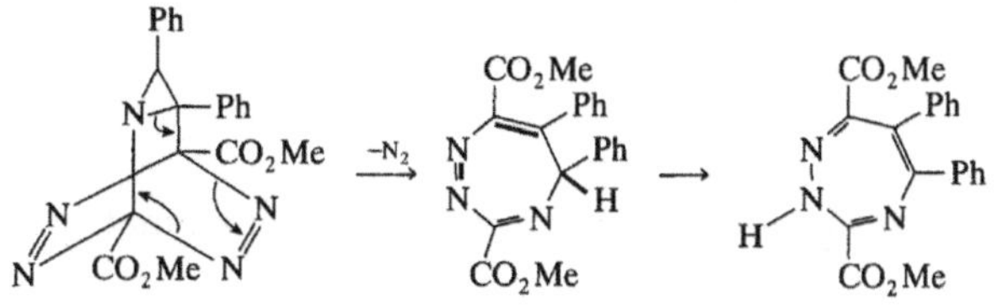

6-22   **6-23**

Another possible decomposition of **6-22**, $[\pi_s^4 + \pi_s^2]$, could give **6-24**, which then could open in a 6π electron electrocyclic reaction to give **6-23**.

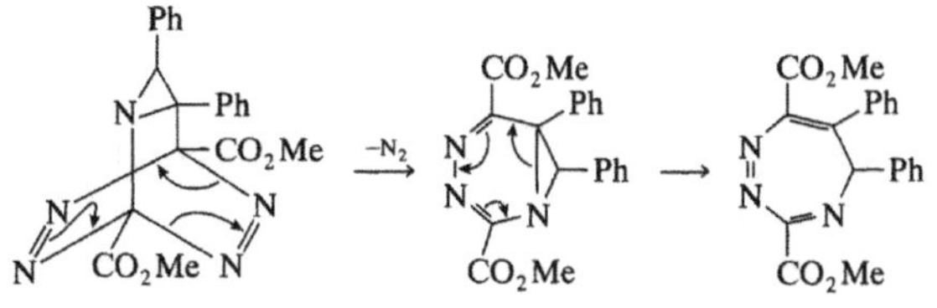

6-22   **6-24**   **6-23**

The direct opening of **6-22** to **6-23** appears more favorable on the basis of immediate relief of the strain of the three-membered ring.

c. This reaction is a $[\pi_s^6 + \pi_s^4]$ intramolecular cycloaddition. The paper cited noted that the reaction is "periselective"; that is, none of the [4 + 2] cycloadduct is produced.

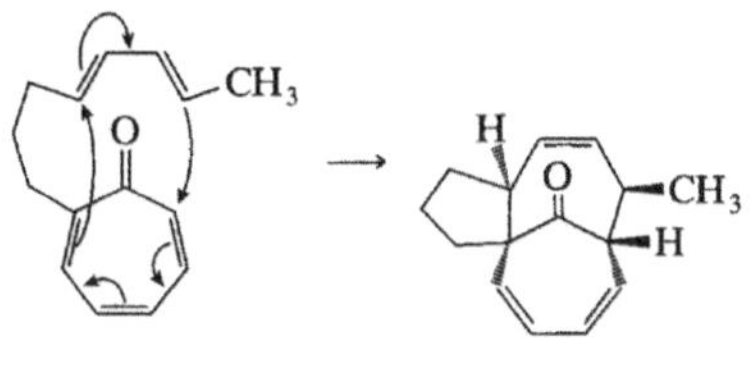

The two alternative $[\pi_a^6 + \pi_a^4]$ modes of reaction are extremely unlikely, because they give products with very high strain energy.

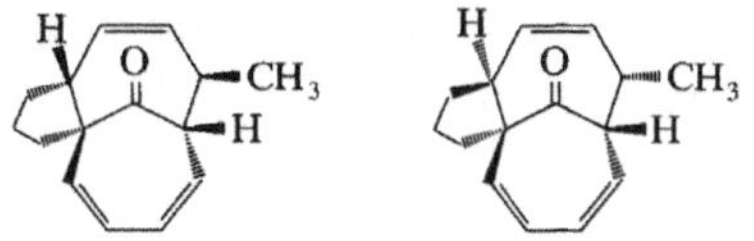

d. The product can be explained as the result of two consecutive [1,5]-sigmatropic shifts.

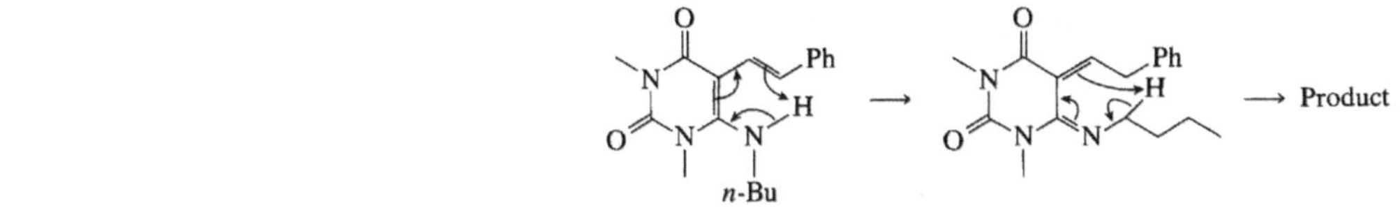

e. The product can be produced by two consecutive [3,3]-sigmatropic shifts.

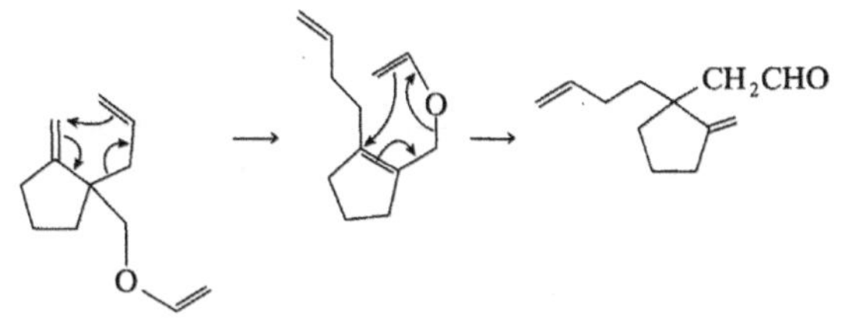

The first reaction is a Cope rearrangement, and the second reaction, rearrangement of a vinyl ether, is a Claisen rearrangement.

f. Step a looks like a [1,3]-sigmatropic shift. However, because a concerted shift would have to be an antarafacial–suprafacial reaction, which is unfavorable sterically, the mechanism probably is not concerted. A radical chain reaction, initiated by a small amount of peroxide or oxygen, is possible.

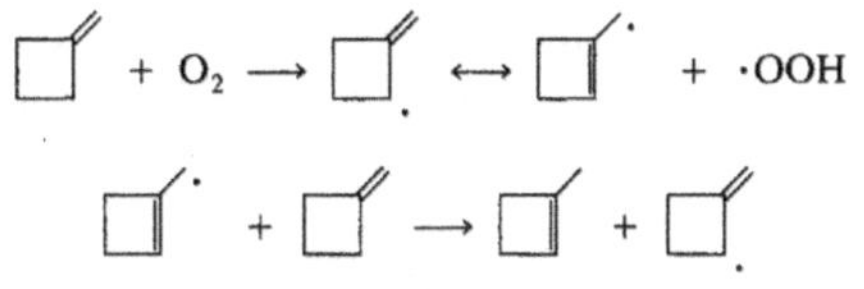

Step b is the ene reaction.

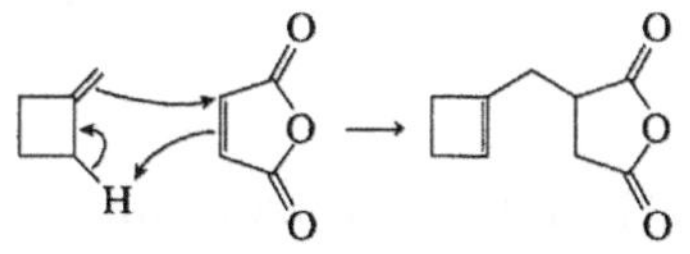

Steps c and d are electrocyclic ring-opening ($[\pi_a^4]$) processes. Because there are only hydrogens on the sp^3-hybridized carbons of the cyclobutene rings, there is no way to observe the stereochemical consequences of ring opening. Therefore, the arrows shown indicate only the rearrangement of electrons.

Step c:

Step d:

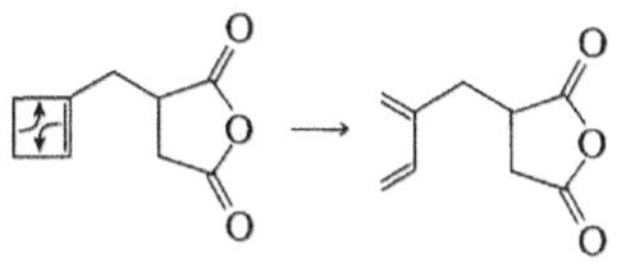

Steps e and f are $[\pi_s^4 + \pi_s^2]$ cycloadditions.

Problem 6.19

In the following equations, R is the allyl group. The first step, electrocyclic ring opening of the four-membered ring, is very common for benzocyclobutenes. There are two possible products for this ring opening (each from a possible conrotatory mode), but only 6-25 has the correct orientation to continue the reaction. The next two steps are both [3,3]-sigmatropic shifts.

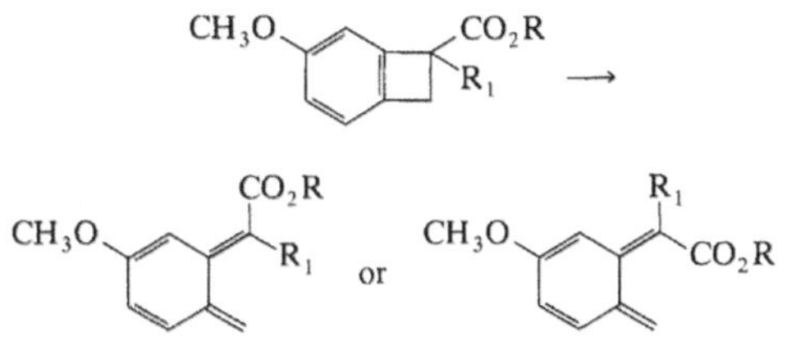

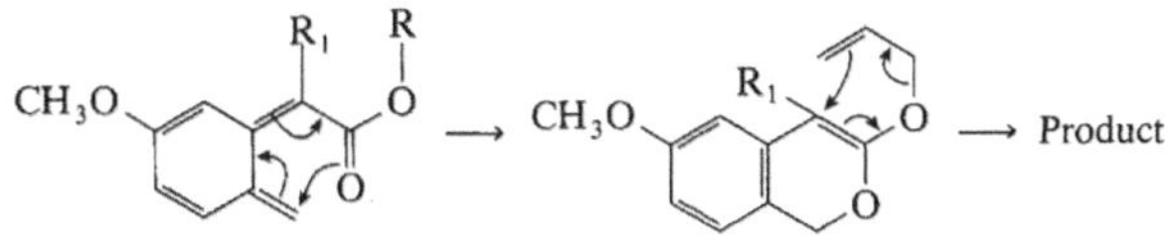

Problem 6.20

The fact that the reaction occurs readily suggests that it is not concerted because a $[\pi_s^4 + \pi_s^4]$ reaction is not a thermally allowed process (Table 6.2). Whereas the stereochemistry of the product shows that anthracene has reacted suprafacially, the presence of oxygen and nitrogen at the ends of the other 4π electron component precludes a determination of the stereochemistry of addition at this component. However, antarafacial reaction of this component is unlikely because it would require a large amount of twist in the transition state. Thus, this transformation probably proceeds through an intermediate, which is likely to be charged. A positive charge could be stabilized by conjugation with two aromatic rings of the anthracene component. A negative charge could be stabilized by N and/or O, as well as by the electron-withdrawing chlorines. Therefore, reaction of anthracene on either oxygen or nitrogen gives a stabilized intermediate:

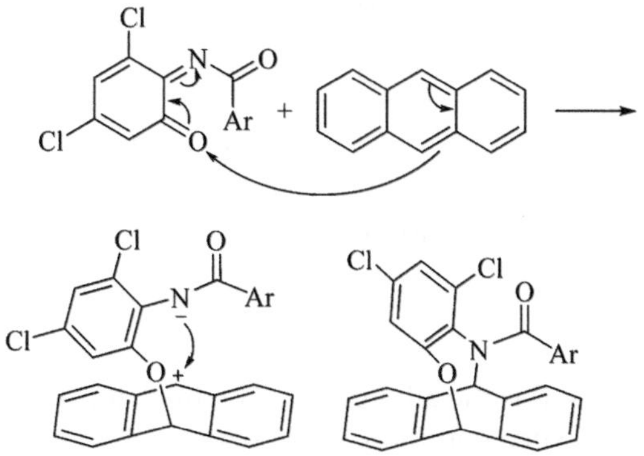

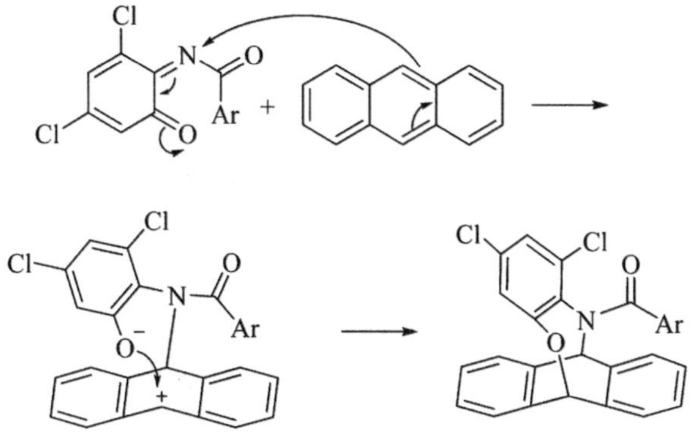

Problem 6.21

a. The first ring opening is a conrotatory process typical of a 4π electron thermal reaction. That is, in **6-26**, there are an anion and a π bond giving a total of four electrons:

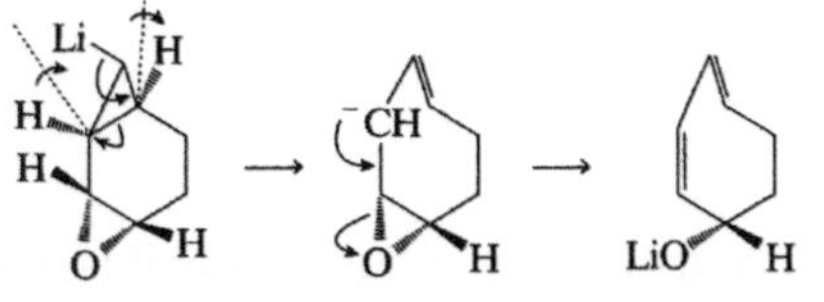

Formation of **6-19** follows from conrotatory ring closure of the heptadiene:

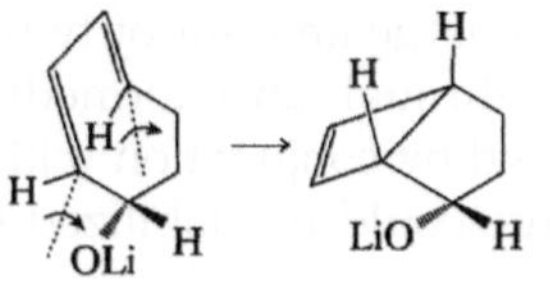

Notice that the conrotatory motion moves the hydrogens up and the carbons down. Ring closure to form **6-19** is favored because of the ring strain associated with a *trans* double bond in a seven-membered ring.

The isobenzofuran acts as a 4π electron component in a $[4+2]$ cycloaddition to trap the intermediate. As usual, both components must react suprafacially. Thus, because the fusion between the six- and seven-membered rings in **6-18** is *trans*, it must be the *trans* double bond in **6-16** that reacts in this trapping reaction.

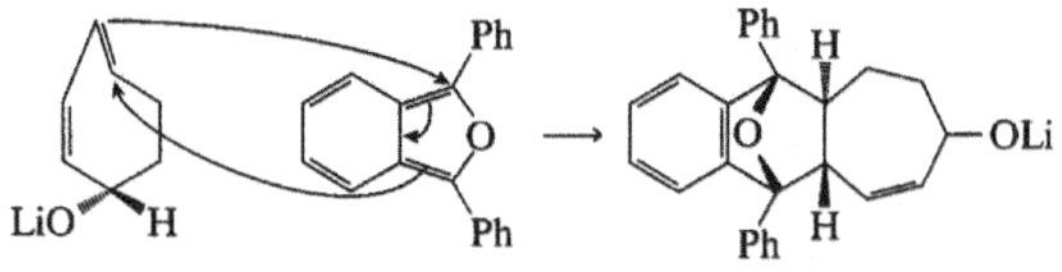

6-18

Problem 6.22

The single node should be located symmetrically, that is, at the center of the system. Because the node is located incorrectly, the orbital is neither symmetric nor antisymmetric.

Problem 6.23

The lowest energy orbital shown has no nodes, but the next higher one has two nodes. There should be an MO of intermediate energy with one node. In addition, the orbitals do not alternate in symmetry. Both of the orbitals have the same symmetry with respect to the σ plane (S) and the C_2-axis (A).

Problem 6.24

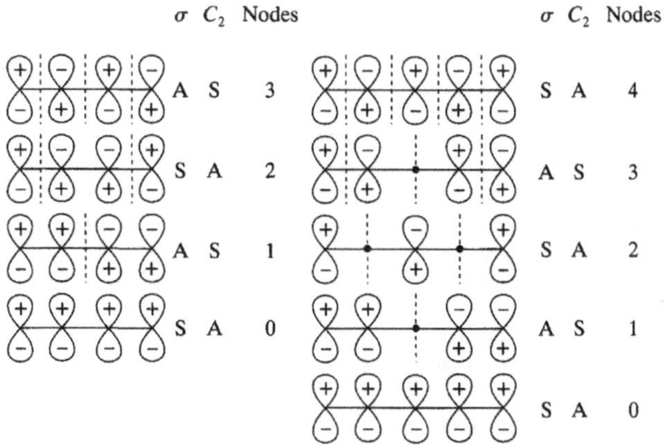

Some of the nodes coincide with nuclei in the five-carbon π system because the nodes must be placed symmetrically with respect to the mirror plane that bisects the system.

Problem 6.25

The pentadienyl cation has 4π electrons; the radical and the anion have 5π and 6π electrons, respectively. To obtain the electronic configuration, fill in the MOs in order, starting with the orbital of lowest energy.

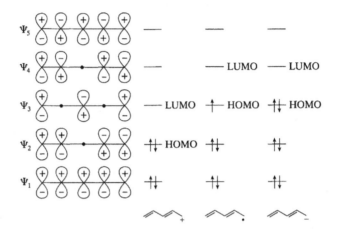

References

Alder, K.; Dortmann, H. A. *Chem. Ber.* **1952**, *85*, 556—565.

Alder, K.; Schmitz-Johnson, R. *Ann.* **1955**, *595*, 1—37.

Bailey, W. J.; Baylouny, R. A. *J. Org. Chem.* **1962**, *27*, 3476—3478.

Baird, M. S.; Lindsay, D. G.; Reese, C. B. *J. Chem. Soc. C*, **1969**, 1173—1178.

Baran, J.; Mayr, H. *J. Am. Chem. Soc.* **1987**, *109*, 6519—6521.

Barrack, S. A.; Okamura, W. H. *J. Org. Chem.* **1986**, *51*, 3201—3206.

Carey, F. A.; Sundberg, R. J. *Advanced Organic Chemistry, Part A: Structure and Mechanism,* 3rd ed; Plenum Press: New York, 1990, 44—46.

Carte, B.; Kernan, M. R.; Barrabee, E. B.; Faulkner, D. J.; Matsumoto, G. K.; Clardy, J. *J. Org. Chem.* **1986**, *51*, 3528—3532.

Coates, R. M.; Last, L. A. *J. Am. Chem. Soc.* **1983**, *105*, 7322—7326.

Conia, J. M.; Robson, M. J. *Angew. Chem. Int. Ed. Engl.* **1975**, *14*, 473—485.

Cookson, R. C.; Dance, J.; Hudec, J. *J. Chem. Soc.* **1964**, 5417—5422.

Curran, D. P.; Jacobs, P. B.; Elliott, R. L.; Kim, B. H. *J. Am. Chem. Soc.* **1987**, *109*, 5280—5282.

Darcy, P. J.; Heller, H. G.; Strydom, P. J.; Whittall, J. *J. Chem. Soc. Perkin Trans. I*, **1981**, 202—205.

de Meijere, A.; Erden, I.; Weber, W.; Kaufmann, D. *J. Org. Chem.* **1988**, *53*, 152—161.

Dolbier, W. R., Jr.; Koroniak, H.; Burton, D. J.; Bailey, A. R.; Shaw, G. S.; Hansen, S. W. *J. Am. Chem.Soc.* **1984**, *106*, 1871—1872.

Evans, D. A.; Andrews, G. C. *Acc. Chem. Res.* **1974**, *7*, 147—155.

Funk, R. L.; Bolton, G. L. *J. Am. Chem. Soc.* **1986**, *108*, 4655—4657.

Grigg, R.; Kemp, J.; Sheldrick, G.; Trotter, J. *J. Chem. Soc., Chem. Commun.* **1978**, 109—111.

Hart, H.; Rodgers, T. R.; Griffiths, J. *J. Am. Chem. Soc.* **1969**, *91*, 754—756.

Hart, J.; Rodgers, T. R.; Griffiths, J. *J. Am. Chem. Soc.* **1969**, *91*, 754—756.

Heine, H. W.; Suriano, J. A.; Winkel, C.; Burik, A.; Taylor, C. M.; Williams, E. A. *J. Org. Chem.* **1989**, *54*, 5926—5930.

Hill, A. E.; Greenwood, G.; Hoffmann, H. M. R. *J. Am. Chem. Soc.* **1973**, *95*, 1338—1340.

Hoffmann, H. M. R. *Angew. Chem. Int. Ed. Engl.* **1969**, *8*, 556—577.

Houk, K. N. *Acc. Chem. Res.* **1975,** 361—369.

Huisgen, R.; Dahmen, A.; Huber, H. *J. Am. Chem. Soc.* **1967,** *89,* 7130—7131.

Huisgen, R. *Angew. Chem. Int. Ed. Engl.* **1963,** *2,* 565—598.

Huisgen, R. *Angew. Chem., Int. Ed. Engl.* **1963,** *2,* 565—598.

Jefford, C. W.; Hill, D. T. *Tetrahedron Lett.* **1969,** 1957—1960.

Johnson, G. C.; Levin, R. H. *Tetrahedron Lett.* **1974,** 2303—2306.

Kametani, T.; Suzuki, Y.; Honda, T. *J. Chem. Soc., Perkin Trans. I,* **1986,** 1373—1377.

Krimse, W.; Rondan, N. G.; Houk, K. N. *J. Am. Chem. Soc.* **1984,** *106,* 7989—7991.

Machiguchi, T.; Hasegawa, T.; Itoh, S.; Mizuno, H. *J. Am. Chem. Soc.* **1989,** *111,* 1920—1921.

Matyus, P.; Zolyomi, G.; Eckhardt, G.; Wamhoff, H. *Chem. Ber.* **1986,** *119,* 943—949.

Miller, B.; Baghdadchi, J. *J. Org. Chem.* **1987,** *52,* 3390—3394.

Minami, T.; Harui, N.; Taniguchi, Y. *J. Org. Chem.* **1986,** *51,* 3572—3576.

Nakai, T.; Mikami, K. *Chem. Rev.* **1986,** *86,* 885—902.

Paquette, L. A.; Wang, T.-Z. *J. Am. Chem. Soc.* **1988,** *110,* 3663—3665.

Reich, H. J.; Yelm, K. E.; Wollowitz, S. *J. Org. Chem.* **1983,** *105,* 2503—2504.

Rondan, N. G.; Houk, K. N. *J. Am. Chem. Soc.* **1985,** *107,* 2099—2111.

Roth, W. R.; Konig, J.; Stein, K. *Chem. Ber.* **1970,** *103,* 426—439.

Sauter, H.; Gallenkamp, B.; Prinzbach, H. *Chem. Ber.* **1977,** *110,* 1382—1402.

Schlatmann, J. L. M. A.; Pot, J.; Havinga, E. *Reel. Trav. Chim. Pays-Bas* **1964,** *83,* 1173—1184.

Shishido, K.; Shitara, E.; Fukumoto, K.; Kametani, T. *J. Am. Chem. Soc.* **1985,** *107,* 5810—5812.

Snider, B. B.; Phillips, G. B. *J. Org. Chem.* **1984,** *49,* 183—185.

Spellmeyer, D. C.; Houk, K. N.; Rondan, N. G.; Miller, R. D.; Franz, L.; Fickes, G. N. *J. Am. Chem. Soc.* **1989,** *111,* 5356—5367.

Takeshita, H.; Sugiyama, S.; Hatsui, T. *J. Chem. Soc., Perkin Trans. II,* **1986,** 1491—1493.

Trost, B. M.; Seoane, P. R. *J. Am. Chem. Soc.* **1987,** *109,* 615—617.

Tsai, C.-Y.; Sha, C.-K. *Tetrahedron Lett.* **1987,** *28,* 1419—1420.

Vedejs, E. *Acc. Chem. Res.* **1984,** *17,* 358—364.

von, R.; Schleyer, P.; Van Dine, G. W.; Schollkopf, U.; Paust, J. *J. Am. Chem. Soc.* **1966,** *88,* 2868—2869.

Vos, G. J. M.; Reinhoudt, D. N.; Benders, P. H.; Harkema, S.; van Hummel, G. J. *J. Chem. Soc. Chem. Commun.* **1985,** 661—662.

Woodward, R. B.; Hoffmann, R. *The Conservation of Orbital Symmetry;* Verlag Chemie: Weinheim, 1970, 38.

Wu, P.-L.; Chu, M.; Fowler, F. W. *J. Org. Chem.* **1988,** *55,* 963—972.

Xu, S. L.; Moore, H. W. *J. Org. Chem.* **1989,** *54,* 6018—6021.

Ziegler, F. E.; Piwinski, J. J. *J. Am. Chem. Soc.* **1982,** *104,* 7181—7190.

Oxidations and Reductions

Oxidations and reductions are very common transformations in organic chemistry. Seemingly simple (merely a change in oxidation state with little change in molecular architecture), the mechanisms are often complex and can still deliver specific molecular change through the influences of subtle and dynamic forces. Even the simple conversion of one carbon center from sp^2 to sp^3 can have a significant impact on the conformation of the whole molecule. The spectrum of oxidizing and reducing reagents and transformation types are vast and it is not our intention to be comprehensive in this chapter. We will cover a broad selection of reagent types and mechanistic pathways and have included a group of oxidations and reductions which we feel are illustrative of a variety of processes and mechanism types that can be applied to the understanding of oxidation, reduction, and other significant processes in synthetic organic chemistry.

1. DEFINITION OF OXIDATION AND REDUCTION

Oxidations are chemical processes where an atom undergoes a net loss of electrons. This is a long-standing definition and can be applicable to both inorganic and organic chemistry. Another definition common to organic chemistry is that an oxidation reaction is a process where an atom gains a bond (or bonds) to a more electronegative element (usually oxygen, a halide, nitrogen, or sometimes a carbon) and/or loses bonds to a less electronegative element (most likely a hydrogen).

Reduction is a process where an atom gains a bond (or bonds) to a less electronegative element (again, most likely hydrogen) and/or loses bonds to a more electronegative element (most likely oxygen, a halide, nitrogen, or sometimes a carbon).

Common parlance is to refer to "the molecule" being oxidized. For our purposes and for the sake of aiding you in applying the models described here, we will refer to the functionality being affected and more specifically, the circumstances of a particular atom. This perspective will ultimately be refocused to describe the situation in the context of the molecules transformation.

A. Determining Changes in Oxidation State

As the first example, we will consider the oxidation of methanol to formaldehyde (Example 7.1). Here, the oxidation of the carbon in methanol leads to the replacement of a

355

Copyright © 2014 Elsevier Inc. All rights reserved.

carbon—hydrogen bond with a second (now a double) bond to oxygen. If formaldehyde is oxidized, another carbon—hydrogen bond is replaced with a carbon—oxygen bond, the result is formic acid.

• Oxidation

Methanol   Formic acid

In contrast, an example of a reduction reaction is the conversion of a ketone to an alcohol. In the example below, the carbon's double bond to oxygen (the ketone) is converted to a single carbon to oxygen bond and a new carbon bond to hydrogen is formed.

• Reduction

These are simple examples used to illustrate the definitions of reduction and oxidation as we will use them here. It should again be pointed out that the situation is often much more complicated than the simple exchange of bonds to hydrogens and oxygens. Challenges such as chemoselectivity, stereochemistry, and driving forces that move the transformation along through to the products need to be considered and will be addressed as part of the reaction mechanisms described below.

HINT 7.1

The change in oxidation level can be followed by tracking the oxidation state or "oxidation number" for the atoms being transformed. The following model can be used to track the changes and aid in identifying the type of process being evaluated as well as understand how specific functional groups relate to one another. Initially, let us consider the carbon atoms being evaluated.

In the first part of the process, identify the carbon atoms that have been changed in the transformation. These are assigned an oxidation level or "score" through the following steps:

1. For the carbon atom being evaluated, assign a score of −1 for the following:
   a. Each bond to an atom more electropositive than carbon (hydrogen for example).
   b. Each negative charge associated with the carbon being evaluated.
2. For the carbon atom being evaluated, assign a score of zero for the following:
   a. Each bond to another carbon.
   b. Each lone pair on the carbon being evaluated.
3. For the carbon atom being evaluated, assign a score of +1 for the following:
   a. Each bond to an atom more electronegative than carbon (oxygen or halide for example).
   b. Each positive charge associated with the carbon being evaluated.

For the second part of the process, add up the values for each atom being evaluated (the corresponding carbons that changed during the transformation) for both the products and reactants. Now determine the difference in states (scores) between the products and reactants.

1. If the difference is a positive number, then the transformation is an oxidation.
2. If there is no net change (the value is zero), then it was not a reduction or an oxidation.
3. If the difference is a negative number, then the transformation is a reduction.

This is really an accounting system that can assist us in understanding the changes that result from the transformations we are evaluating. This system can be complicated by alkylation and hydration reactions. Even so, the system will still provide value in understanding the "relative state" of the atom being evaluated and some specific examples will be presented. To get the most value from this model, remember to consider the functionality that was transformed in the process as well as the specific atom being evaluated.

EXAMPLE 7.1.A. CALCULATING OXIDATION NUMBER

Below we apply our tool to the oxidation sequence described in Example 7.1. Note the change in "oxidation number" at the red carbon as we move from alcohol, to aldehyde to acid.

Below are a few examples showing the different oxidation numbers carbon can have including examples where different molecules have carbon with the same oxidation number. Carbon can exist in nine different states, $+4$ (more oxidized) to -4 (more reduced).

EXAMPLE 7.2. CALCULATING OXIDATION NUMBER

Determine the oxidation number for the red carbon in each structure.

Solutions:

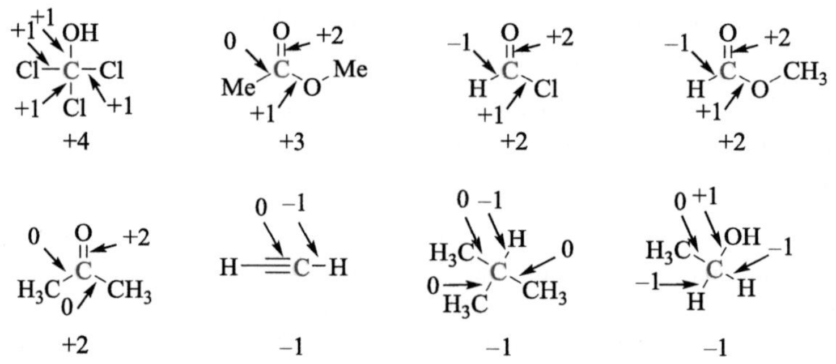

This system is useful for two reasons:

1. It allows us to judge the "relative oxidation level" of a carbon and relate it to what we are trying to achieve.
2. It allows us to track the changes in oxidation state during a transformation to understand how the functionality (and perhaps the whole molecule) has changed.

PROBLEM 7.1

What is the oxidation state for the red atoms in the following molecules?

(A)

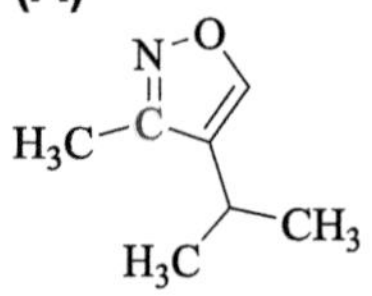

(B)

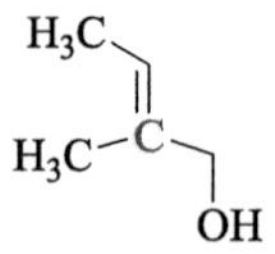

(C)

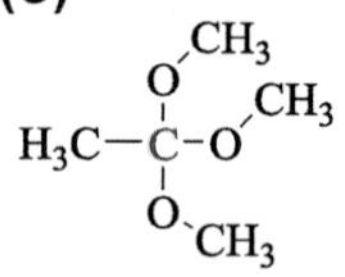

(D)

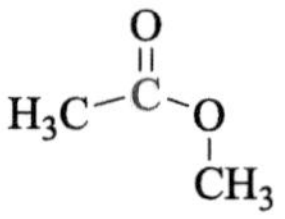

(E)

(F)

H_2O

(G)

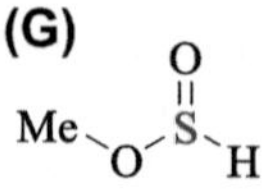

(H)

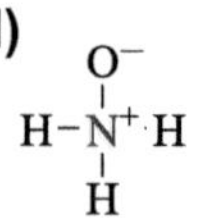

PROBLEM 7.2

Describe the change in oxidation number for the following transformation?

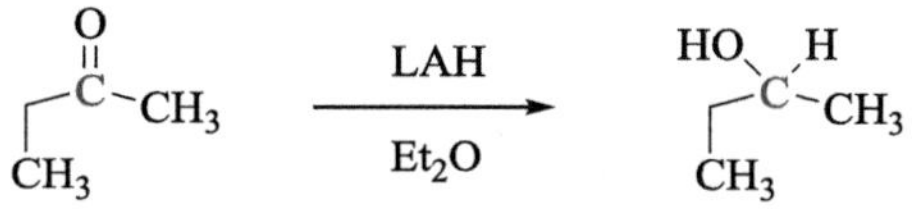

PROBLEM 7.3

Describe the change in oxidation number for the following transformation?

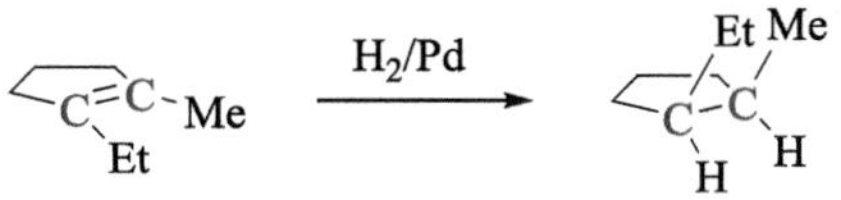

PROBLEM 7.4

Do these transformations represent reductions or oxidations? Explain.

(A)

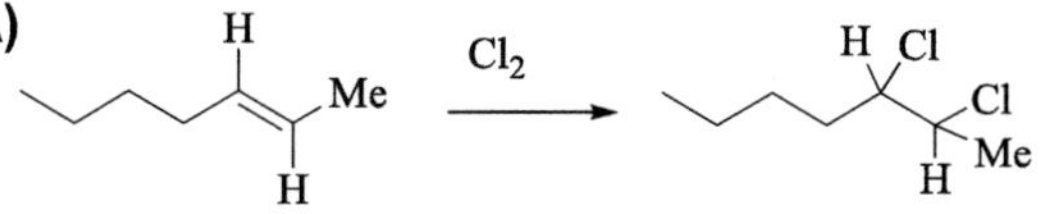

(B)

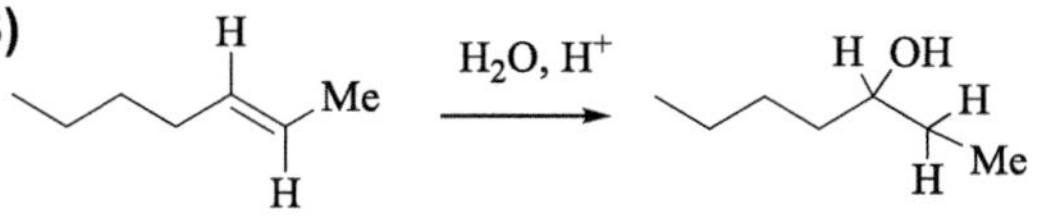

2. OXIDATIONS

A. Olefins

Ozonolysis

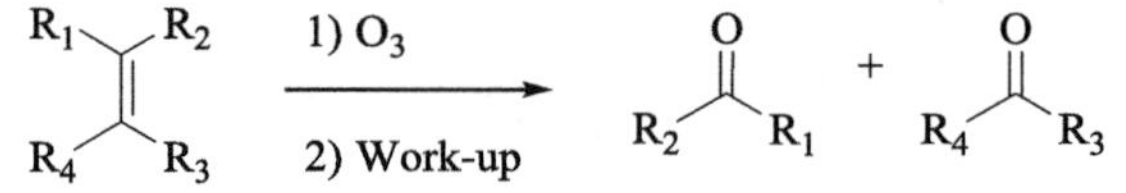

Ozonolysis of olefins is a common tool used in organic chemistry because of its reliability, low cost, relatively safe methods for reagent (ozone) generation and because of the minimal side products or waste associated with the reagents. Ozone will undergo a concerted $[4+2]$ 1,3-dipolar cycloaddition (7-1) leading to the cleavage of the carbon−carbon double bond. The addition of ozone to the olefin initially leads to the primary ozonide 7-2. The ozonide then opens up through a process called cycloreversion to form the (presumably tight) zwitterionic intermediate 7-3 that recombines to form the secondary ozonide intermediate 7-4. The mechanism shown below was established by the work of Criegee in the 1960s and has at various times come into question. A careful investigation utilizing oxygen isotope labels and ^{17}O-NMR has shown that in the examples studied, the results are consistent with this proposed model. The primary ozonide is unstable and not isolated, although there is spectroscopic evidence for its existence. Even though the secondary ozonide is more stable than the primary ozonide and has in some cases been isolated, it is also potentially explosive and is therefore usually reduced in-situ.

The sources are Rubin, (2003); Criegee, (1975); Geletneky and Berger, (1998); Kuczkowski, (1992).

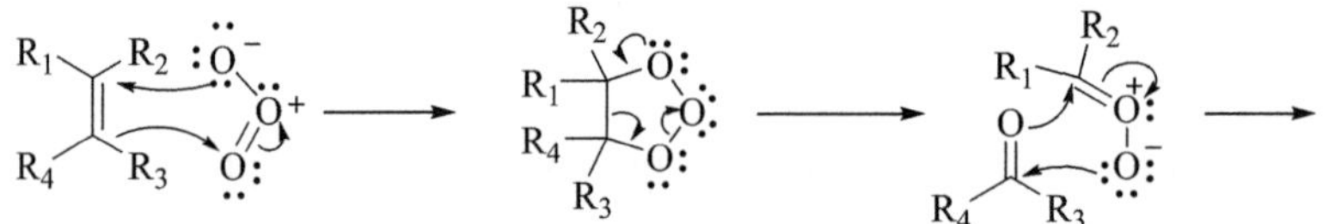

7-1　　　　7-2　　　　7-3

Primary ozonide
also called a molozonide
or the Criegee intermediate

Cycloreversion through the
zwitterionic intermediate

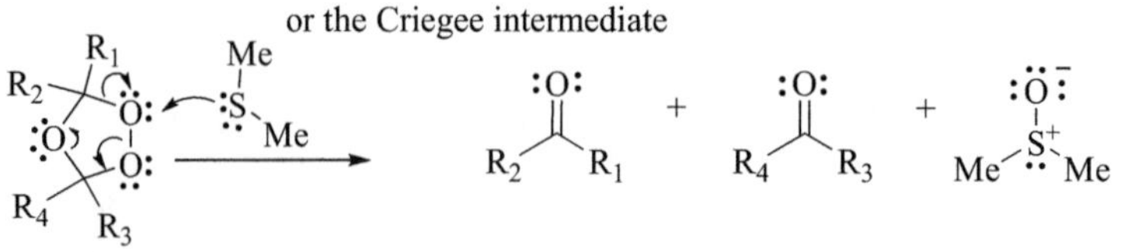

7-4

Secondary ozonide
also called the Staudinger ozonide

During an aqueous workup, there is a strong chance of secondary oxidation resulting from hydrolysis of the secondary ozonide and liberation of hydrogen peroxide. In the scheme above, we describe quenching the ozonide with dimethyl sulfide (a common reductant in this process) that would occur before an aqueous workup to remove the chance of secondary oxidation products. Other reductants that have been used for similar effect include zinc, sodium borohydride, phosphines, and trialkyl phosphites $(RO)_3P$.

The source is Knowles and Thompson, (1960).

A closer examination of the mechanism of the ozone addition to the olefin shows that, among other complexities, if we start with an unsymmetrical olefin, the primary ozonide could potentially open with electron flow in either direction and deliver different intermediates. This is of little consequence if we desire only the final aldehyde, but might provide a valuable clue as to the forces that drive the transformation.

R = H   styrene
R = Methyl *trans* propenylbenzene

A series of experiments were run with styrene and the *trans*-methyl styrene (*trans*-propenylbenzene). When the reaction is run in the presence of methanol, the intermediate zwitterion can be trapped, leading to a mixture of aldehydes and peroxides that result from the zwitterionic intermediates **7-7a** and **7-7b**.

The source is Keaveney et al., (1967).

When styrene is ozonolized in the presence of methanol, the product mixture is close to 50/50, and offers only a slight preference for the phenyl hydroperoxide methyl ether **7-5** (route A). In contrast, when *trans*-propenylbenzene is ozonolized in the presence of methanol, the methyl hydroperoxide methyl ether **7-6** and benzaldehyde are favored (route B) over formaldehyde and the phenylhydroperoxide in a ratio of better than 9:1. The outcome favoring the benzaldehyde and methyl hydroperoxide **7-6** in path B from *trans*-propenylbenzene might at first be thought to be the result of stabilization of the zwitterionic intermediate **7-7b**, but stabilization of the cation through resonance would favor the phenyl peroxide products (route A and zwitterionic intermediates **7-7a**) through structures **7-8**. It has instead been postulated that the inductive effect on the primary ozonide is the dominant factor and that the electronic donation from the methyl group (the push of electron density in the direction shown) facilitates electron flow in the direction shown below 7-9.

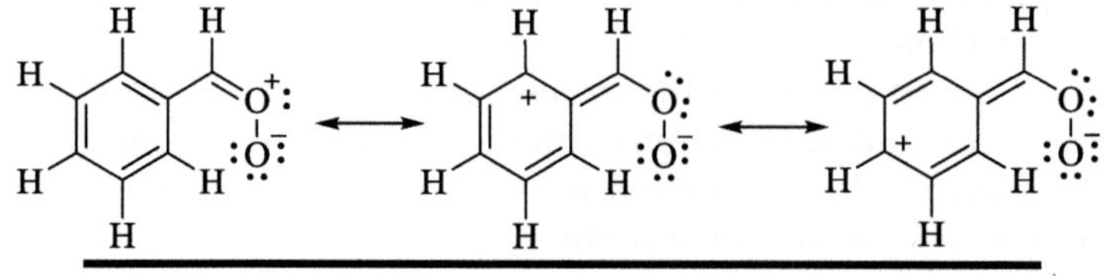

7-8

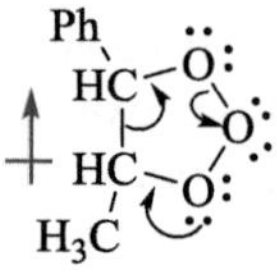

7-9

In the case of the alkyl peroxide **7-6**, which leads to the major product, the phenyl ring has an inductive influence that pulls electron density toward it. Thus, reinforcing the effect on the movement of electrons in the early stages of the transformation.

HINT 7.2

In systems with highly reactive intermediates, like the primary ozonide, the transition state more closely resembles the reactants (the primary ozonide rather than the zwitterionic intermediates) and influences early in the transformation are more significant.

Such subtle factors like these may not be considered important, especially in a case where we will most likely reduce the secondary ozonide to collect the desired carbonyl products and not isolate any intermediate. But these seemingly insignificant circumstances and the clues therein can collectively lead to a greater understanding of the complex nature of the processes we are attempting to influence. By inference, this "understanding" affords us the perspective to make better predictions on the outcomes of other systems in chemistry.

Wacker Oxidation

Originally developed by Wacker Chemie in the 1950s and 1960s for the oxidation of ethylene to acetaldehyde, the Wacker oxidation has found application in a wide variety of other systems. The Wacker process or the Hoechst—Wacker process (named after the chemical companies of the same name) was the first organometallic reaction applied on an industrial scale. In this process, a terminal olefin is oxidized on the proximal side (more substituted end) to yield a ketone. This reaction is carried out under an oxygen atmosphere in the presence of catalytic palladium(II) chloride and copper(I) chloride and is reliable, cost-effective, and amenable to use on an industrial scale.

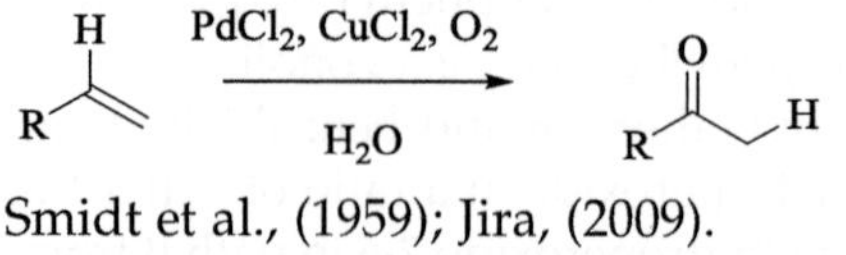

The sources are Smidt et al., (1959); Jira, (2009).

The Wacker oxidation involves a series of redox steps with oxygen serving as the terminal oxidant. When a compound is being oxidized, there is always an accompanying reduction, and the Wacker oxidation is a great example of this principle. In the process, Pd(II) is reduced to Pd(0) upon formation of the oxidized organic product. The Pd(0) is then reoxidized to

Pd(II) by Cu(II) which is acting as a "co-oxidant." The Cu(I) formed from oxidation of the Pd is then reoxidized to Cu(II) by oxygen. In this way, the Pd(II) and Cu(II) are constantly regenerated to the active oxidation state needed to convert olefin to the ketone, and thus the expensive metals can be used in catalytic quantities.

Because the reaction involves catalytic reagents, the mechanism is best represented by the catalytic cycle described below. Palladium (II) coordinates with the *p*-electrons in the olefin bond to form a complex **7-10**. Displacement of chloride by a molecule of water forms **7-11**. This initial coordination is thought to be a very fast process and reversible. Deprotonation of the bound water followed by hydroxide attack onto the bound olefin leads to β-hydroxyl palladium intermediate **7-12**. Simultaneous dissociation of chloride and β-hydride elimination generates the palladium–enol complex **7-13**. Insertion of the bound olefin into the Pd–H bond results in formation of the α-hydroxypalladium intermediate **7-14**. Another β-H elimination step **7-15** and palladium dissociation results in the ketone product **7-16** along with water and HCl.

The sources are Nelson et al., (2001); Keith et al., (2007).

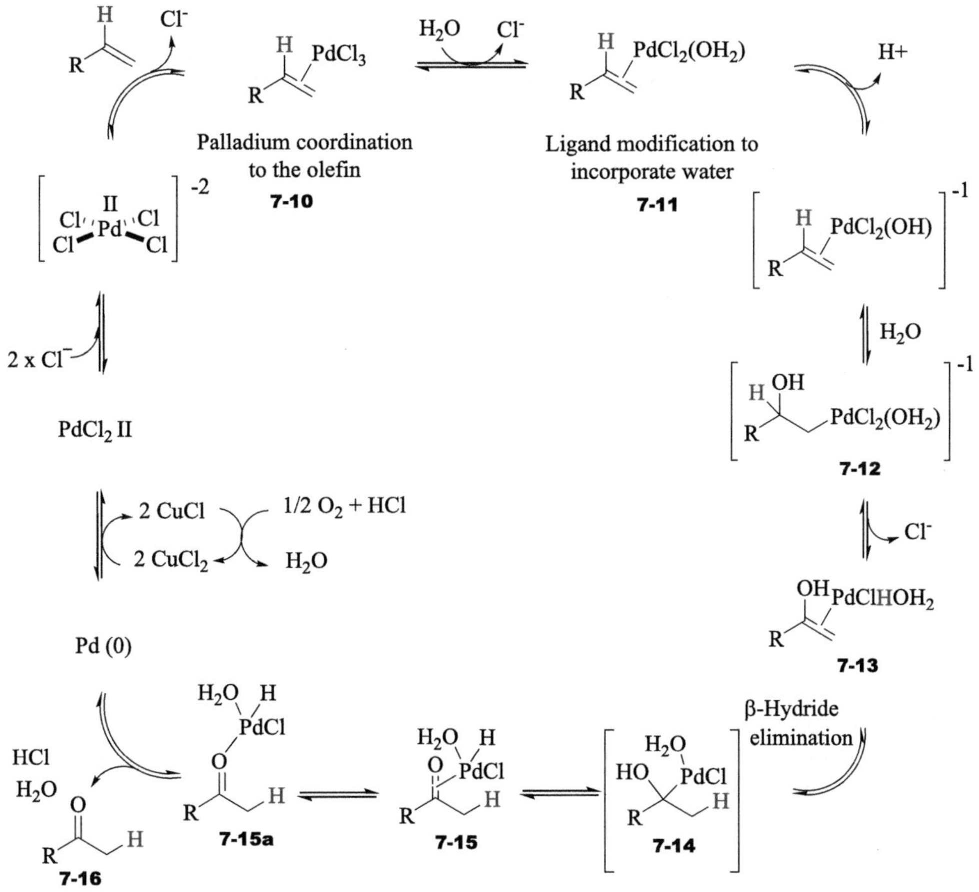

The oxidation system:

Palladium is expended and goes from Pd(II) to Pd(0) as the olefin is oxidized.

$$PdCl_2^{2-} \rightarrow Pd(0) + 2 \times Cl^-$$

The oxidation of Pd(0) to Pd(II) is facilitated by the reduction of Cu(II) to Cu(I).

$$Pd(0) + 2 \times CuCl_2 \rightarrow PdCl_2^{2-} + 2 \times CuCl$$

The copper is reoxidized from Cu(I) to Cu(II) by action of the oxygen.

$$(CuCl)_2 + 2 \times HCl + 1/2 O_2 \rightarrow 2CuCl_2 + H_2O$$

It should also be noted that in the Wacker process, oxygen (O_2) is not directly oxidizing the substrate. The source of oxygen that is incorporated into the substrate in this process is water and oxygen is working to reoxidize the metals. Consider the following example of an intramolecular nitrogen analog "aza-wacker" oxidation where the nitrogen is inserted to form the quinolone.

EXAMPLE 7.3. INTRAMOLECULAR AZA-WACKER OXIDATION REACTION

The Wacker oxidation is known to deliver very consistent results, even on very large scale. The Wacker oxidation has been applied successfully to other systems where different nucleophiles, nitrogen for example, have been used to further expand its utility. Although seemingly more complicated, notice how the aniline nitrogen plays a similar role to the first water in the Wacker oxidation cycle show above.

7-17   **7-18**

In these first steps, palladium coordinates to the olefin (**7-17**) and then coordinates to the nitrogen of the aniline **7-18**. The bridged system is now ready for the new bond to be created between the nitrogen and the proximal end of the olefin.

7-18a   **7-19**   **7-20**

The palladium inserts into the Pd—olefin bond to form the N—C bond (**7-19**); β-hydride elimination generates the Pd—olefin complex **7-20**. Reinsertion of the Pd—H (**7-21**) followed by another β-hydride elimination results in formation of the 4-hydroxyl-3,4 dihydroquinoline **7-22**. Dehydration and aromatization lead to the quinolone product **7-23**.

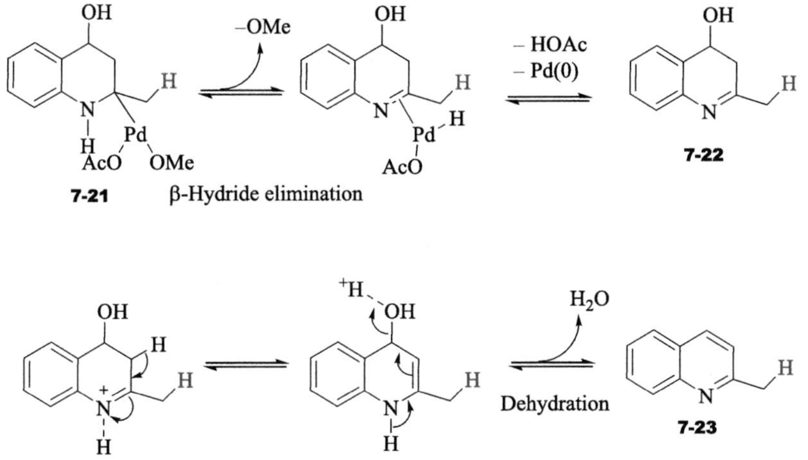

7-21   β-Hydride elimination

Consider the transformation just described. Where has this molecule been oxidized? Where has it been reduced?

Source: Zhang et al., (2008).

Epoxidations of Olefins

PERACID OXIDATIONS

A widely used method for conversion of olefins to epoxides is the application of peroxycarboxylic acid such as *m*-chloroperoxybenzoic acid in a process known as the Prileschajew reaction. Because the transformation proceeds with retention of geometry relative to the starting alkene and the rate of the reaction is not affected by changes in solvent polarity, it has been postulated that this is a concerted process as depicted below.

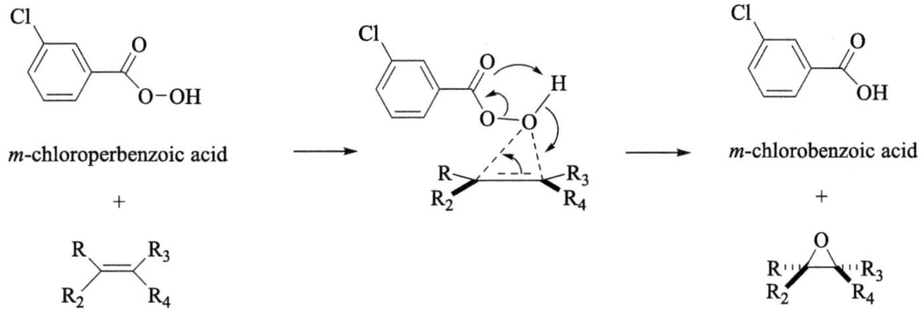

m-chloroperbenzoic acid   *m*-chlorobenzoic acid

The sources are Schwartz and Blumberg, (1964); Porto et al., (2005).

The peracids (also known as peroxyacids) will generally approach from the less hindered side of the olefin in the case of ridged cyclic systems. The reaction is sensitive to electronic and steric effects, thus functionality on or near the olefin can direct the orientation of the incoming reagent. For example, in the case of cyclohexene-ol **7-24**, epoxidation occurs from the same side as the alcohol through the effect of precomplexation of the peracid to the hydroxyl group (**7-25**). The acetate analog **7-26** which lacks the hydrogen bonding coordination provides the other stereochemical result with the epoxide resulting from the opposite (less sterically demanding) face of the olefin to give epoxide **7-27**.

The sources are Henbest and Wilson, (1957); Kwart and Takeshita, (1963).

Electron-rich olefins (where the olefin is functionalized with electron donating substituents) are more reactive in this transformation. Electron-poor peroxyacids (substituted with stronger electron withdrawing substituents) are more active in this reaction. It has thus been postulated that the olefin is acting as a nucleophile and the peracid tends to act as the electrophile in this process.

Dihydroxylation

KMnO$_4$ AND OsO$_4$ OXIDATION OF OLEFINS

Potassium permanganate (KMnO$_4$) and osmium tetroxide (OsO$_4$) can be used to oxidize olefins to the corresponding diols. This occurs through the simultaneous addition of the permanganate or osmylate oxygens and the opening of the olefin bond (**7-28**). The intermediate ester **7-29** is hydrolyzed to reveal diol **7-30**. In the case of permanganate, this transformation is carried out under alkaline conditions to prevent transformation to the corresponding alpha hydroxy ketone and then further oxidation of the carbon–carbon bond. The dihydroxylation of olefins with OsO$_4$ proceeds through a mechanism similar to the KMnO$_4$ process. In the case of OsO$_4$, the process stops at the cyclic osmate ester intermediate which can be isolated or hydrolyzed with a solution of sodium sulfite. Because of the low cost and more manageable toxicity on large-scale processes, permanganate is more commonly used even though the yields are generally less satisfactory.

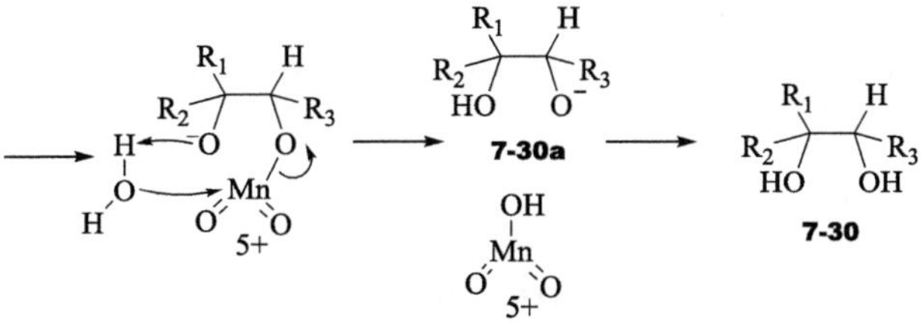

With both $KMnO_4$ and OsO_4, periodate can be added to the reaction to cleave the resultant diol and to reoxidized the osmium or manganese intermediates, which allows for the catalytic use of $KMnO_4$ or OsO_4 reagent. In the case of the osmium tetroxide process, this can be advantageous since the osmium tetroxide is costly and toxic.

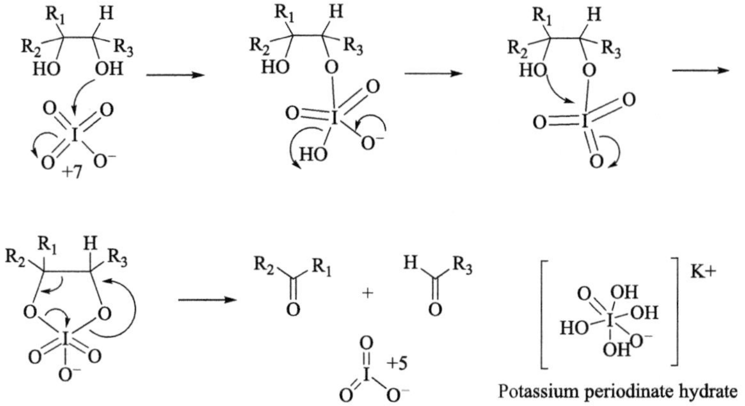

Potassium periodinate hydrate

Together, the mixture of potassium permanganate and potassium periodate ($KMnO_4$–KIO_4) is an effective agent for the oxidation of benzylic and allylic alcohols, as well as aldehydes, to the corresponding acids. For this reason, it is common for the cleavage of an olefin to result in the corresponding acids. The hydrated form of the periodinate is considered to be the active reagent.

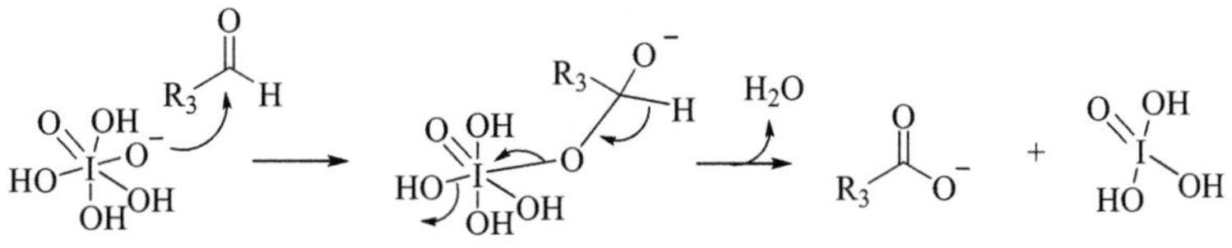

The process is often run deficient in $KMnO_4$, in which case the IO_4 can be used to reoxidize the reduced manganese back to MnO_4.

EXAMPLE 7.4. OXIDATION OF 1,2 DIOLS

Show a potential mechanism for the oxidation of the 1,2 diol to the 1,2 hydroxy ketone under acidic conditions with $KMnO_4$.

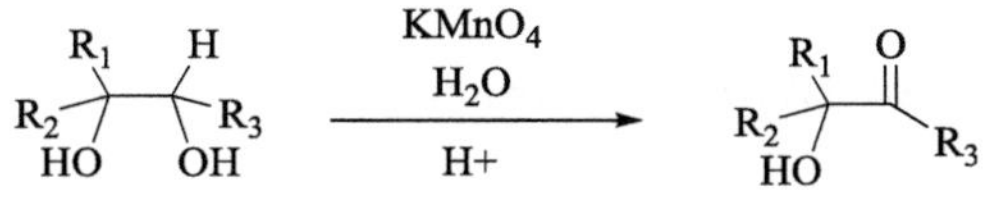

A potential mechanism would begin with the protonation of one of the hydroxyls (**7-31**) and elimination/dehydration leading to enol **7-32**. This could easily be oxidized by the KMnO$_4$, leading to α-hydroxy-ketone (acyloin) moiety **7-33**.

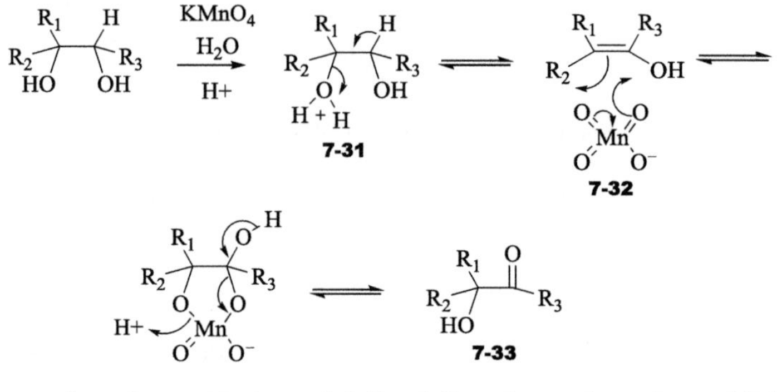

7-31 **7-32**

7-33

Permanganate and osmium oxidations of olefins deliver the product of *syn* addition as a result of the concerted process leading through the five-membered cyclic intermediate; the *trans* product is not possible. In the case of the oxidation of a rigid system, the osmium tetroxide will approach from the less hindered side (**7-34**), providing the corresponding diols.

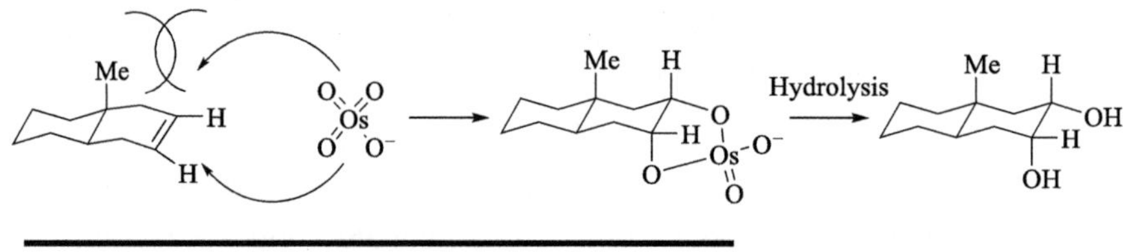

7-34

Source: Schroeder, (1980).

PROBLEM 7.5

Show a mechanism for the following transformation

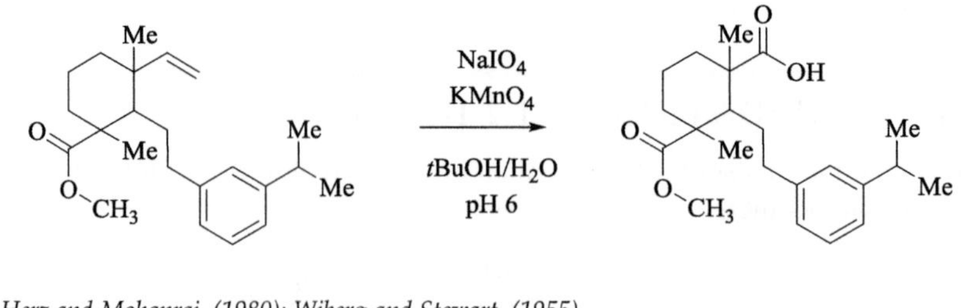

Sources: Herz and Mohanraj, (1980); Wiberg and Stewart, (1955).

B. Oxidation of Alcohols and Carbonyls

Carbonyl Oxidation

SPECIAL TOPIC—MECHANISM ELUCIDATION

THE BAEYER VILLIGER   In Chapter 4, we covered the basics of the Baeyer–Villiger oxidation. This transformation involves the addition of peracids to ketones resulting in the insertion of an oxygen adjacent to the ketone to form the ester. We revisit this oxidation to better understand the nuances of the process and the genesis of our understanding of the mechanism.

The Baeyer–Villiger Transformation

The transformation is thought to proceed through addition of terminal peracid **7-35** oxygen (*m*-chloro perbenzoic acid is commonly used, as well as peracetic acid and the more reactive trifluoro peracetic acid) to the carbonyl. The peroxide dicarbonyl intermediate (**7-36**) decomposes and the alkyl group migrates to the proximal peroxide oxygen (**7-37**). The last step in the mechanism **7-37** (heterolysis of the O—O bond and migration) is concerted and believed to be rate determining. The ability of the peracid ester (**7-36**) to accommodate the charge determines the activity of the peracid in this reaction. Thus, a more electron withdrawing functional group on the original peracid (at R3) will most likely increase its activity (by better accommodating the electrons it gains as it is ejected as the acid) in this transformation. It has also been postulated that in the initial addition step, the peracid with the more electron withdrawing functional group will be more nucleophilic at the terminal oxygen.

Although steric factors come in to play, in cases where the ketone is unsymmetrical, the propensity to migrate is enhanced by an electron releasing group (functionality that is more able to stabilize the positive charge or put another way, holds its electrons less tightly). For example, with para-substituted aryl ketones, the order of ease of migration is

$CH_3O > CH_3 > H > Cl > NO_2 > NH_2$ (Doering and Speers, 1950) and for alkyl groups, the more substituted carbons have a higher propensity to migrate, with the order being:

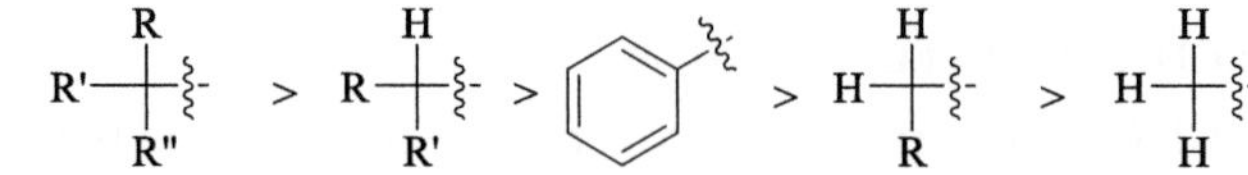

(Friess and Farnham, 1950). Aldehydes tend to yield the corresponding carboxylic acids (the hydrogen shifts) and enolizable ketones such as 1,3 diketones are inert.

Isotopic Labeling and Support for a Plausible Reaction Mechanism

Isotopic labeling has been an effective tool in the elucidation of reaction mechanisms in organic chemistry since the middle of the twentieth century. We saw an example earlier, in **Problem 3.9**, and the utility of isotopically labeling compounds to better understand the mechanism of the reactions can be seen in the large number of examples in the literature. The source is Maitra and Chandrasekhar, (1997).

Until the early 1950s, there were at least three postulated mechanisms for the Baeyer–Villiger oxidation. Doering and Dorfman clarified our understanding of the mechanism when they described the results of a well-planned and executed oxygen labeling experiment in 1953. Although each of the three proposed mechanisms would lead to the same product, their hypothesis was that the different pathways leading to the products could be differentiated if the two oxygens of the ester were tracked.

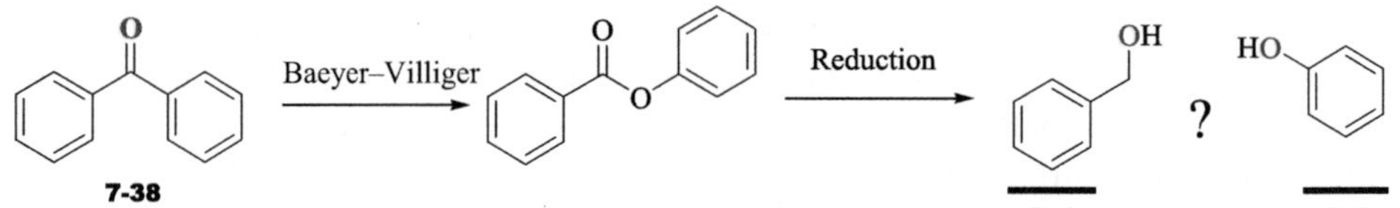

The three possible mechanisms (A, B, and C) are depicted below. In proposed mechanism A, the ketone oxygen adds into the peracid to form a peroxy-carbonyl. The peroxy-carbonyl intermediate rearranges and then enolizes to form the ester where the R2 group has migrated onto what was the carbonyl oxygen. In mechanism B, the terminal peracid oxygen adds into the ketone. The oxygen then adds into the proximal peroxy oxygen to form a dioxirane which can open as the R2 group shifts. This system is symmetrical and would be anticipated to deliver equal amounts of the R2 group shifted onto what was originally the ketone oxygen or to what was the terminal peracid oxygen. The proposed mechanism C moves through the intermediate resulting from peracid addition into the ketone carbonyl as well. But in this mechanism, the new ester carbonyl is formed at this point (without going through the dioxirane) and the R2 group shifts to what was the terminal peracid oxygen to generate the ester. Each of the mechanisms is plausible and leads to the same products. But if the carbonyl oxygen could be differentiated, each mechanism would deliver a unique result.

When the ^{18}O-labeled ketone (below) was exposed, the Baeyer–Villiger conditions and the final product was then reduced with LAH, it was discovered that the R1 alcohol oxygen (derived from the carbonyl) was nearly completely labeled and the phenol (R2) oxygen was not. This result is consistent with mechanism C. The other two routes would have inverted the ratio or resulted in a 50/50 mixture.

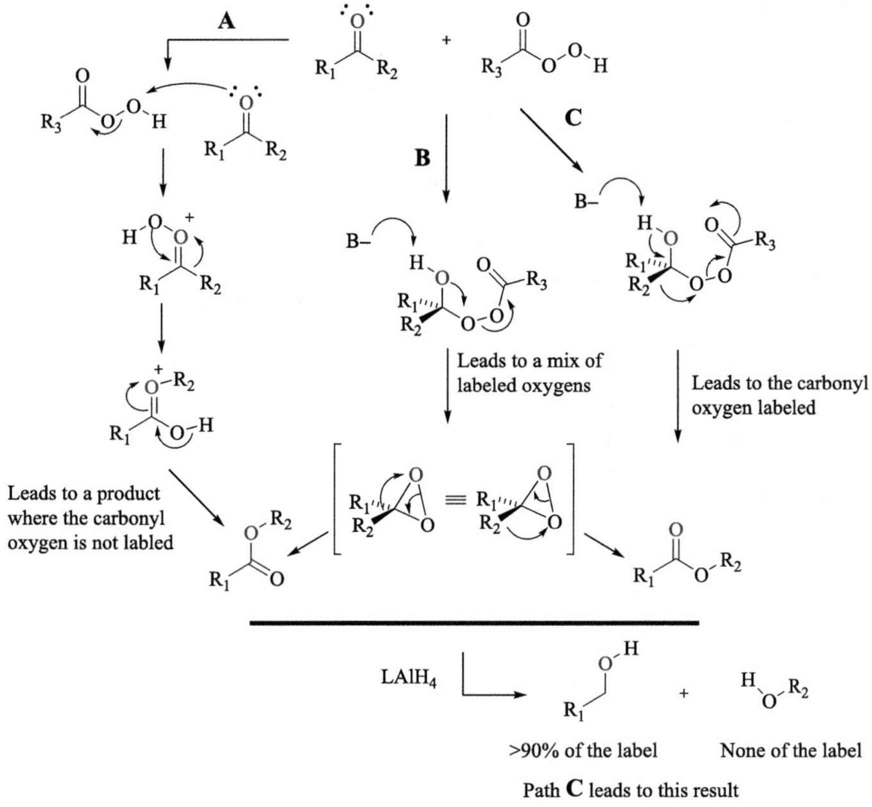

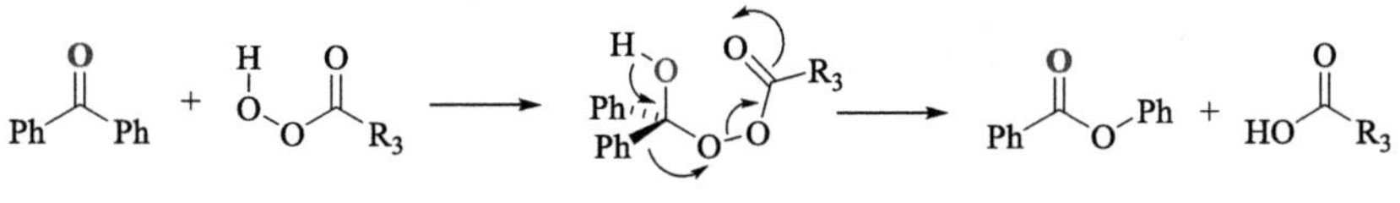

This is a striking example of the power of a well-planned and executed labeling experiment. It should be pointed out that although this is strong support for the mechanism described, it is not proof.

HINT 7.3

We cannot *prove* that a process will follow a particular mechanism, rather we can find supporting data or disprove a mechanism for a particular situation under specific conditions. It is common to see a mechanism described as "supported by the observed results" or "consistent with experimental observations." It should also be pointed out that with enough evidence (like that described above), we can feel comfortable in applying a particular proposed mechanism to predict the behavior and outcome of related transformations.

Alcohol Oxidation

JONES AND OTHER CHROMIUM-BASED OXIDATIONS

Although the metal is toxic, chromium-based oxidants such as chromic acid and chromic anhydride are effective reagents for the oxidation of primary and secondary alcohols and have had a long history of use in organic chemistry. The Jones reagent, 8 N chromic acid in dilute sulfuric acid, has been used extensively. Chromic anhydride and anhydrous pyridine (Sarret's reagent or the Collins reagent) generates a complex of CrO_3 and pyridine. Collins' reagent, when diluted into methylene chloride, allows for the oxidations of alcohols in the presence of sensitive functionality. These reagents are very hygroscopic and have a propensity to catch fire and burn during their preparation. Two commercially available chromium-based alternatives described by E. J. Corey, pyridinium chlorochromate (PCC) and pyridinium dichromate (PDC) are now used as an almost universal replacement for the older Jones or Collins procedures, because they are so safe and effective.

The sources are Cornforth et al., (1962); Corey and Suggs. (1975); Corey and Schmidt, (1979); Piancatelli et al., (1982).

Pyridinium dichromate (PDC)   Pyridinium chlorochromate (PCC)

Oxidations with chromium oxidants proceed through a similar mechanism.

The first step in the process is a fast and reversible addition of the alcohol to the chromium, leading to the pentavalent intermediate (**7-39**) that loses water to form the chromate ester **7-40**. The chromate ester intermediate then decomposes and delivers the newly formed carbonyl compound **7-41**.

The source is Zhao et al., (1998).

Based on a series of deuterium labeling studies, the deprotonation step which occurs in the oxidation **7-40** is hypothesized to be rate determining.

Faster   Slower

Note: *Deuterium is heavier than hydrogen, and thus, the bond it makes with carbon vibrates more slowly. This slower vibration is thought to strengthen the deuterium bond. The energy required to break the stronger C–D bond is referred to as the zero point energy. A change in reaction rate may be observed when breaking the bond to an isotope. This change in rate is referred to as a primary isotope effect.*

The sources are Westheimer and Nicolaides, (1949); Wiberg, (1955); Westheimer, (1961); Awasthy et al., (1967); Kabilan et al., (2002); Atzrodt et al., (2007).

Further, the chromium-based oxidation is effected by the steric demands of the alcohol being oxidized. In situations where the starting alcohol is sterically crowded, due to other surrounding functionality (Example 7.5), it may have a faster reaction rate due to the relief of steric strain as it transitions from an sp^3-hybridized center to an sp^2. This effect can be counterbalanced by the bulk of the groups adjacent to the alcohol, which could offset the rate enhancement by impeding access to that position. In very congested systems, formation of the chromate ester can be slowed or even stopped.

EXAMPLE 7.5. COMPETING STERIC INFLUENCES ON THE OXIDATION OF ALCOHOLS

The initial formation of the chromate ester may be slowed or blocked by a very hindered alcohol.

The rate of oxidation in the example below is faster than expected. The conversion of the alcohol to the ketone results in the oxygen moving away from the axial methyl groups. This change in geometry removes the strain associated with the diaxial repulsion and leads to a relatively favorable (lower energy) structure.

This is a case where the steric hindrance of the alcohol results in increased reaction rates, due to a decrease in steric strain.

Also see Muller, (1970) for examples and discussion of rate changes associated with sterics, support of developing charge, and bond angle strain on the developing sp^2-hybridized carbonyl.

Sources: Albright and Goldman, (1965); Rocek et al., (1962); Lee and Raptis, (1973); Smith, (2011).

The change in reaction rate has an electronic component as well. This electronic effect could be related to the ability of the alcohol carbon to support a positive charge as the hydrogen is removed. Studies showing changes in reaction rates associated with various aryl groups have demonstrated the impact of functionalization at this position.

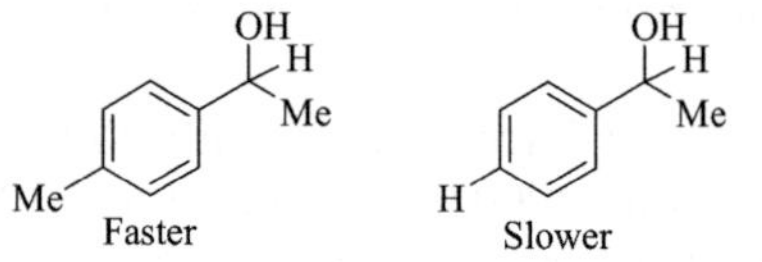

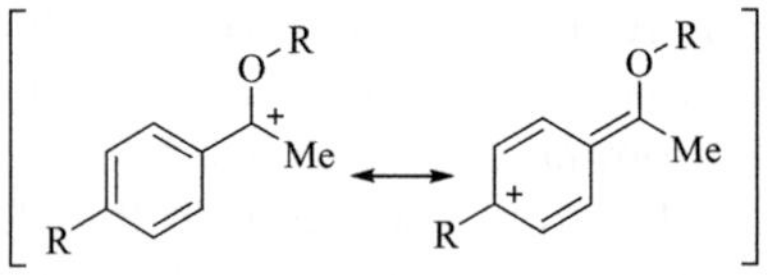

Because this result is not consistent with the previously drawn mechanism, it may indicate that another mechanism is at play.

Alternative Mechanism for the Action of Chromium Trioxide on an Alcohol

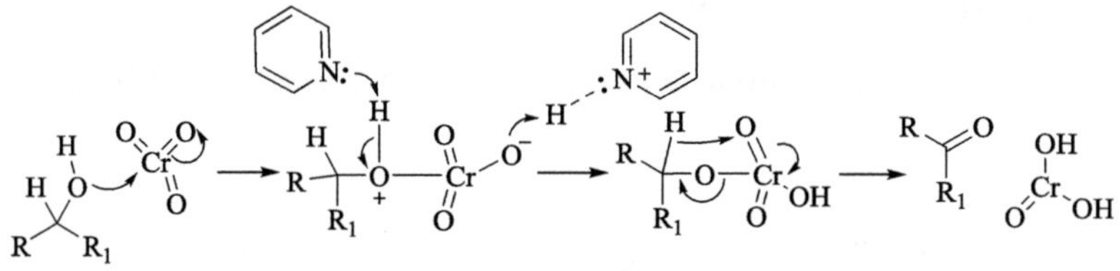

Our understanding of what is happening in this process is still developing. For any mechanism, it may not be possible to clearly represent all the complexities of the transformation through the model we use to describe it. In this case, the first mechanism presented is commonly utilized because it is consistent with most examples to which it has been applied and is generally a dependable predictor of the outcome of this transformation.

PROBLEM 7.6

When primary alcohols are mixed with chromium-based oxidants in systems where water is present, the resulting aldehyde is potentially susceptible to further oxidation. Show a mechanism for this process.

Hint: The process is presumed to proceed through the corresponding hydrate (acetal).

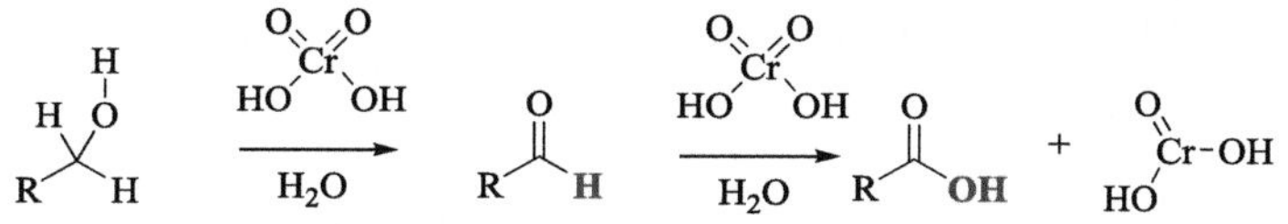

Activated Sulfoxide Oxidations of Alcohols

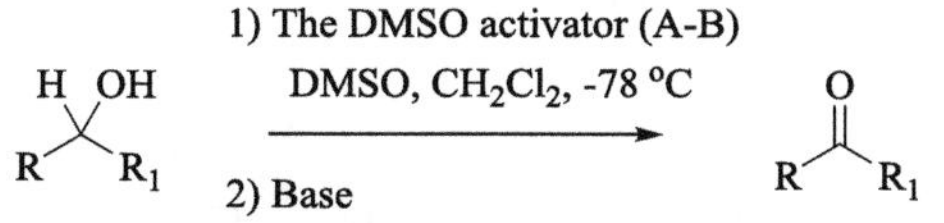

1) The DMSO activator (A-B)

DMSO, CH_2Cl_2, -78 °C

2) Base

There is a whole family of oxidations utilizing *activated sulfoxide* or similar systems to generate carbonyl compounds from the corresponding alcohols. The reaction conditions are relatively mild and, in general, other functionality is unaffected by the process. The activated sulfoxide oxidation does not require the use of heavy metals, but this benefit is offset by the dialkyl sulfide that is produced as a byproduct.

The general mechanism for activated sulfoxide-mediated oxidations starts with the addition of the dimethylsulfoxide (DMSO) oxygen into the activating agent to give **7-41**. A variety of activating agents can be used in the generation of **7-41** (Table 7.1). The alcohol adds into the sulfur, displacing the oxygen bound to the activating group, to give **7-42**, which is deprotonated and leads to the carbonyl product (**7-43**) and the byproduct dialkyl sulfide.

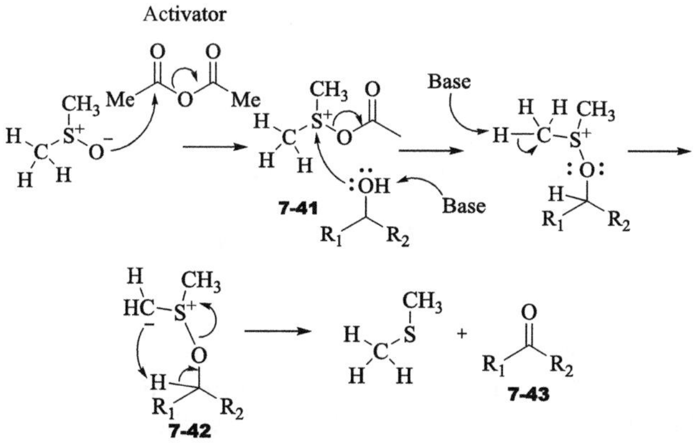

Swern Oxidation

The initial step in the Swern oxidation creates activated chlorodimethyl sulfonium chloride **7-44** by the reaction of dimethyl sulfoxide with oxalyl chloride. The process generates one equivalent each of CO_2, CO (it is often possible to see the mixture bubble as the fast

TABLE 7.1   Activated Sulfoxides

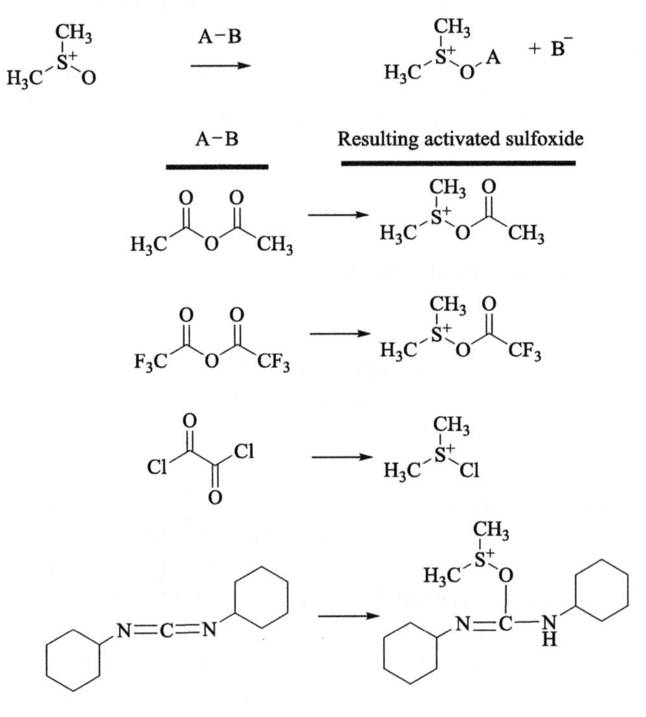

A–B	Resulting activated sulfoxide

decomposition ensues), and chloride ion. The chloride ion attacks the sulfur (**7-45**) to form the sulfonium chloride.

Step 1

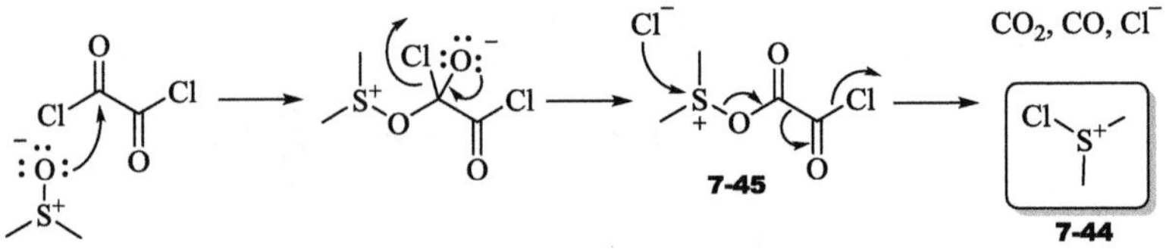

Dimethyl sulfonium chloride

In the second step of the process, the alcohol displaces the chloride (**7-46**) to deliver alkoxy sulfonium intermediate **7-47**. The hydrogens on the carbon adjacent to the sulfonium sulfur are relatively acidic, due to stabilization afforded by the sulfonium salt (pKa ~18.9, comparable to an enolizable proton adjacent to a carbonyl, see appendix C).* The triethyl-amine deprotonates **7-47** to form the sulfur ylide (**7-48**). In a cyclic transformation, the anionic carbon abstracts a proton and the alcohol is oxidized to the carbonyl, as the sulfur—oxygen bond is broken and dialkyl sulfide is released.

The sources are Corey and Kim, (1972); Mancuso and Swern, (1981); Marx and Tidwell, (1984); Tidwell, (1990a); Tidwell, (1990b); McConnell et al., (2008).

Moffat Oxidation

The Moffat oxidation, a method related to the Swern oxidation, proceeds through the same sulfonium intermediate described above, but arrives there through the initial activation of the DMSO by dicyclohexylcarbodiimide (DCC) **7-49**, followed by alcohol addition (**7-50**). Depro-tonation of the acidic proton adjacent to the sulfonium sulfur forms an ylide (**7-52**) and finally abstraction of the proton leads to the formation of the carbonyl and dimethyl sulfide. It has been noted that the oxidation step is not slowed with increased steric demand and that yields improve slightly as the bulk of the substituents at the alcohol carbon center is increased.

Another interesting observation was that when DMSO and DCC were mixed in methylene chloride, there was little evidence of the sulfonium intermediate (implying that the equilibrium lies far to the left), but as soon as the alcohol was added, the transformation was quite rapid.

The sources are Pfitzner and Moffatt, (1965a); Pfitzner and Moffatt, (1965b); Omura and Swern, (1978).

Moffatt Oxidation

*This value for the pKa of a hydrogen adjacent to the sulfonium ion is an estimate. It is based upon data associated with related compounds and correlations made based on similar factors associated with those compounds. For details on how this estimate was made, see Crosby and Stirling, (1970), table 2, item d.

Step 2

An alternate mechanism for the Moffatt oxidation has been postulated. This is a very similar process to that described above, except that the authors describe the transformation proceeding through a transition state containing a quaternary sulfur (**7-53**) that leads directly to the active sulfonium salt and the urea. This second model has been supported by deuterium-labeling studies. When DMSO-d_6 was used, the urea product showed deuterium incorporation (**7-51a**), which would most likely arise from the second (intramolecular) mechanism.

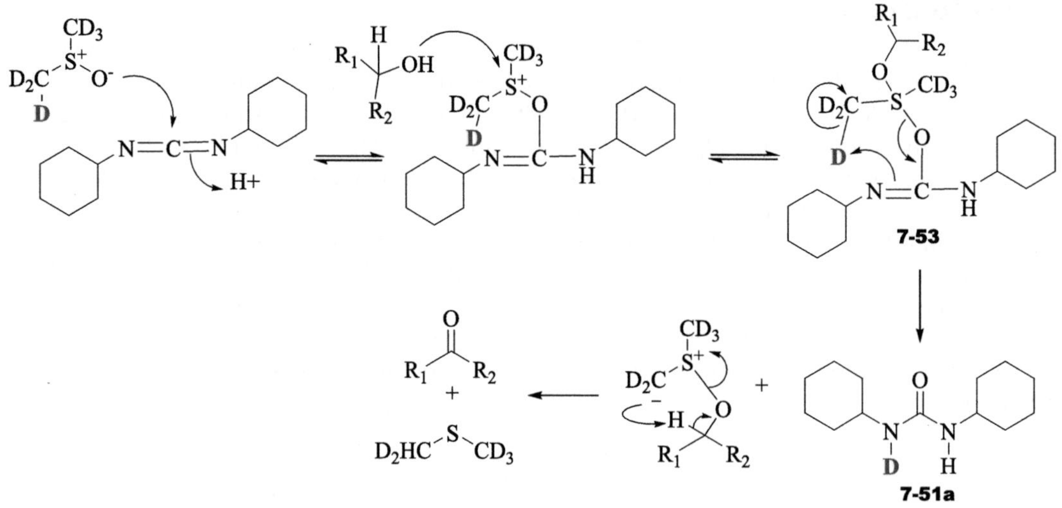

7-51a

Dicyclohexyl urea

Both mechanisms describe the observed product outcomes (except when dimethyl sulfoxide-d_6 was used) and are considered reasonable models for our purposes.

The sources are Fenselau and Moffatt, (1966); Moffatt, (1971).

There have been many industrial accidents associated with the rapid decomposition of DMSO mixtures with other materials. Mixtures of DMSO with acyl chlorides, aryl halides, alkyl halides (at warmer temperatures), perchloric acid, perchlorates, permanganates, periodic acids, phosphorous pentoxide, sodium hydride, and a number of other substances have been found to be incompatible under certain circumstances.

The sources are Rowe et al., (1968); Lam et al., (2006).

PROBLEM 7.7

For the following transformations:

(a) Which of these two steps represent an oxidation?

(b) Show a mechanism for part 2.

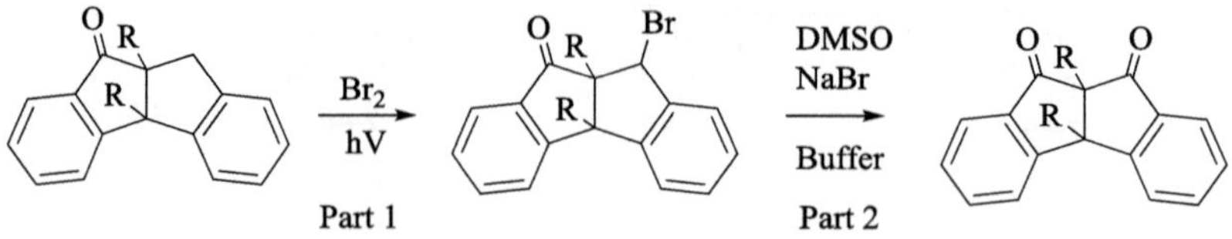

Part 1   Part 2

Source: Paisdor and Kuck, (1991).

Manganese Dioxide Oxidation of Benzylic and Allylic Alcohols

Manganese dioxide is a mild reagent for the oxidation of allylic, alkynic, and benzylic alcohols to ketones without oxidizing alkyl alcohols.

The sources are Fatiadi, (1976a); Fatiadi, (1976b).

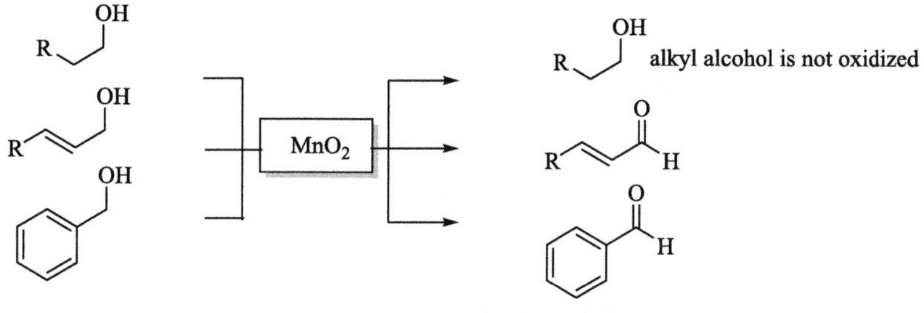

This transformation has been hypothesized to proceed through a radical mechanism, as shown below. The rate of the first step is affected by the ease of abstraction of the C−H hydrogen and is proposed to be the source of the chemoselectivity.

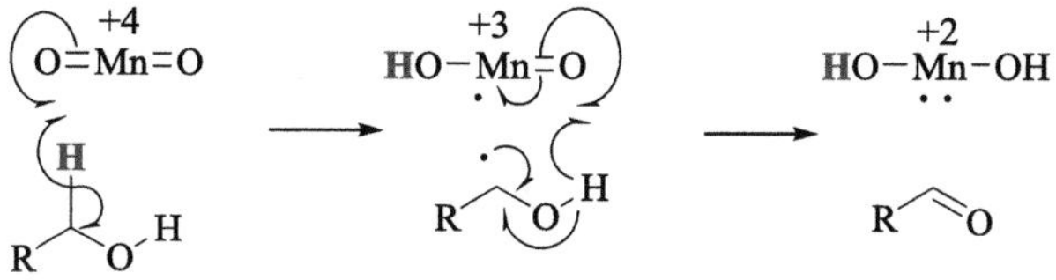

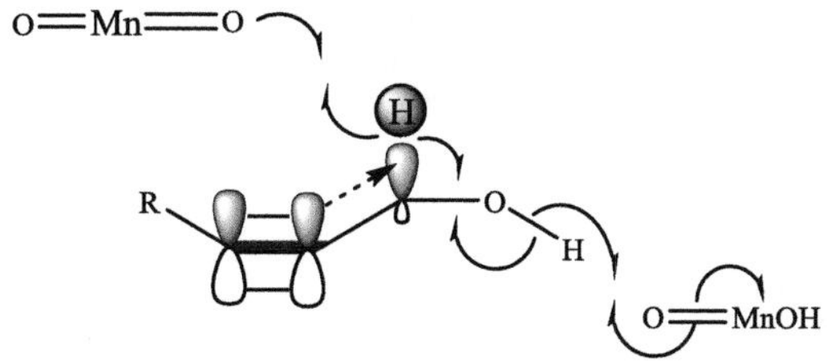

It is often necessary to use a large excess of MnO$_2$ to get satisfactory conversion to product. This process may be dependent upon impurities and poor solubility of the MnO$_2$, as different results from different lots and preparations of MnO$_2$ from the same manufacturer are common (Gritter and Wallace, 1959). These complicating factors and other intricacies associated with the interaction between the MnO$_2$ and the substrate, including the potential ionic nature of the process are still a matter of debate. Additional evidence supporting the mechanism and the relative reactivity of allylic and benzylic systems comes from changes in the reactivity of conformationally strained systems where the bond geometry in the molecule does not allow for the proper olefin orbital overlap.

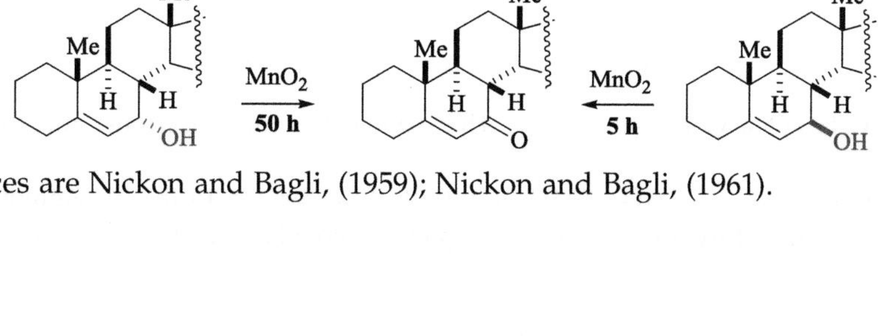

The sources are Nickon and Bagli, (1959); Nickon and Bagli, (1961).

Oxidations with TEMPO (2,2,6,6-tetramethylpiperidin-1-yl)oxidanyl

Since the first preparations of oxoammonium salts in the 1960s, they have been found to be very versatile oxidants. The oxoammonium salt (7-56) has been prepared via a one-electron oxidation of a nitroxyl radical (7-55 TEMPO for example), which in turn is derived from the corresponding amine (7-54).

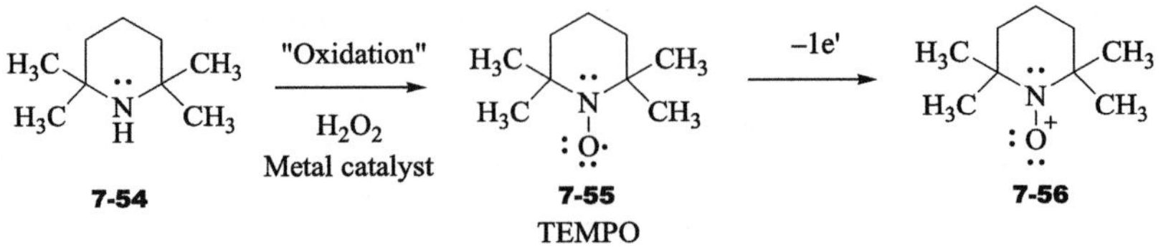

The source is De Nooy et al., (1996).

A simplified view of the redox analogs shows that a one-electron reduction (addition of one electron) of the nitroxide radical (7-55) leads to the aminoxy ion (7-57) (corresponding to the hydroxyl amine 7-58), whereas the one-electron oxidation (removal of an electron) leads to oxoammonium cation 7-56 or resonance structure 7-56b.

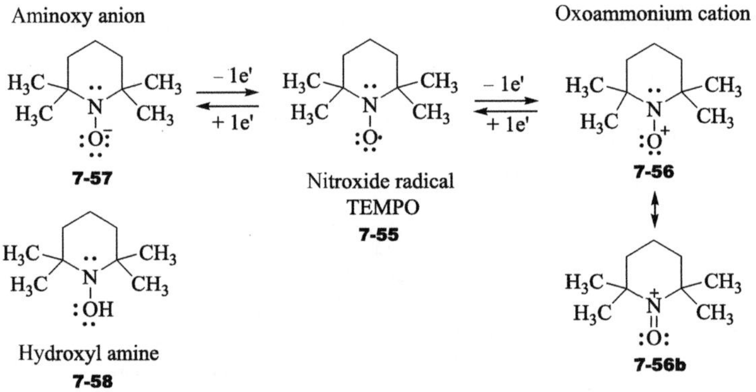

TEMPO is a very practical reagent due to its stability and commercial availability. TEMPO is expensive, but is typically used catalytically. The nitroxide radical in TEMPO is stabilized by resonance and the tetra-alkyl substitution pattern prevents hydrogen abstraction alpha to the nitrogen.

EXAMPLE 7.6. WHICH OF THESE RESONANCE STRUCTURES OF THE NITROXIDE RADICAL IS NOT A VIABLE ALTERNATIVE? WHY?

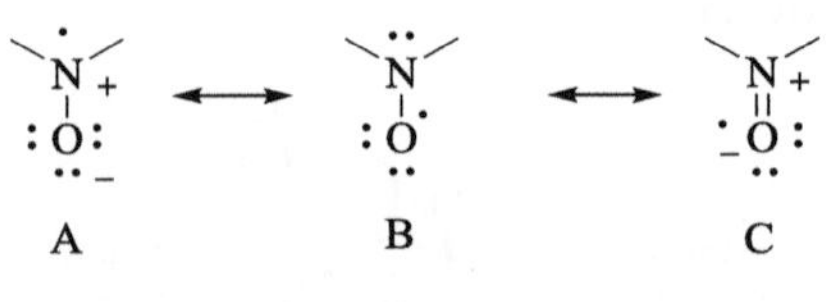

Structure **C** is not a viable alternative. The oxygen has nine electrons, and thus, violates the octet rule.

In the oxidative cycle, TEMPO (7-55) is initially transformed to the oxoammonium cation (7-56) (the active agent) by a cooxidant (examples of cooxidants include hypochlorite, peroxy acid, bromite, and O_2 with copper chloride) via a one-electron oxidation. A common procedure utilizes a biphasic mixture of water and methylene chloride and sodium hypochlorite (bleach), sodium chlorite, potassium bromide, sodium bicarbonate, and TEMPO. The oxoammonium salt (7-56b) then oxidizes the alcohol to deliver the desired carbonyl and hydroxyl amine byproduct 7-59 (Example 7.7). The hydroxyl amine is reoxidized (two electrons, see reoxidation mechanism below) to the oxoammonium anion by the cooxidant and is again ready to repeat the cycle. The process can be run under either slightly acidic or basic conditions. The mechanism below represents the details of the process when run under slightly acidic conditions. In some instances, the HBr is reoxidized by yet another cooxidant, HOCl, derived from NaOCl.

EXAMPLE 7.7. THE OXIDATIVE CYCLE FOR TEMPO OXIDATION MECHANISM

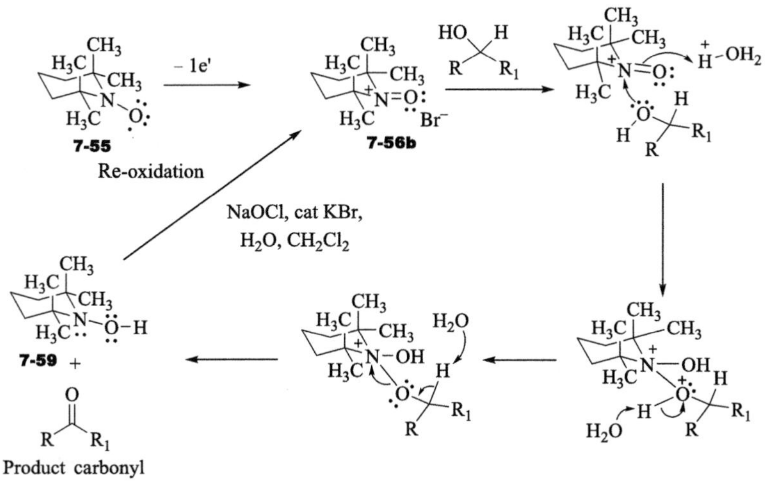

The reoxidation cycle:

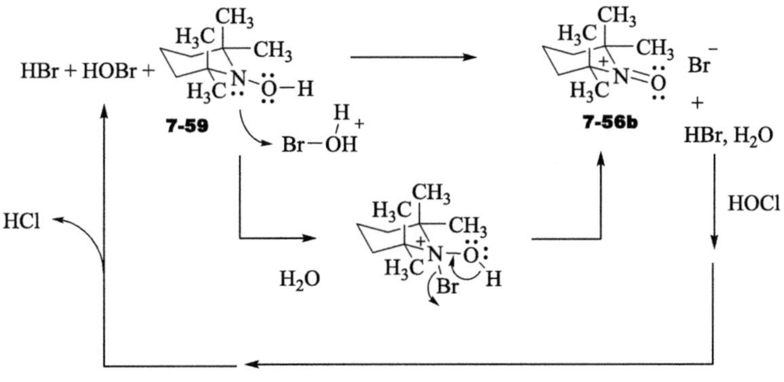

Re-oxidaiton to form the oxoammonium salt

EXAMPLE 7.8. EXAMPLE TRANSFORMATIONS EFFECTED THROUGH THE USE OF TEMPO

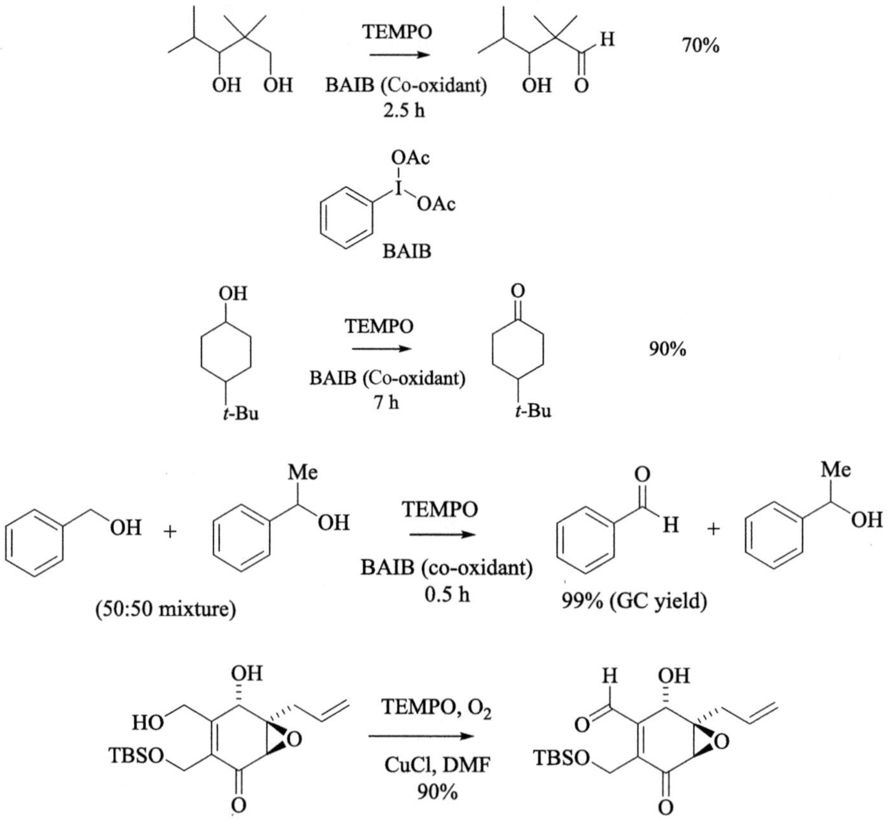

TEMPO oxidations show selectivity for primary alcohols over secondary alcohols, which are oxidized at a slower rate. Secondary alcohols will eventually be oxidized, providing a mild and selective oxidation protocol. Because TEMPO is catalytic, does not rely upon a heavy metal and its byproducts/cooxidants are relatively innocuous, it is considered a "greener" alternative, relative to other oxidation systems.

Sources: De Mico et al., (1997); Mehta and Pan, (2004).

PROBLEM 7.8

What would be the reaction mechanism for the transformation seen in Example 7.7 look like under basic conditions?

Sources: Naik and Braslau, (1998); Bailey et al., (2007).

PROBLEM 7.9

TEMPO is relatively stable, but the α-hydrogen analog is not. Why?
Show a mechanism explaining what could happen.

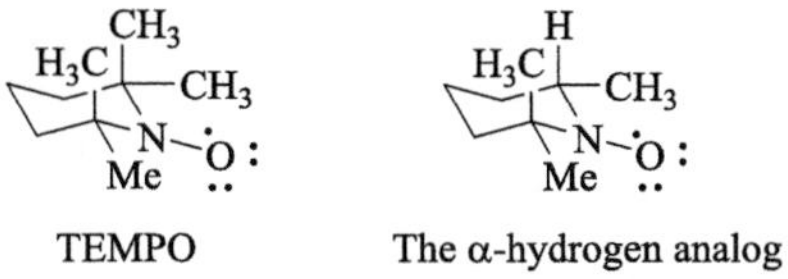

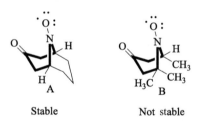

TEMPO   The α-hydrogen analog

Source: Naik and Braslau, (1998).

PROBLEM 7.10

Why is nitroxide A so stable in comparison to other nitroxides with alpha hydrogens like B?

Stable   Not stable

Sources: Naik and Braslau, (1998); Dupeyre and Rassat, (1966).

Oppenauer Oxidation

A classic technique for oxidizing an alcohol to a ketone **7-62** is the Oppenauer oxidation. In this procedure, the alcohol (**7-60**) is mixed with a sacrificial ketone (**7-61**, commonly acetone or cyclohexanone) in the presence of an aluminum alkoxide and the mixture is heated. The reaction is an equilibrium-based process. This transformation can be driven forward in three ways:

1. By using a very reactive sacrificial ketone (**7-61**).
2. By adding an excess of the alcohol (**7-60**), driving the equilibrium by a change in concentration (Le Chatelier's principle).
3. By using an excess of the carbonyl donor (**7-61**) and removing product (or byproduct). The removal of product can be accomplished through techniques such as distillation.

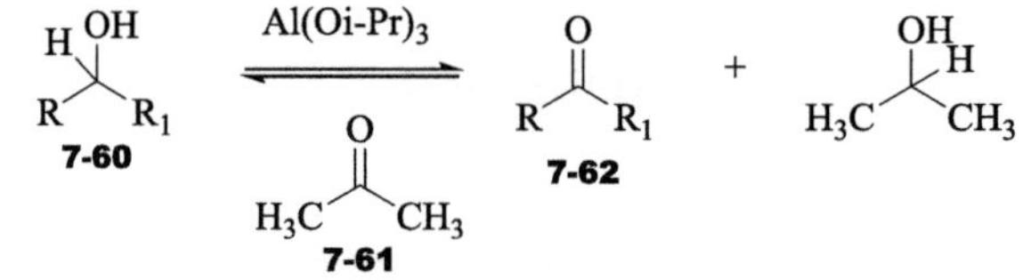

The source is Wilds, (1944).

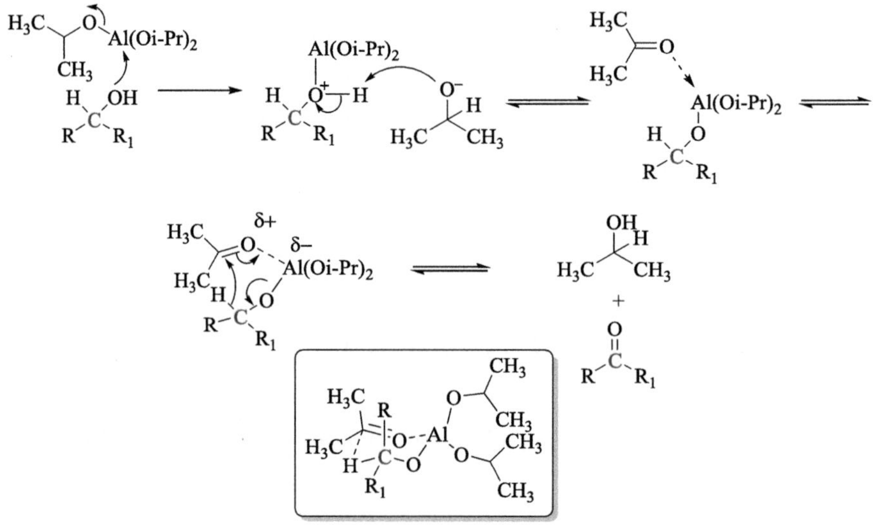

Steric demands of the substrate can impact the reaction performance. Hindered alcohols react less readily and in rigid cyclic systems, such as cyclohexanol derivatives, an equatorial hydroxyl is faster to react than an axial hydroxyl.

EXAMPLE 7.9. MECHANISM OF THE OPPENAUER OXIDATION

Show a mechanism for the following transformation.

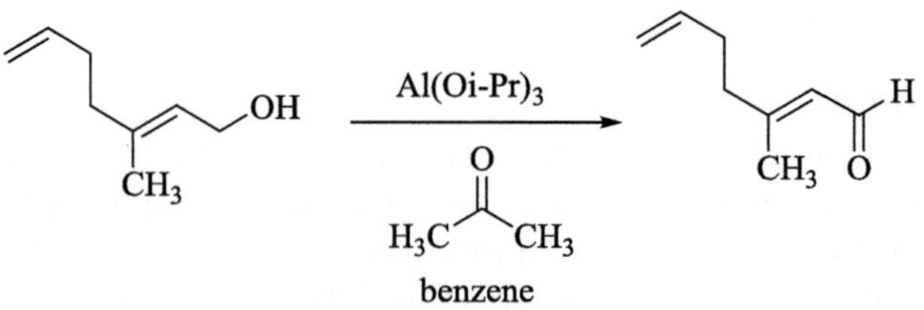

In the proposed mechanism for the Oppenauer oxidation, the alcohol substrate interacts with the aluminum, displacing isopropanol **7-63**. The ketone comes into close proximity to the aluminate complex (**7-64**) and the hydride is transferred to the activated carbonyl in the six-membered transition state **7-65**. The lowest energy chair transition state represented by **7-65a** has been postulated to be the main contributor to the output for this process. The carbonyl **7-66** (the newly formed aldehyde in this example) is released and the aluminum alkoxide is ready for another molecule of the alcohol substrate to be oxidized.

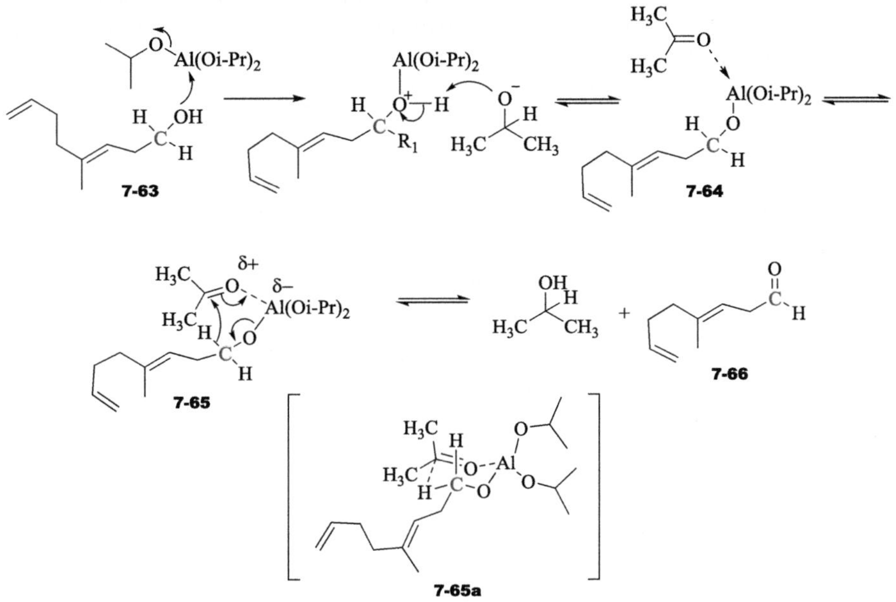

3. REDUCTIONS

A. The Relationship between Oxidation and Reduction

Meerweine–Pondorffe–Verley Reduction

For each oxidation reaction, there is a component of the reaction that was reduced. If we look back at the transformations we have covered in the oxidation section of this chapter, we can pick out the redox partners for each oxidation. A great example is the Meerwein–Pondorff–Verley reduction reaction, which is the reverse of the Oppenauer oxidation. In this reaction, the reduction–oxidation partners have been reversed and the desired product is from carbonyl reduction and not alcohol oxidation.

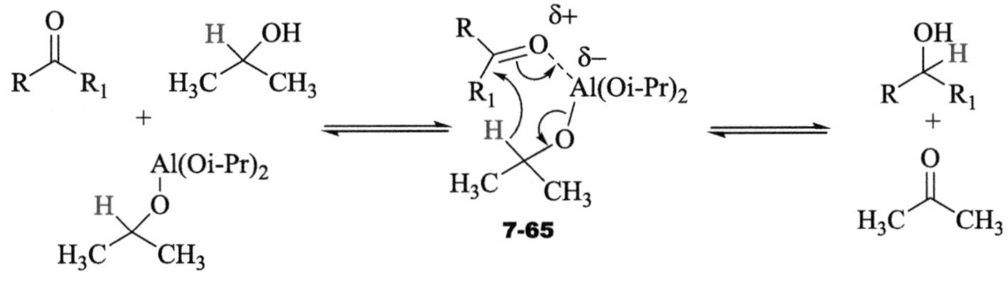

The transformation proceeds through the same cyclic transition state (**7-65**), as seen in the Oppenauer oxidation and is again an equilibrium-dependent process. An excess of isopropyl alcohol drives the reaction to the right and yields are often better than 70%, though the process is sensitive to the steric demands of the substrate. The hydride is delivered from an angle that will keep the large aluminum trialkoxide in a position that minimizes the steric conflict. In cases where the substrate is a rigid six-membered ring system, like the cyclohexanone derivative shown below (**7-67**), the results tend to favor the equatorial alcohol product.

EXAMPLE 7.10. MEERWEIN−PONDORFF−VERLEY REDUCTIONS IN A CONSTRAINED SYSTEM

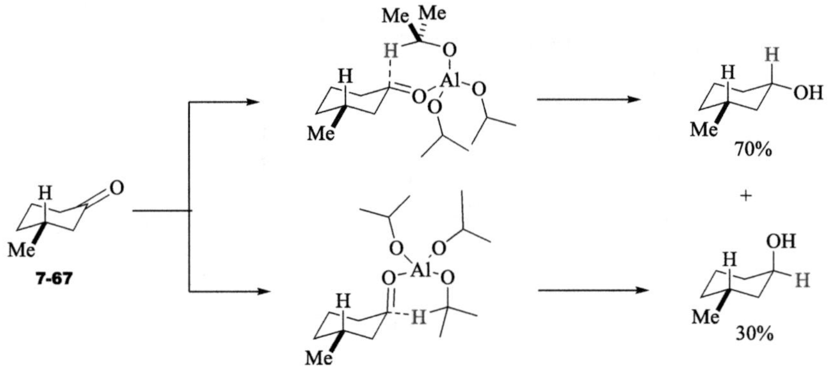

Support for the cyclic transition state, with the direct transfer of a hydride, was obtained from a deuterium-labeling experiment showing that when deuterated water is added to the mixture, no deuterium is incorporated into the product.

Source: Moulton et al., (1961).

PROBLEM 7.11

(a) Show the mechanism for the following transformation.
(b) The process only delivers about 67% of the product and about 33% of the starting material is recovered, why might the authors not be able to push the process further to the right?

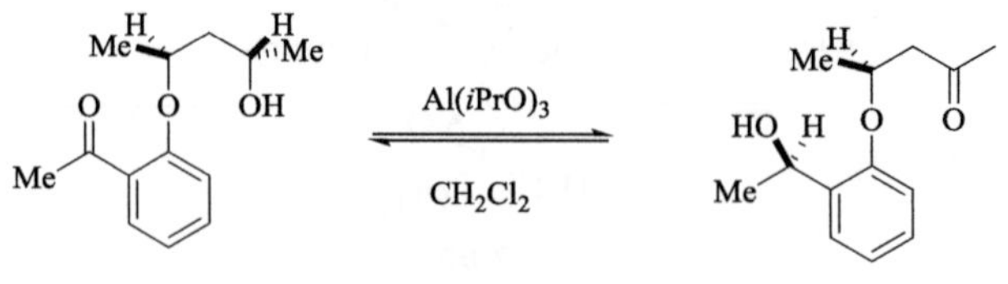

Source: Fujita et al., (1997).

B. Borane Reductions

Hydroborations with borane or alkyl borane derivatives represent an important class of transformations in organic chemistry. Borane (BH_3), which is too unstable to be isolated, exists as a dimer (diborane, B_2H_6) or as a complex with a simple electron donor like THF, diglyme, or dimethyl sulfide. In the literature, it is often referred to as $BH_3 \cdot L$.

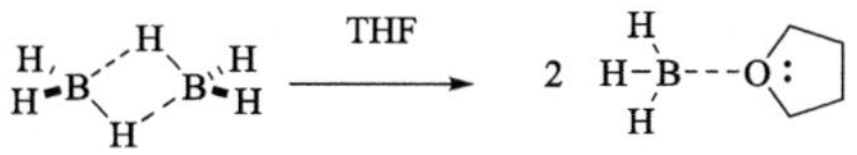

Reduction of Olefins

Boranes are commonly used to affect the hydroboration of olefins. In the first step, borane adds across the olefin in a concerted manner and in an anti-Markovnikov fashion (the hydrogen atom adds to the more substituted carbon center and the boron to the less substituted center). The intermediate (**A**) is then either oxidized with hydrogen peroxide in alkaline water to give the alcohols or worked up through a protonolysis procedure to deliver the alkane.

EXAMPLE 7.11. THE HYDROBORATION INTERMEDIATE REPRESENTS A NET REDUCTION OF BOTH CARBONS OF THE ORIGINAL ALKENE

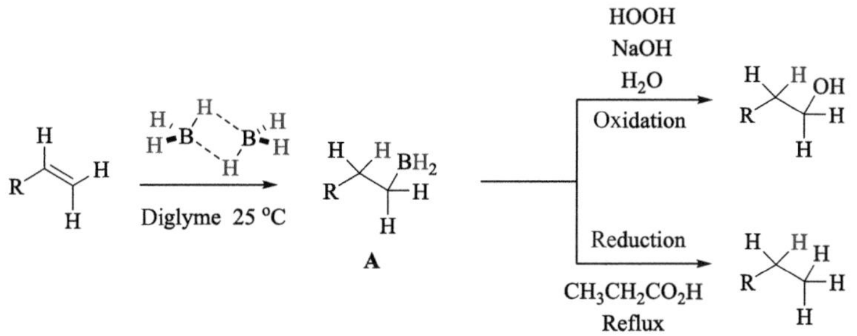

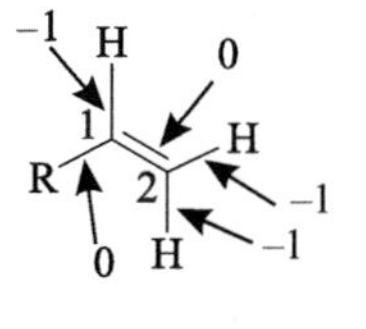

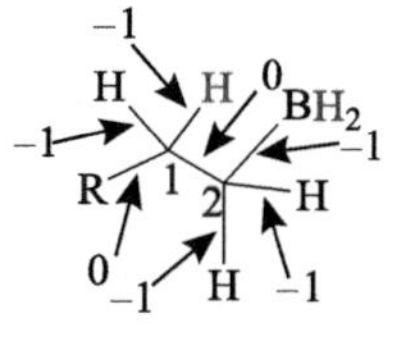

Carbon #1 = $-1 + 0 + 0 = -1$

Carbon #2 = $-1 + (-1) + 0 = -2$

Carbon #1 = $-1 + (-1) + 0 + 0 = -2$

Carbon #2 = $-1 + (-1) + (-1) + 0 = -3$

Change for carbon #1 = -1 (reduction)
Change for carbon #2 = -1 (reduction)

The regiochemical orientation of the borane addition across the olefin is driven by a mixture of steric and electronic effects. The electronegativity of boron (2.0) is less than that of hydrogen (2.1) and leads to an addition product with the opposite regiochemistry compared to addition of water across an olefin (see Table 1.2 in Chapter 1).

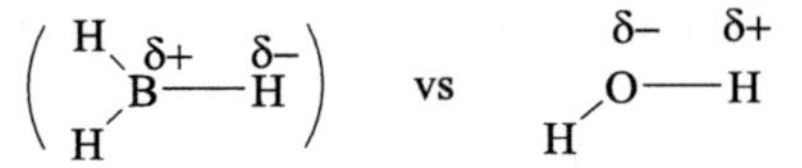

For many of the examples described here, only the intermediate organoborane is shown. We will discuss the further functionalization of the hydroboration intermediate (borane) in more detail later in this chapter.

As borane ads across the olefin, a partial positive charge develops on one of the carbons of the olefin. This positive charge is better stabilized at the more substituted center.

EXAMPLE 7.12. ORIENTATION AS THE BORANE INTERACTS WITH THE OLEFIN (PART 1)

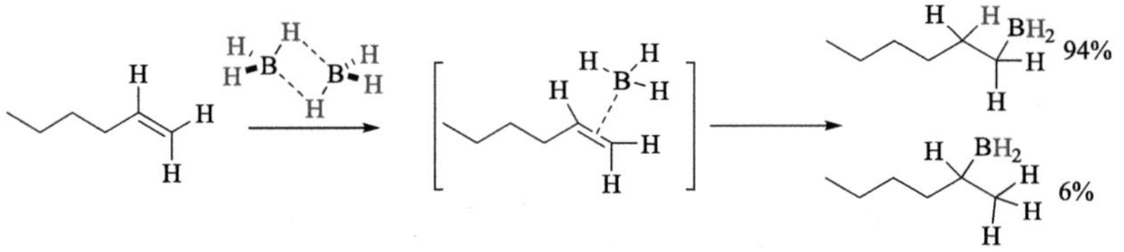

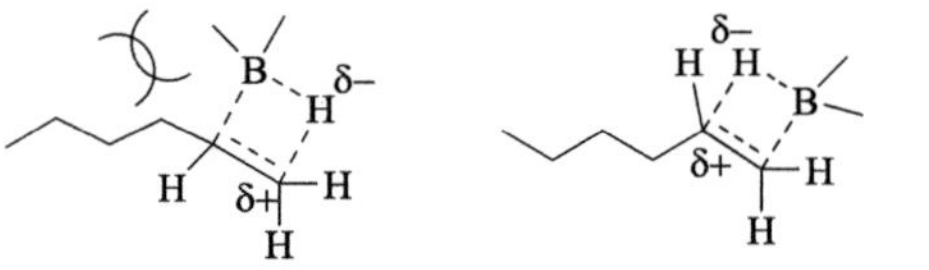

Favored transition state

7-68a   7-68b

The regiochemistry of the hydroboration process is also influenced by the steric demands of both the olefin and the organoborane. For example, even though the reagent borane can deliver all three of its hydrides, the addition of the final hydride across an olefin is slower than addition of the first two. This effect has been attributed to the steric bulk of the borane side chains that result with each hydroboration.

EXAMPLE 7.13. ORIENTATION AS THE BORANE INTERACTS WITH THE OLEFIN (PART 2)

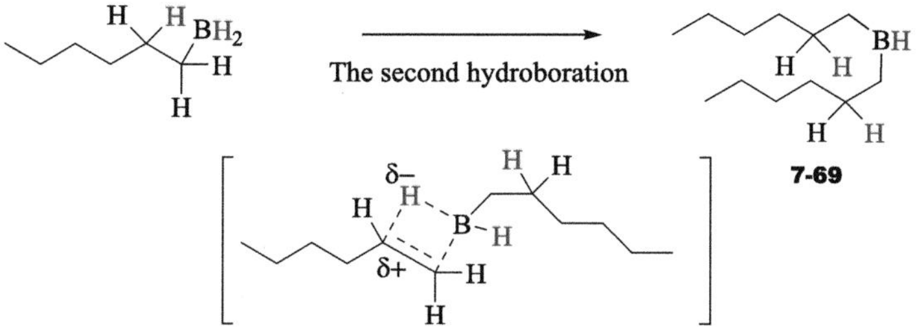

Favored transition state
7-69a

For the second hydroboration, the electronic preference for the developing positive charge on the more substituted carbon is reinforced by the increased bulk on the borane, directing the alkylborane away from the alkane chain (**7-69a**). The steric demands of the transition state result in a slower and more selective process, reinforcing the orientation of the transition state to provide the anti-Markovnikov product (**7-69**) with good selectivity. This effect has been exploited through the use of even more hindered dialkyl borane reagents (Example 7.14).

EXAMPLE 7.14. ORIENTATION AS THE BORANE INTERACTS WITH THE OLEFIN (PART 3)

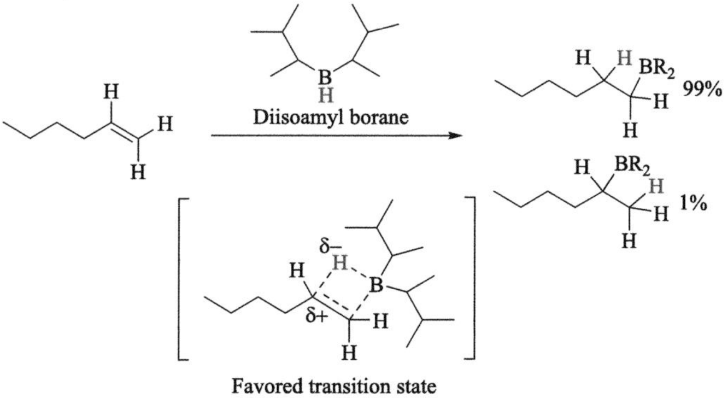

Favored transition state

The impact of competing steric and electronic factors can also be seen in hydroboration of styrene derivatives. The system with an electron-donating substituent on the aryl ring increases the ability of the benzylic carbon to bear a positive charge, but the presence of an electron-withdrawing

TABLE 7.2   The Regiochemical Effect on a Hydroboration Related to Substitution on an Adjacent Aryl Ring

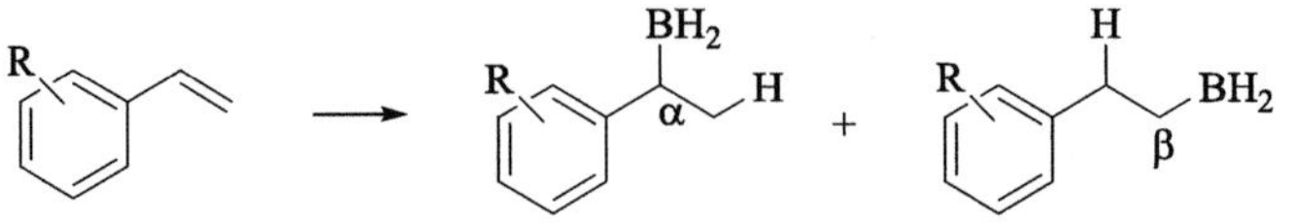

| | Boron Distribution | |
R	α Markovnikov	β Anti-Markovnikov
p-CH$_3$O	7	93
m-CH$_3$O	19	81
H	19	81
p-Cl	26	74
p-CF$_3$	34	66

group destabilizes the carbocation and increases the ratio of Markovnikov product. Note that unlike the para-methoxy styrene, the meta-methoxy styrene is not able to benefit in the same way from resonance stabilization of the cationic center and this leads to a product ratio similar to the parent styrene (Table 7.2).

Source: Brown and Sharp, (1966).

The electronic effects seen in the styrene example demonstrate the impact of the resonance stabilization afforded by the substituent on the aryl group. When propenyl styrene is used, the terminal methyl group can change the resulting regiochemical outcome, through a combination of steric and electronic effects, leading to the reversed regiochemistry.

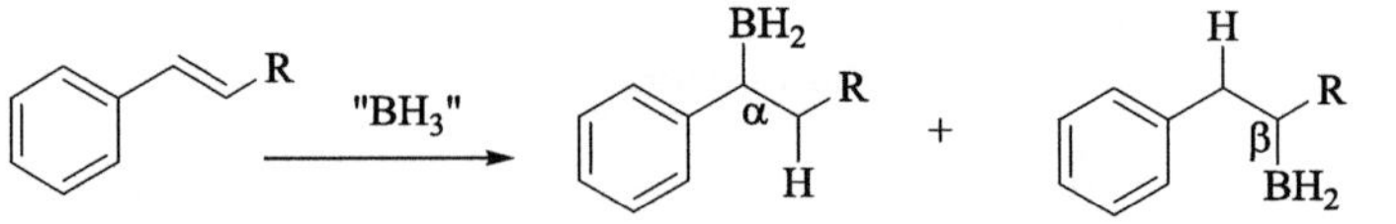

R = H	20%	80%
R = Me	75%	25%

The source is Allred et al, (1960).

The methyl group is electron donating (relative to the aryl ring) and shows a propensity to stabilize the developing carbocation in the opposite direction through induction. In the styrene case, the steric demands of the aryl ring favored having the boron approach at the terminal carbon of the olefin, but the methyl group of the propenyl styrene balances the steric influence the aryl ring has and thus the "β" position is no longer favored.

Styrene Reactivity—Competing Factors

Aryl resonance effect is stabilizing the positive charge at the position adjacent to the ring

Aryl inductive effect is destabilizing the positive charge at the position adjacent to the ring

For Propenyl Styrene—Inductive Effects are Significant

Aryl inductive effect and the alkyl electron push reinforce one another stabilizing the positive charge at the position beta to the aryl ring

Although we have displayed the transition state as "partially charged," the overall transition state has been shown to be relatively nonpolar and the electronic effects seem to be more impactful in determining the lengths of the developing carbon–boron and carbon–hydrogen bonds, and thus, the steric demands as the bonds form. In any case, the steric influences (size of the borane and substitution on the olefin) and the electron density at the olefin are dominant factors in determining the regiochemical outcome of this process.

Hydroboration is a stereospecific reaction. Because of the concerted nature of the addition (the new hydrogen–carbon and boron–carbon bonds being formed at the same time as the boron–hydrogen bond of the borane is being broken), the bonds being formed are on the same face of the olefin (syn addition).

7-70   **7-71**   **7-72**

The source is Burkhardt and Matos, (2006).

The borane approaches from the less sterically demanding face of the olefin (away from the dialkylated bridge, **7-70**) and is oriented with the boron in the less sterically demanding position, the bridged carbon (**7-71**). The developing cation is centered at the more substituted position of the alkene with the boron and hydrogen adding to the same face. Together this provides the regio- and stereochemical outcome observed in **7-72**.

The versatility of the hydroboration is ultimately demonstrated by the variety of functionalities into which intermediate borane **7-73** can be converted. A common workup for the borane intermediate is oxidation with hydrogen peroxide in alkaline water to deliver the alcohol. The mechanism is proposed to proceed as described below. The peroxide is deprotonated and adds into the boron atom (7-74) leading to a subsequent rearrangement (7-75) to establish the carbon–oxygen–boron bond structure (7-76). The final oxidation step occurs with retention of configuration.

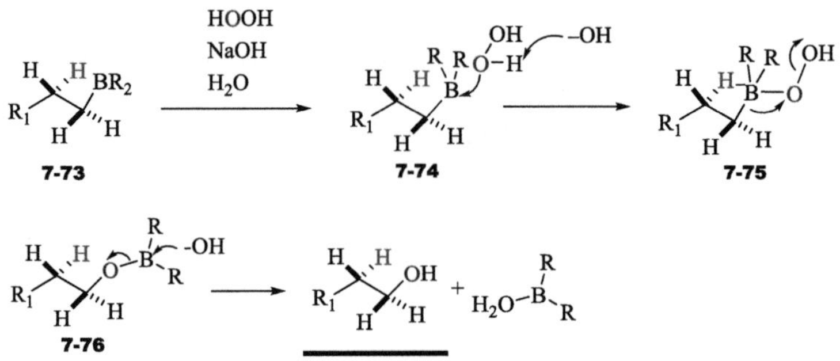

The overall product of the hydroboration and peroxide workup sequence results in a net anti-Markovnikov addition of water across the olefin. This classic reaction has found continued use because of its dependable yields and regio- and stereospecificity.

EXAMPLE 7.15. A COMBINATION OF STEREO- AND REGIODIRECTING EFFECTS IN ONE SYSTEM

Hydroboration of this rigid compound demonstrates both the regio- and stereoselectivity associated with the hydroboration reaction.

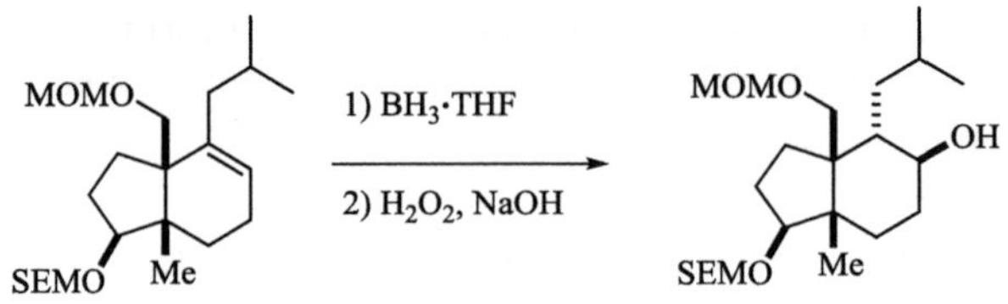

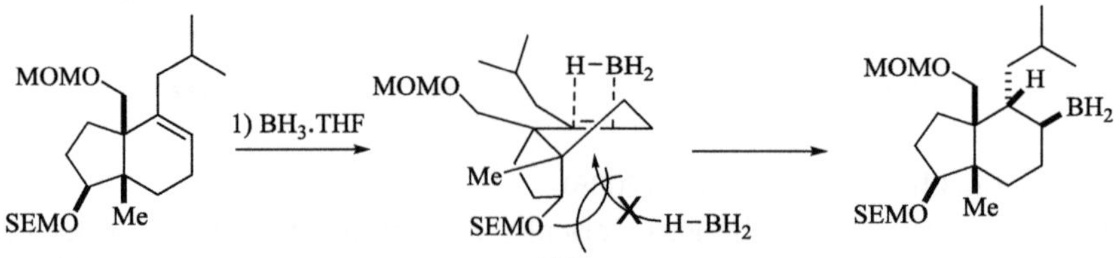

The borane adds to the less sterically encumbered face of the olefin (7-77). The bottom face is blocked, preventing reaction from this side. The steric influences are a result of the adjacent functionality, as well as the overall architecture of the substrate, controlling the orientation of the borane addition across the olefin, as well as the face of attack.

Three key aspects of this transformation are the following:

1. Attack from the top of the system is the result of the ring conformation and functionality of the system.
2. The orientation of the borane is the result of the conformation of the bicyclic ring system and adjacent functionality.
3. The anti-Markovnikov orientation is reinforced by electronic and steric forces.

The borane is worked up with hydrogen peroxide (**7-78**), resulting in addition of the peroxide to the borane, followed by rearrangement (**7-79**) and ultimate loss of the boron (**7-80**), leading to the alcohol product (**7-81**). The other possible regio- and diastereo chemical outcomes were not observed.

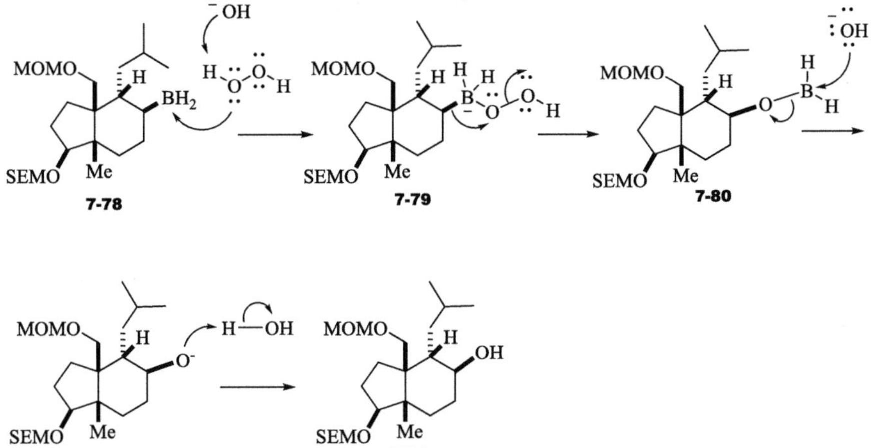

A reductive workup of the organoborane intermediate is possible through interaction with a carboxylic acid to give the alkane. This reduction of olefins is sometimes preferred over direct hydrogenation procedures, due to the stereospecificity of the reaction and the chemoselectivity for the alkene when there are other hydrogenation sensitive functional groups in the molecule. The process has been proposed to be initiated by the formation of a complex between the boron atom and the carbonyl oxygen of the carboxylic acid, followed by a proton transfer to the alkane as the boron−carbon bond is severed (**7-82**).

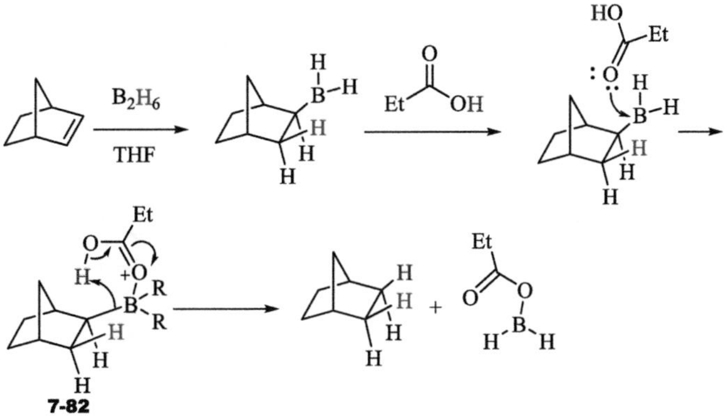

Source: Paquette and Sugimura, (1987).

PROBLEM 7.12

How is the action of sulfuric acid and water on an olefin similar to the two-step hydroboration reaction/oxidation sequence? How are they different? Why?

PROBLEM 7.13

In the transformation below, show the expected product (including stereochemistry) and the intermediate that leads to it.

Sources: Zweifel and Brown, (1964); Brown and Kawakami, (1969).

PROBLEM 7.14

Provide a mechanism describing the following transformation.

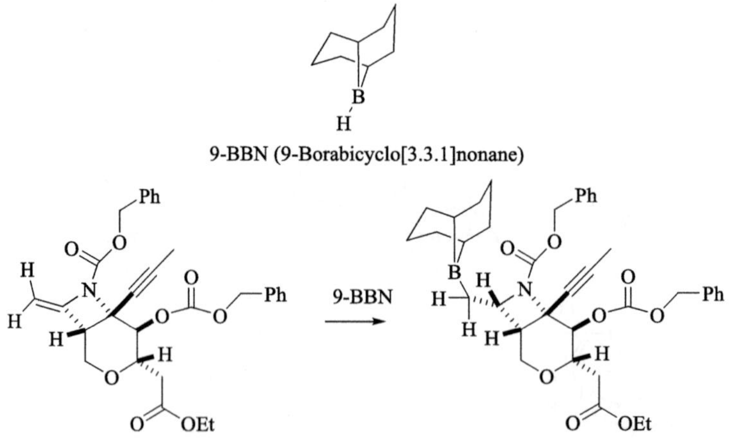

9-BBN (9-Borabicyclo[3.3.1]nonane)

Source: Diethelm and Carreira, (2013).

Sodium perborate is an alternative oxidant to hydrogen peroxide in the workup of hydroborations. In water, the sodium perborate (dimer **7-83**) can deliver small amounts of peroxide at a lower pH than traditional, concentrated hydrogen peroxide solutions. This allows for milder conditions and circumvents the use of unstable and very reactive hydrogen peroxide. It has been postulated that in solution, the sodium perborate forms sodium peroxyborate **7-84** that then goes on to deliver a hydroperoxy species to the substrate.

The sources are McKillop and Sanderson, (1995); Kabalka et al., (1989).

EXAMPLE 7.16. THE OXIDATIVE WORKUP USING PERBORATE AFTER A HYDROBORATION PROCEDURE

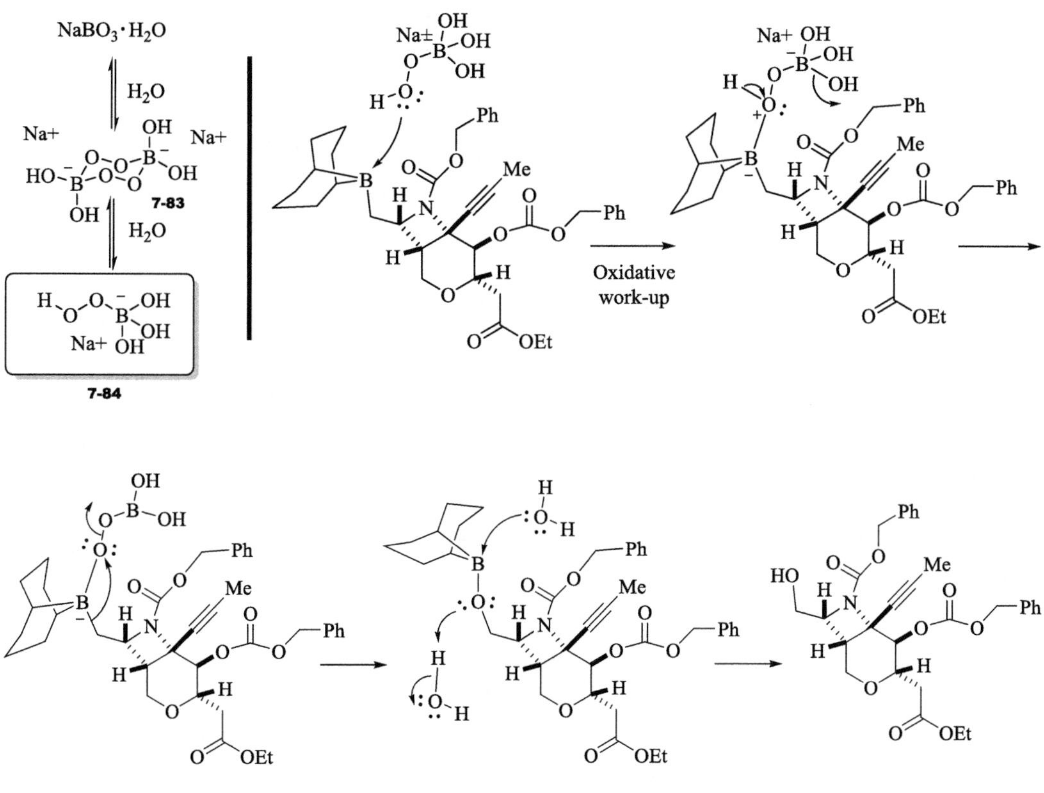

Reduction of Carbonyls

In addition to alkenes, borane can affect the reduction of carbonyls. Borane is electron deficient (a Lewis acid) and is thought to require precomplexation to the carbonyl (**7-85**) for delivery of the hydride to the carbonyl carbon.

The borane acts as a Lewis Acid

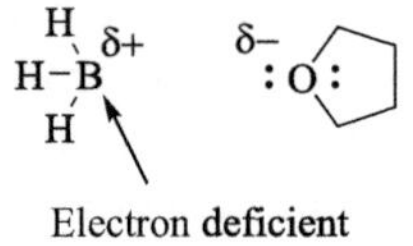

Electron deficient

Once the complex is established, hydride is delivered until the carbonyl is reduced to the hydroxyl or the borane has no more hydrides to deliver. The final products, after workup with water, are the corresponding alcohol and boric acid. Carboxylic acids and carbon nitrogen double and triple bonds are susceptible to reduction with borane. Because of the lack of chemoselectivity including the potential for hydroboration of olefins, the application of borane in this transformation is limited.

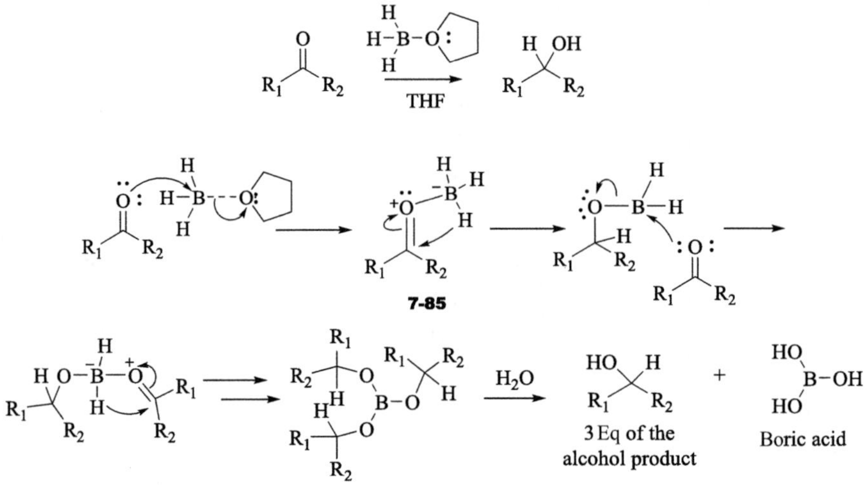

7-85

In the proposed mechanisms for the reduction of an acid with borane, both oxygen atoms interact with the borane, generating hydrogen gas as the transformation occurs. It has been suggested that the diborane acetal is the product before workup. The final hydride transfer can then lead to hydrolysis and deliver the alcohol product and boric acid. Two possible pathways are shown below.

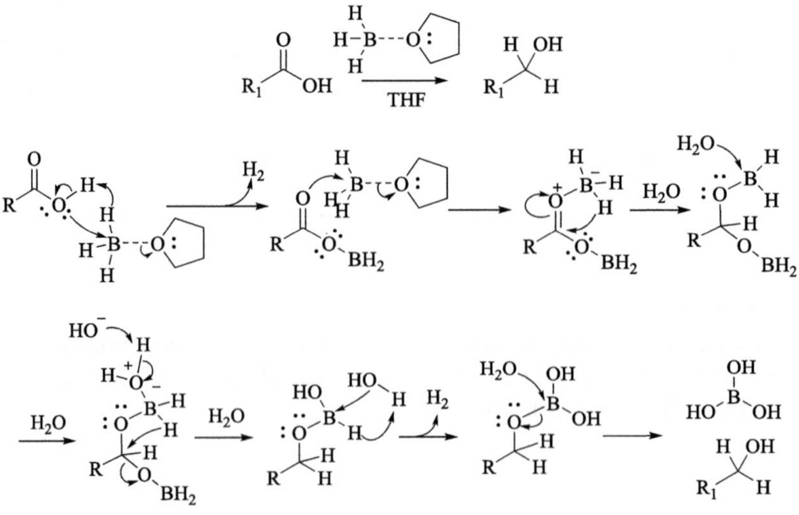

The following mechanism is also plausible.

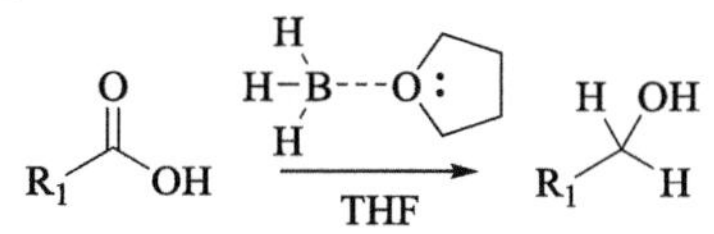

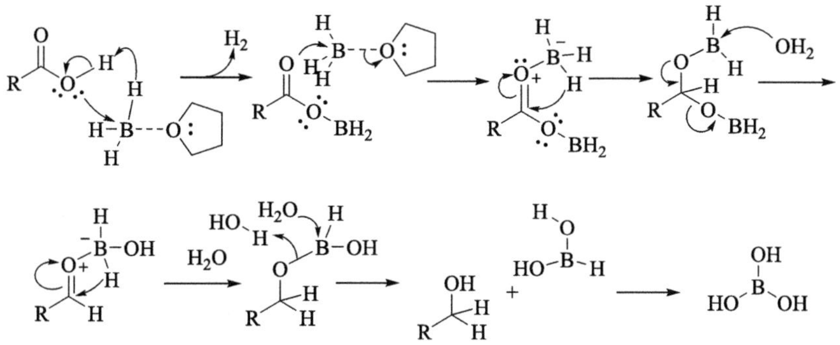

The broad reactivity profile associated with boranes, and certain risk factors as well as the ease of access and use associated with sodium borohydride and lithium aluminum hydride, has led to diminished use of borane relative to sodium borohydride and lithium aluminum hydride for carbonyl reductions.

C. Reduction of Carbonyls with Complex Metal Hydrides

Lithium Aluminum Hydride and Sodium Borohydride Reductions

Related chemical transformations of carbonyls to those seen with borane can be accomplished with sodium borohydride (NaBH$_4$) and lithium aluminumhydride (LiAlH$_4$). Both these reagents are electron rich and have some complimentary properties when compared to borane reagents. NaBH$_4$ and LiAlH$_4$ are effective at reducing aldehydes, ketones, esters, and sometimes carboxylic acids. LiAlH$_4$ is the most reactive of the two and one equivalent will deliver each of its hydrides, up to four equivalents, but with decreasing efficiency as each is delivered. In the case of LiAlH$_4$ for example, the hydrides are delivered without the prior complexation between the substrate carbonyl oxygen and borane, as seen in the mechanisms with diborane.

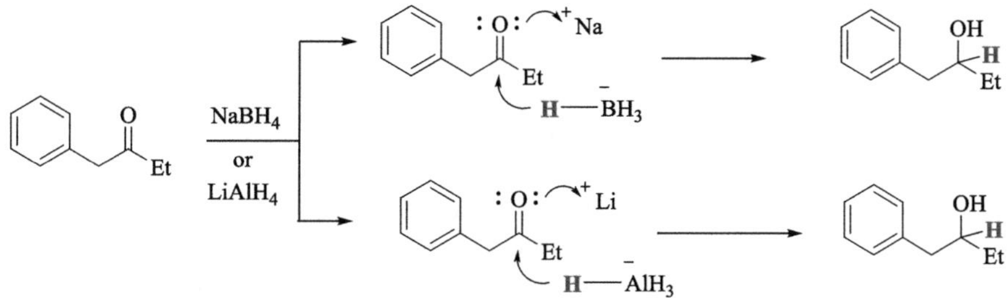

The sources are Brown and Krishnamurthy, (1979); Ashby and Boone, (1976).

Cram's Model

Analogs of LiAlH$_4$ such as lithium tri-*tert*-butoxyaluminum hydride (**7-86**) are less reactive than the parent LiAlH$_4$. These reagents are larger than LiAlH$_4$, and are thus more susceptible to the effects of sterics, and tend to show more selectivity in reductions with hindered carbonyls.

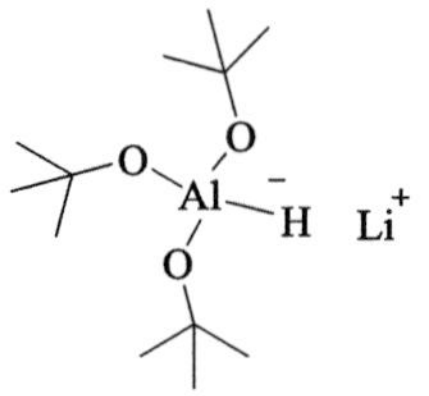

7-86

The selectivity associated with the addition of nucleophiles to ketones that possess adjacent stereo centers has been explained through the application of several different models. Using a "Newman projection" and aligning the largest group on the adjacent center to the carbonyl carbon away from the carbonyl oxygen, it has been predicted that LiAlH$_4$ would be delivered from the side with the smallest functionality on the adjacent center. This, like many models, is a simplification of the situation and is useful only to a point. This early model describing the approach of a nucleophile to a carbonyl was presented by Cram and explained the results seen in many hydride additions to carbonyls. The model requires that the nucleophile approach form the *least* sterically congested side (the side with the "small" group as opposed to the "medium" group). Although this model was based on a simple mix of qualitative and quantitative measures of size and developed from a limited set of historical results, it was a significant breakthrough in predicting the direction of these types of processes. But, there are certain limitations with Cram's model. A major weakness presented by Hugh Felkin in the late 1960s was that the size of the substituent on the carbonyl carbon affected the stereochemical outcome of the reactions. This would make sense because this substituent eclipses the large group in Crams model **7-87**, but this effect is not accounted for in Cram's model.

7-87

The sources are Cram and Elhafez, (1952); Cornforth et al., (1959).

Felkin presented his own model, where the substituents are staggered and the nucleophile approaches from the side opposite the large group. Although this cleared up some of the conflicts, it led to other ambiguities that needed to be resolved. Mainly, it was not clear which configuration should be used since both **7-88a** and **7-88b** (shown below) would allow for the nucleophile to approach from the side opposite the large group, with similar conformational and steric demands as it approached. There may also be some difficulty in predicting the outcome between two different "large/medium" groups of similar size.

Felkin Model

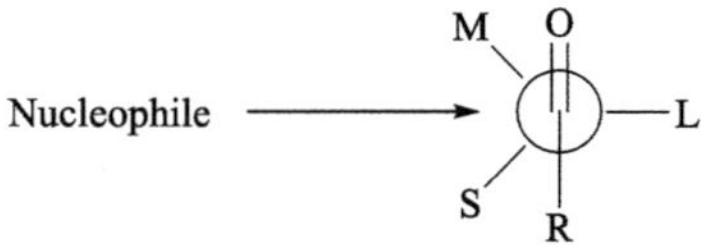

7-88a

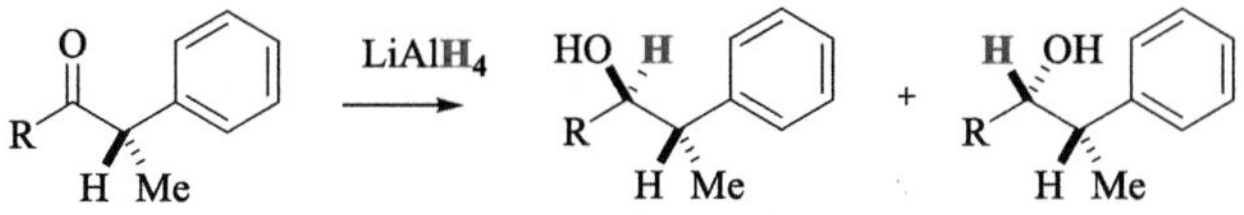

Nucleophile ⟶

R = Et
R = *i*-propyl
R = *t*-butyl

~ 70 : 30
83 : 17
98 : 2

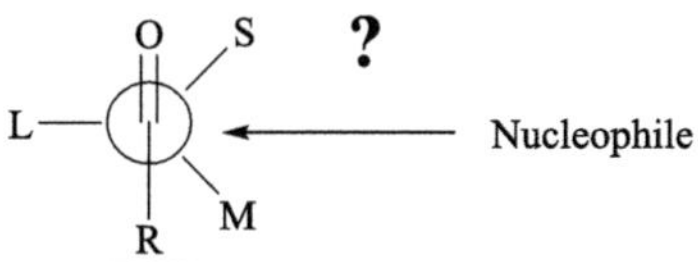

7-88b

Nucleophile

The source is Chérest et al., (1968).

A solution addressing these challenges was proposed by Nguyen T. Anh and O. Eisenstein. In Anh's proposal, the staggered transition state is oriented such that no groups were eclipsed and differentiation between side chains is made based upon the steric and electronic framework of the carbonyl. This approach agreed with some of Felkin's earlier hypotheses. The group that is more electronegative (LG^2) will take up a position parallel and away from the incoming nucleophile. This anti-periplanar arrangement allows for the best overlap with the π-bond of the carbonyl (as shown below). This group is thus designated as the "large" group. The next part of Anh's solution described the true "angle of attack" as the nucleophile approaches the carbonyl carbon. Rather than a straight on approach as we have previously described, Ahn postulated that nucleophiles would approach at an angle of about 107°, the Burgi–Dunitz angle. This is close to the angles in an sp^3-hybridized atom and represents the space occupied by the open orbital position on that carbon atom. More specifically, the nucleophile is approaching at the lowest unoccupied molecular orbital of the carbonyl π-system.

Where LG_2 is the large group and/or most electron withdrawing group, allowing for the best electronic stabilization

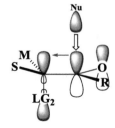

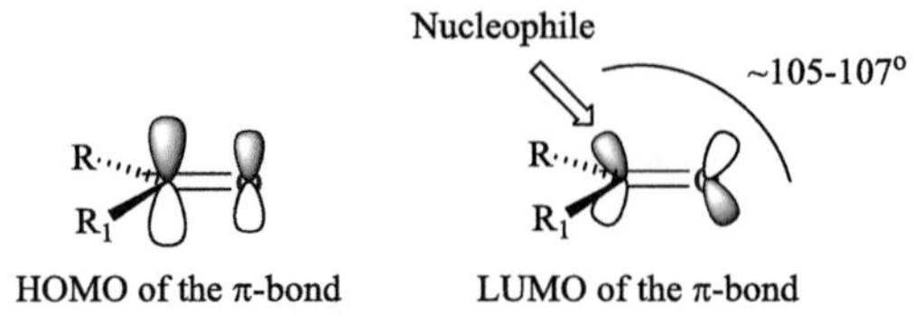

HOMO of the π-bond   LUMO of the π-bond

When we apply this model, there is a clear difference between the two different nucleophile approaches, with one having to come by the small group and one by the medium/large group adjacent to the carbonyl carbon. As the size of the substituent on the carbonyl increases, the approach of the nucleophile from the side with the small substituent is reinforced, as the carbonyl rotates to have the R group away from the medium-/large-sized substituent, as Felkin described.

Felkin–Anh Model

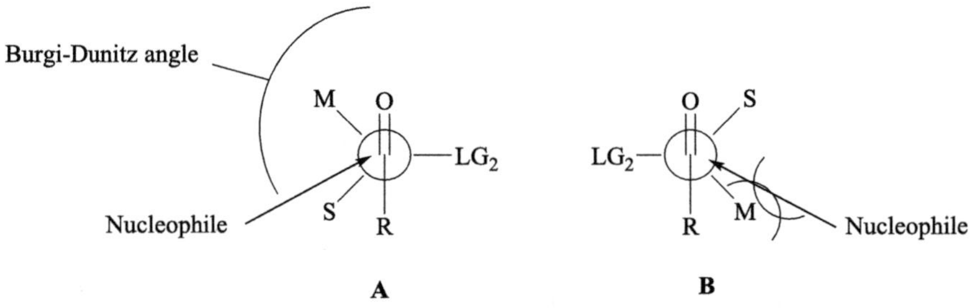

Note that the size of the R group reinforces the benefits of conformation A

The sources are Anh and Eisenstein, (1977); Bürgi et al., (1974); Anh, (1980); Mengel and Reiser, (1999).

These models are complex, have evolved over a long period of time, and are the result of many experimental observations. There has been additional work on these and related systems to further refine the models, expand their utility, and explain specific situations. Good models are useful to us as tools of prediction and explanation. The "Felkin-Anh model" is effective for a wide range of examples, agrees with our understanding of these types of systems, and has proven to be useful in predicting the outcomes of many other transformations.

Using the Felkin–Anh model, we can explain the lithium aluminum hydride reduction of carbonyls. It is postulated that the lithium coordinates to the oxygen as the hydride is delivered by the aluminum hydride and from the appropriate face, as described above.

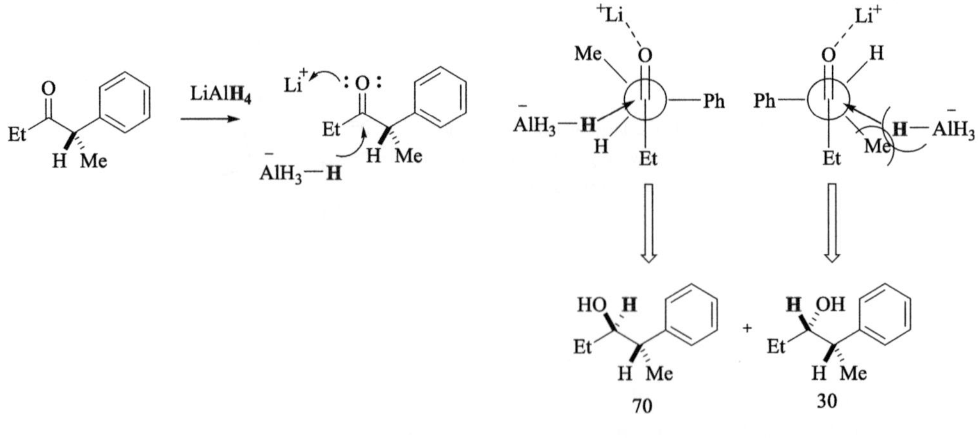

70   30

EXAMPLE 7.17. REDUCTION OF AN ACID WITH LITHIUM ALUMINUM HYDRIDE

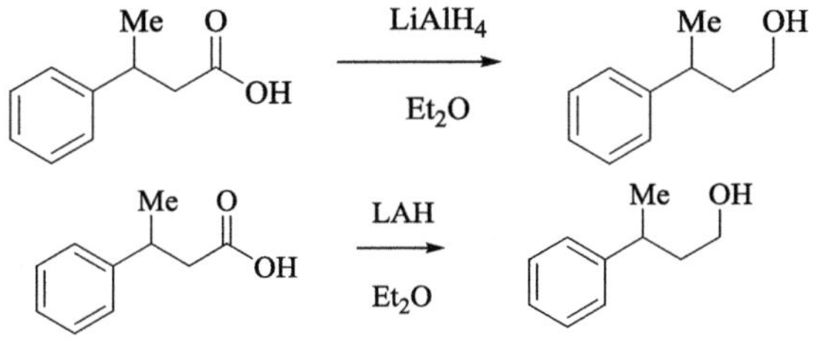

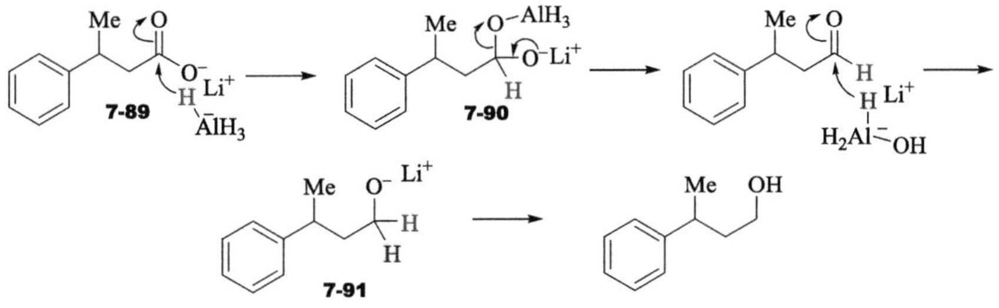

After the first equivalent of hydride deprotonates the acid, and forms hydrogen gas, reduction of the carboxylate (**7-89**) occurs forming aluminate **7-90**. The aluminum alkoxide is released and the resulting aldehyde reduced by another hydride, to deliver the alkoxide (**7-91**), which upon work-up produces the alcohol.

Sodium borohydride is a milder reductant, but its reduction of a carbonyl is influenced in the same way as LiAlH$_4$. NaBH$_4$ is not an effective reagent for the reduction of acids like LiAlH$_4$. To reduce esters, NaBH$_4$ often requires heat and extended reaction times to achieve adequate conversion. However, sodium borohydride is safer and the reaction work-up is less complicated, leading to its popularity.

Source: Kogure and Eliel, (1984).

EXAMPLE 7.18. AN EXAMPLE OF THE APPLICATION OF SODIUM BOROHYDRIDE AND THE MECHANISM FOR THIS TWO-STEP PROCESS

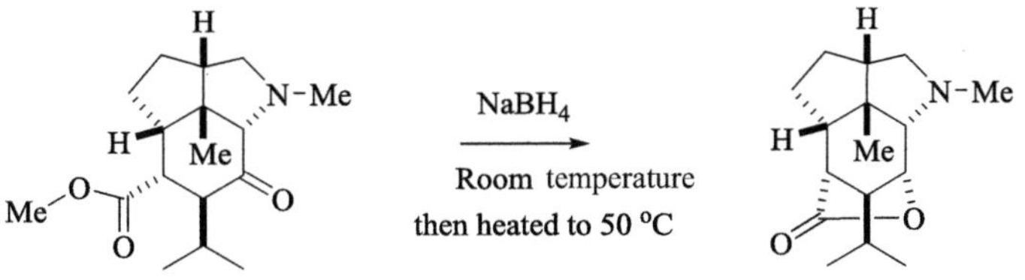

The ketone carbonyl is more susceptible to reduction than the ester, and as it complexes to the sodium ion, the hydride is delivered (**7-92**) to provide alkoxide **7-93**. As represented below, the bottom face of the molecule is more sterically congested, due to the concave shape and the adjacent methylamino moiety (**7-94**). Once reduced, the alkoxide is positioned well for the next step, and upon heating, the system cyclizes to form the product lactone (**7-95**).

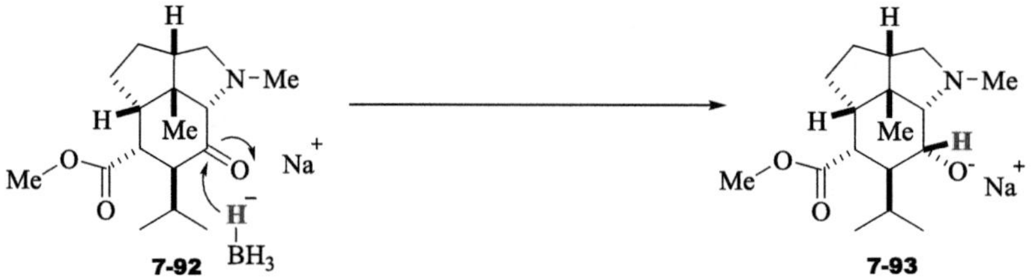

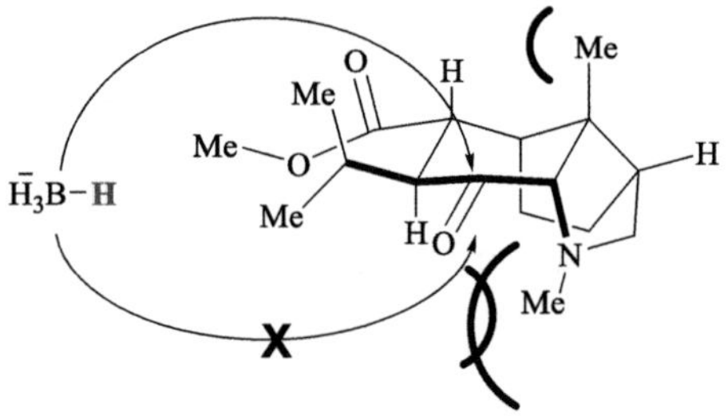

When the substrate possesses an α,β-unsaturated carbonyl, the reduction may occur in a 1,4-manner, leading to reduction of the olefin and not the carbonyl. Boranes reduce α,β-unsaturated ketones in a 1,2-fashion generating the enol as a result of the initial interaction between the Lewis acidic borane and the carbonyl oxygen. The example below demonstrates the mechanism of a hindered, and less reactive (more selective), borane working to reducing an α,β-unsaturated ketone.

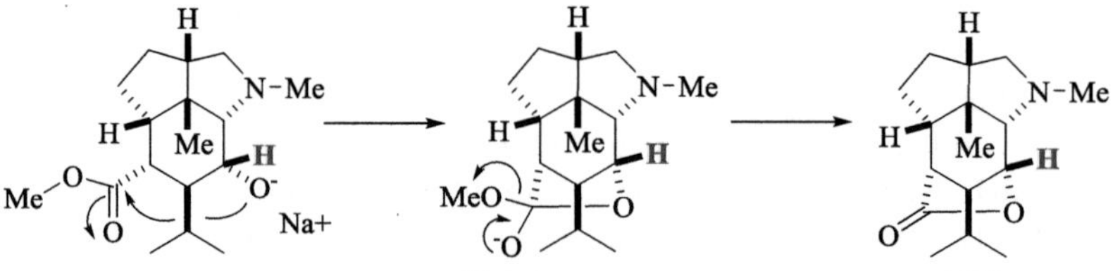

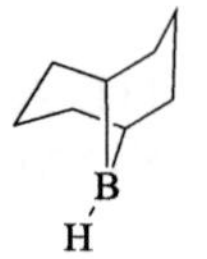

9-BBN (9-Borabicyclo[3.3.1]nonane)

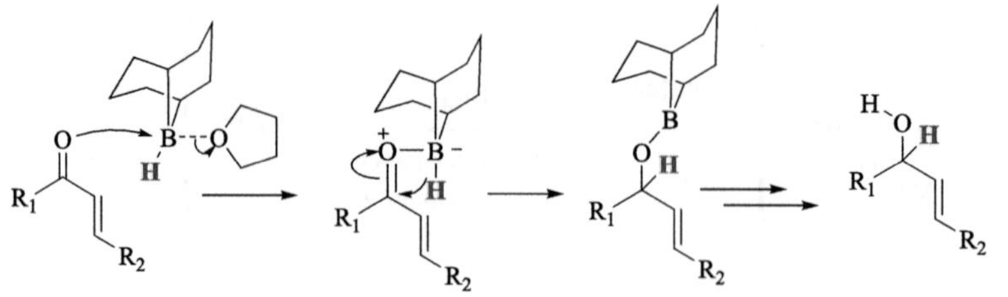

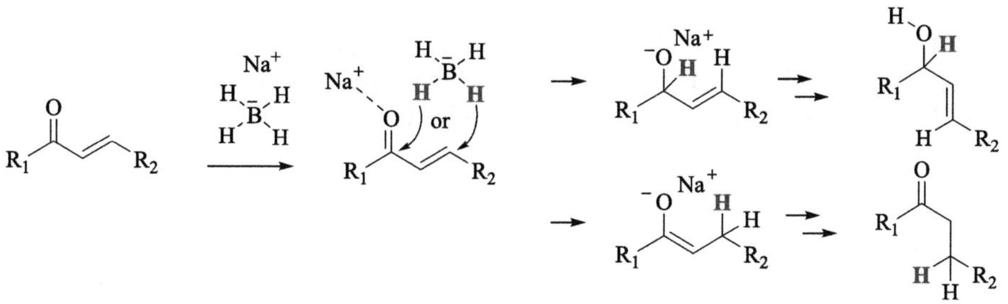

Complex metal hydrides (ionic metal hydrides), sodium borohydride for example, are charged and do not require the same kind of the initial interaction with the carbonyl that the Lewis acidic borane does. So sodium borohydride and lithium aluminum hydride will reduce the α,β-unsaturated carbonyl to produce a mixture of reduced products. Addition of cerium trichloride "Leuche conditions" to a sodium borohydride reduction of an (α,β)-unsaturated ketone, activates the carbonyl toward 1,2-reduction, leaving the olefin intact (Table 7.3).

It has been hypothesized that in the systems with methanol, the lanthanide actually activates the alcoholic solvent to be slightly more acidic, leading to a carbonyl that is more susceptible to 1,2-reduction without activating the olefin.

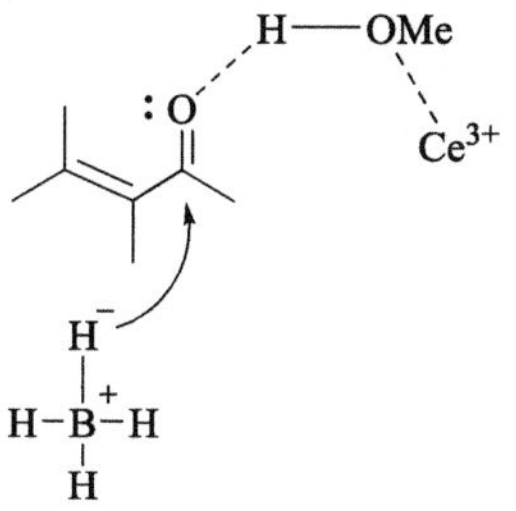

Sources: Kreis and Carreira, (2012); Krishnamurthy and Brown, (1977); Gemal and Luche, (1981).

TABLE 7.3   Regioselection in the Reduction of α,β-Unsaturated Carbonyls

| | | 1,2 | 1,4 | 1,2 : 1,4 Selectivity | |
| | | | | in MeOH | in MeOH with CeCl₃ |
		Reduction			
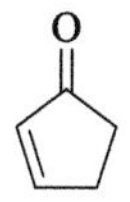	NaBH₄	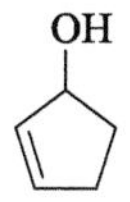	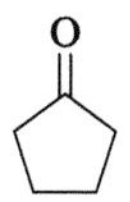	8 : 92	93 : 7
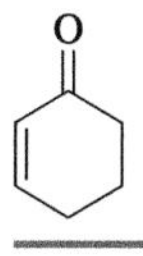		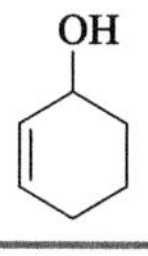	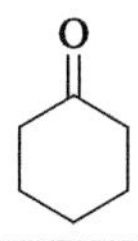	50 : 50	99 : 1

EXAMPLE 7.19. USE OF SODIUM BOROHYDRIDE WITH CERIUM TRICHLORIDE (LEUCHE CONDITIONS) IN A CLASSIC EXAMPLE WORKING TOWARD THE SYNTHESIS OF PROSTAGLANDIN

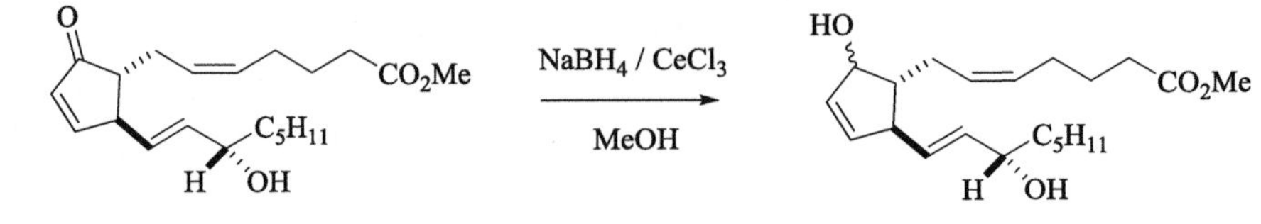

Source: Luche et al., (1978).

D. Reductive Aminations and Imines

The synthesis of new carbon−nitrogen single bonds, derived from the reaction of a carbonyl (ketone or aldehyde) and an amine (primary or secondary) followed by reduction, is called a reductive amination. In the first step of the transformation, the amine adds to the carbonyl to produce hemiaminal **7-96**. The hemiaminal intermediate dehydrates to form the iminium ion (or Schiff base) **7-97**. The process is facilitated by the use of slightly acidic conditions. The resulting iminium ion is then reduced via hydrogenation or by applying a hydride-based reductant, such as lithium aluminum hydride, sodium borohydride, or sodium triacetoxyborohydride Na(OAc)$_3$BH, to provide the amine (**7-98**). Although sodium triacetoxyborohydride will reduce ketones and aldehydes very slowly, it is an effective reagent for the reduction of iminium ions, and thus the reductant can be added directly to the reaction, before the iminium ion has been fully formed.

The source is Abdel-Magid et al., (1996).

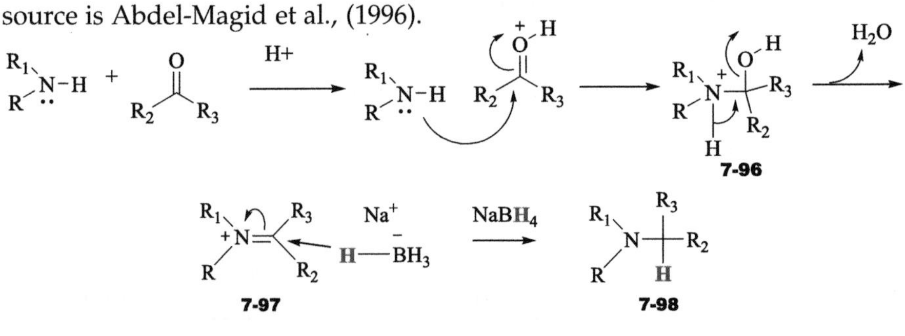

The source is Gomez et al., (2002).

PROBLEM 7.15

Show a possible mechanism for the following transformation.

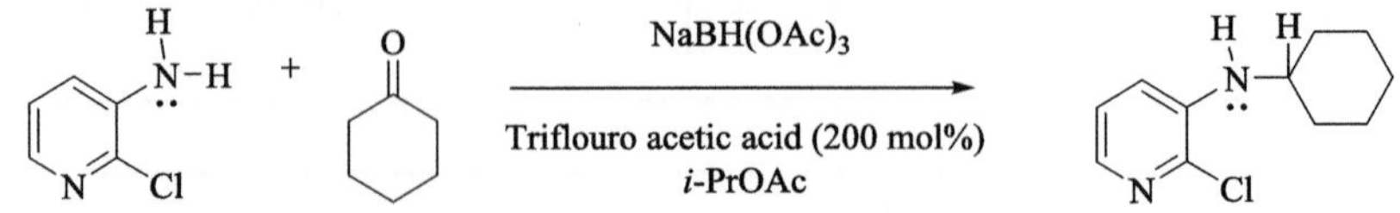

Source: McLaughlin et al., (2006).

E. Hydrogenation of an Olefin

In the presence of certain catalysts, hydrogen can be added across double and triple bonds with great efficiency. Because of the importance of such processes and the extreme conditions required to accomplish this transformation without catalysts, many techniques have been developed to facilitate the execution of this process.

Hydrogenations with Heterogeneous Catalysts

There are a variety of heterogeneous catalysts available to facilitate hydrogenations that tolerate a wide range of conditions and lead to varying selectivity's. The base metal catalysts, such as nickel, copper, and iron, were originally developed for this task, but were supplanted by more active catalysts derived from palladium, platinum, and ruthenium. Typical preparation of the catalysts requires a chemical treatment and the metal to be ground into a fine powder, so that the once shiny metal is black and has the consistency of dust. This provides the most surface area possible, and since the reaction is proposed to occur at the surface of the metal, this turns out to be a very significant factor in a heterogeneous process. The more reactive metals can promote hydrogenations at temperatures below 100 °C and at pressures near 1 atm. These reactions are often carried out in a reactor that can be pressurized and which is mounted in a mechanical rocker to allow for good mixing (an important factor in heterogeneous reactions) of the three phases (H_2 (*gas*), solvent/reactant (*liquid*), and catalyst (*solid*)).

In general, the reactivity trend of the metal catalysts follows the order nickel < rhodium and ruthenium < palladium < platinum. Hydrogenations are effected by other parameters, adding to the versatility of this process. For example, solvent can alter the activity of the catalytic systems. For example, acidic solvents, such as acetic acid or ethanol, are more active than ethyl acetate.

Other functional groups are susceptible to hydrogenation with the following trend in reactivity: acid chlorides, nitro > alkenes, alkynes > nitriles, aldehydes > ketones > imines, arenes. An examination of the energy of activation for hydrogenation reactions reveals the need for extreme conditions or a catalyst and sometimes both. Based upon calculations, 41 kcal/mol are required to convert the carbon—carbon double bond to a single bond and 103 kcal/mol to break the hydrogen—hydrogen bond in a molecule of dihydrogen. This 144 kcal/mol represents a large energy hurdle to overcome. In the end, the net gain in energy resulting from the creation of the two new carbon—hydrogen bonds leads to a favorable (30 kcal/mol) energy surplus. Thus, the process is energetically favorable; it just requires significant energy input or a catalyst to get over the initial (and quite considerable) energy barrier. The catalyst is not changing the relative energies of the starting materials and products; rather it is modifying the activation energy, which results in a change in the rate of the reaction.

C=C	100 kcal
C−C	59 kcal
H−H	103 kcal
C−H	87 kcal

1. 100 kcal/mol − 59 kcal/mol = 41 kcal/mol (the energy cost of going from a double to a single carbon−carbon bond).
2. 41 kcal/mol + 103 kcal/mol = 144 kcal/mol (the combined energy necessary for breaking the hydrogen molecule and taking the carbon−carbon bond to a single bond).
3. 2 × 87 kcal/mol = 174 kcal/mol (the energy payback for creating two new carbon−hydrogen bonds).

Energy cost	144 kcal
Energy payback	174 kcal

Energy surplus 30 kcal/mol *(the transformation is energetically favorable).*

The mechanism of heterogeneous metal catalyzed hydrogenation has been proposed to occur through the following sequence. Initially the hydrogen is adsorbed onto the exposed surface of the metal catalyst. In the second step, the substrate alkene comes into contact with the metal surface and becomes "activated." In the third step, the hydrogen atoms are transferred to the activated alkene, breaking the C−C double bond. Finally, the reduced product dissociates from the metal surface.

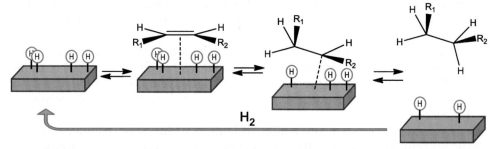

The initial hydrogen transfer step is reversible and can, in situations of incomplete reduction, promote a *cis*-to-*trans* isomerization. It has also been suggested that the interaction between the metal and both the substrate and the hydrogen merely leads to weakened bonds. It is thought that the weakened bonds and the close proximity facilitate a simultaneous transfer of hydrogen atoms as the hydrogen bond is broken. Both mechanisms are sufficient to explain the results seen in the many examples described in the literature.

The sources are Allen and Kiess, (1956); Brandt et al., (2008); Musolino et al., (2010).

PROBLEM 7.16

Why are such harsh conditions necessary for adding hydrogen across and olefin even when the reaction is energetically favorable? As part of the answer, draw a simple energy diagram with labels for the different points along the transformation and show how this could change if a catalyst were applied.

In general, the hydrogenation conditions can be selected to promote complete conversion with little or no reversion. It has been shown that under these conditions, the two hydrogen atoms are delivered from the same face of the alkene, leading to *syn* addition. The proposed syn addition of hydrogen has been supported by a series of experiments, including addition of hydrogen across butynes, leading to *cis*-butene derivatives and the reduction of 1,2-dimethylcyclohexene to yield *cis*-1,2-dimethylcyclohexane. See Example 7.20.

EXAMPLE 7.20. STEREO- AND FACIAL SELECTIVITY IN HYDROGENATION REACTIONS

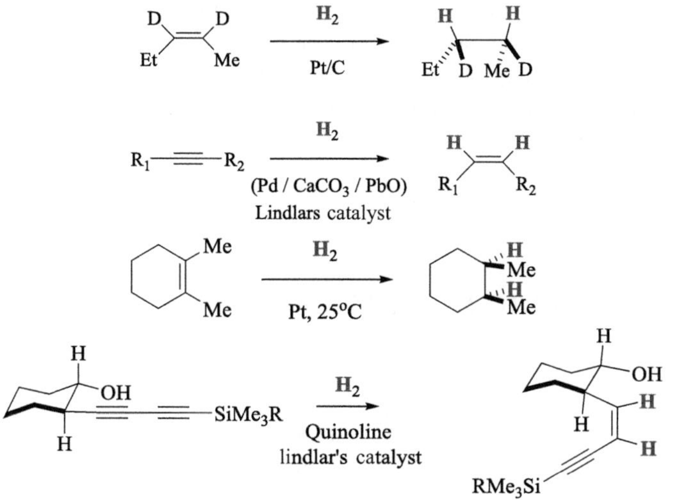

It has also been shown in the cases where the olefins are hindered or in rigid systems that addition occurs from the less hindered face (Example 7.21).

Sources: Cram and Allinger, (1956); Kalsi, (2005); Holmes et al., (1976).

EXAMPLE 7.21. FACIAL SELECTIVITY DUE TO THE STERIC FORCES IN A RIGID SYSTEM

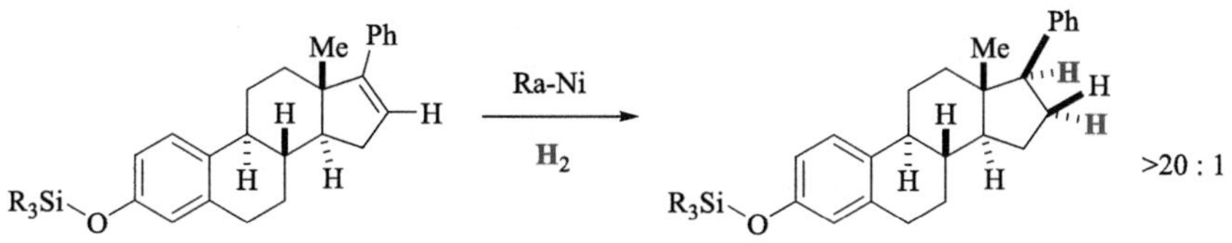

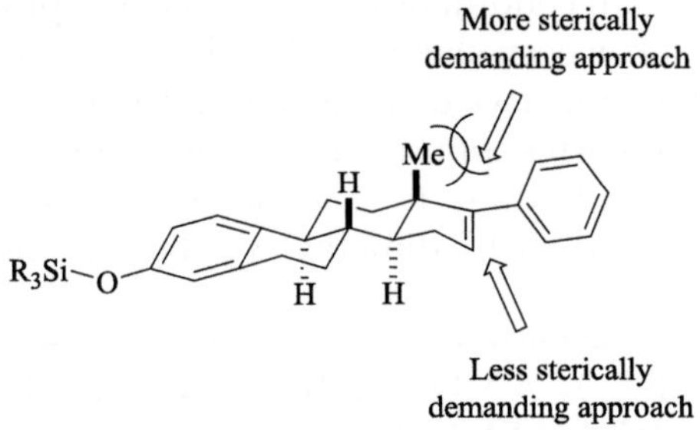

More sterically
demanding approach

Less sterically
demanding approach

Substitution around the olefin can have an impact on the reactivity of the double bond toward hydrogenation. In general, the more substituted the olefin, the less reactive it will be to hydrogenation. Substitution-based chemoselectivity between olefins within the same molecule is possible, but not commonly observed in systems with heterogeneous catalysts (Example 7.22). Notice the selectivity difference between the heterogeneous and homogeneous catalyst promoted transformation. Example 7.22b.

Sources: Thompson, (1971); Jun Shi et al., (2009).

EXAMPLE 7.22. STERIC INFLUENCES AND ACCESSIBILITY IN HYDROGENATION REACTIONS

The least substituted / most accessible olefin in the molecule

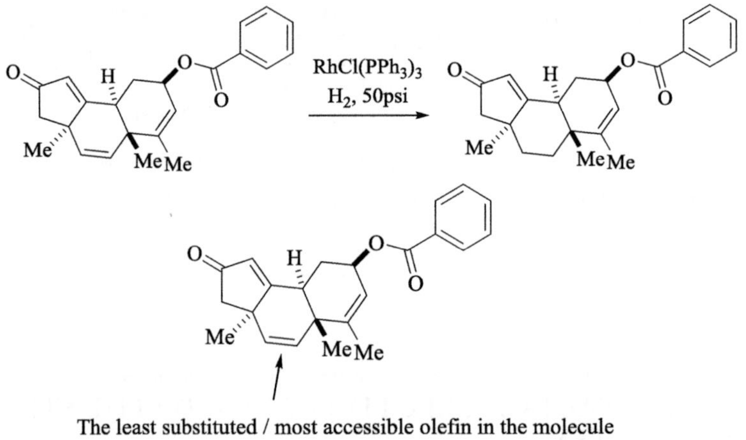

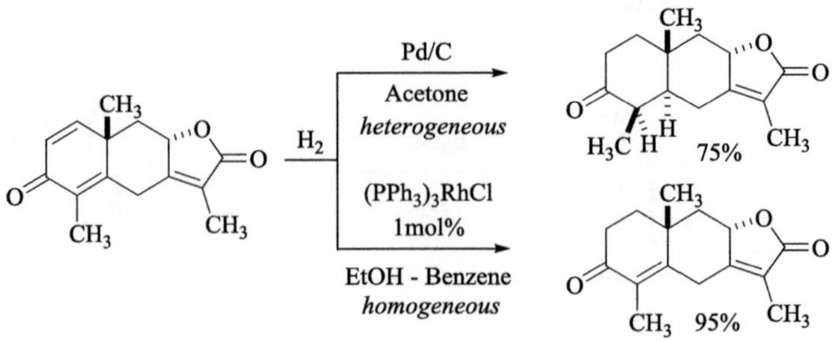

Selectivity differences seen in the heterogeneous and
homogeneous catalysis systems

The functional groups in the molecule can also affect the stereochemical outcome of the transformation. Hydrogenations are, in some cases, influenced by neighboring heteroatoms through a catalyst–substrate interaction that directs the hydrogenation of the olefin. Consider the effect of changing an ester group to a hydroxyl group on the ratio of hydrogenation products in Example 7.23.

Sources: Ward et al., (2000); Blay et al., (1996).

EXAMPLE 7.23. DIRECTIVE FORCES IN RIGID SYSTEMS (PRECOMPLEXATION VS STERICS)

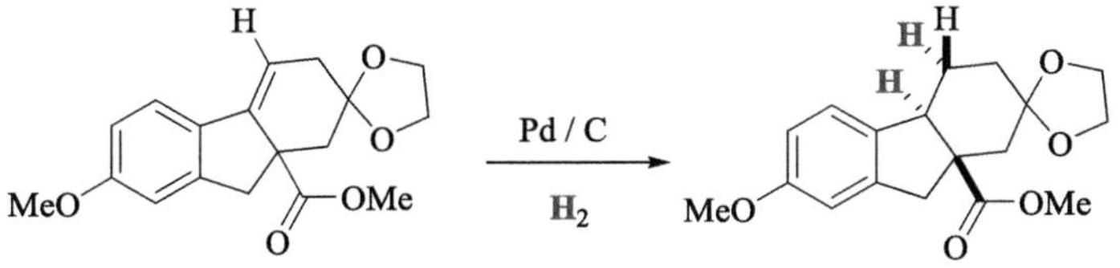

86% of the product has the *anti* relationship between the ester functionality and the newly added hydrogens

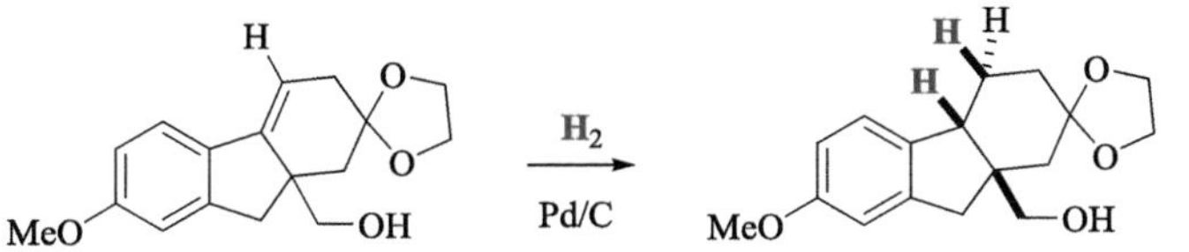

95% of the product has the *syn* relationship

Syn selectivity is the result of precoordination between the alcohol- and the hydrogen-impregnated catalyst.

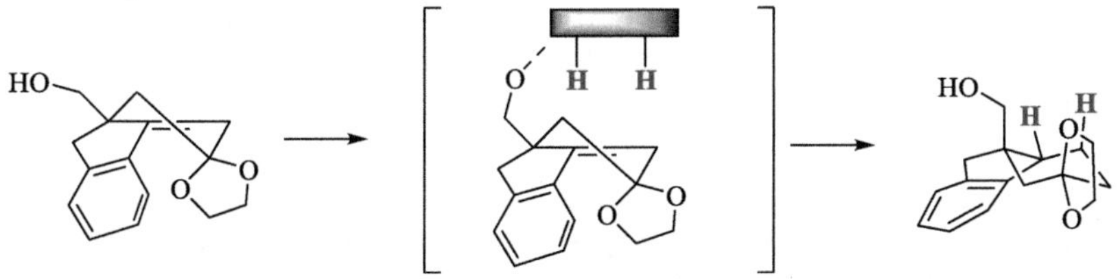

Dimethylformamide will compete for the same binding sites as the directing group and is presumed to block the coordination, leading to hydrogen addition from the less sterically demanding bottom face, further supporting the precoordination hypothesis. Conversely, when a less polar solvent is utilized, the result is a product from the addition of hydrogen to the same (top) face as the directing group.

Source: Thompson, (1971).

PROBLEM 7.17

Describe a mechanism for the following transformation.
The reductive amination reaction discussed earlier in the chapter can be accomplished through the application of metal-catalyzed hydrogenation instead of hydride reduction. Show the mechanism for the transformation below.

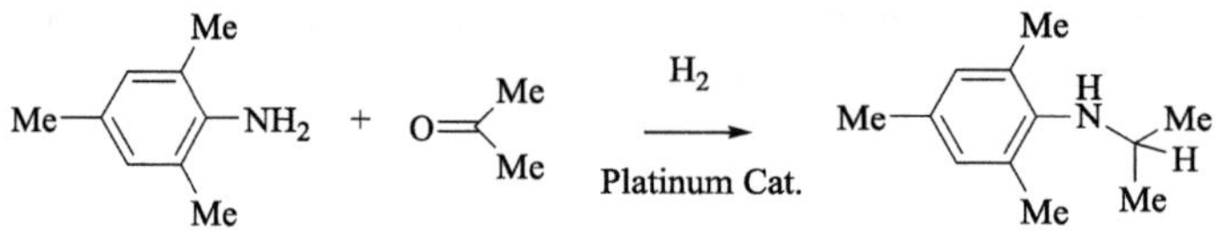

Source: Gomez et al., (2002).

Homogeneous Catalyst Processes

Homogeneous catalysis is a powerful tool for the hydrogenation of olefins. One of the earliest and most efficient catalysts for effecting this transformation is Wilkinson's catalyst, chlorotris-(triphenylphosphine) rhodium, $[(Ph_3P)_3RhCl]$. With this system, hydrogen addition occurs in a *syn* fashion, and isomerization is less prevalent. Wilkinson's catalyst is bulky and sterics play a role in the reactivity of the substrate. In general, *cis*-olefins are reduced faster than *trans*-olefins and terminal olefins are reduced faster than more substituted systems. Trisubstituted and tetrasubstituted olefins tend to be much less reactive than mono- and disubstituted olefins, although trisubstituted cycloalkenes are still suitable substrates. The steric demands of the substrate can also play a significant role in the stereochemical outcome of the hydrogenation as illustrated in Example 7.24. Here, the bias for approach from the less demanding top face of the system is exaggerated when the bottom face is hindered, as in the case of the methyl analog (**Example 7.24a**).

EXAMPLE 7.24. THE IMPACT OF STERICS UPON A HOMOGENEOUS CATALYSIS HYDROGENATION IN A RIGID SYSTEM

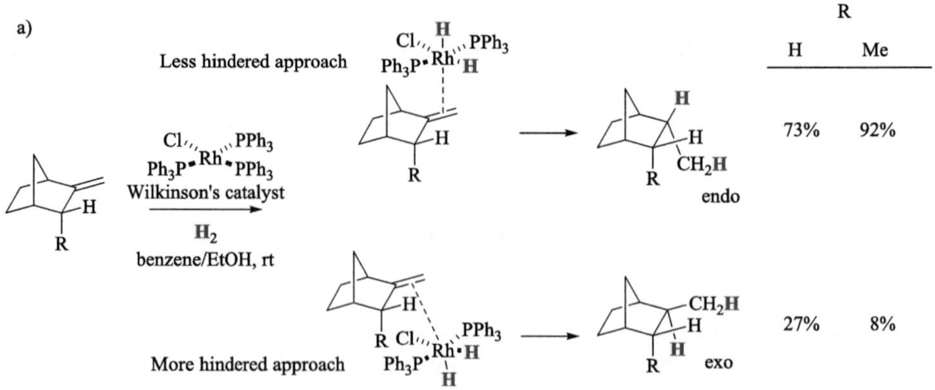

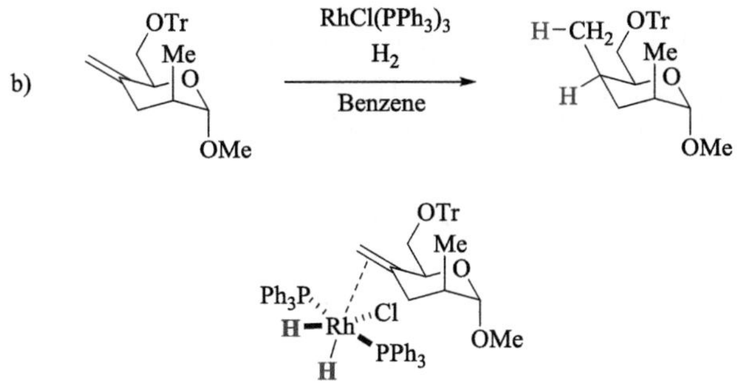

b)

Attack from the less sterically demanding face of the molecule

The catalytic process is postulated to proceed through the cycle described below. Initially, the catalyst dissolves and one of the triphenylphosphine ligands is replaced with a molecule of solvent (**7-99**), which dissociates when the Rh complex undergoes oxidative addition of hydrogen (**7-100**). After olefin coordination, the metal inserts into the double bond (**7-101**) and is reduced through migratory insertion, to form **7-102**, followed by reductive elimination, to release an alkane (**7-103**) and regenerate the catalyst to continue the cycle.

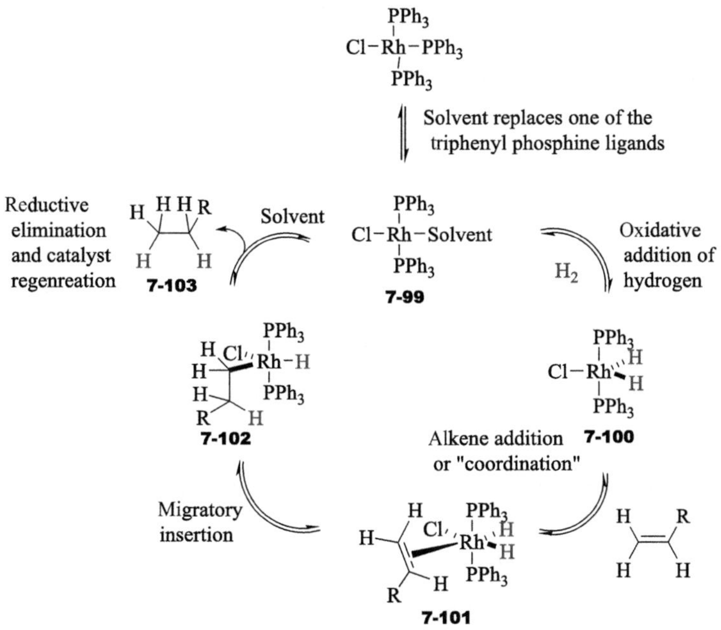

Crabtree's catalyst ([Ir(cod)py(PCy$_3$)]PF$_6$) is a more active homogenous hydrogenation catalyst (at least 100 times more active than Wilkinson's catalyst). It will catalyze the hydrogenation of tetrasubstituted alkenes and has a faster turnover rate than Wilkinson's catalyst. In addition to performing complimentary transformations, Crabtree's catalyst has also been employed to perform "directed hydrogenations" where functionality on the substrate directs the hydrogenation onto a specific face of the alkene (**Example 7.25**). This is thought to be similar to the process described previously for heterogeneous catalysis.

Sources: Jardine, (1981); Sum and Weiler, (1978).

EXAMPLE 7.25. THE EFFECT OF PRECOMPLEXATION ON HYDROGENATION OF OLEFINS IN RIGID SYSTEMS WITH THE USE OF THE CRABTREE CATALYST

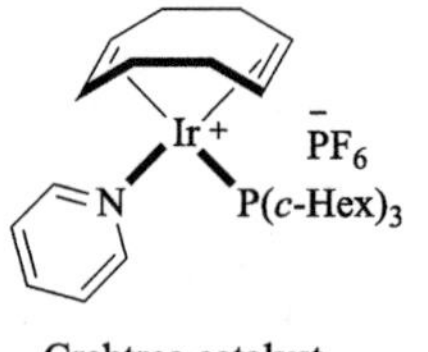

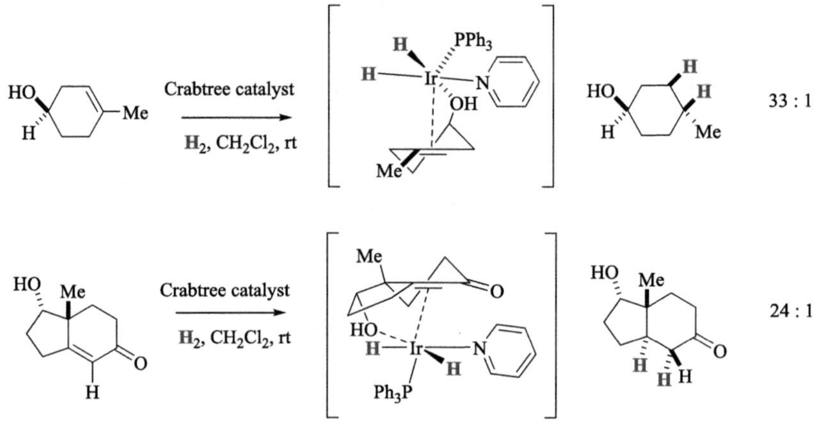

Crabtree catalyst

33 : 1

24 : 1

Sources: Stork and Kahne, (1983); Hoveyda et al., (1993).

PROBLEM 7.18

Show the product and the key intermediate that helps explain the outcome for each of the following transformations.

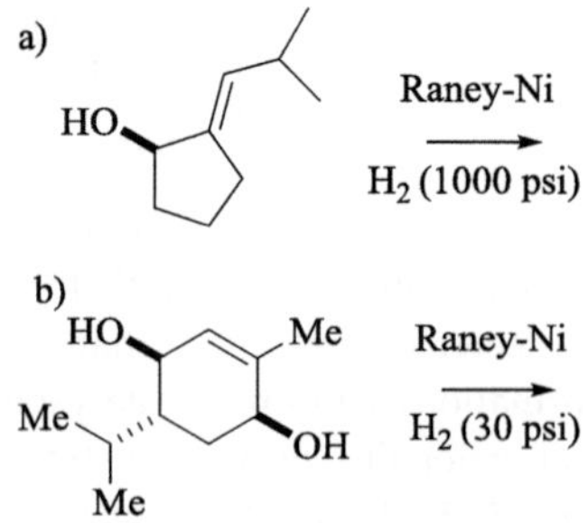

a)

Raney-Ni

H_2 (1000 psi)

b)

Raney-Ni

H_2 (30 psi)

c) 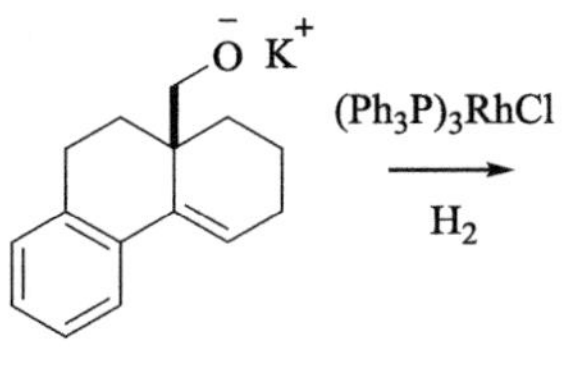 (Ph₃P)₃RhCl, H₂

Sources: Sehgal et al., (1975); Stolow and Sachdev, (1971); Hoveyda et al., (1993).

Transfer Hydrogenations

Diimide (**7-105**) can also affect the reduction of an alkene or alkyne. This process is referred to as a "transfer hydrogenation." Although unstable, diimide can be generated in situ by the decarboxylation of potassium azodicarboxylate **7-104**. The addition of hydrogen by diimide to an olefin is a concerted process and gives *syn* addition of hydrogen and liberates nitrogen gas.

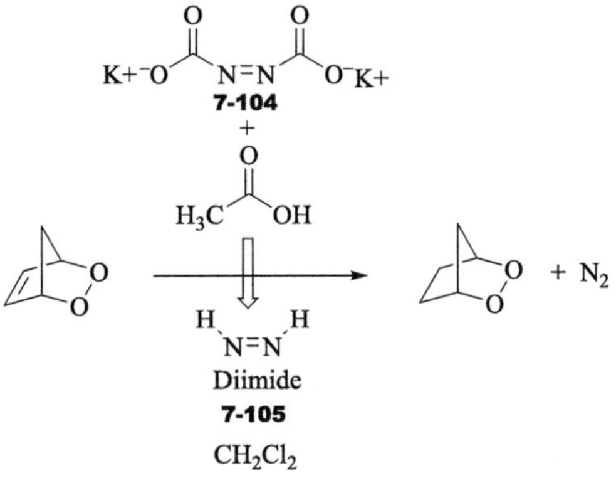

The formation of a gas provides a strong driving force rendering the transformation irreversible. The proposed cyclic transition state has been supported by kinetic studies of a mixture of *cis* and *trans* diimide in the presence of reactive olefins. The diimide is generated as a mixture of *cis*- and *trans*-isomers, but only the *cis*-isomer reacts and isomerization is not observed, further reinforcing the postulated concerted mechanism.

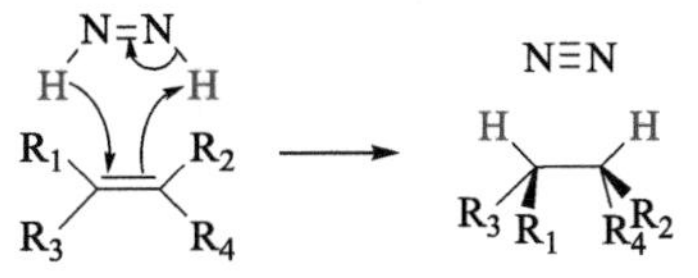

The source is Back, (1984).

Diimide is selective for the less sterically demanding olefin (**Example 7.26**). However, the reactivity of diimide prevents selective reduction of an alkyne to an alkene, leading instead to the alkane as the predominant product.

EXAMPLE 7.26. SELECTIVITY IN DIIMIDE HYDROGEN TRANSFER REACTIONS

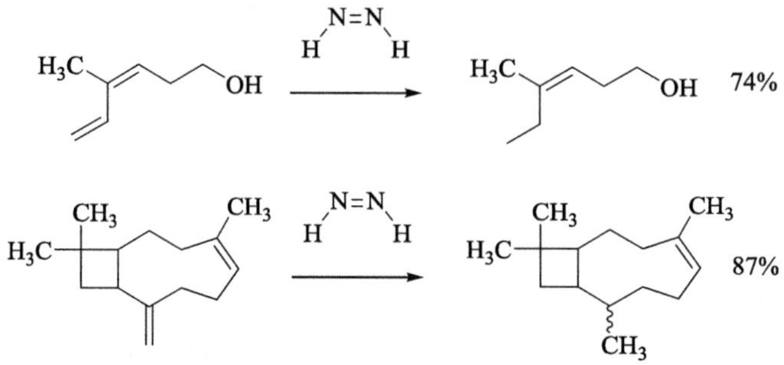

It has also been shown that diimide reductions of olefins can be substrate directed in a similar fashion to metal-catalyzed hydrogenations (Example 7.27).

Sources: Mori et al., (1972); Rao and Devaprabhakara, (1978); Pasto and Taylor, (2004).

EXAMPLE 7.27. HYDROGEN TRANSFER WITH DIIMIDE IS ALSO INFLUENCED BY PRECOMPLEXATION

Hydrogen transfer occurs form the same side as the alcohol tether.

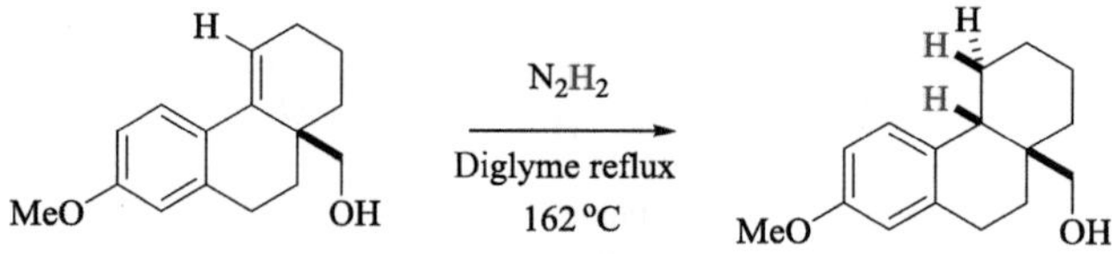

Source: Thompson and McPherson, (1977).

EXAMPLE 7.28. HYDROGEN TRANSFER WITH DIIMIDE GENERATED FROM P-NITROBENZENESULFONYLHYDRAZIDE

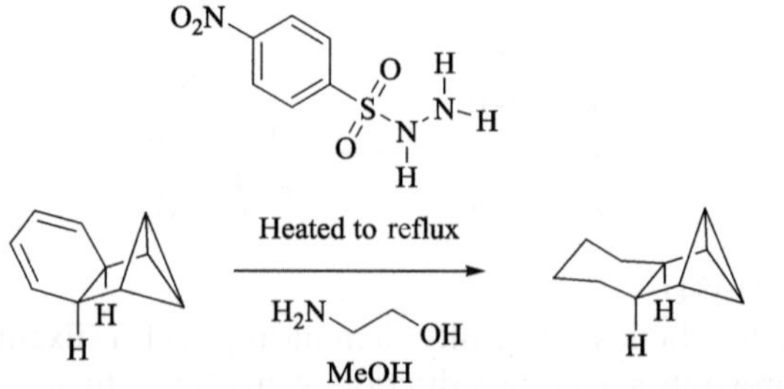

In this example, the diimide is generated by heating the *p*-nitrobenzenesulfonylhydrazide (**7-106**) in a mixture of methanol and ethanolamine. Because the diimide (**7-105**) is not stable under these conditions, the substrate to be hydrogenated is mixed with the diimide precursor before the system is heated, to "trap" the active diimide as it is generated.

Diimide generation

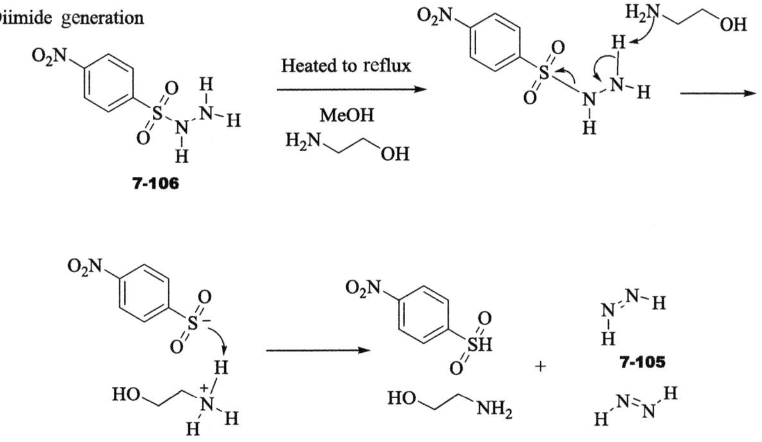

7-106

Heated to reflux

MeOH

7-105

Diimid hydrogenation

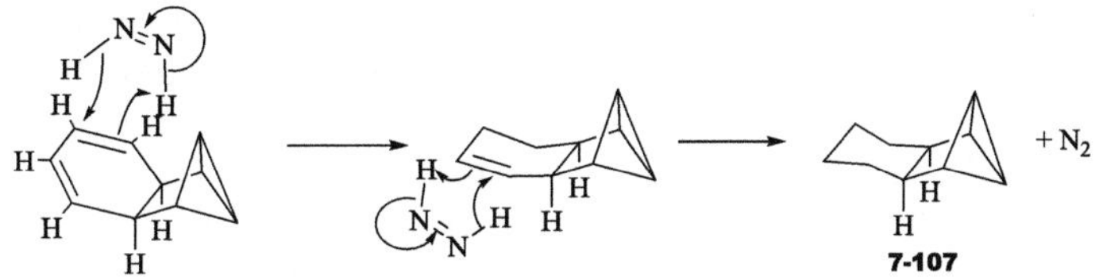

7-107

The diimide transfers hydrogen to both olefins, providing the alkane product (**7-107**).

Source: Pasto and Taylor, (2004).

Oxidations and reductions are very common transformations in organic chemistry but are often considered less significant than the structure-building processes associated with carbon—carbon bond forming reactions. In any case, these oxidation/reduction processes constitute a significant aspect of many synthetic sequences and represent a real opportunity for continued development and improvement in the field of synthetic organic chemistry. As quoted in a survey text of process chemistry:

> "Oxidation of secondary alcohols to ketones can be separated into three categories: those that are dangerous, those that generate large volumes of often hazardous by-products, or those that are dangerous **and** generate large volumes of often hazardous by-products. When confronted with the scale-up of an alcohol oxidation it is advisable to invest time in redesigning the route to avoid oxidation."

The source is Harrington, (2011).

There is room for improvement and through a better understanding of the existing technologies, we can confront the new obstacles that avail themselves, and perhaps a new generation of reductions and oxidations will be developed to meet these future challenges.

ANSWERS TO PROBLEMS

Problem 7.1

The oxidation numbers are solved as described in the model.

(A)

0 + 2 + 0 = +2

(B)

0 + 0 + 0 = 0

(C)

0 + 1 + 1 + 1 = +3

(D)

0 + 1 + 2 = +3

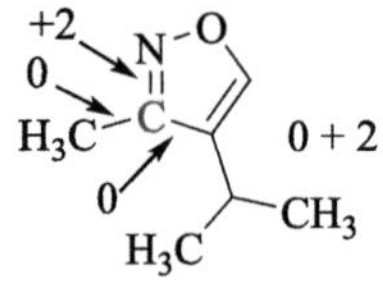

The following examples show how the model can be applied to hetero atoms. The rules are nearly identical as described for carbon. In the case of a heteroatom, it is important to remember that if, for example, you are solving for a nitrogen that is bonded to another nitrogen atom, that bond gets a "score" of 0 (we assume the bond is not polarized). If we are solving for a heteroatom (nitrogen or oxygen for example) and it is bonded to a carbon (both nitrogen and oxygen are more electronegative than carbon (see Table 1.2)) and that would receive a score of −1. Sulfur could be considered a close comparator to carbon, but in this book and as a general guideline, it makes more sense to consider it to be more electronegative relative to carbon and further, to consider it in the same way, oxygen is considered relative to carbon.

(E)

−1 + 0 = −1

(F)

0 + 0 −1 −1 = −2

(G)

−1 + 1 + 2 = +2

(H)

+1 + 1 + 1 −1 −1 = +1

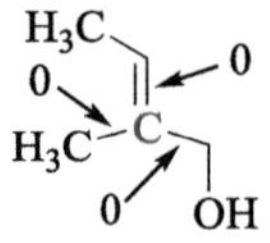

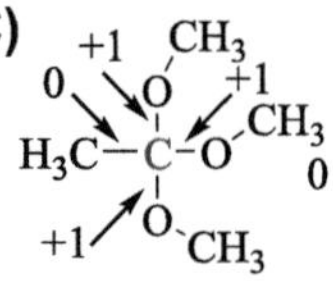

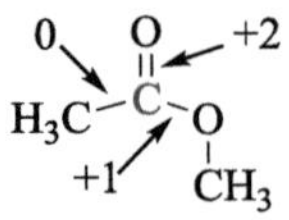

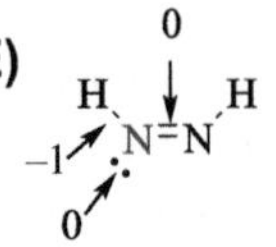

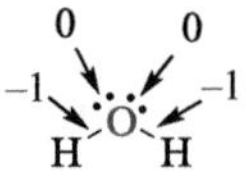

Problem 7.2

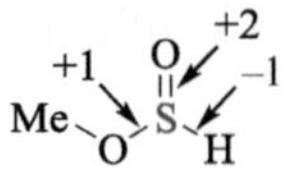

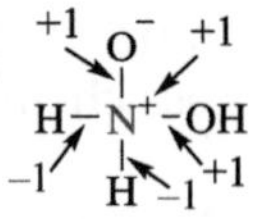

For the following example, the ketone carbon is converted to an alcohol. The ketone carbon has an oxidation number of +2. The resulting alcohol has an oxidation number of 0, leading to a net change of −2. A negative number represents a reduction (reduction in the oxidation number or reduction in oxidation state) for that atom.

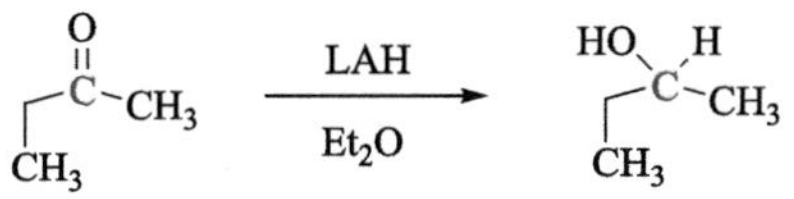

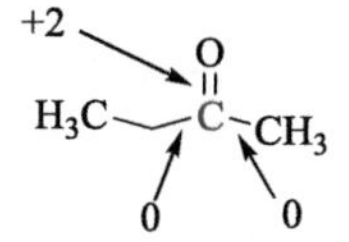

= +2

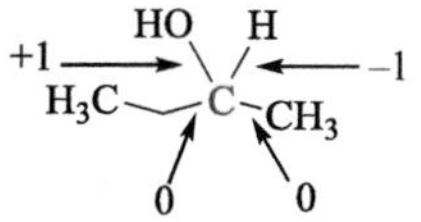

+1 —→ HO H ←— -1 = 0
      H₃C——C–CH₃
         ⁄   ⁄   ⁄ ⁄
        0   0
                        = -2   The net change is -2 so the
                                red carbon was reduced

Problem 7.3

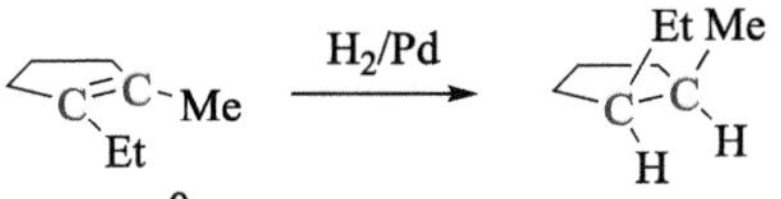

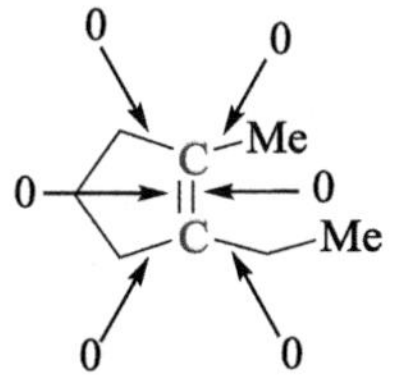

Carbon #1 = 0

Carbon #2 = 0

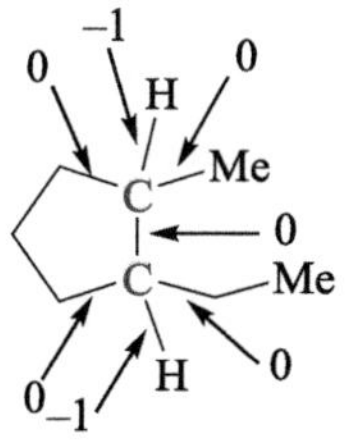

Carbon #1 = -1

Carbon #2 = -1

Carbon #1   0 → -1 = Net change -1 (Reduced)

Carbon #2   0 → -1 = Net change -1 (Reduced)

Both carbons of the original double bond were reduced and it can also be said that the olefin has been reduced.

Problem 7.4

a)

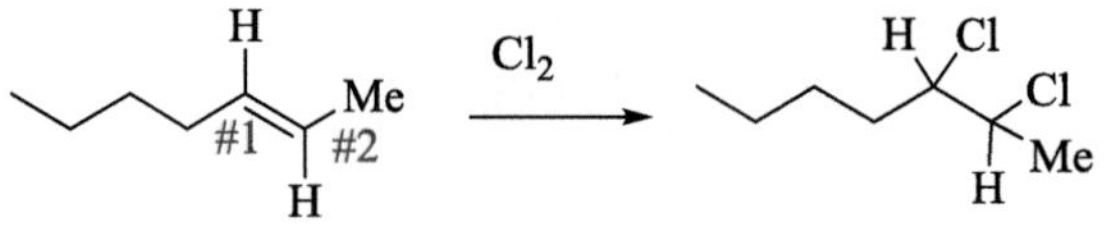

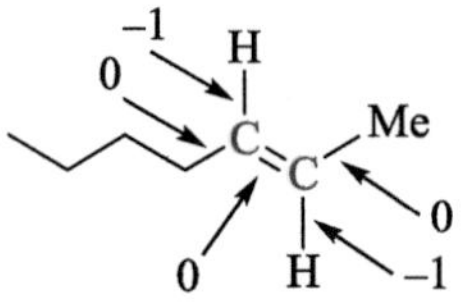

Carbon #1 = −1

Carbon #2 = −1

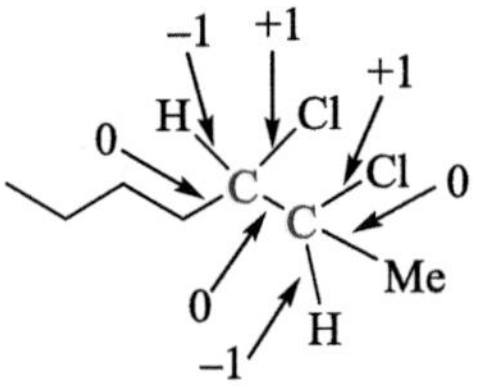

Carbon #1 = 0

Carbon #2 = 0

Carbon #1 −1 ⟶ 0 Net change of +1 (Oxidized)

Carbon #2 −1 ⟶ 0 Net change of +1 (Oxidized)

Both carbons of the original double bond were oxidized and it can also be said that the olefin has been oxidized.

b)

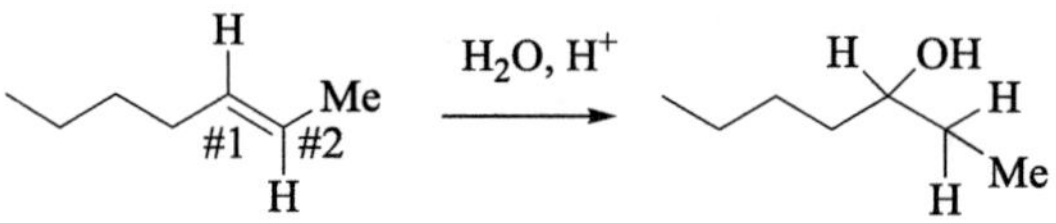

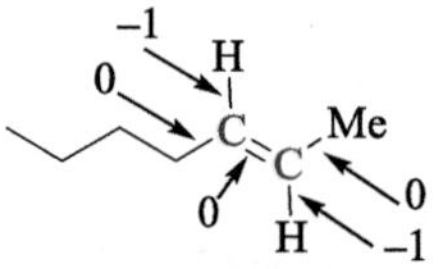

Carbon #1 = −1

Carbon #2 = −1

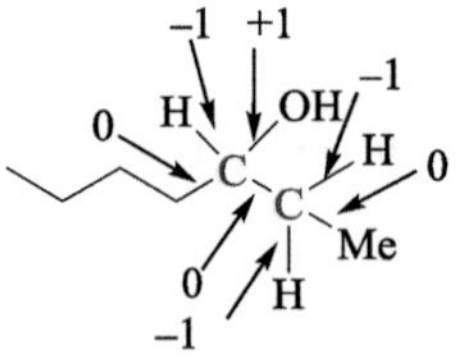

Carbon #1 = 0

Carbon #2 = −2

Carbon #1 −1 ⟶ 0 Net change of +1 (Oxidized)

Carbon #2 −1 ⟶ −2 Net change of −1 (Reduced)

Looking at each atom individually, the carbon–carbon double bond has been transformed into a single bond and carbon #1 has been converted to the alcohol and carbon #2 is now a methylene. Carbon #1 has been oxidized and carbon #2 has been reduced. The olefin functionality overall has not been reduced or oxidized. This is a hydrolysis and it demonstrates that our accounting scheme allows for cases where the oxidation state of individual atoms that are part of the functionality can change oxidation state, with no net modification to the oxidation state of the molecule as a whole. Even though the overall oxidation state of the molecule has not been altered, the functionality has been changed at the atom level and our tool tracks this. The differentiation allows us to better understand subsequent chemistry opportunities at each atom as a result of the transformation.

Problem 7.5

In the initial step of this mechanism, the permanganate reacts with the olefin in a concerted manner (7-108) to form the manganate ester (7-109), which is hydrolyzed to form the diol (7-110). The diol is then further oxidized by the periodinate to form the iodinate ester (7-111), which simultaneously eliminates as it cleaves the carbon–carbon bond to form the aldehyde.

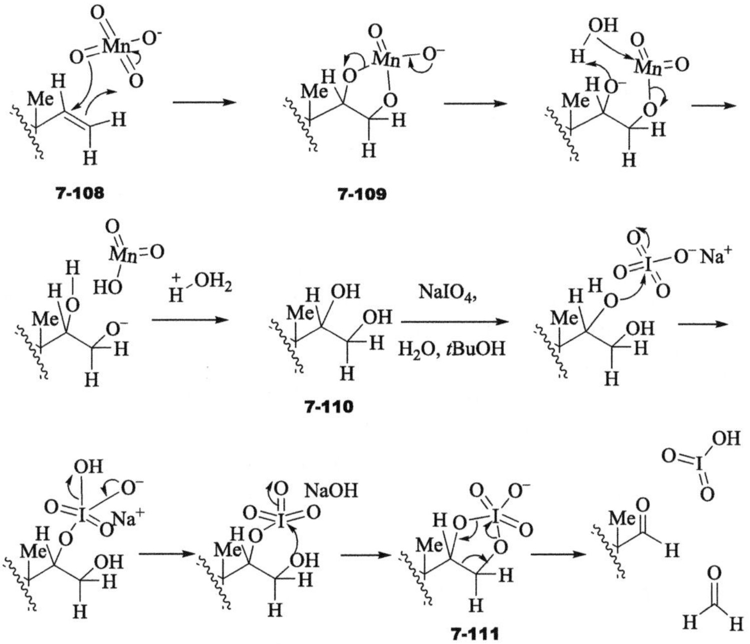

The aldehyde is then further oxidized to the acid through the action of the permanganate (manganese tetroxide) leading to the observed acid product.

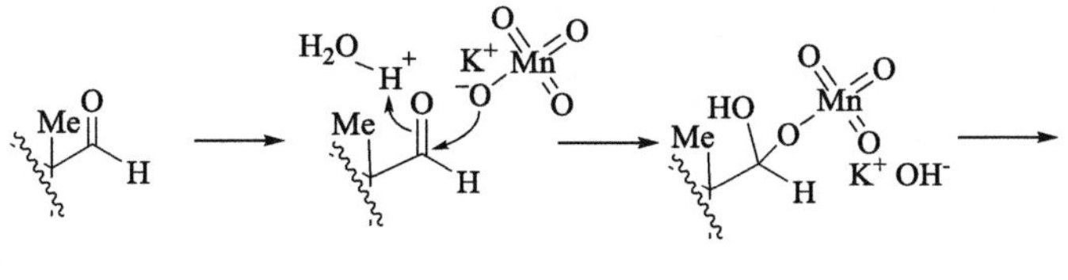

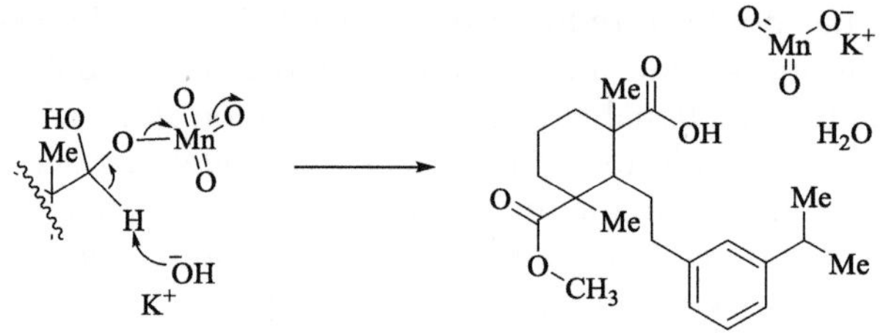

In this process, the aldehyde carbon has gone from an oxidation number of $+1$ to $+3$ for the acid, a change of $+2$ (it was oxidized). The manganese has gone from a $+7$ oxidation state to $+5$, a change of -2... they balance. Often, the mechanism for the $KMnO_4$ oxidation of an aldehyde will show the final form of manganese to be MnO_4 in the $+6$ oxidation state (not the $+5$ for MnO_3 as shown above). This is accomplished through the interaction of two (MnO_3^{-1})'s coming together to create one MnO_4^{-2} and one MnO_2.

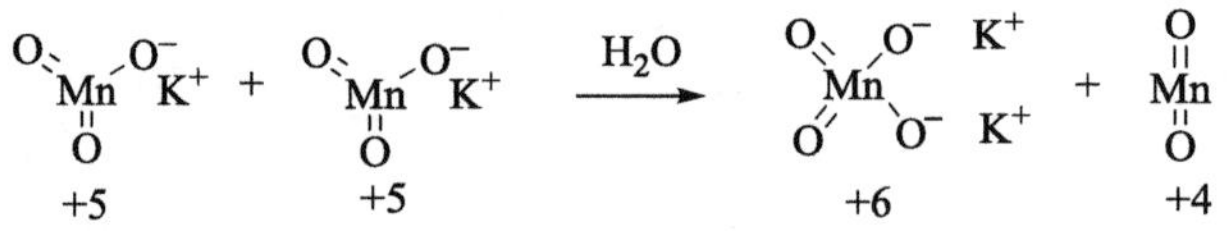

Problem 7.6

The following is a proposed mechanism for the transformation of the aldehyde to the corresponding acid. This represents a common result when affecting the oxidation of a primary alcohol using Jones conditions $(CrO_3/H_2SO_4/H_2O)$. This "over oxidation" of aldehydes that proceeds through the acetal or hydrated aldehyde, can be avoided by (1) using reagents such as PCC and PDC; (2) removing any water in the system (using molecular sieves as a drying agent); or (3) removing the aldehyde by distillation as it is produced.

In this process, the acetal interacts with the chromium to form the chromate ester. Loss of the proton and elimination of the chromium leads to the oxidized product.

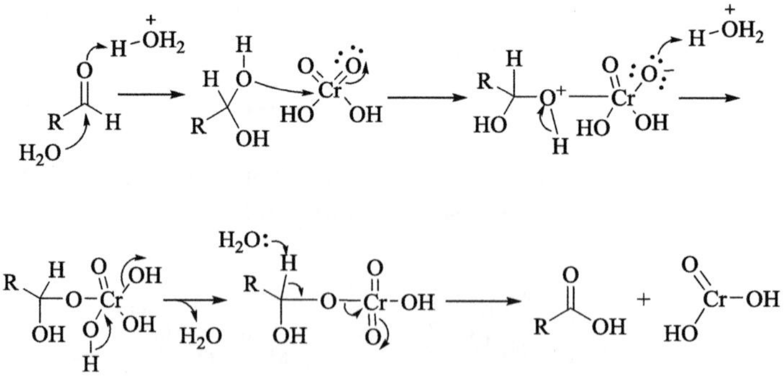

Problem 7.7

(a)

Part 1

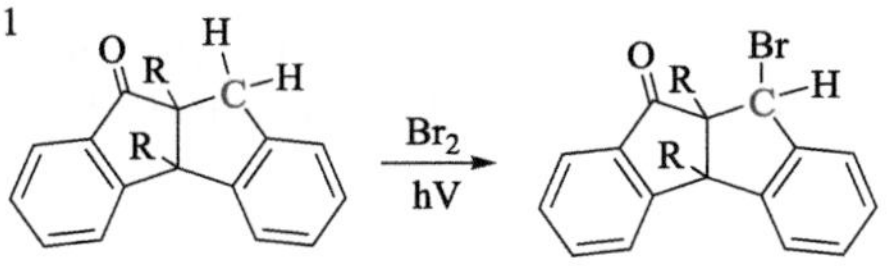

$$\text{Br}_2 \xrightarrow{h\text{V}}$$

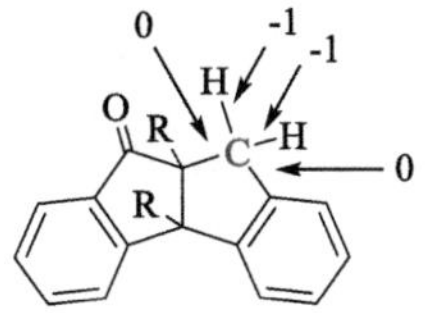

$$0 + (-1) + (-1) + 0 = -2$$

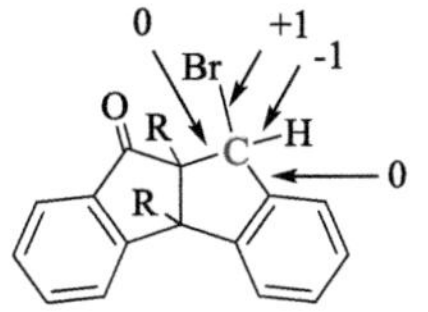

$$0 + 1 + (-1) + 0 = 0$$

There is a net change of +2, thus the red carbon has been oxidized. A halogenation of an alkane is an oxidation, and that carbon is then considered to be at the corresponding "alcohol oxidation state."

Part 2

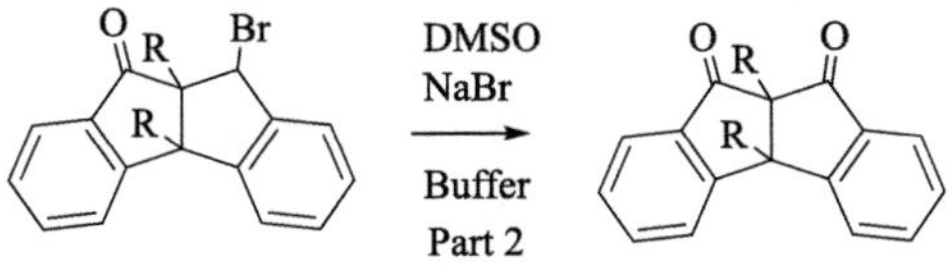

DMSO
NaBr
Buffer
Part 2

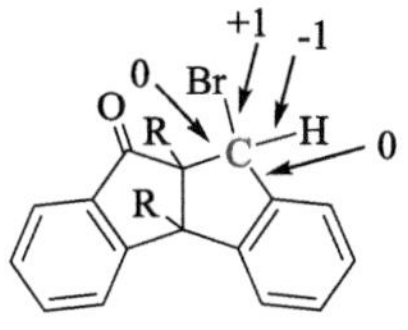

$$0 + 1 + (-1) + 0 = \mathbf{0}$$

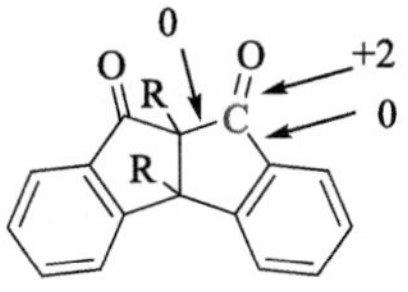

$$0 + 2 + 0 = \mathbf{2}$$

The change at the alkyl bromide is +2, the red carbon was oxidized to the carbonyl. This is a Kornblum oxidation and through this process, it is possible to convert halogenated or tosylated alkanes to the corresponding carbonyl.

The sources are Kornblum et al., (1959); Kornblum et al., (1957).

(b) Mechanism for the Oxidation

The DMSO oxygen adds in and displaces the bromide (**7-112**). As discussed in Chapter 1, the bromide is benzylic, making it fairly susceptible to substitution reactions. The bromite then assists in the loss of the proton from the sulfonium ion, to form the ylide (**7-113**), setting up the intramolecular deprotonation–elimination step leading to the ketone.

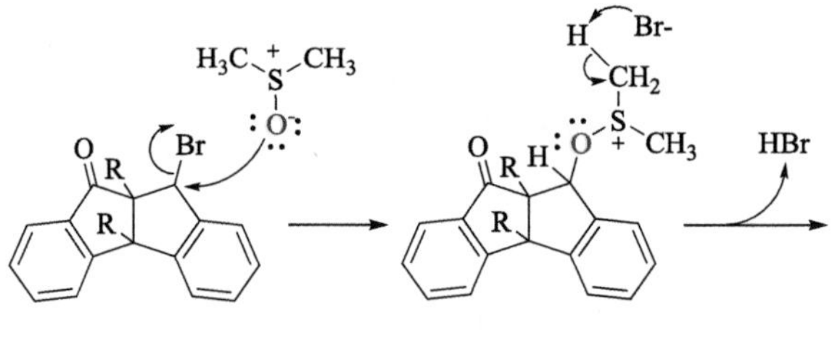

7-112

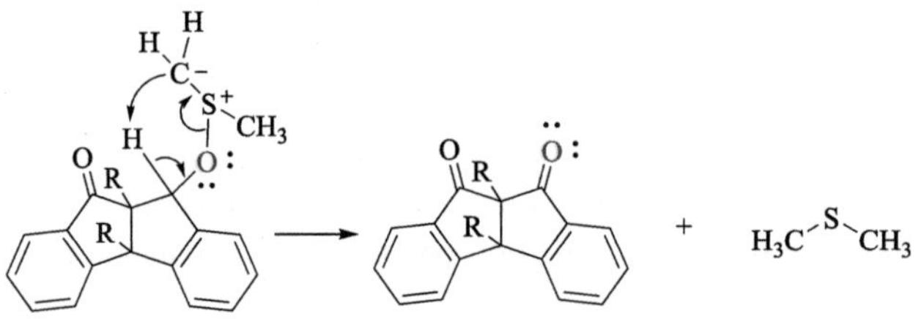

7-113

As expected, the order of reactivity for halogens in this process is I > Br > Cl. In cases where the chloro or bromo compounds are not reactive enough for the initial step, it has been found that the addition of iodine to promote a halogen exchange, via the Finkelstein reaction, has led to improved results.

The source is Dave et al., (1986).

Problem 7.8

Similar to the process that occurs under acidic conditions, it has been shown that primary and less hindered alcohols are more rapidly oxidized under basic conditions. There is little or no difference in the rates of oxidation for primary or secondary alcohols under acidic conditions.

Initially, the alcohol adds into the oxoammonium cation to generate the ylide. This will then reorganize through the five-membered transition state (**7-114**) to deliver the carbonyl and hydroxyl amine **7-115**. The tight five-membered transition state is more susceptible to steric demands of the substrate, leading to the difference in reaction rates for primary alcohols as compared to secondary alcohols under basic conditions.

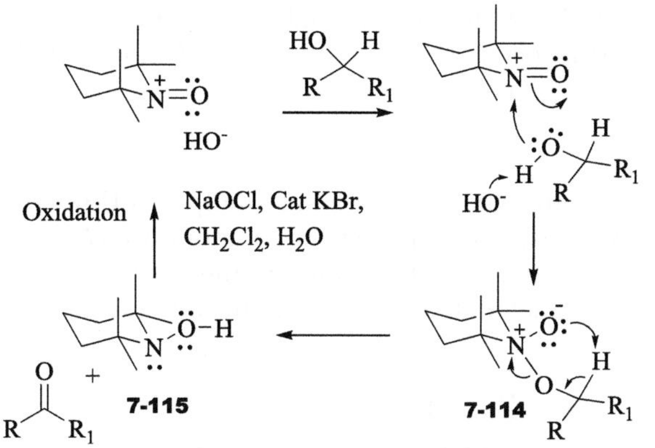

For the TEMPO oxidation under basic conditions, it is also a reasonable to show the more "linear" mechanism, with the only real difference being the intermolecular proton removal (7-114a).

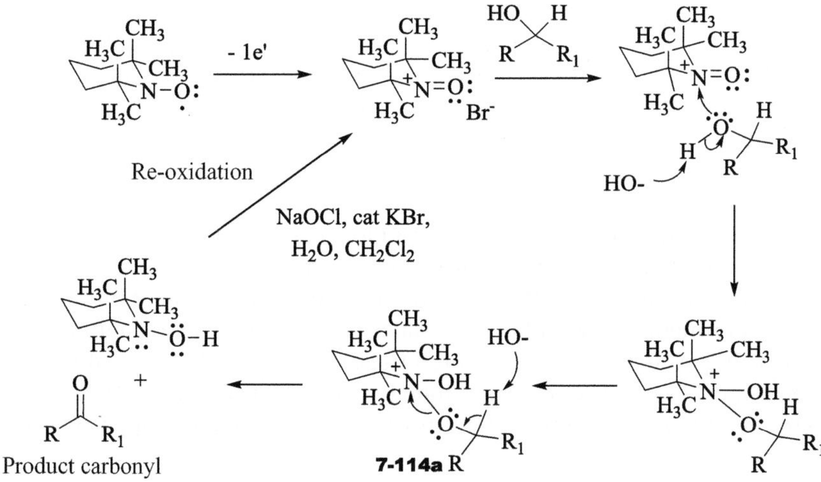

Problem 7.9

The nitroxide radical (7-116) looks very similar to TEMPO and would seem to be just as stable. When evaluating the stability of a molecule, it is necessary to consider the compatibility of the molecule in a bimolecular sense (will it react with another equivalent of itself).

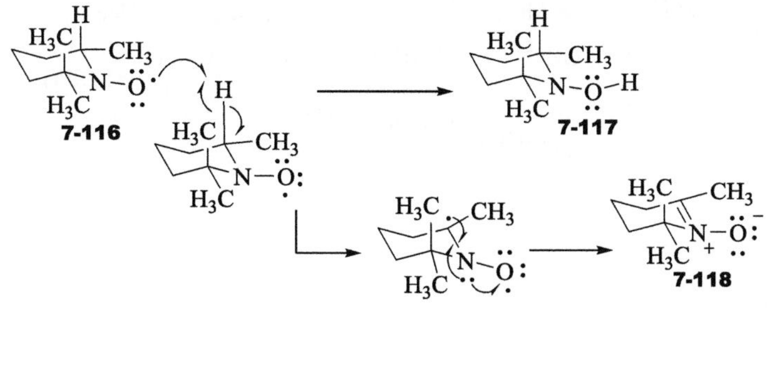

The alpha hydrogen in this system is susceptible to abstraction by another molecule of the nitroxide radical. This leads to a molecule of the hydroxyl amine **7-117** and imine oxide **7-118**.

Problem 7.10

This problem would seem to be related to the example in **Problem 7.9**. Again there is an alpha proton that can be abstracted. But, bicyclic compound **7-120** is stable, while compound **7-116**, as we discussed before, is not. Even though the hydrogens adjacent to the nitroxide radical in **7-120** can be abstracted, that process is unproductive.

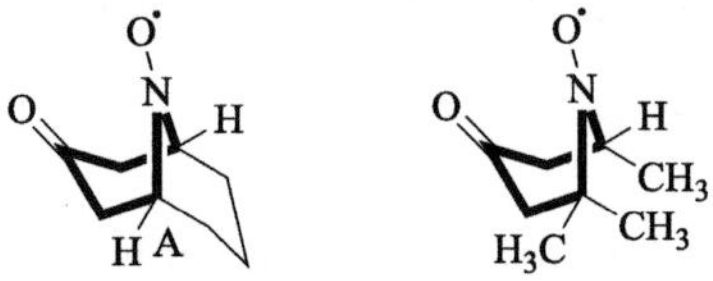

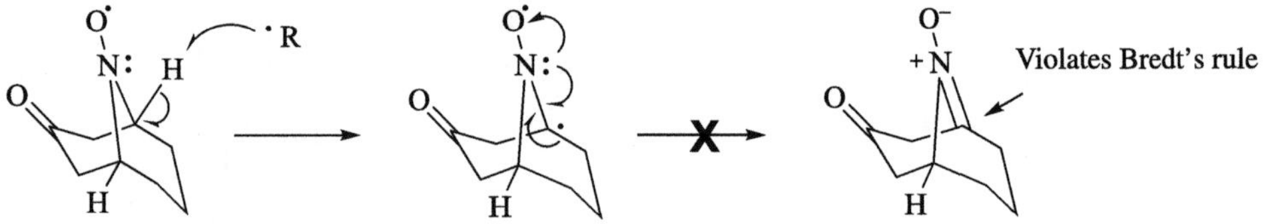

In the case of the bicyclic system, the process of losing the hydrogen would lead to an olefin at the bridgehead and it is not possible to achieve the proper orbital alignment for a π-bond at that position. Named for Julius Bredt, this is an example of a violation of Bredt's rule. This rule is based upon experimental observation and states that a double bond cannot be placed at the bridgehead of a ring system (again, attributed to the unacceptable strain at the bridge head). The hydrogen is most likely abstracted and then the newly formed radical proceeds no further: it either abstracts a hydrogen from another molecule or is quenched upon workup.

Problem 7.11

Is this an Oppenauer oxidation or a Meerwein–Pondorff–Verley reduction? The answer is both, occurring in an intramolecular sense with one part of the molecule being reduced, as another part of the molecule oxidized.

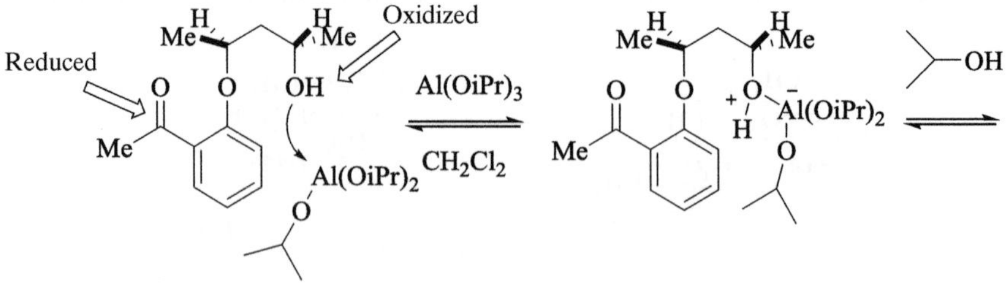

7-121

The stereochemical transfer provides some evidence that the process is potentially influenced by another part of the molecule and is thus consistent with an intramolecular (within the molecule) process. The stereochemical outcome is the result of maintaining all the methyl groups in pseudoequatorial positions and at the same time, the aluminate holding the bridged system in a rigid conformation as the hydrogen transfers (**7-121**).

This particular example does not allow for the removal of the product as it is generated, or addition of more starting material to move the equilibrium toward products as we had seen in previous examples. The levers for pushing the process in one direction or the other are limited. The ratio of starting material and product arises as a result of the relative stability of the product and starting material after a thermodynamic equilibrium is reached.

Problem 7.12

The action of sulfuric acid and water on addition across the olefin is complimentary to the two-step hydroboration–oxidation reaction, as they are both net hydration reactions. Starting with the olefin, one carbon is oxidized and one is reduced. The hydration using sulfuric acid and water is initiated by the activation of the double bond with acid, leaving the more stable carbocation. Water then adds into the carbocation to deliver the alcohol with the Markovnikov regiochemical relationship.

The mechanism for the hydroboration–oxidation sequence proceeds through the proposed mechanism below. Oxidation occurs at the terminal olefin carbon (in an anti-Markovnikov sense) and a reduction of the carbon at the proximal end of the olefin. This is thought to be the result of the decreased steric resistance associated with having the boron at the terminal end of the olefin. To a lesser extent, this stereochemical result may also depend on the electronic structure of the olefin. As the borane approaches, the more electronegative hydrogen of the borane ends up at the position better able to support the developing positive charge (the proximal end of the olefin), leading (after oxidation) to the terminal alcohol.

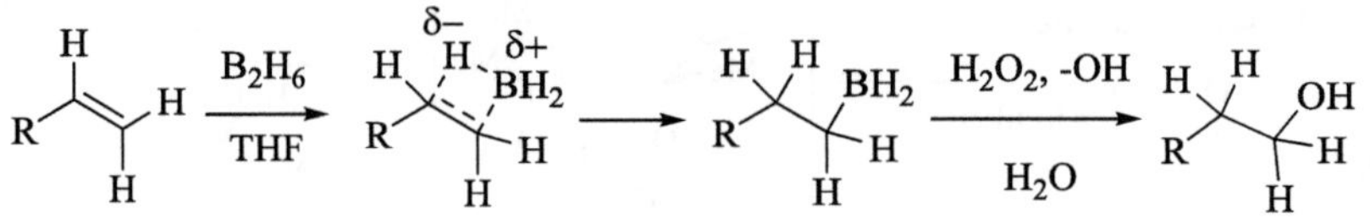

Problem 7.13

The stereochemistry of the hydroboration reaction is the result of the borane approaching the olefin from the less congested bottom face of the molecule (**7-122**). The borane will be oriented with the boron at the terminal end of the olefin and the hydrogen at the more sterically demanding proximal position (**7-123**). The electronic factors would also favor the boron being delivered at the less substituted carbon of the olefin, as it becomes more electron rich in the process.

7-122 **7-123**

Favored approach

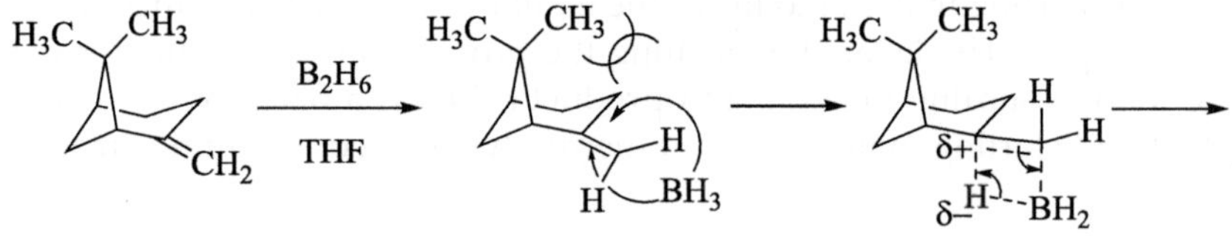

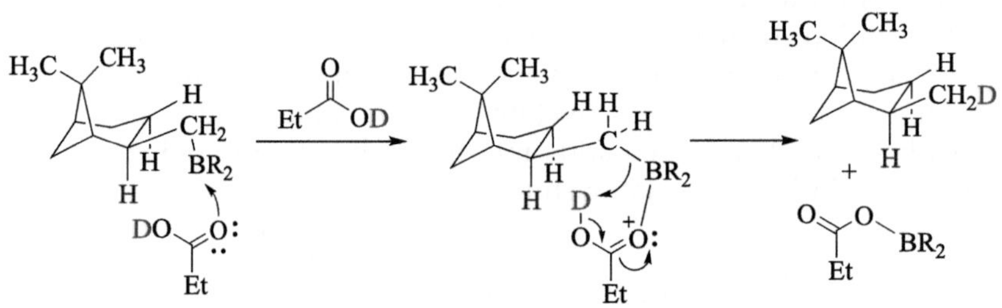

The borane is then reduced with propanoic acid, leading to the borane ether and deuterium label at the terminus of what was the olefin.

Problem 7.14

Hydroboration occurs away from the sterically demanding bottom face of the olefin (**7-124**) and with the boron oriented toward the terminal (less substituted and less sterically demanding) end of the olefin (**7-125**). The bulky borane used in this example is even more susceptible to steric demands as it approaches the olefin.

7-124

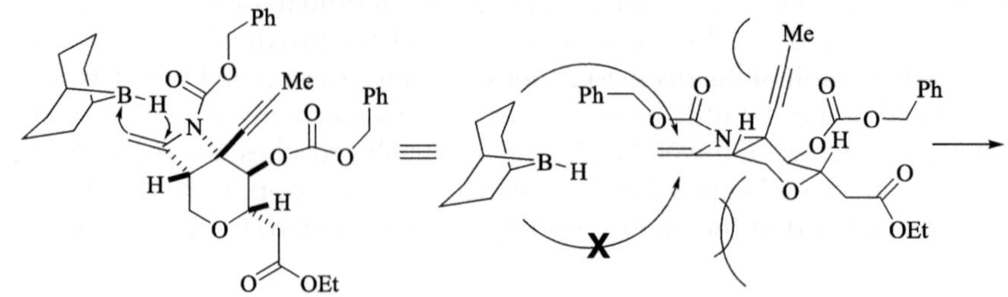

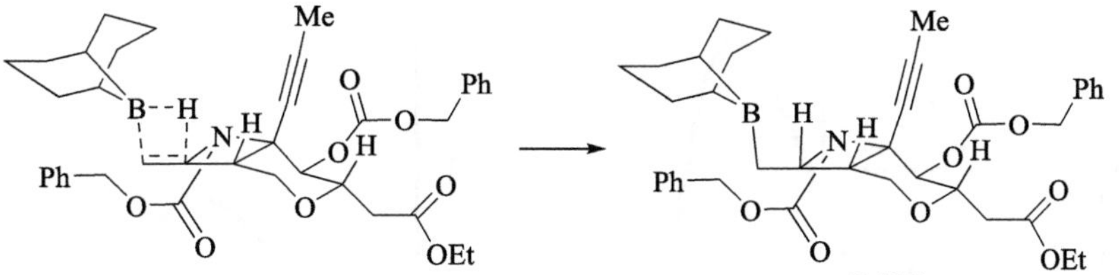

7-125

Problem 7.15

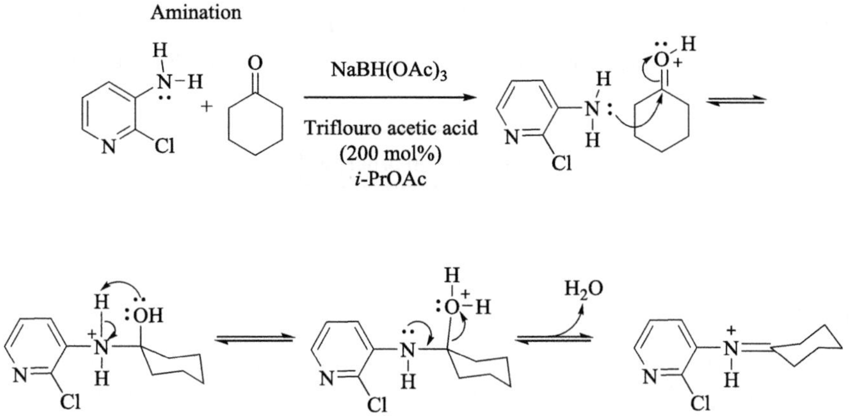

Amination

7-126

The reductive amination of this somewhat less reactive amine is affected through the use of a stronger acid to aid in the initial formation of imine **7-126**. The amine adds into the ketone and is then dehydrated, leading to the imine. The imine is then reduced with sodium triacetoxyborohydride **7-127** to provide the new amine.

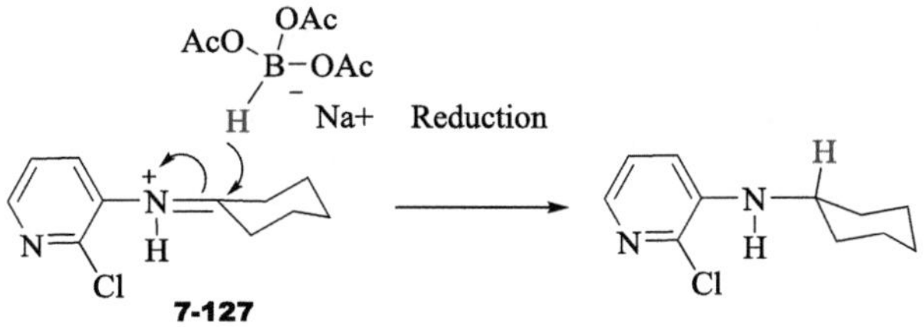

7-127

Problem 7.16

The catalyst changes the reaction activation energy leading to lower energy transition states. The net energy change between the alkene and the alkane (30 kcal/mol) remains the same.

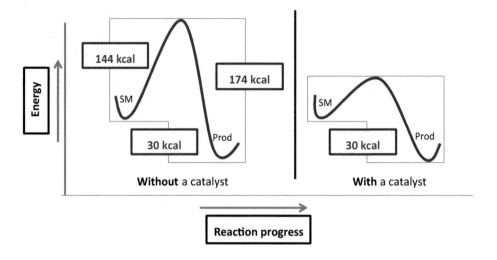

Problem 7.17

This is a reductive amination carried out on a hindered and less reactive amine (an aniline). The initial condensation of the amine and the ketone leads to the imine and loss of water. The imine is then reduced via hydrogenation to provide the amine.

Hydrogenation of the imine

Problem 7.18

In each of these examples, a temporary "tether" is proposed to be formed in the initial step. This places one face of the substrate and the activated/hydrogenated metal in close proximity and provides an explanation consistent with the observed products.

References

Abdel-Magid, A. F.; Carson, K. G.; Harris, B. D.; Maryanoff, C. A.; Shah, R. D. *J. Org. Chem.* **1996**, *61*, 3849–3862.

Albright, J. D.; Goldman, L. *J. Org. Chem.* **1965**, *30*, 1107–1110.

Allen, R. R.; Kiess, A. A. *J. Am. Oil Chem. Soc.* **1956**, *33*, 355–359.

Allred, E. L.; Sonnenberg, J.; Winstein, S. *J. Org. Chem.* **1960**, *25*, 26–29.

Anh, N. T. *Top. Curr. Chem.* **1980**, *88*, 145–162.

Anh, N. T.; Eisenstein, O. *Nouv. J. Chim.* **1977**, *1*, 61–70.

Ashby, E. C.; Boone, J. R. *J. Am. Chem. Soc.* **1976**, *98*, 5524–5531.

Atzrodt, J.; Derdau, V.; Fey, T.; Zimmermann, J. *Angew. Chem. Int. Ed.* **2007**, *46*, 7744–7765.

Awasthy, A. K.; Rocek, J.-V.; Moriarty, R. M. *J. Am. Chem. Soc.* **1967**, *89*, 5400–5403.

Back, R. A. *Rev. Chem. Intermed.* **1984**, *5*, 293–323.

Bailey, W. F.; Bobbitt, J. M.; Wiberg, K. B. *J. Org. Chem.* **2007**, *72*, 4504–4509.

Blay, G.; Cardona, L.; Garcıa, B.; Pedro, J. R.; Sanchez, J. J. *J. Org. Chem.* **1996**, *61*, 3815–3819.

Brandt, B.; Fischer, J.-H.; Ludwig, W.; Libuda, J.; Zaera, F.; Schauermann, S.; Freund, H.-J. *J. Phys. Chem. C* **2008**, *112*, 11408–11420.

Brown, H. C.; Kawakami, J. H. *J. Org. Chem. Soc.* **1969**, *92*, 1990–1995.

Brown, H. C.; Krishnamurthy, S. *Tetrahedron* **1979**, *35*, 567–607.

Brown, H. C.; Sharp, R. L. *J. Am. Chem. Soc.* **1966**, *88*, 5851–5854.

Bürgi, H. B.; Dunitz, J. D.; Lehn, J. M.; Wipff, G. *Tetrahedron* **1974**, *30*, 1563–1572.

Burkhardt, E. R.; Matos, K. *Chem. Rev.* **2006**, *106*, 2617–2650.

Chérest, M.; Felkin, H.; Prudent, N. *Tetrahedron Lett.* **1968**, *18*, 2199–2204.

Corey, E. J.; Kim, C. U. *J. Am. Chem. Soc.* **1972**, *94*, 7586–7587.

Corey, E. J.; Schmidt, G. *Tetrahedron Lett.* **1979**, *20*, 399–402.

Corey, E. J.; Suggs, J. W. *Tetrahedron Lett.* **1975**, *16*, 2647–2650.

Cornforth, J. W.; Cornforth, R. H.; Mathew, K. K. *J. Chem. Soc.* **1959,** 112—127.

Cornforth, R. H.; Cornforth, J. W.; Popjak, G. *Tetrahedron* **1962,** *18,* 1351—1354.

Cram, D. J.; Allinger, N. L. *J. Am. Chem. Soc.* **1956,** *78,* 2518—2524.

Cram, D. J.; Elhafez, F. A. A. *J. Am. Chem. Soc.* **1952,** *74,* 5828—5835.

Criegee, R. *Angew. Chem. Int. Ed.* **1975,** *87,* 745—752.

Crosby, J.; Stirling, C. J. M. *J. Chem. Soc. B* **1970,** 671. table 2, item d.

Dave, P.; Byun, H.-S.; Engle, R. *Synth. Commun.* **1986,** *16,* 1343—1346.

De Mico, A.; Margarita, R.; Parlanti, L.; Vescovi, A.; Piancatelli, G. *J. Org. Chem.* **1997,** *62,* 6974—6977.

De Nooy, A. E. J.; Besemer, A. C.; Bekkum, H. *Synthesis* **1996,** 1153—1174.

Diethelm, S.; Carreira, E. M. *J. Am. Chem. Soc.* **2013,** *135,* 8500—8503.

Doering, W. V.-E.; Speers, L. *J. Am. Chem. Soc.* **1950,** *72,* 5515—5518.

Dupeyre, R. M.; Rassat, A. *J. Am. Chem. Soc.* **1966,** *88,* 3180—3181.

Fatiadi, A. J. *Synthesis* **1976a,** 65—104.

Fatiadi, A. J. *Synthesis* **1976b,** 133—167.

Fenselau, A. H.; Moffatt, J. G. *J. Am. Chem. Soc.* **1966,** *88,* 1762—1765.

Friess, S. L.; Farnham, N. *J. Am. Chem. Soc.* **1950,** *72,* 5518—5521.

Fujita, M.; Takarada, Y.; Sugimura, T.; Tai, A. *Chem. Commun.* **1997,** *17,* 1631—1632.

Geletneky, C.; Berger, S. *Eur. J. Org. Chem.* **1998,** 1625—1627.

Gemal, A. L.; Luche, J.-L. *J. Am. Chem. Soc.* **1981,** *103,* 5454—5459.

Gomez, S.; Peters, J. A.; Maschmeyer, T. *Adv. Synth. Catal.* **2002,** *344,* 1037—1057.

Gritter, R. J.; Wallace, T. J. *J. Org. Chem.* **1959,** *24,* 1051—1056.

Harrington, P. J. *Pharmaceutical Process Chemistry for Synthesis;* John Wiley and Sons: Hoboken, New Jersey, 2011.

Henbest, H. B.; Wilson, R. A. L. *J. Chem. Soc.* **1957,** 1958—1965.

Herz, W.; Mohanraj, S. *J. Org. Chem.* **1980,** *45,* 5417—5419.

Holmes, A. B.; Raphael, R. A.; Wellard, N. K. *Tetrahedron Lett.* **1976,** *17,* 1539—1542.

Hoveyda, A. H.; Evans, D. A.; Fu, G. C. *Chem. Rev.* **1993,** *93,* 1307—1370.

Jardine, F. H. *Prog. Inorg. Chem.* **1981,** *28,* 63—202.

Jira, R. *Angew. Chem. Int. Ed.* **2009,** *48,* 9034—9037.

Jun Shi, J.; Shigehisa, H.; Guerrero, C. A.; Shenvi, R. A.; Li, C.-C.; Baran, P. S. *Angew. Chem. Int. Ed.* **2009,** *48,* 4328—4331.

Kabalka, G. W.; Shoup, T. M.; Goudgaon, N. M. *J. Org. Chem.* **1989,** *54,* 5930—5933.

Kabilan, S.; Girija, R.; Reis, J. C. R.; Segurado, M. A. P.; de Oliveira, J. D. G. *J. Chem. Soc., Perkin Trans.* **2002,** *2,* 1151—1157.

Kalsi, P. S. *Stereochemistry Conformation and Mechanism,* 6th ed.; New Age International, 2005; pp. 444—450.

Keaveney, W. P.; Berger, M. G.; Pappas, J. J. *J. Org. Chem.* **1967,** *32,* 1537—1542.

Keith, J. A.; Nielsen, R. J.; Oxgaard, J.; Goddard, W. A. *J. Am. Chem. Soc.* **2007,** *129,* 12342—12343.

Knowles, W. S.; Thompson, Q. E. *J. Org. Chem.* **1960,** *25,* 1031—1033.

Kogure, T.; Eliel, E. L. *J. Org. Chem.* **1984,** *49,* 576—578.

Kornblum, N.; Powers, J. W.; Anderson, G. J.; Jones, W. J.; Larson, H. O.; Levand, O.; Weaver, W. M. *J. Am. Chem. Soc.* **1957,** *79,* 6562.

Kornblum, N.; Jones, W. J.; Anderson, G. J. *J. Am. Chem. Soc.* **1959,** *81,* 4113—4114.

Kreis, L. M.; Carreira, E. M. *Angew. Chem. Int. Ed.* **2012,** *51,* 3436—3439.

Krishnamurthy, S.; Brown, H. C. *J. Org. Chem.* **1977,** *42,* 1197—1201.

Kuczkowski, R. L. *Chem. Soc. Rev.* **1992,** 79—83.

Kwart, H.; Takeshita, T. *J. Org. Chem.* **1963,** *28,* 670—673.

Lam, T. T.; Vickery, T.; Tuma, L. *J. Therm. Anal. Calorim.* **2006,** *85,* 25—30.

Lee, D. G.; Raptis, M. *Tetrahedron* **1973,** *29,* 1481—1486.

Luche, J.-L.; Rodriguez-Hahn, L.; Crabbe, P. *J. Chem. Soc. Chem. Commun.* **1978,** 601—602.

Maitra, U.; Chandrasekhar, J. *Resonance* **1997,** 23—28.

Mancuso, A. J.; Swern, D. *Synthesis* **1981,** *3,* 165—185.

Marx, M.; Tidwell, T. T. *J. Org. Chem.* **1984,** *49,* 788—793.

McConnell, J. R.; Hitt, J. E.; Daugs, E. D.; Rey, T. A. *Org. Process Res. Dev.* **2008,** *12,* 940—945.

McKillop, A.; Sanderson, W. R. *Tetrahedron* **1995,** *51,* 6145—6166.

McLaughlin, M.; Palucki, M.; Davies, I. W. *Org. Lett.* **2006**, *8*, 3307–3310.

Mehta, G.; Pan, S. C. *Org. Lett.* **2004**, *6*, 3985–3988.

Mengel, A.; Reiser, O. *Chem. Rev.* **1999**, *99*, 1191–1223.

Moffatt, J. G. *J. Org. Chem.* **1971**, *36*, 1909–1912.

Mori, K.; Ohki, M.; Sato, A.; Matsui, M. *Tetrahedron* **1972**, *28*, 3739–3745.

Moulton, W. N.; Van Atta, R. E.; Ruch, R. R. *J. Org. Chem.* **1961**, *26*, 290–292.

Muller, P. *Helv. Chim. Acta* **1970**, *53*, 1869–1873.

Musolino, M. G.; Caia, C. V.; Mauriello, F.; Pietropaolo, R. *Appl. Catal., A* **2010**, *390*, 141–147.

Naik, N.; Braslau, R. *Tetrahedron* **1998**, *54*, 667–696.

Nelson, D. J.; Li, R.; Brammer, C. *J. Am. Chem. Soc.* **2001**, *123*, 1564–1568.

Nickon, A.; Bagli, J. F. *J. Am. Chem. Soc.* **1959**, *81*, 6330.

Nickon, A.; Bagli, J. F. *J. Am. Chem. Soc.* **1961**, *83*, 1498–1508.

Omura, K.; Swern, D. *Tetrahedron* **1978**, *34*, 1651–1660.

Paisdor, B.; Kuck, D. *J. Org. Chem.* **1991**, *56*, 4753–4759.

Paquette, L. A.; Sugimura, T. *J. Am. Chem. Soc.* **1987**, *109*, 3017–3024.

Pasto, D. J.; Taylor, R. T. *Organic Reactions, Reductions with Diimide* **2004**, *Vol. 40; pp. 91–155 (chapter 2).*

Pfitzner, K. E.; Moffatt, J. G. *J. Am. Chem. Soc.* **1965a**, *87*, 5661–5670.

Pfitzner, K. E.; Moffatt, J. G. *J. Am. Chem. Soc.* **1965b**, *87*, 5670–5678.

Piancatelli, G.; Scettri, A.; D'Auria, M. *Synthesis* **1982**, 245–258.

Porto, R. S.; Vasconcellos, M. L. A. A.; Ventura, E.; Coelho, F. *Synthesis* **2005**, 2297–2306.

Rao, V. V. R.; Devaprabhakara, D. *Tetrahedron* **1978**, *39*, 2223–2227.

Rocek, J.-V.; Westheimer, F. H.; Eschenmoser, A.; Moldovanyi, L.; Schreiber, J. *Helv. Chim. Acta* **1962**, *45*, 2554–2567.

Rowe, J. J. M.; Gibney, K. B.; Yang, M. T.; Dutton, G. G. S. *J. Am. Chem. Soc.* **1968**, *90*, 1924.

Rubin, M. B. *Helv. Chim. Acta* **2003**, *86*, 930–940.

Schroeder, M. *Chem. Rev.* **1980**, *80*, 187–213.

Schwartz, N. N.; Blumbergs, J. H. *J. Org. Chem.* **1964**, *29*, 1976–1979.

Sehgal, R. K.; Koenigsberger, R. U.; Howard, T. J. *J. Org. Chem.* **1975**, *40*, 3073–3078.

Smidt, J.; Hafner, W.; Jira, R.; Sedlmeier, J.; Sieber, R.; Rüttinger, R.; Kojer, H. *Angew. Chem. Int. Ed.* **1959**, *71*, 176–182.

Smith, M. B. *Organic Chemistry: An Acid-Base Approach;* CRC Press, 2011; p. 818.

Stolow, R. D.; Sachdev, K. *J. Org. Chem.* **1971**, *36*, 960–966.

Stork, G.; Kahne, D. E. *J. Am. Chem. Soc.* **1983**, *105*, 1072–1073.

Sum, P.-E.; Weiler, L. *Can. J. Chem.* **1978**, *56*, 2700–2702.

Thompson, H. W. *J. Org. Chem.* **1971**, *36*, 2577–2581.

Thompson, H. W.; McPherson, E. *J. Org. Chem. Soc.* **1977**, *42*, 3350–3353.

Tidwell, T. T. *Synthesis* **1990a**, *10*, 857–870.

Tidwell, T. T. *Org. React.* **1990b**, *39*, 297–572.

Ward, D. E.; Gai, Y.; Qiao, Q. *Org. Lett.* **2000**, *2*, 2125–2127.

Westheimer, F. H. *Chem. Rev.* **1961**, *61*, 265–273.

Westheimer, F. H.; Nicolaides, N. *J. Am. Chem. Soc.* **1949**, *71*, 25–28.

Wiberg, K. B. *Chem. Rev.* **1955**, *55*, 713–743.

Wiberg, K. B.; Stewart, R. *J. Am. Chem. Soc.* **1955**, *77*, 1786–1795.

Wilds, A. L. *Org. React.* **1944**, *2*, 178–223.

Zhang, Z.; Tan, J.; Wang, Z. *Org. Lett.* **2008**, *10*, 173–175.

Zhao, M.; Li, J.; Song, Z.; Desmond, R.; Tschaen, D. M.; Grabowski, E. J. J.; Reider, P. J. *Tetrahedron Lett.* **1998**, *39*, 5323–5326.

Zweifel, G.; Brown, H. C. *J. Am. Chem. Soc.* **1964**, *86*, 393–397.

Additional Problems

This chapter includes additional problems related to the materials in Chapters 3–7. Some of the mechanisms are mixed; for example, there might be a pericyclic reaction followed by hydrolysis in either acid or base. If a reaction appears to be pericyclic, be sure to determine whether the reaction is symmetry-allowed or symmetry-forbidden under the reaction conditions. If it is symmetry-forbidden, a nonconcerted reaction pathway through radical or charged intermediates will be the most likely mechanism. There are also a few problems with multiple steps requiring you to use material from different chapters of the book.

PROBLEM 8.1

Choose one principle or mechanism from each of Chapters 3–7, and find a reaction from the recent (past 5 years) literature that illustrates that principle or mechanism. Write a detailed step-by-step mechanism for each reaction. Journals to look at include *J. Org. Chem., Tetrahedron Lett., Synthesis,* and *Synthetic Comm.* Sometimes a synthetic sequence can be found that answers the entire question!

PROBLEM 8.2

Show how the following transformations could occur.

a.

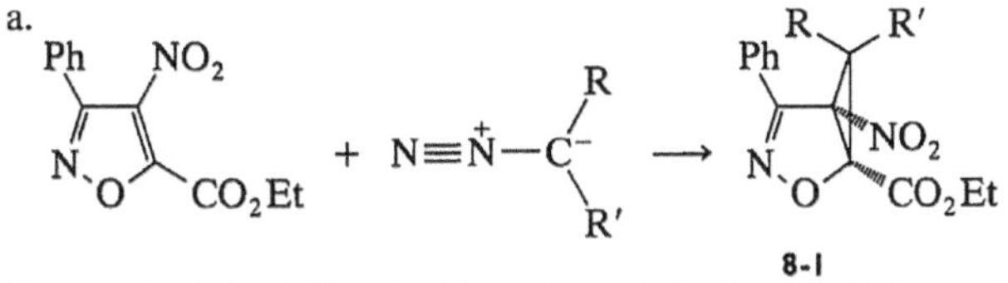

8-1

Copyright © 2014 Elsevier Inc. All rights reserved.

b.

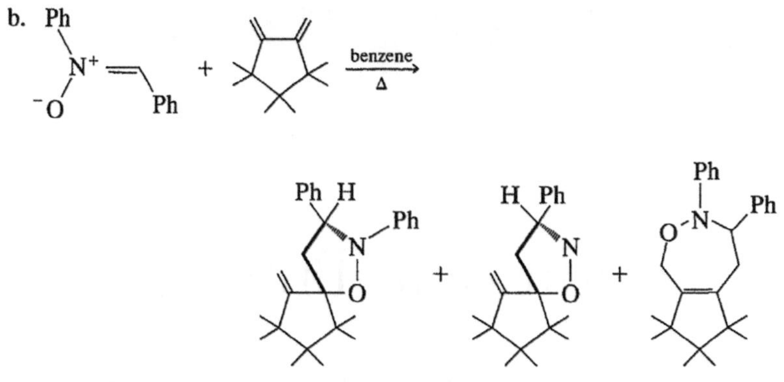

8-2 + **8-3** + **8-4**

c.

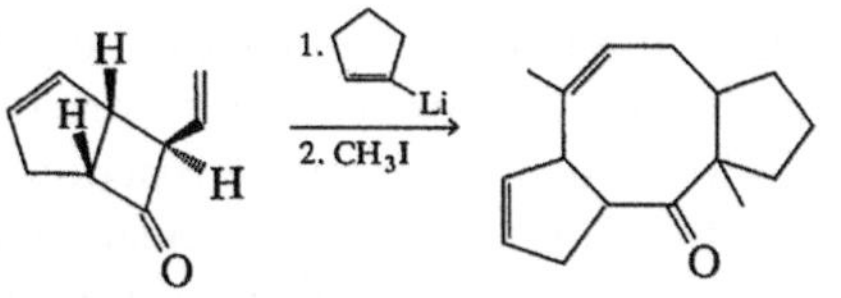

8-5   **8-6**

Sources: Nesi et al., (1989); Baran and Mayr, (1987); Paquette et al., (1983).

PROBLEM 8.3

What is the designation for the sigmatropic shift in the transformation of **8-7** to **8-8** in the following reaction sequence?

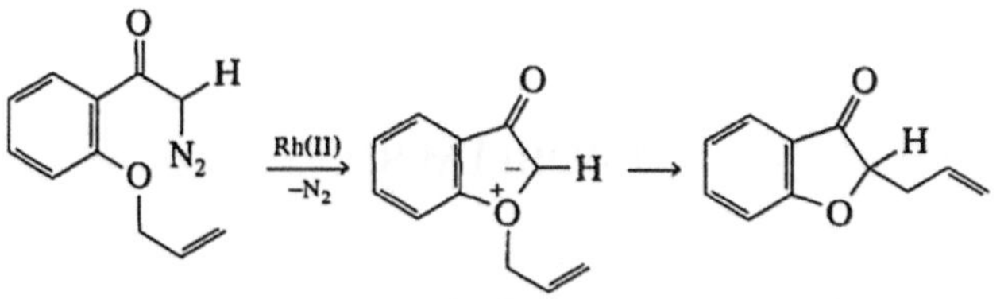

8-7   **8-8**

Source: Pirrung and Werner, (1986).

PROBLEM 8.4

Write a step-by-step mechanism for the following transformation. Other products for this reaction are covered by Problem 4.17.b.

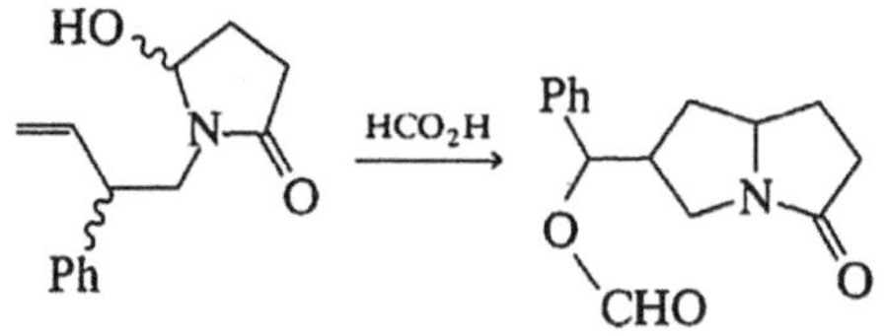

Source: Ent et al., (1986).

PROBLEM 8.5

Problem 5.14 showed the transformation of dibenzyl sulfide to a number of products by reaction with *n*-butyllithium, followed by treatment with methyl iodide. Some tetramethylenediamine (TMEDA) was included in the reaction mixture to coordinate the lithium. Another product **8-9** is produced by this reaction. Write a reasonable mechanism for its formation.

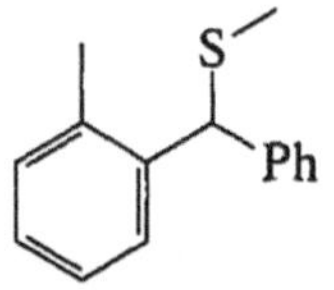

8-9

PROBLEM 8.6

By using mechanistic principles as well as the molecular formulas, determine the structures of **8-10** and **8-11**.

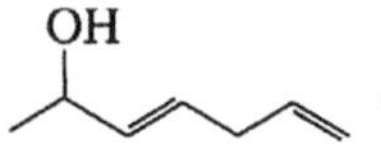

1. KH/THF
2. Cl_3CCN
3. trace MeOH

→ $C_9H_{12}Cl_3NO$ 93% **8-10**

xylene, Δ → $C_9H_{12}Cl_3NO$ 72% **8-11**

Source: Roush et al., (1987).

PROBLEM 8.7

a. Write a step-by-step mechanism for the formation of **8-14**.
b. Propose a reasonable structure for the isomeric product **8-15**.

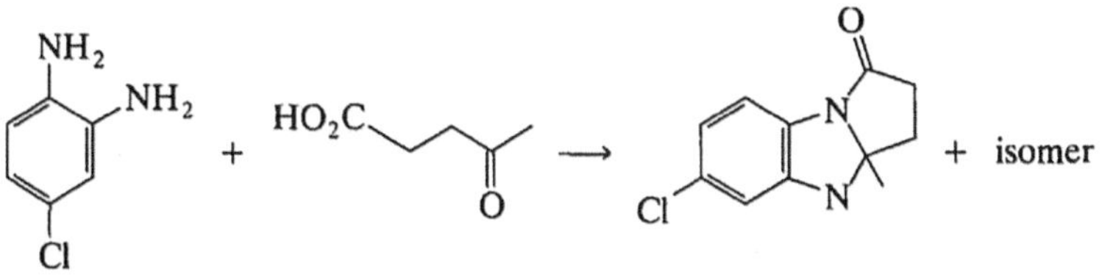

8-12   8-13   8-14   8-15

The reaction is run in refluxing toluene with azeotropic removal of water.

Source: Chimirri et al., (1988).

PROBLEM 8.8

Write a mechanism for the following reaction, a reduction rearrangement in which zinc is the reducing agent. The initial steps of the mechanism could involve protonation of the nitrogen and reaction with zinc.

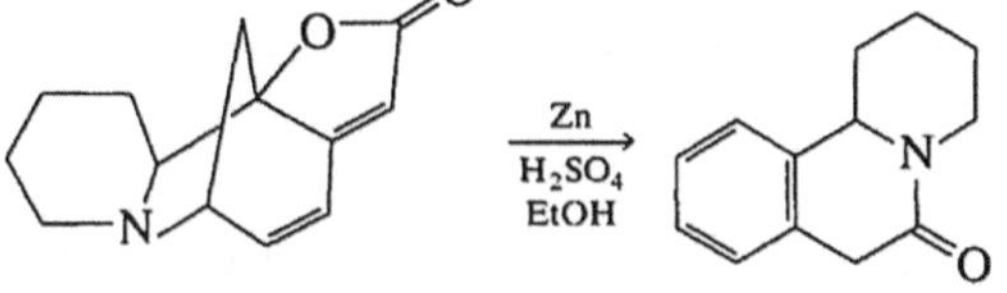

PROBLEM 8.9

Write mechanisms for the formation of both products derived from the following degradation of the fungal neurotoxin, verrucosidin.

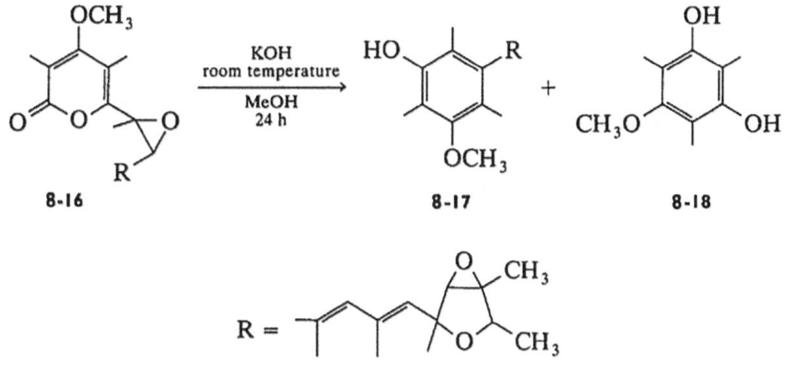

8-16   8-17   8-18

Source: Ganguli et al., (1984).

PROBLEM 8.10

Propose mechanisms for the steps shown in the following synthetic sequence. The overall process involves reduction by the transition metal reagents, as well as cyclization.

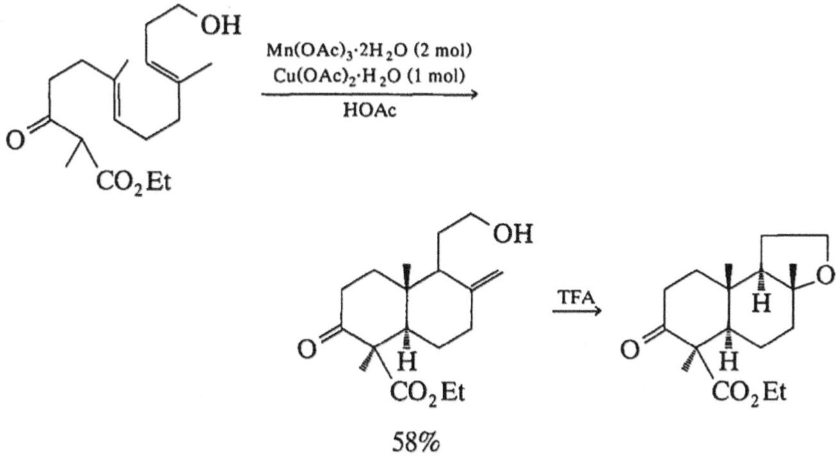

58%

Source: Zarecki et al., (1998).

PROBLEM 8.11

Propose a mechanism detailing the steps involved in the following transformation. Keep track of the stereochemical relationships of the groups involved in this process.

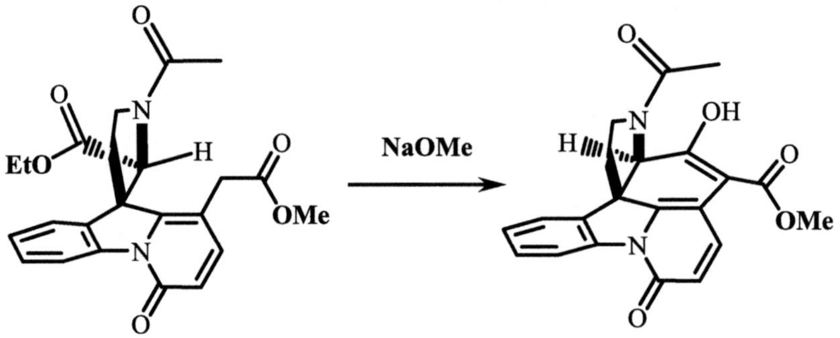

Source: Woodward et al., (1963).

PROBLEM 8.12

Propose a mechanism that describes the following transformation.

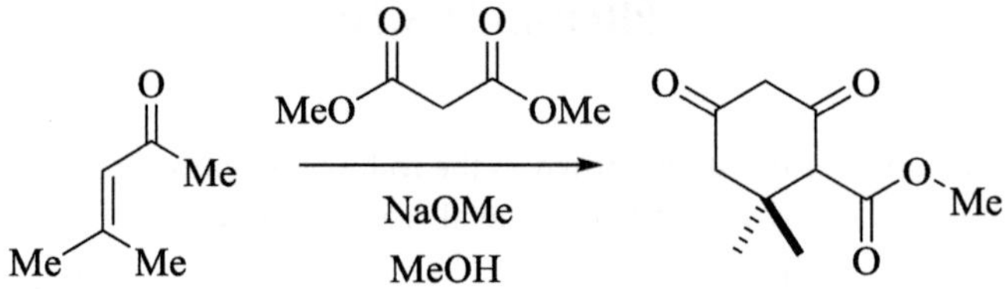

Sources: Cha et al., (2011); Steiner and Willhalm, (1952).

PROBLEM 8.13

Propose a mechanism detailing the steps involved in the following transformation.

1) H+ / H₂O

2) ⟶ MgBr

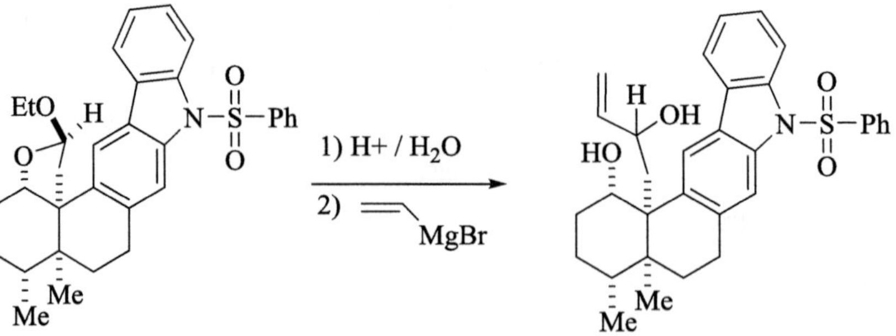

Source: Bian et al., (2012).

PROBLEM 8.14

a. Describe the mechanism including the initiation and chain propagation steps
b. What other related ring systems could have been generated? Is this the result you would expect? Explain.

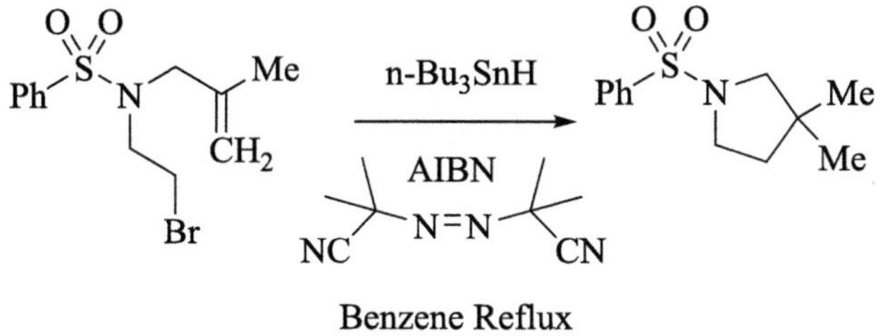

Benzene Reflux

Source: Padwa et al., (1985).

PROBLEM 8.15

Describe the process for the following transformation and show the stereochemical outcome at position A and explain why that is the result.

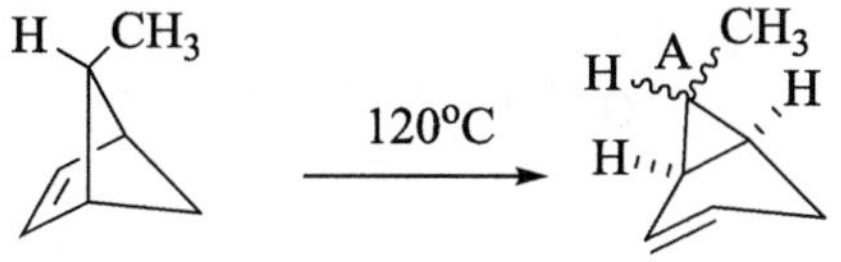

Source: Roth and Friedrich, (1969).

PROBLEM 8.16

What is the oxidation number of the red carbons?

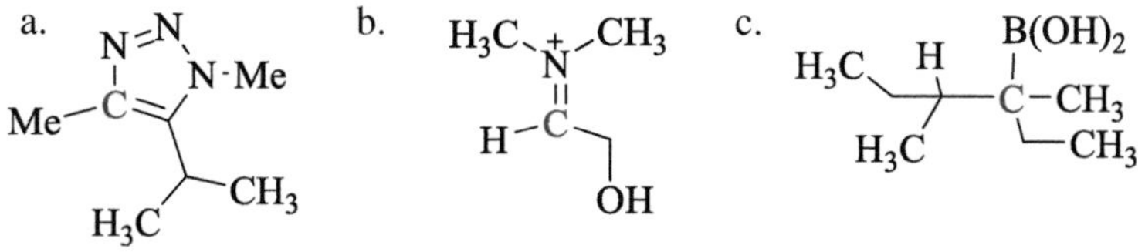

PROBLEM 8.17

For the highlighted atom, determine if this transformation is an oxidation, reduction, or neither?

a)

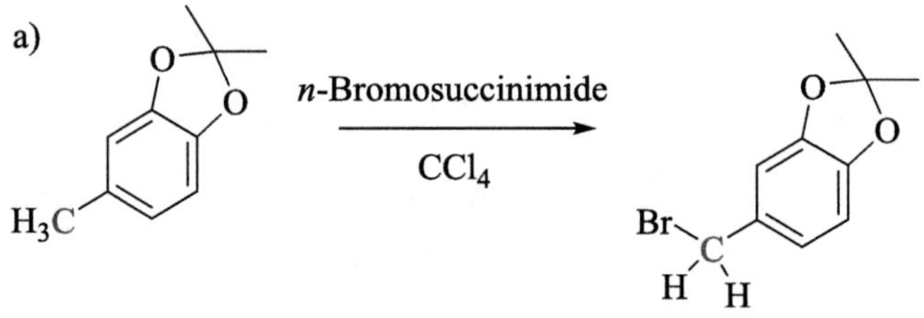

n-Bromosuccinimide

CCl_4

b)

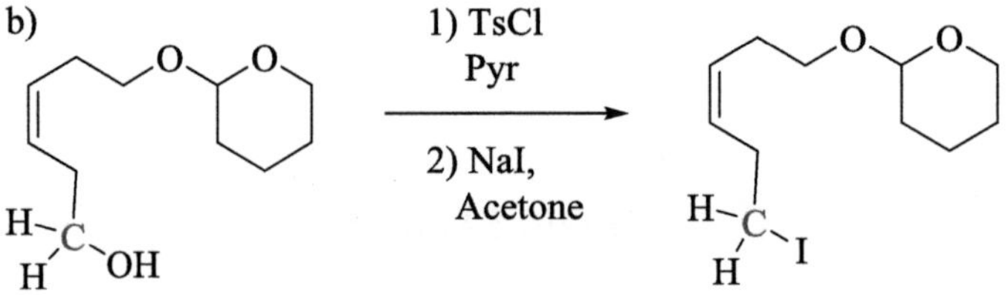

1) TsCl
   Pyr

2) NaI,
   Acetone

c)

Catalytic Sulfuric acid

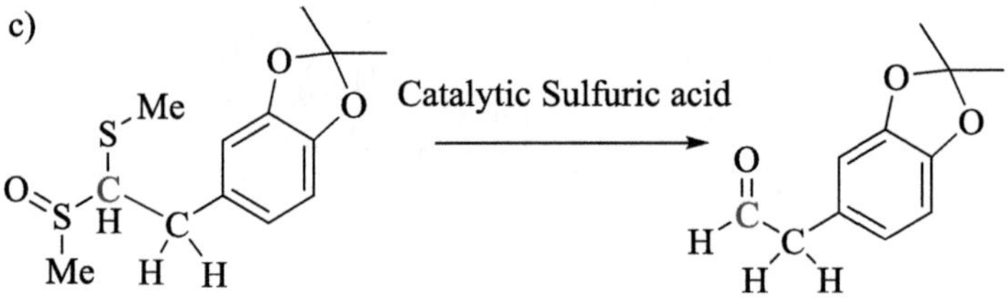

Sources: Ogura and Tsuchihashi, (1971); Carvalho and Prestwich, (1984); Ogura and Tsuchihashi, (1971).

PROBLEM 8.18

The conversion of phenyl acyl bromides to glyoxals has been effected using the Kornblum procedure as shown below. In certain circumstances, it is also possible to get the side product shown. Describe a mechanism leading to the desired product and to the side product.

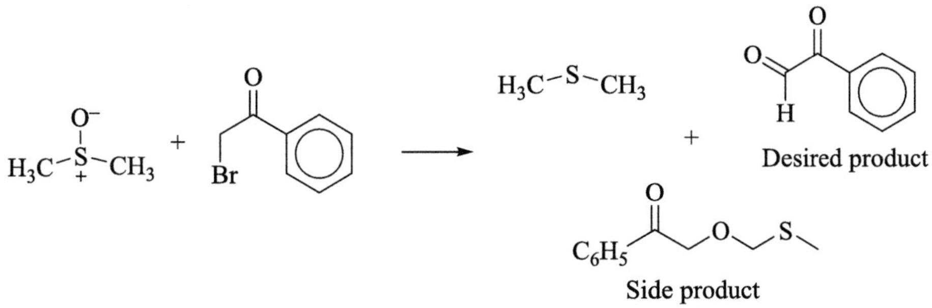

Desired product

Side product

Source: Kornblum et al., (1957).

PROBLEM 8.19

Describe the mechanism for these transformations and identify compounds A and B.

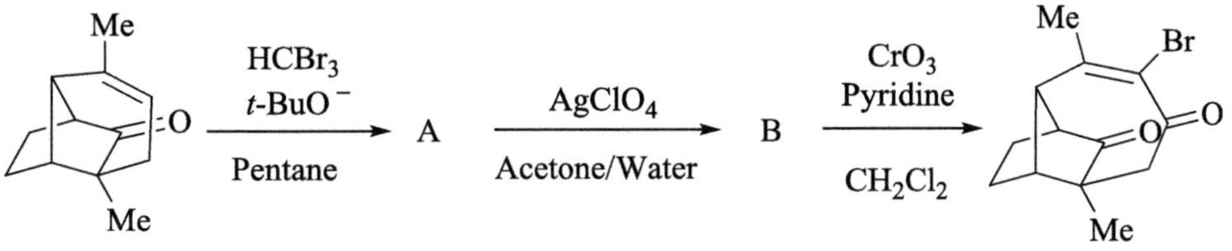

Source: McMurry and Isser, (1972).

PROBLEM 8.20

Describe the mechanism for each step leading to the product shown.

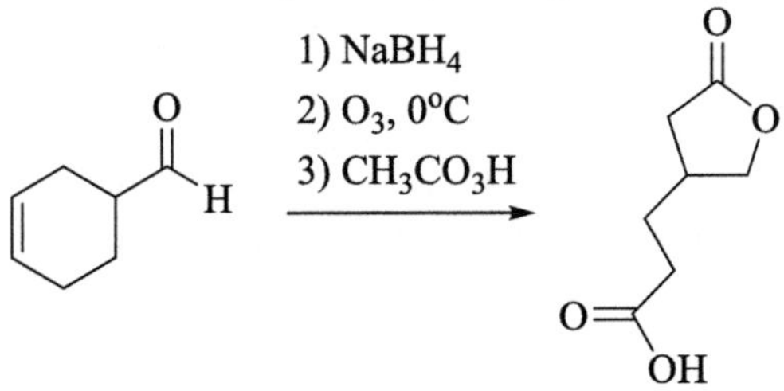

1) NaBH₄
2) O₃, 0°C
3) CH₃CO₃H

Source: White et al., (1982).

PROBLEM 8.21

Describe a mechanism for the two steps that lead to the product shown.

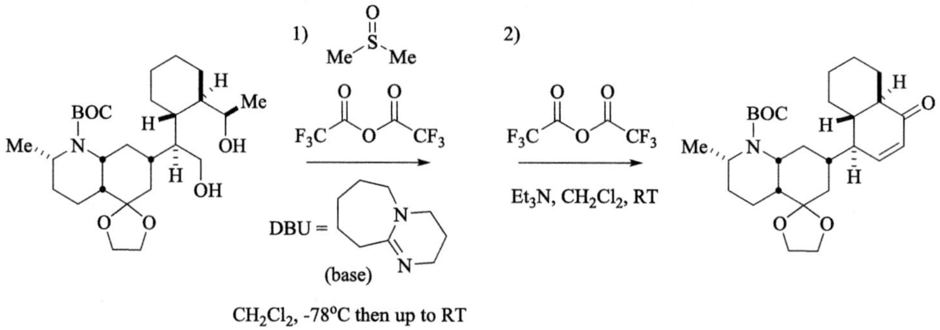

Source: Zi et al., (2010).

PROBLEM 8.22

A short sequence representing a combination of steps—protection, oxidation, and reduction.

Describe the mechanism for the three-step sequence shown below.

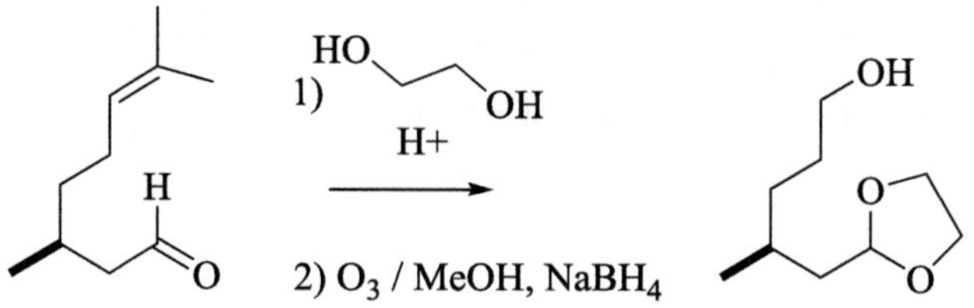

Source: Deng et al., (2012).

ANSWERS TO PROBLEMS

Problem 8.2

a. There are several reasons why it is unlikely that loss of nitrogen from the diazo compound, followed by direct addition of the carbene to the double bond, is the mechanism of the reaction. (1) Even though the carbene has two alkyl substituents, it is still quite electrophilic and would not be expected to react with one of the double bonds to the exclusion of the other. (2) There is no catalyst or ultraviolet light to aid in the generation of the carbene.

A more probable mechanism could be 1,3-dipolar addition to the electrophilic double bond to give **8-19**, followed by loss of nitrogen from the intermediate to give the product.

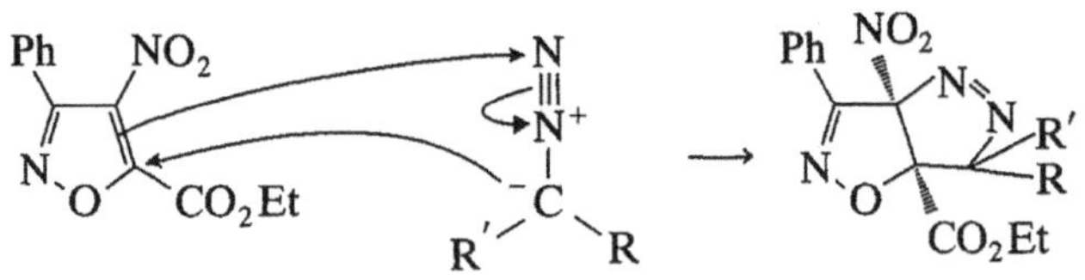

8-19

The loss of nitrogen might occur with the intermediate formation of ions or radicals. In the heterolytic process, a negative charge develops on carbon because of resonance stabilization by the nitro group and the nitrogen in the oxazole ring.

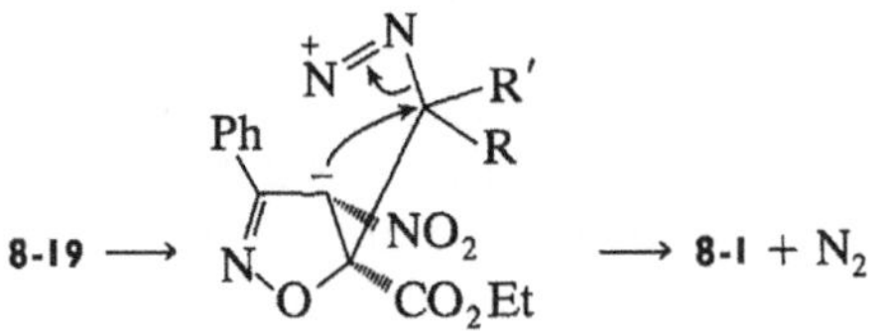

$$\textbf{8-19} \longrightarrow \longrightarrow \textbf{8-1} + N_2$$

b. The formation of **8-2** and **8-3** can result from 1,3-dipolar cycloaddition of the nitrone to one double bond. This is a $[\pi_s^4 + \pi_s^2]$, thermally allowed reaction. The four-electron component is the nitrone, and the two-electron component is the alkene.

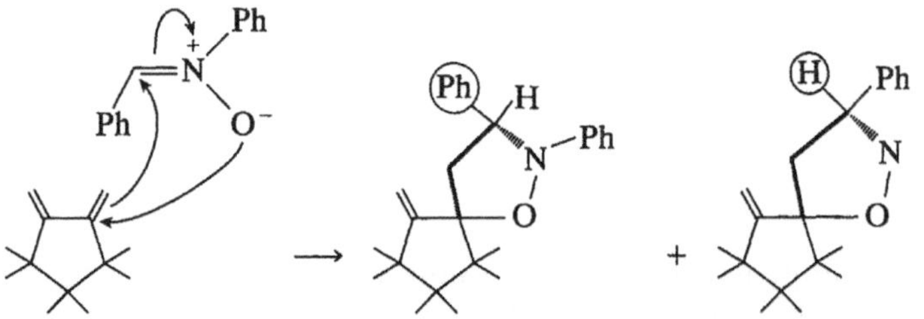

8-2   **8-3**

Representing the reaction in this way does not clearly indicate how the two isomers arise. A better appreciation of what is going on is obtained by considering the orientation required for the molecular orbitals of the two components to interact in a pericyclic process. The most favorable orientation has the faces of the π systems parallel to one another, as shown in the following structures.

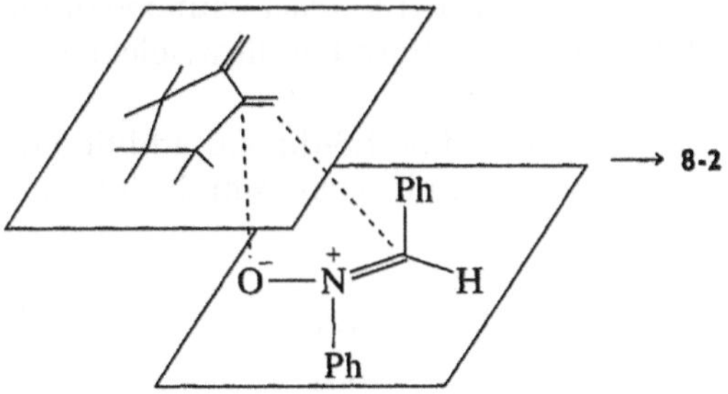

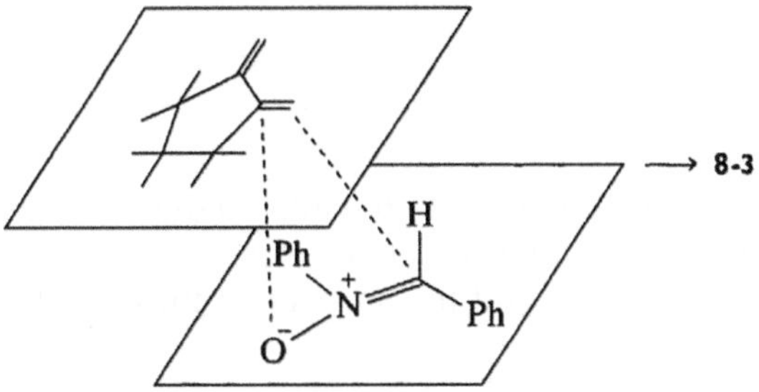

The formation of the stereoisomeric products **8-2** and **8-3** is analogous to the formation of *exo* and *endo* adducts in the Diels–Alder reaction. One way to keep track of the stereochemical relationships is to note that in **8-2**, the circled phenyl group and the *exo*-methylene group in the cyclopentane ring are on the same side of the oxazole ring (*cis*), whereas in **8-3**, it is the circled hydrogen that is *cis* to the *exo*-methylene group.

Concerted formation of **8-4** would be a $[\pi_s^4 + \pi_s^4]$ cycloaddition, which is not thermally allowed. Thus, formation of this product probably involves a diradical intermediate.

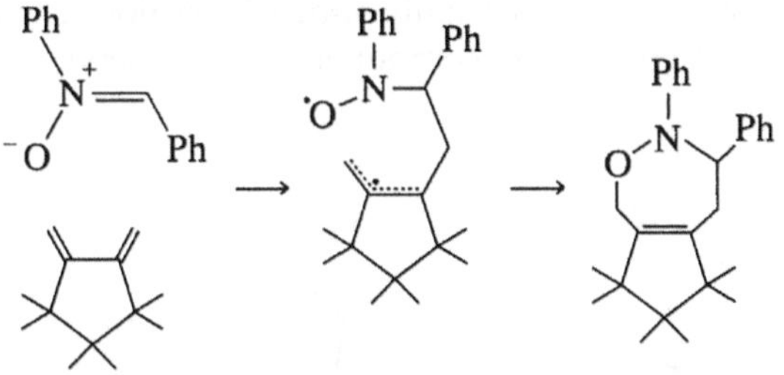

Another possibility is formation of an intermediate ion pair, **8-20**. However, in a nonpolar solvent like benzene, radical intermediates, which require less solvation than ions, seem more likely.

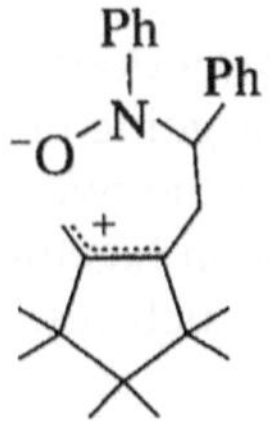

Formation of **8-2** and **8-3** also is possible through these intermediates.

The following reaction mechanism was ruled out by the authors cited: [4 + 2] cycloaddition of the diene with the nitrone to give tertiary amine oxide, **8-21**, which then thermally rearranges to the product. (Thermal rearrangement of a tertiary amine oxide to an alkylated hydroxylamine is called *Meisenheimer rearrangement*.)

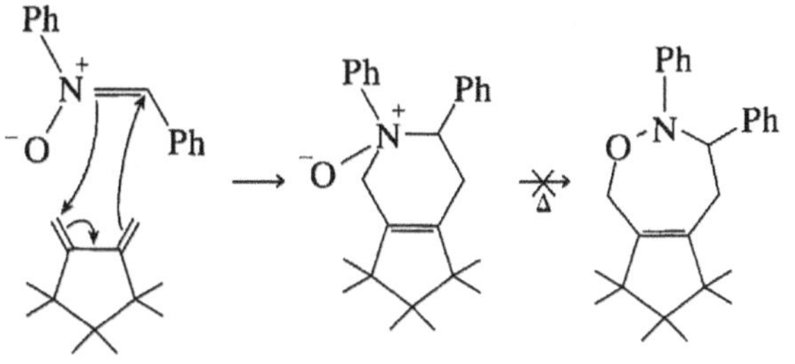

8-21   **8-4**

When **8-21**, synthesized by an alternate route, was subjected to the reaction conditions, it gave **8-22**, not **8-4**. Therefore, **8-21** cannot be an intermediate in the mechanism of **8-4**.

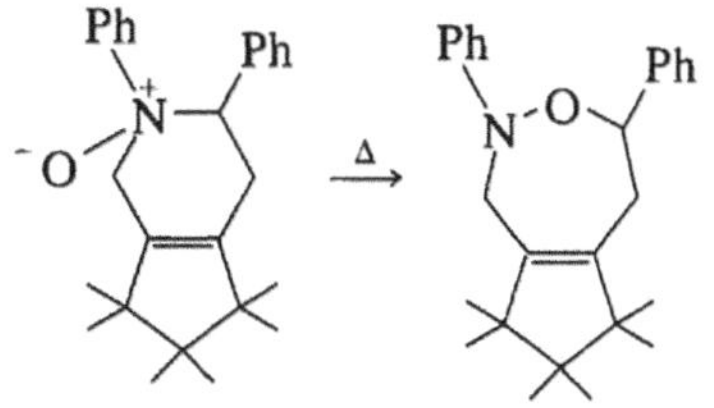

8-21   **8-22**

c. The lithium reagent must react with the carbonyl carbon of **8-5** to give **8-23** as an intermediate. Because neither of the double bonds in the starting material is conjugated with the carbonyl, 1,4-addition of the lithium reagent is not possible.

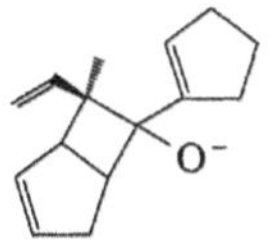

8-23

Compare the structures of the starting material **8-5**, intermediate **8-23**, and the product **8-6**. Note the following:

(1) The double bond in the five-membered ring of the starting ketone, **8-5**, is retained in the product.

(2) The other two double bonds in **8-23** have rearranged. The allyl double bond and the double bond in the new cyclopentenyl group are situated suitably for a [3,3]-sigmatropic shift.

(3) There is one methyl group in the starting material and two methyl groups in the product. The second methyl group must be introduced by an S$_N$2 reaction of methyl iodide. One of the methyls in the product is bonded to a carbon α to the carbonyl group. Thus, the nucleophile in the S$_N$2 reaction most likely is an enolate ion.

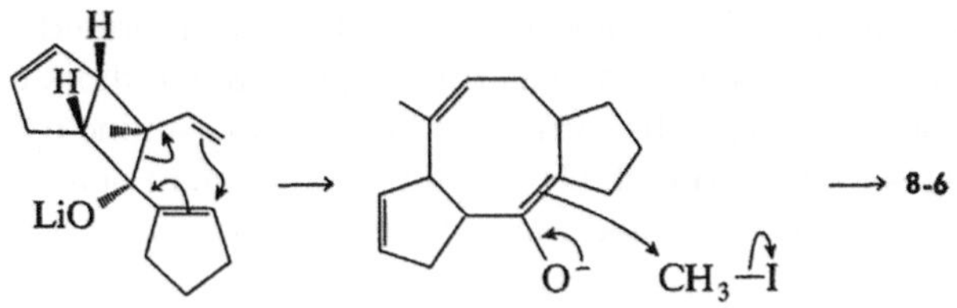

Ring opening to allyl anion **8-25** would be less likely because **8-25** would be considerably less stable than enolate anion **8-24**, which is resonance stabilized.

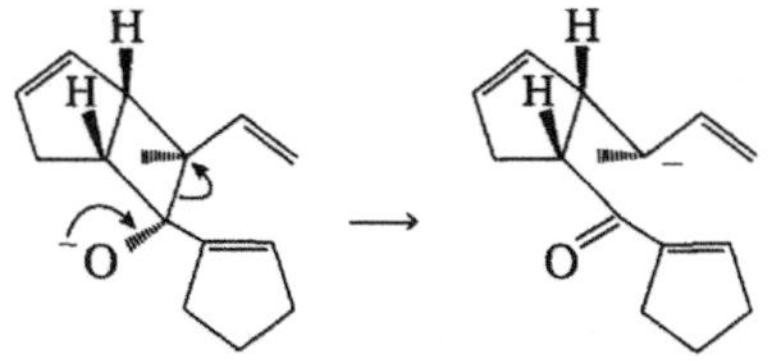

Problem 8.3

The usual numbering scheme shows that this is a [2,3]-sigmatropic shift involving a total of six electrons.

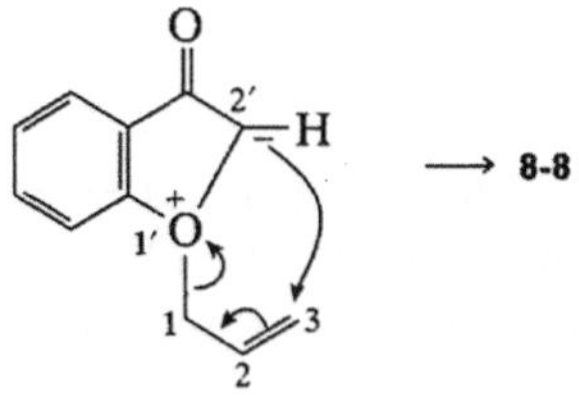

Another potential mechanism would be a [1,2]-sigmatropic shift of the allyl group. However, if concerted, this would be a $[\pi_s^2 + \sigma_s^2]$ reaction, which is not thermally allowed. For purposes of illustration *only*, suppose that a [1,2]-sigmatropic shift takes place. There are two imaginable ways to write the flow of electrons for this reaction. The usual arrow-pushing technique cannot be correct because it leaves a pentavalent, 10-electron carbon with a negative charge, which is not possible.

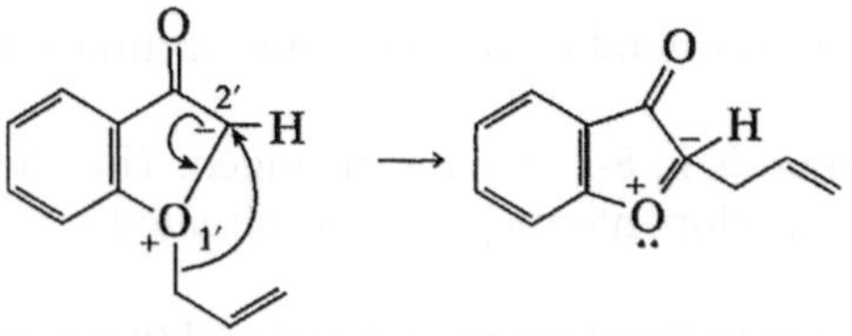

The method of writing electron flow, as we have been doing it in this book, *actually means the shift of a single electron from one atom to another*. That is, in the preceding mechanism, the two electrons that were on C-2′ are shared with oxygen in the product. C-2′ now is perceived as having just one of those original two electrons; that is, a single electron has shifted to the oxygen.

The other way to write the flow of electrons for a hypothetical [1,2]-shift would be transfer of two electrons from carbon-2′ to the oxygen, as illustrated in the following. But, as just reemphasized, reaction mechanisms are not written in this way.

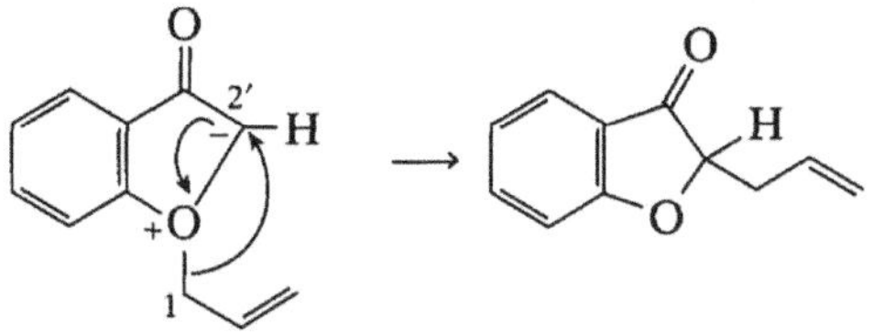

For further reading on this concept of the single-electron shift, see Pross, (1985).

Problem 8.4

Protonation of the hydroxyl group followed by loss of water gives an intermediate, **8-26**, which can undergo a [3,3]-sigmatropic shift to **8-27**. It is also reasonable to show neighboring group assistance in the loss of water leading to the intermediate **8-26a**, and then proceeding forward from this point.

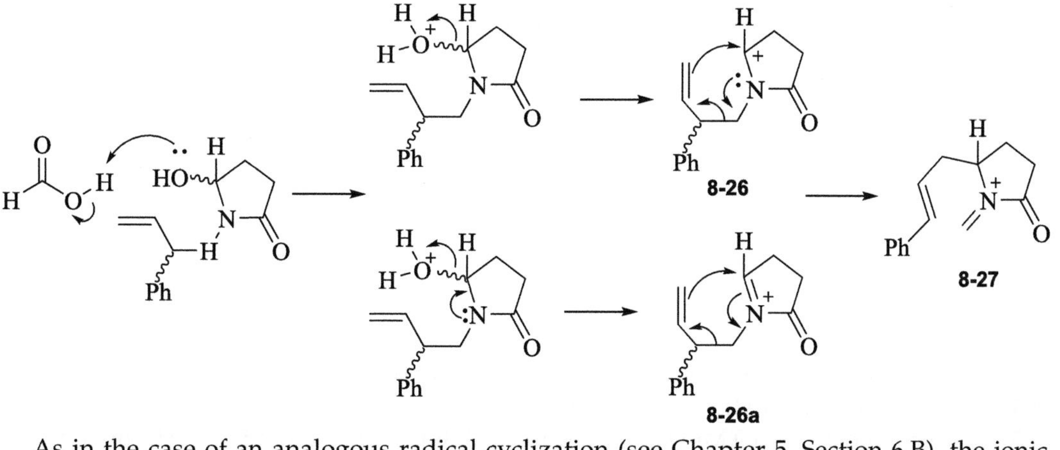

As in the case of an analogous radical cyclization (see Chapter 5, Section 6.B), the ionic cyclization of **8-27** to generate the five-membered ring is favored here. The reasons for this preference are subtle. With other substitution patterns, there might be steric and/or electronic factors that would favor cyclization to the six-membered ring. For example, it can be postulated that the phenyl ring stabilized carbocation is one factor contributing to the observed product.

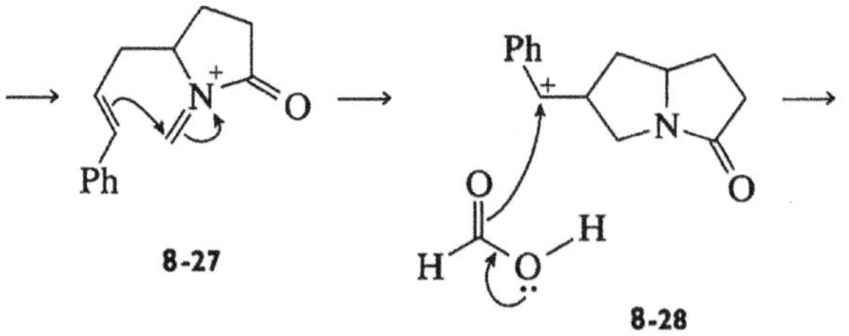

In the nucleophilic reaction of the carboxylic acid with the cation, **8-28**, the carbonyl oxygen, not the C—O oxygen, acts as the nucleophile because only then is the resulting intermediate stabilized by resonance.

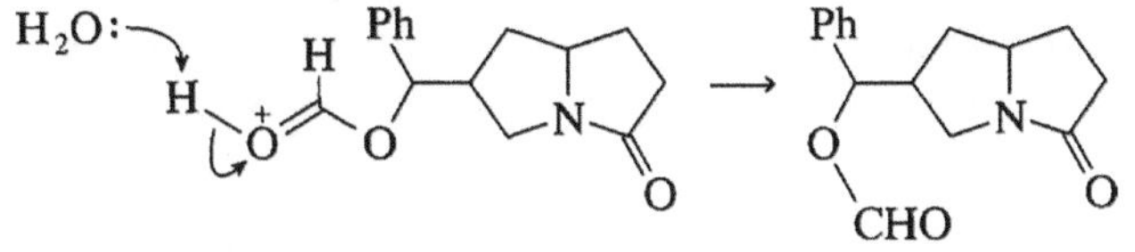

Problem 8.5

The first step in the reaction is removal of the most acidic proton in dibenzyl sulfide, **8-29**, by *n*-butyllithium to give anion **8-30**. This anion then undergoes a [2,3]-sigmatropic rearrangement to **8-31**. This rearrangement utilizes a σ-bond, the anionic electrons, and one π bond of one of the aromatic rings.

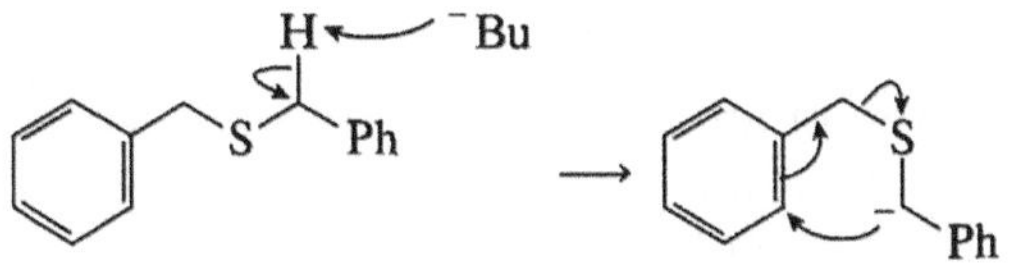

8-29   **8-30**

The intermediate, **8-31**, can be converted to the more stable benzylic anion, **8-32**, by a tautomerization.

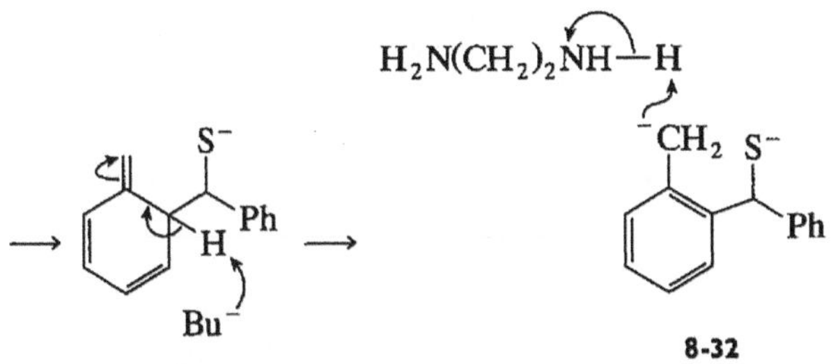

The base, which removes the proton from **8-31**, could be butyllithium, a molecule of anion **8-32**, or the anion from TMEDA. Anion **8-32** then can remove a proton from a molecule of **8-31** or from a molecule of TMEDA to produce **8-33**. It is unlikely that **8-32** picks up a proton from butane because of the high pK_a value of butane relative to the amine or **8-31**. (Compare the following pK_as from Appendix C: diisopropylamine, 36 or 39; ethane, 50; toluene, 43; and ammonia, 41.)

A concerted [1,3]-thermal shift of hydrogen to convert **8-31** to **8-32** is not a reasonable mechanism because such shifts are ruled out by the selection rules for pericyclic reactions (Chapter 6).

The last step is an S_N2 reaction of the nucleophilic sulfur anion with methyl iodide to give the methylated sulfur.

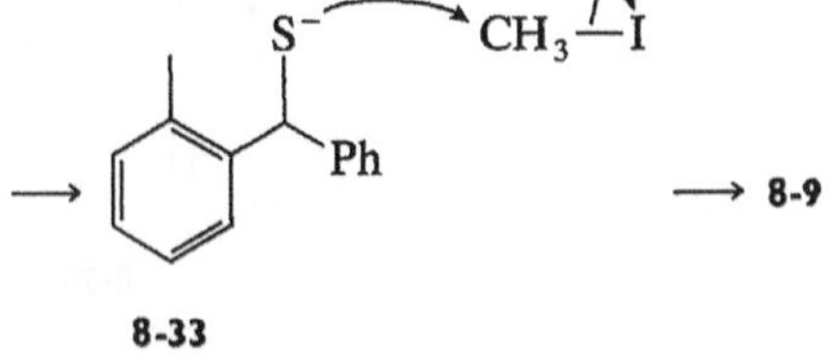

8-33

Problem 8.6

The molecular formula of **8-10** indicates that trichloroacetonitrile has reacted with the starting material. By using Hint 1.2, three rings and/or π bonds are calculated. Because the starting alcohol has two π bonds and the nitrile has two π bonds, a nucleophilic reaction with the nitrile is likely. Potassium hydride should remove the most acidic proton from the starting alcohol, the proton on the oxygen.

There are two possible condensation reactions for the nitrile. One is nucleophilic reaction of the oxyanion with the carbon of the nitrile functional group. This carbon is activated by nitrogen and by the strongly electron-withdrawing trichloromethyl group. The other possibility, S_N2 displacement of chloride, is ruled out because there are three chlorines and four π bonds in the product. After the nucleophilic reaction, a trace of methanol is needed to form a neutral product by protonation of the anion. The second reaction is isomerization. That is, the molecular formula of **8-11** is the same as that of **8-10**. Because xylene is an unreactive aromatic hydrocarbon, a reasonable assumption is that it functions only as the solvent. Therefore, the first thing to look for is a concerted reaction. A [3,3]-sigmatropic reaction is possible (Look for two double bonds separated by three single bonds).

8-10

8-10   8-11

Problem 8.7

a. Two types of condensations occur. One is the reaction of an amine with a carboxylic acid, and the other is the reaction of an amine with a ketone. The latter should occur more readily (and is written first in the following mechanism) because the resonance effect of the hydroxyl of the carboxylic acid group reduces the electrophilicity of its carbonyl group compared to that of a ketone. Two rings are formed in this reaction. Either the smaller five-membered ring or the larger eight-membered one could be formed first. Entropy factors would favor more rapid formation of the five-membered ring.

The first step is protonation of the ketone carbonyl group by the carboxyl group. This might be either an intermolecular or an intramolecular reaction.

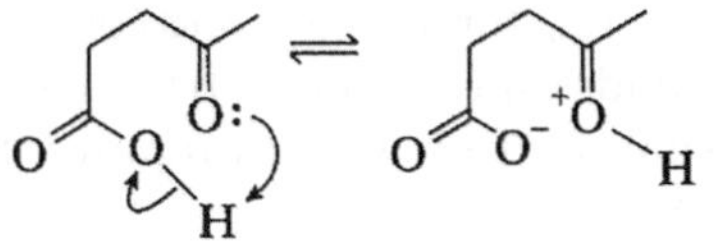

Then the nitrogen of the aniline can react, as a nucleophile, at the electrophilic carbon of the protonated carbonyl group:

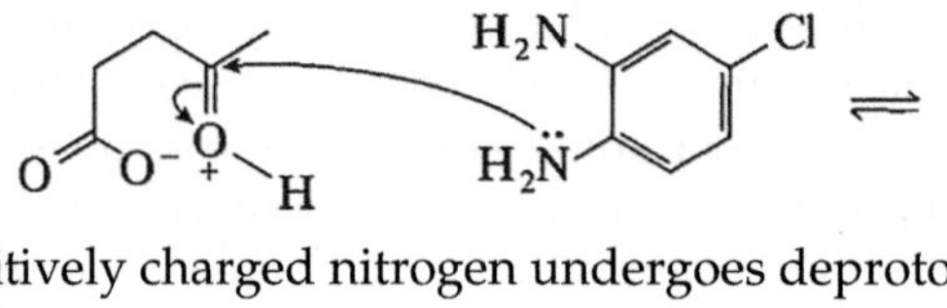

Subsequently, the positively charged nitrogen undergoes deprotonation, followed by protonation of the hydroxyl group.

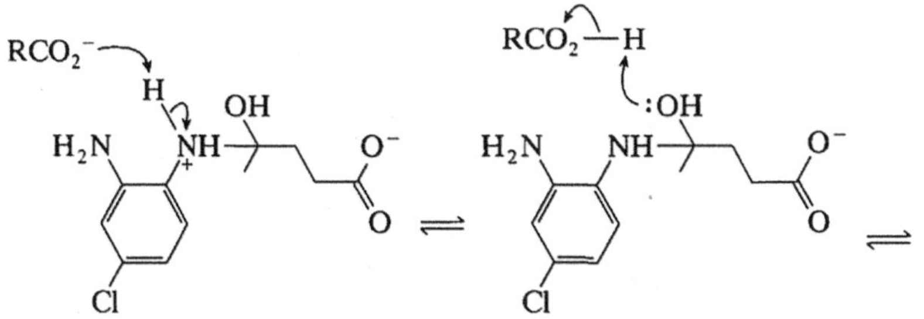

Loss of a molecule of water leads to an iminium ion, **8-35**, which is susceptible to nucleophilic reaction at carbon:

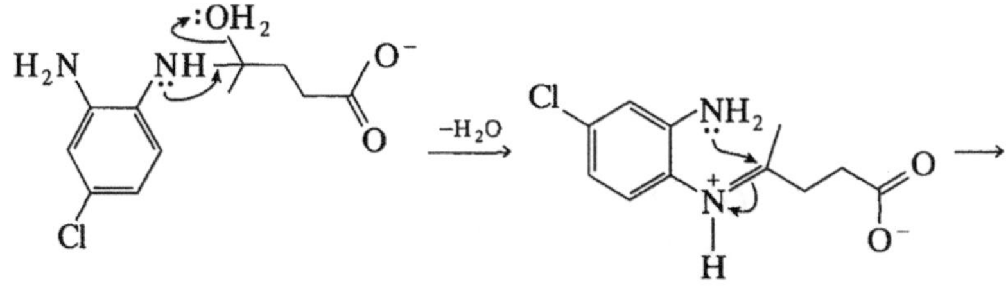

8-34   8-35

After deprotonation of the positively charged nitrogen, one of the nitrogens acts as a nucleophile with the electrophilic carbon of the neutral carboxy group. In the structures that have been written up to this point, the carboxylate anion has been indicated. However, this anion and the neutral carboxylic acid can be written interchangeably because they both would be present under the reaction conditions. (As stated previously, entropy favors the initial formation of the five-membered ring because of the low probability for productive collision of the ends of an eight-atom chain.)

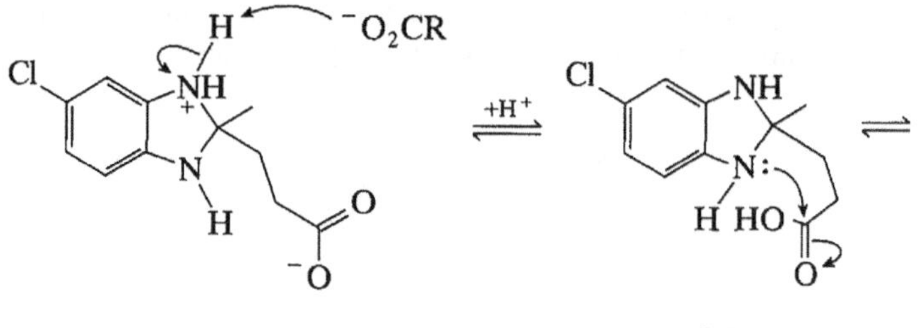

8-36

Protonation of oxygen and deprotonation of nitrogen occur.

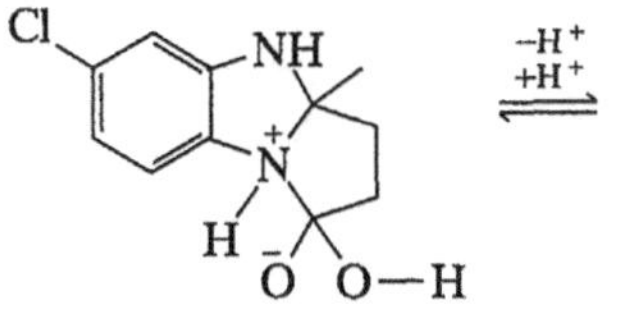

Then, protonation of one of the hydroxyl groups, elimination of water, and deprotonation give the product.

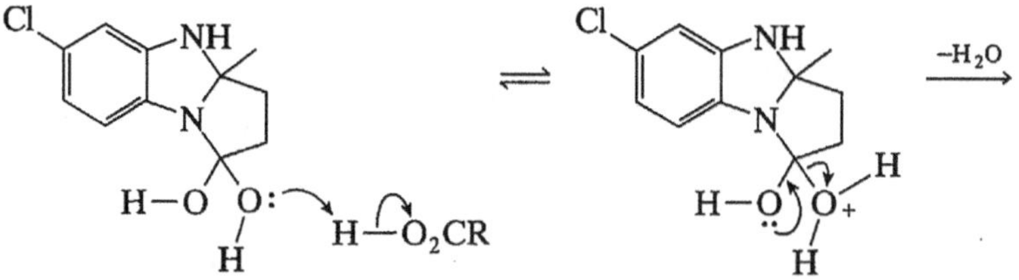

During the formation of **8-14**, water is lost in two separate steps. Both steps are reversible, but the iminium ion **8-35** is hydrolyzed much more readily than the amide group in **8-14**. Azeotropic removal of water by reflux in toluene drives both reactions toward completion.

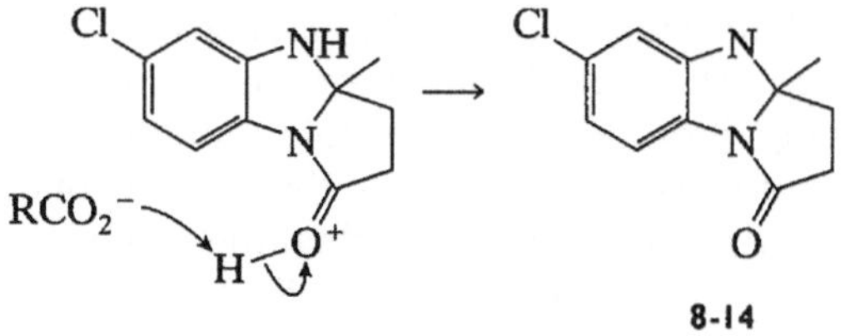

8-14

Additional Comments

(1) The following equilibrium, which undoubtedly occurs under the reaction conditions, is unproductive. That is, it is not a reaction of the amine that is on a pathway to the product.

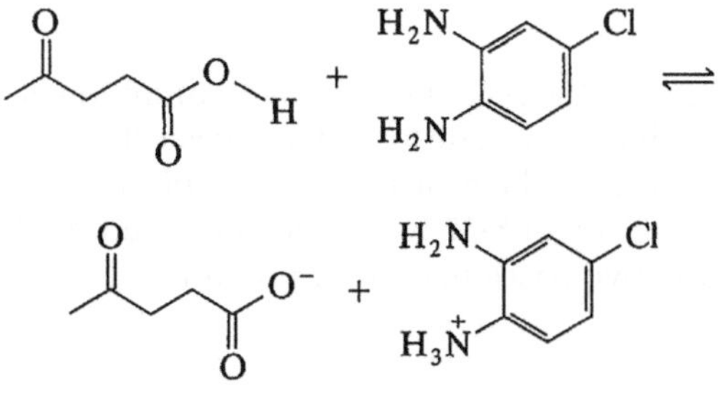

(2) There are several possible mechanisms for the loss of water from the protonated aminol, **8-34**. A partial structure representing the possibility depicted earlier is as follows:

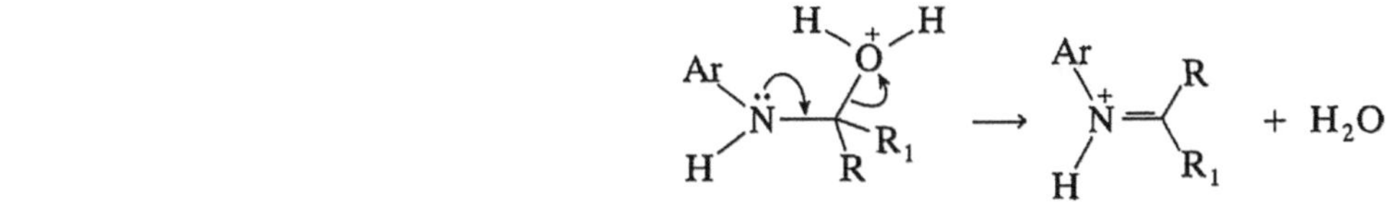

8-34   **8-35**

Another possibility is the simultaneous loss of a proton from the nitrogen:

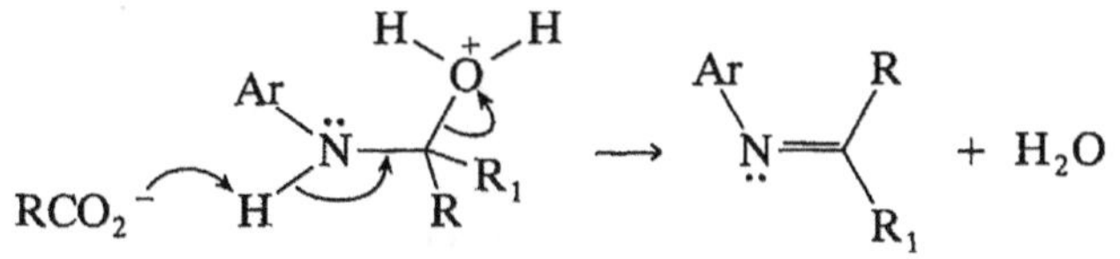

Finally, the proton might be removed from the nitrogen before elimination of water takes place:

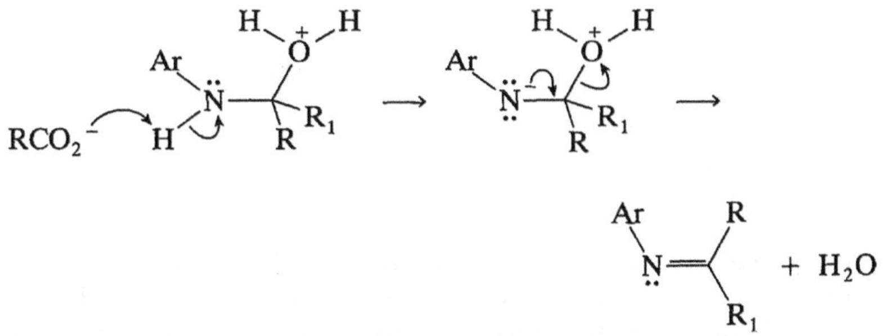

This last mechanism is an example of an E_{1CB} elimination, which is not common unless the intermediate anion is especially stabilized by strongly electron-withdrawing groups. Although the anion produced in this reaction is resonance-stabilized by the ring, it probably is not stable enough to be produced without the driving force of simultaneous formation of the double bond. Also, the bases usually used in E_{1CB} reactions are much stronger than the carboxylate anion.

(3) Estimations of the basicity of carboxylate ion **8-37**, formed from **8-13**, and the starting aniline, **8-12**, suggest that either one may be used as the base in the mechanisms written.

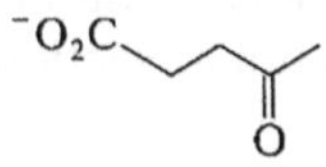

8-37

From Appendix C, the pK_a of p-chloroanilinium ion is 4.0, and that of acetic acid is 4.76. This means that the acetate ion is more basic than p-chloroaniline, but by less than a factor of 10. However, the basicity of **8-12** would be enhanced by the second amino group, and the basicity of **8-37** would be decreased by the keto group. Thus, the basicities of these two bases may be approximately the same.

(4) The anion **8-38**, which is formed by removal of a proton from one of the amino groups, would not function as a base in this reaction because the equilibrium to form **8-38** is completely on the side of starting material. This can be calculated from the values given in Appendix C. The hydronium ion, H_3O^+, ($pK_a = -1.7$) is at least 10^{20} times more acidic than the substituted aniline (pK_a of m-chloroaniline is 26.7).

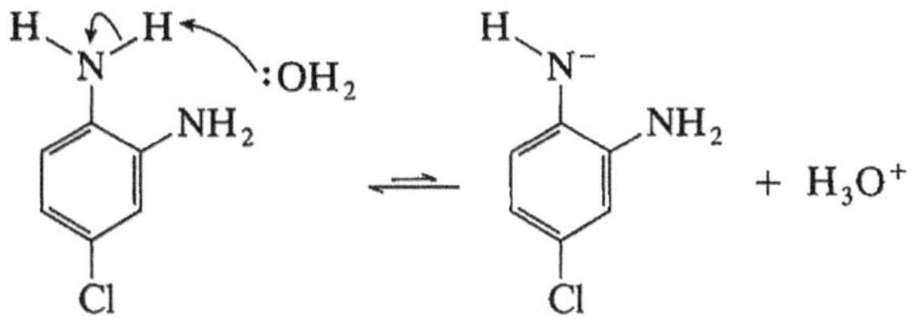

8-38

b. Formation of the isomeric product **8-15**.

If the other nitrogen in **8-36** reacts with the carboxyl carbon, the following isomeric product would be formed:

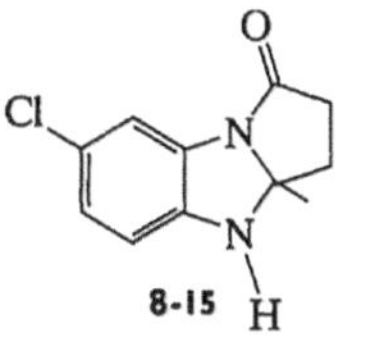

8-15

The nitrogens in the starting amine are not identical, so that **8-15** also could be produced by the nitrogen *meta* to chlorine, acting as the original nucleophile. The mechanism then would be written just as it was in part a. See the paper cited for experimental evidence that supports just one of these two possibilities.

Problem 8.8

Protonation of the nitrogen is followed by reduction by zinc. This is a two-electron reduction, in which zinc is oxidized to Zn^{2+}. The remainder of the mechanism involves acid-catalyzed reactions.

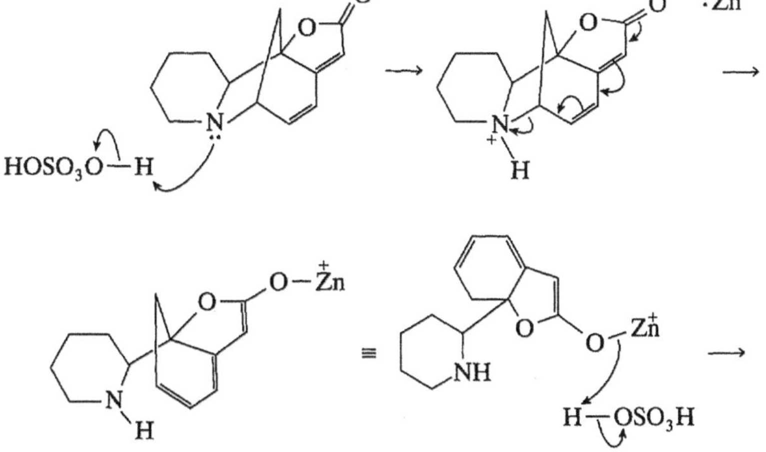

8-39

→ Product

An alternative route from **8-39** involves formation of the remaining six-membered ring before the five-membered ring opens:

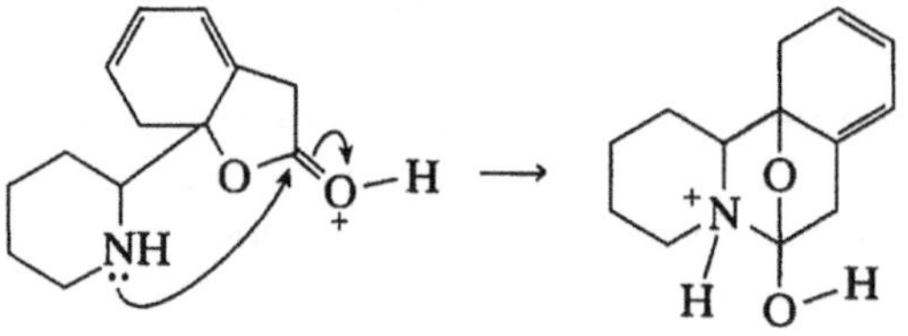

8-39

Subsequent acid-catalyzed steps will also give the product.

If you failed to use zinc in this reaction, a reduction would not occur. The product of acid-catalyzed reactions of the starting material then would contain two less hydrogens, as in **8-40**.

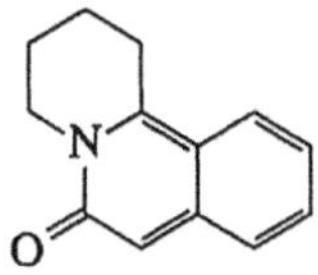

8-40

Problem 8.9

Application of Hint 2.15 may help to organize your thoughts about how the products might form. Formation of the first product involves the loss of one carbon. It appears most likely that this would be CO_2 or CO. It is probably CO_2 because CO usually is lost only in strong acids or, under thermal conditions, in reverse Diels–Alder reactions or free radical decarbonylation of aldehydes.

Once the atoms in the starting material, **8-16**, have been numbered, the carbons attached to the R group and the methoxy group in product **8-17** can be numbered.

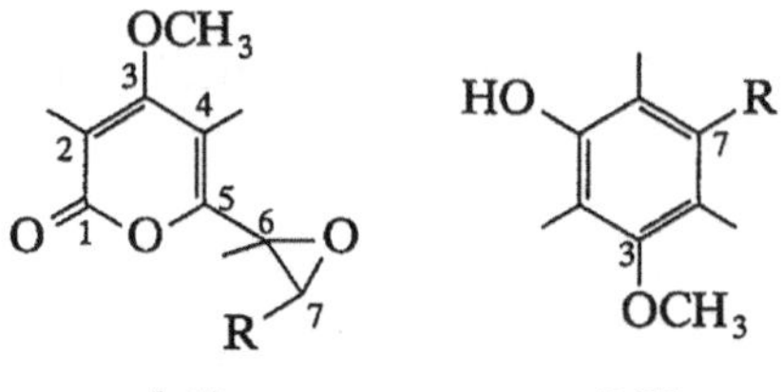

8-16   **8-17**

There are two possible numbering schemes for the remaining atoms in **8-17**:

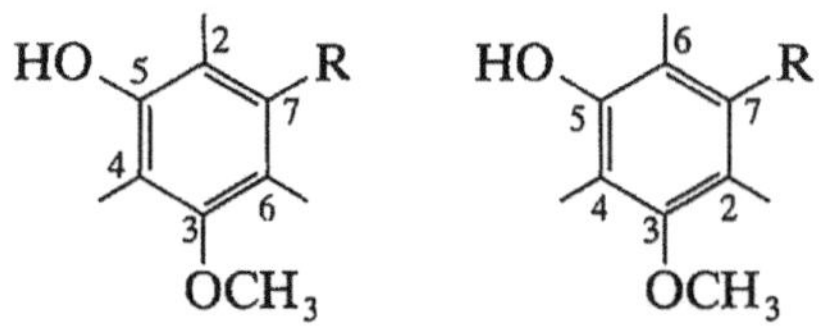

8-41   **8-42**

Getting to **8-42** involves much less bond making and breaking than getting to **8-41**. Thus, application of Hint 2.17 suggests that **8-42** is a better guide to what is occurring than **8-41**. That is, C-2 becomes bonded to C-7, and C-1 is lost. All other carbons remain attached to the same atoms.

Because the reaction conditions are basic, the next thing to do is to search for protons in the molecule that are acidic and for carbons susceptible to reaction with a nucleophile. At this point, we will look at possible nucleophilic reactions and will return to acidic protons later in the discussion. Two obvious electrophilic carbons are those of the epoxide. Moreover, if the C-2—C-7 bond is formed by nucleophilic reaction of an anion at C-2 on carbon C-7, ring opening of the epoxide would also occur in this step. This ring opening would be a driving force for the reaction. Thus, we should search for steps that would lead to an anion at C-2. This is not likely to be obvious from initial inspection of the product. In addition to the epoxide carbons, electrophilic positions in **8-16** include the carbonyl carbon (C-1) and the positions β and δ to it (C-3 and C-5).

One other comment should be made before we start writing a mechanism. The acidities of methanol ($pK_a = 15.5$) and water ($pK_a = 15.7$) are practically identical, so that methoxide or hydroxide ion as base and methanol or water as acid can be used interchangeably in the mechanism.

$$^-OH \ + \ HOCH_3 \ \rightleftharpoons \ HOH \ + \ ^-OCH_3$$

We need to start trying the various nucleophilic reactions suggested previously. We will start by adding hydroxide ion to the carbonyl group. Then a ring-opening reaction can occur.

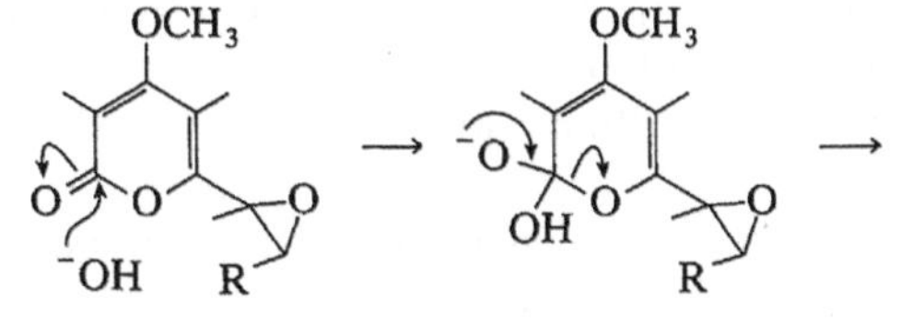

8-16

This ring-opening reaction leads to **8-43**, a resonance-stabilized anion:

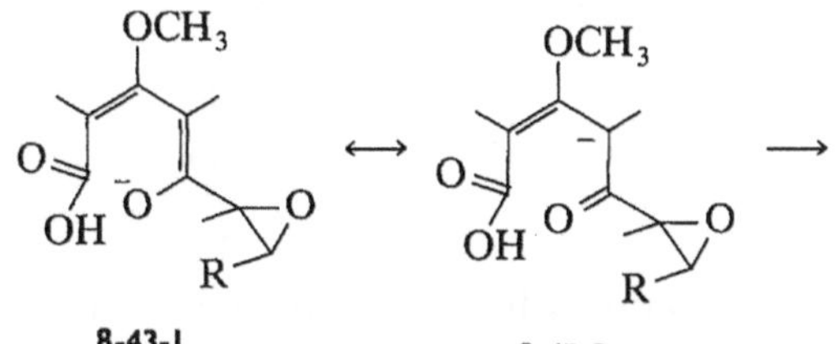

8-43-1　　　　　　　　　　**8-43-2**

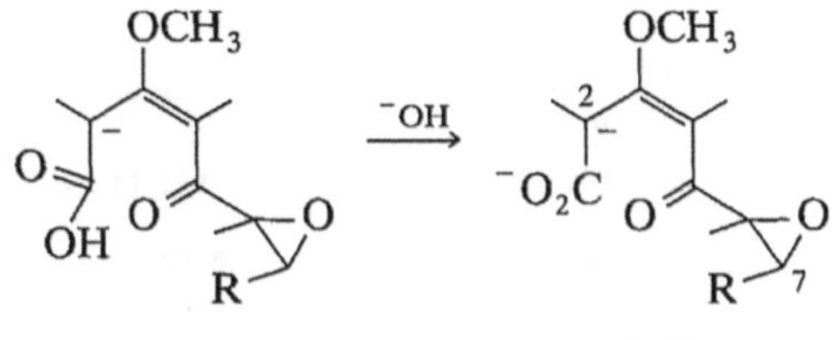

8-43-3　　　　　　　　　　**8-44**

In the basic medium, the carboxylic acid will be converted rapidly to the carboxylate anion, **8-44**. Rotation in this intermediate brings C-2 and C-7 close enough to react with each other. After the ring opening of the epoxide, the resulting oxyanion can remove a proton from the solvent to give the corresponding alcohol.

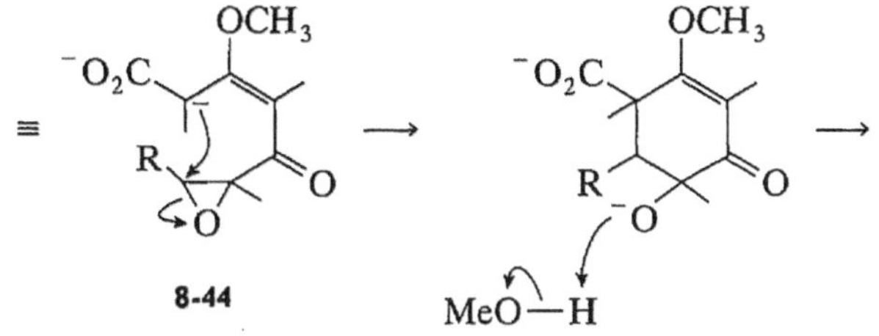

8-44   MeO—H

The relationship of the carboxylate ion to the ketone carbonyl makes it possible to write a mechanism analogous to the decarboxylation of acetoacetic acid. This suggests that decarboxylation should occur readily.

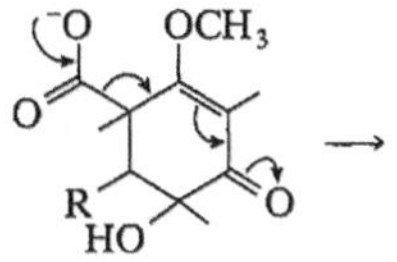

The resulting anion then undergoes base-promoted elimination of water to give the anion of a phenol, **8-45**. This adds a proton, upon workup, to give the neutral product, **8-17**. The driving force for this elimination is the formation of an aromatic system. Hence, this reaction occurs with facility, even though the leaving group is a poor one.

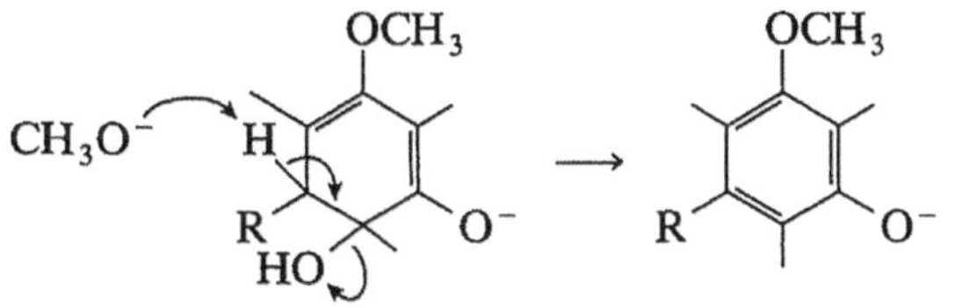

8-45

Another possible mechanism for transforming **8-44** to product involves a different sequence of reactions. Thus, **8-44** might pick up a proton to give **8-46**.

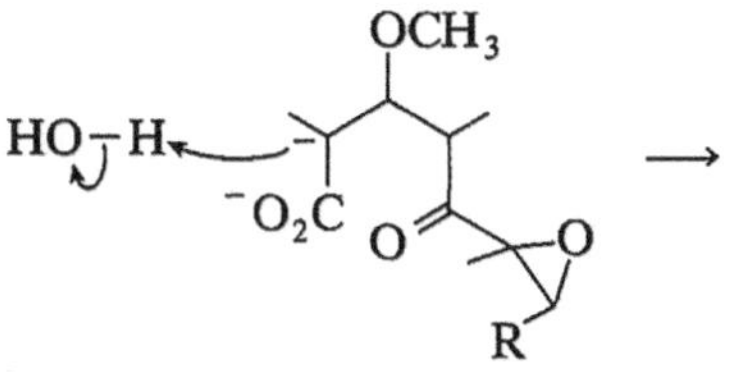

8-44

This intermediate can then decarboxylate to give an anion, which can then ring close with the epoxide.

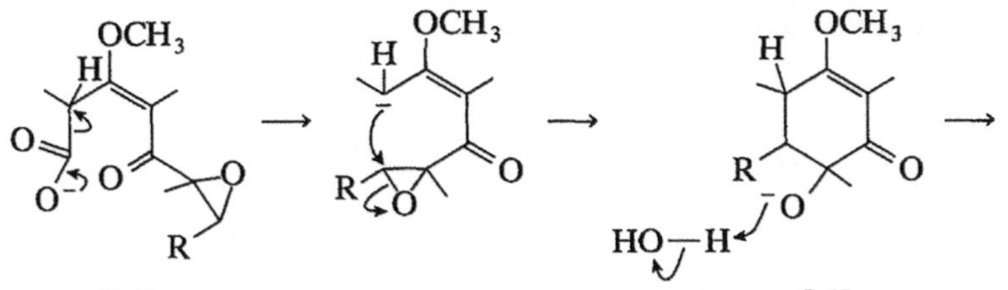

8-46   **8-47**

Protonation of **8-47** by water or methanol gives an intermediate that can undergo base-promoted elimination of water and tautomerization to give product **8-17**.

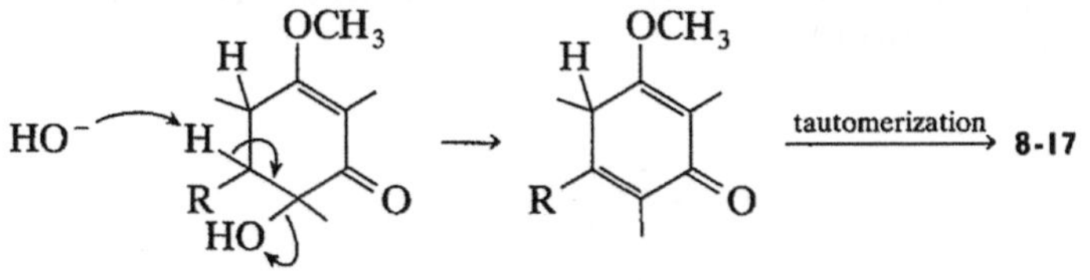

Another possible mechanism for the formation of **8-17** begins with nucleophilic addition of hydroxide to C-5 of **8-16**. The resulting anion reacts with the epoxide ring, which opens to give **8-48**.

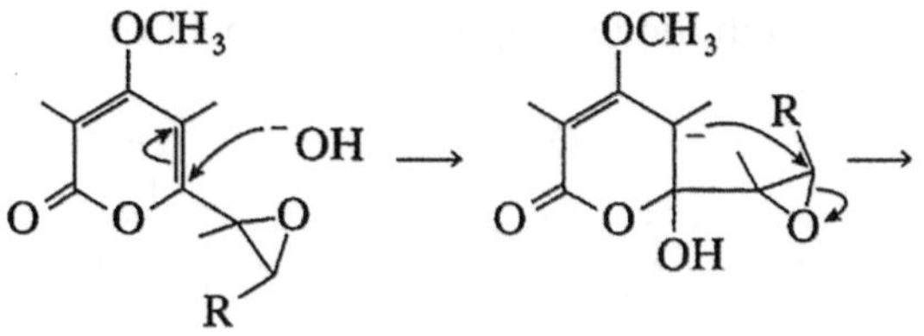

8-16

Intermediate **8-48** picks up a proton on one oxygen and loses a proton from another oxygen.

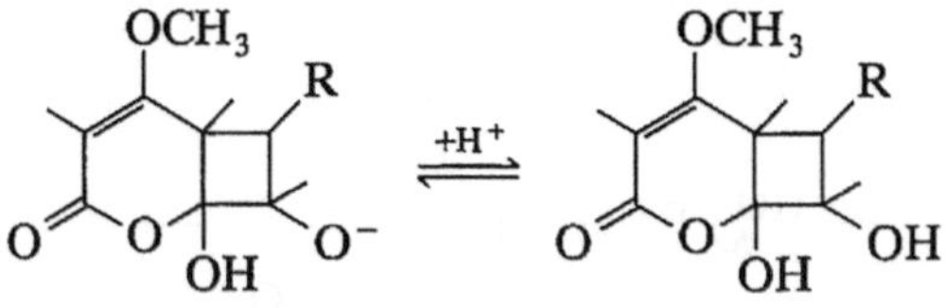

8-48

The six-membered ring of the new alkoxide ion, **8-49**, then opens with decarboxylation, to give an anion, **8-50**. Cyclization then occurs by reaction of the anion with the carbonyl group to give **8-51**.

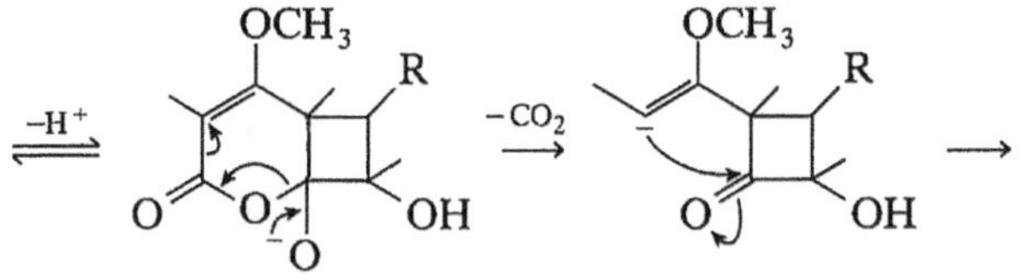

The remaining steps from **8-51** to product are protonation, electrocyclic ring opening, and elimination of water.

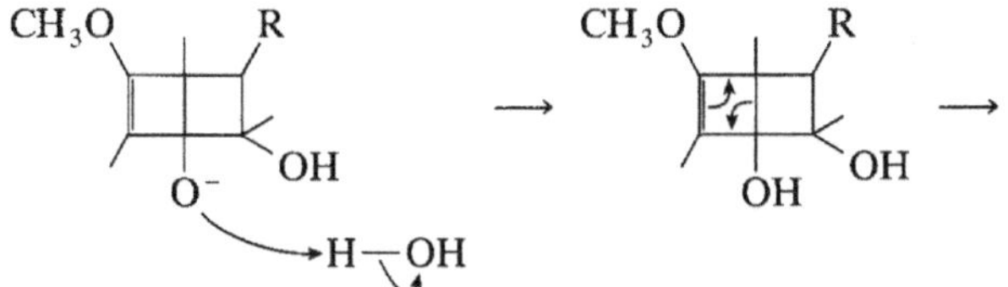

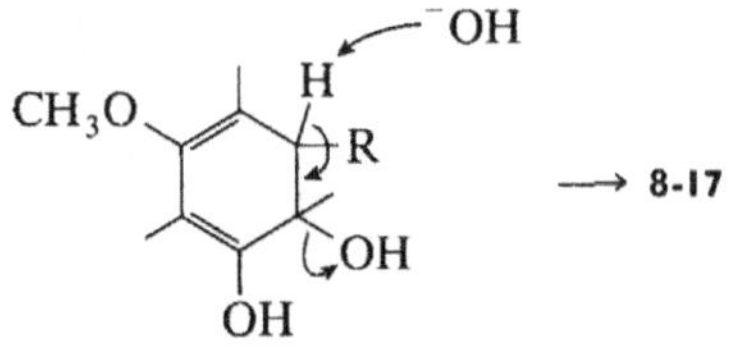

This mechanism has several disadvantages when compared to those given previously. Intermediate carbanion **8-50** is relatively unstable because its negative charge is localized on an sp^2-hybridized carbon. In contrast, the carbanions of the other mechanisms are stabilized by delocalization. Another intermediate of this mechanism, the bicyclo[2.2.0]hexenyl system, is not very stable because of high strain energy.

Mechanism for the formation of **8-18**.

Use of Hint 2.15 to analyze the relationship between **8-16** and **8-18** gives the following:

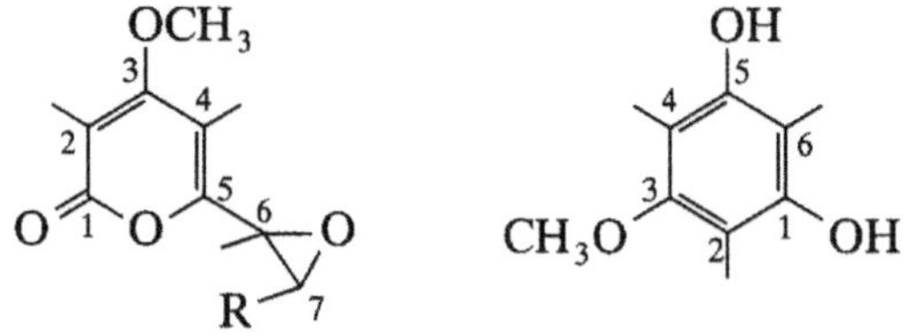

This numbering scheme in the product gives the fewest changes in bond making and bond breaking and shows that the only new connection is between C-6 and C-1. C-7 and the R-group attached to it are not present in the product. The paper cited shows the following mechanism for this transformation:

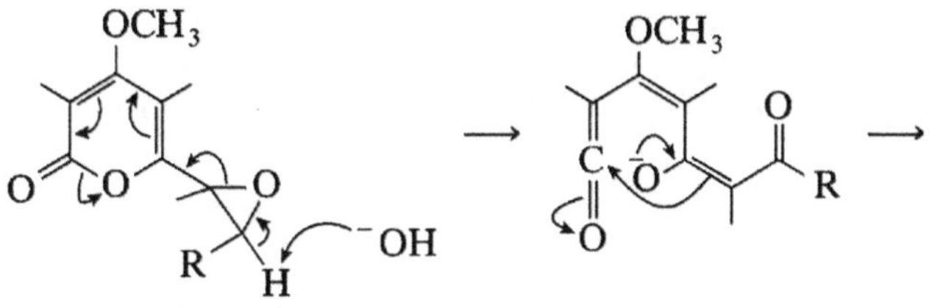

8-16

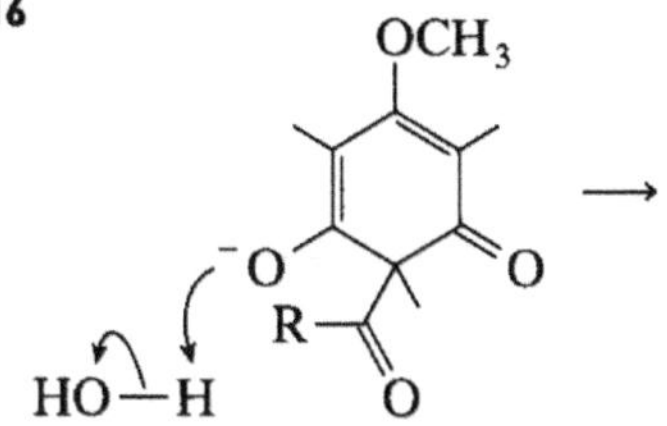

8-52

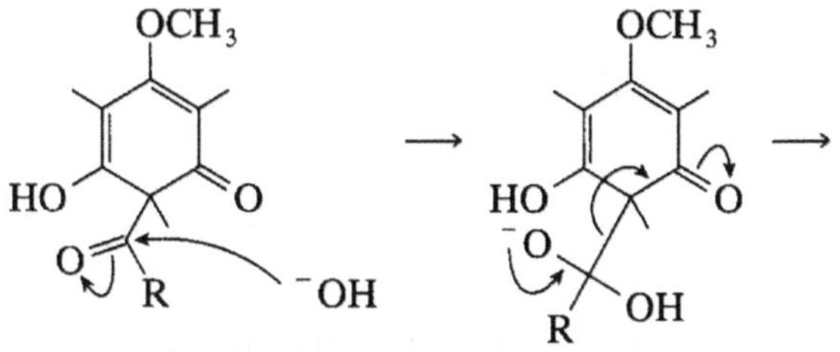

The phenolate then can pick up a proton during acidic workup to give the product phenol.

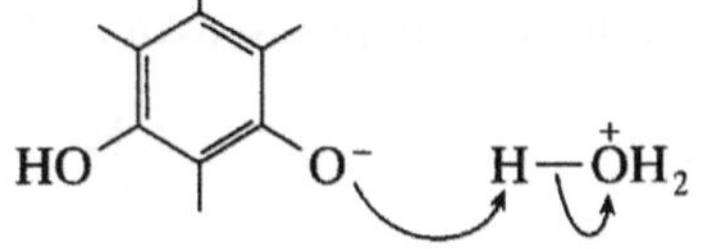

→ 8-18

An alternative mechanism for the formation of intermediate **8-52** might be nucleophilic addition of hydroxide to C-5 of **8-16**, followed by ring opening and subsequent modification of the epoxide:

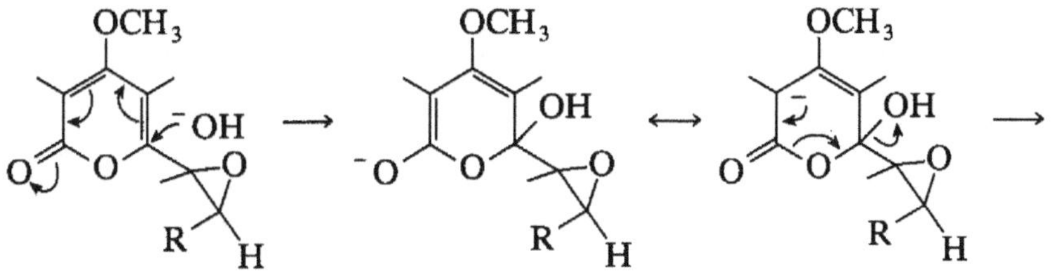

8-16

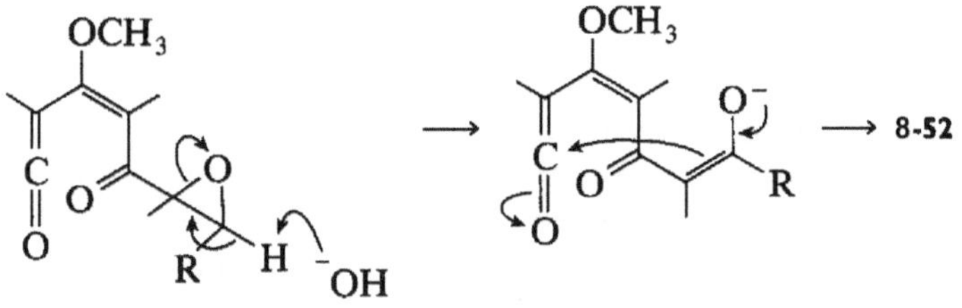

The following reaction was proposed by a student as part of a mechanism for the reaction of **8-16**.

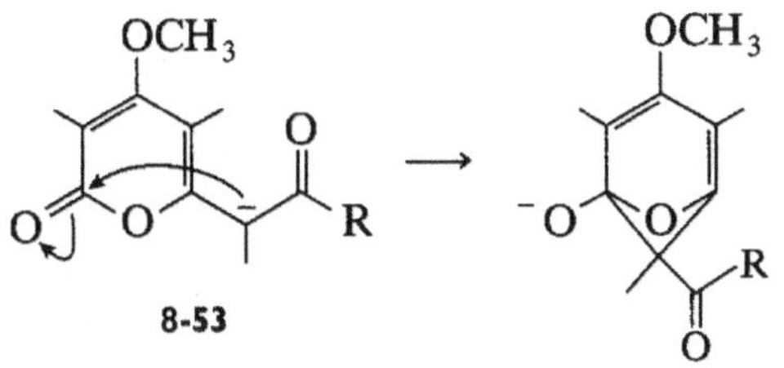

8-53

8-54

This mechanistic step suffers from three problems. First, anion **8-53** is attacking itself! This anion is a resonance hybrid; one of the resonance forms, **8-53-2**, shows that there already is considerable electron density at the carbonyl group.

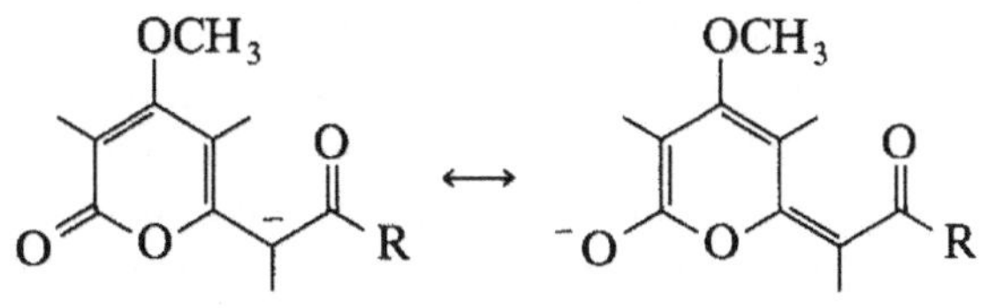

8-53-1

8-53-2

Second, all the atoms involved in resonance stabilization must be coplanar. Thus, the exocyclic carbanion is not in a position geometrically favorable for reaction with the carbonyl group. Third, in **8-54**, there is a double bond at a bridgehead that involves considerable strain energy.

Another student wrote the following as the first step in the reaction:

8-16   8-55

Anion **8-55** is not stabilized by resonance in any way. The acidity of the proton ring, relative to that of normal hydrocarbons, is enhanced by the higher s character in the C—H bond and by the adjacent oxygen. Nonetheless, neither of these factors enhances its acidity enough to bring it close to the pK_a's of methanol or water. The mechanism given in the paper cited starts with removal of this proton, but makes it part of a concerted reaction in which the epoxide ring is opened. In this case, the release of strain energy and the stability of the resulting resonance-stabilized anion provide considerable driving force.

Problem 8.10

In analyzing the first cyclization, it may be useful to ask ourselves what kind of intermediate is involved. We note that 2 mol of the manganese(III) compound and 1 mol of the copper(II) compound are used in the reaction. Manganese has stable oxidation states from +2 to +7, so that manganese(III) could function as either an oxidizing or a reducing agent. Copper is stable in the +1 and +2 states, so that the copper(II) compound would be expected to function as an oxidizing agent. By comparing the formulas of the starting material and product, we can see that the product has two less hydrogen atoms than the starting material. How is the hydrogen lost? As the first step, we could conceive of loss of a proton, a hydrogen atom, or a hydride ion. In acidic solution, loss of a proton to form a carbanion is not likely to occur. Unassisted loss of a hydride ion is an unlikely process, but loss of hydride by transfer to one of the transition metals (especially manganese) is a possibility. However, there are no positions in the starting material where transfer of a hydride could lead to a reasonably stable carbocation. On the other hand, abstraction of a hydrogen atom from the β-ketoester group leads to a radical that is stabilized by resonance with both the keto and ester groups. The authors of the article cited suggest that this radical (**8-56**) is an intermediate in the cyclization. This suggestion seems reasonable, especially because it is known that heating manganese(III) acetate produces the radical $\cdot CH_2CO_2H$, which could abstract a hydrogen atom from the starting material.

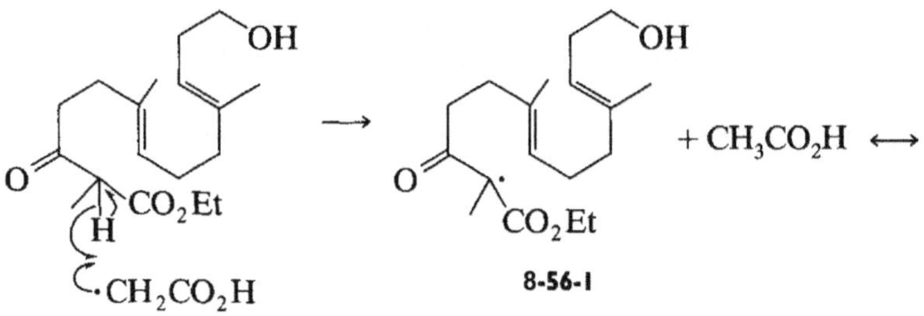

8-56-1

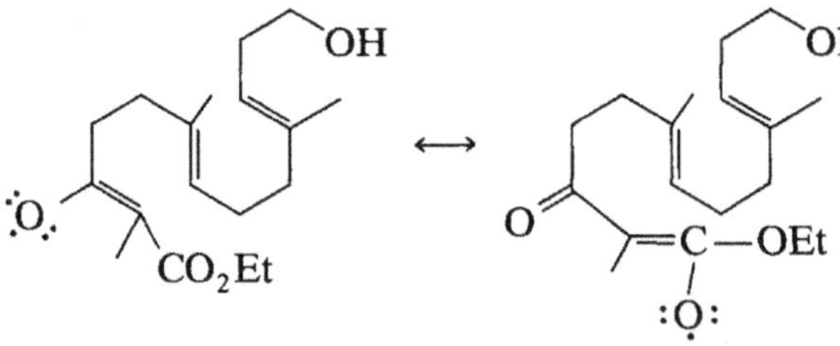

8-56-2   8-56-3

The radical **8-56** could then cyclize as shown to form the ring system of the tricyclic intermediate.

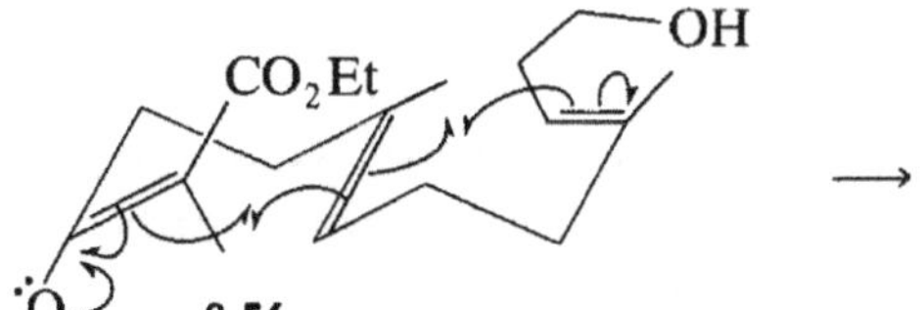

8-56

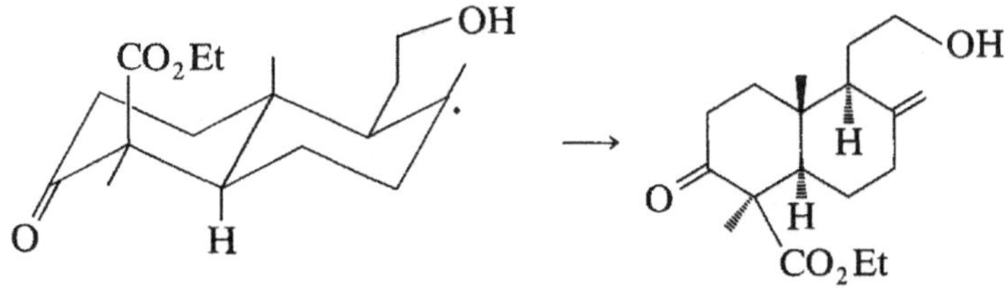

8-57

8-58

Notice that although there are four asymmetric centers in the molecule, one isomer is formed in 58% yield. This points to a concerted cyclization step, in which the stereochemistry incorporated in the C=C double bonds is transferred to the developing asymmetric centers in the product. (This phenomenon is analogous to the well-known stereospecific formation of the steroid skeleton by cationic cyclizations of similar olefinic systems with defined stereochemistry at the C=C double bonds.)

Although a radical intermediate can be written for this transition-metal-mediated reaction, this is not a radical chain reaction because the transition metal reagents are employed in stoichiometric amounts. In addition, if the reaction were to proceed by a chain reaction, one might reasonably expect to obtain a different product (**8-59**) as a result of hydrogen abstraction.

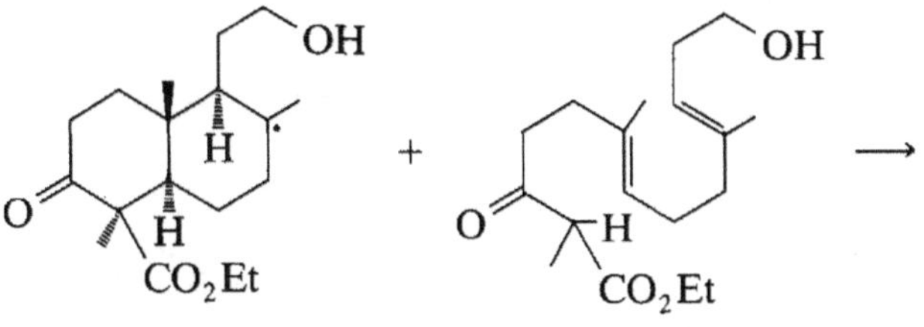

8-57

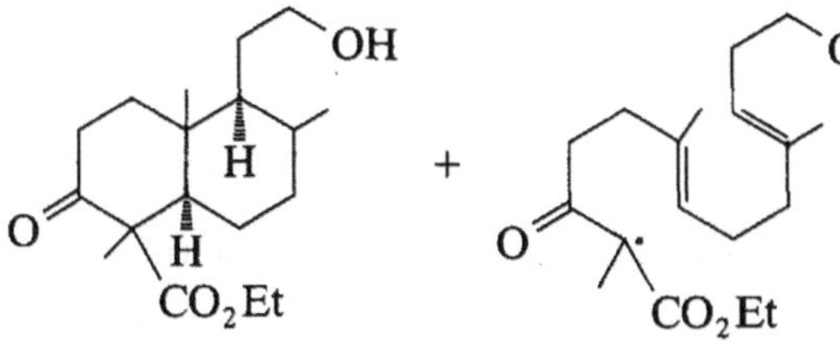

8-59   8-56

How can we get from the cyclized radical **8-57** to the product? One possibility is to transfer an electron to the copper(II) acetate to form a tertiary carbocation, **8-60**, which then loses a proton to form the exocyclic olefin.

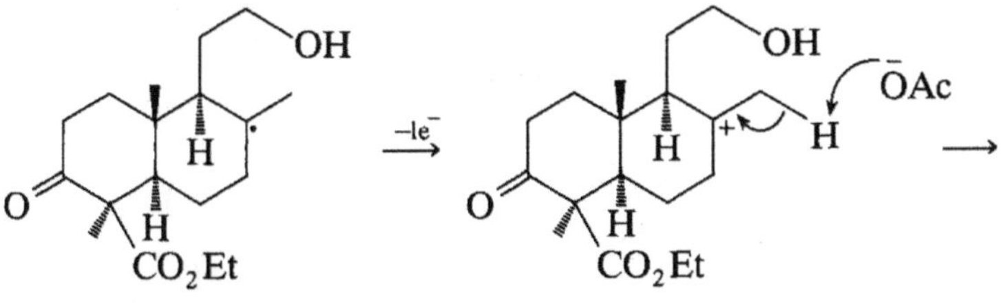

8-57   8-60

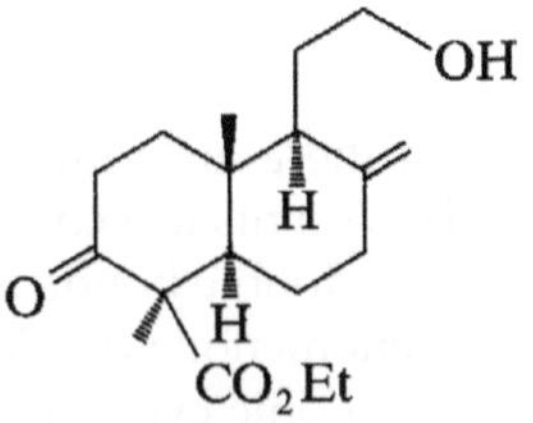

8-58

The second step of the synthetic sequence is relatively straightforward. The exocyclic olefin is protonated in trifluoroacetic acid to form the tertiary carbocation, which cyclizes by an intramolecular S_N1 reaction to give the final product.

8-60   8-61

If the carbocation **8-60** is an intermediate in the first reaction, why does it not immediately cyclize to form the final product? To answer this, it may help to look at the reaction conditions. The first cyclization is carried out in acetic acid, whereas the final cyclization uses trifluoroacetic acid. To form the exocyclic olefin, the loss of a proton from **8-60** to form the exocyclic methylene group in **8-58** must be a rapid process compared to reaction with a nucleophile, even by the intramolecular hydroxyl group. Once **8-58** is formed, acetic acid is not a strong-enough acid to form an appreciable concentration of the tertiary carbocation **8-60**. However, in trifluoroacetic acid (a much stronger acid), **8-58** can be protonated to reform **8-60** as many times as necessary to give the hydroxyl group time to react, and the reaction can proceed to the more stable product, **8-61**.

Problem 8.11

In this classic transformation, the authors were confronted with a situation that may have seemed unworkable. The ester carbon center is initially out of reach for the cyclization reaction required to get to product. The necessary stereochemical relationship is in fact unfavored due to the steric demands encountered when the ester is pressed into the space necessary for the bond-forming step **8-64**.

The participating carbon centers are
not in proximity with one another

8-64

The transformation works because as the aminoester epimerizes and the two esters come into close proximity **8-64a** (the more sterically demanding orientation), there is a chance that the bond will be formed and locked in. This bond-forming step **8-65** is an intramolecular Claisen reaction known as the Dieckmann condensation and once the new bond has been formed, the ketal intermediate decomposes and methanol is ejected. The acidic proton of the newly formed beta-keto ester **8-66** is then lost providing the enol ester **8-17**.

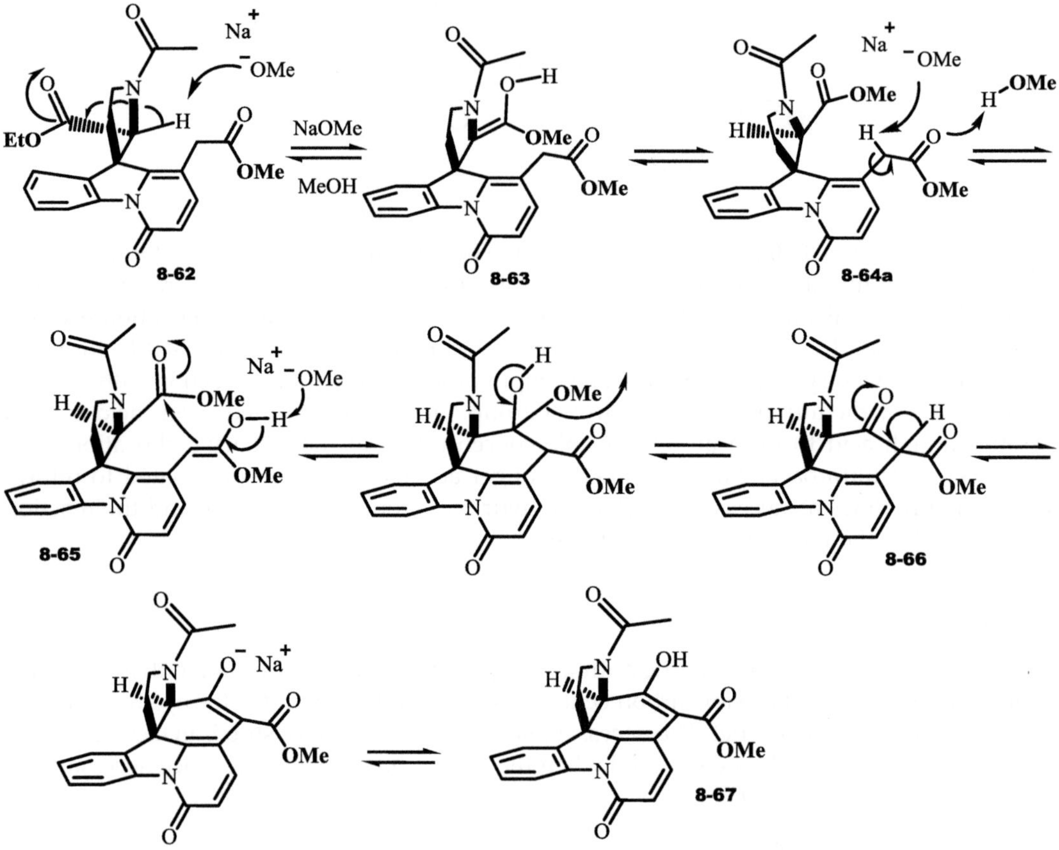

This mechanism demonstrates that although the equilibrium may lay far to the left, perhaps better than 99:1 of the "wrong" stereochemical relationship, every time the center epimerizes and the ester comes into proximity with the second bond forming center, there is a chance to form the new bond. The driving forces are in favor of the products (ring-forming reaction leading to small leaving group (methanol) and the creation of strong structural components (the enol ester)) and this drives the transformation forward.

Problem 8.12

The symmetrical starting diester **8-68** can exist in the enolester form and acts as a nucleophile. The enol ester adds to the dimethyl enone in a 1,4-fashion (a Michael reaction) to provide the symmetric enol-diester intermediate **8-69**. After enolization, this system then cyclizes and ejects methoxide to form the diketoester product **8-70**.

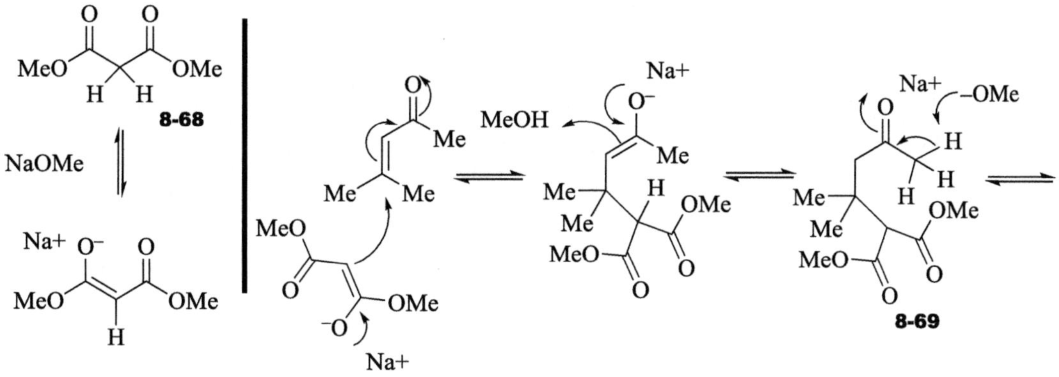

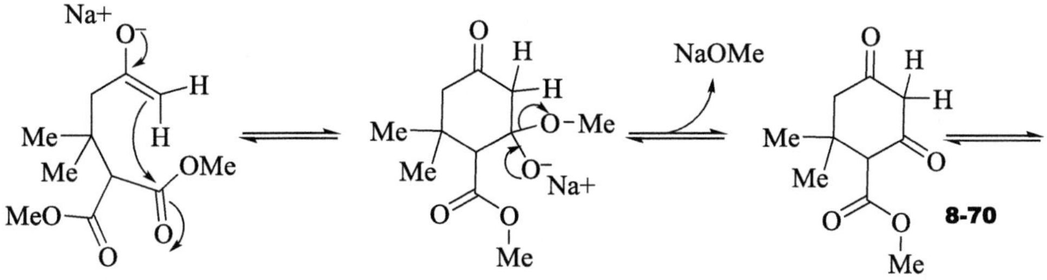

The ester-di-ketone can go on to enolize and exist in one or more of the forms shown below.

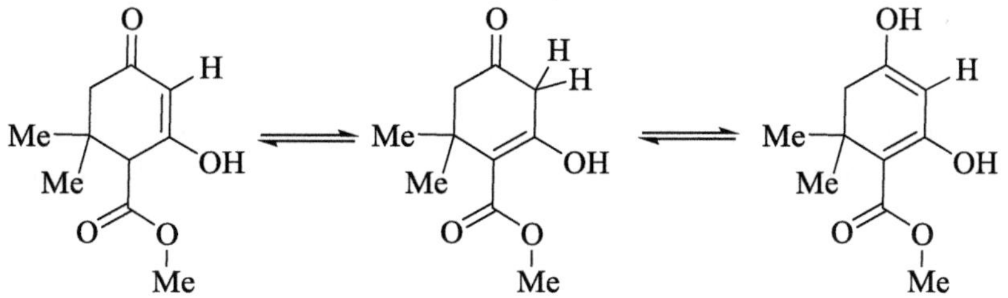

Problem 8.13

The starting material is treated with acid and water to hydrolyze the cyclic acetal and provide the parent alcohol–aldehyde **8-71**.

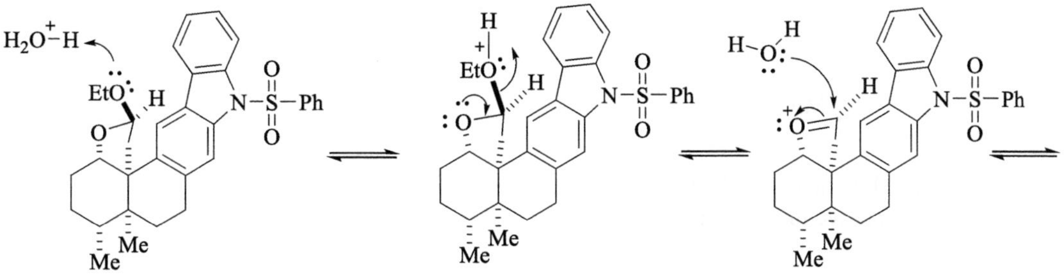

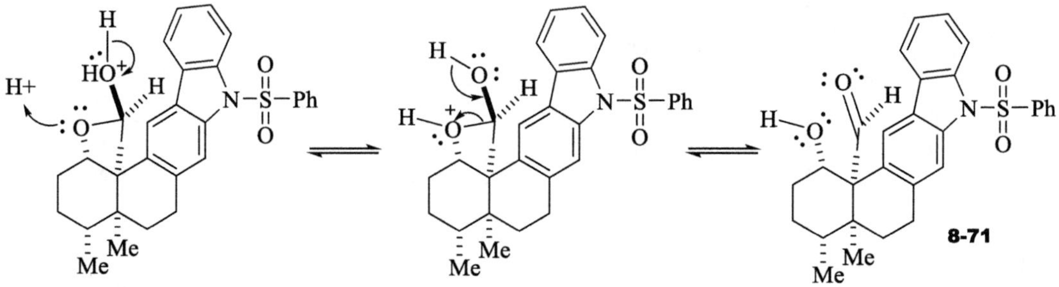

In the second step, the Grignard is added to the aldehyde and after workup, provided the desired diol.

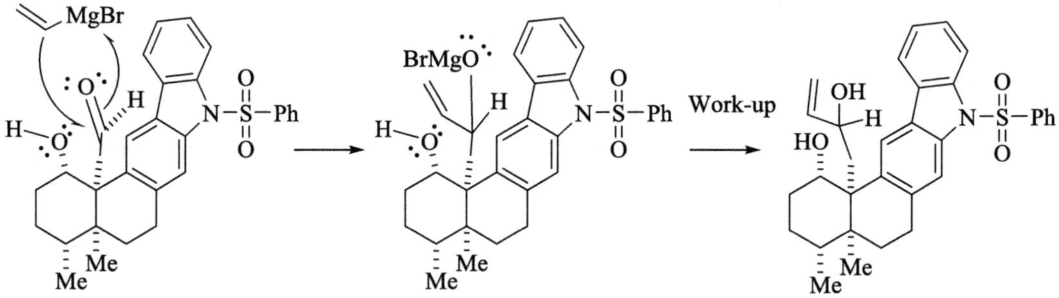

This addition of the vinyl Grignard does not deliver the high yields hoped for, even when run in toluene at 60°C for 5 h. It is presumed that the aldehyde-alcohol (8-71) is in equilibrium with the lactol (aka hemiacetal), with the equilibrium weighed heavily in the direction of the lactol.

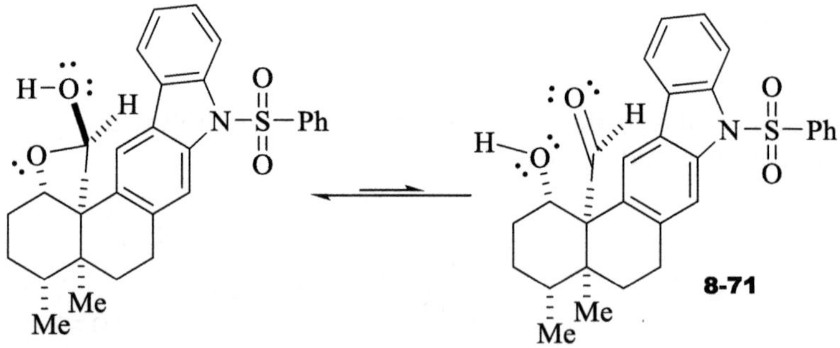

The hemiacetal or lactol is
unreactive to the Grignard

Problem 8.14

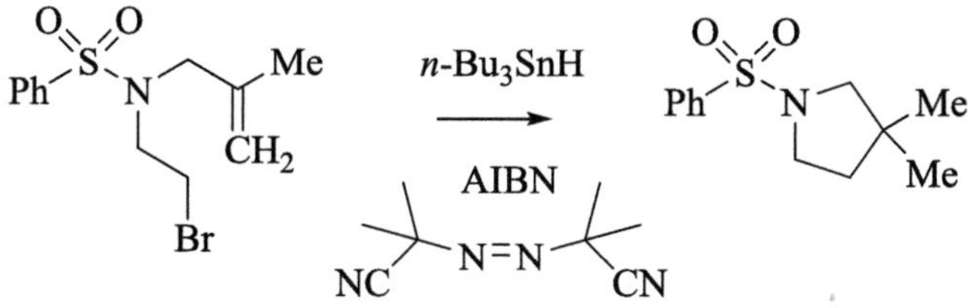

Benzene Reflux

This reaction is initiated by heating the mixture to benzene reflux which causes the Azobisisobutyronitrile (AIBN) to decompose. The AIBN breaks down into nitrogen gas and two equivalents of the cyanoprop-2-yl radicals **8-72**. The cyanopropyl radicals abstract hydrogen from the tin hydride generating tri-*n*-butyl tin radicals **8-73** that create the alkyl radical that then goes on to cyclize in an *exo* fashion **8-74** leading to the five-membered ring with a terminal radical center **8-75**. This terminal alkyl radical then abstracts a hydrogen creating another tri-*n*-butyl tin radical which propagates the reaction and leads to the observed product **8-76**.

Initiation

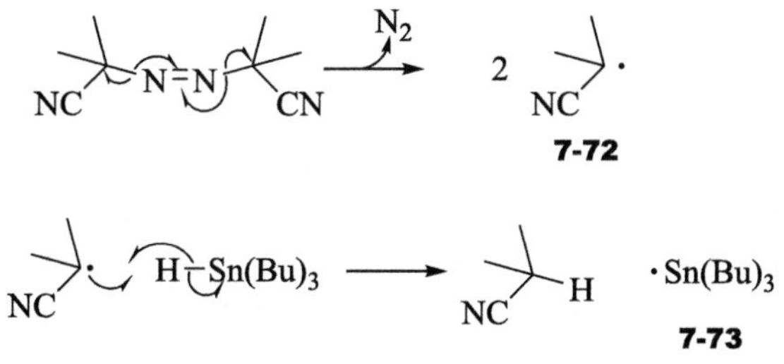

Propogation and product generation steps

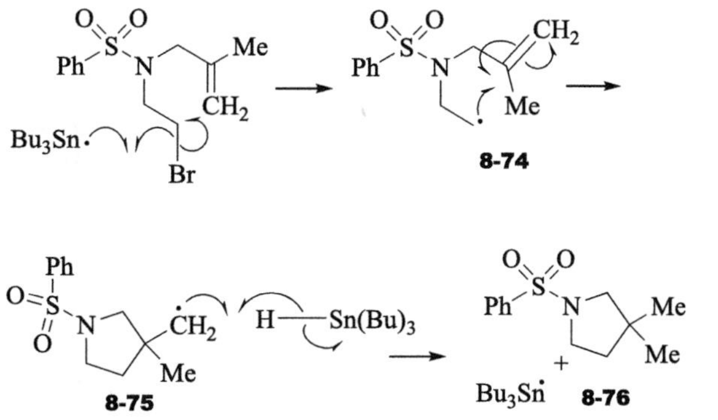

b. The other product that could have been generated via "endo" cyclization is **8-77**, which would generate **8-78**.

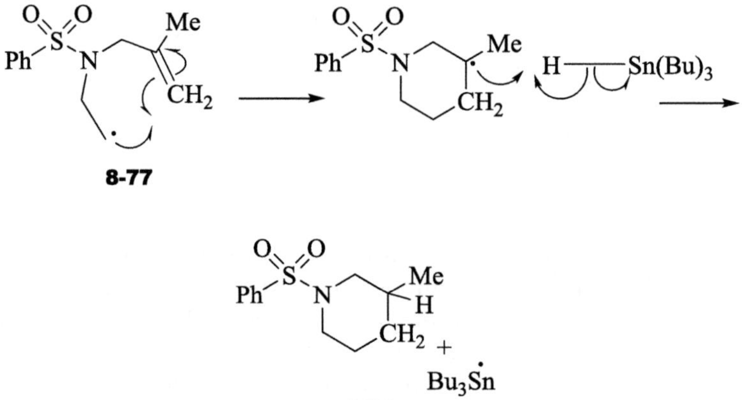

8-77

8-78

   The 1,6-endo cyclized product leading to the six-membered ring was the expected result based upon literature precedent for this substitution pattern. The authors believe that stereoelectronic factors associated with the sulphonamide may play a particularly strong role in blocking the required ring configuration necessary for endo cyclization.

   In other situations, the five-membered ring formation is faster (as was discussed in Chapter 5, Section B) than the six-membered ring (even though the six-membered ring would be thermodynamically favored). Radical ring closures generally proceed through a kinetic manifold (fastest products are formed and the process does not generate a thermodynamic mix of products) and this may also be considered a reasonable explanation for the observed result.

   The sources are Beckwith, et al., (1985); Spellmeyer et al., (1987).

Problem 8.15

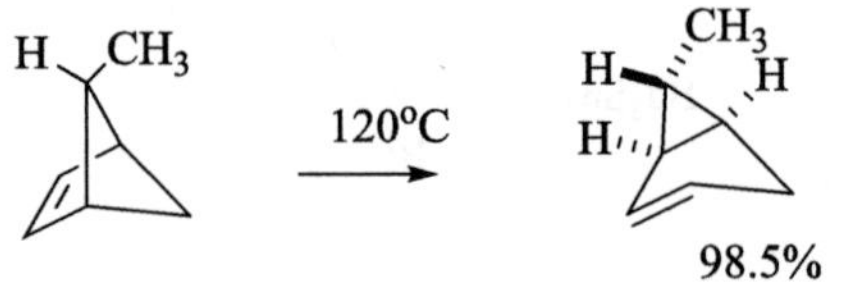

   This is an example of a 1,3-thermally promoted shift. This is a forbidden transformation if a hydrogen is migrating, but in this case, the back lobe of the carbon atom σ-bond can interact with the π-bond orbital in an antarafacial manner. This leads to inversion of the migrating center to deliver the product with very little of the other diastereomer.

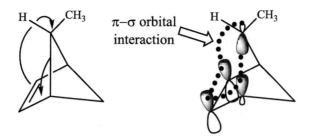

$$\sigma 2a + \pi 2s \text{ or } \sigma 2s + \pi 2a$$

1,3-antarafacial rearrangement with inversion

The corresponding transformation for the compound possessing the other stereochemical arrangement on the top of the bridge requires higher temperatures and does not deliver the inverted stereochemical result in the same high ratio.

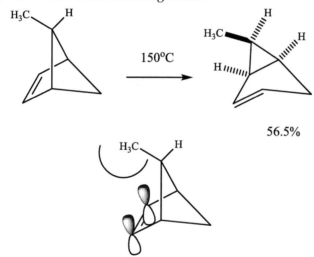

It is postulated that part of the reason that the same ratio cannot be attained in this system is that the methyl group partially blocks the required orbital overlap. This provides further support of the postulated mechanistic rationale for the result we see in the first system.

Problem 8.16

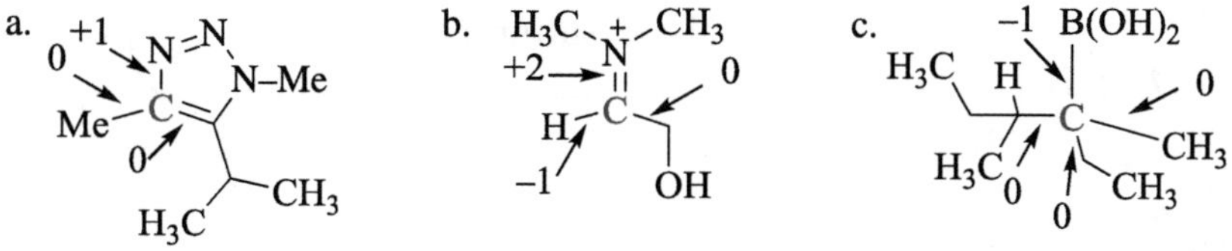

a.

+1 + 0 + 0 = **+1**

b.

+2 + (−1) + 0 = **+1**

c.

−1 + 0 + 0 + 0 = **−1**

Problem 8.17

Oxidation, reduction or neither?

a.

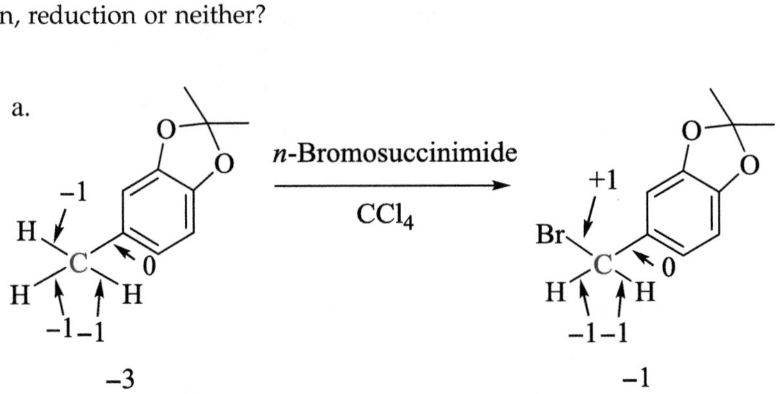

n-Bromosuccinimide

CCl_4

Oxidation: The halogenation brings the methyl group up to the same oxidation state as the corresponding alcohol.

b.

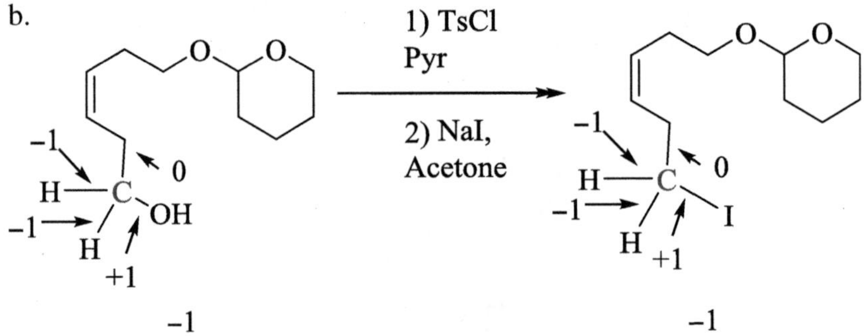

1) TsCl
   Pyr

2) NaI,
   Acetone

Neither: The alcohol and iodo analog are at the same oxidation state. This is the Finkelstein reaction which is used to convert compounds to iodo analogs or to generally convert one group to a better leaving group.

c.

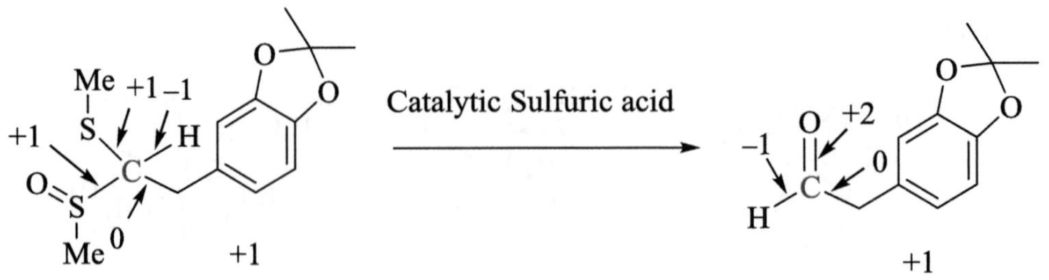

Catalytic Sulfuric acid

Neither: The carbon is converted from a di-thio-acetal to an aldehyde. This does not constitute a change in oxidation state since both are considered to be at the aldehyde oxidation state.

Problem 8.18

The direct oxidation of a halide (at the alcohol oxidation state) to the ketone is the Kornblum oxidation. This transformation proceeds through the mechanism described below, with the sulfoxide oxygen displacing the bromide. The resulting sulfonium intermediate can pull a proton leading to the ejection of dimethyl sulfide and leading to the desired glyoxal (alpha keto-aldehyde) 8-79. The side reaction is called the Pummerer reaction and proceeds through the sulfoxide species to the alkoxy anion 8-80 that attacks back at the activated sulfonium ion species to deliver the mixed oxygen-sulfur acetal side product 8-81.

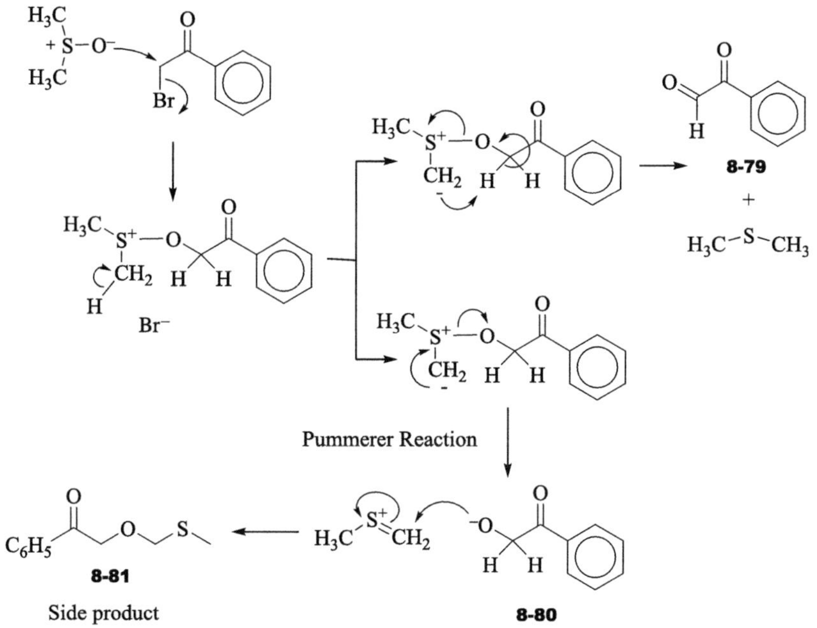

Problem 8.19

In this classic example, the carbene is formed as a result of mixing bromoform and t-butoxide with the olefin. The first step leads to the dibromocyclopropane A. The dibromocyclopropane A is then exposed to silver perchlorate. Silver is a halophile and leads, through the removal of one of the bromines, to the opening of the strained cyclopropane in the direction that provides to the more stable, tertiary carbocation intermediate **8-82**.

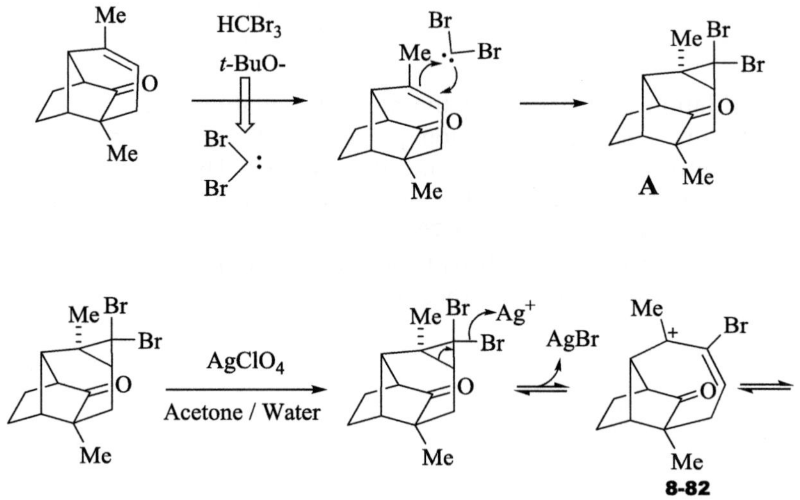

Water then adds in and the olefin shifts to provide compound **B** with the tetrasubstituted olefin and the alcohol in the allylic position. The alcohol is then oxidized with chromium trioxide–pyridine, to form the chromate ester. The system and then loses the alpha proton as it is oxidized and ejects the chromium to provide the product diketone **8-83**.

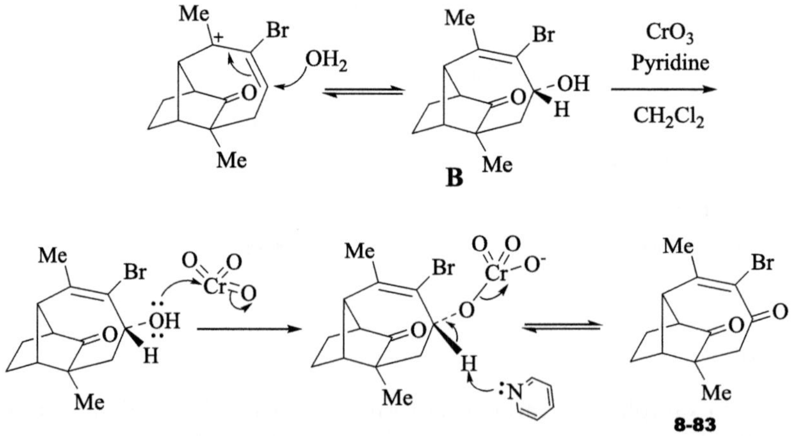

Problem 8.20

In this example, the aldehyde is reduced with sodium borohydride to produce the cyclohexene alcohol. The olefin is then exposed to ozone to first deliver the primary ozonide **8-84**, which then rearranges through the intramolecular zwitterionic transition state to provide the secondary ozonide **8-85**.

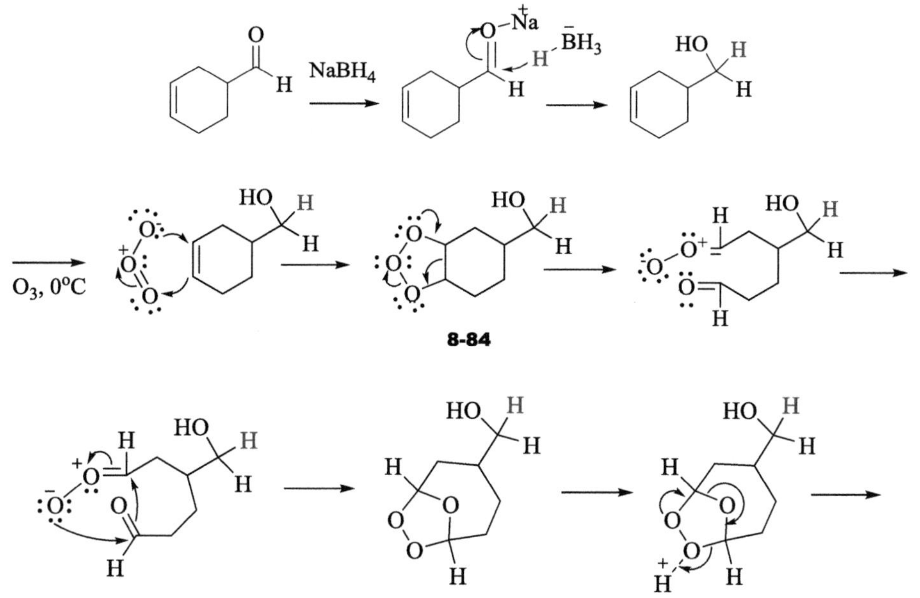

8-84

8-85

This material is worked up under oxidative conditions with peroxiacetic acid to deliver the alcohol di-acid **8-86** through the proposed mechanism described below. The alcohol di-acid can then go on to form the five-membered lactone **8-87** through the process described below. (Can you identify the steps where the carbon centers are oxidized?)

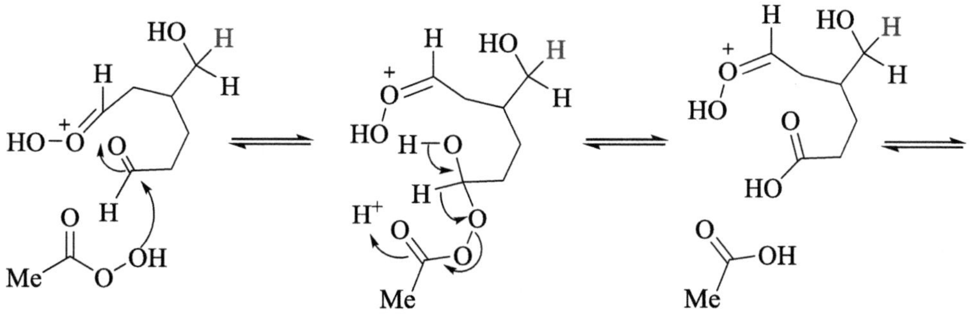

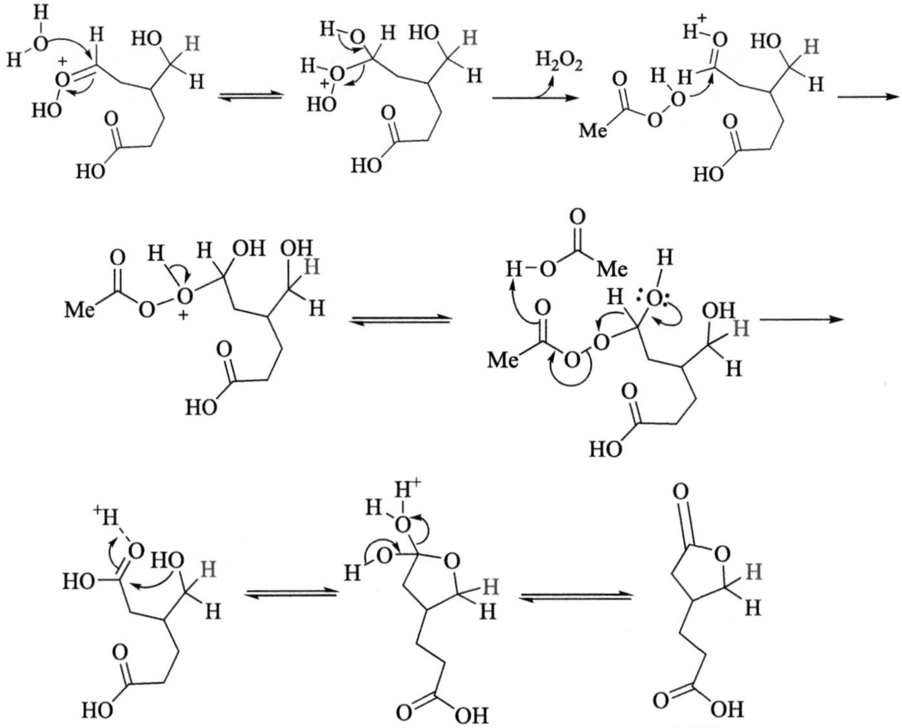

8-86   **8-87**

The Baeyer-Villiger type oxidation with the hydrogen shifting has been observed in many examples in the literature. But, it is also plausible that the alcohol di-acid could be formed through the intramolecular step shown below.

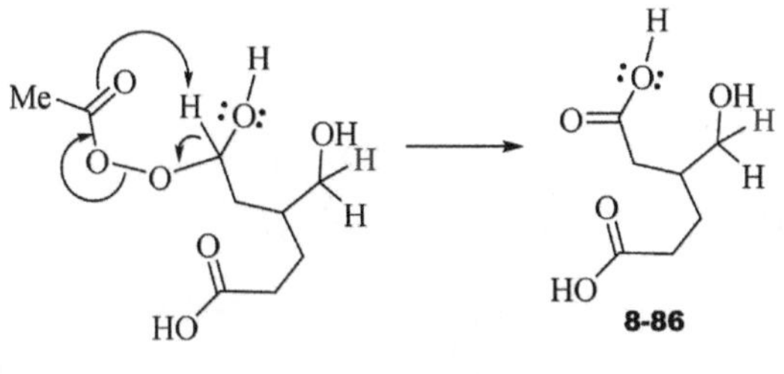

8-86

Problem 8.21

This is a sequence of two Swern or "activated sulfoxide" oxidations. In the first step, the dimethyl sulfoxide is mixed with trifluoroacetic anhydride to generate the activated sulfoxide species.

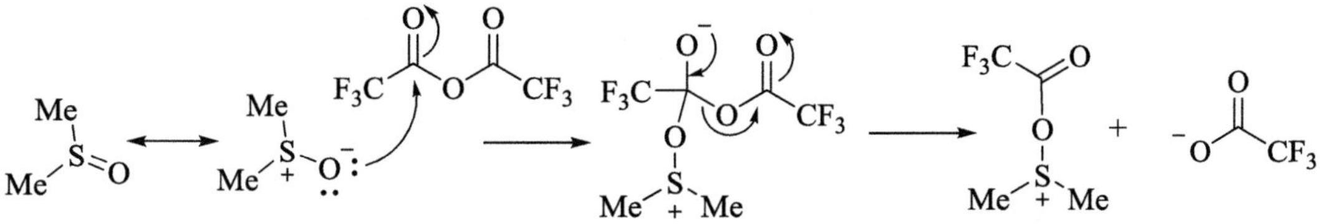

Activated Species

The substrate, a diol, is then added and the primary and secondary alcohols are oxidized to the corresponding aldehyde and ketone through the process shown below.

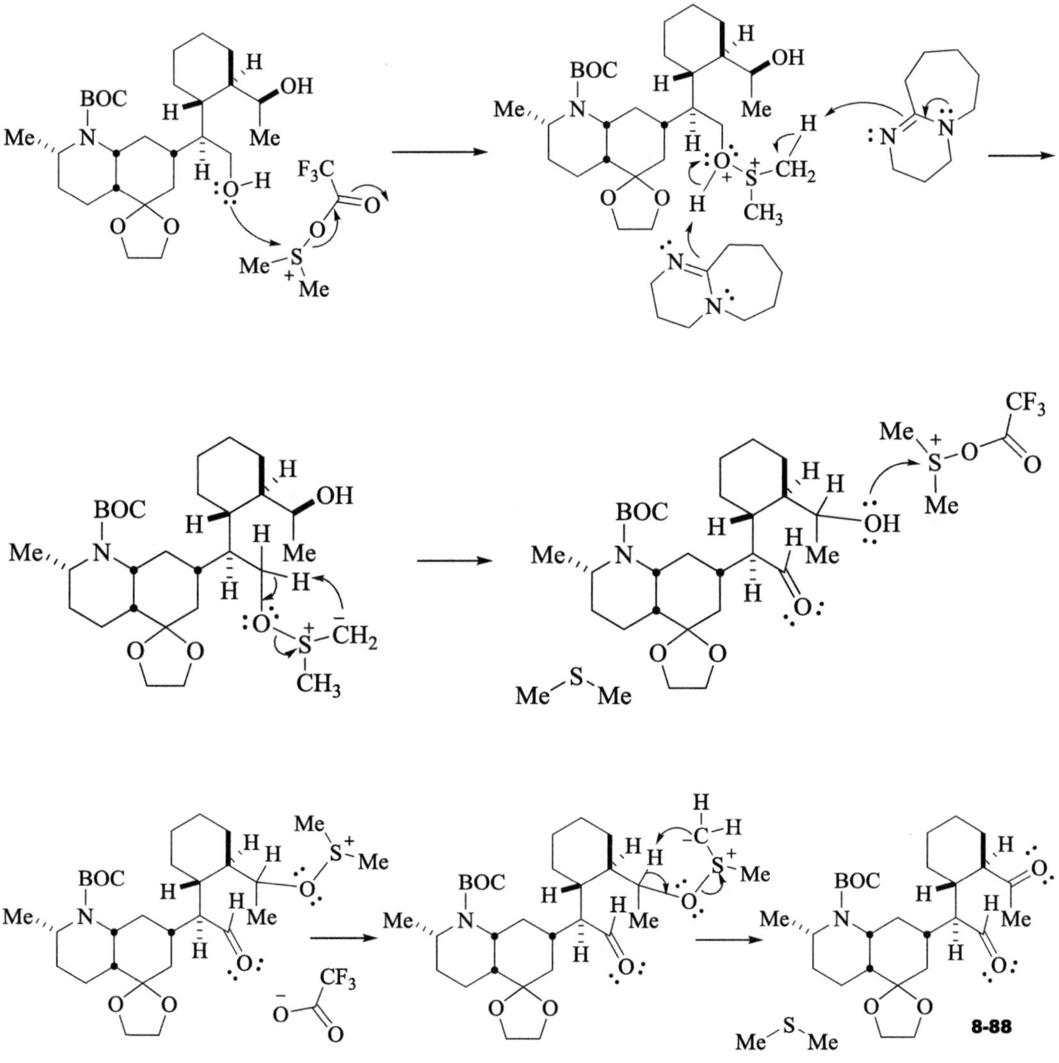

8-88

With the aid of the amidine base DBU (*1,8 Diazabicyclo [5.4.0]undec-7-ene*), the ketoalde-hyde intermediate **8-88** cyclizes through an aldol condensation to provide the β-hydroxy ketone **8-89**.

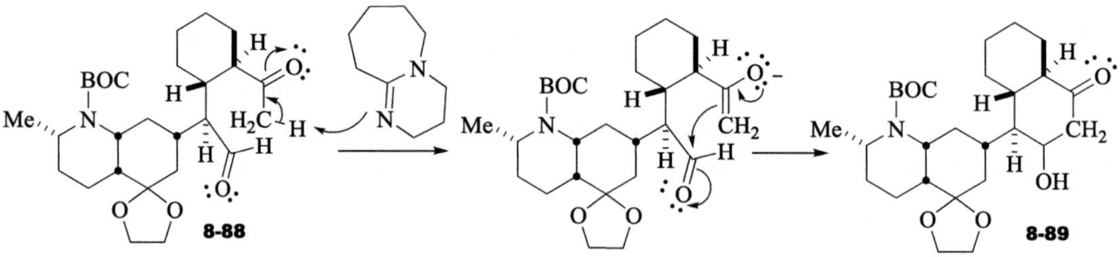

Step 2

The β-hydroxy ketone **8-89** was treated with more of the hexafluoroacetic anhydride to form the trifluoroacetate. The trifluoroacetate is a good leaving group and is eliminated as shown below to provide the product enone **8-90**.

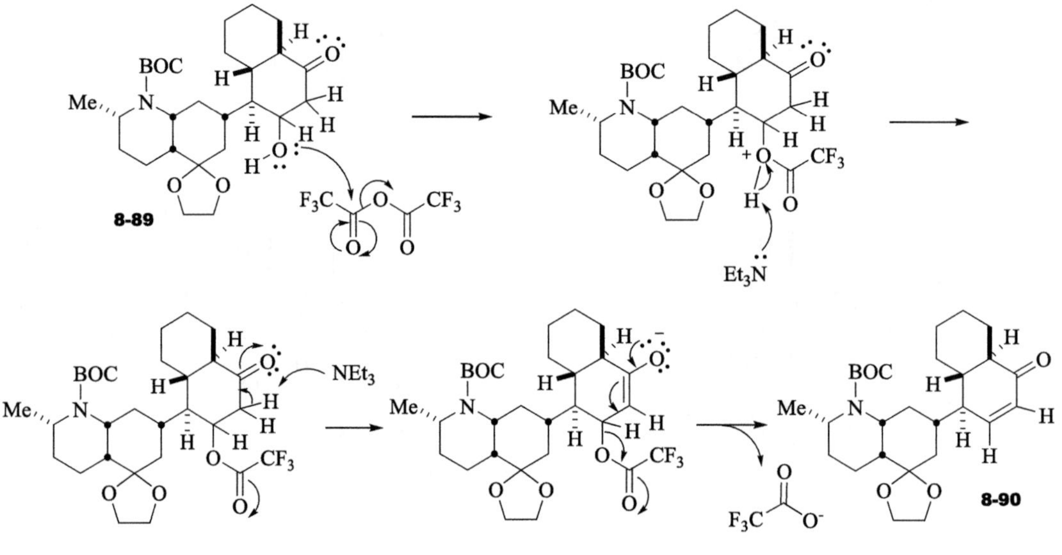

Problem 8.22

In this sequence, the authors needed to protect an aldehyde in preparation for an olefin cleavage and aldehyde reduction. In the first step the acetal is generated from the aldehyde using ethylene glycol. Initially one of the glycol oxygens adds into the acid activated alde-hyde. Under these conditions, the alcohol hydroxyl can be activated to leave by the acidic conditions by the glycol ether oxygen. The tethered hydroxyl of the glycol then adds in to create the cyclic acetal **8-91**.

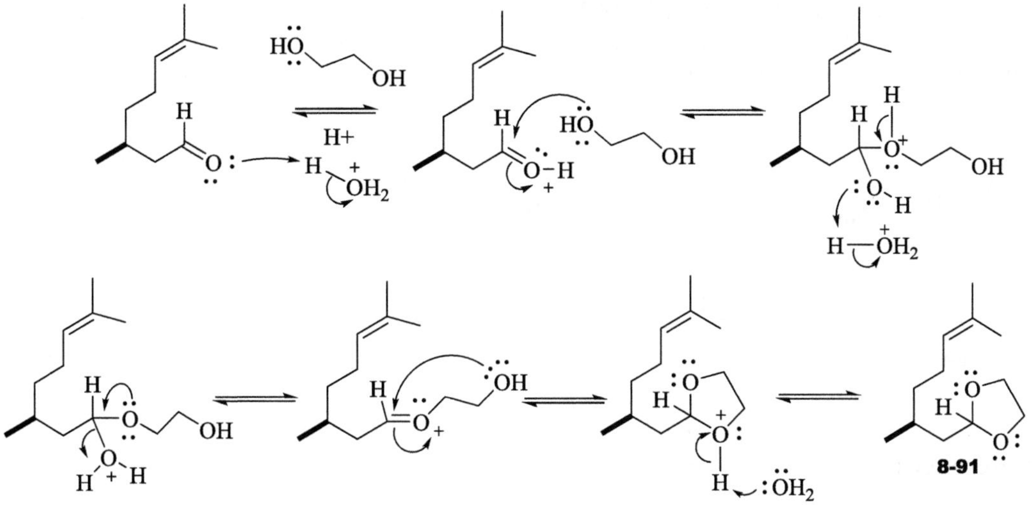

The compound was then exposed to ozone, which adds in a concerted fashion, to form the primary ozonide **8-92**. The primary ozonide reorganizes through the zwitterionic intermediate to form the secondary ozonide **8-93**. This material is not isolated; rather it is reduced directly with sodium borohydride. The NaBH$_4$ adds into one of the peroxy oxygens leading to the reduction of the ozonide and the aldehyde intermediate **8-94**.

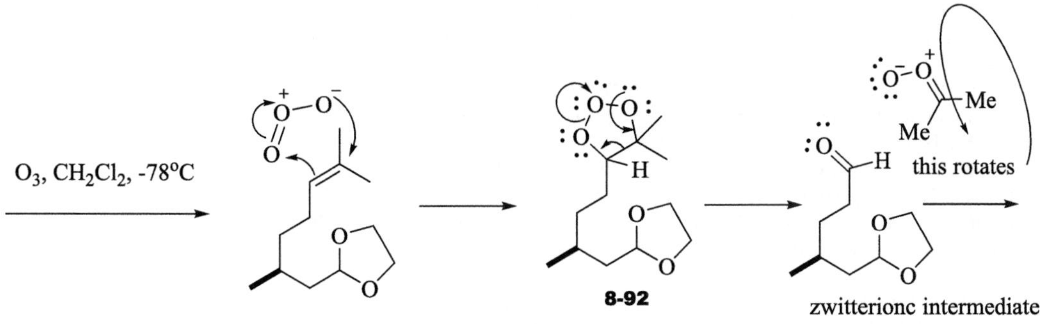

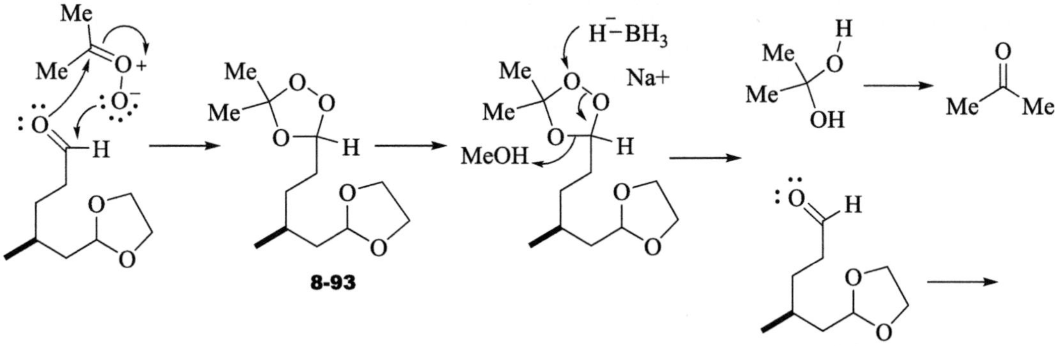

The intermediate product aldehyde **8-94** is not isolated, for under the conditions of the reaction, the aldehyde is further reduced by sodium borohydride to the alkoxy anion, which is ultimately quenched to provide the alcohol **8-95**.

References

Baran, J.; Mayr, H. *J. Am. Chem. Soc.* **1987**, *109*, 6519–6521.

Beckwith, A. L. J.; Schiesser, C. H. *Tetrahedron* **1985**, *41*, 3925.

Bian, M.; Wang, Z.; Xiong, X.; Sun, Y.; Matera, C.; Nicolaou, K. C.; Li, A. *J. Am. Chem. Soc.* **2012**, *134*, 8078–8081.

Carvalho, J. F.; Prestwich, G. D. *J. Org. Chem.* **1984**, *49*, 1251–1258.

Cha, J. Y.; Yeoman, J. T. S.; Reisman, S. E. *J. Am. Chem. Soc.* **2011**, *133*, 14964–14967.

Chimirri, A.; Grasso, S.; Monforte, P.; Romeo, G.; Zappala, M. *Heterocycles* **1988**, *27*, 93–100.

Deng, J.; Zhu, B.; Lu, Z.; Yu, H.; Li, A. *J. Am. Chem. Soc.* **2012**, *134*, 920–923.

Ent, H.; de Koning, H.; Speckamp, W. N. *J. Org. Chem.* **1986**, *51*, 1687–1691.

Ganguli, M.; Burka, L. T.; Harris, T. M. *J. Org. Chem.* **1984**, *49*, 3762–3766.

Kornblum, N.; Powers, J. W.; Anderson, G. J.; Jones, W. J.; Larson, H. O.; Levand, O.; Weaver, W. M. *J. Am. Chem. Soc.* **1957**, *79*, 6562–6562.

McMurry, J. E.; Isser, S. J. *J. Am. Chem. Soc.* **1972**, *94*, 7132–7137.

Nesi, R.; Giomi, D.; Papelao, S.; Bracci, S. *J. Org. Chem.* **1989**, *54*, 706–708.

Ogura, K.; Tsuchihashi, G. -I. *Tetrahedron Lett.* **1971**, *34*, 3151–3154.

Ogura, K.; Tsuchihashi, G. -I. *Tetrahedron Lett.* **1971**, *34*, 3151–3154.

Padwa, A.; Nimmesgern, H.; Wong, G. S. K. *J. Org. Chem.* **1985**, *50*, 5620–5627.

Paquette, L. A.; Andrews, D. R.; Springer, J. P. *J. Org. Chem.* **1983**, *48*, 1147–1149.

Pirrung, M. C.; Werner, J. A. *J. Am. Chem. Soc.* **1986**, *108*, 6060–6062.

Roth, W. R.; Friedrich, A. *Tetrahedron Lett.* **1969**, *31*, 2607–2610.

Roush, W. R.; Straub, J. A.; Brown, R. J. *J. Org. Chem.* **1987**, *52*, 5127–5136.

Spellmeyer, D. C.; Houk, K. N. *J. Org. Chem.* **1987**, *52*, 959.

Steiner, U.; Willhalm, B. *Helv. Chim. Acta.* **1952**, 1752–1756.

White, W. L.; Anzeveno, P. B.; Johnson, F. *J. Org. Chem.* **1982**, *47*, 2379–2387.

Woodward, R. B.; Cava, M. P.; Ollis, W. D.; Hunger, A.; Daeniker, H. U.; Schenker, K. *Tetrahedron* **1963**, *19*, 247–288.

Zarecki, P. A.; Fang, H.; Ribiero, A. A. *J. Org. Chem.* **1998**, *63*, 4779.

Zi, W.; Yu, S.; Ma, D. *Angew. Chem. Int. Ed.* **2010**, *49*, 5887–5890.

Lewis Structures of Common Functional Groups

Structure	Name	Structure	Name
$-\overset{\displaystyle |}{\underset{\displaystyle |}{C}}-$	Aliphatic carbon	$\overset{\displaystyle :O:}{\underset{\displaystyle -C-\ddot{N}H_2}{\|}}$	Amide
$\underset{\diagup}{\diagdown}C=C\underset{\diagdown}{\diagup}$	Alkene	$:\ddot{O}-\ddot{O}:$	Molecular oxygen (triplet)
$-C\equiv C-$	Alkyne	$-:\ddot{O}:-H$	Hydroxy
$\underset{\diagup}{\diagdown}C=\ddot{O}$	Carbonyl	$-\ddot{O}-$	Ether
$\overset{\displaystyle :O:}{\underset{\displaystyle -C-\ddot{O}H}{\|}}$	Carboxylic acid	$-\ddot{O}-\ddot{O}-H$	Hydroperoxy
$-\ddot{O}-\ddot{O}-$	Peroxy	$-\ddot{N}-\ddot{N}=\ddot{O}$	Nitrosamine
$R-\ddot{X}:$	Halide (X = F, Cl, Br, I)	$R-\ddot{N}=\ddot{N}-R$	Azo
$R_3N:$	Amine (tertiary)	$\underset{\diagup}{\diagdown}C=\overset{+}{N}=\ddot{N}:^-$	Diazo
$\underset{\diagup}{\diagdown}C=\ddot{N}\underset{\diagdown}{}$	Imine	$-\ddot{N}=\overset{+}{N}=\ddot{N}:^-$	Azide
$\underset{\diagup}{\diagdown}C=C\overset{\displaystyle H}{\underset{\displaystyle \ddot{N}R_2}{}}$	Enamine	$-\ddot{S}-$	Sulfide
$-C\equiv N:$	Nitrile	$-\ddot{S}-H$	Thiol
$-N\equiv C:$	Isonitrile	$-\ddot{S}-\ddot{S}-$	Disulfide
$R_3\overset{+}{N}-\ddot{O}:^-$	Amine oxide	$\overset{\displaystyle :O:}{\underset{\displaystyle -S-}{\|}}$	Sulfoxide
$\underset{\diagup}{\diagdown}C=\ddot{N}-\ddot{O}-H$	Oxime	$\overset{\displaystyle :O:}{\underset{\displaystyle :O:}{\overset{\|}{\underset{\|}{-S-}}}}$	Sulfone

$-\ddot{\text{N}}=\ddot{\text{O}}$	Nitroso	$R_3P:$	Phosphine
$-N\overset{\ddot{\text{O}}}{\underset{\cdot\ddot{\text{O}}:}{\diagup}}$	Nitro	$R_3P=\ddot{\text{O}}$	Phosphine oxide
$-\ddot{\text{N}}\text{H}\ddot{\text{N}}\text{H}_2$	Hydrazine	$(RO)_3P:$	Phosphite ester
$\underset{/}{\overset{\backslash}{}}\text{C}=\ddot{\text{N}}-\ddot{\text{N}}\text{H}_2$	Hydrazone	$(RO)_3P=\ddot{\text{O}}$	Phosphate ester

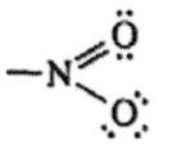

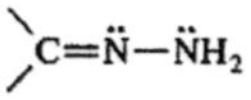

B

Symbols and Abbreviations Used in Chemical Notation

Symbol	Meaning
$\longrightarrow$	Reaction
$\rightleftharpoons$ or $\rightleftharpoons$	Equilibrium
$\longleftrightarrow$	Resonance structures
⌣	Movement of an electron pair
⌣ or ⌣	Movement of an unpaired electron
↑ or ↓	An electron (in molecular orbital diagrams)
: or |	Electron pair
.	Unpaired electron
	Circle inside a polygon denotes aromaticity
	Dotted line indicates delocalized π system. It may or may not be aromatic.
Me	CH_3-, methyl group
Et	CH_3CH_2-, ethyl group
i-Pr	$(CH_3)_2CH-$, isopropyl group
t-Bu	$(CH_3)_3C-$, *t*-butyl group
Ph or ϕ	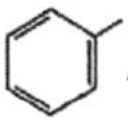 , phenyl group
Ar	Aryl, any aromatic residue, e.g., 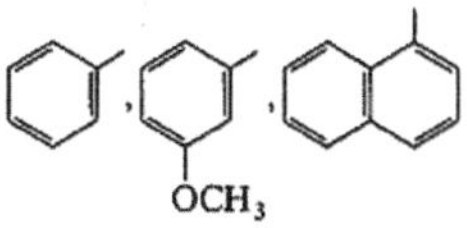
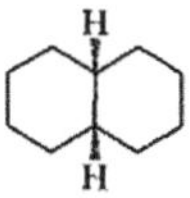	Dotted wedge indicates that the bond is directed below the plane of the page. The structure shown has a *cis* ring junction.

Symbol	Meaning
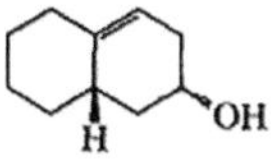	Solid wedge indicates that the bond is directed toward the viewer (above the plane of the page). The hydrogen and hydroxy groups are *trans* to one another.
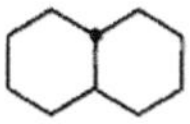	Solid circle indicates a hydrogen atom above the plane of the page. The ring junction is *trans*.
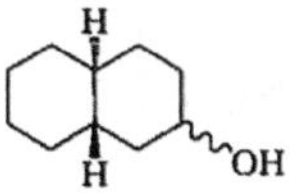	Wavy bond indicates that the stereochemistry is unspecified.
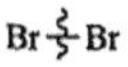 Br $\overset{\xi}{\rightarrow}$ Br	Wavy line across bond indicates homolytic cleavage.
*	Excited state

C

Relative Acidities of Common Organic and Inorganic Substances[a]

Acid	Solvent	pK_a	Conjugate base	References[h]
HI	Extrapolated[b]	-10	I^-	1
HBr	Aqueous H_2SO_4	-9	Br^-	2
HCl	Aqueous H_2SO_4	-8	Cl^-	2
$(CH_3)_2\overset{+}{S}H$	Aqueous H_2SO_4	-6.99	$(CH_3)_2S$	3
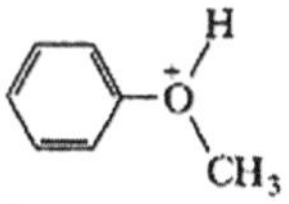 PhÖH(CH₃) (protonated anisole)	Aqueous H_2SO_4	-6.5	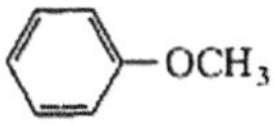 Ph—OCH_3	4
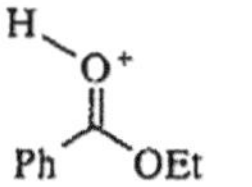 protonated PhCO₂Et	Aqueous H_2SO_4	-6.2	$PhCO_2Et$	5
CF_3SO_3H	Estimate[c]	$-5.1(-5.9)$	$CF_3SO_3^-$	3, 6
$HClO_4$	Estimate[c]	-5.0	ClO_4^-	6
FSO_3H	Estimate[c]	$-4.8(-6.4)$	FSO_3^-	3, 6
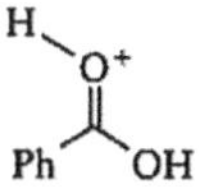 protonated PhCO₂H	Aqueous H_2SO_4	-4.7	$PhCO_2H$	5
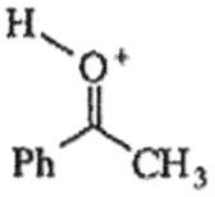 protonated PhCOCH₃	Aqueous H_2SO_4	-4.3	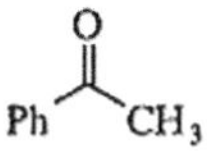 Ph—CO—CH_3	5
$PhCH{=}\overset{+}{O}H$	Aqueous H_2SO_4	-3.9	$PhCH{=}O$	5
$(CH_3)_2\overset{+}{O}H$	Aqueous H_2SO_4	-3.8	$(CH_3)_2O$	7
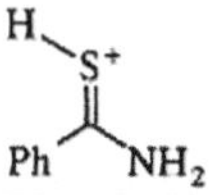 protonated PhC(=S)NH₂	Aqueous H_2SO_4	-3.20	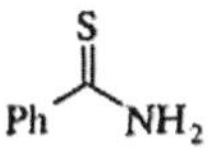 Ph—C(=S)—NH_2	8

Acid	Solvent	pK_a	Conjugate base	References[h]
$(CH_3)_2C=\overset{+}{O}H$	Aqueous H_2SO_4	−2.85	$(CH_3)_2C=O$	3
$PhSO_3H$	Estimate[c]	−2.8	$PhSO_3^-$	6
H_2SO_4	Estimate[c]	−2.8	HSO_4^-	6
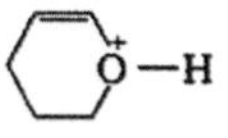	Aqueous H_2SO_4	−2.8	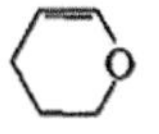	7
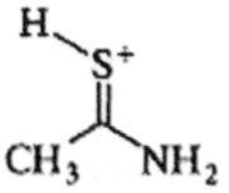	Aqueous H_2SO_4	−2.51	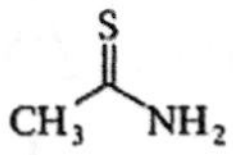	8
$CH_3\overset{+}{O}H_2$	Aqueous H_2SO_4	−2.5	CH_3OH	9
CH_3SO_3H	Extrapolated[d]	−1.9	$CH_3SO_3^-$	1
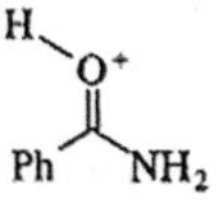	Aqueous H_2SO_4	−1.74	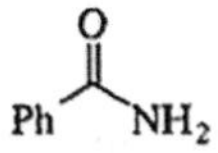	10
H_3O^+	Estimate[c]	−1.7	H_2O	6
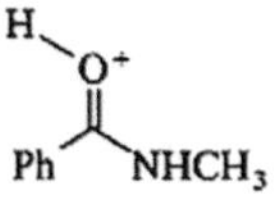	Aqueous H_2SO_4	−1.7	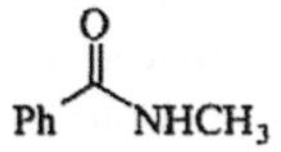	10
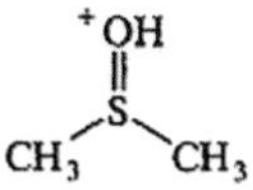	Aqueous H_2SO_4	−1.5	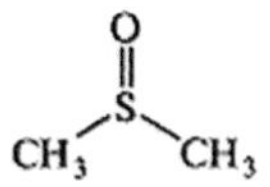	8
HNO_3	Estimate[c]	−1.3	NO_3^-	6
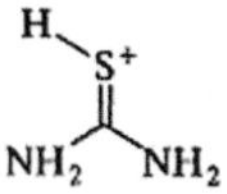	Aqueous H_2SO_4	−1.26	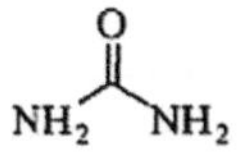	11
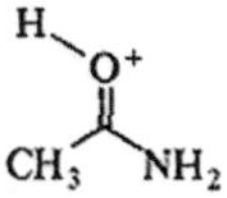	Aqueous H_2SO_4	−0.6	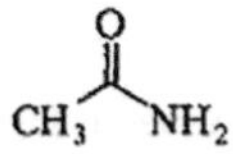	10
CF_3CO_2H	H_2O	−0.6	$CF_3CO_2^-$	1
Cl_3CCO_2H	H_2O	−0.5	$Cl_3CCO_2^-$	1
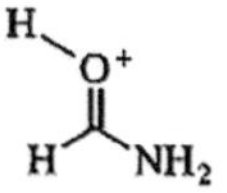	Acetic acid[e]	−0.48	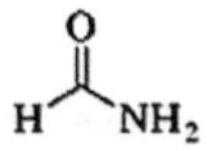	12
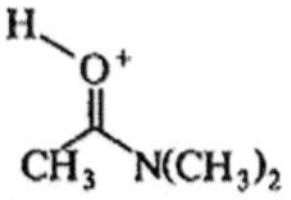	H_2O, CH_3NO_2	0.1	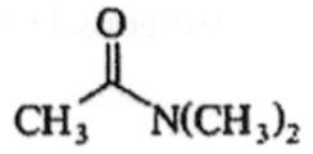	13
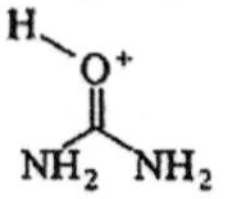	Acetic acid[e]	0.5	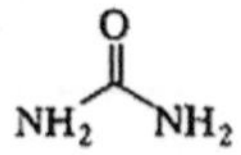	12

Acid	Solvent	pK_a	Conjugate base	References[h]
$(Ph_2)\overset{+}{N}H_2$	H_2O	0.8	$(Ph)_2N$	14
$PhSO_2H$	H_2O	1.2	$PhSO_2^-$	1
HO_2CCO_2H	H_2O	1.25	$HO_2CCO_2^-$	1
Cl_2CHCO_2H	H_2O	1.35	$Cl_2CHCO_2^-$	1
$PhCH{=}\overset{+}{N}HOH$	H_2O	2.0	$PhCH{=}NOH$	1
H_3PO_4	H_2O	2.1	$H_2PO_4^-$	14
CH_3SO_2H	H_2O	2.3	$CH_3SO_2^-$	1
$\overset{+}{N}H_3CH_2CO_2H$	H_2O	2.35	$\overset{+}{N}H_3CH_2CO_2^-$	1
FCH_2CO_2H	H_2O	2.6	$FCH_2CO_2^-$	1
$ClCH_2CO_2H$	H_2O	2.86	$ClCH_2CO_2^-$	1
HF	H_2O	3.2	F^-	2
HNO_2	H_2O	3.4	NO_2^-	6
CH_3COSH	H_2O	3.4	CH_3COS^-	15
O_2N—C$_6$H$_4$—CO_2H	H_2O	3.44	O_2N—C$_6$H$_4$—CO_2^-	1
H_2CO_3	H_2O	3.7	HCO_3^-	16
HCO_2H	H_2O	3.75	HCO_2^-	17
$HOCH_2CO_2H$	H_2O	3.8	$HOCH_2CO_2^-$	17
Cl—C$_6$H$_4$—$\overset{+}{N}H_3$	H_2O	4.0	Cl—C$_6$H$_4$—NH_2	18
O_2N—C$_6$H$_3(NO_2)$—OH	H_2O	4.1	O_2N—C$_6$H$_3(NO_2)$—O^-	17
$PhCO_2H$	H_2O	4.2	$PhCO_2^-$	17
C$_6$H$_5$—$\overset{+}{N}H_3$	H_2O	4.6	C$_6$H$_5$—NH_2	18
CH_3CO_2H	H_2O	4.76	$CH_3CO_2^-$	6
$PhCH_2\overset{+}{N}H_2OH$	H_2O	4.9	$PhCH_2NHOH$	1
NH_2—C$_6$H$_4$—CO_2H	H_2O	4.92	NH_2—C$_6$H$_4$—CO_2^-	14
$Ph\overset{+}{N}H(CH_3)_2$	H_2O	5.1	$PhN(CH_3)_2$	19
C$_5$H$_5$$\overset{+}{N}H$ (pyridinium)	H_2O	5.2	C$_5$H$_5$N (pyridine)	19
$CH_3\overset{+}{N}H_2OH$	H_2O	6.0	CH_3NHOH	1
$^+NH_3OH$	H_2O	6.0	NH_2OH	1
C$_6$H$_5$—SH	H_2O	6.5	C$_6$H$_5$—S^-	20
H_2S	H_2O	7.0	HS^-	14

Acid	Solvent	pK_a	Conjugate base	References[h]
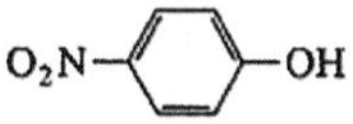 O_2N–C$_6$H$_4$–OH (4-nitrophenol)	H_2O	7.2	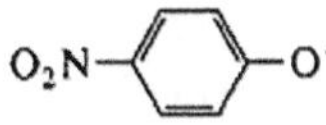 O_2N–C$_6$H$_4$–O$^-$	17
$^+NH_3OH$	H_2O	8.0	NH_2OH	45
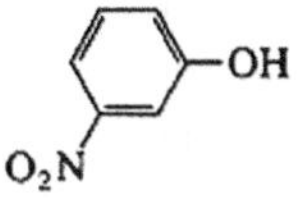 3-nitrophenol (O_2N–C$_6$H$_4$–OH)	H_2O	8.3	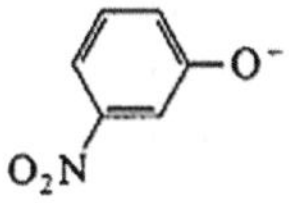 3-nitrophenoxide (O_2N–C$_6$H$_4$–O$^-$)	17
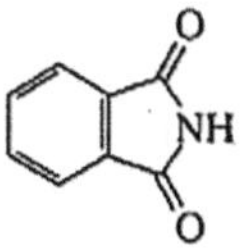 phthalimide (NH)	H_2O	8.3	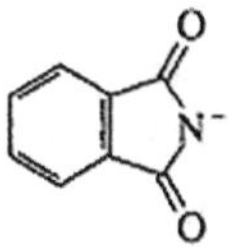 phthalimide anion (N$^-$)	17
$CH_3\,COCH_2COCH_3$	H_2O	9.0	$CH_3COCHCOCH_3$	21
NH_4^+	H_2O	9.2	NH_3	9
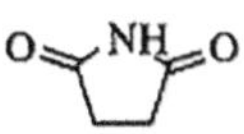 succinimide (O=C–NH–C=O)	H_2O	9.6	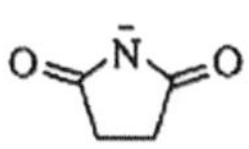 succinimide anion (O=C–N$^-$–C=O)	22
$^+NH_3CH_2CO_2^-$	H_2O	9.8	$NH_2CH_2CO_2^-$	14
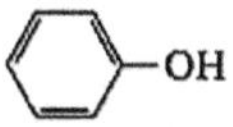 C$_6$H$_5$–OH (phenol)	H_2O	10.0	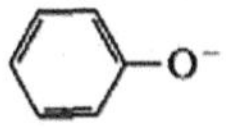 C$_6$H$_5$–O$^-$	17
CH_3NO_2	H_2O	10.0	$^-CH_2NO_2$	2
HCO_3^-	H_2O	10.2	CO_3^{2-}	14
PhSH	DMSO	10.3	PhS$^-$	23
CH_3CH_2SH	H_2O	10.6	$CH_3CH_2S^-$	1
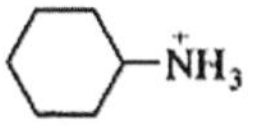 cyclohexyl–$\overset{+}{N}H_3$	H_2O	10.7	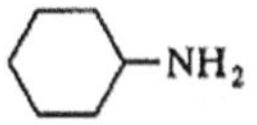 cyclohexyl–NH_2	19
$(CH_3CH_2)\overset{+}{N}H$	H_2O	10.8	$(CH_3CH_2)_3N$	19
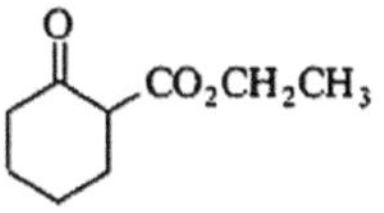 2-carbethoxycyclohexanone ($CO_2CH_2CH_3$)	H_2O	10.9	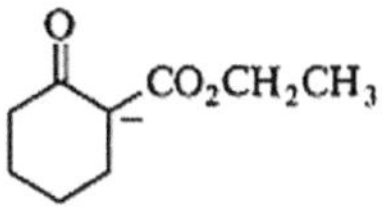 2-carbethoxycyclohexanone anion ($CO_2CH_2CH_3$)	17
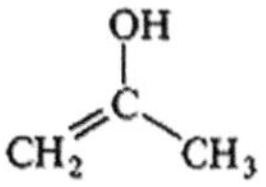 $CH_2{=}C(OH)CH_3$	H_2O	11.0	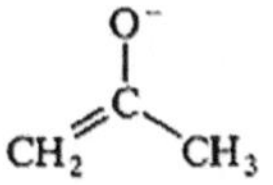 $CH_2{=}C(O^-)CH_3$	24
$PhCO_2H$	DMSO	11.0	$PhCO_2^-$	2
$(CH_3CH_2)_2\overset{+}{N}H_2$	H_2O	11.0	$(CH_3CH_2)_2NH$	19
$CH_2(CN)_2$	DMSO	11.1	$CH(CN)_2^-$	25
 piperidinium ($\overset{+}{N}H_2$)	H_2O	11.1	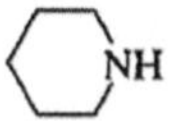 piperidine (NH)	19
$CH_2(CN)_2$	H_2O	11.4		26
HOOH	H_2O	11.6	HOO$^-$	27

Acid	Solvent	pK_a	Conjugate base	References[h]
(2-indanone)	H_2O	12.2	(2-indanone anion)	28
$PhCH_2NO_2$	DMSO	12.3	$Ph\bar{C}HNO_2$	2
CH_3CO_2H	DMSO	12.3	$CH_3CO_2^-$	2
CF_3CH_2OH	H_2O	12.4	$CF_3CH_2O^-$	29
$CH_2(SO_2CH_3)_2$	H_2O	12.5	$\bar{C}H(SO_2CH_3)_2$	47
$NCCH_2CO_2CH_3$	DMSO	12.8	$NC\bar{C}HCO_2CH_3$	30
$CH_2(COCH_3)_2$	DMSO	13.4	$\bar{C}H(COCH_3)_2$	25
guanidinium ($\overset{NH_2}{\underset{\overset{+}{N}H_2\;NH_2}{C}}$)	H_2O	13.4	guanidine ($\overset{NH_2}{\underset{HN\;NH_2}{C}}$)	1
$CH_3COCH_2CO_2CH_2CH_3$	DMSO	14.2	$CH_3CO\bar{C}HCO_2CH_2CH_3$	30
imidazole (N $\smile$ NH)	H_2O	14.5	imidazolide (N $\smile$ N$^-$)	31
CH_3OH	H_2O	15.5	CH_3O^-	29
H_2O	H_2O^f	15.7	HO^-	32
CH_3CH_2OH	H_2O	15.9	$CH_3CH_2O^-$	32
2-nitro-4-nitroaniline (O_2N–⬡(NO_2)–NH_2)	DMSO	15.9	(O_2N–⬡(NO_2)–$\bar{N}H$)	33
Ph–CH_2–CO–CH_3	H_2O	15.9	Ph–$\bar{C}H$–CO–CH_3	28
CH_3CHO	H_2O	16.5	$\bar{C}H_2CHO$	34
$(CH_3)_2CHNO_2$	DMSO	16.9	$(CH_3)_2\bar{C}NO_2$	31
$(CH_3)_2CHOH$	H_2O	17.1	$(CH_3)_2CHO^-$	32
CH_3NO_2	DMSO	17.2	$\bar{C}H_2NO_2$	35
$(CH_3)_3COH$	H_2O	18	$(CH_3)_3CO^-$	32
cyclopentadiene	DMSO	18.1	cyclopentadienide	25
CH_3CSNH_2	DMSO	18.5	$CH_3CS\bar{N}H$	36
CH_3COCH_3	H_2O	19.2	$\bar{C}H_2COCH_3$	24
O_2N–⬡–NH_2	DMSO	20.9	O_2N–⬡–$\bar{N}H$	33
$PhCH_2CN$	DMSO	21.9	$Ph\bar{C}HCN$	35
Ph_2NH	H_2O/DMSO	22.4	$Ph_2\bar{N}$	37
fluorene	DMSO	22.6	fluorenide	35

Acid	Solvent	pK_a	Conjugate base	References[h]
Ph_2NH	DMSO	23.5	Ph_2N^-	33
$CHCl_3$	H_2O^g	24	$^-CCl_3$	46
CH_3COPh	DMSO	24.7	$\bar{C}H_2COPh$	35
$HC{\equiv}CH$		25	$HC{\equiv}C^-$	38
CH_3CONH_2	DMSO	25.5	$CH_3CO\bar{N}H$	36
Ph–C(=O)–CH(CH₃)₂ (isobutyrophenone)	DMSO	26.3	Ph–C(=O)–C⁻(CH₃)CH₃	35
CH_3COCH_3	DMSO	26.5	$\bar{C}H_2COCH_3$	35
3-chloroaniline (Cl–C₆H₄–NH₂)	DMSO	26.7	3-chloroaniline anion (Cl–C₆H₄–N̄H)	33
NH_2CONH_2	DMSO	26.9	$NH_2CO\bar{N}H$	3
$CH_3CH_2COCH_2CH_3$	DMSO	27.1	$CH_3CH_2CO\bar{C}HCH_3$	35
$Ph-C{\equiv}CH$	DMSO	28.8	$Ph-C{\equiv}C^-$	35
CH_3SO_2Ph	DMSO	29.0	$\bar{C}H_2SO_2Ph$	35
4-chloroaniline (Cl–C₆H₄–NH₂)	DMSO	29.4	4-chloroaniline anion (Cl–C₆H₄–N̄H)	35
CH_3CO_2Et	DMSO	30.5	$\bar{C}HCO_2Et$	30
$(Ph)_3CH$	DMSO	30.6	Ph_3C^-	35
$PhNH_2$	DMSO	30.7	$Ph\bar{N}H$	33
1,3-dithiane (ring, –H)	DMSO	30.6	1,3-dithiane anion	2
2-phenyl-1,3-dithiane (ring, –Ph)	DMSO	30.7	2-phenyl-1,3-dithiane anion (=Ph)	2
$CH_3SO_2CH_3$	DMSO	31.1	$\bar{C}H_2SO_2CH_3$	35
CH_3CN	DMSO	31.2	$\bar{C}H_2CN$	25
$(Ph)_2CH_2$	DMSO	32.3	$(Ph)_2\bar{C}H$	25
$CH_3CON(CH_2CH_3)_2$	DMSO	34.5	$\bar{C}H_2CON(CH_2CH_3)_2$	30
$[(CH_3)_2CH]_2NH$	THF	35.7	$[(CH_3)_2CH]_2\bar{N}$	39
$[(CH_3)_2CH]_2NH$	THF	39	$[(CH_3)_2CH]_2\bar{N}$	40
NH_3	35	41	$\bar{N}H_2$	25
$PhCH_3$	DMSO	43	$Ph\bar{C}H_2$	25
benzene (C₆H₆ ring)	CHA	43	phenyl anion	41
$CH_2{=}CH_2$	35	44	$CH_2{=}\bar{C}H$	42

Acid	Solvent	pK_a	Conjugate base	References[h]
$CH_3CH=CH_2$	35	47.1–48.0	$\bar{C}H_2CH=CH_2$	43
CH_3CH_3	35	~50	$\bar{C}H_2CH_3$	44
CH_4	35	58 ± 5	$\bar{C}H_3$	43

[a]Abbreviations: DMSO, dimethyl sulfoxide; THF, tetrahydrofuran; CHA, cyclohexylamine. Most acidities were measured at 25 °C. Some are extrapolated values while others are values from kinetic studies. Errors in some cases are several pK units. The farther the pK value is from 0 to 14, the larger the errors because of estimates and assumptions made when water is not the solvent. Values of pK's for the same substance in different solvents differ because of differences in solvation. Although the acids' actual structures are listed in this Appendix, not all references do this. Thus, you may find lists of the pK_a values for organic amines that refer to the pK_a of the protonated amine rather than the amine itself. A good rule of thumb is that if the pK_a value given for an amine is less than 15, it must be the pK_a of the protonated amine rather than the amine itself.

[b]Calculated from vapor pressure over a concentrated aqueous solution extrapolated to infinite dilution.

[c]Estimated from model kinetic studies extrapolated to aqueous media.

[d]Highly concentrated solutions extrapolated to dilute aqueous media.

[e]Titrated in acetic acid and corrected to H_2O at 20 °C.

[f]Corrected from 14 because H_2O concentration is 55 mol/liter.

[g]Acidities of very weak acids are measured and/or calculated by a variety of indirect methods and may contain large errors.

[h]References: 1. Stewart, R. *The Proton: Applications to Organic Chemistry*; Academic Press: New York, 1985. 2. Bordwell, F. G. *Acc. Chem. Res.* **1988**, *21*, 456–463. 3. Perdoncin, G.; Scorrano, G. *J. Am. Chem. Soc.* **1977**, *99*, 6983–6986. 4. Arnett, E. M.; Wu, C. Y. *Chem. Ind.* **1959**, 1488. 5. Edward, J. T.; Wong, S. C. *J. Am. Chem. Soc.* **1977**, *99*, 4229–4232. 6. Guthrie, J. P. *Can. J. Chem.* **1978**, *56*, 2342–2354. 7. Arnett, E. M.; Wu, C. Y. *J. Am. Chem. Soc.* **1960**, *82*, 4999–5000. 8. Lemetais, P.; Charpentier, J.-M. *J. Chem. Res. (Suppl.)* **1981**, 282–283. 9. Deno, N. C.; Turner, J. O. *J. Org. Chem.* **1966**, *31*, 1969–1970. 10. Yates, K.; Stevens, J. B. *Can. J. Chem.* **1965**, *43*, 529–537. 11. Janssen, M. J. *Reel. Trav. Chim. Pays-Bas* **1962**, *81*, 650–660. 12. Huisgen, R.; Brade, H. *Chem. Ber.* **1957**, *90*, 1432–1436. 13. Adelman, R. L. *J. Org. Chem.* **1964**, *29*, 1837–1844. 14. *CRC Handbook of Chemistry and Physics*; Weast, R. C., Ed.; CRC Press: Boca Raton, FL, 1982–1983. 15. Kreevoy, M. M.; Eichinger, B. E.; Stary, F. E.; Katz, E. A.; Sellstedt, J. H. *J. Org. Chem.* **1964**, *29*, 1641–1642. 16. Bell, R. P. *The Proton in Chemistry*, 2nd ed.; Cornell Univ. Press: Ithaca, NY; 1973. 17. Bell, R. P.; Higginson, W. C. E. *Proc. R. Soc. (London)* **1949**, *197A*, 141–159. 18. Biggs, A. E.; Robinson, R. A. *J. Chem. Soc.* **1961**, 388–393. 19. Perrin, D. D. *Dissociation Constants of Organic Bases in Aqueous Solution*; Butterworths; London; 1965. 20. Liotta, C. L.; Perdue, E. M.; Hopkins, H. P., Jr. *J. Am. Chem. Soc.* **1974**, *96*, 7981–7985. 21. Pearson, R. G.; Dillon, R. L. *J. Am. Chem. Soc.* **1953**, 75, 2439–2443. 22. Pine, S. H. *Organic Chemistry*, 5th ed.; McGraw-Hill: New York; **1987**. 23. Bordwell, F. G.; Hughes, D. J. *J. Am. Chem. Soc.* **1985**, *107*, 4737–4744. 24. Chiang, Y.; Kresge, A. J.; Tang, Y. S.; Wirz, J. *J. Am. Chem. Soc.* **1984**, *106*, 460–462. 25. Bordwell, F. G.; Bartness, J. E.; Drucker, G. E.; Margolin, Z.; Matthews, W. S. *J. Am. Chem. Soc.* **1975**, *97*, 3226–3227. 26. Hojatti, M.; Kresge, A. J.; Wang, W. H. *J. Am. Chem. Soc.* **1987**, *109*, 4023–4028. 27. Everett, A. J.; Minkoff, G. J. *Trans. Faraday Soc.* **1953**, *49*, 410–414. 28. Ross, A. M.; Whalen, D. L.; Eldin, S.; Pollack, R. M. *J. Am. Chem. Soc.* **1988**, *110*, 1981–1982. 29. Ballinger, P.; Long, F. A. *J. Am. Chem. Soc.* **1959**, *81*, 1050–1053. 30. Bordwell, F. G.; Fried, H. E. *J. Org. Chem.* **1981**, *46*, 4327–4331. 31. Walba, H.; Isensee, R. W. *J. Am. Chem. Soc.* **1956**, *21*, 702–704. 32. Murto, J. *Acta Chem. Scand.* **1964**, *18*, 1043–1053. 33. Bordwell, F. G.; Algrim, D. J. *J. Am. Chem. Soc.* **1988**, *110*, 2964–2968. 34. Guthrie, J. P. *Can. J. Chem.* **1979**, *57*, 1177–1185. 35. Matthews, W. S.; Bares, J. E.; Bartmess, J. E.; Bordwell, F. G.; Cornforth, F. J.; Drucker, G. E.; Margolin, Z.; McCallum, R. J.; McCollum, G. J.; Vanier, N. R. *J. Am. Chem. Soc.* **1975**, *97*, 7006–7014. 36. Bordwell, F. G.; Algrim, D. J. *J. Org. Chem.* **1976**, *41*, 2507–2508. 37. Dolman, D.; Stewart, R. *Can. J. Chem.* **1967**, *45*, 911–925, 925–928. 38. Cram, D. J. *Fundamentals of Carbanion Chemistry*; Academic Press: New York; 1965. 39. Fraser, R. T.; Mansour, T. S. *J. Org. Chem.* **1984**, *49*, 3442–3443. 40. Chevrot, C.; Perichon, *J. Bull. Soc. Chim. Fr.* **1977**, 421–427. 41. Streitwieser, A., Jr.; Scannon, P. J.; Neimeyer, H. H. *J. Am. Chem. Soc.* **1972**, *94*, 7936–7937. 42. Maskornick, M. J.; Streitwieser, A., Jr. *Tetrahedron Lett.* **D**, 1625–1628. 43. Juan, B.; Schwar, J.; Breslow, R. *J. Am. Chem. Soc.* **1980**, *102*, 5741–5748. 44. Streitwieser, A., Jr.; Heathcock, C. H. *Introduction to Organic Chemistry*, 3rd ed.; Macmillan: New York; 1985. 45. Bissot, T. C.; Parry, R. W.; Campbell, D. H. *J. Am. Chem. Soc.* **1957**, *79*, 796–800. 46. Margolin, Z.; Long, F. A. *J. Am. Chem. Soc.* **1973**, *95*, 2757–2762. 47. Hine, J., Philips, J. C.; Maxwell, J. I. *J. Org. Chem.* **1970**, *35*, 3943.

Index

Note: Page numbers followed by f indicate figures; t, tables; b, boxes.

A

Abnormal products, 191, 195
Acetal, hydrolysis and formation, 179–180, 179b–180b
Acetic acid, esterification with methanol in strong
    acid, 62b–63b
Acetyl chloride, hydrolysis in water, 63b
Acid-base equilibrium
    equilibrium constant calculation, 31b, 52
    tautomers, 26b
Acid catalysis
    acetic acid esterification in strong acid, 62b–63b
    carbonyl compounds
        1,4-addition, 181–182, 181b–182b, 213–215
        hydrolysis of carboxylic acid derivatives
            amide hydrolysis, 176b–178b, 209–211
            ester hydrolysis, 178b, 211–213
            steps, 176–178
        hydrolysis and formation of acetals, ketals and
            orthoesters, 179–180, 179b–180b
        nitrile hydrolysis, 211–213
        phenylhydrazone synthesis, 110b–111b
    zinc as reducing agent, 436b, 453–455
Acidity, see pKₐ
Acrolein, electrophilic addition of hydrochloric acid,
    181b–182b, 214
Acylium ion, intermediate in electrophilic aromatic
    substitution, 218–219
Addition-elimination mechanism, nucleophilic
    substitution at aromatic carbons, 101–103,
    102b–103b, 138–139
Addition reactions, see Carbene, see Cycloaddition
    reactions, see Electrophilic addition,
    see Nucleophilic addition, see Radical
1,4-Additions
    carbon nucleophile to carbonyl compounds,
        113–121, 114b, 119b, 121b, 146–148
    electrophilic, 181–182, 181b–182b, 213–215
AIBN, see Azobis(isobutyronitrile) (AIBN)
Alcohol–aldehyde reaction, 438b, 467–468
Alcohol oxidation
    activated sulfoxide oxidations, 375, 376t, 442b,
        476–478
    chromate ester intermediate, 372
    chromium-based oxidants, 372

chromium trioxide, 374–375
    Collins' reagent, 372
    deuterium, 373
    electronic effect, 374
    Jones reagent, 372
    over oxidation, 374b–375b, 420
    steric strain, 373–374, 373b–374b
Aldol condensation, carbon nucleophile addition to
    carbonyl compounds, 112–113, 112b–113b,
    144–146, 286
Alkyllithium reagents, addition reactions
    aldehydes and ketones, 107–108, 107b–108b
    carboxylic acid derivatives, 108–110, 108b–109b
Amide hydrolysis, leaving groups, 73b–74b
Anions, representation, 12
Aromaticity
    antiaromatic compounds, 24–25, 24b
    aromatic carbocycles, 22–23, 22b–23b
    aromatic heterocycles, 23–24, 23b–24b
    classification of compounds, 25b, 47
Arrows
    bond-making and bond-breaking, 58–61, 58b–60b,
        81–82
    electron density redistribution, 58b, 446–447
    radical reaction representation, 237
Aryne mechanism, nucleophilic substitution at
    aromatic carbons, 103–104, 103b–104b,
    139–140, 155, 156
Atom numbering, writing reaction mechanisms, 76b,
    79b, 87–88, 123b–125b, 134–135, 166b,
    201–202, 455
Azide, intramolecular 1,3-dipolar cycloaddition,
    311b
Azobis(isobutyronitrile) (AIBN), 239, 269–270, 284

B

Baeyer–Villiger rearrangement, electron-deficient
    oxygen, 196t, 198b–199b, 225
Baeyer–Villiger transformation, 369
Balancing equations
    atoms, 56b
    charges, 57b, 81
    criteria in organic chemistry, 56b
Baldwin rules, radical cyclization, 249

Barton nitrite photolysis, 251b, 271–272
Basicity
    comparison with nucleophilicity, 33
    determination with resonance structures, 31b, 49–50
    leaving group ability, inverse relationship to base
        strength, 73
    solvent effects, 78b
BDE, see Bond dissociation energy (BDE)
Beckmann rearrangement, electrophilic nitrogen, 195,
        195b, 222–223
Benzilic acid rearrangement, 122–123, 122b–123b,
        149–150
Benzoic acid, Birch reduction, 262b–263b
Benzophenone oxime, Beckmann rearrangement,
        195b
Benzylic and allylic alcohols, manganese dioxide
        oxidation, 379
BHT, see Butylated hydroxytoluene (BHT)
Bicyclo[3.1.0]hexane system, geometric constraint to
        disrotatory ring opening, 302b
Bimolecular elimination, see E2 elimination
Birch reduction
    benzoic acid, 262b–263b
    mechanism, 263b, 279
    solvent, 262–263
Bond dissociation energy (BDE)
    determining feasibility of radical reactions, 244–246,
        246b, 268
    table of values, 245t
Bond number
    carbon, 3–4
    estimation, 2, 36–39, 44–47
    hydrogen, 3
    nitrogen, 4
    oxygen, 4–5
    phosphorus, 5
    radicals, 10b
    sulfur, 5
Brønsted acid and base, 28
Butadiene, cyclobutene interconversion and
        correlation diagrams, 331b–333b
Butene, bromination, 173b
Butylated hydroxytoluene (BHT), 243
t-Butyl hypochlorite, radical chain halogenation
    energetics, 241b–243b, 268
    mechanism, 246b, 268

C
Carbene, 191b–192b, 441b, 473–474
    addition reactions
        butene, addition of singlet dichlorocarbene, 189b
        stereospecificity, 189–190
        carbenoid, 187, 192b
    formation
        alkyl halides in base, 187b–188b
        diazo compounds as starting compounds, 188b
        Simmons–Smith reagent, 188b
    insertion reactions, 192
    reactivity, 186
    rearrangements, 192–193
    singlet carbene, 186–187
    substitution reactions, 190–192, 190b, 221
    triplet carbene, 186–187
Carbocation, see also Electrophilic addition
    fates, 164–165
    formation
        alkyl halide reaction with Lewis acid, 164
        electrophile addition to π bond, 163–164,
            163b–164b
        ionization, 162
    rearrangement
        alkyl shift, 166b–168b, 200–203
        dienone-phenol rearrangement, 168–169,
            168b–169b
        hydride shift, 166b
        overview of pathways, 165–168
        pinacol rearrangement, 170–171, 170b–171b,
            202–207
        stability, 165
    resonance stabilization in electrophilic aromatic
        substitution, 183t, 183b–184b, 215–217
    stability, factors affecting, 161
Carbon, bond number, 3–4
Carbonyl oxidation, 369
Carbonyls reduction
    carboxylic acids and carbone nitrogen, 396
    lewis acid, 395–396
    mechanism, 397
    pathways, 396
Chain process, see Radical
Chemical notation, symbols and abbreviations,
        485t–486t
2-Chlorobutane, reaction with aluminum trichloride,
        164b
Claisen rearrangement, 342–343, 348
Conrotatory process, electrocyclic transformations,
        296b–297b, 297, 343–344, 350–351
Cope rearrangement, 293–294, 312, 312b–313b,
        342–343, 348
Correlation diagrams, pericyclic reaction analysis
    classification of relevant orbitals, 331–332
    orbital phase correlations, 333
    principle, 331
    symmetry characteristics of reaction, 333
    symmetry correlations between bonding orbitals of
        reactants and products, 332

Crabtree catalyst, 412b
Cram's model, 398
C_2 symmetry, molecular orbital theory, 326,
  329b–330b, 351
Curtius rearrangement, electrophilic nitrogen, 196,
  196t
Cycloaddition reactions
  allyl cations and 1,3-dipoles, 309–312, 309b–312b,
    341–342
  atom number in classification, 304–305, 305b
  electron number in classification, 303, 303b
  frontier orbital theory, 334–335, 335b
  orbital symmetry, 322b, 349–350
  overview, 293, 302
  selection rules, 305–308, 306t
  stereochemistry
    allowed stereochemistry, 306b–307b
    antarafacial process, 303–304
    classification of reactions, 340
    exo:endo ratio in Diels–Alder reaction,
      308–309
    suprafacial process, 303–304
Cyclobutene, butadiene interconversion and
  correlation diagrams, 331b–333b
Cyclooctatetraene, thermal cyclization,
  298b–299b
Cyclopentanone, Baeyer–Villiger oxidation, 198b, 225
Cyclopropyl cation, *see* Electrocyclic transformations
Cyclopropyl tosylates, solvolysis, 301b

D

Dewar benzene, geometry, 20–22
Di-*t*-butyl nitroxide
  radical inhibition, 244
  stability, 238
Diazoacetophenone, Wolff rearrangement, 193b
Diels–Alder reaction, 293, 303b, 304, 308–309, 319,
  346, 443–444, 455
Dienone–phenol rearrangement, 168–169,
  168b–169b
Dinitrobenzene, radical trapping, 20b, 43, 244
1,4-Dinitrotoluene, radical inhibition, 244
1,3-Dipolar cycloaddition, 309–312, 309b–312b,
  341–342, 442–443
Dipole
  direction, 16b, 40
  relative dipoles in common bonds, 15b
Disrotatory process, electrocyclic transformations,
  297–300, 302b, 337–339
Driving forces, chemical reactions
  leaving groups, 73b–74b
  overview, 73b
  small stable molecule formation, 74

E

E_{1CB} elimination, 452
E2 elimination
  aldol condensation, 112b–113b, 286
  concerted process, 105
  leaving groups, 105–106
  stereochemistry, 105
Ei elimination
  pyrolytic elimination from a sulfoxide, 106b
  stereochemical restrictions, 107b, 140–141
  transition states, 106–107
Electrocyclic transformations
  cyclooctatetraene, thermal cyclization, 298b–299b
  cyclopropyl cation reactions, 300–302, 301b–302b,
    339–340
  frontier orbital theory, 334
  intermediates, 349
  overview, 295, 293
  selection rules, 295–296, 296b, 298b
  stereochemistry
    conrotatory process, 296b–297b, 297, 343–344,
      350–351
    disrotatory process, 297–300, 302b, 337–339
    effect of reaction conditions, 339
Electronegativity
  periodic table, trends, 14–16
  polarity of bonds, 15b
  relative values of elements, 15t
Electrophile, *See also specific electrophiles*
  common types, 34t
  definition, 32
  identification of centers, 36b, 52–53
Electrophilic addition
  1,4-addition, 181–182, 181b–182b, 213–215
  regiospecificity, 172
  steps, 172
  stereochemistry
    anti addition, 172–173, 173b
    nonstereospecific addition, 174–176,
      174b–175b
    syn addition, 173–174, 174b
  temperature effects, 175b–176b, 207–209
Electrophilic substitution
  intermediate carbocations and resonance
    stabilization, 183t, 183b–184b, 215–217
  mechanisms, 185b–186b, 217–221
  metal-catalyzed intramolecular reaction,
    184b–185b
  nitrenium ion intermediate, 226–227, 231
  substituent influence in aromatic substitution, 183t,
    183b
  toluene, electrophilic substitution by sulfur trioxide,
    183b

Elimination, *see* E$_{1CB}$ elimination, *see* E2 elimination, *see* Ei elimination
Elimination-addition mechanism, nucleophilic substitution at aromatic carbons, 103—104, 103b—104b, 139—140, 155, 156
Ene reactions
  intramolecular reactions, 319b—320b, 345
  overview, 294, 319—323
Equilibrium constant, calculation, 31b, 52
Ester, hydrolysis in acid, 178b, 211—213
Ethyl 2-chloroethyl sulfide, neighboring group effect in hydrolysis, 98b

F
Favorskii rearrangement, 121—122, 122b—123b, 149
Felkin—Anh model, 400—404
Felkin model, 399—400
Formal charge
  calculation, 6—11, 6b, 11b, 36—39, 97b
  dimethyl sulfoxide, 9b—10b
Frontier orbital theory, pericyclic reactions
  cycloaddition reactions, 334—335, 335b
  electrocyclic reactions, 334, 334b
  overview, 294, 334
  sigmatropic rearrangements, 335—337, 335b—337b

G
Grignard reagents, addition reactions
  aldehydes and ketones, 107b—108b, 110b, 141—142
  esters, 109b
  nitriles, 108b, 110—111

H
Hexanamide, Hofmann rearrangement to pentylamine, 196b
Highest occupied molecular orbital (HOMO), pericyclic reaction analysis, 330—331, 330b—331b, 352
Hoechst—Wacker process, *see* Wacker oxidation
Hofmann rearrangement, electrophilic nitrogen, 196b
Homolytic bond cleavage, radical formation, 237—238, 268, 288
Hückel's rule
  aromatic carbocycles, 22—23, 22b—23b
  aromatic heterocycles, 23—24, 23b—24b
Hybrid orbitals, representation, 12—14, 13t, 14b, 39
Hydrogen
  bond number, 3
  sigmatropic shifts, 92, 293—294, 312b—313b, 315—317, 315b—316b, 342—344, 348
Hydrogen abstraction
  incorporation in mechanisms, 464
  radical formation, 239
  rates of abstraction and radical stability, 239

Hydrogenations, olefin
  double and triple bonds, 405
  heterogeneous catalysts
    activated alkene, 406
    base metal catalysts, 405
    functional group, 405—406
    cis-to-*trans* isomerization, 406
    metal-catalyzed hydrogenation, 409b—410b, 428
    precomplexation *vs.* sterics, 409b
    reaction activation energy, 406b, 427—428
    stereo- and facial selectivity, 407b
    steric force, rigid system, 407b—408b
    steric influences and accessibility, 408b—409b
  homogeneous catalyst process
    Crabtree catalyst, 412b
    sterics impact, 410b—411b
    temporary "tether,", 412b—413b, 429
    Wilkinson's catalyst, 410—413
  transfer hydrogenations
    carbon—carbon bond forming reactions, 415
    concerted process, 413
    diimide hydrogen transfer reactions, 414b
    cis- and *trans*-isomers, 413
    P-nitrobenzenesulfonylhydrazide, 414b—415b
    precomplexation, 414b

I
Indene, chlorination, 174b
Induction/inductive effects, 161, 207, 361—362, 390—391
Intermediate
  resonance stabilization, 21b—22b, 31b, 43—47, 50—52, 72b—73b, 89—90, 462—463
  stability required in mechanism writing, 70—73, 70b—71b
  tautomer stabilization, 448
Intramolecular aza-Wacker oxidation reaction, 364b—365b
Intramolecular elimination, *see* Ei elimination
Intramolecular zwitterionic transition, 441b, 475—476

K
Ketal, hydrolysis and formation, 179—180, 179b—180b

L
Leaving group
  ability
    common groups, 73, 95t
    inverse relationship to base strength, 73
    solvent effects, 95
  amide hydrolysis reactions, 73b—74b
  E2 elimination, 105—106
  S$_N$2 reactions, 94—96
Lewis acid
  alkyl halide reactions, 97b, 98

carbonyl compound reactions, 96b
definition, 161
Lewis structure, *see also* Resonance structures
 acetaldehyde, 5b—6b
 bond number estimation, 3—6, 3b—5b
 common functional groups, table, 483t—484t
 dimethyl sulfoxide (DMSO), 9b—10b
 drawing, 3—6, 3b—5b, 7b—8b, 11b, 22b, 36—39,
  44—47
 formal charge calculation, 6—11, 6b—7b
Lone pairs, representation, 11—12
Lowest unoccupied molecular orbital (LUMO),
  pericyclic reaction analysis, 330—331,
  330b—331b, 352

M

Markovnikov, 172, 215, 387, 389, 390t, 392, 393, 425
Meerwein—Pondorff—Verley reduction, 385—386,
  386b, 424—425
Meisenheimer rearrangement, 445—446
Methyl acetate, hydrolysis in strong base, 62b
3-Methyl-2-cyclohexen-l-one, hybridization and
  geometry of atoms, 14b
exo-6-Methylbicyclo[3.1.0]hexenyl cation, sigmatropic
  shifts, 317b—318b
Michael reaction, carbon nucleophile addition to
  carbonyl compounds, 114b
Moebius—Huckel theory, pericyclic reactions, 294
Moffat oxidation, 377—378
Molecular orbital theory, pericyclic reaction analysis
 C_2 symmetry, 326, 329b—330b, 351
 correlation diagrams
  classification of relevant orbitals, 331—332
  orbital phase correlations, 333
  principle, 331
  symmetry characteristics of reaction, 333
  symmetry correlations between bonding orbitals
   of reactants and products, 332
 frontier orbital theory, 294, 334—337, 334b—337b
 highest occupied molecular orbital, 330—331,
  330b—331b, 352
 lowest unoccupied molecular orbital, 330—331,
  330b—331b, 352
 mirror plane, 324b, 325
 nodes, 324b, 325t, 328b, 330b, 351
 π orbitals
  allyl system, 329b, 351
  basis set, 326b—327b
  bonding system in chemical reactivity, 324
  energy levels, 327b—329b, 351
  ethylene, 326b—327b
  types, 327b
 wavefunctions, 324, 326b—327b

N

Naphthalene, resonance structures, 17b—18b
Neighboring group effect, nucleophilic substitutions,
  98b
Nitrene
 features, 193
 synthesis, 194b
Nitrenium ion
 features, 194
 intermediate in electrophilic substitution, 226—227,
  231
 synthesis, 194b
p-Nitroanisole, resonance structures, 18b—19b
Nitrogen
 bond number, 4
 electron deficient, rearrangements
  Beckmann rearrangement, 195, 195b,
   222—223
  Curtius rearrangement, 196, 196t
  Hofmann rearrangement, 196b
  Schmidt rearrangement, 196, 196t, 224
 positively-charged species, 72b—73b
 valence shell accommodation of electrons,
  71b
Normal products, 191
Nucleophile, *See also specific nucleophiles*
 common types, 33t
 definition, 32
 diester, 438b, 466—467
 hydroxide ion, 62b
 identification of centers, 36b, 52—53
 nucleophilicity
  comparison with basicity, 33
  ranking of nucleophiles, 33—34, 35t
  solvent dependence, 34—35
  substrate structure effects, 34
Nucleophilic addition
 addition followed by rearrangement, 123b—125b,
  150—151
 carbonyl compounds
  carbon nucleophiles, reactions with carbonyl
   compounds
   1,4-additions, 113—121, 114b, 119b, 121b,
    146—148
   aldol condensation, 112—113, 112b—113b,
    144—146
   Michael reaction, 114b
  nitrogen-containing nucleophiles, reactions with
   aldehydes and ketones
  overview, 110
  phenylhydrazone formation mechanism,
   110b—111b, 142
  steps in mechanism, 111b, 142—144

Nucleophilic addition (*Continued*)
  organometalic reagents to
    aldehydes and ketones, 107b—108b, 110b,
      141—142
    carboxylic acid derivatives, 108—110
    esters, 109b
    nitriles, 108b
    overview, 107—110
    reversibility of additions, 107
  combination addition and substitution reactions,
    125b—126b
  overview, 93
Nucleophilic substitution
  aromatic carbon substitution
    addition-elimination mechanism, 101—103,
      102b—103b, 138—139
    elimination-addition mechanism, 103—104,
      103b—104b, 139—140
  carbonyl group substitution
    ester hydrolysis in base, 99—101, 99b
    examples of steps in substitution, 100b—101b,
      132—137
    resonance-stabilized intermediates, 133, 137
    sp^2 versus sp^3-hybridized centers, susceptibility to
      substitution, 100
    tautomers in mechanisms, 133, 136
  combination addition and substitution reactions,
    125b—126b
  overview, 93
  proton abstraction preference *vs.*
    substitution, 68b
  S_N2 reactions
    alcohol protonation, 96b, 129
    features, 94
    leaving groups, 94—96
    neighboring group effect, 98
    phenolic oxygen alkylation, 97b, 129
    reactivities of carbons, 94
    stereochemistry, 96—97
    writing of mechanisms, 97b—98b, 129—132

O

Occam's razor, simplicity in writing reaction
  mechanisms, 78, 135
Olefin, cationic polymerization, 165b
Olefin oxidation
  dihydroxylation
    1,2 diols, 367b—368b
    $KMnO_4$ and OsO_4 oxidation, 366—368
    manganate ester, 368b, 419—420
  hydrogenations
    Crabtree catalyst, 412b
    double and triple bonds, 405

heterogeneous catalysts, 405—410, 406b, 427—428
  homogeneous catalyst process, 410—413,
    410b—411b
  metal-catalyzed hydrogenation, 409b—410b, 428
  precomplexation *vs.* sterics, 409b
  stereo- and facial selectivity, 407b
  steric force, rigid system, 407b—408b
  steric influences and accessibility, 408b—409b
  temporary "tether,", 412b—413b, 429
  transfer hydrogenations, 413—415, 414b
ozonolysis
  alkyl peroxide, 362
  1,3-dipolar cycloaddition, 360
  trans-propenylbenzene, 361—362
  secondary ozonide, 360
  transition state, 362b
  unsymmetrical olefin, 361
  zwitterionic intermediate, 361
peracid oxidations, 365—366
Wacker oxidation
  co-oxidant, 362—363
  copper reoxidization, 364
  deprotonation, 363
  intramolecular aza-Wacker oxidation reaction,
    364b—365b
Oppenauer oxidation, 383—385, 384b—385b
Orthoester, hydrolysis and formation, 179—180,
  179b—180b
Oxidations, 442b, 472—473, 478—480
  alcohol oxidation
    activated sulfoxide oxidations, 375, 376t, 442b,
      476—478
    chromate ester intermediate, 372
    chromium-based oxidants, 372
    chromium trioxide, 374—375
    Collins' reagent, 372
    deuterium, 373
    electronic effect, 374
    Jones reagent, 372
    over oxidation, 374b—375b, 420
    steric strain, 373—374, 373b—374b
  benzylic and allylic alcohols, manganese dioxide
    oxidation, 379
  carbon—carbon double bond, 359b, 418—419
  carbon—hydrogen bond, 355—356
  carbonyl oxidation, 369
  definition, 355
  DMSO oxygen, 378b, 422
  level/score, 356b—357b
  Moffat oxidation, 377—378
  olefins, *see* Olefin oxidation
  Oppenauer oxidation, 383—385, 384b—385b
  oxidation number, 356b—357b, 359b

hetero atoms, 358b, 416
  ketone carbon, 359b, 416—417
oxidation state, 378b, 421
plausible reaction mechanism, 370—371, 371b
products and reactants, 356b—357b
relative state, 356b—357b
Swern oxidation, 375—377
2,2,6,6-tetramethylpiperidin-1-yl)oxidanyl (TEMPO),
  380—383, 382b—383b, 422—424
  Bredt's rule, 383b, 424
  nitroxide radical, 380b—381b, 383b, 424
  oxidative cycle, 381b
  transformations, 382b
Oxygen
  Baeyer—Villiger rearrangement, 196t, 198b—199b,
    225
  bond number, 4—5
  positively-charged species, 26b
Ozonolysis
  alkyl peroxide, 362
  1,3-dipolar cycloaddition, 360
  trans-propenylbenzene, 361—362
  secondary ozonide, 360
  transition state, 362b
  unsymmetrical olefin, 361
  zwitterionic intermediate, 361

P

Pericyclic reactions
  concerted nature, 293
  cycloadditions
    allyl cations and 1,3-dipoles, 309—312, 309b—312b,
      341—342
    atom number in classification, 304—305, 305b
    electron number in classification, 303, 303b
    frontier orbital theory, 334—335, 335b
    orbital symmetry, 322b, 349—350
    overview, 293, 302
    selection rules, 305—308, 306t
    stereochemistry
      allowed stereochemistry, 306b—307b
      antarafacial process, 303—304
      classification of reactions, 340
      exo:endo ratio and secondary factors, 308—309
      suprafacial process, 303—304
  electrocyclic transformations
    cyclooctatetraene, thermal cyclization, 298b—299b
    cyclopropyl cation reactions, 300—302, 301b—302b,
      339—340
    frontier orbital theory, 334
    intermediates, 349
    overview, 293, 295
    selection rules, 295—296, 296b, 298b

  stereochemistry
    conrotatory process, 296b—297b, 297, 343—344,
      350—351
    disrotatory process, 297—300, 302b, 337—339
    effect of reaction conditions, 339
  ene reactions
    intramolecular reactions, 319b—320b, 345
    overview, 294, 319—323
  molecular orbital theory
    C_2 symmetry, 326, 329b—330b, 351
    correlation diagrams
      classification of relevant orbitals, 331—332
      orbital phase correlations, 333
      principle, 331
      symmetry characteristics of reaction, 333
      symmetry correlations between bonding orbitals
      of reactants and products, 332
    frontier orbital theory, 294, 334—337, 334b—337b
    highest occupied molecular orbital, 330—331,
      330b—331b, 352
    lowest unoccupied molecular orbital, 330—331,
      330b—331b, 352
    mirror plane, 324b, 325
    nodes, 324b, 325t, 328b, 330b, 351
    π orbitals
      allyl system, 329b, 351
      basis set, 326b—327b
      bonding system in chemical reactivity, 324
      energy levels, 327b—329b, 351
      ethylene, 326b—327b
      types, 327b
    wavefunctions, 324, 326b—327b
  selection rules, theory, 294
  sigmatropic rearrangements
    Claisen rearrangement, 342—343, 348
    Cope rearrangement, 293—294, 312, 312b—313b,
      342—343, 348
    frontier orbital theory, 335—337, 335b—337b
    overview, 293—294, 312
    selection rules
      alkyl shifts, 317—319, 317t, 317b—318b
      hydrogen shifts, 315—317, 315b—317b, 343—344
      terminology, 312—314, 312b—314b, 342—343
  symmetry-allowed reactions, 294, 331—333, 338—339
  symmetry-forbidden reactions, 294, 331—333, 337
  writing mechanisms, 320b—321b, 345—349
Phenyl acyl bromides to glyoxals conversion, 441b,
  473
Phenylhydrazone, synthesis, 110b—111b
Phosphorus, bond number, 5
π bond
  electrophile addition, 163—164, 163b—164b
  estimation of number, 2—3

π orbital
  allyl system, 329b, 351
  basis set, 326b–327b
  bonding system in chemical reactivity, 324
  energy levels, 327b–329b, 351
  ethylene, 326b–327b
  types, 327b
Pinacol rearrangement, 170–171, 170b–171b,
    202–207
pK_a
  approximation from related compounds, 123b–125b,
    152, 453
  calculation, 31b–32b, 52
  carbonyl groups, 152
  definition, 28
  values for common functional groups, 29t–30t,
    487t–493t
Proton removal
  epimerization of reactants, 125b–126b
  nucleophilic substitution, competition with, 67,
    67b–68b
  susceptibility of specific protons, 65b–67b, 83–87
  writing reaction mechanisms
    condensation reactions, 449, 451–453
    rationale, 73b–74b
    strong base, 62b
    weak base, 63b
Protonation
  carbonyl groups, 163b
  nitrogen, 232–233
  olefins, 163b
  susceptibility of specific centers, 66b, 86–87
  writing reaction mechanisms
    condensation reactions, 450–451, 449
    rationale, 73b–74b
    strong acid, 62b–63b
    weak acid, 63b

R
Radical
  addition reactions
    intermolecular addition, 247–248, 247b–248b,
      268–269
    intramolecular cyclization, 249–252, 249b–252b,
      269–273
  Birch reduction, 262–263, 262b–263b
  bond dissociation energies in determining feasibility
    of reactions, 244–246, 246b, 268
  chain process
    balancing of equations, 240
    coupling of radicals, 241b, 268
    disproportionation, 241b–242b
    halogenation by t-butyl hypochlorite, 241b–242b

    initiation, 240, 241b–242b, 248b, 250b–251b,
      252–253, 268–271, 282–284, 289, 439b, 469–470
    propagation, 240, 241b–242b, 248b, 250b–252b,
      253, 269–271, 282–283, 285, 291, 439b, 469–470
    termination, 240, 241b–242b
  definition, 237
  depicting mechanism, 237
  formation from
    functional groups, 239–240
    homolytic bond cleavage, 237–238, 238b, 268,
      288
    hydrogen abstraction, 239
  fragmentation reactions, loss of small molecules
    addition followed by fragmentation, 254b–255b
    CO, 253–255
    CO_2, 252–253
    ketone, 253
    N_2, 253–255
    writing of mechanisms, 251b–252b, 272–273
  inhibitors, 243–244
  rearrangements
    alkyl migration, 166b–168b, 200–203
    apparent alkyl migration, 258b–259b, 273–276
    aryl migration, 256b–257b
    halogen migration, 257b–258b
    mechanisms in anion rearrangement, 264, 264b,
      280–281
    non-migrating groups, 256b
  resonance stabilization, 268
  $S_{RN}1$ reaction
    enolate reaction with aromatic iodide, 260b–261b
    features, 259–260
    identification of reactions, 262b, 277–279, 281–282
    initiation, 260, 285
    propagation, 260, 286
  stability, 238–239
  stereochemistry of reactions, 237
Rearrangement, see also Sigmatropic rearrangements
  Baeyer–Villiger rearrangement of electron-deficient
    oxygen, 196t, 198b–199b, 225
  base-promoted rearrangements
    benzilic acid rearrangement, 122–123, 122b–123b,
      149–150
    Favorskii rearrangement, 121–122, 122b–123b,
      149
    nucleophilic addition followed by rearrangement,
      123b–125b, 150–151
    stereochemistry, 123b, 149–150
  carbenes, 192–193
  carbocation
    alkyl shift, 166b–168b, 200–203
    dienone-phenol rearrangement, 168–169,
      168b–169b

hydride shift, 166b
overview of pathways, 165—168
pinacol rearrangement, 170—171, 170b—171b, 202—207
stability, 165
electrophilic nitrogen
Beckmann rearrangement, 195, 195b, 222—223
Curtius rearrangement, 196, 196t
Hofmann rearrangement, 196b
Schmidt rearrangement, 196, 196t, 224
radicals
apparent alkyl migration reactions, 258b—259b, 273—276
aryl migration, 256b—257b
halogen migration, 257b—258b
mechanisms in anion rearrangement, 264, 264b, 280—281
non-migrating groups, 256b
Reductions, 442b, 472—473, 478—480
aminations and imines, 404, 404b, 427
borane reductions
carbonyls reduction, *see* Carbonyls reduction
electronic effects, 390
hydroboration, 387b—388b, 394b, 426—427
olefins reduction, 387—390
orientation, 388b—390b
perborate hydroboration, 395b
propenyl styrene, 391—395
stereo- and regiodirecting effects, 392b—393b
stereochemistry, 394b, 426
styrene reactivity, 391
two-step hydroboration reaction/oxidation sequence, 394b, 425—426
definition, 355
lithium aluminum hydride and sodium borohydride reductions, 401b—402b, 403t, 403b—404b
vs. borane reagents, 397
Cram's model, 398
Felkin—Anh model, 400—404
Felkin model, 399—400
Meerwein—Pondorff—Verley reduction, 385—386, 386b, 424—425
olefin hydrogenation, *see* Hydrogenations, olefin
Reimer—Tiemann reaction, 190b, 221
Resonance effects
basicity, 31b, 49—50
carbocation stability, 161
protonation, 66b, 86—87
proton removal, 65b—67b, 83—87
radical stability, 268
Resonance structures
cyclooctatetraenyl anion, 18b—19b

definition, 16
distinguishing from tautomers, 26b—28b, 48—49, 153
drawing, 16—20, 19b, 21b—22b, 40—47
naphthalene, 17b—18b
p-nitroanisole, 18b—19b
rules, 20—22
stability, 21b—22b, 31b, 43—47, 50—52, 72b—73b, 89—90, 462—463
Ring number, estimation, 2—3

S
Schmidt rearrangement, electrophilic nitrogen, 196, 196t, 224
Sigmatropic rearrangements, 344—345
Claisen rearrangement, 342—343, 348
Cope rearrangement, 293—294, 312, 312b—313b, 342—343, 348
frontier orbital theory, 335—337, 335b—337b
overview, 293—294, 312
selection rules
alkyl shifts, 317—319, 317t, 317b—318b
hydrogen shifts, 315—317, 315b—317b, 343—344
terminology, 312—314, 312b—314b, 342—343
Simmons—Smith reagent, synthesis, 188b
S_N2 reactions
alcohol protonation, 96b, 129
features, 94
leaving groups, 94—96
neighboring group effect, 98
phenolic oxygen alkylation, 97b, 129
reactivities of carbons, 94
stereochemistry, 96—97
writing of mechanisms, 97b—98b, 129—132
Solvent, effects
basicity, 78b
leaving group ability, 95
mechanism of reaction, 77—78
nucleophilicity, 34—35
radical stability, 444
sp hybridization, overview, 12, 13t, 39
sp^2 hybridization
nucleophilic substitution at aliphatic carbon, 99—101, 99b—101b, 132—137
overview, 12—13, 13t, 14b, 39
sp^3 hybridization, overview, 12—13, 13t, 14b, 39
$S_{RN}1$ reaction
enolate reaction with aromatic iodide, 260b—261b
features, 259—260
identification of reactions, 262b, 277—279, 281—282
initiation, 260, 285
propagation, 260, 286
Stereochemistry, 290, 438b—439b, 465—466, 470—471

Stereochemistry (*Continued*)
  base-promoted rearrangements, 123b, 149–150
  cycloaddition reactions
    allowed stereochemistry, 306b–307b
    antarafacial process, 303–304
    classification of reactions, 340
    exo:endo ratio and secondary factors, 308–309
    suprafacial process, 303–304
  E2 elimination, 105
  electrocyclic transformations
    conrotatory process, 296b–297b, 297, 343–344, 350–351
    disrotatory process, 297–300, 302b, 337–339
    effect of reaction conditions, 339
  electrophilic addition
    anti addition, 172–173, 173b
    nonstereospecific addition, 174–176, 174b–175b
    syn addition, 173–174, 174b
  epimerization of reactants, 125b–126b
  pinacol rearrangement, 170b–171b, 203
  radical reactions, 237
  S_N2 reactions, 96–97
cis-Stilbene, bromination in acetic acid, 174b–175b
Sulfur, bond number, 5
Swain–Scott equation, nucleophilicity calculation, 33–34
Swern oxidation, 375–377, 442b, 476–478
Symbols, chemical notation, 485t–486t

T
Tautomer
  acid-base equilibrium, 25b
  definition, 25
  distinguishing from resonance structures, 26b–28b, 48–49, 153
  drawing of structures, 27b, 48

  enolization, strong acids and bases as intermediates, 63b–64b
  equilibrium, 25–28
  ketones, 26b
  stabilization of intermediates, 448
Toluene, electrophilic substitution by sulfur trioxide, 183b
Triflate, spontaneous ionization, 162b
Trifluoroiodomethane, photochemical addition to allyl alcohol, 247b
Trimolecular reaction
  breaking down into several bimolecular steps, 69b–70b, 79b–80b, 89, 158
  rarity, 69

U
Unproductive step, 67b

V
Verrucosidin, degradation products, 437b, 455–462
Vitamin D_2, synthesis, 316b–317b

W
Wacker oxidation
  co-oxidant, 362–363
  copper reoxidization, 364
  deprotonation, 363
  intramolecular aza-Wacker oxidation reaction, 364b–365b
Wilkinson's catalyst, 410–413
Wolff rearrangement
  nitrogen analogues, 196–197, 196t, 196b
  overview, 192–193, 193b

Z
Zinc, reducing agent in acid catalysis, 436b, 453–455

Printed in the United States by Baker & Taylor Publisher Services

Printed and bound by CPI Group (UK) Ltd, Croydon, CR0 4YY

03/10/2024

01040318-0002